2nd International Conference on Materials Technology and Energy (ICMTE 2019)

IOP Conference Series: Materials Science and Engineering
Volume 943

Miri, Malaysia
6 – 8 November 2019

Editors:

Sukanta Roy
Yam Ke San

ISBN: 978-1-7138-4777-9
ISSN: 1757-8981

Printed from e-media with permission by:

Curran Associates, Inc.
57 Morehouse Lane
Red Hook, NY 12571

proceedings.com

Some format issues inherent in the e-media version may also appear in this print version.

Institute of Physics
Dirac House, Temple Back
Bristol BS1 6BE UK

Phone: 44 1 17 929 7481
Fax: 44 1 17 920 0979

techtracking@iop.org

Additional copies of this publication are available from:

Curran Associates, Inc.
57 Morehouse Lane
Red Hook, NY 12571 USA
Phone: 845-758-0400
Fax: 845-758-2633
Email: curran@proceedings.com
Web: www.proceedings.com

TABLE OF CONTENTS

Preface

Curtin University Malaysia hosted 2nd International Conference on Materials Technology & Energy (ICMTE) 2019 at its campus from 6 to 8 November 2019. The conference was co-organized with The Institution of Engineers Malaysia (IEM) Miri Branch and The Institution of Materials Malaysia (IMM) Miri Chapter with support from the Sarawak Convention Bureau (SCB) and Serba Dinamik Holdings Berhad, Malaysia.

ICMTE 2019 provided a unique platform to exchange ideas on innovations, trends and future challenges related to materials technology, structures, construction, green energy, and industrial technologies. The conference hosted over 100 delegates, academics, researchers and industry professionals from Malaysia, Australia, Brunei, India, Indonesia, Singapore, UAE, Dubai, and South Africa to showcase their knowledge and expertise in different technologies focused on the above topics.

The conference featured seven keynote speakers, such as Prof. Mohd. Nasir Tamin (University Teknologi Malaysia) who spoke on 'Hybrid Experimental-Computational Characterisation of Materials for Reliability and Life Prediction Models', Assoc. Prof. Yang Wenming (National University of Singapore) whose topic was 'Fuel Design and its Impact on the Performance and Emissions of Internal Combustion Engines', and Assoc. Prof. Ir. Chong Wen Tong (University of Malaya) who delivered a talk on 'Innovative Wind Turbine Designs and Technologies'. Assoc. Prof. Sujan Debnath (Curtin University Malaysia) gave a presentation on 'Mechanical Performance of Natural Fiber Polymer Composites Using Palm Oil Biomass' while Dr. Suvash Saha (University of Technology Sydney) spoke on 'Heat Transfer and Thermal Storage in Different Types of Heat Exchangers'. Dr. Ir. Wahyudi Sutopo (Universitas Sebelas Maret) talked about 'Intervening for Accelerating a Technology Commercialisation in University using Early Supply Chain Integration', and Assoc. Prof. Chithirai Pon Selvan (Curtin University Dubai), spoke about 'Investigation of Process Parameters for Depth of Cut in Abrasive Waterjet Machining of Materials'.

Apart from the keynote talks, 8 parallel sessions were hosted presenting the papers. A total of 88 papers have been submitted to the conference, out of which 66 papers have been selected for publication in the conference proceedings. The delegates found the conference an excellent opportunity for domestic and international participants to share their research findings and initiate collaboration in areas of recent technical advancements and innovation with regards to materials technology and energy. The conference ended with a half-day technical tour to the Sarawak BioValley Pilot Plant at the Curtin Malaysia campus, and a half-day master class on the interpretation of industrial welding codes.

The organizers hope that these proceeding papers will be a good source of information for the relevant Scientific community from academia and industry.

Editor(s)

Dr. Sukanta Roy
Curtin University Malaysia

Dr. Yam Ke San
Curtin University Malaysia

IOP Conf. Series: Materials Science and Engineering **943** (2020) 011002 doi:10.1088/1757-899X/943/1/011002

COMMITTEE

Patron (s)	Prof. Jim Mienczakowski, Curtin University Malaysia Ir. Prof. Lau Hieng Ho, Institute of Engineers Malaysia
General Chair (s)	Prof. Beena Giridharan, Curtin University Malaysia Ir. Assoc. Prof. Edwin Jong Nyon Tchan, Institution of Materials Malaysia
International Scientific and Advisory Board	A. Prof. Vincent Lee Chieng Chen, Curtin University Malaysia A. Prof. Ian Davies, Curtin University, Australia A. Prof. Chithirai Pon Selvan, Curtin University, Dubai A. Prof. Yang Wenming, National University of Singapore, Singapore Prof. Ir. Amir Azam Khan, Universiti Malaysia Sarawak, Malaysia Prof. Mohd Nasir Tamin, Universiti Teknologi Malaysia Prof. Ujjwal K Saha, Indian Institute of Technology Guwahati, India Ir. Paul Chew Lik Ing, Institution of Materials Malaysia A. Prof. Chong Wen Tong, University of Malaya, Malaysia A. Prof. Akshay Dvivedi, Indian Institute of Technology Roorkee, India A. Prof. Ranjan Das, Indian Institute of Technology Ropar, India Dr. Hubert Branger, IRPHE/CNRS/Aix Marseille University, France Dr. Suvash Saha, University of Technology Sydney, Australia Dr. Alokesh Pramanik, Curtin University, Australia Dr. Noman Ahsan Khandoker, Central Queensland University, Australia Dr. Denni Kurniawan, University of Technology Brunei Dr. Kristofeer Mckee, Curtin University, Australia
Chairperson	Dr Yam Ke San, Curtin University Malaysia
Co-Chairperson (s)	Dr. Henry Foo Chee Yew, Curtin University Malaysia Ir. Andrew Ling, Shell Sarawak Berhad
Organizing Secretary	Dr. Sukanta Roy, Curtin University Malaysia
Treasurer Committee	Ir Lim Chye Ing, Curtin University Malaysia Ir. Meheron Selowara Joo, Curtin University Malaysia Dr. Fethma M Nor, Curtin University Malaysia Dr. Law Kah Haw, Curtin University Malaysia Dr. Tan Inn Shi, Curtin University Malaysia Ir Dr. Vera Loo Hui, Curtin University Malaysia Dr. Sung Aun Naa, Curtin University Malaysia

IOP Conf. Series: Materials Science and Engineering **943** (2020) 011002 doi:10.1088/1757-899X/943/1/011002

Thomas Ting, Akal Solutions Sdn. Bhd.

Editor (s)

Dr. Sukanta Roy, Curtin University Malaysia
Dr. Yam Ke San, Curtin University Malaysia

Technical Committee

Prof. Muhammad Ekhlasur Rahman, Curtin University Malaysia
A. Prof. Ravi Kumar Gupta, Manipal University Jaipur, India
A. Prof. Sujan Debnath, Curtin University Malaysia
A. Prof. John Lau Sie Yon, Curtin University Malaysia
A. Prof. John Lau Sie Yon, Curtin University Malaysia
A. Prof. Law Ming Chiat, Curtin University Malaysia
A. Prof. Mohan Reddy Moola, Curtin University Malaysia
A. Prof. Garenth Lim King Hann, Curtin University Malaysia
A. Prof. Stephanie Chan, Curtin University Malaysia
Dr. Ehsan Pashajavid, Curtin University, Perth, Australia
Dr. Vinod Kushvaha, Indian Institute of Technology Jammu, India
Dr. Bhaskor J. Bora, RGIPT Energy Institute, Bangalore, India
Dr. Le Ling Yong, Universiti Teknologi Malaysia
Dr. Naveen Kumar Elumalai, Curtin University Malaysia
Dr. Agnimitra Biswas, National Institute of Technology Silchar, India
Dr. Kishore Debnath, National Institute of Technology Meghalaya, India
Dr. Teoh Yi Peng, Universiti Tunku Abdul Rahman, Malaysia
Dr. Ooi Zhong Xian, Universiti Tunku Abdul Rahman, Malaysia
Dr. Ziad Bennour, Curtin University Malaysia
Ir. Dr. Wong Wei Kitt, Curtin University Malaysia
Dr. Mahmood Anwar, Curtin University Malaysia
Dr. Jung Hyul Chyul, Curtin University Malaysia
Dr. Law Kah Haw, Curtin University Malaysia
Dr. Kelvin Yong Sheng Chek, Curtin University Malaysia
Dr. Tina Ting, Curtin University Malaysia
Dr. Daniel Tang Kuok Ho, Curtin University Malaysia
Dr. Jibrail Kansedo, Curtin University Malaysia

2nd International Conference on Materials Technology and Energy (ICMTE 2019)

KEYNOTE SPEAKERS

Prof. Mohd Nasir Tamin earned his doctoral degree in Mechanical Engineering and Applied Mechanics from the University of Rhode Island, USA in 1997. He has been with the Faculty of Mechanical Engineering, Universiti Teknologi Malaysia since 1984. He had his fair share of administrative positions as the Head of the Applied Mechanics Department, Head of the Materials Engineering Department, Director of UTM Center for Composites, and the Deputy Dean (Research and Innovation) of the faculty. He is currently leading his research team with 11 doctoral candidates, a post-doctoral researcher, and a project manager at the Computational Solid Mechanics Laboratory (CSMLab) UTM, which he founded in 2006. Prof. Tamin's research team activities focus on the development of constitutive and damage models for ductile metals and fibrous composite laminates. He leads his research team on few successful research collaborations with industries including Intel Technology on the development of a validated methodology for reliability prediction of microelectronic BGA packages and through-silicon via (TSV) interconnects, with Kiswire (Korea) for fatigue life improvement of steel wire ropes, and with Airbus and Aerospace Malaysia Innovation Center (AMIC) for damage detection in FRP composite laminates using the digital image correlation technique (DIC). Prof. Tamin has been invited as a visiting researcher at Sophia University, Tokyo (Japan), a visiting professor at the Institut Supérieur de l'Automobile et des Transport, Nevers (France) and Dongguk University, Seoul (Korea), and currently a visiting research professor at the University of Southampton (Malaysia Campus). He is keen in promoting the university-industry research collaboration among colleagues across the globe.

Dr. Yang Wenming is currently an Associate Professor with tenure in the department of Mechanical Engineering, National University of Singapore. His research interests include development of Lower Temperature Combustion (LTC) Technologies, fuel design, combustion and emissions control of internal combustion engines fueled by various fuels, microscale combustion and micro power generators. He has authored/co-authored more than 280 papers in international peer-reviewed journals and conferences, of which, more than 210 papers are SCI index, and most of his papers are published in top 10% journals. His papers have been cited by more than 7500 times with a H-index of 47. He is currently the Principal Investigator or Co-PI for 14 competitive research grants funded by National Funding Agencies and Ministry of Education of Singapore, worth more than $30 million US dollars. He has won a series of awards including 4 times of scientific progress award (2nd prize) by the Ministry of Education of China and the Society of Mechanical Engineering of China; two best paper awards by Applied Energy (Elsevier) in 2014 and 2016; He is also the regular reviewer for many peer-reviewed journals such as Applied Energy, Energy Conversion and Management, Fuel, Progress in Energy and Combustion Sciences etc.

Dr. Chong Wen Tong is presently working in University of Malaya as Associate Professor and Head of Centre for Energy Sciences. He is the Editorial Board Member of two SCI-Q1 journal (top 10%), i.e. the International Journal of Energy Research (IJER) and International Journal of Precision Engineering and Manufacturing-Green Technology (IJPEM-GT). He has co-authored >110 ISI technical articles (h-index = 27). He is also the recipient of the University of Malaya Excellence Award 2016, Excellent Lecturer (Science & Technology). He has secured more than RM 6 million research grants, completed 18 research projects (renewable energy & green technology) as principal investigator and successfully supervised >35 post-graduate graduates. He has filed 12 intellectual property rights and won the First Prize of National Intellectual Property Award 2016 (Cross-Axis-Wind-Turbine). He is graduated from the Faculty of Mechanical Engineering, UTM (Mechanical-Aeronautics). He was awarded the Best Student Award for his academic achievement; and obtained his Ph.D. from the same university. Dr. Chong joined the Department of Mechanical Engineering, UM (# 33, QS-Subjects-Ranking-2016/2017) after spending 10 years in various industries. Before joining UM, he was the Head of Technology and Mechanical Team in Dyson Malaysia, and has been responsible for research, design and development works on vacuum cleaner, hand dryer and bladeless fan. He is a Chartered Engineer registered by the Engineering Council (UK). He served as the Programme Coordinator (Mechanical) for about 5 years (2011-2015) and teaches Renewable Energy at the M.Eng. level in UM. He is the Conference Chair of International Conference on Sustainable Energy and Green Technology (SEGT 2018 & 2019).

Dr. Suvash C. Saha is working at the University of Technology Sydney, Sydney, Australia as a Senior Lecturer of Mechanical Engineering. He has also worked as a postdoctoral research fellow for more than 6 years at the School of Chemistry, Physics and Mechanical Engineering of Queensland University of Technology, Brisbane, Australia. Dr. Saha received his PhD in Computational Fluid Dynamics applications to heat transfer in buildings from the School of Engineering and Physical Sciences of James Cook University, Australia in 2009. His current research interests are Thermal storage using Phase Change Materials (PCM); Solar thermal energy technology; Natural convection heat transfer in buildings and other confined geometries; Scale analysis for the transient flow; Computational biomechanical engineering etc. Dr. Saha has more than 160 publications including 104 journal papers mostly published in high impact journals. Total citations of Dr. Saha is 1318 (Google Scholar) and 872 (Scopus) with h-Index of 19 (Google Scholar) and 16 (Scopus). He has more
than $1.2m funding including 2 Australian Research Council Linkage grants and is involved in two National Natural Science Foundation of China (NSFC) funding.

Dr. Ir. Wahyudi Sutopo, ST., M.Si, IPM is hold an engineering professional degree (Ir) from Study Program of Professional Engineer – Universitas

Sebelas Maret (UNS) at 2019. He also registered as a Professional Engineer (IPM – Insinyur Profesional Madya) with The Institution of Engineers Indonesia – IEI since 2016 (https://pii.or.id). He obtained his Doctorate in the field of Industrial Engineering and Management (Dr) from Institut Teknologi Bandung (Indonesia)at 2011, Master of Science in Management (M.Si) from Universitas Indonesia at 2004 and Bachelor of Engineering in Industrial Engineering (ST) from Institut Teknologi Bandung at 1999. He currently is the Head of Department of Industrial Engineering, UNS, FY 2015-2019. In UNS, he is Asset Auditor for Internal Audit Unit (FY 2013-now) and Coordinator of Industrial Engineering and Techno-Economy (RITE) Research Group (FY 2016-now). He also as General Chairman of BKSTI (Agency for Cooperation of Higher Education of Industrial Engineering in Indonesia) FY 2017-2020. His research interests are Logistics & Supply Chain Management, Engineering Economy & Cost Analysis, and Technology Commercialization. He has published 3 monographs and 7 book chapters. He has published sevral research articles in reputed journal and conference proceedings and obtained 30+ research grants.

Dr. Sujan (CEng MIMechE) is an Associate Professor at Department of Mechanical Engineering in Curtin University Malaysia. He had served as the HOD of Mechanical Engineering from October 2014 to December 2017 in Curtin Malaysia. Dr. Sujan obtained his PhD Degree from the University of Science Malaysia in 2006 majoring applied mechanics and heat transfer with specific research focuses on interfacial thermal mismatch stress analysis in electronic packaging. Over the years, he has been working in the area of thermal management in electronic packaging, green composite materials using bio-waste, nano polymer composites and nano-enhanced MQL Machining. Dr. Sujan has more than 100 publications in reputable international journals and conference proceedings. Dr. Sujan is a Registered Level 1 supervisor in Curtin University, Australia for Doctoral and Master Thesis.

Dr. Chithirai Pon Selvan obtained his Bachelor in Production Engineering, Master in Computer Aided Design and Ph.D. in Mechanical Engineering. Dr. Pon Selvan has over twenty years of experience in teaching, educational assessment, classroom management and student relations. He currently works as Associate Professor at Curtin University, United Arab Emirates and serves as the Head of School of Science and Engineering. He has published 100+ research articles in international journals and conferences. He has been invited and honored as keynote speaker, session chair, resource person and technical committee member in various conferences in UAE, India, Thailand, Malaysia, UK, Germany, and Italy. He is in the editorial board of more than 50 international journals. His research interests are in the areas of machine design, optimization techniques and manufacturing practices, particularly non-traditional manufacturing methods.

IOP Conf. Series: Materials Science and Engineering **943** (2020) 011005 doi:10.1088/1757-899X/943/1/011005

Acknowledgement of Sponsor: ICMTE2019

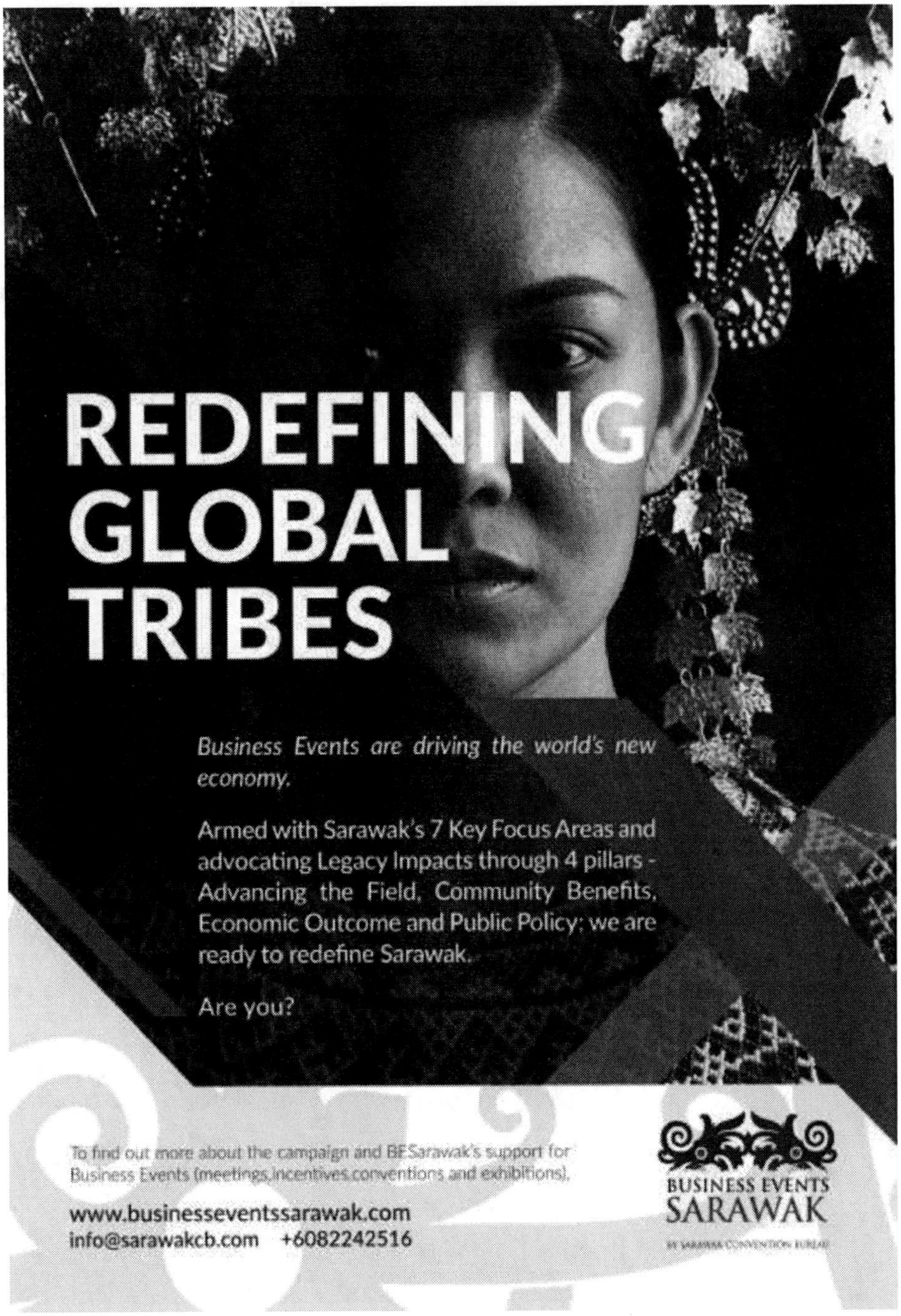

IOP Conf. Series: Materials Science and Engineering **943** (2020) 011005 doi:10.1088/1757-899X/943/1/011005

Acknowledgement of Sponsor: ICMTE2019

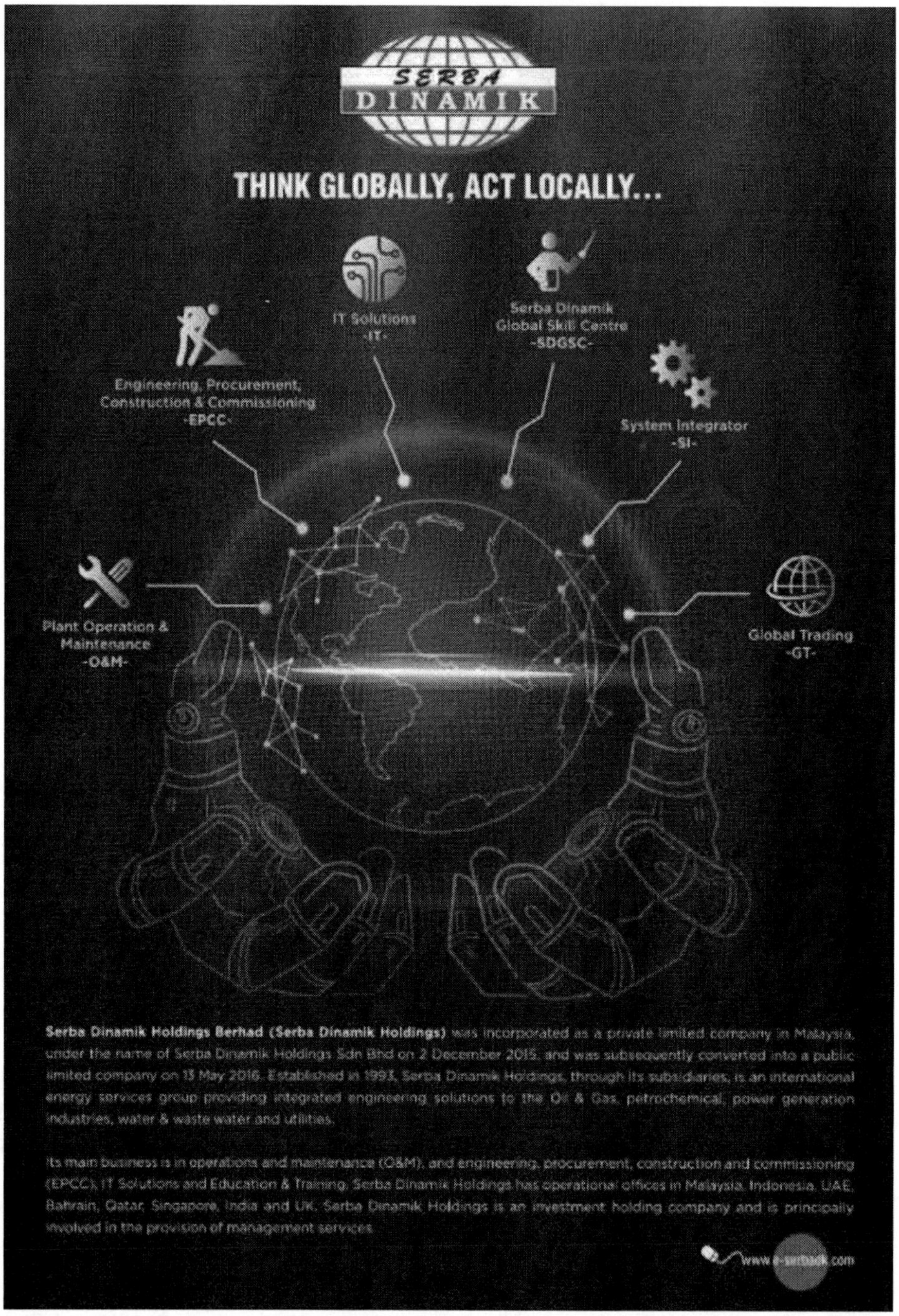

IOP Conf. Series: Materials Science and Engineering **943** (2020) 011006 doi:10.1088/1757-899X/943/1/011006

Peer review declaration

All papers published in this volume of IOP Conference Series: Earth and Environmental Science have been peer reviewed through processes administered by the Editors. Reviews were conducted by expert referees to the professional and scientific standards expected of a proceedings journal published by IOP Publishing.

Type of peer review: Other
Open, double peer review by expert technical committee. Minimum qualification for reviewer is Doctoral degree in the relative field.

Conference submission management system:
The abstracts and full-length papers were submitted through online management system. Upon successful submission, the papers were sent to at least 2 reviewers per paper. Based on the comments received from the reviewers, the papers were either (a) accepted as original submission, (b) accepted with minor corrections, (c) accepted with major corrections or (d) rejected.

Number of submissions received: 88

Number of submissions sent for review: 88

Number of submissions accepted: 66

Acceptance Rate (Number of Submissions Accepted / Number of Submissions Received X 100): 75%

Average number of reviews per paper: 2 reviews

Total number of reviewers involved: 35

Any additional info on review process:
The review process was carried out in accordance with the instructions given to the reviewer.

Contact person for queries: Dr. Sukanta Roy, Curtin University, Malaysia, sukanta.roy@curtin.edu.my

Weed Pixel Level Classification Based on Evolving Feature Selection on Local Binary Pattern With Shallow Network Classifier

Basil Andy Lease, WK Wong, L Gopal, and W. R. Chiong

Curtin University Malaysia, CDT 250, Miri 98009, Sarawak, Malaysia

E-mail: basil.andy.lease@postgrad.curtin.edu.my

Abstract. This paper proposes an evolving feature selection on a rotation-invariant Local Binary Pattern (LBP) with genetic algorithm (GA) and a non-linear classifier to perform pixel-wise classification on biomass pixels. Early true leaf growth of carrots and weeds consists of 60 images provided by a public benchmark Crop/Weed Field Image Dataset (CWFID) [1] was used. The GA encodes LBP feature parameters generated from normalized difference vegetation index (NDVI) image slices as genome consisting of number of neighbours, radius, size of image slice, number of LBP combinations. LBP is a lightweight rotation invariant texture feature descriptor which encodes discriminative local texture information between crops and weeds. The study evaluated multiple ensemble models evolved by GA, where GA evolves the LBP feature parameter selection, and the number of LBP features as input variables used. The classifier can handle crop and weed overlaps, which are often present in commercial fields. Weeds often grow in close proximity and overlap crops and are similar in size, which adds complexity in discriminating them. Our experiments have shown that combinations with a back-propagation neural network with a symmetrical two hidden-layer network achieved the best classification accuracy when compared to other non-linear classifiers. By utilizing a single type of feature (LBP texture feature), our resulting Artificial Neural Network (ANN) ensemble classifier is able to achieve a classification accuracy of 83.5 %.

1. Introduction

Certain parts of the world are currently experiencing a food crisis. This is due to the high food production demands unmet with ever-growing world population. The agricultural field needs improvements in agriculture production technology to meet such demands. Thus, precision agriculture and smart agriculture is currently an important field of research. Freshly sown field with young crops are often plagued with naturally growing weeds which compete for living resources. The early growth stages are critical as it will affect crop yield. The conventional method of farmers to eliminate weeds is by blanket spraying chemical herbicides using a tractor as it is cost-effective and easy to use. However, they pose environmental, and health concerns as excessive use may lead to groundwater contamination [2], river contamination [3], and food contamination [4]. The current advancement of technology allows for computer vision-enabled autonomous robots to manage and control weeds in the field. This would enable far greater precision than blanket spraying as robots would be able to target and spray weeds with the correct dose effectively. Thus, reduce waste and excessive use of chemical herbicides. To achieve this level of autonomy, an accurate weed discrimination system is required. The main challenge is often related to occlusion and complex natural environment of weeds among the crop. Thus, the system needs to understand the discriminative features between weeds and crops.

This paper focuses on the development of a weed classification system in the natural complicated environment setting of crops and weeds. This paper suggests a strategy that does not involve image segmentation of crop and weeds. Instead, we propose that the texture characteristics can separate weeds and plants through Local Binary Pattern (LBP) obtained from the picture windows centred around each pixel containing crop data. LBP is selected as it has desirable characteristics such as lightweight, invariant of rotation, capable of encoding pattern replication and providing data on local texture. The image window would encode the texture data of neighbouring crops as a function. Our model utilises genetic algorithm to select the optimal combinations of texture features and its feature parameters to best describe the differences of crops and weeds . A non-linear classifier will then evaluate the chosen features to predict the label for each pixel.

2. Literature Review

Features are often used to describe the property of an image and encode it in numerical vectors for image classification problems. Classification models then use the encoded data to understand relationships and patterns within the encoded data to distinguish between the classes marked. In many classification tasks, Local Binary Pattern (LBP) is often used as a visual descriptor for texture. Some popular uses in LBPs are face detection [5], facial recognition [6], visual inspection [7], motion analysis [8] and texture analysis [9]. LBP encodes the local textures of each surrounding pixel as a feature vector. The literature has demonstrated that LBP to be a robust texture descriptor in many applications, thus its potential should be studied in smart agriculture applications.

Current technology makes the solution to weed control tasks in the field of computer vision possible. Different kinds of features are frequently used in literature as the input to a classifier model. Ahmed et al. [10] proposed the approach to using the supporting vector machine (SVM) classifier to discriminate against weeds. They chose shape, colour and moment as the feature vector in their work. The SVM model was for six plant species. This technique averaged 97.3 % for the resultant classification model. The writers note that the technique could be more robust if more pictures were used. They also noticed that the model was affected by image noise and plant holes.

Hemming [11] used colour and biological morphology features for carrot, cabbage and weed. The classifier is based on weighted fuzzy classification. Colour features were shown to distinguish plants from the soil, and the classifier shows an improvement in classification accuracy when using colour features in addition to morphology features. Their technique has led to a classification rate of 51-95 %, which varies based on weed density. They have observed that this method is prone to occlusion of the leaves.

McCool et al. [12] suggested using lightweight DCNN models for pixel-level crop-weed classification tasks. The authors suggested that the classification model be trained in three phases. The first phase utilizes a pre-trained DCNN model that delivers cutting-edge performance at the cost of high computational requirements. This model is not suited for mobile automated field robotics as the frame rate is relatively low at 0.12 frames per second (fps). In the second stage, therefore, the pre-trained DCNN model is used as a teacher model to form a lightweight DCNN student model. This technique is known as model compression resulting in reduced accuracy but significantly reduced complexity. The third phase includes various lightweight designs coupled as a mix of models resulting in greater classification performance. The classification accuracy of the mixture of lightweight models is 90 %. The research discovered that while cutting-edge DCNN model could attain 93.9 % ranking accuracy, it could only operate low frame rates. Whereas, the lightweight DCNNs combination was able to compute with fewer

Figure 1. An image from the crop/weed field image dataset by Haug et al. [1].

parameters at a faster framerate of about 1.07 to 1.83 fps.

Haug et al. [13] suggested a less complex design for plant classification without the use of segmentation techniques. Crops and weeds often grow in close proximity and overlap are prevalent. The authors, therefore, suggested the use of a Random Forest classifier to classify pixels instead of segments. Pixel classification eliminates difficulties combined with segmentation techniques like overlapping and challenging background. The classifier predicts the position of the spatial pixel based on the characteristics of overlapping neighbourhood pixels. Markov Random Field was used to merge similar predictions. The model achieved an average classification accuracy of 93.8 %.

Current literature has outlined problems related to plant holes, occluded leaves, and high complexity models. Therefore, we suggest a model that seeks to reduce certain problems by combining an effective texture descriptor, LBP, in an ensemble classification model with the evolving feature selection in order to classify each pixel without segmentation as inspired by Haug et al. [13].

3. Classification Model

3.1. Image Dataset

The images used are from a crop/weed field image dataset by Haug et al. [1]. The dataset includes 60 images of top-down images of row crops on organic carrot farms. The images were captured with a multi-spectral camera, which contains red and near-infrared visual information, mapped to 3 colour channels. Channel 1 represents red, while channel 2 represents near-infrared, and channel 3 is a repetition of channel 1. A sample image is shown in Figure 1. The complicated spatial composition of overlapping and close growth of the weeds and crops are also apparent in the figure. The data set provides labelled images, in which a human expert annotated each pixel. The labels are "1" for weed and "0" for crops, which simplifies the task to a binary classification.

3.2. Background Removal

The original image includes soil and other visual information that is not useful in discriminating weeds. Thus, removing non-contributing image information is essential. The image includes red and near-infrared channels, so a Normalized Difference Vegetation Index (NDVI) was adopted to calculate the differences between plant and soil variations in spectral reflection. The NVDI

Figure 2. Masked NDVI vegetation image after background removal.

used to obtain soil vegetation mask is calculated, as shown in Equation 1. Besides separating the plant from the soil, the method also simplifies the image to a single-channel greyscale image.

$$NDVI = \frac{NIR - Red}{NIR + Red} \tag{1}$$

Most background information is filtered out by NDVI. Otsu's thresholding method[14] further filters out the remaining background pixels, which produces a binary vegetation mask which represents the vegetation pixels in 1 (white) and the background as 0 (black). The binary mask is used to filter the soil pixels on the NDVI image, resulting in a picture shown in Figure 2 where in NDVI space, only the vegetation pixels are present. The resulting image, therefore, includes only contributing plant information in a single channel NDVI space. Reduction to a single channel is essential because the feature extractor requires it.

3.3. Feature Extraction

The image is divided into sub-images, which contains plant pixels. This paper will use the notion of the image window to represent the sub-images. Every image window, as shown in Figure 3 comprises of the same pixel length and height where the centre pixel is the label class. The image window size is determined in Equation 2.

$$W_{length} = (n * 2) + 1 \tag{2}$$

Where n represents the half length of the image window.

From the image window, a rotation-invariant Local Binary Pattern (LBP) [15] is used as a feature extractor encoding local pixel texture patterns in the to create a feature vector. LBP compares its neighbouring pixels to a reference pixel situated in the middle of the image window to form a binary vector. LBP can be defined as follows in terms of equations:

$$LBP_{P,R}(x_c, y_c) = \sum_{p=0}^{p-1} = s(i_p - i_c) * 2^p \tag{3}$$

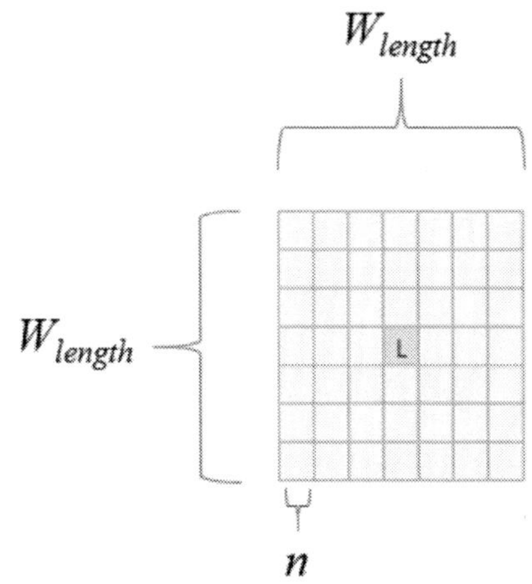

Figure 3. The image window representation where L is the label to labeled middle pixel and n is the number of pixels.

$$s(v) = \begin{cases} 1 & : v \geq 0 \\ 0 & : v < 0 \end{cases} \tag{4}$$

From Equation 3, P is the number of neighbours, and R is the radius from the centre. The subscript c denotes the centre and p denotes the current pixel neighbour. i_c represents the pixel value of the center reference pixel i_p, which is located at (x_c, y_c). The LBP process requires the image window to be divided into cells. Equation 5 calculates the number of cells.

$$numCells = floor\left(\frac{W_{length}}{CellSize}\right)^2 \tag{5}$$

Where W_{length} represents the length of the image window and $CellSize$ represents the chosen cell size.

The LBP extraction method can be summarized as follows:

(i) Locate the plant pixel.
(ii) The plant pixel will correspond to a label based on the annotated image.
(iii) The plant pixel will be the centre of the image window.
(iv) Divide the image window into cells.
(v) The middle pixel is set as the reference, compare every other N neighbouring pixels through an R radius circular manner.
(vi) For pixels which are larger than the middle will be valued as 1, otherwise 0 in a continuous sequence given by Equation 4.
(vii) This will form a binary sequence of N number of neighbours which is converted to decimal given by Equation 3.
(viii) From the obtained number, a histogram is computed over each cell, and the frequency is added to each bin.
(ix) L2 normalization is applied to normalize the histogram.
(x) The resulting feature vector represents the normalized summation of the histogram of each cell for the particular image window.
(xi) Stride and repeat for all plant pixels.

IOP Conf. Series: Materials Science and Engineering **943** (2020) 012001 doi:10.1088/1757-899X/943/1/012001

Table 1. Table of parameters

LBP Parameters	Lower Boundary	Upper Boundary
Image Window Size (S)	3x3	101x101
Radius (R)	1	5
Number of Neighbours (N)	8	16

3.4. Feature Selection

Feature selection is a significant component in the model where most contributing features are chosen to allow the classifier to learn the differences between the two classes. As the search space is large, genetic algorithm is used to search for the ideal feature parameters in the ensemble. In parameter tuning, the LBP extraction method enables tuning of window size S, reference pixel R radius, and number of neighbours N. The cell size is set to 1. A list of LBP features with varying S, R, N parameters were generated for timely optimization. The image windows S size ranges from 3x3 to 101x101 pixels with 2 pixel increments for each dimension. Zero paddings are applied to the image window for out of bound regions. The radius of the neighbouring pixels R range from 1 to 5. The number of neighbours N ranges from 8-16 determines the number of pixels to be included for the binary pattern calculation. The LBP features produced are invariant to rotation. Therefore, all feasible combinations are produced for feature selection with the stated parameter boundaries.

Neural networks have been chosen as the classifier of choice since no prior knowledge of data processes is necessary, and they can establish relationships from the features given. The neural network classification structure is comprised of a model of three layers, namely, an input layer, 2 hidden layers and an output layer. The input layer provides one type of LBP feature from the GA selection. Each of the two hidden layers has 10 interconnected neurons. Finally, the output layer consists of a 2-class output, although it is a binary classification task, a 2-class output is needed to generate certainty scores of each class. At the output layer, tangent sigmoid was used as the activation function, and the model loss uses the mean squared error of the classification accuracy. This structure forms a simple NN unit that is used in the ensemble.

The ensemble consists of several simple stacked NN units. Contrary to the traditional stacked neural networks, this ensemble model does not supply all simple NN units with the same input. The output of all simple NN units is combined with a sum average of each output class. The highest value would determine the selection of the predicted class.

3.5. Training of Model

3.6. Neural Network Classifier

The divisions for train and test are 40 images and 20 images respectively. The training consists of 2 stages, which are feature selection by GA and NN training through backpropagation. The full ensemble classification model is shown in Figure 4. For the first stage, genetic algorithm (GA) optimization was used to find the optimal feature parameter combinations. The GA starts with a randomly generated population with varying feature parameters. Each generation will consist of a GA population size of 20 individuals where each individual will its NN units trained, and its fitness evaluated. Fitness is defined as the classification accuracy given by the ensemble NN units. Stopping criteria will be met when there is no improvement of over 50 generations. The features selected by GA will be the inputs of each NN unit and the outputs are the 2 class certainty scores (i.e. crop or weed). The 2 class certainty scores from each NN units are combined with all the outputs in the ensemble model through sum average. The resultant score represents the classifier accuracy. This classification accuracy is then used by the GA to determine the best combinations of parameters and features to be used.

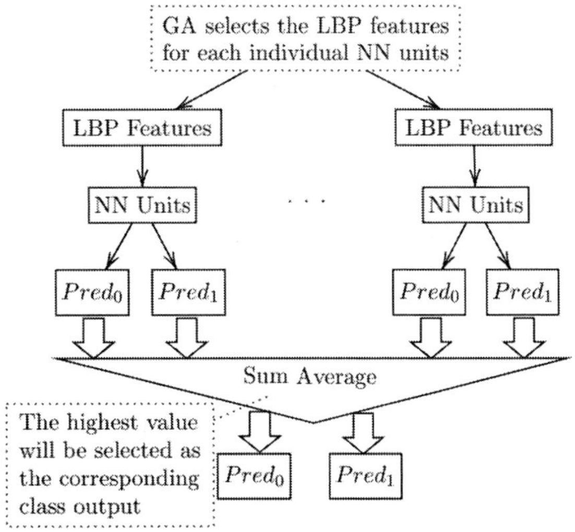

Figure 4. The ensemble classification model

4. Results and Discussion

Evaluation metrics provide a standard for discussing and comparing each model's strengths and weaknesses. The standard metrics are accuracy, accuracy, recall and F1-score from Equations (6)-(9) respectively. Accuracy indicates how well weeds and plants can be properly classified in the classification model. Precision refers to how well classification of the weed class can be determined without a crop being misclassified as a weed. The precision metric is particularly essential in order to avoid accidental crop spraying. The recall metric, or sometimes called sensitivity, will determine how much weeds are sprayed considering the actual number of weeds in the field. The F1 score is the weighted average of recall and precision. The F1 metric takes into account the imbalanced dataset and measures recall and precision as one measure.

$$\text{Accuracy} = \frac{TP + TN}{TP + FP + FN + TN} \tag{6}$$

$$\text{Precision} = \frac{TP}{TP + FP} \tag{7}$$

$$\text{Recall} = \frac{TP}{TP + FN} \tag{8}$$

$$\text{F1 Score} = \frac{2 * (\text{Recall} * \text{Precision})}{\text{Recall} + \text{Precision}} \tag{9}$$

Where T is "True", F is "False", P is "Positive", and N is "Negative".

Our proposed method used a combination of simple NN units in an ensemble. The number of simple NN units in the ensemble are 5, 10, and 15. From the results shown in Table 2, a simple NN unit with a single LBP feature at best could only achieve an accuracy of 80.8 %. The LBP as a feature extractor shows potential in weed classification tasks as even the single

Table 2. Comparison of classification scores from literature studies on CWFID [1] dataset

Methods	Accuracy	Precision	Recall	F1-score
Random Forest Tree [1]	85.9 %	79.6 %	80.8 %	80.2 %
PWDS [16]	85.8 %	91.1 %	86.8 %	88.8 %
ANN with 15 units in ensemble	83.5 %	83.5 %	97.8 %	90.1 %
ANN with 10 units in ensemble	83.0 %	82.8 %	98.2 %	89.9 %
ANN with 5 units in ensemble	82.3 %	84.0 %	95.8 %	89.5 %
ANN without ensemble	80.8 %	84.5 %	91.8 %	88.0 %

type of LBP feature combination paired with NN could achieve relatively high precision, recall and F1-score when compared to Random Forest Tree [1]. In terms of classification accuracy, our best ensemble model with 15 NN units at 85.9 % did not surpass in terms of accuracy when compared to Random Forest Tree [1] of 85.9 % and PWDS [16] of 85.8 %. The ensemble model was able to increase the overall accuracy by 2.7 % from 80.8 % to 83.5 % with 15 units in the ensemble. Even though a higher number of ensemble increase the overall accuracy, it can be seen that the precision maintained relatively the same, averaging around 83 %. The increase in accuracy is due to the increasing recall metric. Recall, as mentioned previously, is the sensitivity or hit rate. This means that the classifier is very good at determining the targeted weed class, but as reflected in accuracy, it does not perform well in classifying the outlier crop class.

The limitation of our current proposed work is at the method by which the ensemble outputs are combined. As the predicted classes are a sum average of all the NN units, the ensemble model is constrained to equal weights between each class and each NN unit. This would limit the flexibility of the ensemble model as it would not be able to allocate more weights to the appropriate contributing class and unit. In other words, the current ensemble model weighs all predictions from each output class of the NN units the same and does not boost the weak class well.

Our method achieved the highest F1-score of 90.1 % from 15 units in the ensemble when compared with Random Forest Tree [1] and PWDS [16]. This indicates that the classifier performs quite well in regards to a well-balanced precision and recall. Thus, improvements could be made towards the classification of the crop class to improve the classification accuracy further.

5. Conclusion

This paper demonstrated that LBP could potentially be used to identify weeds and crops in the agriculture field. LBP, as a feature used in this weed classification task, appears to perform quite well in classifying weeds as reflected by the recall score. Improvements could be further made to the structure of the ensemble to improve the classification of the weaker crop class further. For future work, we would propose the use of a weighted sum average to the ensemble paired with an evolving optimizer such as particle swarm optimization (PSO) to find the optimal weights. We hypothesize that certain combinations of LBP feature parameters may be able to discriminate a particular class better. Thus, with a weighted ensemble, the classification of the weaker class as the weaker units could be boosted by the weights.

References

[1] Haug S and Ostermann J 2014 A crop/weed field image dataset for the evaluation of computer vision based precision agriculture tasks *European Conference on Computer Vision* (Springer) pp 105–116
[2] Guzzella L, Pozzoni F and Giuliano G 2006 *Environmental Pollution* **142** 344–353
[3] Pereira W E and Hostettler F D 1993 *Environmental science & technology* **27** 1542–1552

[4] Landrigan P J and Benbrook C 2015 *New England Journal of Medicine* **373** 693–695
[5] Jin H, Liu Q, Lu H and Tong X 2004 Face detection using improved lbp under bayesian framework *Third International Conference on Image and Graphics (ICIG'04)* (IEEE) pp 306–309
[6] Liu L, Fieguth P, Zhao G, Pietikäinen M and Hu D 2016 *Information Sciences* **358** 56–72
[7] Tajeripour F, Kabir E and Sheikhi A 2008 *EURASIP Journal on Advances in Signal Processing* **2008** 60
[8] Zhao G and Pietikainen M 2006 Local binary pattern descriptors for dynamic texture recognition *18th International Conference on Pattern Recognition (ICPR'06)* vol 2 (IEEE) pp 211–214
[9] Mäenpää T and Pietikäinen M 2005 Texture analysis with local binary patterns *Handbook of pattern recognition and computer vision* (World Scientific) pp 197–216
[10] Ahmed F, Al-Mamun H A, Bari A H, Hossain E and Kwan P 2012 *Crop Protection* **40** 98–104
[11] Hemming J and Rath T 2001 *Journal of agricultural engineering research* **78** 233–243
[12] McCool C, Perez T and Upcroft B 2017 *IEEE Robotics and Automation Letters* **2** 1344–1351
[13] Haug S, Michaels A, Biber P and Ostermann J 2014 Plant classification system for crop/weed discrimination without segmentation *Applications of Computer Vision (WACV), 2014 IEEE Winter Conference on* (IEEE) pp 1142–1149
[14] Otsu N 1979 *IEEE transactions on systems, man, and cybernetics* **9** 62–66
[15] Ojala T, Pietikäinen M and Mäenpää T 2002 *IEEE Transactions on Pattern Analysis & Machine Intelligence* 971–987
[16] Umamaheswari S, Arjun R and Meganathan D 2018 Weed detection in farm crops using parallel image processing *2018 Conference on Information and Communication Technology (CICT)* (IEEE) pp 1–4

IOP Conf. Series: Materials Science and Engineering **943** (2020) 012002 doi:10.1088/1757-899X/943/1/012002

Influence of Azodicarbonamide on Physical and Mechanical Behaviours of Thermoplastic Blend

Ruey Shan Chen and Sahrim Ahmad

Materials Science Program, Faculty of Science and Technology, Universiti Kebangsaan Malaysia, 43600 UKM Bangi, Selangor, Malaysia

Email: chen@ukm.edu.my

Abstract. Foaming technology is promising in producing foamed materials which can offer benefits such as lowering the cost and weight of materials as well as their utility in many applications. In this study, the thermoplastic blend of recycled high density polyethylene and recycled polyethylene terephthalate was foamed with azodicarbonamide chemical foaming agent (CFA) at 0.5-2.0 phr. The blending of thermoplastic was carried out via extrusion and the foaming process occurred during hot/cold pressing. The cell morphology, density and mechanical properties were examined. Field emission scanning electron microscope (FESEM) micrograph showed that the thermoplastic blend foamed with 1.5 phr CFA had the cellular morphology structure with non-collapsed cells and better cell distribution. As the CFA content increased, the density of blend foams decreased accordingly. The minimum density was achieved at 1.00 g/cm3 for 2.0 phr ADC. Tensile and bending properties generally decreased with the content of CFA added. The impact strength of blend foams increased when the CFA added up to 1.0 phr and then reduced with further CFA content. It can be concluded that the fabricated foamed blend can give value added to the recycled polymers.

1. Introduction

The public awareness on environmental pollution and the increasing cost of raw materials have attracted industries' attention in exploring ways to fabricate lightweight products with desirable properties by using less material [1]. In this context, foamed materials have been investigated for replacing the solid counterparts as foaming technology can reduce the materials' consumption, and hence lead to the reduction of weight and density as well as the production cost [1, 2]. Polymeric foams, so-called cellular polymers are defined as two-phase materials where a gas dispersed in a continuous macromolecular phase [3, 4]. Practically, polymer foams are applied in various fields, such as packaging materials, automobile industry, sports equipment and building materials, owing to the low cost and weight, good mechanical properties as well as energy insulation and high energy absorption [5].

Chemical blowing agent or chemical foaming agent (CFA) can be defined as a chemical compound that is used to promote foaming process in which the gas is generated via its thermal decomposition [4]. In literature, it was established that there are various CFA used for different types of thermoplastics. Chang et al. has investigated the mechanical and surface properties of polypropylene (PP) blown with 1-9 wt% endothermic CFA masterbatch, and they demonstrated that the best cell uniformity with the smallest average cell was at 5 wt% CFA [1]. Sodium bicarbonate is one of the endothermic CFAs which has been added into natural rubber at 4-12 phr; the findings showed that the amount of decomposed carbon dioxide gas increased with the concentration of foaming agent and hence increasing the number of cells per unit volume with smaller average cell sizes [6]. On the other hand, for exothermic CFA, azodicarbonamide is one of the interesting chemical compounds because it decomposes at a relatively low temperature and produces closed-cell foam [7] for neat polymers and composites. Stange and Münstedt reported that with the use of azodicarbonamide, a remarkable enhancement of the foaming behavior where a higher expansion ratio and more homogeneous cell size distribution with a lower amount of connected cells are found with the increasing content of the long-chain branched-PP up to 50 wt% [8]. In the study of Bledzki et al., an exothermic CFA was used to foam PP/wood fibre composites and it was found that 2 wt% CFA produced a finer cellular structure and the density was reduced up to 30% [9].

In spite of numerous researches on polymer foaming, very few reports on the polymer foams based on polymer blending were found. In this research, the recycled polymer blend made from post-consumer high-density polyethylene and polyethylene terephthalate was used for foaming. The focus of this work was to investigate the effect of azodicarbonamide CFA content varying from 0 to 2 phr on the morphology, density and mechanical properties of recycled thermoplastic blend.

2. Experimental Section

2.1 Materials

Recycled high-density polyethylene (RHDPE) and recycled polyethylene terephthalate (RPET) used in this study were supplied by Bio Composites Extrusion Sdn. Bhd. The RHDPE has a melt flow index of 0.072 g/10 min (at 190°C) and a density of 923 kg/m^3. The Lotader AX8840 type of ethylene glycidyl methacrylate (EGMA) with a melt index of 5 g/10 min (190°C, 2.16 kg) and a glycidyl methacrylate content of 8% was used as compatibilizer. Chemical foaming agent (CFA) used in this study is azodicarbonamide foaming agent, and zinc oxide (ZnO) foaming agent decomposition stabilizer were provided by Sigma Aldrich.

2.2 Method

Prior compounding, the RPET was oven-dried at 90 °C. Figure 1 shows the experimental section for fabrication of recycled thermoplastic blend foams. The RHDPE, RPET and EGMA were melt- compounded in a co-rotating twin screw extruder (Thermo Prisma TSE 16PC) with an extrusion temperature profile of 250-270-240-190 °C and a screw speed of 30 rpm. The composition of RHDPE/RPET/EGMA was fixed at 75/25/5 (wt/wt/phr). The extrudates were taken for second extrusion with CFA at 170-215-210-195 °C at the same screw rotation speed. The CFA content was varied at 0.0, 0.5, 1.5 and 2.0 phr. After extrusion, the extruded granules were compression molded at 200 °C under 1000 psi using a hot/cold pressing machine (LP50, LABTECH Engineering Company LTD). The periods of hot press and cold press were set to 10 min and 5 min, respectively.

For structural observation, field emission scanning electron microscopy (FESEM) was performed using LEO 1450 VP with an operation accelerating voltage of 14 kV. The density measurement of thermoplastic blend (either solid or foamed ones) was determined manually as: density = mass per unit volume. The tensile testing was done on 3 mm thickness dumbbell specimens (ASTM D638-03 (type I)) using a universal tensile machine (Model Testometric M500-50CT) at 5 mm/min crosshead speed. The Izod impact testing was performed using Ray-Ran Universal Pendulum Impact System (ASTM D256-05), with a velocity of 3.46 ms-1. Bending measurement was determined using the same machine as tensile test, but the sample dimension was prepared follow the standard ASTM D790-03.

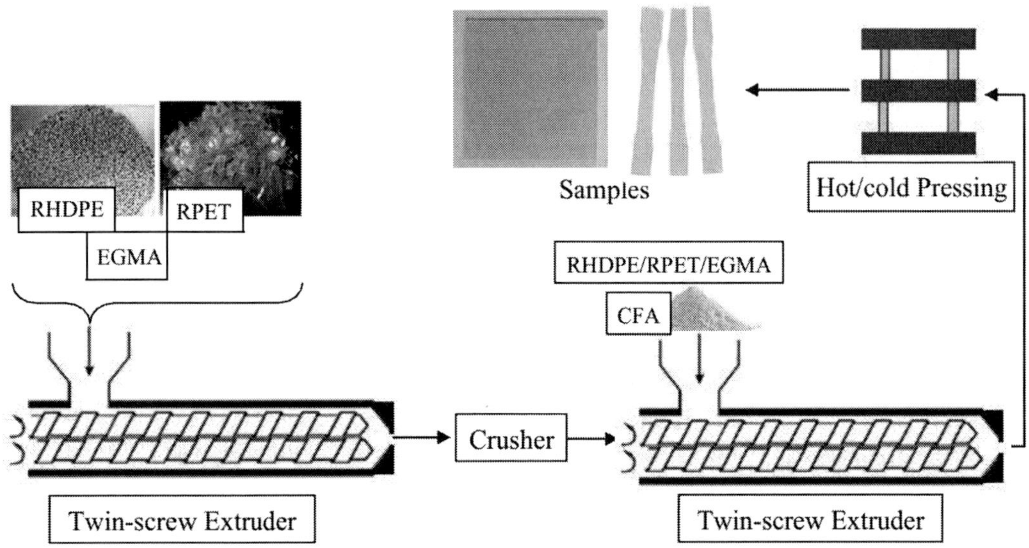

Figure 1. Schematic experimental section for fabrication of recycled thermoplastic blend foams.

3 Results and Discussion

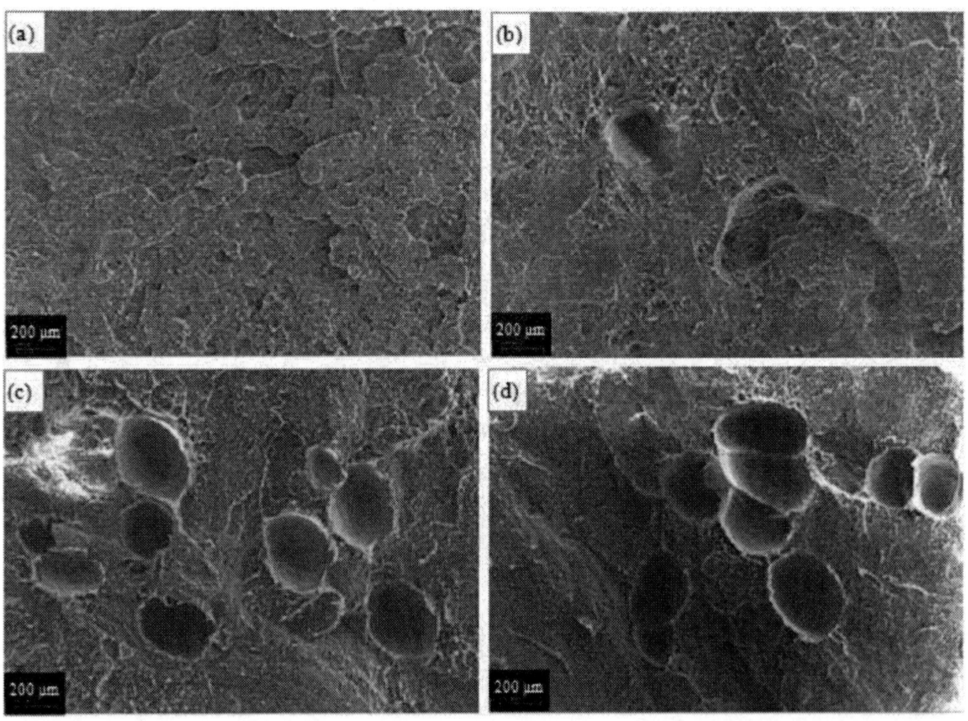

Figure 2. FESEM micrograph of thermoplastic blend foams with (a) 0.0 phr, (b) 0.5 phr, (c) 1.5 phr and 2.0 phr of CFA.

Figure 2 shows morphological structure of thermoplastic blend foamed with different content of CFA at (a) 0, (b) 0.5, (c) 1.5 and (d) 2.0 phr. In Figure 1(a), without CFA, the solid thermoplastic blend displayed a compact structure. By adding CFA at low content (0.5 phr), the surface morphology of the specimen showed the effect of foaming as proven by the presence of the dents. In this case, the foaming effect seems not stable and sufficient as the foam cells formed are not perfect circular shaped. This indicates the content of CFA at 0.5 phr was insufficient in this blend. When the CFA was added at higher content, the foam cells are more stabilized. Amongst all the CFA contents, the blend with 1.5 phr CFA displayed the most satisfied structure with having perfect spherical foam cells with no coalescence and less variety of cell sizes, whereas, the blend with 2.0 phr CFA had an inhomogeneous distribution and collapsing of the foam cells. The coalescence between cells (bubbles) could be due to the high amount of gas produced from CFA [7]. The observations in FESEM micrograph are tally with the density measurement shown in Figure 3. In which the density results showed the downward trend of density with the increasing CFA content. The reduction of density is ascribed to the increasing number of cells generated (Figure 2). However, the reduction of density was insignificant (up to only 6%) and this is possibly due to the low uniformity of the foam cells in the blend. As shown in FESEM micrograph, the cells not fully distributed throughout the entire surface.

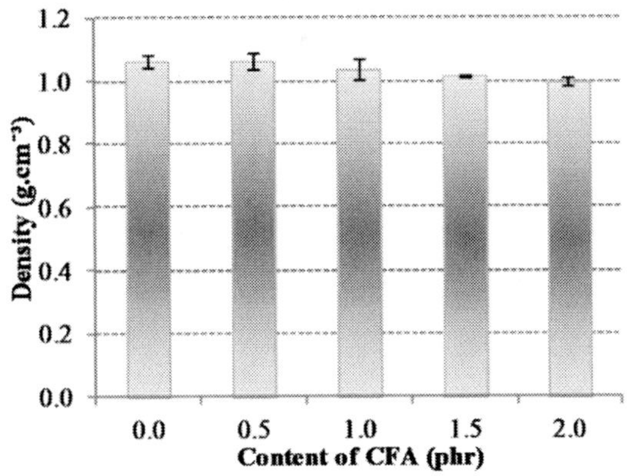

Figure 3. Density measurement of thermoplastic blend foams with different content of CFA.

Figure 4 presents the mechanical properties of thermoplastic blend as a function of CFA. The model of Gibson and Ashby stated that the tensile modulus of foams are determined by the factors of foam density, tensile modulus of the material used and foam structure [10]. In general, it is found that as the density of composite decreased upon the addition of CFA, the tensile and bending strength and modulus also decreased. For the solid blend, the mechanical properties were recorded at 19.1 MPa for stress at yield (tensile strength), 351.5 MPa for Youngs Modulus (tensile modulus), 29.9 MPa for bending strength, and 1166.6 MPa for bending modulus. A gradual linear reduction was shown for tensile and bending properties (except for Youngs Modulus) and the lowest properties (maximum reduction by about 8-18 %) were exhibited by 2.0 phr CFA. This phenomenon has been similarly reported in previous research regarding the neat polymer foams and polymer composite foams [9, 11]. This is expected as the voids resulting from foaming process could adversely affect the mechanical properties [11], especially in the case of non-uniform and non-homogenous cell sizes and distribution in the structure. The tensile and bending strength and modulus decreased with the rising of CFA content. In fact, the foamed blend with higher content of CFA generates larger voids which act as crack creation sites that may severely deteriorate mechanical properties (i.e. stiffness) [1]. For the modulus, especially tensile modulus (Figure 4 (b)), the negative effect by CFA content was inconsistent. Nevertheless, the lowering of modulus of the blend resulted in the higher impact strength and the softness of the final foam product [10].

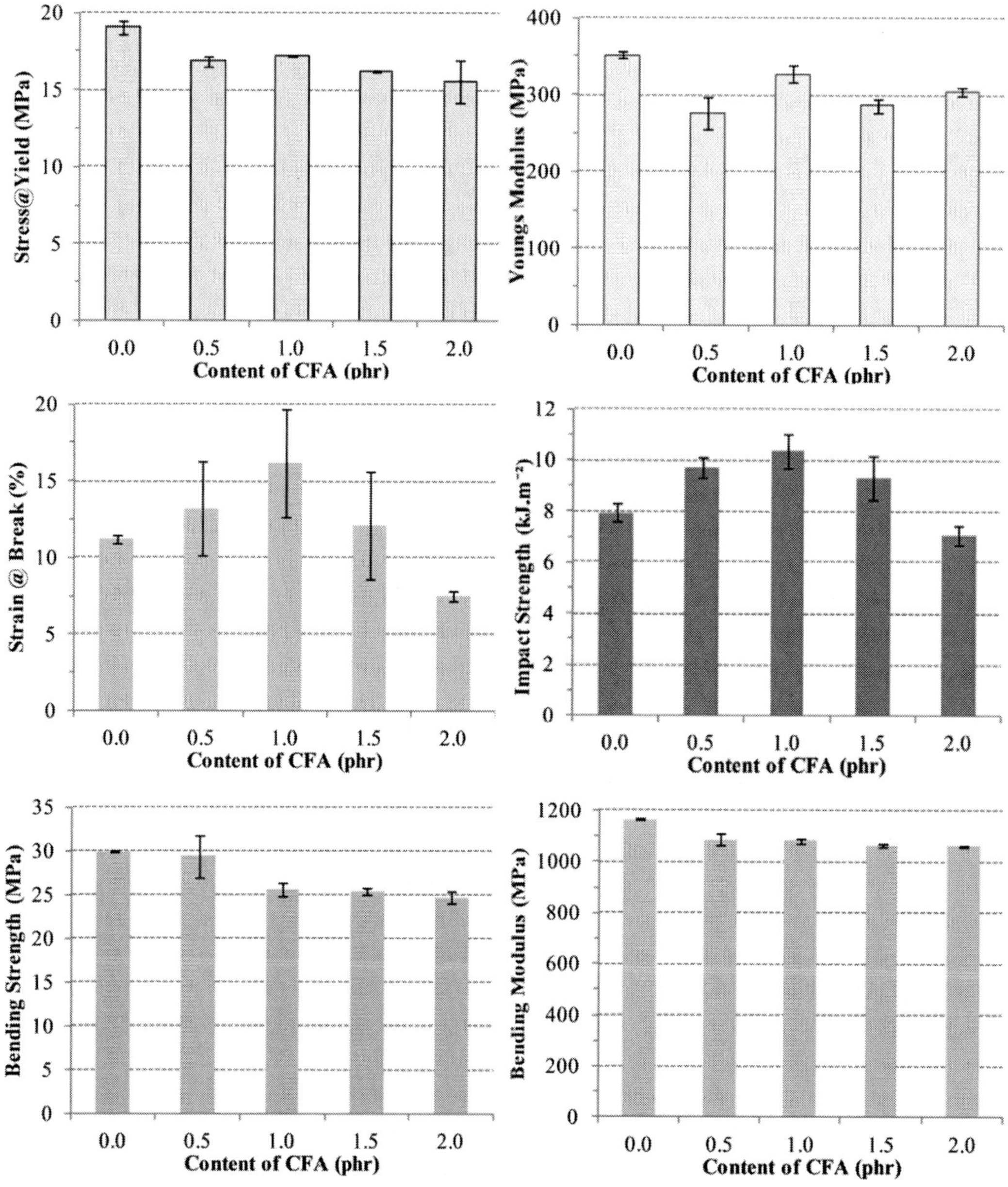

Figure 4. Mechanical properties of thermoplastic blend foams with different content of ADC.

Interestingly, the strain at break and impact strength showed some improvements with the addition of less than 1.5 phr CFA. The maximum strain at break (16.2 %) and impact strength (1.4 kJ/m2) were shown for samples with 1.0 CFA. According to Chang et al., the impact strength is governed by several parameters like cell size, skin thickness and void fraction [1]. The trend of earlier increase in strains at break and impact strength has been published in the previous work on polypropylene foam with 5 phr CFA [1]. The unique properties of foamed material are owing to the outstanding energy-absorbing characteristics induced by the presence of the gaseous phase [6]. The latter reduction implies the weakened ability of the foam cells to absorb impact energy as a result of increasing number of cells per unit volume which led to poor crosslink density. It is stated that the properties of the foam product will be deteriorated when cells with large size are present. This is because these cells will have lower cell wall thickness and the cell wall is more likely to collapse and rupture when a force is applied [6]. This can be obviously observed in foamed blend with 2.0 phr CFA.

4. Conclusions

The solid and foamed RHDPE/RPET/EGMA (75/25/5) thermoplastic blends were prepared via melt-blending technique and followed by compression molding. The effect of exothermic azodicarbonamide content varied from 0.5 to 2.0 phr on the microstructure, density and mechanical behaviours of blend foams was investigated. By increasing azodicarbonamide content, the amount of cells formed in the blend structure increased, thereby led to the reduction of blend density. With the presence of 2.0 phr azodicarbonamide, the cells structure collapsed as shown in FESEM micrograph. From tensile and bending results, the strength reduced by 17-19% and modulus by 8-14% for foamed thermoplastic blends with respect to azodicarbonamide content. Whilst, the strain at break and impact strength of the blend foams showed an optimum increment at 1.0 phr before decreasing trend with higher azodicarbonamide content. The findings of this study preserve a good preliminary research in polymer foaming for further improvement in weight reduction while maintaining or even improving the material performance.

5. Acknowledgements

The authors are grateful for the supports provided by Electron Microscopy Unit of Universiti Kebangsaan Malaysia (UKM) and UKM Research Grant Scheme (GGPM-018-061).

Reference

[1] Chang E, Mahmud M B, Li X, Mohebbi A, and Park C B. *Optimizing chemical blowing agent content in foam injection molding process of polypropylene*. in *SPE-ANTEC meeting, Society of Plastics Engineers, Anaheim, CA, USA*. 2017.
[2] Hassan N A A, Ahmad S, Chen R S, Zailan F D, and Shahdan D 2019 *Materials Today: Proceedings*.vol p 601-606
[3] Saiz-Arroyo C, de Saja J A, Velasco J I, and Rodríguez-Pérez M Á 2012 *J Mater Sci*.vol p 5680-5692
[4] Lopez-Gonzalez E, Salmazo L O, Lopez-Gil A, and Rodriguez-Perez M A 2019 *Polymer Engineering & Science*.vol p 791-798
[5] Heidari A and Fasihi M 2019 *eXPRESS Polymer Letters*.vol p 429-442
[6] Najib N, Ariff Z, Manan N, Bakar A, and Sipaut C 2009 *Journal of Physical Science*.vol p 13-25
[7] Charoeythornkhajhornchai P, Samthong C, Boonkerd K, and Somwangthanaroj A 2017 *Journal of Cellular Plastics*.vol p 287-303
[8] Stange J and Münstedt H 2006 *Journal of cellular plastics*.vol p 445-467
[9] Bledzki A K and Faruk O 2006 *Journal of Cellular Plastics*.vol p 63-76
[10] Reichelt N, Stadlbauer M, Folland R, Park C B, and Wang J 2003 *Cell Polym*.vol p 315-327
[11] Jeoung S K, Hwang Y J, Lee H W, Kwak S B, Han I-S, and Ha J U 2016 *AIP Conference Proceedings*.vol p 100002

IOP Conf. Series: Materials Science and Engineering **943** (2020) 012003 doi:10.1088/1757-899X/943/1/012003

Analysis of Wind Energy Possibilities of Different Sites in India.

P Sarmah [1] , M Z Hussain[2], S Khan [3]
[1]Department of Mechanical Engg., Dibrugarh University, PIN - 786004, India.
[2,3]Department of Mechanical Engg., Jamia Millia Islamia, New Delhi, PIN 110025,India.
E-mail: pksnit07@gmail.com

Abstract. Wind energy is one of the most cost effective renewable energy resources. The purpose of this study is to investigate wind energy potential at different sites in India from year 2008-2017 for six month basis (March to August). The average wind speed data for different sites indicates that the average wind speed for site Tiruchirappalli is highest. The normal probability plot shows that wind speed data are normally distributed except for the site Kolkata. Weibull cumulative probability plot shows that the higher probability of occurrence of wind energy was at higher average wind speed for each site. Weibull probability density function shows that the probability density function is minimum at higher average wind speed. It is also found that the scale parameter is an important parameter compare to shape parameter that affect wind speed. The analysis of power density shows that the average power density for site Tiruchirappalli is highest and lowest for site Agartala. The analysis also shows that higher wind energy possibility is in Tiruchirappalli followed by Coimbatore.

1. Introduction

Energy is very essential input for the growth of human development as it is directly correalated to the consumption of energy. The need of renewable energy to satisfy human needs and economic development of a country is increasing rapidly day by day [1]. Due to industrialization, development of technology and increasing population of a country, fossil fuel reserves got depleted and increase environmental pollution. The cost of fossil fuels is also increasing trends in recent years [2]. Therefore, all country needs to search for other sources of energy to overcome the shortage of conventional energy. Renewable energy can serve human need of energy. Renewable energy is energy which is collected from resources like bio-mass, sunlight, wind, tides, and geothermal heat. Accordingly, REN21's 2017 report, renewable energy resources contribute 19.3% to global energy consumption and 2.2% energy is collected from resources: wind, solar, geothermal, and biomass [3]. In India, the demand for energy has increased for satisfying people needs and India has ranked third for the consumption of energy after China and United States, and also second rank for coal consumption after China [4]. Wind energy is one of the most cost effective non-conventional energy sources which are non-polluting, cost effective, clean, and never-ending resource compared to conventional energy resources which can satisfy energy demand. Therefore the transformation of wind power to electrical energy has been developing very rapidly [5]. Wind energy met ~4% of global

IOP Conf. Series: Materials Science and Engineering **943** (2020) 012003 doi:10.1088/1757-899X/943/1/012003

energy demand in 2015. In recent years, wind power generation in India has significantly increasing. The wind generation capacity for state Tamil Nadu, India is ~29% of total capacity and a total installed wind capacity is 8,197 MW, followed by Gujarat [6].

Temitope R Ayodele et al. investigated wind power potential of Port Elizabeth using statistical Weibull parameters [7]. The results show that the average wind power density is maximum during spring season and lowest during autumn season. Hussain et al. studied the wind energy potential at different cities Assam, India and found that most suitable city for wind energy generation was Dibrugarh [8]. Kamran Abbas et al. developed best fitting distribution of wind speed in Islamabad on a daily basis from year 2001-2003 using different probability models: [9]. They found that Gamma, Lognormal and Burr distributions was more consistent compare to other models considered. Serap AKCAN compared time series methods for wind energy at nine stations in Turkey [10]. He found that the country Turkey was rich wind energy potential to compensate energy needs. Mahyoub H. Al Buhairi investigated possibility of renewable energy in country Yemen using statistical methods [11]. He found that the Weibull distribution showed better power density estimations compare to Rayleigh distribution. Djamal Hissein Didane et al. studied monthly and yearly wind speed data, and wind power density for different stations in Chad by using Weibull function [12]. The yearly averaged wind data revealed that the most suitable metrological station in Chad for generation of wind energy. Rajat Gupta and Agnimitra Biswas investigated the wind energy potential of district Silchar, Assam, India over a period of five years by using Weibull and Rayleigh's distribution functions [13]. The average wind velocity data showed that the region Silchar was considered as low wind power generation region. Sengupta et al. studied wind energy potential of Silchar (Assam, India) by using different models over a period of four years from 2008-2011 [14]. They found that the Weibull model best fitted compare to Rayleigh's model to measure the probability density distribution.

Therefore, the geographical position of wind speed and distribution of wind speed analysis by using the statistical model is essential for selection of a particular region for wind energy generation. So; in this article, an investigation has been made for possibility of wind energy potential for different region in India. The selected sites in this study were Dibrugarh Airport, Kolkata, Agartala, Tiruchirappalli, Coimbatore and Jai Prakash Narayan airport, Patna.

2. Location of Sites

The meteorological sites used in our study are Dibrugarh Airport (Dibrugarh Assam); Kolkata (West Bengal); Agartala (Tripura); Tiruchirappalli (Tamil Nadu); Coimbatore (Tamil Nadu); Jay Prakash Narayan Airport (Bihar). Dibrugarh Airport is located at a distance of ~15 km from Dibrugarh city, Assam, India. The airport is located at 110 m above sea level. The latitude and longitude of site Dibrugarh Airport is 27.4728° N and 94.9120° E respectively. The next site has been selected Kolkata which is known as capital of West Bengal, India. The latitude and longitude of site Kolkata are 22°33'45.47"N and 88°21'46.94"E respectively. Agartala, the capital of Tripura located at 23° 30' 0" N, 91° 30' 0" E. The geographic coordinates of Tiruchirappalli having latitude: 10°48′55″ N, longitude: 78°41′47″ E and 79 m elevation above sea level: 79 m. Coimbatore is a major site of Tamil Nadu, India. The latitude of Coimbatore is 11.0168° N, and the longitude is 76.9558° E. Coimbatore site elevation is 420 meters height above sea level. Jay Prakash Narayan Airport is located in Patna, the state capital of Bihar in India. The Latitude and longitude of Jai Prakash Narayan airport Patna is 25.5947° N and 85.0908° E respectively.

3. Data Collection

The necessary wind speed data for different sites for the month of March to August from year 2008-2017 have been collected from website www.weatheronline.in. The average wind speed data for six months for sites is shown in table 1.

IOP Conf. Series: Materials Science and Engineering **943** (2020) 012003 doi:10.1088/1757-899X/943/1/012003

Table 1. Wind Speed Data (m/s), different sites of India

Site no.	Site	2008	2009	2010	2011	2012
1	Dibrugarh	1.59	1.36	2.06	1.50	1.74
2	Kolkata	0.54	0.85	0.93	0.84	0.82
3	Agartala	1.38	1.34	1.57	1.09	1.15
4	Tiruchirappalli	4.78	4.82	3.43	3.39	4.03
5	coimbatore	3.19	3.31	3.38	4.06	4.35
6	Patna	1.97	2.30	2.47	2.78	1.95
Site no.	**Site**	**2013**	**2014**	**2015**	**2016**	**2017**
1	Dibrugarh	1.82	1.78	1.53	2.18	2.41
2	kolkata	0.81	2.15	2.38	2.48	2.64
3	Agartala	1.17	1.17	1.01	1.03	0.98
4	Tiruchirappalli	4.37	4.07	3.43	3.61	3.28
5	coimbatore	4.14	4.05	3.36	3.22	3.85
6	Patna	2.93	2.48	2.35	2.99	2.24

4. Results and Discussion

4.1 Analysis of Average Wind Speed variation

The variation of average wind speed (six month basis) of different sites in India from year 2008 to 2017 is shown in figure 1. It has been observed from figure 1 that the average wind speed is maximum for the site Tiruchirappalli is highest in 2009, followed by site Coimbatore in 2012. The lowest wind speed has been observed in the site Kolkata in 2008 followed by site Agartala in the year 2017. The highest and lowest average wind speed for (six months average) for site Tiruchirappalli is 4.82 m/s in the year 2009 and 3.28 m/s in the year 2017 respectively. For site Coimbatore 4.35 m/s in 2012 and 3.19 m/s in 2008, for site Patna 2.99 m/s in 2016 and 1.95 m/s in the year 2012 respectively. The monthly average wind speed in all the years revealed that the wind speed is maximum for site Tiruchirappalli~5.78 m/s in july and minimum for site Agartala in march~0.9916 m/s. The average wind speed indicates that the site Tiruchirappalli is best suitable for wind power generation followed by site Coimbatore. The grand average wind speed in years considered shows that the grand average wind speed is highest 4.39 m/s for site Tiruchirappalli, followed by coimatore 4.12 m/s. The lowest grand average wind speed for the years considered is found in the city Agartala about 1.14 m/s. The analysis also revealed that the site Tiruchirappalli is best suitable for wind power generation followed by site Coimbatore.

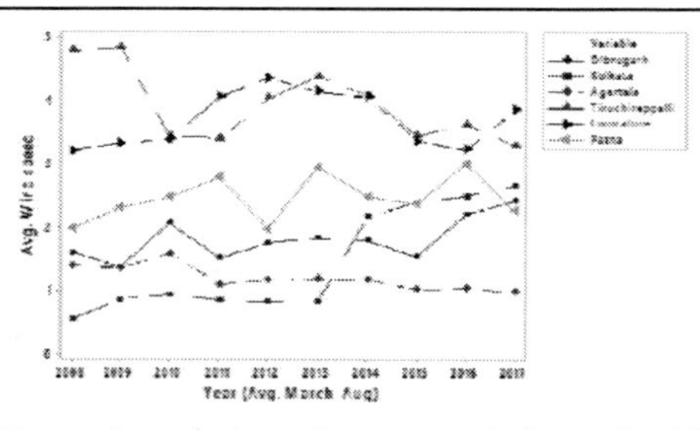

Figure 1. variation of average wind speed with different sites from year 2008-2017.

4.2 Frequency distribution against Average. Wind Speed

The frequency distribution of averages wind speed from the year 2008-2017 (for six months/year) as shown in figure 2. The figure revealed that the frequency of average wind speed for city Dibrugarh lies in between 1.3 m/s to 2.5 m/s and maximum frequency lies in between 1.5 m/s-1.9 m/s. For site Tiruchirappalli, the frequency distribution lies in between average wind speed 3.1 m/s- 4.9 m/s and maximum frequency of three for average wind speed lies 3.3 m/s-3.5m/s. The mean and standard deviation of wind speed for 10 years (March-August) for site Tiruchirappalli are 3.926m/s and 0.5825 respectively, which are maximum values compare to other sites considered. The mean wind speed and standard deviation for site Agartala are minimum compare to other sites. The frequency of average wind speed for each year six month basis is minimum at higher wind speed for all sites from year 2008-2017.

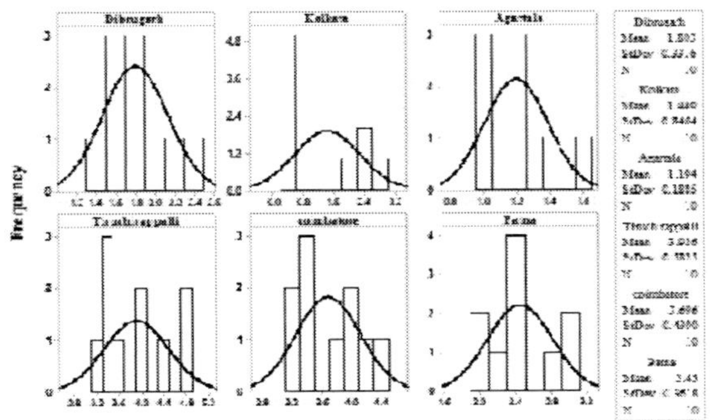

Figure 2. Frequency Disribution of Avgerage Wind Speed for sites, 2008-2017(March-August)

4.3 Normal probability plot

Figure 3 shows probability plot of average wind speed for considered sites from 2008-2017. It is observed from the figure that the data are normally distributed for all sites except the site Kolkata at 95% confidence interval since p-value >.05. The figure shows that the speed variation is minimum for site Agartala as standard deviation is minimm followed by site Dibrugarh. The figure also revealed that the variation of wind speed is maximum for site Kolkata followed by site Tiruchirappalli and Coimbatore.

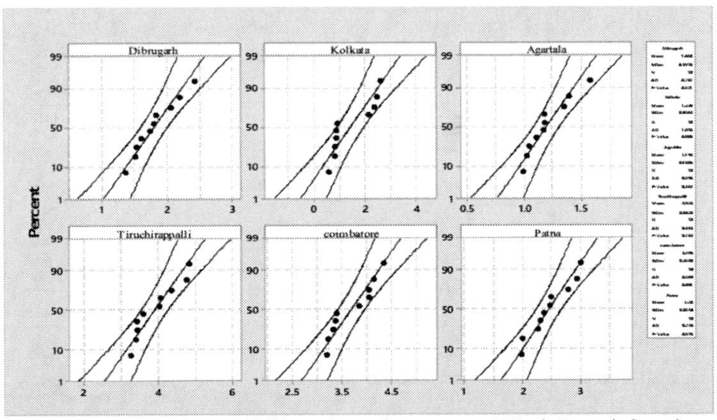

Figure 3. Normal probability plot of Average Wind Speed for sites, 2008-2017(March-August)

4.4 Weibull cumulative probability plot

Probability distributions and functions are used to analyze wind speed data. In the articles, Weibull plot is widely used for this purpose [15-16]. Table 2 shows that the cumulative probability of average wind speeds from year 2008-2017. The table 2 revealed that the higher probability of occurrence of wind power at higher average wind speed for site Dibrgarh. The occurrence of maximum average wind speed (2.41667 m/s) for site Dibrugarh is 97.6071% in 2017. Similarly, the occurrence of maximum average wind speed for site 95.0277% in the year 2009.

Table 2. Weibull cumulative probability of different sites in India.

Dibrugarh		kolkata	
Weibull with shape = 5.981 and scale = 1.939		Weibull with shape = 1.947 and scale = 1.646	
x	P(X ≤ x)	x	P(X ≤ x)
1.59722	0.269162	0.5463	0.110223
1.36574	0.115669	0.85648	0.244438
2.06019	0.762371	0.93519	0.282958
1.50926	0.200246	0.84722	0.23999
1.74537	0.413158	0.82407	0.22896
1.82407	0.500372	0.81481	0.224585
1.78704	0.458695	2.15278	0.814815
1.53241	0.217099	2.38426	0.872214
2.18981	0.873816	2.48148	0.891813
2.41667	0.976071	2.64352	0.919166
Tiruchirappalli		coimbatore	
Weibull with shape = 7.543 and scale = 4.174		Weibull with shape = 9.942 and scale = 3.885	
x	P(X ≤ x)	x	P(X ≤ x)
4.78241	0.938635	3.19907	0.134937
4.8287	0.950277	3.31944	0.188824
3.43056	0.203664	3.38889	0.22671
3.39815	0.191034	4.06019	0.787832
4.03241	0.53736	4.35648	0.955972
4.375	0.759693	4.14352	0.850043
4.07407	0.565237	4.05093	0.78032
3.43981	0.20738	3.36111	0.210931
3.61574	0.287215	3.22685	0.146119
3.28241	0.150608	3.85185	0.600817
Patna		Agartala	
Weibull with shape = 7.856 and scale = 2.602		Weibull with shape = 6.653 and scale = 1.274	
x	P(X ≤ x)	x	P(X ≤ x)
1.97685	0.109064	1.38889	0.830708
2.30093	0.316539	1.34259	0.757679

2.47222	0.487791	1.57407	0.983165
2.78241	0.816075	1.09722	0.309375
1.95833	0.101701	1.15741	0.410251
2.93519	0.923988	1.17593	0.443938
2.48148	0.497903	1.1713	0.43542
2.35648	0.368111	1.01389	0.196564
2.99074	0.949502	1.03241	0.218754
2.24537	0.269552	0.98148	0.16165

The cumulative probability plot for different sites from year 2008-2017 is shown in figure 4(a)-(f). The Weibull shape parameter (k) and scale parameter (c) for different sites were evaluated by using MINITAB version 18 as shown in figure 3. It is observed from figures that the two parameters are high at higher average wind speed. The scale parameter for site Tiruchirappalli is maximum about 4.174 m/s, followed by site Coimbatore. The minimum value of scale parameter for site Agartala is 1.646 m/s. From the analysis it is observed that the scale parameter is an important parameter compare to shape parameter that affect wind speed.

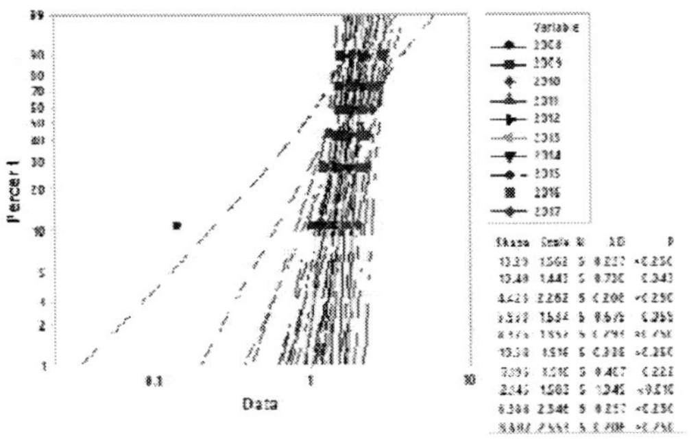

(a) Weibull Probability plot for Dibrugarh from year 2008-2017

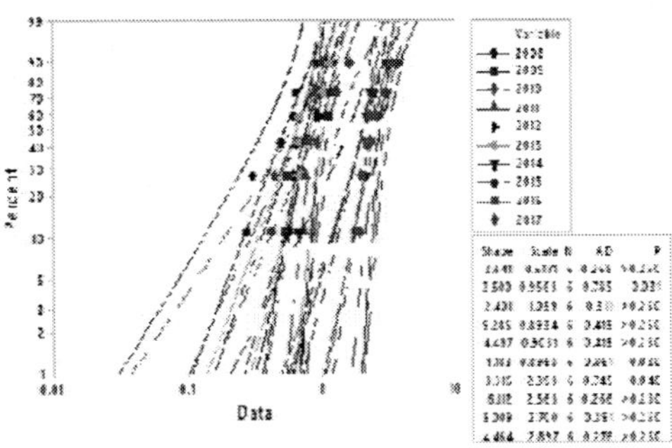

(b) Weibull Probability plot for Kolkata from year 2008-2017

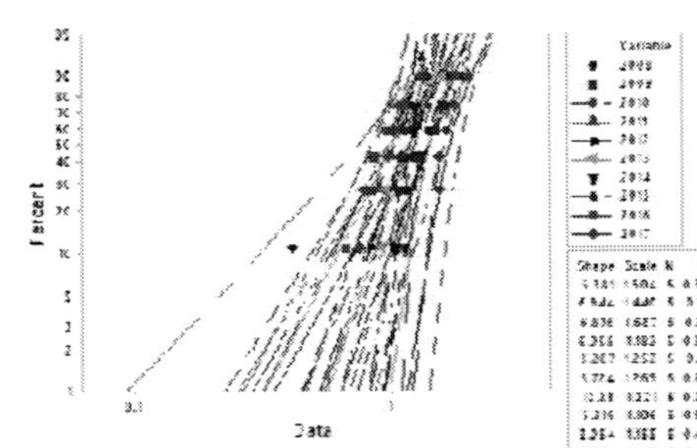

(c) Weibull Probability plot for Agartala from year 2008-2017

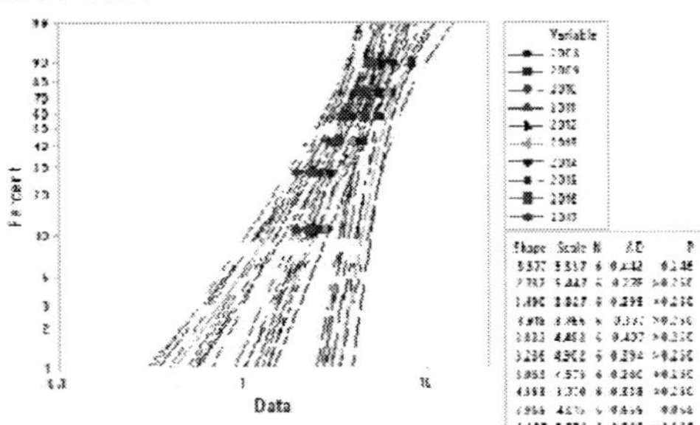

(d) Weibull Probability plot for Tiruchirappalli from year 2008-2017

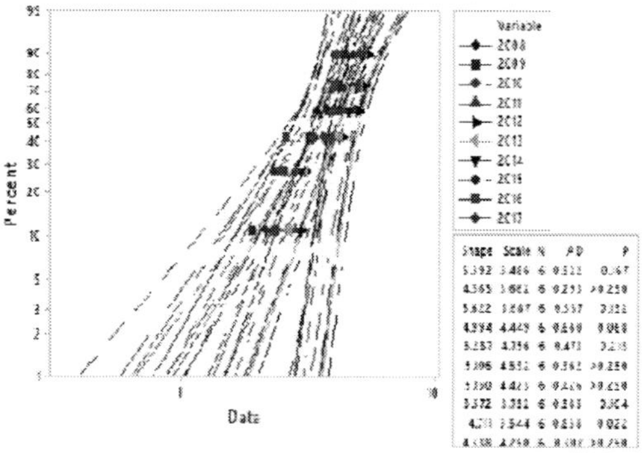

(e) Weibull Probability plot for Coimbatore from year 2008-2017

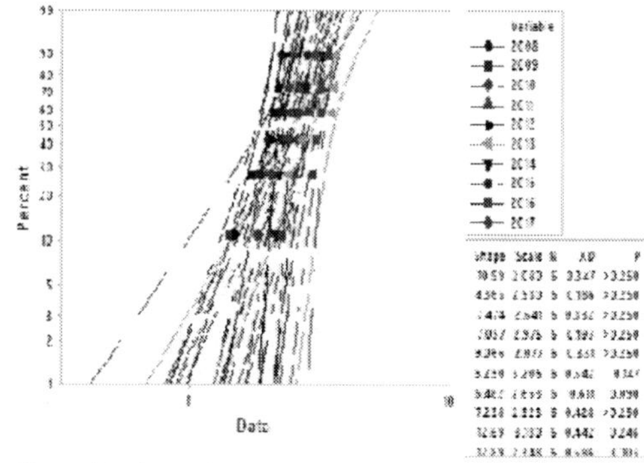

(f) Weibull Probability plot for Patna from year 2008-2017

Figure. 4(a)-(f) Weibull Probability plot for different sites

4.5 Probability density function plot

The Weibull probability density function plot for different sites from year 208-2017 is shown in figure 5. From the figure 4, it is observed that the probability density function is minimum for site Tiruchirappalli at higher average wind speed and maximum for site Agartala at lower average wind speed. This indicates that at higher average wind speed, probability density function is minimum.

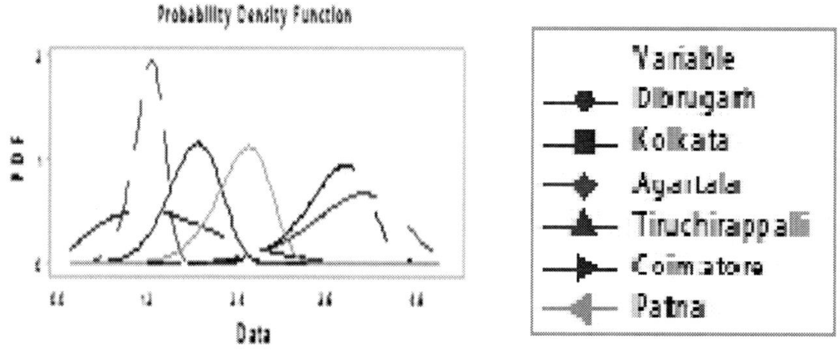

Figure. 5 Weibull Probability plot for different sites from year 2008-2017

4.6 Wind Power Density

The wind power density based on Weibull distribution function is calculated using the following equation [17]:

$$\frac{P}{A}=\frac{1}{2}\rho c^{3}\Gamma\left(\frac{k+3}{k}\right)$$

Where,

k is Weibull shape parameter

c is Weibull scale parameter

ρ is density of air =1.225 kg/m^3

Table 3 shows the average power density for different sites for six month basis. The average wind power density for site Dibrugarh is highest in the year 2017 and equal to 9.084104 W/m^2. The average wind power density for site Tiruchirappalli is 103.493 W/m^2 in the year 2009. The average wind power density for site Coimbatore is 58.6611W/m^2 in the year 2012. The grand all-year average power density for site Tiruchirappalli is highest (52.66388 W/m^2), followed by Coimbatore (38.00481 W/m^2), Patna (10.428 W/m^2), Kolkata (4.413379 W/m^2), Dibrugarh (4.38688 W/m^2) and Agartala (1.24900 W/m^2). The variation of Average Wind Power Density with respect to years, 2008-2017 as shown in figure 6 and the figure revealed that the most suitable site for wind power generation is site Tiruchirappalli, followed by Coimbatore, Jai Prakash Narayan airport Patna, Kolkata, Dibrugarh, Agartala.

Table 3. Average Wind Power Density for different sites of India

Year	Dibrugarh	Kolkata	Agartala	Tiruchirappalli	Coimbatore	Patna
2008	2.563727	0.153338	1.85338	87.73091	23.07605	4.97145
2009	1.655031	0.503871	1.62038	103.493	27.26018	8.98949
2010	6.41341	0.82394	2.60511	32.56471	27.09907	10.0098
2011	2.368662	0.390568	0.89626	30.09393	48.19865	14.2918
2012	3.463323	0.410012	1.07013	51.11856	58.6611	4.90227
2013	3.873912	0.402221	1.11054	70.07961	50.86569	17.9757
2014	3.819065	8.086015	1.01168	58.15169	47.25627	10.1287
2015	3.623622	9.136266	0.73826	29.58649	30.49365	8.72109
2016	7.003945	10.77062	0.9368	36.49443	24.48805	17.0785
2017	9.084104	13.45694	0.64749	27.32547	42.64933	7.21613
Average	4.38688	4.413379	1.249	52.66388	38.00481	10.428

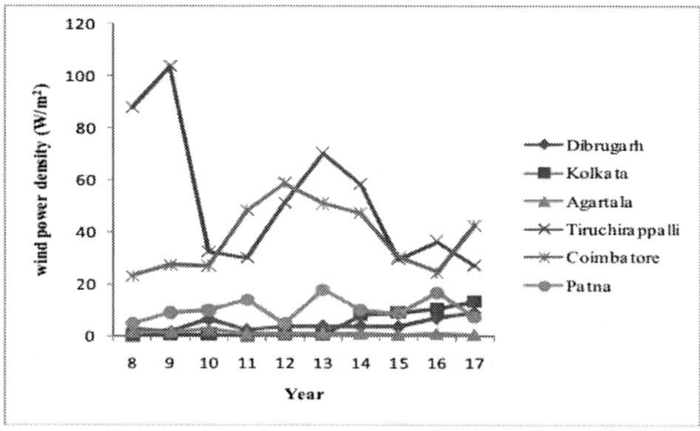

Figure 6. Variation of Average Wind Power Density vs. years 2008-2017.

5. Conclusion

In this article an investigation has been made to evaluate the possibility of wind energy potential at different sites in India by using statistical tools. The variation of average wind speed of different sites in India revelled that the average wind speed for the site Tiruchirappalli is the highest compare to other sites considered and lowest average wind speed is for site Agartala. The frequency of average wind speed for each year is minimum at higher wind speed for all sites. The normal probability plot shows that the wind data are normally distributed except for site Kolkata and the variation of average wind speed is maximum for site Kolkata. Weibull cumulative probability plot revealed that the higher probability of occurrence of wind power at higher average wind speed. The Weibull probability density function shows that at higher average wind speed, probability density function is minimum. It is also found that the scale parameter is an important parameter compare to shape parameter that affect wind speed. The most important factor of the wind energy is the wind speed. The six-month average power density from year 2008-2017 for site Tiruchirappalli is highest (52.66388 W/m^2), followed by Coimbatore (38.00481 W/m^2) and minimum for site Agartala (1.24900 W/m^2). Therefore, high wind power generation possibility is in the region Tiruchirappalli followed by Coimbatore. So, an accurate evaluation of probability distribution of wind speed data is necessary to find wind energy possibility at different sites.

References

[1] Owusu, P.A. and Asumadu-Sarkodie, S., 2016. A review of renewable energy sources, sustainability issues and climate change mitigation. Cogent Engineering, 3(1), p.1167990.

[2] Patel, N.K. and Shah, S.N., 2015. Biodiesel from Plant Oils. In Food, energy, and water (pp. 277-307). Elsevier.

[3] Renewables 2017 Global Status Report [Electronic resource]. URL: http://www.ren21.net/wpcontent/uploads/2017/06/178399_GSR_2017_Full_Report_0621_Opt.pdf (accessed: 02.07.2019)

[4] Website of U.S. Energy information Administration Country Analysis Brief: India, 2016 [Electronic resource] (accessed: URL: http://https://www.eia.gov/beta/international/analysis.php?iso=IND&src=home-b6.

[5] Aras, H., Yilmaz, V. and Celik, H.E., 2003, July. Estimation of Monthly Wind Speeds of Eskişehir, Turkey. In The First International Exergy, Energy and Environment Symposium, Hotel Princess, Izmir, Turkey , pp. 13-17.

[6] Website of the Indian wind turbine manufacturer's association. http://www.indianwindpower.com/pdf/Installed-Wind-Capacity.pdf.

[7] Ayodele, T.R., Jimoh, A.A., Munda, J.L. and Agee, J.T., 2012. Statistical analysis of wind speed and wind power potential of Port Elizabeth using Weibull parameters. Journal of Energy in Southern Africa, 23(2), pp.30-38.

[8] Hussain, M.Z., Roy, D.K., Khan, S., Sharma, P.K. and Talukdar, R., 2018. Wind Energy Potential at Different Cities of Assam Using Statistical Models. International Journal, 6(1), pp.38-43.

[9] Abbas, K., Alamgir, K.S., Ali, A., Khan, D.M. and Khalil, U., 2012. Statistical analysis of wind speed data in Pakistan. World Applied Sciences Journal, 18(11), pp.1533-1539.

[10] AKCAN, S., 2017. Wind speed forecasting using time series analysis methods. Cukurova Univ J Fac Eng Archit, 32(2), pp.161-172.

[11] Al Buhairi, M.H., 2006. A statistical analysis of wind speed data and an assessment of wind energy potential in Taiz-Yemen. Ass. Univ. Bull. Environ. Res, 9(2), pp.21-33.

[12] Didane, D.H., Rosly, N., Zulkafli, M.F. and Shamsudin, S.S., 2017. Evaluation of wind energy potential as a power generation source in Chad. International Journal of Rotating Machinery, 2017.

[13] Gupta, R. and Biswas, A., 2010. Wind data analysis of Silchar (Assam, India) by Rayleighs and Weibull methods. Journal of Mechanical Engineering Research, 2(1), pp.010-024.

[14] Sengupta A.R., Biswas A., Gupta R. 2016 An Analysis of Wind Energy Potential of Silchar (Assam, India) By using different models. IJEMS, 7 **(2)** ,100-107.

[15] Tuller, S.E. and Brett, A.C., 1984. The characteristics of wind velocity that favor the fitting of a Weibull distribution in wind speed analysis. Journal of Climate and Applied Meteorology, 23(1), pp.124-134.)

[16] Xie, M., Tang, Y. and Goh, T.N., 2002. A modified Weibull extension with bathtub-shaped failure rate function. Reliability Engineering & System Safety, 76(3), pp.279-285.

[17] Keyhani, A., Ghasemi-Varnamkhasti, M., Khanali, M. and Abbaszadeh, R., 2010. An assessment of wind energy potential as a power generation source in the capital of Iran, Tehran. Energy, 35(1), pp.188-201.

IOP Conf. Series: Materials Science and Engineering **943** (2020) 012004 doi:10.1088/1757-899X/943/1/012004

CCSDS-MHC on Raspberry Pi for Lossless Hyperspectral Image Compression

N A A Samah[1], N R M Noor[1,*], E A Bakar[2] and M K M Desa[1]

[1]School of Electrical & Electronic Engineering, Universiti Sains Malaysia, Engineering Campus, 14300, Penang, Malaysia
[2]School of Aerospace Engineering, Universiti Sains Malaysia, Engineering Campus, 14300, Penang, Malaysia

Email: nrmn@usm.my

Abstract. This paper is about Consultative Committee for Space Data System for Lossless Multispectral and Hyperspectral Image Compression (CCSDS-MHC) algorithm that is implemented on Raspberry Pi 3 Model B+ system using Open Multi-Processing (OpenMP). CCSDS-MHC algorithm along with the full prediction mode is opted due to its best compression ratio (CR) performance. The issue of Hyperspectral Image Compression is the loss of data when compress, thus CCSDS algorithm is used in this research. Besides, with current technologies that require low-power but high-performance devices, Raspberry Pi was chosen to be tested in terms of its performances while being compared to other platforms. AVIRIS (airborne) and Hyperion (spaceborne) are used to test the performance of the system. OpenMP is introduced to simplify the computational operation through parallelization to take advantage of the multi-core architecture of the hardware system. In term of execution time, CCSDS-MHC algorithm when parallelize using OpenMP gave the best performances about 69.4% for AVIRIS 1997 dataset, 69.3% for AVIRIS 2006 dataset and 67.7% for Hyperion dataset. The execution time of performance CCSDS-MHC in Raspberry Pi 3 Model B+ comparing with other different multicore platform is validate and the mean parallelize is measured. The result of comparison proves that faster performing task gives the best result due to higher speed of CPU-cores performances. From this research, it is proven that Raspberry Pi which is a low-power embedded platform able to compress the hyperspectral images at optimized speed with the implementation of OpenMP through CCSDS algorithm. Therefore, this research will be useful for further studies in lossless of hyperspectral images.

1. Introduction

In this era, our planet is continuously being observed by airborne and satellite sensors that acquires tons of data to be processed and analyzed daily. One of the types of remote sensing is hyperspectral sensing, also referred to as hyperspectral imaging. Hyperspectral imaging has been widely employed in real life applications industrial, biomedical, and biometric applications such as face detection and recognition [1].

Hyperspectral images are represented in a forms of three-dimensional (x, y, λ) hyperspectral data cube for processing and analysis, where x and y represent two spatial dimensions of the scene, and λ represents the spectral dimension [2]. The hyperspectral cubes are generated from airborne sensors like

NASA's Airborne Visible/Infrared Imaging Spectrometer (AVIRIS), or from satellites like NASA's EO-1 with its hyperspectral instrument Hyperion.

The images need to be transmitted to the ground from the sensors for specific applications. However, the size of these images is typically enormous in quantity due to the bulky number of capturing wavelengths and the length of compressed images that vary from image to image. Thus, the efficient compression techniques are required to fit the available bandwidth for reducing the transmission time [3].

Due to the limitation that impacts the loss of data, there is a need to improve the algorithm in order to compress a lossless image. Therefore, CCSDS-MHC algorithm is developed. The CCSDS-MHC algorithm is a set of reformulations at the prediction stage [4]. However, in this project all parameters of hyperspectral image compression is taken from the papers of international conference of data compression [5] and has been set out into specific declaration.

This research focuses on the lossless compression of AVIRIS images and Hyperion images by using CCSDS-MHC algorithm. Since the aim of the project is a lossless compression of hyperspectral imaging using CCSDS-MHC algorithm, only full predictions will be used.

Dividing the hyperspectral image into blocks and efficiently distributing the prediction and encoding workload to independent tasks running in parallel can improve the compression speed significantly [6]. Therefore, the parallelization of OpenMP yields a large impact to the program execution time [7].

This research is related to lossless hyperspectral image where the need only lies in wanting to compress the image on the satellite which could be time saving with low memory performances that is relatable by the article [8]. Embedded systems are often used in the recent trends. Embedded system has a combination of a computer's hardware and software function where it is capable of being programmable to be designed in a specific manor or function. Research in [9] proves that FPGA is one of the low-memory and high-performance architectures in the implementation of CCSDS. Graphical Processing Unit (GPU) implementation has been performed such as in [10] but can only be suitable for terrestrial application due to high power consumption of a GPU [11] and low power requirement of satellite-based remote sensing [12]. A compact low-cost and portable spectral imaging system for general purposes through Raspberry Pi has been implemented [13].

Based on the research, Raspberry Pi 3 Model B+ is one of the best low-cost, low-power consumption embedded platforms architecture to implement CCSDS-MHC algorithm for lossless hyperspectral image compression.

2. Methodology

In this research flow, there are nine main part subsystem that combines together to form into one system. The main part for each block in the flowchart Figure 1 will be discussed in detail in the following sections.

The hardware setup consists of four primarily important ports namely micro SD Card port, Ethernet port, HDMI port, and micro USB port that are being used as a power supply. HDMI cable, micro USB cable, Local Area Network (LAN) cable and micro SD Card with 32GB memory were prepared for the connection of the whole hardware setup. Raspberry Pi 3 Model B+ board is used as the main device to implement the CCSDS-MHC algorithm in C language.

For software development, Codeblocks Software is used to operate the CCSDS-MHC algorithm of C language and is run under Raspbian operating system (OS). To establish an interface connection between Mac OS and Raspbian OS, VNC Viewer is used through internet network to control the desktop of the computer in real-time between laptops or another computer as remote control.

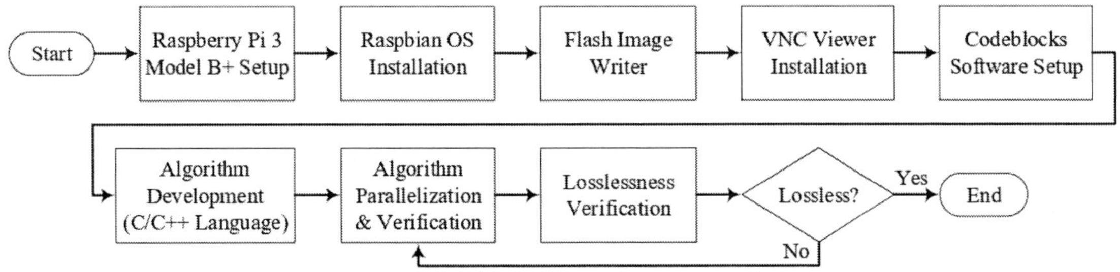

Figure 1. Flowchart of the overall process

2.1 CCSDS-MHC Algorithm

In CCSDS-MHC algorithm, hyperspectral image which consist of three-dimensional array of integer is set as input to the compressor in an image. Due to the differences type of image content, the length of compressed images will vary from image to image which means the compressed image is variable-length. Therefore, to compress the hyperspectral images without any loss of information there are two types of functional parts named predictor and encoder as shown in Figure 2.

In the predictor, adaptive linear prediction method is used to predict the value of each image sample that was performed sequentially in a single pass. The prediction residual which is the differences between the predicted and actual sample values will be mapped on to an unsigned integer that can be represented using the same number of bits as the input data sample.

There are two ways to perform the prediction of the neighbouring samples which is by using *column-oriented* (i.e. reduced-prediction mode) or *neighbour-oriented* local sums (i.e. full-prediction mode, FP) that can be applied. The use of reduced-prediction mode in a combination with column-oriented local sums tends to yield smaller compressed image data volumes for raw (uncalibrated) input images from push-broom imagers. The use of full-prediction mode in a combination with neighbour-oriented local sums tends to yield smaller compressed image data volumes for whiskbroom imagers, frame imagers, and calibrated imagery.

In encoder, a compressed image consists of a header that encodes image and compression parameters followed by a body, that produced by an entropy coder which losslessly encodes the mapped prediction residuals. Entropy coder parameter are adaptively adjusted during this process to adapt changes of the mapped prediction residuals. Encoder encodes the mapped prediction residual by Sample or Block based adaptive entropy coding. The block-adaptive encoding approach was included as an option in the standard so that implementers could take advantage of existing space-qualified hardware implementations of this encoder.

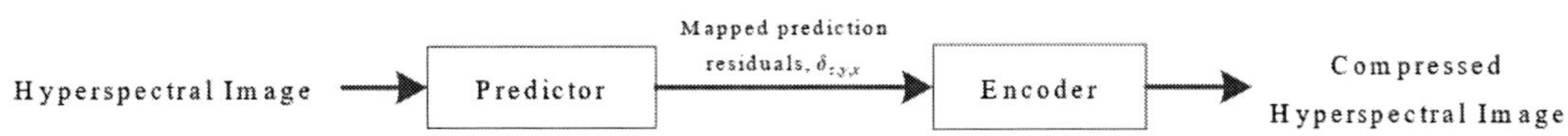

Figure 2. CCSDS-MHC Compressor Schematic

2.2 CCSDS-MHC Coding

In CCSDS-MHC, input hyperspectral image is chosen at the first place with a number of datasets from AVIRIS and Hyperion images. Once the image has been chosen, the code will compress the image through the encoded binary image form. The CCSDS-MHC algorithm focus only on the performance of an OpenMP parallelization in compression part. While the decompression part is taken to check the losslessness in hyperspectral image. However, Raspberry Pi can only perform compression operation due to memory performance whereby decompression part is inapt but can be performed with other multicore system. Overall data flow of Hyperspectral Image Compression is shown in Figure 3.

IOP Conf. Series: Materials Science and Engineering **943** (2020) 012004 doi:10.1088/1757-899X/943/1/012004

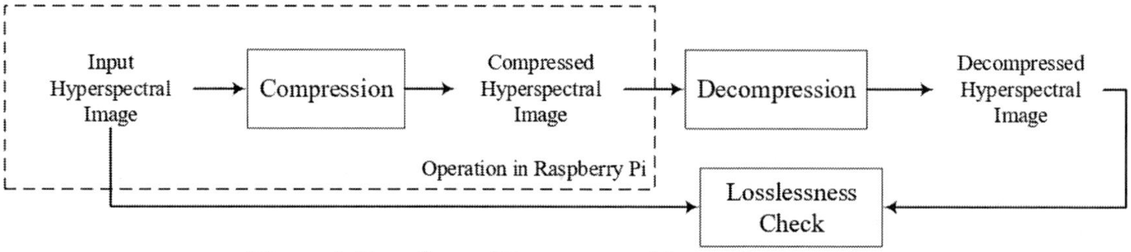

Figure 3. Data flow of Hyperspectral Image Compression

2.3 Coding Structure

In Listing 1, there is eight important function in developing an OpenMP code:

1. #pragma omp parallel for
2. prepareNPreviousBands
3. ec_init
4. prepareBands
5. predictor_compress
6. codeSample
7. ecupdate
8. numBitsWritten

```
1.  #pragma omp parallel for schedule(static,band/omp_get_max_threads())
    reduction (+:numBitsWritten)
2.  for(z = 0; z < band; z++){
3.    prepareNPreviousBands(z, omp_get_max_threads(), ...);
4.    ec_init(z, ...);
5.    prepareBands(z, ...);
6.    for(y = 0; y < row; y++){
7.      for(x = 0; x < col; x++){
8.        outputMappedResidual[x][y][z] = predictor_compress(x, y, z, ...);
9.        numBitsWritten+=codeSample(outputMappedResidual[x][y][z], x, y, z, ...);
10.       ecupdate(outputMappedResidual[x][y][z], x, y, z, ...);
11.   }}}
12.  totalNumBitsWritten += numBitsWritten;
```

Listing 1. CCSDS-MHC Parallelization

#pragma omp for only delegates portions of the loop for different threads in the current team. A team is the group of threads executing the program. At program start, the team consists only of a single member. To create a new team of threads, the parallel keyword must be specified. Therefore, syntax of #pragma omp parallel for in line 1 is used as the compiler directive to connect the communication to the compiler, making the next for loop (line 2) a parallel region.

For schedule clause, the directive parallel for schedule is used to assign contiguous range of interaction that are normally called chunks in each processor. There are four type of scheduling which is in dynamic, guided, runtime and static. In this project, schedule clause for static is used, specifying the size of the chunk in the processor. Therefore, the syntax of #pragma omp parallel for schedule (static,band/omp_get_max_threads) is declared.

prepareBands is the function that prepares the prediction neighbourhood by bands shifting. Before commencing the execution, the number of bands to perform in this operation is already being specified in the declaration section. The execution time for prepareNPreviousBands is too long compared to prepareBands. Therefore, in this project, schedule clause is performed to reduce the time execution that can combine both functions together at one time.

ec_init is the implementation that supports the accumulator initialize constant. In this project, the accumulator constant is specified in the declaration sections. predictor_compress is declared in this function to compress the data read from the prepareBands. predictor_compress function consists of mathematical calculation to compress the data so that the execution time will be lower.

On CCSDS-MHC, `codeSample` is declared from the number of prediction bands where the data was taken from the compression bands. Data samples have a fixed-size of dynamic range of D bits where D should be the integer in the range $2 < D < 16$ [14]. If the image data is read and the dynamic range is too small according to the range, error sign will appear in the program terminal.

Along the process to compress the bands, `ecupdate` is important to read back the fixed size of original sample data. `ecupdate` is in charge of rescaling the counter size of compression bands. Lastly, `numBitsWritten` is the data update for the image of compressed size after performing some mathematical calculation on a dynamic range value. All the important functions in developing an OpenMP code are as shown in Listing 1.

2.4 Execution Time

In this project, the execution time is used to measure and analyse the performance of CCSDS-MHC. The execution time consist of CPU time and real-world time. The I/O execution time which is known as CPU time is ignored because the focus on this research is only on the time for reading/writing the hyperspectral/compressed image from/to the memory which is known as real-world time.

For the execution time performance parameter, it is simply evaluated in seconds. When the computer is multitasking, the real time for each program is determined separately, and depends on how the microprocessor allocates resources among the programs. Therefore, there is a difference between real-world time and CPU time in an embedded system. To get the real time to be used by a task within a C application, `omp_get_wtime` function is needed to measure a multi-threaded program as shown in Figure 4.

The `omp_get_wtime` function returns a double precision value equal to the number of seconds since the initial value of the operating system is the real-time clock. The initial value is guaranteed unchanging during the execution of the program. The value returned by the `omp_get_wtime` function is not consistent across all threads in the team [15].

2.5 Compression Ratio

Compression ratio is a term used to evaluate the reduction in data-representation size produced by CCSDS-MHC algorithm. Compression ratio is calculated as shown in Table 1 and is defined in equation (1).

$$Compression\ Ratio, CR = \frac{Size\ of\ Original\ Image}{Size\ of\ Compressed\ Image} \quad (1)$$

The compression ratio in Table 1 is used as guide to verify the correct data of parallelization in CCSDS-MHC for hyperspectral image compression. The lossless compression performance through the bit-rate of the compressed image, in the number of bpppb. Note that both the original AVIRIS and Hyperion datasets have bit-rates of 16 bpppb [16]. The bit-rate (in bpp) of the compressed data are defined in equation (2).

$$bpp = \frac{Original\ image\ bit\text{-}rate}{Compression\ Ratio} \quad (2)$$

Table 1. Compression ratio for different hyperspectral images using CCSDS-MHC

IOP Conf. Series: Materials Science and Engineering **943** (2020) 012004 doi:10.1088/1757-899X/943/1/012004

Hyperspectral Image		CCSDS-MHC-FP
AVIRIS 1997 dataset	Cuprite1	3.2795
	Jasper1	3.2439
	Low1	2.9894
	Low5	2.9963
	Lunar1	3.2206
AVIRIS 2006 dataset	YSCal0	4.0982
	YSCal3	4.2042
	YSCal10	4.8597
	YSCal11	4.4667
	YSCal18	4.0071
Hyperion dataset	Atturbah	2.4858
	Benoni	2.4601
	Boston	2.0027
	Coolamon	2.4740
	Dubbo1	2.4751
	Edenton	1.9988
	Greenland	2.3501
	Maizhokunggar	2.4824
	Okha	2.4558
	Portobago	2.4729

3. Result & Discussion

In this project, the parallelized CCSDS-MHC algorithm using OpenMP and without OpenMP that implement inside Raspberry Pi Model 3 B+. Section 3.1 is the result measured and verification of the lossless in terms of execution time. Raspberry Pi only perform the operation through the compression part meanwhile, the decompression part is not the interest in this project. Section 3.2 shows the result of the execution time when running in C/C++ language with OpenMP algorithm whereby different kinds of desktop consist of multicore system and data is executed based on the performances.

3.1 CCSDS-MHC using OpenMP

High data of hyperspectral images with a multithreaded function can increase the execution time for the system to read and write, especially upon performances in a low-power computer. To optimize the execution time for the system to operate in a low-power computer with high data, OpenMP in CCSDS-MHC algorithm is implemented.

Therefore, Table 2 show the execution time when running the CCSDS-MHC algorithm with and without OpenMP with hyperspectral image. Table 2 proves that there is a higher difference of execution time when OpenMP is applied in CCSDS-MHC algorithm. This is because OpenMP divides a task among the thread so that each thread executes its own allocated part of the code. This OpenMP is achieved by the parallelism task and data on CCSDS-MHC algorithm.

The speed performance and efficiency of CCSDS-MHC algorithm using the parallelization of OpenMP through its implementation inside Raspberry Pi 3 Model B+ is reported as average and percentage values. Running the CCSDS-MHC algorithm in Raspberry Pi with 4-threaded CPU-core without OpenMP for AVIRIS 1997 dataset was able to occupy the average of 289.0340 seconds with 88.4737 seconds using the OpenMP. The execution time for AVIRIS 1997 dataset increased to about 69.4% optimization when OpenMP is implemented.

For AVIRIS 2006 dataset, the average reading for CCSDS-MHC algorithm without OpenMP is 289.0340 seconds and 88.8774 seconds is achieved with OpenMP. It shows that AVIRIS 2006 dataset execution time is optimized to 69.3% when using the OpenMP.

For Hyperion image dataset, the average of 59.6981 seconds is recorded when the CCSDS-MHC algorithm is not implemented using OpenMP meanwhile, 19.2643 seconds is obtained when it is implemented with the OpenMP, optimizing the execution time to about 67.7%.

To summarize, it is proven that OpenMP can be exploited in CCSDS-MHC algorithm and compute faster results to about half of the execution time efficiency in Raspberry Pi 3 Model B+ with a low memory embedded system.

Table 2. Execution time with and without OpenMP in Raspberry Pi 3

Hyperspectral Image		CCSDS-MHC algorithm without OpenMP (seconds)	CCSDS-MHC algorithm with OpenMP (seconds)	Improvement (%)
AVIRIS 1997 dataset	Cuprite1	287.8375	88.3160	69.3
	Jasper1	285.5075	88.0509	69.2
	Low1	290.5157	88.8951	69.4
	Low5	290.2677	88.5578	69.5
	Lunar1	290.0447	88.5488	69.4
	Average	**288.8346**	**88.4737**	**69.4**
AVIRIS 2006 dataset	YSCal0	288.8649	89.1358	69.1
	YSCal3	290.2122	89.3906	69.2
	YSCal10	288.2797	88.8543	69.2
	YSCal11	289.3909	88.0401	69.6
	YSCal18	288.4221	88.9664	69.2
	Average	**289.0340**	**88.8774**	**69.3**
Hyperion dataset	Atturbah	60.5935	18.8207	68.9
	Benoni	60.4675	19.5144	67.7
	Boston	58.9004	19.0454	67.7
	Coolamon	59.4605	19.1911	67.7
	Dubbo1	60.1562	18.9905	68.4
	Edenton	60.7881	19.4695	68.0
	Greenland	58.2046	19.2856	66.9
	Maizhokunggar	59.0472	19.6949	66.7
	Okha	59.9177	19.6098	67.3
	Portobago	59.4457	19.0207	68.0
	Average	**59.6981**	**19.2643**	**67.7**

3.2 Performances with Other Implementations

This project consists of four types of different platform used which are Raspberry Pi 3 Model B+, Desktop with 4 threads, Desktop with 8 threads and Desktop with 16 threads. For specifications, Raspberry Pi 3 Model B+ processor consists of Broadcom BCM2837B0, Cortex-A53 (ARMv8) 64-bit @ 1.4GHz quad-core processor, Intel® Core™ i5-3210M CPU @ 2.50GHz, 2-core processor for the Desktop with 4 threads, Intel® Core™ i7-2600S for the Desktop with 8 threads and Intel® Xeon® CPU E5-1660 v4 @ 3.20GHz for the Desktop with 16 threads.

OpenMP compiler directives and without any major source code modifications can easily be ported to run in parallel on any shared-memory, multicore systems with reasonably good computational performance. The shared-memory of the multicore architecture is the state-of-the-art hardware environment for modern microprocessors, including computational nodes of workstation, personal desktop and laptop [17].

The execution time performances and efficiency of lossless hyperspectral image compression that implements the OpenMP in CCSDS-MHC algorithm with low memory computerization using Raspberry Pi 3 Model B+ and high memory computerization through various threads of desktop are shown in Table 3.

Performance and efficiency of OpenMP parallel implementation depend on the hardware system as well whereby all of the maximum thread in the system is used. Time execution is measured by implementing CCSDS-MHC algorithm on different processor performance. The performance and efficiency of CCSDS-MHC algorithm using the parallelization of OpenMP that is implemented inside Raspberry Pi 3 Model B+ and with desktop of 16 threads, 8 threads as well as 4 threads of time execution is measured and reported in terms of average and percentage. The result of the performances (in percentage) is obtained from the value of the average execution time for each Desktop which is made in comparison with the Raspberry Pi 3 Model B+ data. It is noted that, all the processor threads are used to run the CCSDS-MHC algorithm for lossless hyperspectral image compression.

When running the CCSDS-MHC algorithm for AVIRIS 1997 image dataset, the Raspberry Pi 3 B+ managed to be executed with the average reading of 88.4737 seconds whereas 18.0756 seconds for the Desktop with 4 threads, 10.7640 seconds for the Desktop with 8 threads and 4.7184 seconds for the Desktop with 16 threads. By running the CCSDS-MHC algorithm on a Desktop of 4 threads with 2 CPU-core, the mean parallel efficiency is 79.6%. In contrast to 4-thread, the 8-thread with 4 CPU-core has the mean parallel efficiency of 87.8%. When 16 threads with 8 CPU-core are used, the parallelization of CCSDS-MHC algorithm has improved by 94.7% as compared to the 4 CPU cores of Raspberry Pi 3 B+.

For AVIRIS 2006 image dataset, the average that was obtained from Raspberry Pi 3 is 88.8774 seconds, 17.9008 seconds for the 4-thread Desktop, 10.6580 seconds for the 8-thread Desktop and 4.7870 seconds for the Desktop with 16 threads. By running the CCSDS-MHC algorithm on a Desktop of 4 threads with 2 CPU-core, the mean parallel efficiency is 79.9%. In contrast to the 4 threads, the 8 threads with 4 CPU-core has the mean parallel efficiency of 88.0%. When 16-thread Desktop with 8 CPU-core are used, the parallelization of CCSDS-MHC algorithm has accelerated by 94.6% compared to the 4 CPU cores of Raspberry Pi 3 B+.

However, using Hyperion image dataset that were cropped to a size of 256 × 256 pixels, only 196 out of 242 bands were classified on the execution time for Hyperion images whereby it is lower compared to AVIRIS images. Therefore, the average of the execution time obtained for Raspberry Pi 3 B+ is 19.2643 seconds, 3.7439 seconds for the Desktop with 4 threads, 1.5257 seconds for the Desktop 8 threads and 0.6143 seconds for the Desktop with 16 threads. Comparing the mean performances with Raspberry Pi 3 B+ execution time, the 4-thread with 2 CPU-cores obtained 80.6% and 92.1% for the 8-thread with 6 CPU-cores. When the 16-thread with 8 CPU-cores are used, the parallelization of CCSDS-MHC algorithm has accelerated by 96.8% compared to the 4 CPU cores of Raspberry Pi 3 B+.

The optimizing performance for the 16-thread Desktop with 8 CPU-cores is increased to 96.8% which is the percentage of the performance that furnishes a large difference compared to AVIRIS 1997 dataset which has a difference of 2.1% and a difference of 2.2% for AVIRIS 2006 dataset.

The differences in performances of the execution time for every different embedded platform are due to the CPU type, CPU-core processor, difference in CPU thread and number of RAM. It can be concluded that, the number of processor threads available gives a large effect on the execution time of CCSDS-MHC algorithm in performing the task.

Table 3. Time execution with other implementation

Hyperspectral Image		Desktop (16 threads)[a]	Desktop (8 threads)[b]	Desktop (4 threads)[c]	Raspberry Pi 3 B+ (4 threads)[d]
AVIRIS 1997 dataset	Cuprite1	4.5000s	10.8100s	17.8910s	88.3160s
	Jasper1	4.6400s	10.6390s	18.2290s	88.0509s
	Low1	4.9060s	10.8110s	18.1640s	88.8951s
	Low5	4.8740s	10.8420s	18.2350s	88.5578s
	Lunar1	4.6720s	10.7180s	17.8590s	88.5488s
	Average	**4.7184s**	**10.7640s**	**18.0756s**	**88.4737s**
	Comparison*	**94.7%**	**87.8%**	**79.6%**	
AVIRIS 2006 dataset	YSCal0	4.7180s	10.7170s	17.9640s	89.1358s
	YSCal3	4.7960s	10.5930s	17.9660s	89.3906s
	YSCal10	4.8740s	10.6230s	17.8330s	88.8543s
	YSCal11	4.7810s	10.6550s	17.8540s	88.0401s
	YSCal18	4.7660s	10.7020s	17.8870s	88.9664s
	Average	**4.7870s**	**10.6580s**	**17.9008s**	**88.8774s**
	Comparison*	**94.6%**	**88.0%**	**79.9%**	
Hyperion dataset	Atturbah	0.6100s	1.5290s	3.7400s	18.8207s
	Benoni	0.6250s	1.5130s	3.7460s	19.5144s
	Boston	0.6250s	1.6070s	3.7520s	19.0454s
	Coolamon	0.6250s	1.5130s	3.7970s	19.1911s
	Dubbo1	0.6100s	1.5440s	3.7210s	18.9905s
	Edenton	0.6100s	1.4980s	3.7090s	19.4695s
	Greenland	0.6100s	1.5130s	3.7190s	19.2856s
	Maizhokunggar	0.6090s	1.4970s	3.7190s	19.6949s
	Okha	0.6090s	1.4980s	3.7630s	19.6098s
	Portobago	0.6100s	1.5450s	3.7730s	19.0207s
	Average	**0.6143s**	**1.5257s**	**3.7439s**	**19.2643s**
	Comparison*	**96.8%**	**92.1%**	**80.6%**	

*Comparison between [a], [b], [c] with respect to [d]

4. Conclusion

With the support of OpenMP in CCSDS-MHC algorithm, each thread was executed independently by the parallel section of the code which was found to be more efficient and allowed higher level of constructs into a simplified form. Table 2 proved that the parallelization of CCSDS-MHC algorithm using OpenMP was optimized to approximately half of the execution time compared to the CCSDS-MHC algorithm without OpenMP. This proved that in this research, CCSDS-MHC algorithm by using OpenMP was successfully developed and can be implemented inside the Raspberry Pi 3 Model B+.

The execution time performances with others implementation results in Table 3 shown that performing CCSDS-MHC algorithm with OpenMP was able to obtain the averages of 88 seconds for AVIRIS and 19 seconds for Hyperion images. Besides, when implemented using Desktop with different types of multicore, the execution time for hyperspectral image compression decreases depending on the Desktop specification itself such as processor speeds and architectures as well as RAM sizes. This proved that the validation and measurement of the performance for losslessness hyperspectral image compression algorithm using CCSDS-MHC was successful and the objectives of this study are achieved.

Acknowledgement
This work was supported by Universiti Sains Malaysia Short Term Grant 304/PELECT/6315003.

Reference

[1] D. A. Landgrebe, *Signal Theory Methods in Multispectral Remote Sensing*. New Jersey: John Wiley & Sons, 2003.

[2] G. McNamara, J. M. Larson, S. A. Schwartz, and M. W. Davidson. (2019, 14th. Oct.). *Spectral Imaging and Linear Unmixing*. Available: https://www.microscopyu.com/techniques/confocal/spectral-imaging-and-linear-unmixing

[3] M. I. Afjal, M. A. Mamun, and M. P. Uddin, "Band reordering heuristics for lossless satellite image compression with 3D-CALIC and CCSDS," *Journal of Visual Communication and Image Representation*, vol. 59, pp. 514-526, 2019/02/01/ 2019.

[4] G. Lopez, E. Napoli, and A. G. M. Strollo, "FPGA implementation of the CCSDS-123.0-B-1 lossless Hyperspectral Image compression algorithm prediction stage," in *2015 IEEE 6th Latin American Symposium on Circuits & Systems (LASCAS)*, 2015, pp. 1-4.

[5] J. E. Sanchez, E. Auge, J. Santalo, I. Blanes, J. Serra-Sagrista, and A. Kiely, "Review and Implementation of the Emerging CCSDS Recommended Standard for Multispectral and Hyperspectral Lossless Image Coding," presented at the Proceedings of the 2011 First International Conference on Data Compression, Communications and Processing, 2011.

[6] M. Olaru and M. Craus, "Lossless multispectral and hyperspectral image compression on multicore systems," in *2017 21st International Conference on System Theory, Control and Computing (ICSTCC)*, 2017, pp. 175-179.

[7] H. Du, W. Zhang, N. Guan, and W. Yi, "Scope-aware data cache analysis for OpenMP programs on multi-core processors," *Journal of Systems Architecture*, vol. 98, pp. 443-452, 2019/09/01/ 2019.

[8] S. Balakrishnan, D. Langerman, E. Gretok, and D. A. D. George, "Deep Learning for Hyperspectral Image Classification on Embedded Platforms," in *2018 IEEE International Conference on Image Processing, Applications and Systems (IPAS)*, 2018, pp. 187-191.

[9] N. Kefalas and G. Theodoridis, "Low-memory and high-performance architectures for the CCSDS 122.0-B-1 compression standard," *Integration*, 2018/04/10/ 2018.

[10] R. L. Davidson and C. P. Bridges, "GPU accelerated multispectral EO imagery optimised CCSDS-123 lossless compression implementation," in *2017 IEEE Aerospace Conference*, 2017, pp. 1-12.

[11] B. Hopson, K. Benkrid, D. Keymeulen, and N. Aranki, "Real-time CCSDS Lossless Adaptive Hyperspectral Image Compression on Parallel GPGPU & Multicore Processor Systems," in *NASA/ESA Conference on Adaptive Hardware and Systems (AHS), 2012*, Erlangen, Germany, 2012, pp. 107-114.

[12] S. Wang, S. Zhang, X. Huang, J. An, and L. Chang, "A Highly Efficient Heterogeneous Processor for SAR Imaging," *Sensors*, vol. 19, no. 15, p. 3409, 2019.

[13] N. Lopez-Ruiz, F. Granados-Ortega, A. Carvajal Miguel, and A. Martinez-Olmos, "Portable multispectral imaging system based on Raspberry Pi," *Sensor Review*, vol. 37, no. 3, pp. 322-329, 2017.

[14] CCSDS, *Lossless Multispectral & Hyperspectral Image Compression - Recommended Standard (CCSDS 123.0-B-1)* (no. 1). Washington DC, USA: CCSDS Secretariat, NASA, 2012, p. 52.

[15] Saisang *et al.* (2019, 14th. Oct.). *Run-time library functions*. Available: https://docs.microsoft.com/en-us/cpp/parallel/openmp/3-run-time-library-functions?view=vs-2019

[16] N. R. Mat Noor, "Karhunen-Loève Transform based Lossless Hyperspectral Image Compression for Space Applications," PhD eThesis, Department of Engineering, University of Leicester, Leicester, 2015.

[17] X. Xu, "OpenMP parallel implementation of stiffly stable time-stepping projection/GMRES(ILU(0)) implicit simulation of incompressible fluid flows on shared-memory, multicore architecture," *Applied Mathematics and Computation*, vol. 355, pp. 238-252, 2019/08/15/ 2019.

IOP Conf. Series: Materials Science and Engineering **943** (2020) 012005 doi:10.1088/1757-899X/943/1/012005

A review of chemical stabilisation in road construction

E H Tan[1], E M M Zahran[2], S J Tan[2]

[1] Civil Engineering Programme Area, Universiti Teknologi Brunei (UTB), Brunei Darussalam
[2] Centre for Transport Research, Universiti Teknologi Brunei (UTB), Brunei Darussalam

E-mail(s): angelteh86@gmail.com, elsaid.zahran@utb.edu.bn, soonjiann.tan@utb.edu.bn

Abstract. This paper presents a literature review of the application of cement, lime and polymer emulsion in road construction. The focus is mainly on the soil candidates, binding mechanism, the reported strengths and weaknesses for each chemical stabiliser used in soil stabilisation. While it is generally agreed that cement is best suited for sand-sized to gravel-sized soil mixtures (well-graded), lime is mainly used for wet heavy-clay soil mixtures. Polymer emulsion is more suitable for silt-sized to fine gravel-sized soil mixtures (well-graded). Organic soils are the most difficult soil type to stabilise, and for any kind of chemical stabiliser used, the strength gain is relatively minimal. Cement undergoes hydration reaction which is rapid, and results in high compressive strength, while lime undergoes a slower pozzolanic reaction is slower, and provide lower compressive strength. Polymer emulsion depends on the rate of evaporation for the film formation to complete, and results in ductile polymer-soil matrix. Like lime, polymer does not provide significant compressive strength but does provide great flexural strength that prevents crack propagations. The outcomes from this paper provide an insight into recent and upcoming research trend using chemical soil stabilisers on local soil mixtures to improve the engineering properties of pavement layers made of these soil mixtures.

1. Introduction

Soil stabilisation is an important work package in road construction. It is performed when the sub-grade soils are soft and have low bearing capacity, when the local natural aggregates are of poor quality and importing them is costly, and/or when excavating the sub-grade soils and replacing them with granular materials are expensive. Soft and weak soils such as peats and soft clays typically represent the most problematic spectrums of soils encountered in road construction as they are not sufficiently strong to support the overlying road structure; causing structural damages and premature failures. Expansive soils that contain smectite clay minerals are known to be significantly problematic due to the swelling-shrinking behaviour caused by moisture variability [1, 2]. Peat soils deform greatly when subjected to loads on the account of the soils' exceptionally high water and organic contents [3, 4]. Roads built on raised embankment over soft and weak soils (compressible soils) will give rise to instability problems during construction, and long-term and persistent settlement after construction [5].

In some geological regions, the locally sourced natural aggregates of poor-quality lack the hardness and strength to withstand loads, thus may deform excessively. Stabilising the poor-quality aggregate allow these regions to use their local resources to reduce the dependency of importing expensive high-quality aggregate. Groundwater and floodwater are known to cause moisture-related damages to the unbound and bound layers of the roads, especially when the road is submerged frequently. The road will eventually experience serviceability problems and structural failures. Soil stabilisation, when performed optimally to the underlying granular layers and/or sub-grade soils, offers significant support to the road, and protection against damages caused by traffic loading, and exposure to climatic and ground conditions.

Although there are many different types of soil stabilisation methods, the choice of method are governed by several important factors such as the soil and aggregate types (e.g. clayey soils, silty soils, sandy soils, siliceous aggregate or calcareous aggregate) and their respective gradations (i.e. well-graded, poorly-graded and gap-graded), the geotechnical layer to be stabilised (i.e. sub-grade, sub-base or base), the degree of improvement of soil properties, the climate conditions (e.g. tropical, dry and mild) and ground conditions (e.g. sandy soils underlain by very soft clay and alluvial deposits overlain by thick peat). Generally, in order to improve the soil's gradation, workability and plasticity, small amount of chemical stabiliser will be sufficient. When it is desired to improve the soil's strength and durability, larger amount of chemical stabiliser will be needed. Generally, fine-grained soils have smaller load bearing capacity than coarse-grained soils, while the former also have greater degree of compressibility [6].

The objective of this paper is to present a literature review of three types of chemical stabiliser, i.e., cement, lime and polymer emulsions used in stabilising the soil or soil-aggregate defined by its effective sizes. The primary binding mechanism for each of the three chemical stabilisers is also reviewed in this paper. The outcome from the paper is expected to lead to an improved insight into recent and future research trend related to the application of the chemical stabilisers on the local soil or soil-aggregate and subsequently, the development of recommendation, guideline and/or general specification for local soil or soil-aggregate stabilisation in road construction.

2. Chemical soil stabilisation

The main purpose of soil stabilisation is to alter the geotechnical properties of natural soils to meet specific engineering purposes. The soil's geotechnical properties can be improved by enhancing the shear strength parameters [7], increasing the tensile strength [8], and improving stiffness [9]. It also aims to reduce plasticity [10], and minimise volumetric changes in expansive or fine-grained soils caused by moisture variability [1]. Chemical stabilisation through the use of chemical compounds or mixtures is very common. It is regarded as a cost-effective, performance-efficient and environmental-friendly method of soil stabilisation [11, 12]. Chemical modification involves soil particles bound together through the use of chemical compounds or mixtures, and the changes in the geotechnical properties normally more complex than mechanical modification. Chemical modification, nonetheless, is the preferred choice when roads are to be constructed under the adverse prevailing traffic loading, climatic and ground conditions.

The most frequently used traditional chemical stabilisers are the calcium-based powdered cement and lime. The common issues when using these stabilisers are the lengthy curing times and large quantity required for soil strength improvement [13]. Another major concern is the brittle behaviour of calcium-treated soils [14], which can cause cracks when subjected to dynamic loading. Soils containing a high level of sulphate ions (SO_4^{2-}) treated with calcium-based stabilisers will cause swell/heave problems as well as having a deleterious impact on the cementitious stabilisation mechanisms [15]. In spite of the limitations, traditional stabilisers remain widely used in countries where the supply is consistent and the procurement cost is low.

The non-traditional type of chemical stabilisers can be categorised into various groups [13], but the most popular ones are polymer emulsions. Soil amendment using polymer emulsion is a growing industry and it receives great interest in applications for improving various soil properties. Polymer emulsions have an advantage as their chemical formulations to treat a wider range of soil compositions.

They also have faster curing period, and lesser quantity is required [13]. Water-soluble polymers have the advantage of easier access to the surfaces of soil particles than the powdered stabilisers; and water-soluble polymers do not produce excessively viscous liquids [16]. They also do not require a solvent carrier and do not pose environmental concerns when used in bulk [17]. In contrast to calcium-based stabilisers, polymer-stabilised soils exhibit flexible [18] or ductile [14] behaviour.

2.1. Cement stabilisation

Cement is the single largest calcium-based soil stabiliser used in road construction. It contains 21.9% silicon oxide (SiO), 6.9% aluminium oxide (Al_2O_3), 3.9% iron oxide (Fe_2O_3) and 63% calcium oxide (CaO) [19].

Cement is the preferred choice for granular and sandy soils [20, 21] but its applications have been extended to include clayey soils [22, 23] and peat soils [3, 24]. Acceptable soil candidates for cement stabilisation are poorly cohesive granular soils [25] and/or soils whose fines fraction is less than 0.074mm but plastic soils should be avoided [26]. Generally, well-graded granular soils that possess sufficient fines that produce a homogeneous mixture (floating aggrege matrix) are best suited for cement stabilisation [26]. Reference [21], on the other hand, stated that the optimum proportion of soils for cement stabilisation is 75% sand with 25% silt and clay. Clay soils that are suitable for cement stabilisation are kaolinitic and illitic clay soils [23], and cement is found to be effective in stabilising low-plasticity clay under wet test conditions [13]. Though peat soils lack the silica (SiO_2) and alumina (Al_2O_3) that are found in clay soils to allow for pozzolanic reaction to take place, the primary hydration reaction is sufficient to create the cementitious bond for peat soils. While it is almost apparent that all types of soils can be stabilised with cement, there are several exceptions. Some types of organic soils, high plastic clays, soils with medium to high levels of SO_4^{2-} and poorly reacting sandy soils are not suitable for cement stabilisation [27]. Organic soils can retard the primary hydration reaction due to the high organic content, and the low content of pozzolans and silts makes cement alone insufficient to provide the desired stabilisation strengths, unless cement is added in large dosage [28, 29]. Soils containing organic matters greater than 2%, and with pH less than 5.5, are considered unacceptable soil candidates for cement stabilisation [27] as lower pH values will negatively affect the rate of hydration reaction, resulting in slower strength gain [29]. Soils with 10% of clay content (an effective minimum for soils to be cohesive) may require only 5% cement for stabilisation, but the cement content will increase to 10% or more when the clay content increases to 30% [30]. Cement content between 6% to 10% would start to cause shrinkage cracking [31], and cement content greater than 10% is not economical [30]. Shrinkage cracking relates to the drying of the cement paste at atmospheric temperature and pressure drying.

Cementitious hydration is a chemical process primary and unique to cement. The phase compounds found in the cement are tricalcium silicate (C_3S), dicalcium silicate (C_2S), tricalcium aluminate (C_3A) and tetracalcium aluminoferrite (C_4AF) [32]. When C_3S and C_2S react with water, the hydration reaction produces tricalcium silicate hydrate (CSH) and crystallised calcium hydroxide (CH); it is these hydration products that give the hardened cement paste (HCP) most of its engineering properties such as strength and stiffness [32]. The rate of reaction of C_3S is fast and the slowest being C_2S [33]. C_3A is particularly important in the early stage of the hydration as it reacts very violently with water resulting in an immediate stiffening of the cement paste [32], and it reacts faster than C_3S. This reaction, albeit important, has be to delayed so gypsum is added to prevent the immediate stiffening [32]. C_3A and gypsum react with water to produce ettringite. When gypsum is exhausted, C_3A reacts with water to produce tricalcium aluminate hydrate (CAH) [32]. The CSH and CAH form a gel network that serves as binder to the soil particles, thus resulting in a soil-cement matrix [22], and when the cement paste hardened the strength is established. As a result, the HCP becomes tough and insoluble, which gives the soils the required compressive strength. Strength gain occurs from day 1 to day 27 [22]; after that, the strength gain gradually slows down until the cement grains are completely hydrated. The rate of cement hydration is influenced by cement fineness, water-to-cement ratio and curing temperature [34].

Leaching of calcium is minimal when cement is used [22]. However, the single use of cement in soil stabilisation is highly discouraged, and this is mainly due to the negative environmental impacts associated with cement production in terms of high energy consumption and emission of carbon dioxide (CO_2) [35-37].

2.2. Lime stabilisation

After cement, lime is the next most used calcium-based soil stabiliser. The two types of lime used in soil stabilisation are quicklime (CaO) and hydrated lime (Ca(OH): calcium hydroxide). Apart from the difference in chemical compositions, their reactivities are also different. CaO is denser, and more readily reactive than $Ca(OH)_2$. Quicklime appears to be a more effective soil stabiliser than hydrated lime, and quicklime in a slurry state produces higher strength in soils than when it is a dry state [38]. Quicklime has also a higher drying effect on wet soils, but quicklime is very caustic, posing health risks [39].

Lime stabilisation is common for clayey soils as lime works better with the clay minerals [30]. A good soil candidate for lime stabilisation is generally fine-grained clayey soils (reactive clays) with a minimum of 25% by mass passing the 74μm sieve, and a PI of greater than 10 [40, 41]. In order words, low-plasticity soils are not suitable. Soils of medium to high plasticity can also be treated with lime to decrease plasticity, increase workability, reduce swell and increase strength [26]. Lime stabilisation is less suitable for granular soils (non-plastic soils) [10, 20, 30, 38, 42]. The efficacy of lime in low-cohesion or cohesionless soils is poor without the addition of pozzolans [43]. Lime is use only as construction expedient to improve the workability of wet clayey soil with little regards given to time and strength gain [39]. The assumption that lime is more suitable than cement to stabilise clay soils is not true, and cement is as good as lime to stabilise clay soils with PI value between 20 and 50, if not better as cement offers higher strength gain [22]. In order words, lime does not contribute strength to the soils. Hence, it is relatively common to see that lime is mixed with cement. The success of the lime-cement stabilisation depends on the properties of soils, the ratios of lime to cement and curing time.

Cation exchange and flocculation-agglomeration are a two-phase chemical process unique to lime. These reactions cause textural changes within the soils which subsequently improve workability and moisture stability [40]. A small amount of lime will increase the PI, but lime has less pronounced effect on the liquid limit (LL) [39]. They are considered the short-term benefits of lime stabilisation. Most clays are reactive and when stabilised with lime, the soils' strengths may triple or quadruple, and in some unique instances, the strengths improved by an order of ten or more [40]. Clay is a pozzolan, and is a source of silica (SiO_2) and alumina (Al_2O_3) for the pozzolanic reaction, and is regarded as the long-term benefit of lime stabilisation. Pozzolanic reaction is slower than cement hydration [33]. The soluble clay-silica and clay-alumina react with the calcium ions (Ca^{2+}) and hydroxide ions (OH^-) to produce CSH and CAH [40]. They are essentially the same cement hydration products. The pH of water saturated with lime is 12.45 at 25°C, and as long as there is sufficient residual Ca^{2+} in the soil system to react with clay-silica and clay-alumina, the pozzolanic reaction will continue [40]. The continuing pozzolanic reaction can result in self-healing of cracks [39]. The high alkalinity of lime can neutralise acidic soils [39]. The pozzolanic reaction is temperature-dependent, and for temperature between 25°C to 30°C, there is a large increase in strength for every degree increase, thus, making lime stabilisation attractive than cement in countries with warm climate [39]. It is worthwhile to note that CaO consumes considerably larger amount of water when it hydrates in an exothermic reaction [22].

The greatest concern of lime stabilisation is calcium leaching, and it negatively impacts the durability of lime-treated soils. It is resulted from cyclic wetting and drying conditions that are caused by periods of wet and dry seasons, fluctuating groundwater tables and water runoffs followed by dry spells [40].

When clayey soils containing high SO_4^{2-} content are treated with calcium-based stabiliser, the soils will experience heaving problems, and SO_4^{2-} interfere with the pozzolanic reaction [15]. For the same mass of lime and cement and same, the calcium oxide in quicklime (100%) is more than cement (63%), thus quicklime will cause more heaving than cement, and this have been validated by [44]. The heave-induced failure to the road is caused by the highly expansive crystalline ettringite and thaumasite, which are formed when SO_4^{2-} react with Ca^{2+} [45, 46]. The pressure exerted by the calcium-sulphate-aluminate-hydrates can reach to approximately 241 MPa within the crystal [40]. In principle, calcium-sulphate-aluminate-hydrates is not a problem if they are form before compaction, but when they are formed after compaction, and the road is in service, the resulting heave can be destructive [40]. Shrinkage cracking is common in both lime and cement stabilisation, but the effect is less pronounced in lime stabilisation [39].

2.3. Polymer emulsion stabilisation
The use of polymer emulsion for soil stabilisation has been studies as early as the late 1960's [47]. Polymer latexes/dispersion/emulsions consist of very small (0.05μm to 5μm in diameter) polymer particles dispersed in water, and are usually produced by emulsion polymerisation [48]. In general, polymer emulsions are copolymer systems of two or more different monomers, and the total solid particles are between 40% to 50% by mass [48]. However, not all monomers can be copolymerised as there must be compactability among the monomers to form copolymer [49]. Polymer emulsions are regarded as a successful stabiliser for the following reasons: strong adhesive strength with soil surface, strong cohesive strength, low shrinkage during setting, low-water susceptibility after cure, resistance to solar radiation, atmospheric oxygen (oxidation) and biological degradation, non-toxic, non-polluting and inexpensive [49]. The elastomeric or thermoplastic water-based polymers form a continuous polymer film when water completely evaporated off [48]. As water has higher latent heat of vaporisation, it takes a longer time and higher temperature to evaporate, and give the final cured product [49].

The typical polymer-based admixtures are vinyl acetate (homo and copolymer latexes), acrylic-based copolymer latexes and styrene-butadiene (SB) copolymer latex [48]. In general, polymer emulsions can stabilise a greater range of soils as the formulations can be altered. Nonetheless, a majority of published papers has reported on the success of using polymers on granular soils, but the use of polymers of fine-grained soils has mixed results [50]. This could be resulted from the interaction between clay minerals and polymers as the water provides a medium in which clay and polymer react chemically to form clay-polymer complex [49]. References [13] and [51] used proprietary polymers to stabilise clay soils and silty sand respectively, and Reference [52] found that soil stabilisation using polymer emulsion is suitable for granular soils; and is less effective for fine-grained soils due to reduced mixing efficiency and insufficient ability to coat the soil particles, both resulted from high specific surface area. Little or no improvement would be observed if non-plastic crush rocks and poorly-graded soils containing large amount of gravel-sized soils are used with polymer emulsion, and this is simply due to the presence of intergranular air voids even after compaction [53]. A well-graded soil with sufficient fines is the ideal gradation for polymer emulsion to effectively form a soil-polymer matrix [54]. Polymer emulsion is also used to treat peat soils to improve strength but the increase may not be as significant [55, 56].

Cement is commonly added to polymer to further enhance the engineering properties of natural soils. The polymer-cement matrix reduces the propagation of microcracks that are prevalent in cement-stabilised soils [57]. The polymer-cement matrix also provides the stabilised soils with increased flexural and tensile strength and fracture toughness [57]. Similarly, polymer-lime matrix provides the stabilised soils with decreased PI and increased UCS [58]. The commonly used SB polymer is less water sensitive, and has better adhesive property than polyvinyl acetate and acrylic polymers [18]. SB polymer is also suitable for treating non-cohesive sandy soils without significantly reducing the permeability of the treated soils [18]. The non-cohesive sandy and silty sands can be better treated with carboxylated

SB polymer [18], but when either powdered or liquid carboxylated SB is used with cement, there was a significant retardation of the cement hydration reaction as the anionic SB particles strongly retard or suppress the silicates and aluminates in the hydration reaction [59]. Similarly, when cement is added with SB polymer to treat granular soils, the rigid styrene chains and flexible butadiene chains provides the treated soils with good mechanical properties, waterproofing properties and chemical resistance [57, 60]. The flexural strength and modulus of rupture of soils treated with SB polymer and cement [61]. The addition of SB polymer, however, creates a polymer film on the surface of the C_3S that retards or slows down the hydration reaction of silicates to produce CSH [62] and $Ca(OH)_2$ [63]. SB polymer is also used with hydrated lime to stabilise lateritic residual soils to reduce the PI values [58].

When polymer emulsion is used alone to stabilise soils, the binding mechanism involves film formation. When polymer emulsion and cement are added together to the soils, the modification is governed by two processes: cement hydration and film formation [57]. Hydration reaction precedes film formation, except on surfaces where loss of water by evaporation cause the film to form faster than hydration reaction [57]. The process by which the polymer film forms is known as coalescence. With water loss due to the hydration reaction and evaporation, the polymer particles are gradually confined in the capillary pores [57]. As more water is used during hydration reaction and evaporation, the polymer particles flocculate to form a continuous closed-packed film on the surface of the hydration products and unhydrated cement [57]. The film ultimately adheres to all surfaces that it comes in contact with, resulting in a polymeric monolithic network that binds the unhydrated cement, hydration products and soils together. Poorly-formed film will cause weak link or bond between the soil particles, and will soon break apart [49]. The factors that affect the film formation are time, temperature, size and shape of the polymer particles and the physical state of the polymer emulsion [49].

The use of polymer emulsion and cement are often comparable in cost, and both are less expensive than bitumen [53]. In an experimental investigation by Reference [64] shown that 2% to 4% of (acrylic) polymer and 20% to 40% of cement were required respectively to achieve comparable compressive strength of the stabilised sandy soils. This indicates that lesser polymer is required, and the investigation also revealed that the construction cost of 2% polymer content is lower than 20% cement content. Polymer emulsion used in soil stabilisation are often non-toxic and non-polluting [49].

3. Discussion and conclusions

A literature review of three types of chemical stabilisers, i.e., cement, lime and polymer emulsion used for stabilising soil or soil-aggregate defined by its effective size, and their primary binding mechanisms have been presented in the paper. The following are the conclusions drawn from the literature review:

- For cement stabilisation, the ideal soil mixture should be well-graded, consisting of sand-sized to gravel-sized soil particles. If silts and clays are present in sands, they should be less than 25%. Organic soils, high plastic clays and soils with medium to high levels of SO_4^{2-} are the least suitable for cement stabilisation. The primary binding mechanism is hydration reaction, whereby silicates, aluminates and gypsum react with water to from CSH, CAH and CH. CHS and CAH are the main hydration product, and is responsible for the compressive strength in cement-stabilised soils.
- For lime stabilisation, soils with high content of clay (clay rich soils) are the best suited as soluble clay-silica and clay-alumina react with the calcium ions (Ca^{2+}) and hydroxide ions (OH^-) to produce CSH and CAH. Lime is not well suited to treat granular soils, soils with low plasticity and organic soils. When quicklime is added to soils containing high SO_4^{2-}, heaving is greater, though is has good drying effect on wet soils. Though lime is an effective soil stabiliser, but when compared to cement, lime does not provide the much-required strength. Hence, it is not uncommon to find cement added to lime in order to achieve strength.
- For polymer stabilisation, a well-graded soil mixture with sufficient fines is the ideal to ensure a well-established soil-polymer matrix. Soil mixture that contains too much gravel-sized soil

particles are not suitable for polymer stabilisation as high air void content prevents proper coating. Polymer emulsion have negligible chemical reaction with soils as the main binding mechanism is film formation. Similar to lime, polymer emulsion does not provide compressive strength but gives the soil-polymer flexural strength, which is missing for cement-stabilised soil.

While the majority of past researches had focused on cement and lime in soil stabilisation, there has been considerably fewer reported work that investigated the application of polymer emulsion in soil stabilisation. Cement, especially, is easier to procure and it can be used across a wide range of soil mixture than lime. The use of polymer emulsion in soil stabilisation is quickly replacing traditional soil stabilisers due to their lower costs, easy handlings, formula variations and environmental benefits. Nonetheless, for any kind of chemical stabiliser used to stabilise a soil mixture, the target objectives must be clearly stated to obtain optimum results. There will always be future innovative research related to cement and lime stabilisation, while polymer emulsion is expected to receive heightened interest compared to other types of non-traditional chemical stabilisers.

Acknowledgement
The authors are grateful to Universiti Teknologi Brunei for their funding support, and would also express appreciation to Pg. Dr. Hj. Ismail, Mr. Chua Min Chu and Mr. Tang Kum Chuen (SWEE Sdn. Bhd.), Mr. Dhammika Hewage (Meraih Jaya Sdn. Bhd.) and Mr. Mahri Matahir (Terratech Earth Solution Ltd.) for their contribution towards technical and resources supports.

Reference
[1] Holland J E and Richards J 1982 Road pavements on expansive clays *Aust. Road Res.* **12** 173-7
[2] Ruedrich J, Bartelsen T, Dohrmann R and Siegesmund S 2010 Moisture expansion as a deterioration factor for sandstone used in buildings *Environ. Earth Sci.* **63** 1545-64
[3] Zambri N M and Ghazaly Z M 2018 *Peat soil stabilization using lime and cement Int. Conf. Civil Environ. Eng.*, (France)
[4] Zhang L and O'Kelly B C 2013 Constitutive models for peat - a review *Int. Conf. on Comp. Plast. - Fundam. Appl.*, (Barcelona, Spain: International Center for Numerical Methods in Engineering) p 1294–304
[5] Huat B B 1998 An investigation of an embankment failure in soft clay *Int. Conf. Case Hist. Geotech. Eng.*, (Missouri, USA:) p 430-3
[6] Jain V K, Dixit M and Chitra R 2015 Correlation of plasticity index and compression index of soil *Int. J. Innv. Eng. Tech.* **5** 263-70
[7] Afrin H 2017 A review on different types soil stabilization techniques *Int. J. Trans. Eng. Tech.* **3** 19-24
[8] Eme D B and Agunwamba J C 2014 Tensile strength of natural and lime stabilized clay soil in Rivers State using one dimensional indirect tensile testing technique (splitting test) *Int. J. Eng. Sci.* **3** 38-45
[9] Prasad S 2016 Feasibility study on cement treated base and sub-base layers of service roads - a case study on Khed Sinnar NH 50 Project *Int. Res. J. Eng. Tech.* **3** 1455-60
[10] Amadi A A and Okeiyi A 2017 Use of quick and hydrated lime in stabilization of lateritic soil: comparative analysis of laboratory data *Int. J. Geo. Eng.* **8**
[11] Saeid A, Amin C and Hamid N 2012 A review on the lime and fly ash application in soil stabilization *Int. J. Bio. Eco. Environ. Sci.* **1** 124-6
[12] Seyrek E 2018 Engineering behavior of clay soils stabilized with Class C and Class F fly ashes *Sci. Eng. Comp. Mater.* **25** 273-87
[13] Tingle J S and Santoni R L 2003 Stabilization of clay soils with nontraditional additives *J. Trans. Res. Board* **1819** 72-84
[14] Md Zahri A and Zainorabidin A 2019 An overview of traditional and non traditional stabilizer for soft soil *GEOTROPIKA 2019*, (IOP Publishing)

[15] Jones D, Rahim A, Saadeh S and Harvey J 2010 *Guidelines for the stabilization of subgrade soils in California*. (California, USA: University of California)

[16] Naeini S A, Naderinia B and Izadi E 2012 Unconfined compressive strength of clayey soils stabilized with waterborne polymer *KSCE J. Civil Eng.* **16** 943-9

[17] Newman J K and Tingle J S 2004 *Emulsion polymers for soil stabilization*. (New Jersey, USA: U.S. Army Engineer Research and Development Center)

[18] Siddiqi R A and Moore J C 1981 Polymer stabilization of sandy soils for erosion control *J. Trans. Res. Board* **827** 30-4

[19] Dunuweera S P and Rajapakse R M G 2018 Cement types, composition, uses and advantages of nanocement, environmental impact on cement production, and possible solutions *Adv. Mater. Sci. Eng.* **2018** 1-11

[20] Kennedy T W, Smith R, Holmgreen R J and Tahmoressi M 1987 *An evaluation of lime and cement stabilization*. (Texas, USA: University of Texas)

[21] Department of Economic and Social Affairs 1964 *Soil-cement: its use in building* (New York: United Nations)

[22] Prusinski J R and Sankar B 1999 Effectiveness of Portland cement and lime in stabilizing clay soils *J. Trans. Res. Board* **1652** 215-27

[23] Bell F G 1995 Cement stabilization and clay soils, with examples *Environ. Eng. Geosci.* **1** 139-51

[24] Boobathiraja S, Balamurugan P, Dhansheer M and Adhikari A 2014 Study on strength of peat soil stabilised with cement and other pozzolanic materials *Int. J. Civil Eng. Res.* **5** 431-8

[25] Beeghly J H 2003 *Recent experiences with lime-fly ash stabilization of pavement sub-grade soils, base and recycled asphalt Int. Ash Util. Symp.*, (Kentucky, USA)

[26] Guyer J P 2011 *An introduction to soil stabilization for pavements*. (New York, USA: Continuing Education and Development, Inc.)

[27] ACI Committee 230 2009 *ACI 230.1R-09 report on soil cement*. (Farmington Hills, Miami: American Concrete Institute)

[28] Zulkifley M T M, Ng T F, Raj J K, Hashim R, Bakar A F A, Paramanthan S and Ashraf M A 2013 A review of the stabilization of tropical lowland peats *Bull. Eng. Geo. Environ.* **73** 733-46

[29] Skels P, Bondars K and Korjakins A 2013 Unconfined compressive strength properties of cement stabilized peat *Int. Conf. Civil Eng. Construc. Mater.*, (Riga, Latvia:) p 202-6

[30] Norton J 1986 *Building with earth: a handbook* (UK: Intermediate Technology)

[31] Gafar H S H and Qing W D 2002 *Review of chemical stabilisation technologies and applications for public roads in Brunei Darussalam The Regional Seminar on Quality Roads - The Way Forward*, (Bandar Seri Begawan, Brunei Darussalam)

[32] Domone P and Illston J 2010 *Construction materials - their nature and behaviour* (UK: Spon Press)

[33] Reddi L N and Inyang H I 2000 *Geoenvironmental Engineering: Principles and Applications* (New York, USA: Marcel Dekker, Inc.)

[34] Alesiani M, Pirazzoli I and Maraviglia B 2007 Factors affecting OPC early-age hydration studied by NMR: fineness, water-to-cement ratio and curing temperature *Appl. Magn. Reson.* **32** 1-10

[35] Jegandan S, Liska M, Osman A A M and Al-Tabbaa A 2010 Sustainable binders for soil stabilisation *Proceedings of the Institution of Civil Engineers - Ground Improvement*, (Institution of Civil Engineer) p 53-61

[36] Cristelo N, Glendinning S, Fernandes L and Pinto A T 2012 Effects of alkaline activated fly ash and Portland cement on soft soil stabilisation

[37] Firoozi A A, Guney Olgun C, Firoozi A A and Baghini M S 2017 Fundamentals of soil stabilization *Int. J. Geo. Eng.* **8**

[38] Bell F G 1993 *Engineering treatment of soils* (London, UK: E & FN Spon)

[39] Robinson R and Thagesen B 2004 *Road Engineering for Development* (UK: Spon Press)
[40] Little D N 1995 *Handbook for stabilization of pavement subgrades and base courses with lime* (Texas, USA: Lime Association of Texas)
[41] Negi A S, Faizan M, Siddharth D P and Singh R 2013 Soil stabilization using lime *J. Innv. Res. Sci. Eng. Tech.* **2** 448-53
[42] Petry T M and Lee T-W 1988 *Comparison of quicklime and hydrated lime slurries for stabilization of highly active clay soils*. (Texas, USA: University of Texas)
[43] Ibrahim A, Abdullah M S, Jamalludin D and Apo M 2007 Comparison between hydrated lime dry powder and slurry on peat soil stabilization *Acad. J, UiTM P. Penang* **3** 87-100
[44] Knopp J and Moormann C 2016 *Ettringite swelling in the treatment of sulfate-containing soils used as subgrade for road constructions Int. Conf. Transp. Geotech.*, (Guimarães, Portugal) p 128-37
[45] Mitchell J K and Dermatas D 1992 *Clay soil heave caused by lime-sulfate reactions*. (Philadelphia, USA: American Society for Testing and Materials)
[46] Puppala A J, Talluri N, Gaily A and S C C 2013 Heaving mechanisms in high sulfate soils *Int. Conf. Soil Mech. Geotech. Eng.*, (Paris: Presses des Ponts) p 3125-8
[47] Fungaroli A A and Prager S R 1969 Evaluation of some acrylic polymers as soil stabilizers *Industrial & Engineering Chemistry Product Research and Development* **8** 450-3
[48] Ohama Y 1998 Polymer-based admixtures *Cem. Concr. Compos.* **20** 189-212
[49] Siddiqi R A 1978 *Cost, effectiveness and utility of polymer soil stabilizers*. (New York, USA: Oklahoma State University)
[50] Onyejekwe S and Ghataora G S 2014 Soil stabilization using proprietary liquid chemical stabilizers: sulphonated oil and a polymer *Bull. Eng. Geo. Environ.* **74** 651-65
[51] Santoni R L, Tingle J S and Webster S L 2002 Stabilization of silty sand with nontraditional additives *Journal of the Transportation Research Board* **1787** 61-70
[52] Tingle J S, Newman J K, Larson S L, Weiss C A and Rushing J F 2007 Stabilization mechanisms of nontraditional additives *J. Trans. Res. Board* **1989** 59-67
[53] Rodway B 2001 *Polymer stabilisation of clayey gravels ARRB Conference*, (Melbourne, Australia)
[54] Rizenbergs R L, Burchet J L and Davis L E 1973 *Pavement roughness - measurement and evaluation*. (Kentucky, USA: University of Kentucky)
[55] Izzati Norazam P N F, Bakar I, Lim A J M S and Shah Herman H 2017 Stabilization of peat soil by using Envirotac *MATEC Web of Conferences* **103** 07014
[56] Mohd Razali S N, Zainorabidin A, Bakar I and Mohamad H M 2018 Strength changes in peat-polymer stabilization process *Int. J. Integr. Eng.* **10** 37-42
[57] Walters D G 1988 Latex hydraulic cement additives *J. Trans. Res. Board* **1204** 71-6
[58] Snodi L N, Atemimi Y and Ahmad F b 2018 *Effect of stabilizers on the shear strength of residual soil MATEC Web of Conferences*, (p 01019
[59] Baueregger S, Perello M and Plank J 2015 Influence of carboxylated styrene–butadiene latex copolymer on Portland cement hydration *Cem. Concr. Compos.* **63** 42-50
[60] Baghini M S, Ismail A and Firoozi A A 2016 Physical and mechanical properties of carboxylated styrene-butadiene emulsion modified portland cement used in road base construction *J. Appl. Sci.* **16** 344-58
[61] Baghini M S and Ismail A 2015 Short-term effects of applying carboxylated styrene butadiene emulsion-portland cement mixture on road base construction *J. Appl. Sci.* **15** 1266-77
[62] Yue X B and Wang R 2013 Influence of SBR latex on the formation of C-S-H in C3S paste *Adv. Mater. Res.* **687** 329-34
[63] Wang R and Wang P M 2011 Application of styrene-butadiene rubber in cement-based materials *Adv. Mater. Res.* **306-307** 588-93
[64] Verma N 2013 *Effectiveness of using polymers and cement for soil stabilization*. (Punjab, India: Thapar University)

IOP Conf. Series: Materials Science and Engineering **943** (2020) 012006 doi:10.1088/1757-899X/943/1/012006

Synthesis and Size Control of Aluminum Nanoparticles using Solution Plasma Process

B Y Y Liew, M S Ranjit, M R M Julaihi, P Y N Yek and T Shibayama

Faculty of Science, Engineering and Technology, Swinburne University of Technology, Sarawak Campus, Jalan Simpang Tiga, 93350 Kuching, Sarawak, Malaysia

Email: 100062699@students.swinburne.edu.my

Abstract. This research was conducted to investigate the various control techniques to vary the size of aluminum nanoparticles through solution plasma process, specifically the submerged glow-discharge plasma process. Aluminum nanoparticles have received many attention due to their unique combustion, resistance and superhydrophobic properties. A lab-scale based submerged glow-discharge setup has been constructed to produce the nanoparticles. Different concentration of potassium carbonate electrolyte (0.1M – 0.5M) and cathode submerged length (50 mm – 100mm) were used in this study. The results were viewed and analyzed using scanning electron microscopy. As the major results showed that the diameter size distribution ranges from 80nm to 2μm. Higher concentrations have shown to produce smaller nanoparticles due to the overlapping of electron beams on the cathode surface. Also, shorter cathode submerged lengths have resulted in larger-sized nanoparticles.

1. Introduction

The field of nanotechnology deals with matters in the range of 10^{-9} m scale. Although small, this field is of high importance as many large applications that span through all discipline of knowledge depends on it. The main component that this field deals with are nanoparticles. Generally, nanoparticles exhibit unique physicochemical properties that are absent in their bulk (>500 nm) counterparts [1]. Due to their high surface area to volume ratio, it is able to enhance the overall outcomes of many applications. This is seen especially in the case of aluminum nanoparticles.

Aluminum nanoparticles especially its oxide counterpart, which is alumina, has high demands in the global market. Figure 1 shows these demands in different application fields in recent years.

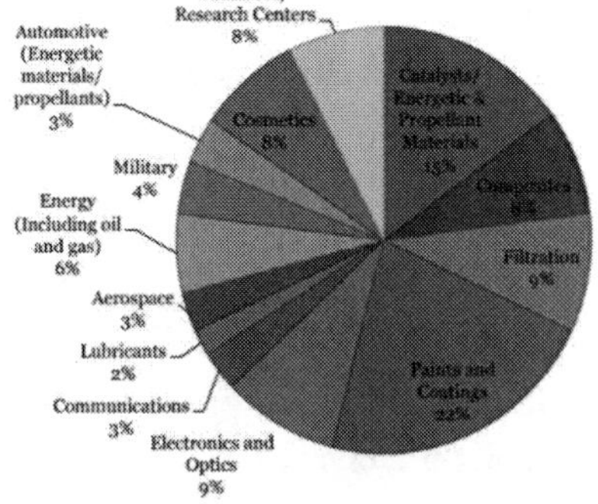

Figure 1. Market demand for aluminum nanoparticles [2].

To keep up with these growing demands, many methods and techniques to synthesize nanoparticles has been discovered such as the chemical reduction, chemical vapor, physical vapor deposition and pulsed laser ablation methods. Each method has the capability of producing different quantities of nanoparticles. The production rate is dependent on the parameters which are supplied, for example, the temperature or voltage supplied. Even with these existing methods, studies to control the size of aluminum nanoparticles is still on the verge of development. Current situation conveys that the aluminum nanoparticles which are used in the different field of applications are not size specific. Hence, the main purpose of this report is to further develop the current state of size control for the synthesis of aluminum nanoparticles.

Among all of the available methods, the submerged glow-discharge plasma via solution plasma method was the chosen method to be used in this research study. There are many advantages of using this method in the synthesis of aluminum nanoparticles, such as requiring a shorter processing time, lower energy and cost to run the experiments [3]. Other than that, it has a simple experimental setup and it does not emit any harmful gases nor require any chemical addition. It can also be conducted under atmospheric pressure in an open system [3]. Size control can be managed easily by controlling the experimental parameters.

The rationale of this research is to develop a viable size control technique for the synthesis of aluminum nanoparticle. Being able to do so would enhance many applicable outcomes of these nanoparticles. This paper discussion the generation mechanism of glow discharge plasma for the synthesis of aluminum nanoparticles. The parameters which were manipulated were the concentration and cathode submerged length.

2.0 Literature Review

2.1 Stages of Submerged Glow Discharge Plasma

The process of a complete submerged glow-discharge plasma can be summarized in four main stages. The first stage is the electrolysis stage. This is the stage where Ohm's Law is obeyed. When there is an increase in voltage, current rises as well. The current flow in the electrolytic cell in this stage slowly heats up the cathode. As the voltage is continually increased over time and the cathode's temperature eventually reaches the electrolyte's temperature, a thin film of water vapor starts to surround the cathode surface. This stage is known as the vaporization point [4].

The supply of more voltage passed the vaporization point causes the film of water vapor to cover more surface area on the cathode. When this film grows wider and thicker, some of the free radical flow in the electrolyte gets hindered. As an implication, the current of the cell drops. This stage is called the transition stage. As this progresses, the gas film eventually covers the entire cathode and the gases starts to continuously ionize with respect to increasing voltage. From here, due to continuous discharge, light emissions can be seen. This is the localized plasma stage [4]. These four stages can be seen in Figure 2. In this figure, Region 1 shows the electrolysis stage. Point 2 is the vaporization point. Region 3 is the transition stage. Region 4 and 5 are the localized plasma stage but in Region 4 indicates where the initiation of the plasma will start and Region 5 indicates where the glow-discharge plasma covers the entire cathode [5].

The formation of nanoparticles through SGDP method occurs during the localized plasma stage. When it is viewed microscopically, current concentration spots forms randomly at the cathode's surface. At these spots, a rise in temperature induces more current to flow in those spots and this causes the temperature to rise even more. When this temperature exceeds the melting point of the material, the cathode starts to melt and hence, forming the nanoparticles. From here, the temperature at this emission point drops until an even temperature diction exist in the area due to heat diffusion. As a consequence, the current concentration spot disappears. While this is happening, new spots starts to form the previous emission point because it has a higher surface temperature than the surrounding. This even is illustrated in Figure 3[6].

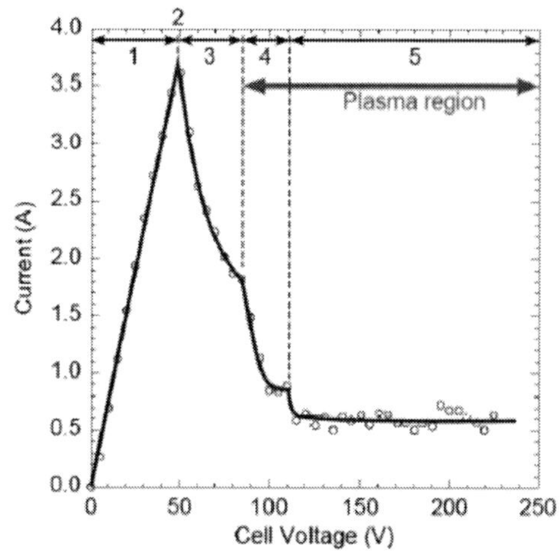

Figure 2. Stages of SGDP Method with Respect to the Current-Voltage Graph [4].

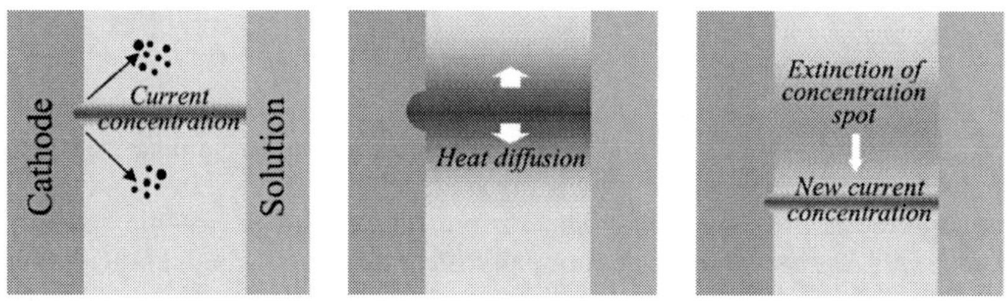

Figure 3. The formation of nanoparticles [6].

2.2 Breakdown Voltage, V_B, Mid-point Voltage, V_D & plasma intensity.

Based on Figure 2, the voltage which corresponds to Point 2 is known as the breakdown voltage, V_B. This voltage is where a gaseous film forms around the cathode. Also, the current at this point is at the maximum. Any increase in voltage would cause a steep drop in the current, indicating that the experiment has moved into the transition stage [7]. The voltage which corresponds to the point between Region 4 and 5 is called the discharge voltage or mid-point voltage, V_D. In most experiments, when the voltage surpasses V_D, the current starts to rise gradually and this is an indicator that full glow-discharge plasma is about to form. The region in between V_B and V_D is where the initiation of the glow-discharge plasma would happen.

Plasma intensity correlates to the excitation temperature, which is the temperature of the plasma at the cathode surface. There is a close relationship between V_D, plasma intensity and the size of the nanoparticles produced. If V_D is low, the plasma intensity decreases and thus, the average nanoparticle size would be small [8].

2.3 Effects of Electrolyte Concentration & Cathode Submerged Length on the Nanoparticle Size

The electrolyte concentration is proportional to the size of the nanoparticles produced [9]. A higher concentration means that the electrolytes are able to dissociate into more free radicals per unit volume. As an implication, the I-V curve which corresponds to the concentration skews to the left. Due to a higher conductivity of the electrolytic cell, a small V_D is adequate enough to initiate the glow-discharge plasma. Subsequently, since V_D is small, the plasma intensity decreases and the resulting nanoparticle size form are small [3]. This is illustrated in Figure 4.

In terms of cathode submerged length, when the cathode is submerged deeper, the I-V curve which corresponds to the submerge length skews slightly to the right. With this, the amount of V_D required to initiate the plasma is slightly more. The plasma intensity with regards to V_D becomes a little bit higher as well [3]. However, this small change in V_D may not be large enough to observe a significant change in the size of the nanoparticles. To obtain a size difference for the nanoparticles, the difference in each cathode submerged lengths in the experiments must be more than one centimeter. This is illustrated in Figure 5.

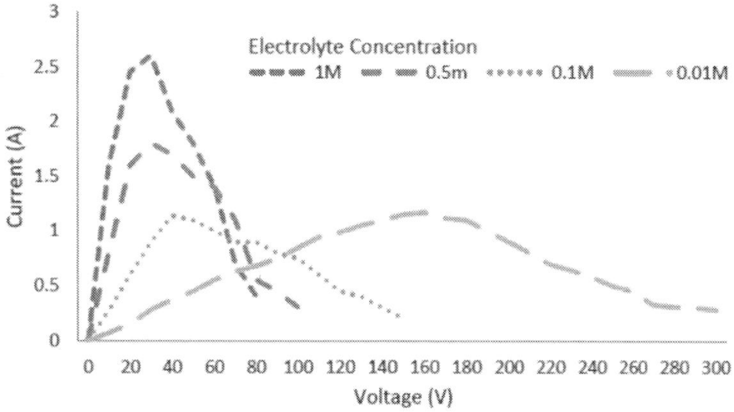

Figure 4. I-V curve for varying electrolyte concentration [3].

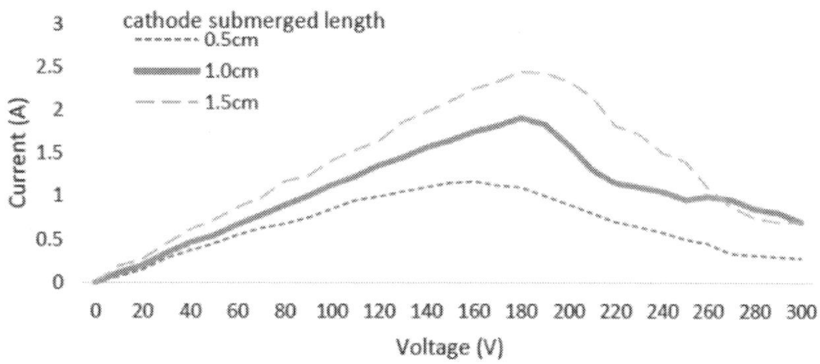

Figure 5. I-V curve for varying cathode submerged length [3].

2.4 Amount of Electric Power to Initiate & Sustain Glow-Discharge Plasma

There are two important terms which are involved in the initiation of the glow-discharge plasma and there are the electric power density for vapor formation, W_V and the power density, W_m required to sustain the plasma. To start the initiation process, W_m has to be greater than W_V, otherwise higher electric power and current are required to start the formation of plasma. W_v can be determined from the maximum current at the vaporization point on an I-V curve. W_v decreases when the temperature of the electrolyte increases. This is due to a decrease in the electrolyte's conductivity at higher temperatures [10]. W_m is determined by calculating the product of the average current and voltage during electrolysis:

$$W_m(W/cm^2) = \frac{Average\ current\ (A) \times Voltage\ (V)}{Surface\ area\ (cm^2)} \quad (1)$$

2.5 Characteristics of Aluminum & Alumina Nanoparticles

Aluminum nanoparticles have a greyish-black appearance and are typically spherical in shape. The nanoparticles vary in different sizes. Commonly, these nanoparticles would range from 10 to 30 nm with a specific surface area of 30 to 70 m^2/g. Those that range from 70 to 100 nm possess a specific are of 5 to 10 m^2/g (*Aluminum Nanoparticles* 2018). Due to these large specific surface areas combined with its high reactivity, Aluminum have very combustibility. As a consequence, Aluminum nanoparticles are normally used as additives in propellants formulations to generate higher and faster energy release [12]. Aluminum and alumina nanoparticles are resistant against corrosion, scratch and wear. Also, these nanoparticles possess high thermal barrier and superhydrophobic properties. Due to these characteristics, they are highly used in the coating sector [2]. Besides that, their application in composites have resulted in improvements in durability, fire retardancy, stiffness, thermal fatigue resistance, fracture toughness, creep resistance and were resistance [2].

3. Materials & Methodology

3.1 Precursor Preparation

Aluminum rods of 5 mm in diameter were cut into 8 cm long pieces using a hacksaw. The oxide layers were removed using a sandpaper and rinsed with water to remove any unwanted bits. Potassium carbonate electrolytes were diluted into 0.1 M, 0.2 M, 0.3 M, 0.4 M and 0.5 M concentration. This is done by adding 4.15 g, 8.29 g, 12.44 g, 16.58 g and 20.73 g of potassium carbonate power into a 400 mL beaker and subsequently filling them up with distilled water up to 300 mL. From here, the electrolytes were placed on a heater plate and stirred until all of the potassium carbonate powders have dissolved.

3.2 Experimental Setup

Figure 6 shows the schematic diagram of the experimental setup. Two different experiments are ran based on varying electrolyte concentration and cathode submerged length. For the varying concentration experiments, the cathode was submerged at a fixed length of 1 cm. On the other hand, for the varying cathode submerged length experiments, the electrolyte concentration was kept at 0.1 M. Table 1 shows the summary of the experimental setups.

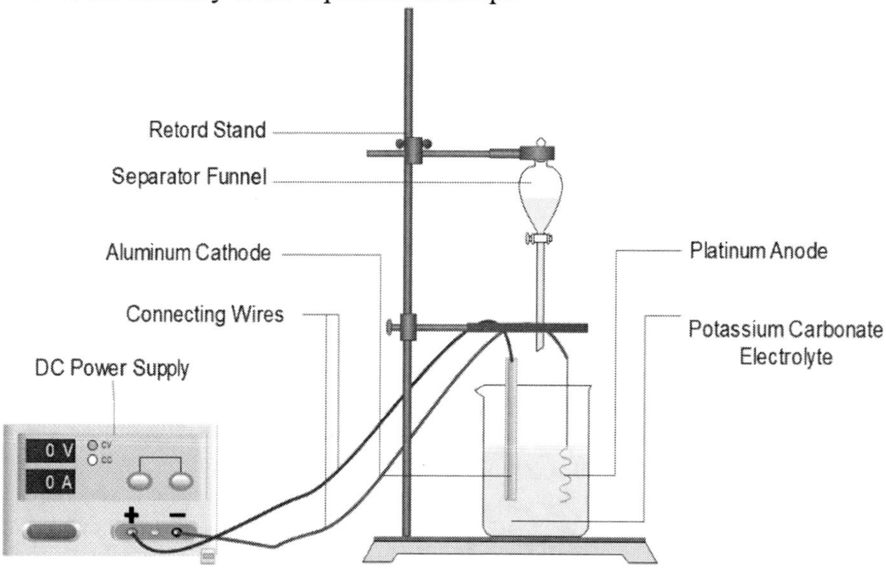

Figure 6. Schematic diagram for the experimental setup.

Table 1. Summary of experimental setup.

	Varying Parameter				
Electrolyte concentration (M)	0.1	0.2	0.3	0.4	0.5
Cathode Submerged Length (cm)	0.50	0.75	1.00	-	-

3.3 Experimental Procedure

After setting up the experiment as based on Figure 6 and Table 1, the DC power supply was switched on and the initial current reading was recorded. Then the voltage was increased by 5 V/ 30 seconds and

IOP Conf. Series: Materials Science and Engineering **943** (2020) 012006 doi:10.1088/1757-899X/943/1/012006

the current reading was recorded for at every increment. This was continually repeated until the glow-discharge plasma has been fully formed. When this happens, the voltage was reduced by 30 V to prevent the melting of the aluminum cathode and the experiment was left to run for 30 minutes to allow the synthesis of the nanoparticles. Distilled water was added regularly to maintain the volume of the electrolyte at 300 mL. When the experiment was completed, the beaker was stored in a box and left overnight to ensure that the nanoparticles were sedimented to the bottom of the beaker. Then, some of the excess electrolyte was removed from each beaker and the samples were transferred into large centrifuge tubes.

3.4 Centrifuging of the Samples

Using a dropper, the remaining electrolyte was removed in each sample until the respective volume is less than 100 ml. The samples were then mixed properly by swirling the dropper around. Subsequently, all of the mixture of each sample was transferred into 4 smaller centrifuge tubes where the volumes is filled up to 12ml. The caps of each tubes were closed, tighten and the samples were shaken. These tubes were then placed into the slots in the centrifuge machine. The speed of the machine was set to 6,000 rpm and the duration of the centrifuge at 15 mins. The centrifuge is then started. After the first cycle is done, a dropper is used to remove the remaining electrolyte in each tube. Next, the tubes were refilled with distilled water up to 12ml and the centrifuge is repeated. After this second cycle, the entire process is repeated for another time.

3.5 Results Reviewing

The size of the aluminum nanoparticles were analyzed through a scanning electron microscope and SEM images were obtained for each samples. Using ImageJ, the images were processed and the sizes of the nanoparticles were measured.

4. Results

4.1 I-V Curves of Submerged Glow-Discharge Plasma

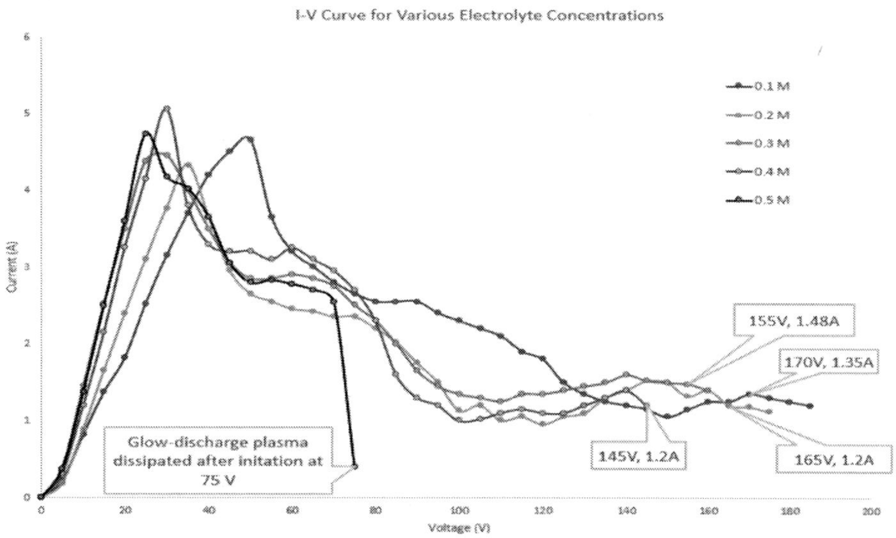

Graph 1. I-V curve for various electrolyte concentration.

IOP Conf. Series: Materials Science and Engineering **943** (2020) 012006 doi:10.1088/1757-899X/943/1/012006

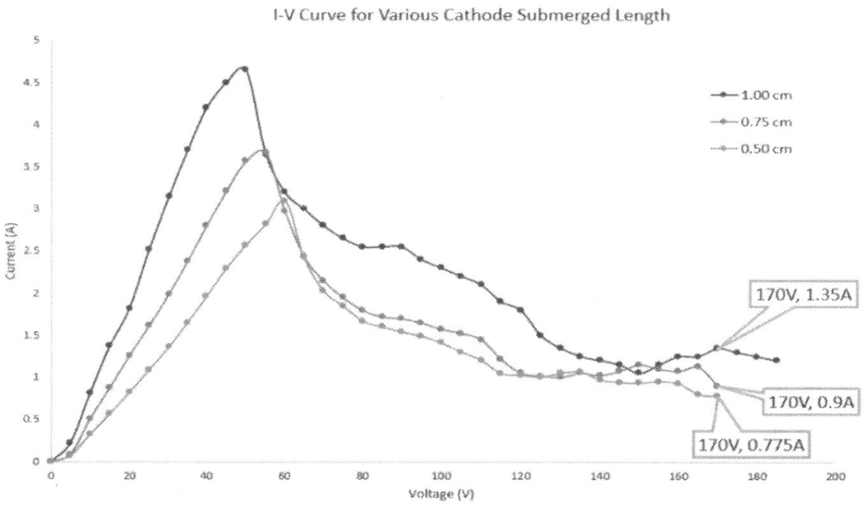

Graph 2. I-V curve for various cathode submerged length.

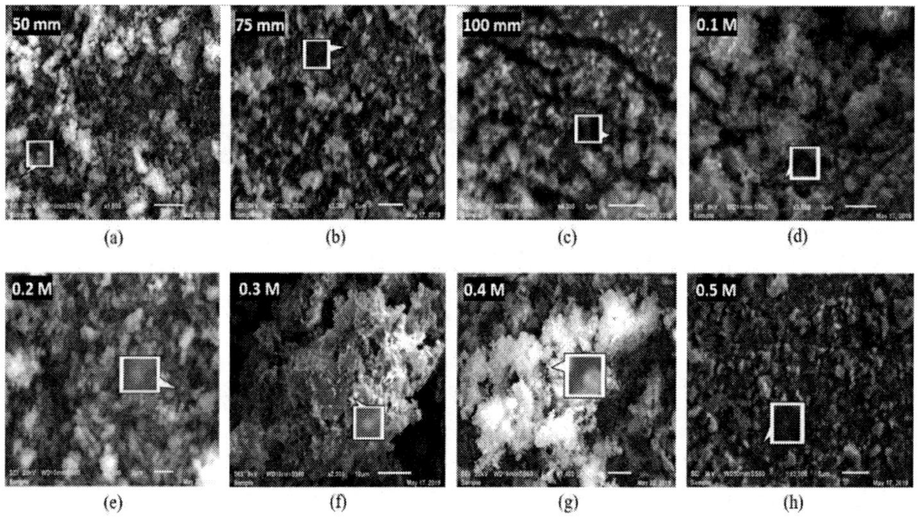

Figure 7. SEM micrograph of aluminum nanoparticles produced from varying submerged length of (a) 50mm, (b) 75mm, (c) 100mm and varying electrolyte concentration of (d) 0.1M, (e) 0.2M, (f) 0.3M, (g) 0.4M and (h) 0.5M.

Table 2. Experimental results for varying electrolyte concentrations.

Electrolyte Concentration (M)	Glow Discharge Voltage (V_D)	Power to initiate glow discharge plasma (W)	Average Size of aluminum nanoparticles (nm)	Standard deviation of aluminum nanoparticles (nm)
0.1	170	229.50	140.00	15.09
0.2	165	198.00	131.90	9.93
0.3	155	229.40	119.00	9.38
0.4	145	174.00	105.60	14.62
0.5	-	-	167.60	32.30

Table 3. Experimental results for varying cathode submerged lengths

Cathode Submerged Length (mm)	Glow Discharge Voltage (V_D)	Power to initiate glow discharge plasma (W)	Average Size of aluminum nanoparticles (nm)	Standard deviation of aluminum nanoparticles (nm)
50	170	229.50	154.90	29.30
75	170	153.00	144.50	9.61
100	170	131.75	140.00	15.09

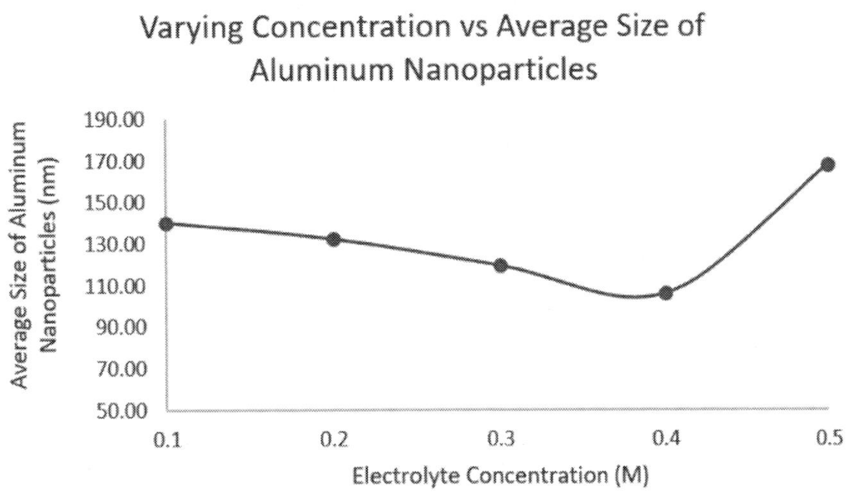

Graph 3. Relationship of electrolyte concentration and the average size of aluminum nanoparticles.

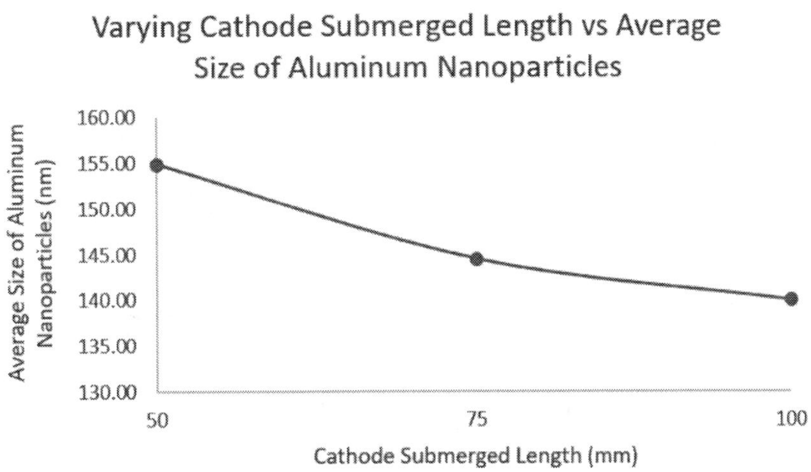

Graph 4. Relationship of cathode submerged length and the average size of aluminum nanoparticles.

Table 4. Relationship between W_v and W_m for varying electrolyte concentrations

Electrolyte Concentration (M)	Total submerged surface area (cm^2)	Power density for vapor formation (W_v/cm^2)	Power density to sustain plasma (W_m/cm^2)
0.1	2.36	97.36	99.02
0.2	2.36	84.00	89.39
0.3	2.36	97.32	107.59
0.4	2.36	73.82	99.72
0.5	2.36	12.73	11.83

Table 5. Relationship between W_v and W_m for varying cathode submerged length

Cathode Submerged Length (mm)	Total submerged surface area (cm^2)	Power density for vapor formation (W_v /cm2)	Power density to sustain plasma (W_m/cm^2)
50	0.98	134.15	135.62
75	1.62	94.41	98.17
100	2.36	97.36	99.02

5. Discussion

5.1 Relationship between glow discharge plasma intensity on the nanoparticles size

The relationship between the glow discharge plasma intensity and the nanoparticle size is seen in this research. A low plasma intensity indicates that the glow discharge is able to be reached faster due to a high conductivity in the electrolytic cell. Subsequently, more current concentration spots are formed on the cathode surface [6]. These spots may overlap with one another to form smaller nanoparticles. At low plasma intensity, the number of spots is not sufficient enough to overlap with one another. As an implication, the size of the nanoparticles formed is larger.

5.2 Influence of electrolyte concentration on the nanoparticles size

Based on Table 2, the discharge voltage drops as the electrolyte concentration increases. This indicates that at a higher concentration, there are more free moving radicals in the electrolyte, thus making the electrolytic cell more conductive. As a result, glow-discharge plasma was seen at a low plasma intensity [3]. Subsequently, when the gaseous sheath forms around the cathode, it is exposed to a higher density of electron bombardments [6]. This implies that more electron beams are targeted towards the cathode. Hence, overlapping of the current concentration spots do occur in this research and the overall size of the aluminum nanoparticles formed are smaller in size with increasing electrolyte concentration [8]. This here is illustrated in Graph 3. In short, at higher electrolyte concentrations, the resulting aluminum nanoparticle size formed is smaller.

In the experiments conducted under varying concentration, it was noticed that the experiment based on 0.5M of potassium carbonate was the upper limit. At any upper limit, the initiation of the plasma is difficult to achieved. As discussed in the literature review, in order to initiate and sustain the plasma, W_m has to be greater than W_v. However, the W_v for the experiment conducted with 0.5M of potassium carbonate, was found to be larger than its corresponding W_m as seen in Table 4. Although no plasma was initiated, some nanoparticles were still formed as a result. This indicates that the temperature on the cathode surface was higher than the aluminum cathode's melting temperature especially when the voltage is increased over time. As a result, nanoparticles were formed in the electrolysis stage but the overall size are larger.

5.3 Influence of cathode submerged length on the nanoparticle size

According to Table 3, the discharge voltage remains the same for all three experiments. However, the resulting size becomes smaller as the cathode submerged length decreases as seen in Graph 4. At a shorter submerged length, the total submerged surface area exposed to the overlapping of electron beams decreases [3]. Hence, for the 50 mm cathode submerged length experiment, it is inferred that more large-sized nanoparticles were produced than smaller-sized ones. When averaging for the size was done for this sample, the resulting average size was larger than the other two samples. An upper limit was found at 125mm of cathode submerged length. This can be seen at 100mm as its W_m approaches closer to W_v based on Table 5. Overall, a longer cathode submerged length results in smaller-sized aluminum nanoparticles.

6. Conclusion

This research aims to create a size control method for the aluminum nanoparticles by varying the electrolyte concentration and the cathode submerged length. Five different electrolyte concentrations were used and experimented to investigate its impact on the average size of the nanoparticles produced. Due to the presence of more electron beams on the cathode surface at higher concentrations, the plasma intensity decreases. Correspondingly, the electron beams tend to overlap more and thus resulting in smaller-sized nanoparticles. Three different cathode submerged lengths were uses to study the impact on the nanoparticle size. A shorter submerged length has a smaller total submerged surface area. This reduces the chance of electron beams to overlap on the cathode surface. Hence, the average nanoparticle size formed is larger.

The size control of aluminum nanoparticle enables them to be used in various applications. These nanoparticles can be used as an additive to biodiesel blends for diesel engines as its required size ranges from 60nm to 160nm. The implementation of this nanoparticles reduces the brake specific fuel consumption by 6% as well as reducing NO_x and CO_x emissions by 6% and 19% respectively. Other application includes being a catalyst for rocket fuels due to its unique combustion properties. Since there are many factors that controls the size of aluminum nanoparticles, the research topic can be extended further such as using different cathode shapes or varying the distance between the cathode and the anode. Besides that, the study on the effects of different materials' melting temperature on the size should be investigated as well since it was noticed that aluminum nanoparticles were still formed even at the electrolysis stage. Therefore, the effects of various electrolyte concentration and cathode submerged lengths reported here can be applied in the field of nanotechnology and nanomaterials.

References

[1] A. K. Singh, "Introduction to Nanoparticles and Nanotoxicology," *Eng. Nanoparticles*, pp. 1–18, 2016.

[2] "The global market for aluminium oxide nanoparticles," *Futur. Mark. Inc*, no. 76, p. 48, 2013.

[3] P. N. Y. Yek, M. R. M. Julaihi, M. S. Osman, T. C. Tiong, W. H. Lee, C. L. Lee, N. Mohamed Noor, and A. W. Azhari, "Submerged Glow-Discharge Plasma: An Economical Approach to Convert Construction Scrap Metal into Nanomaterials," *E3S web Conf.*, vol. 34, 2018.

[4] Muhammad Rafiq Mirza bin Julaihi, "Synthesis of metallic oxide nanoballs via submerged glow-discharge plasma and their photocatalytic effect," *Hokaido Univ.*, 2015.

[5] S. Horikoshi and N. Serpone, "In-liquid plasma: A novel tool in the fabrication of nanomaterials and in the treatment of wastewaters," *RSC Adv.*, vol. 7, no. 75, pp. 47196–47218, 2017.

[6] G. Saito, Y. Nakasugi, and T. Akiyama, "High-speed camera observation of solution plasma during nanoparticles formation," *Appl. Phys. Lett.*, vol. 104, no. 8, 2014.

[7] S. K. Sen Gupta, "Contact glow discharge electrolysis: Its origin, plasma diagnostics and non-faradaic chemical effects," *Plasma Sources Sci. Technol.*, vol. 24, no. 6, p. 63001, 2015.

[8] G. Saito and N. Sakaguchi, "Solution plasma synthesis of Si nanoparticles," *Nanotechnology*,

vol. 26, no. 23, pp. 1–8, 2015.

[9] P. Y. and M. J. MHS Al Anbouri, BWM Yii, "Effect of Electrolyte Concentration during Solution Plasma on Copper Nanoparticle Size," 2018.

[10] G. Saito, Y. Nakasugi, and T. Akiyama, "Generation of solution plasma over a large electrode surface area," *J. Appl. Phys.*, vol. 118, no. 2, 2015.

[11] "Aluminum Nanoparticles," *American Elements*, 2018. [Online]. Available: https://www.americanelements.com/aluminum-nanoparticles-7429-90-5.

[12] M. J. Meziani, C. E. Bunker, F. Lu, H. Li, W. Wang, E. A. Guliants, R. A. Quinn, and Y. P. Sun, "Formation and properties of stabilized aluminum nanoparticles," *ACS Appl. Mater. Interfaces*, vol. 1, no. 3, pp. 703–709, 2009.

IOP Conf. Series: Materials Science and Engineering **943** (2020) 012007 doi:10.1088/1757-899X/943/1/012007

Manipulation of Electrolyte Concentration and Cathode Morphology to Control the size of the Carbon Nanoparticles

M S Ranjit*, B L Y Yang and M R M Julaihi

Swinburne University of Technology Sarawak, Jalan Simpang Tiga, 93350 Kuching Sarawak, Malaysia

Email: 100073374@students.swinburne.edu.my

Abstract. The research aims to study the effects of concentration of the electrolyte and morphology of the cathode rod on the synthesis and size control of the carbon nanoparticles. The research was conducted using the submerged glow discharge plasma (SGDP) method. Potassium carbonate a strong electrolyte was used to maintain the accuracy of the results. The glow discharge voltage and any instabilities were recorded and monitored to obtain trends during the synthesis of carbon nanoparticles. Moreover, the shapes and sizes of the carbon nanoparticles formed were also observed and analyzed. The SGDP method is both eco-friendly and cheap to conduct. Carbon nanoparticles range from 1 – 100 nm. The small size of these nanoparticles gives it a high surface area to its volume ratio over sheet and bulk metals. The nanoparticles have different mechanical properties with respect to their sizes. The ability to control the sizes of the nanoparticles will enable manufacturers to obtain the mechanical properties they desire. The research conducted was able to determine the factors which enables the control of the sizes of carbon nanoparticles formed. At higher electrolyte concentration and higher cathode thickness, the size of the carbon nanoparticles decreased.

1. Introduction

A considerable amount of research is required in the synthesis and size control of carbon nanoparticles. Controlling the sizes of the carbon nanoparticles will have an effect on the mechanical properties of the carbon nanoparticles. The SGDP method was selected as it is easy to set up, requires no hazardous materials and cost effective to carry out **[1]**. The two main parameters controlled during the experimentation are *manipulating the concentration of the electrolyte* and *manipulating the morphology of the cathode* **[2]**.

In this paper, the electrolyte concentration was varied to observe the relationship in the size of the carbon nanoparticles formed with respect to the number of free moving electrons in the solution. The morphology of the cathode is varied through manipulating the thickness of the carbon rod used. This is to investigate the relationship between the surface area and the size of the carbon nanoparticles formed.

2. Background

2.1 Submerged Glow Discharge Plasma Method (SGDP)

The submerged glow discharge plasma method (SGDP) is an electrolysis of water-based experiment. This method comprises of four main stages; electrolysis stage, vaporization stage, transition stage and plasma stage. The formation of the nanoparticles revolves around the collision between the electrons and neutral atoms on the surface of the cathode. The collision causes a spark which ionizes, dissociates and excites the electrons with the emission of light called the "glow phenomenon". The nanoparticles are formed during the plasma stage through an electrical discharge in liquid **[3]**.

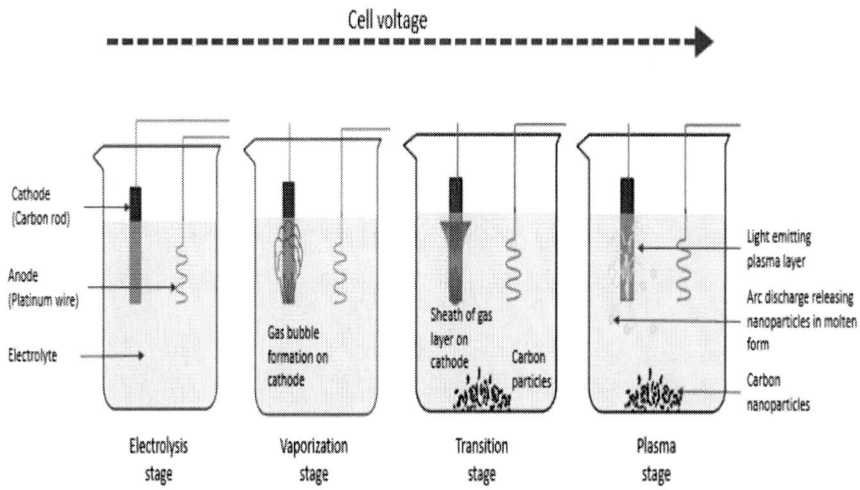

Figure 1. SGDP method reaction at different voltage value.

During the transition phase, vapor bubbles are produced on the surface of the cathode separating the layer between the cathode surface and liquid. When plasma is formed, a pressure difference will exist between these two layers. High pressure on the surface of the cathode enables the nanoparticles to be discharged out in molten form. These particles in molten form falls to the bottom of the beaker and when cooled will turn into nanoparticles.

2.2 Breakdown Voltage (V_b), Glow Discharge Voltage (V_d) and Plasma Intensity

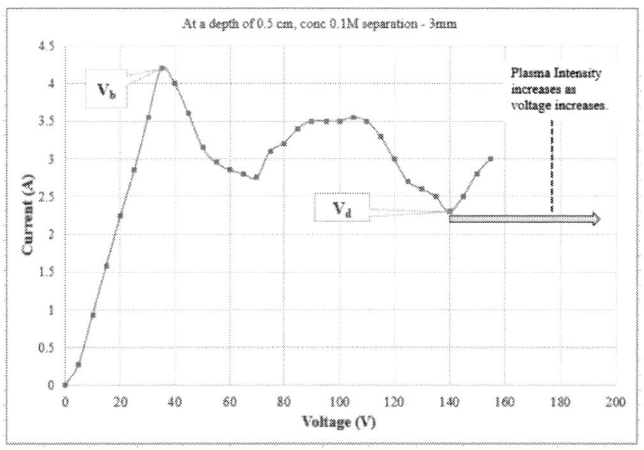

Figure 2. Current against voltage curve. V_b and V_d correspond to the breakdown and discharge voltage respectively. Plasma formation occurs at point V_d.

The breakdown voltage, V_b is the maximum point at which the curve obeys Ohm's law. When the voltage is increased at this point, gaseous bubbles begins to form with a current drop to signify a breakdown of normal electrolysis. As the voltage increase the gaseous sheath around the cathode will slowly stabilize. The current will continue to decrease until light emission is detected. This voltage point is called the glow discharge voltage, current onwards will slowly start to increase again **[4]**. Plasma formation occurs at the glow discharge voltage, nanoparticles are discharged out during this phase. At a higher applied voltage, the nanoparticles formed will increase in size due to the plasma intensity **[5]**. The plasma intensity is proportional to the excitation of electron temperature which occurs during the initiation of plasma on the cathode.

2.3 Size Control of Carbon Nanoparticles

In this research, the formation of the carbon nanoparticles occurs during the plasma stage when the glow discharge voltage, V_d is achieved. This shows that at different V_d values the shape and size of the nanoparticles formed would vary from one another.

With potassium carbonate (K_2CO_3) as the electrolyte, manipulating the concentration of electrolyte will vary the number of free-flowing electrons in the solution. A higher concentration would have more free-flowing electrons in the solution which will increase the conductivity in the solution. This will decrease the glow discharge voltage as the collision intensity between the electrons and neutral atoms increases **[6]**. The current concentration spots on the cathode will increase, thus will cause the discharge of smaller nanoparticles.

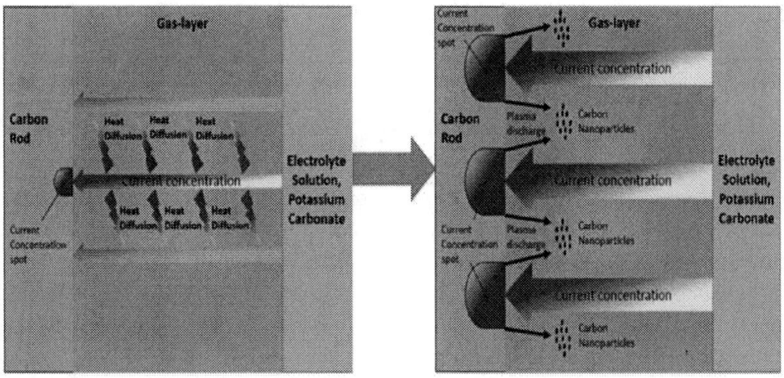

Figure 3. Current concentration spots increases with electrolyte concentration and voltage.

The morphology of the carbon nanoparticles will be studied through varying the thickness of the cathode. This will increase the surface area of the cathode exposed to the electrolyte. Under fixed condition where by the concentration of the electrolyte and depth of the cathode are kept constant, a larger amount of current is now required to initiate the plasma formation **[7]**. The glow-discharge voltage is expected to change depending on the amount of surface area exposed to the electrolyte. To reach plasma stage, requires higher amounts of energy in the form of heat for thicker rods.

2.4 Electrode Reactions

The SGDP method consist of a combination of both physical and chemical reaction occurring simultaneously during the synthesis of the carbon nanoparticles. The reaction takes place on the cathode surface.

$$H_2O + e^- \rightarrow OH^- + H^+ + e^-$$

The formation of the nanoparticles is due to the reaction between the metal ion and the hydrogen radicals which are high energy state **[8]**. At the plasma zone, the high voltage supplied to the aqueous solution during the electrolysis phase will release the hydrogen radicals. The voltage excites the hydrogen radicals cause them to accelerate towards the cathode. The acceleration of the hydrogen radicals towards the cathode is what causes the bombardment of electrons on the wall of the cathode. This bombardment causes the temperature to increase due to the immense heat from the collision on the cathode surface. The plasma layer is formed with a high temperature and electro-conductivity region. There is an electro-thermal instability which occurs due to the current concentrated at the cathode **[9]**. This will form the carbon nanoparticles through the difference in pressure between the cathode surface and the solution.

2.5 Mechanism of the SGDP, Hydrodynamic Instabilities and Plasma Turbulence

From figure 2, the point at which V_b occurs gives the maximum current required for the vapor formation. The power density for the formation of the vapor layer is denoted by W_v, (W/cm^2). An increase in the temperature of the solution will result in a decrease in the conductivity of the solution. Beyond this point the current will slowly decrease as the temperature in the solution increases **[4]**. Upon plasma formation, the electric power is reduced to enable the plasma to self – sustain. Moreover, when the solution has high amounts of resistance the power density required will be higher. Once vapor is formed, current will stabilize and the glow – discharge plasma can be obtained stably. The amount of power W_m (W/cm^2) required to form the plasma can be obtained from the product of the average current (I) and voltage (V) per surface area, A (cm^2) of the cathode exposed to the electrolyte at which the glow – discharge is maintained at.

$$W_m = I \times \frac{V}{A} \tag{1}$$

If the W_m obtained is less than the W_v will suggest that higher amounts of power is required to initiate the plasma formation.

The morphological aspects in the evolution of gas is distinguished between the electrode potential and current. During normal electrolysis, separate bubbles are formed when the temperature in the solution is increased with respect to the voltage. When V_b is achieved at the transition phase, the electrode will be covered by a gaseous blanket moving in a violent motion. This results in a drop in the current and the electrode potential is increased **[10]**. The cathode will be enclosed by a continuous gaseous envelope (plasma layer). The stability of the plasma layer is dependent on the relative speed between the counter flowing gas and liquid streams. If the relative speed is too high, rapid small disturbance will grow and break the contiguous streams at the boundary layer. This small disturbance is directly related to the plasma turbulence.

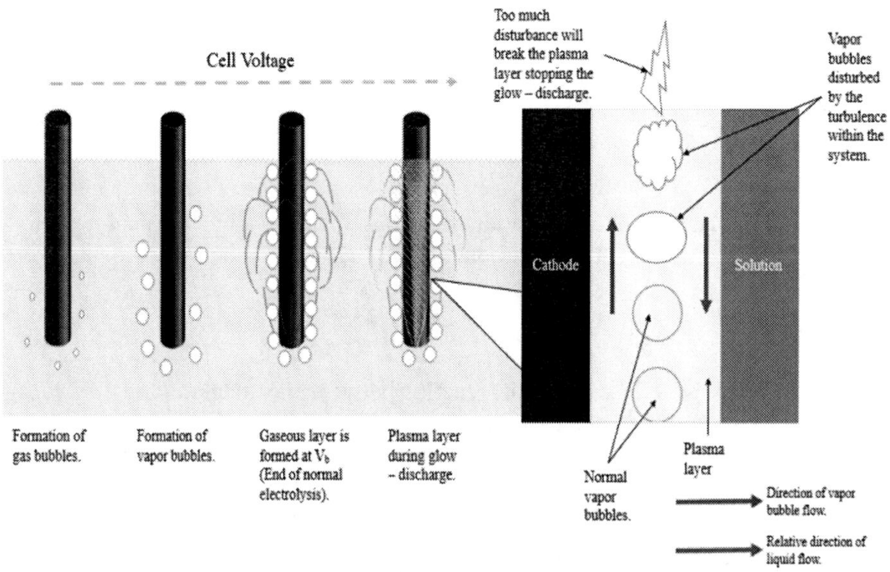

Figure 4. Instability during the glow – discharge.

2.6 Carbon Nanoparticles

Carbon nanoparticles are carbon particles that are nanosized. These particles have a diameter of less than 100nm. The carbon nanoparticles consist of an intermediate crystalline structure of graphite and an amorphous state. This can be described as a turbostatic structure that has a different bulk graphite crystal in an ABAB order **[11]**.

3. Methodology

The submerged glow discharge plasma (SGDP) method was used to synthesis the carbon nanoparticles.

3.1 Experimental Setup

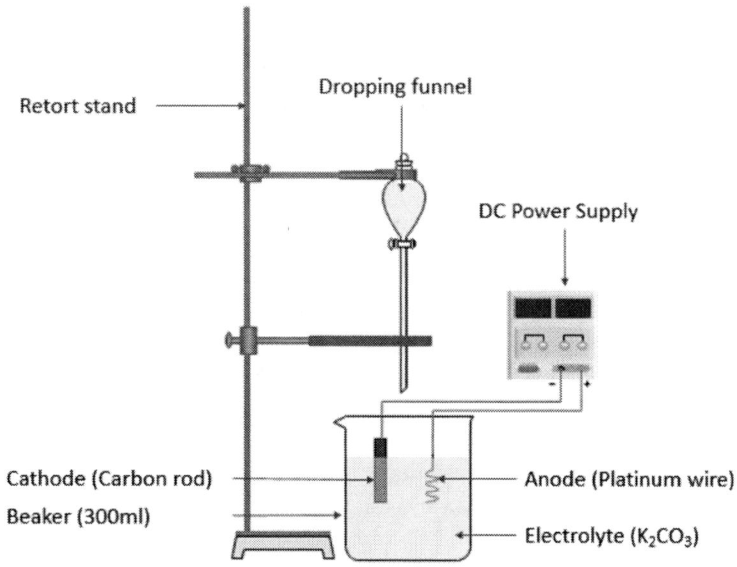

Figure 5. Experimental setup for submerged glow discharge plasma

3.2 Experimental Procedure

a. Preparation of electrolyte

The mass of K_2CO_3 required by the system was calculated using the formula:

$$Mass\ of\ K_2CO_3\ (g) = Volume\ (l) \times Concentration\ \left(\frac{mol}{l}\right) \times Molar\ mass\ K_2CO_3\left(\frac{g}{mol}\right)$$

$$Mass\ of\ K_2CO_3\ (g) = 0.3l\ \times 0.05\left(\frac{mol}{l}\right) \times 138.205\ \left(\frac{g}{mol}\right)$$

$$Mass\ of\ K_2CO_3\ (g) = 2.073g$$

The mass of K_2CO_3 in powdered form was weighted to obtain the mass calculated. The powder form K_2CO_3 was transferred to a beaker and distilled water filled up the beaker to 300 ml. The beaker was then heated on a hot plate before continuing the experiment. The steps were repeated to obtain mass for electrolyte concentration of 0.1M, 0.2M and 0.3M.

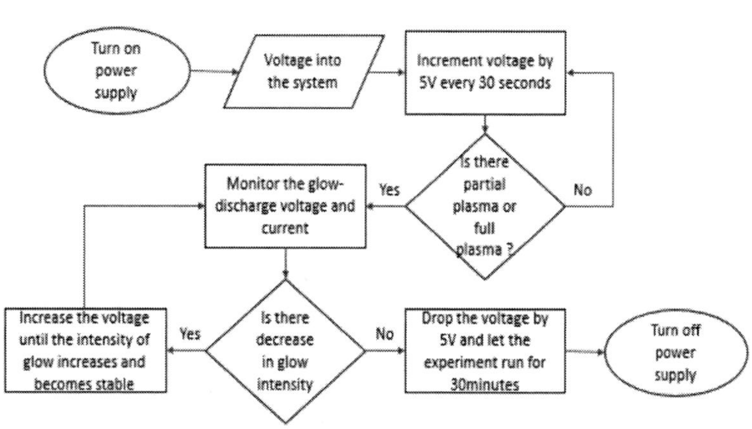

Figure 6: Flow chart to conduct the experiment

b. Manipulating the electrolyte concentration

The apparatus was set as shown in Figure 4. A DC power supply was used to supply an amount of electrical current to the system. Carbon rods of 0.6cm was used as the cathode at a depth of 1.0cm in the electrolyte and a 10cm long platinum wire was used as the anode. The concentration of the K_2CO_3 electrolyte was set to 0.05M. The power supply was switched on and the voltage and current values were recorded. The voltage follows an increment of 5V every 30 seconds. The cathode rod was observed closely at every voltage increment until the formation of sparks (partial plasma). The voltage was further increased until the formation of full plasma. The voltages for both partial plasma and full plasma were recorded down as the glow-discharge voltage. At the glow-discharge voltage, the experiment was left to run for 30 minutes to produce more nanoparticles. The experiment was repeated with electrolyte concentration of 0.1M, 0.2M and 0.3M.

c. Manipulating the morphology of the cathode

The experimental setup is the same as experiment in (b). The changes are the concentration of electrolyte was fixed to 0.1M and the thickness of the cathode was changed to 0.5mm. The power supply was switched on and the voltage and current values were recorded. The voltage follows an increment of 5V every 30 seconds. The cathode rod was observed closely at every voltage increment until the formation of sparks (partial plasma). The voltage was further increased until the formation of full plasma. The voltages for both partial plasma and full plasma were recorded down as the glow-discharge voltage. At the glow-discharge voltage, the experiment was left to run for 30 minutes to produce more nanoparticles. The experiment was repeated with cathode thickness of 1.0mm, 2.0mm and 10.0mm. For both (b) and (c) if the solution volume decreases due to evaporation, a dropping funnel is used to add distilled water into the solution to maintain the solution volume.

d. Centrifuging

After conducting the experimental procedures in (b) and (c), the nanoparticles in the beaker were left overnight inside a covered box for sedimentation. The next day, the solution in the beaker was slowly reduced to around 50ml. The nanoparticles inside the remaining solution were transferred into a centrifuge tube and placed into the centrifuge machine. The machine was set to 6000rpm for 15 minutes per cycle. Three cycles were carried out and at the end of every cycle the remaining electrolyte solution were removed and refilled with distilled water. The nanoparticles were then transferred to another centrifuge tube and sent for SEM viewing.

e. Size analysis of carbon nanoparticles through ImageJ

The carbon nanoparticles obtained from SEM imagery were measured using the ImageJ software. Based on the given scale on the SEM image, the software helps to measure the size of the nanoparticles on the image by adjusting the scale to fit the line drawn on the diameter of selected particle on the actual image.

4. Results

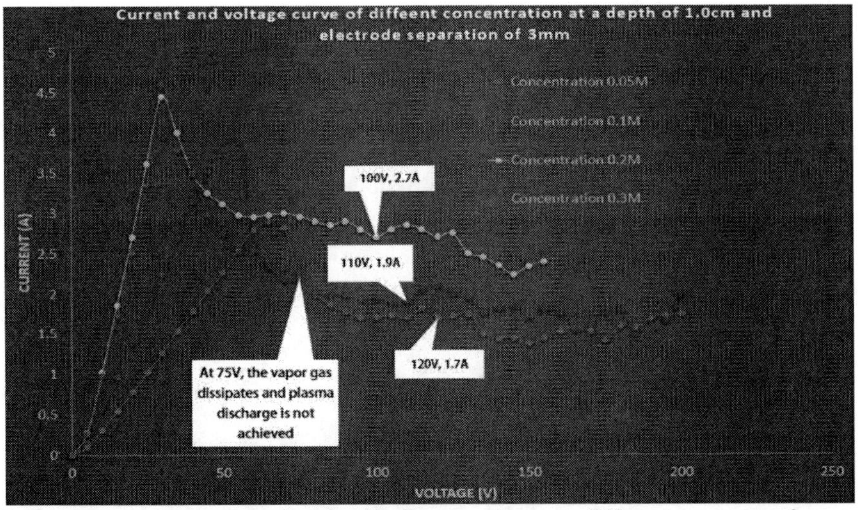

Figure 7. Current - voltage curve for the discharge voltage at different concentration at constant cathode thickness of 0.6mm.

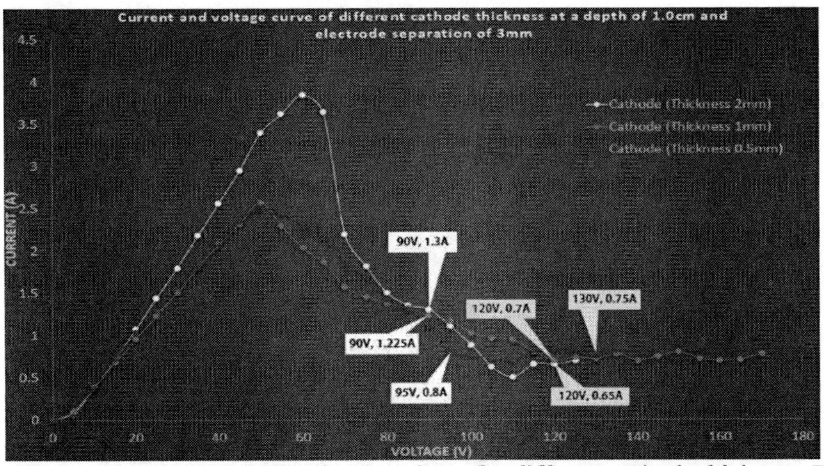

Figure 8. Current - voltage curve for discharge voltage for different cathode thickness. The boxes in white indicate formation of partial plasma and boxes in yellow indicate formation of full plasma

IOP Conf. Series: Materials Science and Engineering **943** (2020) 012007 doi:10.1088/1757-899X/943/1/012007

Figure 9. SEM imagery from a) to c) indicate the CNP's formed through change in electrolyte concentration. SEM imagery from d) to f) for indicates the CNP's formed through change in cathode thickness.

Table 1. Size of nanoparticles with respect to concentration at constant cathode thickness of 0.6cm

Concentration (M)	Power required to initiate glow - discharge, P (W)	Surface area of the cathode, A (cm2)	Power to form the plasma per unit area, Wm (W/cm2)	Average size of Carbon nanoparticles, CNPs (nm)	Standard deviation of sizes of CNPs (nm)
0.05	204	2.4504	83.3	121	26.53
0.1	209		85.3	98.7	22.21
0.2	270		110.2	80.6	12.78
0.3	Unable to form		-	-	-

Table 2. Size of carbon nanoparticles with respect to cathode thickness at constant electrolyte concentration of 0.1M

Thickness, t (mm)	Power required to initiate glow - discharge, P (W)	Surface area of the cathode, A (cm2)	Power to form the plasma per unit area, Wm (W/cm2)	Average size of Carbon nanoparticles, CNPs (nm)	Standard deviation of sizes of CNPs (nm)
0.5	76	0.16	472	128.7	28.49
1.0	110	0.33	334	123.4	29.17
2.0	117	0.69	169	110.9	23.83

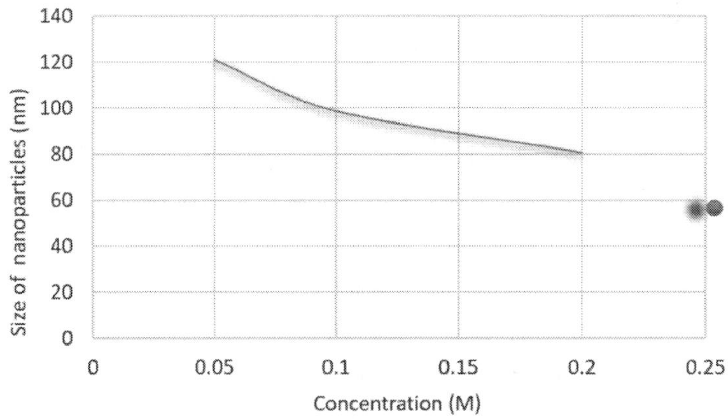

Figure 10. Size of CNP's formed at different electrolyte concentration.

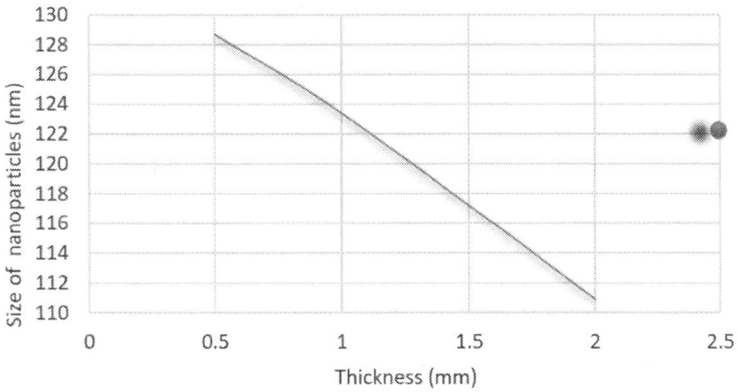

Figure 11. Size of CNP's formed with respect to different cathode thickness.

5. Discussion

5.1 Current – voltage curve relationship for change in electrolyte concentration and size of carbon nanoparticles

The results obtained from Figure 7 shows the initiation of glow – discharge plasma for varying electrolyte concentration. Due to the volatility of the system and to prevent injuries from explosion of carbon rod, the nanoparticles were left under partial plasma condition. At higher concentration the glow – discharge voltage, V_d is lower. The amount of power required to initiate the glow – discharge also increases with the increase in the concentration of the electrolyte as seen in Table 1. This is due to the increase in the number of free moving electrons in solution. More electrons within system would require higher amounts of energy to excite the electrons to bombard neutral atoms on the wall of the cathode (Carbon). The higher energy required to initiate the glow - discharge, the more vigorous the movement of vapor bubbles in the plasma layer.

From the results obtained in Figure 10, the sizes of the carbon nanoparticles formed decreases when the concentration of the electrolyte increases. In addition to that, the glow – discharge voltage was found to be lower when the sizes of the carbon nanoparticles discharged out decreases **[6]**. However, the drop-in voltage corresponds to a higher current output indicating a higher amount of output power. The current concentration spots around the cathode increases with the increase in the number of free – moving electrons within the system. When the number of current concentration spots around the cathode increases, the tendency for these current concentration spots to overlap on the surface of the cathode increases. Due to this, there will be a higher amount of pressure difference on the surface of the cathode where the current concentration spot overlaps compared to other section of the cathode where there is no overlap in the current concentration spots. The higher amounts of pressure difference coupled with the overlapping current concentration spots causes the discharge of smaller carbon nanoparticles. The intensity and frequency of the current concentration spots determines the sizes of the carbon nanoparticles formed. This also relates to an observation made during the experimentation whereby the intensity of the plasma discharge was observed to be more intense at higher electrolyte concentration.

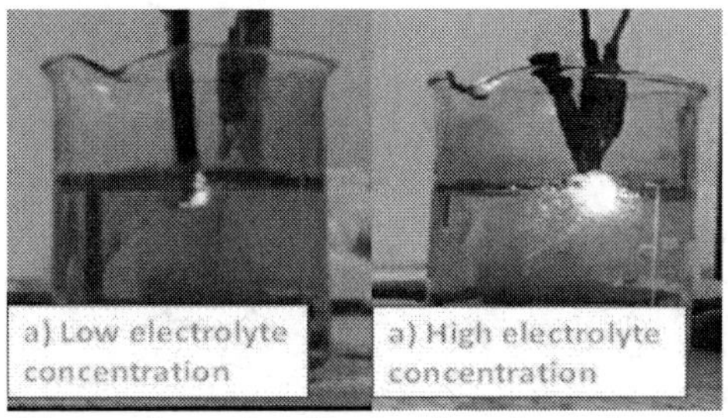

Figure 12: Plasma intensity for varying electrolyte concentration

At 0.3M in Figure 7, shows the dissipation in the formation of the vapor gas. The dissipation in the vapor gas prevents the glow – discharge from occurring. In order to form the glow – discharge, a plasma layer must first be developed from the vapor gasses which is supposed to form gaseous sheath on the surface of the cathode.

5.2 Current – voltage curve relationship for change in the morphology of the cathode and size of the carbon nanoparticles

The results obtained from Figure 8 shows the variation in the initiation of glow – discharge when the morphology of the cathode in terms of its thickness was manipulated. Unlike the experiments conducted for 0.6cm carbon rod, the 0.05cm, 0.1cm and 0.2cm carbon rods were able to achieve full plasma discharge. The system was discovered to be less vigorous compared to the experiment with the 0.6cm carbon rod. The glow - discharge voltage at the initiation of plasma and full plasma stage in Figure 8 shows little variation in terms of when the formation of carbon nanoparticles occurs. However, there is a variation in the power required to initiate the glow – discharge. From Table 2 an increase in the thickness of the cathode will result in higher power required to form the glow – discharge. For a fixed number of free moving electrons in the solution, when the surface area of the cathode exposed to the solution is increased more energy is required to excite and drive the electrons to collide against the neutral atoms at the wall of the cathode. During the experiment, the vigorousness of the vapor bubbles in the system increases with respect to the amount of power required for the formation of the glow – discharge. The amount of power required to form the glow – discharge is dependent on the surface area exposed to the solution at constant electrolyte concentration.

From the graph plotted in Figure 11 the size of the carbon nanoparticles formed decreases when the thickness of the cathode increase. The amount of power required for the initiation of plasma also increases with respect to the thickness of the cathode. The high amount of power is enough to drive electrons around the surface of the cathode. There is a high amount of energy around the current concentration spots formed from the high energy levels within the system which causes heat to diffuse around the current concentration spot. Higher amounts of heat energy when the cathode thickness increases would cause the current concentration spot to increase and overlap one another on the cathode surface. The overlap in the current concentration spots is expected to not be as intense when compared to the experiment conducted with varying electrolyte concentration. This can be seen through comparing the amount power required to initiate the plasma discharge in Table 1 and 2. The power required to initiate the partial plasma discharge corresponds to the amount of plasma intensity and the heat energy from within the system.

5.3 Plasma turbulence, formation and shape of carbon nanoparticles

Not all the carbon nanoparticles formed observed in Figure 9 have a spherical shape. The shapes of some of these nanoparticles are irregular. The irregularity in the shape of the carbon nanoparticles formed is due to the presence of plasma turbulence. These turbulences exist in the plasma layer located in between the cathode surface and solution. It was noticed from the SEM imagery in Figure 9, the frequency of being able to obtain perfectly spherical carbon nanoparticles decreases with respect to both the increase in electrolyte concentration and increase in the thickness of the cathode.

The amount of power required to initiate the partial plasma formation increases with the increase in the electrolyte concentration and cathode thickness. The increase in power indicates an increase in the plasma intensity and heat energy within the system. When the system is experiencing a drop in current, a gaseous blanket will form around the cathode **[10]**. When the amount of energy in the system increase, the intensity and vigorousness of the movement of vapor bubbles increases as well. As mentioned, the stability of the plasma layer is highly dependent on the speed of the counter flowing gas and the relative liquid flow stream. If the system can maintain a stable plasma layer, the carbon nanoparticles discharged out from the cathode will be more spherical in shape. When the energy in the system is high, the movement of the counter flowing gas against the relative liquid flow stream will be high. This creates disturbance (turbulence) within the plasma layer. The consistency of the current concentration spots will decrease as the disturbance begins to grow with time. The pressure difference on the surface of the cathode will vary from each location on the cathode which cause the discharge irregularly shaped carbon nanoparticles. Being able to form perfect spherical carbon nanoparticles at both higher concentration and cathode thickness is due to the plasma layer being highly stable at the start of the experiment. But as the energy from the increase in power of the system increase, the disturbance begins to grow and

eventually breaking the plasma layer thus stopping the submerged glow – discharge plasma from forming anymore carbon nanoparticles.
The experiment conducted at 0.3M in Figure 7 is unable to form the plasma layer as the speed of the vapor gas flow is flowing too high causing the high amount of disturbance which prevents the plasma layer from even forming the gaseous sheath.

5.4 Size of the carbon nanoparticles and cluster formations
The sizes of the carbon nanoparticles vary from one another through a standard deviation of around 12.78nm and 29.17nm. This shows that submerge glow – discharge plasma method is able to control size of the carbon nanoparticles. However, it was noticeable that there were a lot of carbon cluster formation from the SEM images in Figure 9. This is due to carbon being less dense compared to water which causes the nanoparticles to float around the surface of the solution. If left unattended these carbon nanoparticles will form into carbon sheaths thus making it difficult to measure individual spherical carbon nanoparticles. The clusters of carbon if exposed to sunlight will form carbon crystals.
From Figure 9, the carbon nanoparticles have formed agglomerates and fractal structure. The formation of the fractal structures is due to the presence of attractive van der Waals forces. The carbon nanoparticles will attract one another with the aid of the van der Waals forces forming the agglomeration and reducing the accuracy in terms of the sizes of the carbon nanoparticles formed. The presence of agglomerates can reduce the effectiveness of the carbon nanoparticles in terms of its electrical conductivity **[12]**.

6. Conclusion
The synthesis and size control of the carbon nanoparticles was able to be verified through the research conducted. It was found that at higher electrolyte concentration and higher cathode thickness the size of the carbon nanoparticles formed is smaller. The increase in the power required to initiate the partial plasma formation will affect the size of the carbon nanoparticles formed as the nanoparticles was found to be smaller in size. In short, the lower the glow – discharge voltage, the smaller carbon nanoparticles formed. Controlling the size of the carbon nanoparticles brings about enhancing the mechanical properties of materials which would be beneficial in the future for material manufacturing industry.

References

[1] M. Ishigami, J. Cumings, A. Zettl, and S. Chen, “A simple method for the continuous production of carbon nanotubes,” *Chem. Phys. Lett.*, 2000.
[2] P. N. Y. Yek, M. R. M. Julaihi, M. S. Osman, T. C. Tiong, W. H. Lee, and C. L. Lee, “Submerged Glow-Discharge Plasma: An Economical Approach to Convert Construction Scrap Metal into Nanomaterials,” in *E3S Web of Conferences*, 2018.
[3] M. R. M. bin Julaihi, S. Yatsu, M. Jeem, and S. Watanabe, “Synthesis of stainless steel nanoballs via submerged glow-discharge plasma and its photocatalytic performance in methylene blue decomposition,” *J. Exp. Nanosci.*, 2015.
[4] S. K. Sen Gupta, “Contact glow discharge electrolysis : its origin , plasma diagnostics and non-faradaic chemical effects,” *Plasma Sources Sci. Technol.*, p. 63001, 2015.
[5] G. Saito and N. Sakaguchi, “Solution plasma synthesis of Si nanoparticles,” *Nanotechnology*, vol. 26, no. 23, pp. 1–8, 2015.
[6] P. N. Y. Y. and M. R. M. J. M H S Al Anbouri, B W M Yii and Swinburne, “Effect of Electrolyte Concentration during Solution Plasma on Copper Nanoparticle Size,” *Eff. Electrolyte Conc. Dur. Solut. Plasma Copp. Nanoparticle Size*, 2018.
[7] G. Saito, Y. Nakasugi, and T. Akiyama, “Generation of solution plasma over a large electrode surface area,” *J. Appl. Phys.*, vol. 118, no. 2, 2015.
[8] P. Pootawang, N. Saito, and S. Y. Lee, “Discharge time dependence of a solution plasma process for colloidal copper nanoparticle synthesis and particle characteristics,” *Nanotechnology*, 2013.
[9] M. R. M. bin Julaihi, “Synthesis of metallic oxide nanoballs via submerged glow-discharge plasma and their photocatalytic effect,” *Synth. Met. oxide nanoballs via Submerg. glow-discharge plasma their photocatalytic Eff.*, no. March, 2015.
[10] B. Mazza, P. Pedeferri, and G. Re, “Hydrodynamic instabilities in electrolytic gas evolution,”

Electrochim. Acta, vol. 23, no. 2, pp. 87–93, 1978.
[11] J. Kang, O. L. Li, and N. Saito, "Synthesis of structure-controlled carbon nano spheres by solution plasma process," *Carbon N. Y.*, 2013.
[12] X. Deng, Z. Huang, W. Wang, and R. N. Davé, "Investigation of nanoparticle agglomerates properties using Monte Carlo simulations," *Adv. Powder Technol.*, 2016.

IOP Conf. Series: Materials Science and Engineering **943** (2020) 012008 doi:10.1088/1757-899X/943/1/012008

Characterization and evaluation of copper-doped akermanite ceramic

M M Htun[1], A F M Noor[1], M Kawashita[2], and Y M B Ismail [1, *]

[1] Biomaterials Niche Group, School of Materials and Mineral Resources Engineering, Engineering Campus, Universiti Sains Malaysia, 14300 Nibong Tebal, Penang, Malaysia
[2] Department of Inorganic Biomaterials, Institute of Biomaterials and Bioengineering, Tokyo Medical and Dental University, 2-3-10 Kanda-Surugadai, Chiyoda-ku, Tokyo 101-0062, Japan

Email: yannymarliana@usm.my

Abstract. The aim of this study is to investigate the influence of copper (Cu^{2+}) on structural and mechanical properties of akermanite ceramic synthesized via high-energy wet planetary ball-milling route. Based on structural analysis, partial substitution of copper ions did not affect the phase stability of host akermanite. Moreover, Cu^{2+} suppressed the grain growth, thus considerably improved the sinterability and mechanical properties of akermanite ceramic. Vickers microhardness values of doped samples increased to five- or six-times compared to that of pure akermanite at 1200°C. Our study indicated that the Cu^{2+}-doped akermanite ceramic could potentially be used as an alternative material for bone replacement in biomedical applications.

1. Introduction

Silicate bioceramics, such as wollastonite ($CaSiO_3$), diopside ($CaMgSi_2O_6$), akermanite ($Ca_2MgSi_2O_7$), bredigite ($Ca_7MgSi_4O_{16}$) and merwinite ($Ca_3MgSi_2O_8$) have been developed in the past ten years. It has been shown bioceramics containing CaO and SiO_2 possess bone-like apatite-forming ability and exhibit good bioactivity for bone implant materials [1].

An alkaline earth silicate-based compound of akermanite ($Ca_2MgSi_2O_7$), a member of CaO-MgO-SiO_2-based bioceramics, belongs to a tetragonal crystal system, space group $P\text{-}\bar{4}2_1m$ and Z = 2 [2]. Akermanite has garnered considerable interest in biomedical applications due to its dual angiogenesis/osteogenesis stimulation, controllable biodegradability, cytocompatibility, as well as *in vitro* bioactivity [3]. Additionally, akermanite possessed superior mechanical properties to those of stoichiometric hydroxyapatite (HA), beta-tricalcium phosphate (β-TCP) and bioglass [4]. Previously, Wu and co-workers prepared akermanite ceramics at 1370°C for 6 h. They reported that akermanite ceramics induced apatite formation after 10 days of immersion in stimulated body fluid (SBF) [5], which reflects the substrate is suitable in enhancing bone formation both *in vitro* and *in vivo* [1]. Even though akermanite retains both excellent *in vitro* and cell biological behaviours, its low fracture toughness limited its use in load-bearing tissue regeneration [6]. So far, the improvement is still a need.

Copper (Cu^{2+}), an essential inorganic ion in human body, is of interest due to its participation in blood vessel formation (angiogenesis) [7]. Sufficient copper level supports bone growth, but prevents osteoporosis [8]. Earlier researchers demonstrated that the combination of Cu^{2+} and Si^{4+} ions resulted in stronger stimulatory effects of human umbilical vein endothelial cells (HUVECs) on vascularization in comparison to pure $CaSiO_3$ bioceramics [9]. In addition, Željko R et al. highlighted that Cu^{2+} could improve the mechanical properties of calcium phosphate-based bioceramics synthesized hydrothermally [10]. Therefore, the synergic effects of Cu^{2+} and akermanite ceramics might be an interesting approach to achieve better mechanical characteristics than intrinsic counterparts for bone repair and regeneration applications.

IOP Conf. Series: Materials Science and Engineering **943** (2020) 012008 doi:10.1088/1757-899X/943/1/012008

Previously, calcium silicate-based materials were prepared through four main methods such as chemical precipitation, sol-gel, hydrothermal and mechanical activation methods [1]. In our work, high-energy mechanical activation route was applied as a processing technique for nano-structured powders preparation due to its simplicity of experiment, low production cost and rapid synthesis of the product [11].

To the authors' best knowledge, no study has been reported on the physical and mechanical properties of pure and Cu^{2+}-doped akermanite ceramic for bone tissue engineering applications. Hereby, akermanite is doped with different Cu^{2+} contents (2.5, 3.5 and 4.5 mol%) through high-energy planetary ball milling. Phase analysis, morphology and physico-mechanical properties were compared between Cu^{2+}-doped akermanite and akermanite ceramics alone.

2. Material and methods

2.1. Sample preparation

The akermanite and Cu-doped akermanite powders were synthesized via high-energy mechanical activation route. Briefly, the reagents calcium oxide (reagent grade, Sigma-Aldrich, USA), magnesium oxide (99%, Alfa Aesar, USA), silicon dioxide (99%, Sigma-Aldrich, USA), and copper (II) oxide (99%, Sigma-Aldrich, USA) were milled according to stoichiometric compositions of $Ca_{2-x}Cu_xMgSi_2O_7$, (where x = 0, 0.25, 0.35, 0.45) in a planetary ball mill (PM 100, Retsch GmbH, Germany) for 4 h at ambient conditions. The rotational speed of main disc was set at 500 rpm. The ball-to-powder weight ratio used was 10:1. Deionized water was used as suspension media. Afterwards, the resultant powders were uniaxially pressed at 150 MPa in an evacuable metal die into 13 mm diameter discs. Consequently, sintering was performed at 1200°C for 3 h in ambient air.

2.2. Material characterizations

The structural analysis was conducted using X-ray diffractometer (XRD; D2 Phaser, Bruker AXS Inc., USA) fitted with Cu–K_α radiation (λ = 0.154 nm) in a scan range of $20° \leq 2\theta \leq 60°$, with an integration time 1.25 sec/0.02° step. The crystallite size was estimated using the Scherrer's formula.

Morphological observation was conducted on the gold-coated akermanite ceramics using Field Effect Scanning Electron Microscope (FESEM; Gemini Zeiss Supra™ 35 VP, Germany). The average grain size of samples was evaluated using an image analysis software (ImageJ 1.51k, USA).

For diameter shrinkage evaluation, dimensions of ceramic bodies before and after sintering were measured with the use of a digital vernier calliper (Mitutoyo 500-197-30 AOS Absolute, Japan). Then, the shrinkage values were calculated using the equation reported by Hana N et al. [12].

Relative density of the pellets was determined in accordance with ASTM B962-17 by applying Archimedes' relations [13].

Vickers hardness (H_V) was assessed on polished sintered pellets in accordance with ASTM (E-384) procedure [14] applying a digital Vickers microhardness tester (FV-810, Future-Tech Corp., Japan) with 5 kilograms-force (kg.f) load and a dwelling time of 10 secs. Fracture toughness (K_{IC}) was estimated by using Avans and Charles' relation [15].

$$K_{IC} = 0.0824 \frac{F}{C^{3/2}} , \quad (1)$$

where, F is the applied load (N), and C is radial crack length (m) measured from the center point of the indented area.

3. Results and discussion

3.1. Structural analysis

XRD patterns of pure and Cu^{2+}-doped akermanite are illustrated in Figure 1. The resultant diffraction peaks of $Ca_{2-x}Cu_xMgSi_2O_7$ were well matched with akermanite standard pattern (ICDD: 35-592). As an analogue with akermanite in component, a minor presence of diopside ($CaMgSi_2O_6$; ICDD: 41-1370), a bioactive phase, was also detected.

Cu^{2+} doped in akermanite structure slightly altered the lattice parameters, which may be induced by lattice distortion. As evidenced by the enlarged spectra between 25° and 35°, Cu^{2+}-doping somewhat engendered an insignificant shift of diffraction peaks towards the higher angles in term of dopant content This attributed to the partial substitution of Cu^{2+} into Ca^{2+} site of akermanite host. Herein, the explanation for site occupancy of Cu^{2+} in akermanite structure could be made as follows.

A silicate compound akermanite ($Ca_2MgSi_2O_7$), belonging to $X_2T^{(1)}T^{(2)}{}_2O_7$ melilite family of minerals with tetragonal crystal symmetry. The akermanite structure was reported to possess three independent cation sites ($X = Ca^{2+}$, $T^{(1)} = Mg^{2+}$, $T^{(2)} = Si^{4+}$). The Ca^{2+} ion is coordinated by six or eight neighbouring oxygen (O^{2-}) ions. Mg^{2+} and Si^{4+} ions were documented to occupy in the tetragonal sites [16]. Considering the effective ionic radii (r) and charge balance of the cations corresponding to coordination number (CN), the ionic radius of Cu^{2+} (r = 0.73 Å, CN = 6) is close to that of Ca^{2+} (r = 1.00 Å, CN = 6) rather than Mg^{2+} (r = 0.57 Å, CN = 4) and Si^{4+} (r = 0.26 Å, CN = 4). Therefore, from a crystallographic viewpoint, it could be predicted that the accommodation of Cu^{2+} ions in Ca^{2+} sites of $Ca_2MgSi_2O_7$ host is more possible than the Mg^{2+} and Si^{4+} sites.

Adopting Scherrer's relation, the crystallite size was then estimated from the (111), (021), (121), (130), (122), and (132) planes of akermanite. The crystallite size was decreased from 53.981 to 52.524, 51.075 and 49.321 nm in terms of dopant content. The unit cell volume ($V_{cell} = a^2 \times c$) were decreased from 307.374 cm³ to 307.320, 307.254, 307.175 cm³ for AK, 2.5Cu-AK, 3.5Cu-AK and 4.5Cu-Ak respectively, proving lattice contraction [17]. Any additional phase formation was not detected in the profiles, clearly proving the structural stability with doping. Notably, the increase in monoclinic phase with further increase in Cu^{2+} dopant was expected.

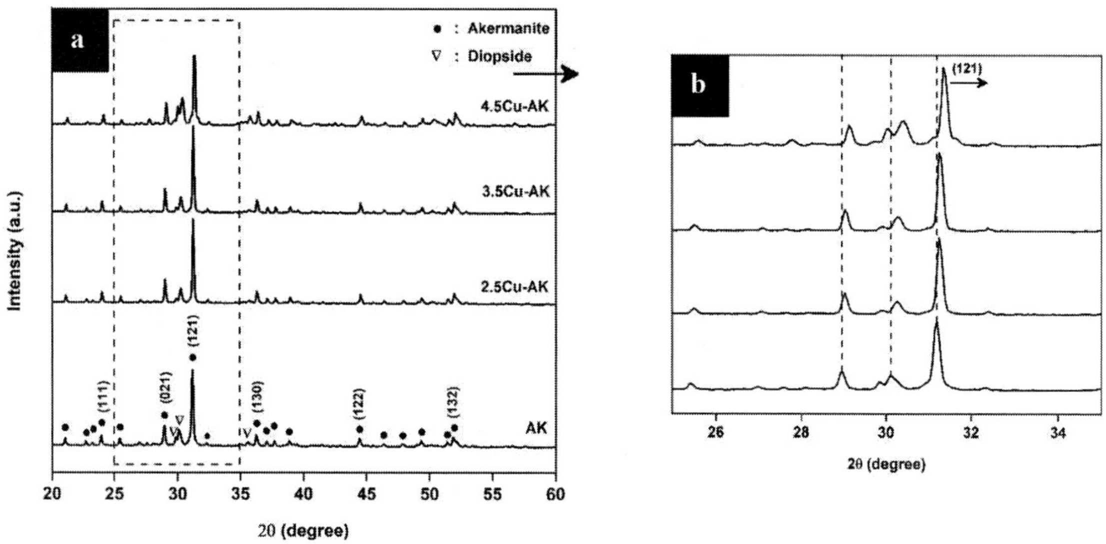

Figure 1. XRD patterns of $Ca_{2-x}Cu_xMgSi_2O_7$ samples: (a) 20° to 60°, and (b) partial magnification from 25° to 35°.

3.2. Morphological and physical evaluations

Figure 2 compares the grain growth behaviour of outermost and fractured surfaces of pure and Cu^{2+}-doped samples at 1200°C. As observed from microstructural analyses, the different amounts of Cu^{2+} dopant had a direct effect on the grain growth. The grains were in irregular geometry. The grains grew slowly and the average granule size range (1.05 μm −1.37 μm) remained a non-significant difference. A possible explanation of this result is that the introduction of smaller Cu^{2+} ions imposed an effective restriction on grain growth of $Ca_2MgSi_2O_7$ by reducing the movement of grain boundaries during sintering.

Further fracture studies signified that the grains were fractured in transgranular and intergranular modes, especially with large fraction of transgranular after doping. It could be assumed that smaller Cu^{2+} ions substituted into akermanite structure reinforced the grain boundary strength leading to a transgranular-prominent fracture. Consequently, this fracture mode restricted crack propagation [18].

In addition, the physical study revealed that dopant changed the density of sintered samples (Table 1). Diameter shrinkage and relative density of ceramics remarkably increased after doping. The shrinkage increased from 15.34% to 26.61% as well as relative density rose from 62.67% to 94.25% with Cu^{2+} content. This result indicated that copper accelerated sinterability of akermanite ceramics. Therefore, the current results were well consistent with the hypothesis that the densification rate was influenced by the radii of ions doped; i.e., the smaller the ionic radii, the better the densification rate [19]. Hence, our study highlighted that Cu^{2+} affects the solid-state reactions and lowered the required densitification temperature of akermanite ceramics. From an economic standpoint, this property is useful for less time- and energy- consumption reduction in producing engineering ceramics.

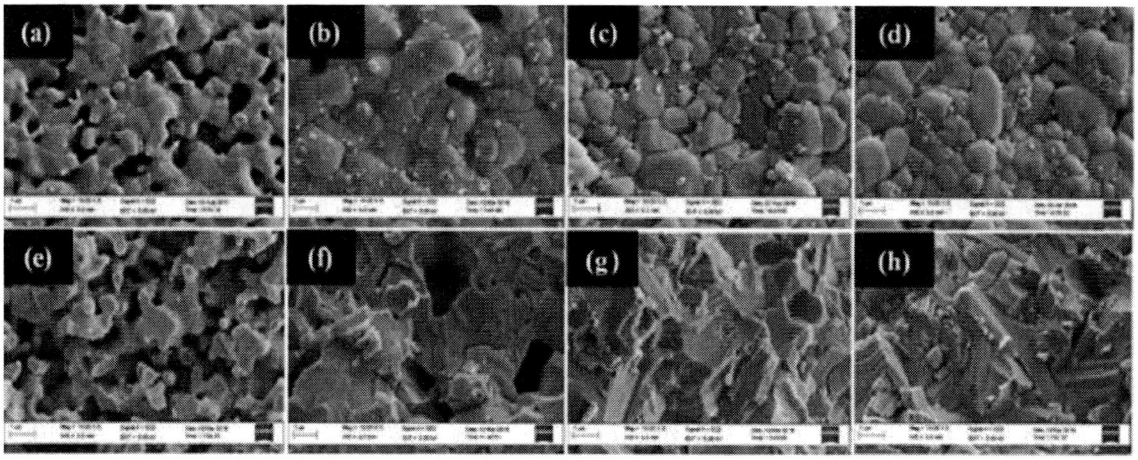

Figure 2. FESEM micrographs of outermost and fractured surfaces: (a, e) AK, (b, f) 2.5Cu-AK, (c, g) 3.5Cu-AK, and (d, h) 4.5Cu-AK ceramics at 1200°C.

Table 1. Physical properties of akermanite ceramics

Sample	Diameter shrinkage (%)	Relative density (%)	Grain size (μm)
AK	15.34	62.67	1.05
2.5Cu-AK	25.25	95.28	1.26
3.5Cu-AK	26.63	98.07	1.27
4.5Cu-AK	26.62	97.12	1.37

3.3. Mechanical evaluations

Hardness and fracture toughness are the important parameters required for the prediction of the mechanical performance of brittle materials for bone implant applications.

The SEM images of resultant impressions are shown in Figure 3 (a-d). As observed from the micrographs, the doped specimens definitely possessed higher densification and toughness than pure ones. The cracking feature around Vickers indentations was surface radial cracks. The indentation depth and size on AK was obviously deeper and larger than that of doped samples. As seen in Table 1, H_V improved from 0.76 GPa to 5.48 GPa, 6.172 GPa and 6.06 GPa for AK, 2.5Cu-AK, 3.5Cu-AK, and 4.5Cu-AK, respectively. As seen, the enhanced in H_V values were obtained in denser doped samples. This could be explained that since H_V technique is a surface technique, the H_V of ceramics mainly depends on the microstructure (i.e., surface porosity) of ceramics [20]. The resultant values were higher than that of hydroxyapatite (1.35 GPa) [21] and human cancellous bone (0.35 GPa) [22] in comparison. A slight reduced H_V in 4.5Cu-AK was assumed to be its microstructure tended towards a combination of crystalline and glassy phases, indicating the margin of Cu^{2+} content.

In a similar trend with H_V, fracture toughness (K_{IC}) increased from 0.75 to 2.06, 2.37, and 2.30 MPa.m$^{1/2}$ for AK, 2.5Cu-AK, 3.5Cu-AK and 4.5Cu-AK, respectively. These K_{IC} values were higher than the previously reported data of pure akermanite (0.35 MPa.m$^{1/2}$) [22], wollastonite (0.67 MPa.m$^{1/2}$) [23],

HA (0.70 MPa.m$^{1/2}$) and β-TCP (0.46 MPa.m$^{1/2}$) [24]. The percentage increment in H_V was went up to 174% (2.5Cu-AK), 216% (3.5Cu-AK), 206% (4.5Cu-AK) in current work, as well as 114% (AK), 11% (wollastonite), 7% (HA), and 63% (β-TCP) compared to the previous works. Thus, densification of akermanite ceramic has a direct effect on its microhardness and fracture toughness as well as agreed well with the earlier microstructural study provided (Figure 2).

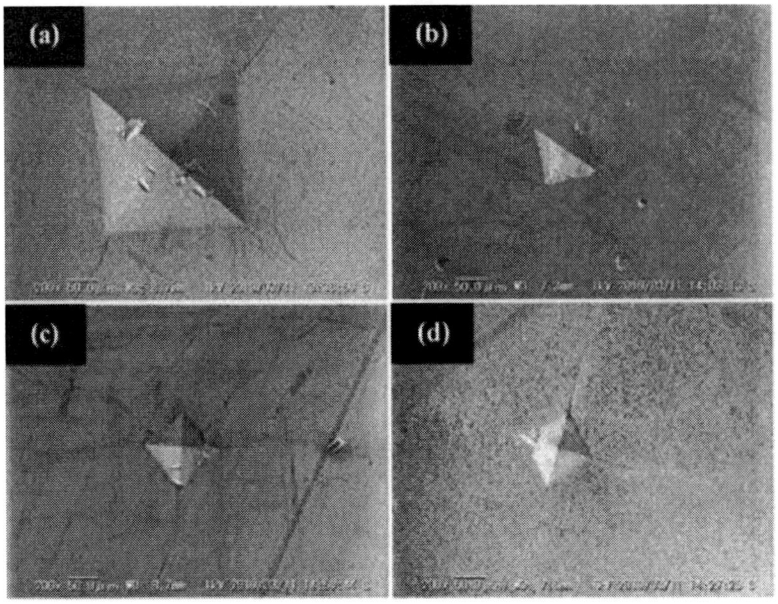

Figure 3. Typical microhardness indented areas at 1200°C: (a) AK, (b) 2.5Cu-AK, (c) 3.5Cu-AK, and (d) 4.5Cu-AK ceramics.

4. Conclusions

A range of Cu^{2+}-doped akermanite ceramics have been successfully prepared through high-energy milling and mixing method, with subsequent sintering at 1200°C. According to the structural analysis, akermanite phase was not disrupted by Cu^{2+} doping. The development of materials with reduced densifying sintering temperature could be beneficial for less time- and energy-consumption. Moreover, this reduced temperature due to Cu^{2+} additive could be advantageous for supressing grain growth, accompanying the effective improved sinterability (i.e., diameter shrinkage and relative density) and mechanical properties (i.e., Vickers hardness and fracture toughness). Therefore, the optimized Cu^{2+} ion-modified akermanite ceramic can meet the requirements for bone replacement materials in biomedical applications. For future studies, the doping effects of Cu^{2+} ions on the biological properties of akermanite ceramics will be conducted.

Acknowledgements
The authors acknowledge AUN/SEED-Net (Grant no. 304.PBAHAN.6050369) and MOHE-FRGS (6071376) for the financial support, as well as Universiti Sains Malaysia and Tohoku University for the facilities.

References

[1] Wu C and Chang J 2013 *Biomed. Mater.* **8** 032001.
[2] Chun-Kuei C and Teng-Ming C 2007 *Appl. Phys. Lett.* **90** 2–5.
[3] Huang Y, Jin X, Zhang X, Sun H, Tu J, Tang T, Chang J, Dai K 2009 *Biomaterials.* **30** 5041–5048.
[4] Wu C, Chang J, Ni S, Wang J. 2006 *J. Biomed. Mater. Res. - Part A.* **76** 73–80.
[5] Chengtie W and Jiang C 2006 *J. Biomater. Appl.* **21** 119–129.
[6] Pei F, Chengde G, Cijun S and Shuping P 2015 *RSC Adv.* **5** 3498–3507.

[7] Bejarano J, Caviedes P and Palza H 2015 *Biomed. Mater.* **10** 025001.
[8] Razmandeh R, Nasli-Esfahani E, Heydarpour R, Faridbod F, Ganjali MR, Norouzi P, Larijani B, Khoda-Amorzideh D 2014 *J. Diabetes Metab. Disord.* **13** 1–6.
[9] Ni K, Kaili L, Haiyan L, and Jiang C 2014 *J. Mater. Chem. B.* **2** 1100–1110.
[10] Željko R, Đorde V, Lidija R, Ilmars Z, Eriks P, Rada P and Đorde J 2018 *Process. Appl. Ceram.* **12** 268–276.
[11] Fariborz T, Caleb A Z, Abbas F, Anoosha F and Daniel J 2019 *Ceram. Int.* **45** 7871–7877.
[12] Hana N, Adel M and Mohamed E M 2014 *J. Adv. Ceram.* **3** 17–30.
[13] ASTM B962–17 2010 *ASTM Int. West Conshohocken* 1–6.
[14] ASTM E384–99 1999 *ASTM Int. West Conshohocken* **14** 1–24.
[15] Antony G E and E Alasdair C 1976 *J. Am. Ceram. Soc.* **59** 371–372.
[16] Erkul K and Bekir K 2015 *Indian J. Chem. - Sect. A Inorganic, Phys. Theor. Anal. Chem.* **54A** 1394–1401.
[17] Vojislav S, Suzana D, Jelena A S, Miodrag M, Bojan J, Ilija B P and Slavica R 2010 *Appl. Surf. Sci.* **256** 6083–6089.
[18] Keisuke T, Yoshiaki A and Robert P W 1986 *Eng. Fract. Mech.* **24** 803–819.
[19] Seyed F S S, Samira G, Hendrik S C M, Bahman N T, Hooman Y, Mahi A and Noor N A O 2015 *RSC Adv.* **6** 190–200.
[20] Umit K, Oguzhan G, Sabri S, Lutfite S O, Simeon A and Ahmet Z S 2010 *BIODENAL* 1271–1274.
[21] Sumit P, Avinash K A and K N Rai 2005 *Trends Biomater. Artif. Organs.* **19** 46–51.
[22] Xianchun C, Jun O, Yun W, Zhongbing H, Yunqing K and Guangfu Y 2010 *J. Mater. Sci. Mater. Med.* **21** 1463–1471.
[23] Sun H, He S, Wu P, Gao C, Feng P, Xiao T, Deng Y, Shuai C 2016 *Materials (Basel).* **9** 1–12.
[24] Hesaraki S, Safari M, Shokrgozar MA 2009 *J. Biomed. Mater. Res. B Appl. Biomater.* **1** 459–469.

IOP Conf. Series: Materials Science and Engineering **943** (2020) 012009 doi:10.1088/1757-899X/943/1/012009

Study on size control of TiO_2 nanoparticles synthesized using solution plasma process

B W M Yii, P Y N Yek, M S Ranjit, B Y Y Liew and M R M Julaihi*

Swinburne University of Technology Sarawak, Jalan Simpang Tiga, 93350 Kuching Sarawak, Malaysia

Email: mjulaihi@swinburne.edu.my

Abstract The research was conducted to study the size control of TiO_2 nanoparticles synthesised. The TiO_2 nanoparticles was formed using the solution plasma process. This method is an electrolysis of water-based experiment. When compared to other methods, the solution plasma method offers both a shorter synthesis time as well as a simple setup. Researches conducted on nanomaterials have proven that these materials tend to have varying properties when compared to bulk metals. With this, being able to vary the sizes of the TiO_2 nanoparticles meant being able to study its possible future applications. One being using these nanoparticles for the fabrication of solar cells. Varying the sizes of these nanoparticles could bring a difference in the reactivity which can help enhance the performance of the solar cells. In order to vary the sizes of the nanoparticles, the concentration of the electrolyte was varied. During the research, it was found that the sizes of the TiO_2 nanoparticles decreases with the increase in the concentration of the electrolyte.

1. Introduction

There are multiple ways to synthesize nanoparticles which includes methods such as sol-gel method, high energy milling and pulse laser ablation [1]. Although these methods can synthesize and control the size of the nanoparticles, there are limitations for the methods mentioned as different researcher is situated in a different environment. One of the few limitations is that the methods mentioned would require specialty equipment such as vacuum chambers and high-powered lasers [1]. Nanomaterials have the tendency to exhibits different properties from their bulk counterparts. Such properties include a change in thermal, optical, reactivity etc. An example of varying properties can be seen from the copper when bulk copper is opaque and copper nanoparticles exhibits transparency. By exploiting the unique properties of nanomaterials, it can be utilized in fields such as medical, aerospace, biotechnology etc. [2].

For this paper, the study revolves around the use solution plasma process for synthesizing titanium dioxide (TiO_2) nanoparticles and size control of the particles. Some benefits of solution plasma over other methods are that it has a simpler setup and can be conducted in small sized laboratories. In addition, the preparation time is shorter than any other methods stated above. The novelty of this research is to device a sustainable technique of producing various size ranges of TiO_2 nanoparticles, which consequently, improve the overall outcomes of its application, otherwise, it is able to provide new ways to synthesis different applicable outcomes. For example, TiO_2 nanoparticles can be used as substitute for the fabrication of solar cells and its reactivity in terms of its plasmonic resonance, helps to enhance the efficiency of the cells.

2. Theory

2.1 Correlation between the size and properties of nanoparticles

As particles micronize, the behavior of the particles tends to be affected in some manner and exhibits different physical properties. The variation in the physical properties is caused by the change of the molecular and atomic bonding states which construct the particles [3]. One of the corresponding effects of reducing the particle size is the change of the number of atoms present on the surface of the particle. This could increase the reactivity of the atoms as the higher number of exposed atoms available for reactions directly corresponds to the higher reactivity. The relationship between particle size and percentage of surface atoms are illustrated in Figure 1.

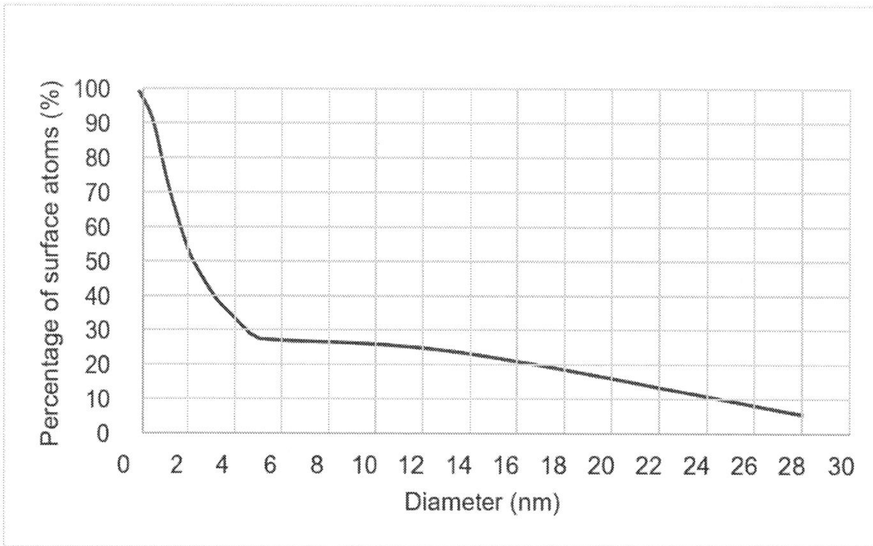

Figure 1. Percentage of surface atoms vs diameter of the particles [3]

Apart from reactivity, the optical and thermal properties are also affected when the size of the particles were reduced. For optical properties, nanoparticles tend to absorb light with a specific wavelength and emit the energy as the electrons returns to its original state. Such property could be exploited and used as pigmentation and light scattering applications because nanoparticle paint pigments is reported to be 100 times greater than ordinary pigments.

As for thermal properties, when metal particles are reduced to the nano spectrum, the melting point greatly reduces as the size gets smaller [3]. This phenomenon is called the melting point depression as nanoparticles have a larger surface-are-to-volume ratio which enables the particles to receive more energy and result in a lower melting point. The melting point depression is shown in Figure 2.

2.1 Method of synthesis – Hydrothermal method

Some common methods to synthesize metallic nanoparticles are shown here. One common method to synthesize fine particles of metal oxides is the hydrothermal method. It involves condensation and hydrolysis process to produce ultrafine particles. The condensation process involves the condensation of metal salts in a high temperature and pressure atmosphere while in contact with a hydrothermal medium.

The high temperature and high-pressure atmosphere are normally achieved in an autoclave which it could produce a controlled pressure and temperature condition where the reaction could occur. Due to this elevated pressure and temperature, the hydrothermal medium could exhibit properties such as viscosity and dielectric constant when compared to normal atmospheric conditions. These changes in properties could lead to better solubility of the metal complexes and thus enables better particle formation and growth [4].

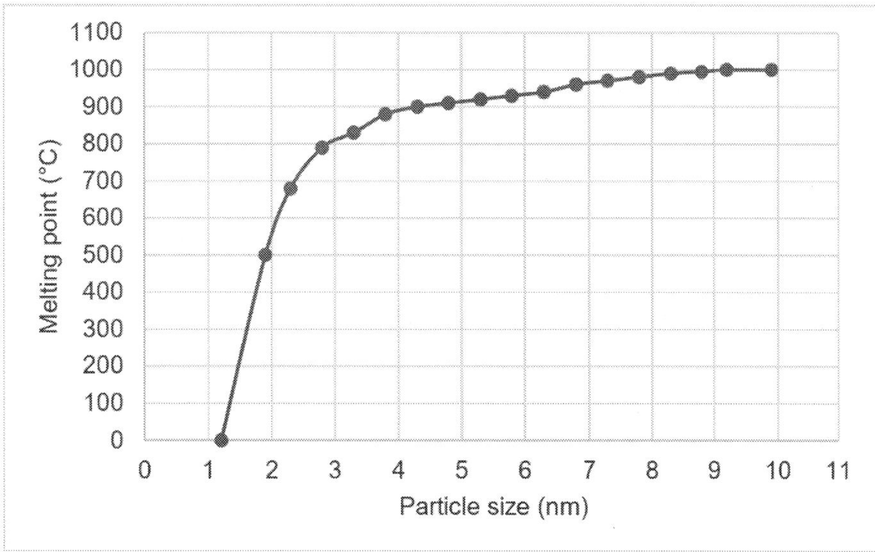

Figure 2. Melting point vs particle size (nm); Melting point depression [3]

2.2 Method of synthesis – Electrochemical method
Another method of nanoparticle synthesis is the electrochemical method. The electrochemical method involves an external circuit and chemical reactions similar to electrolysis process.

The external circuit contains the anode and cathode. The anode contains the bulk metal that will be oxidized and the metal cations would be migrated to the cathode. The electrochemical method causes a reduction along with the formation of metal oxides which are normally present in the zero oxidation states [5]. An additive is normally added to prevent the agglomeration of the metal powders. In addition to that, this method of synthesis is normally conducted in a special nitrogen filled atmosphere and therefore a need for a special gas chamber arises in order to synthesize nanoparticles via electrochemical method.

2.3 Solution plasma process
The solution plasma process, also known as submerged glow discharge plasma, is the focus of this research paper as it was used to synthesize and control the size of TiO_2 nanoparticles. As the name suggests, the solution plasma process utilizes plasma in a solution to synthesize nanoparticles. The generated glow discharge plasma acts upon the cathode and thus melting the surface of the cathode and releasing particles of molten metals into the electrolyte [6].

In solution plasma process, the electrolyte acts as a conductor and coolant. When the molten metal particles are released, the surface tension of the electrolyte forces the particles to be spherical in shape while solidifying it.

The advantages of solution plasma as a method for nanoparticles synthesis can be summarized as the time taken and the simplicity of the experimental setup. Electrochemical and sol-gel methods requires several hours of preparation prior to the synthesis process. Solution plasma process requires up to five minutes for experimental setup, and up to one hour for nanoparticles synthesis depending on the amount of particles desired.

The simple setup of solution plasma process is also a major advantage compared to other methods of nanoparticles synthesis. As a comparison, hydrothermal method requires an autoclave while the electrochemical method requires a special nitrogen atmosphere. Other methods of synthesis such as pulse laser ablation technique require a high-powered laser system whereby not every laboratory houses such complex equipment. On the other hand, solution plasma process only requires a simple setup without the need for any specialty equipment and can be set up in any small sized laboratory.

IOP Conf. Series: Materials Science and Engineering **943** (2020) 012009 doi:10.1088/1757-899X/943/1/012009

3. Methodology

3.1 Required materials and apparatus

1) Electrolyte (Potassium carbonate electrolyte, K_2CO_3)
2) Platinum wire coil
3) Titanium wire (Grade 1 titanium wire, uncoated)
4) DC power supply
5) 300ml beaker

3.2 Experimental setup

The setup is similar an electrolysis setup with an external circuit and an electrolyte. The external circuit consists of a cathode and an anode, and both electrodes were submerged into the electrolyte. The cathode holds the desired nanoparticle material, which in this case was the titanium wire and the anode was the platinum coil which acts as the counter electrode. The platinum wire used in this case was 1000mm in length [6]. As for the electrolyte, potassium carbonate with concentration of 0.1M was used initially. Both electrodes were connected to a DC power supply and both electrodes were held in place away from each other to prevent accidental short circuits.

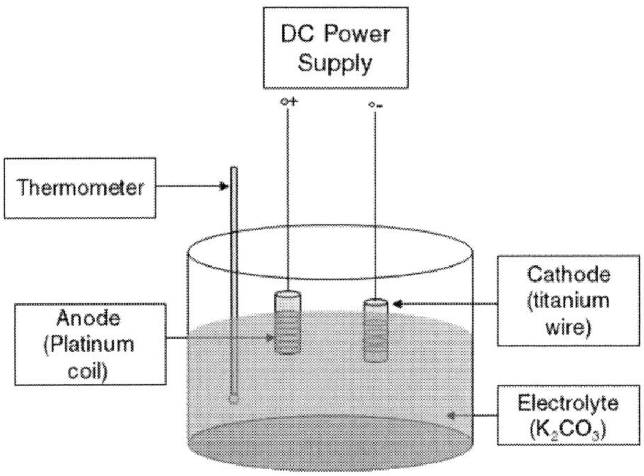

Figure 3. Schematic representation of the setup for solution plasma process

The variable that was set for this experiment was the concentration of K_2CO_3 used. The initial concentration was set to 0.1M and was increased in an increment of 0.2M per iteration. A total of 5 iterations were performed, resulting in a final concentration of 1.0M. The electrolyte concentration has an effect on the plasma formation on the cathode surface. For that reason, the electrolyte concentration was chosen as a variable in this research. In addition, a graph that illustrates the relationship between nanoparticle size and electrolyte concentration was plotted.

3.3 Experimental procedure

The experiment is set up accordingly and the power supply was turned on. Prior to turning on the power supply, the cathode and anode were sufficiently separated from each other or touching each other directly. This was to prevent any short circuits that could result in a minor explosion and other damages to the equipment.

After the power supply was turned on, the voltage was increased slowly at a rate of 5V per minute. This was to enable the electrolyte to reach functional temperature by slowly warming up the electrolyte. The current reading on the DC power supply was constantly observed to ensure that the current did not exceed 4A. Exceeding this current would cause the electrodes to melt prematurely and could lead to dangerous exposure to molten metals.

The voltage was gradually increased until the formation of the discharge plasma. Immediately the voltage was reduced by 5-10V to ensure the stability of the formed plasma discharge and to sustain the plasma. The generated plasma was then sustained for one hour in order for the plasma to synthesize nanoparticles. The electrolyte volume was continuously replenished due to its evaporation

IOP Conf. Series: Materials Science and Engineering **943** (2020) 012009 doi:10.1088/1757-899X/943/1/012009

during plasma discharge.

After the experiment was completed, the nanoparticle suspension was left to cool for 24 hours to allow for the nanoparticles to settle at the bottom of the beaker. After 24 hours, the excess electrolyte was carefully decanted, and distilled water was added to wash away the remaining electrolyte. The sample was then stored in a glass or plastic vial along with the distilled water. To obtain images of the nanoparticles, scanning electron microscopy (SEM) analysis was conducted.

4. Results

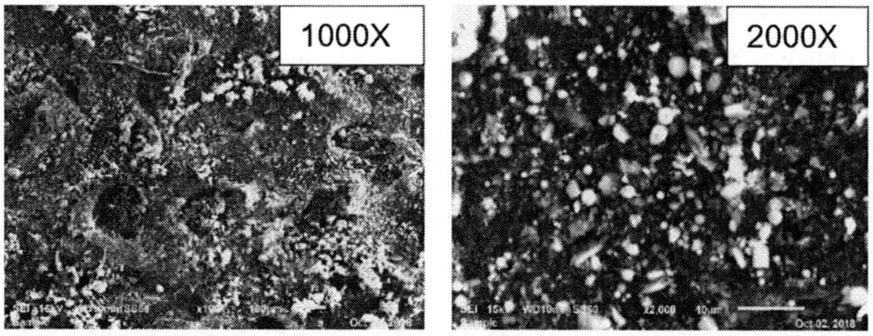

Figure 4. SEM images of TiO_2 for 0.1M concentration.

Figure 5. SEM images of TiO_2 for 0.3M concentration.

Figure 6. SEM images of TiO_2 for 0.5M concentration.

IOP Conf. Series: Materials Science and Engineering **943** (2020) 012009 doi:10.1088/1757-899X/943/1/012009

Figure 7. SEM images of TiO_2 for 0.7M concentration.

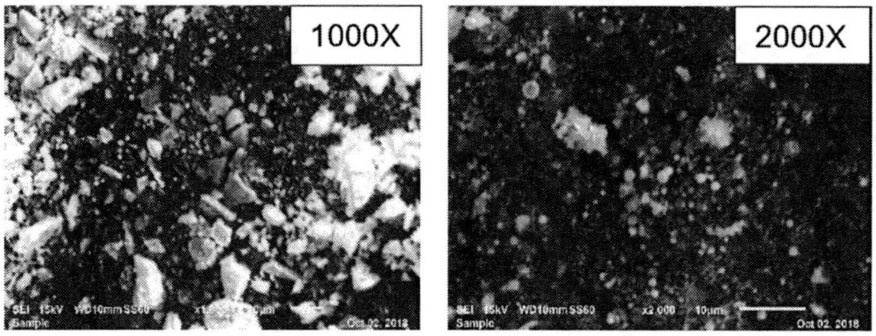

Figure 8. SEM images of TiO_2 for 1.0M concentration

Table 1. Average particle size with the concentration and sustaining voltage

Electrolyte concentration (M)	Average particle size (nm)	Sustained plasma voltage (V)
0.1	538.6	120
0.3	839.5	85
0.5	869.1	80
0.7	991.4	80
1.0	1250.0	76

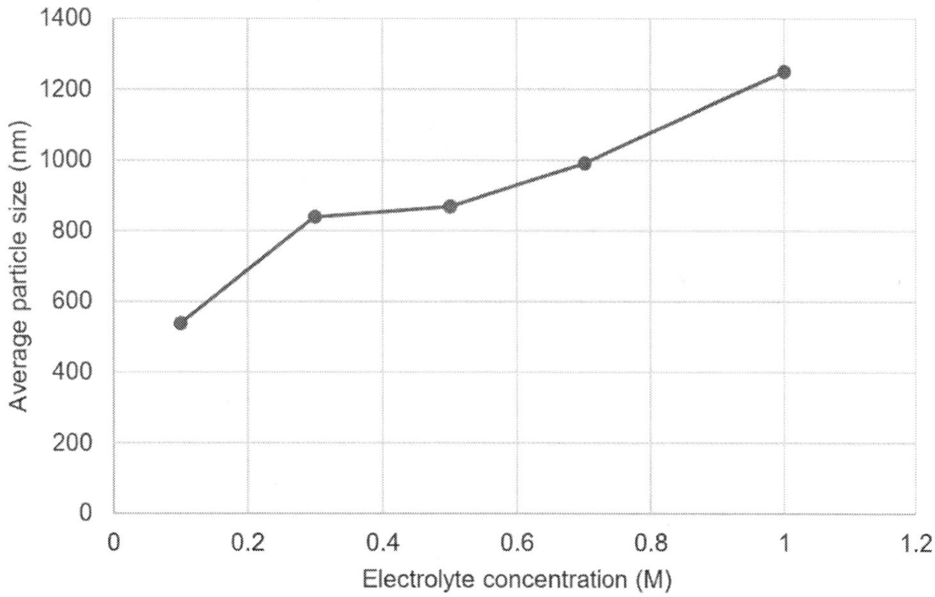

Figure 9. Graph of average particle size vs electrolyte concentration

5 Discussion

5.1 General discussion on the observed size and shape of the particles

Figures 4 to Figure 8 were the obtained SEM images of the synthesized nanoparticles. Observations of size and shape changes were made prior to size measurement by using GATAN Microscopy Suite software. Table 2 shows the observations obtained from the SEM images.

Table 2. Electrolyte concentration and the observation of the images

Electrolyte concentration (M)	Magnification	Observation	Sustained Voltage for Plasma (V)
0.1	2000X	Large numbers of spherical shaped particles	120
0.3	2000X	Large and small sized particles along with irregularly shaped clumps	85
0.5	2200X	Large numbers of irregularly shaped clumps along with fine spherical particles	80
0.7	3300X	Large clumps with small amounts of fine particles	80
1	2000X	Large number of dispersed particles with only small amounts of irregularly shaped clumps	76

There were large clumps of irregularly shaped particles present in every conditions of the experiment. This phenomenon can be explained as an agglomeration of particles. Under normal circumstances, the nanoparticle samples would undergo filtration and are placed in a centrifuge to separate the particles from the excess electrolyte and washed with distilled water. The filtration process would filter out the large clumps of particles and leave behind the finer particles.

For this research experiment, the filtration process was skipped and due to this, there was a presence of irregularly shaped particles in all of the specimens. A comparison of SEM images between filtered and non-filtered specimens are shown below.

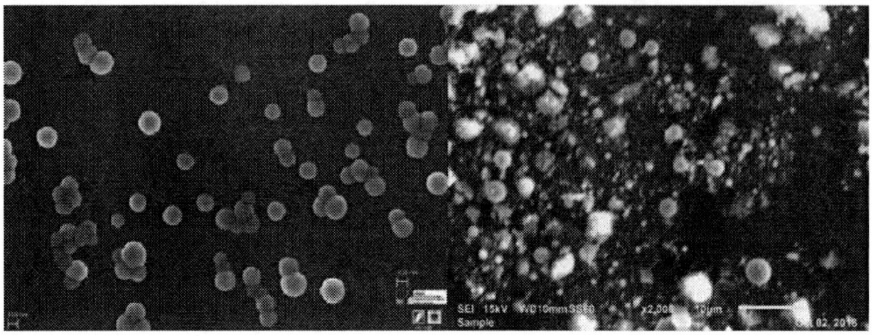

Figure 10: Comparison between filtered (left) and non-filtered (right) nanoparticle samples

The agglomeration of the particles likely occurred during the synthesis process rather than during the post-processing. During the synthesis, molten metals particles were discharged into the electrolyte. The surface tension of the electrolyte forces the molten metal to solidify into spherical balls. In addition, the strong intermolecular forces between the particles act upon each other and therefore causing agglomerations which are very difficult to separate.

5.2 Correlation between the electrolyte concentration and the sustaining voltage for plasma generation

For each electrolyte concentration, the size of the nanoparticles was measured by determining the diameters of twenty particles from the SEM images, and then calculating the average diameter. According to Table 3, it can be observed that the size of the nanoparticles increased along with the concentration of the electrolyte. It is also found that the higher the concentration of the electrolyte, the lower the sustaining voltage for plasma generation.

A higher electrolyte concentration represents a higher number of dissolved ions per fixed volume of the solution. The reasoning behind the decreasing sustaining voltage for plasma generation as the concentration goes up was that the glow discharge plasma was reached when the energy filled electrons produces visible light from repeated collision of photons [6]. The sustaining voltage for the plasma generation would be lower when the concentration is higher due to the solution already yielding high amounts of dissolved ions and did not need as much energy to ionize the electrons.

5.3 Relationship between the electrolyte concentration and nanoparticles size

Table 3. The relationship of the experimental aspects

1	**Electrolyte concentration increases**	**Amounts of ions in the solution increases**
2	Amounts of ions in the solution increases	Sustaining voltage for plasma generation decreases
3	Sustaining voltage for plasma generation decreases	Plasma particle intensity decreases
4	Plasma particle intensity decreases	Average Particle size increases

As the electrolyte concentration increased, the sustaining voltage decreased. However, there were still other relationships to be determined such as the plasma intensity and the size trend of the nanoparticles synthesized.

The relationship can be further discussed by noting that the plasma intensity was higher when the sustaining voltage was higher. A more intense plasma formation would yield a finer nanoparticle sizes due to the more uniform melting of the cathode surface. The increasing uniformity in the melting pattern could be further explained with the size of the plasma particles. By using a high-speed camera, it has been previously observed that a higher sustaining voltage would produce more plasma particles on the cathode surface [6]. The plasma particles produced were finer and contribute to a more uniform melting of the cathode.

6 Conclusion

This research aims to control the size of TiO_2 nanoparticles by changing the concentration of the electrolyte during plasma discharge. It was found that the average size of the nanoparticles increases along with the concentration of the electrolyte. This was mainly due to the plasma intensity that reduced as the electrolyte concentration increased. At the same time, the plasma discharge sustaining voltage was decreased as the electrolyte concentration increased. Subsequently, the plasma intensity was low at high electrolyte concentrations and low sustaining voltages.

Several issues can be considered for future work. One of it is to investigate the influence of electrolyte selection on the trend for nanoparticle size change. For example, in the case of plasma synthesis of CuO nanoparticles, it was found that the average size of nanoparticle decreased as the concentration of NaOH electrolyte increased, while the opposite occurred when using K_2CO_3. The electrolyte pH and the ionic species formed during plasma synthesis could have a major role in the resulting nanoparticles sizes.

In addition to the recommendation above, further research on controlling the plasma particles acting upon the surface of the cathode has the potential to control the yield of nanoparticles. Also, the applications of synthesized TiO_2 nanoparticles could be researched upon as they exhibit different light scattering effect in correlation to the size of the particles. The synthesized nanoparticles can then be utilized in many applications such as pigments, UV blockers and photocatalysts. It would be of great interest to study the varying properties in relation to the size of the nanoparticles as different sizes exhibit different properties.

References

[1] J. Yu, J. Nan, and H. Zeng, "Size control of nanoparticles by multiple- pulse laser ablation," Appl. Surf. Sci., vol. 402, pp. 330–335, 2017.

[2] M H S Al Anbouri et al 2018 IOP Conf. Ser.: Mater. Sci. Eng. 429 012084

[3] H. Masuda, M. Suzuki, K. Ehara, M. Fuji, T. Fukui, H. Suzuki, J. Tatami, K. Hayashi, and K. Toda, Basic Properties and Measuring Methods of Nanoparticles. 2008.

[4] N. K. V. Nadimpalli, R. Bandyopadhyaya, and V. Runkana, "Thermodynamic analysis of hydrothermal synthesis of nanoparticles," Fluid Phase Equilib., vol. 456, pp. 33–45, 2018.

[5] P. Anandgaonker, G. Kulkarni, S. Gaikwad, and A. Rajbhoj, "Synthesis of TiO_2 nanoparticles by electrochemical method and their antibacterial application," Arab. J. Chem., pp. 0–7, 2015.

[6] M. R. M. bin Julaihi, S. Yatsu, M. Jeem, and S. Watanabe, "Synthesis of stainless steel nanoballs via submerged glow-discharge plasma and its photocatalytic performance in methylene blue decomposition," J. Exp. Nanosci., vol. 10, no. 12, pp. 965–982, 2015.

IOP Conf. Series: Materials Science and Engineering **943** (2020) 012010 doi:10.1088/1757-899X/943/1/012010

Application of variable geometry turbine turbochargers to gasoline engines- A review

S Dasgupta[1] P Sarmah[1*], P P Borthakur [1]
[1]Department of Mechanical Engg., Dibrugarh University, PIN - 786004, India.

Email: pksnit07@gmail.com

Abstract. An ideal engine today is considered to be one with low emission, low fuel consumption and high specific power output. Engine downsizing by turbocharging provides an answer toall the three. Though the traditional problems of poor low end torque and inadequate transient response have been overcome to a great extent using variable geometry turbine turbocharger in diesel engine, the application of variable geometry turbine turbocharger is in quite a nascent stage when it comes to gasoline engine. This article primarily describes all the important aspects of application of VGTs to gasoline engines right from when VGT's were restricted to only very expensive research prototypes to the very recent application in cost sensitive production engines. Also at the same time,it highlights the need and development in the flow range of the turbocharger compressor, addressing problems of choke and surge in the turbo chargers compressor.

1. Introduction

Owing to the emphasis on environmental issues, improvement in emission quality and reduction in fuel consumption have become the main driving forces of engine development and research. As per modern day standards, an ideal engine is supposed to have both high power and a low fuel consumption. Although turbocharging historically started as a means to increase the power output of an engine, it is today serving greatly as a means to achieve the emission standards along with lower fuel consumption [1]. The European Union has always been the front runner in achieving emission standards. Figure 1 shows the historical development and future targets for emission standards in the EU [2]. Similarly, most of the large economies have specified converging emission targets for new vehicles. Figure 2 explains the global scenario a bit further. To achieve these emission targets, engine downsizing via forced induction is quite inevitable.

The objective of obtaining more power from an existing engine size, can also be achived by increasing the thermal efficiency by increasing the compression ratio, especially in case of a CI engine. However, an increase in compression ratio is also accompanied by an increase in maximum cylinder pressure and the rate of increase in maximum cylinder pressure is comparatively less than the rate of increase in break mean effective pressure. This, therefore, implies that, for a given maximum allowable cylinder pressure, greater power can be obtained by increasing the pressure at intake as compared to that obtained by increasing the compression ratio [3].

Owing to the aforementioned briefing, turbocharging has been the primary method of engine downsizing. Moreover, from an environmental point of view, turbocharging and lean boosting reduces carbon Dioxide emissions [4] and also leads to lower fuel consumptions [5, 6, and 7], not to mention the reduction of engine weight, thereby leading to lower production cost.

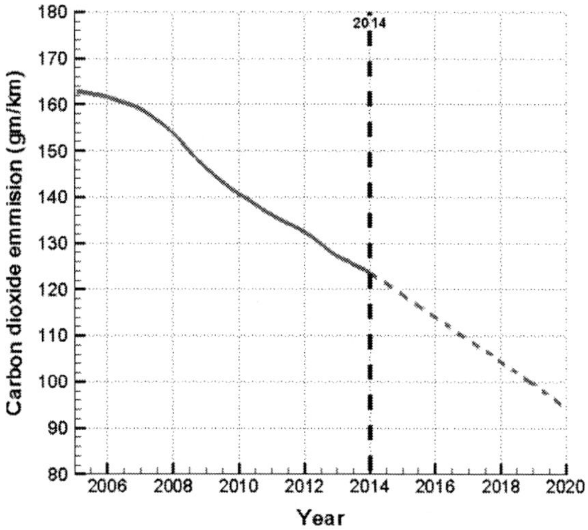

Figure 1. Historical development and future targets for CO2 emission levels of new passenger cars in the EU. [2]

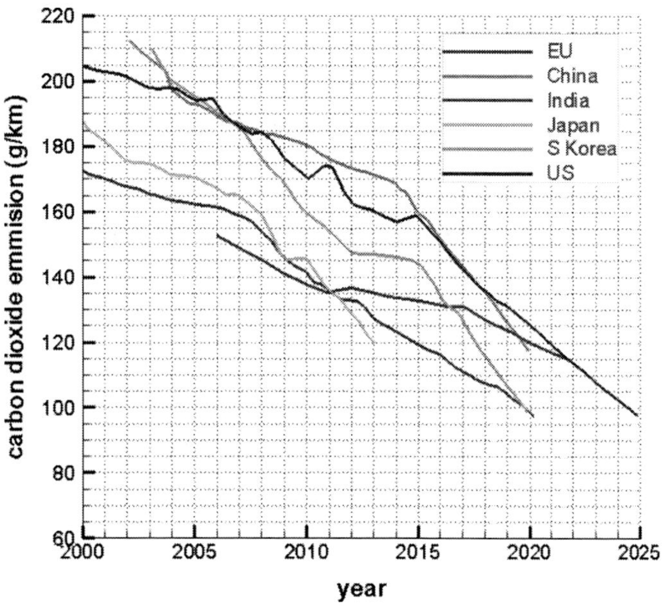

Figure 2. Comparison of global CO2 regulations for new passenger cars. [2]

The degree of turbocharging of a SI gasoline engine is primarily limited by knocking rather than thermal or mechanical loads (a major constraint case in a diesel engine) [8]. Furthermore, SI engines use the throttle to control load in contrast to its CI counterpart, which controls load by governing the amount of fuel injected. This, therefore makes turbocharging of a gasoline engine much more challenging, as the intake manifold pressure increased during turbocharging plays a vital role in load control.

Traditionally, the main problem with turbocharging, be it a gasoline or a diesel engine, has been inadequate low end torque and poor transient response (or 'turbo lag' as it is more popularly known). This lag is due to the additional inertia added to the induction system by the turbocharger. This makes a turbocharged system, with a fixed geometry turbine turbocharger less responsive than its naturally aspirated counterpart. The problems of transient response and low end torque have been addressed by two stage turbo charging or by altering the flow dynamics using different types of variable geometry turbocharger [9, 10] or by using the assistance of an electric motor [11] at low engine speeds. Casing treatment has also proved beneficial in improving the choke and surge limit of the compressor [12, 13]. Apart from these, various synergies of the mentioned techniques have been used to obtain optimum results as per the desired engine requirement. A classic example of these synergetic techniques is the Volkswagen 1.4 liter TSI engine which uses a two stage boosting system using a supercharger along with a fixed geometry turbine (FGT) turbocharger. However, Volkswagen plans to replace the 1.4 liter TSI engine with the new 1.5 liter EA211 TSI evo gasoline engine [34, 35], which makes use of a variable geometry turbine turbocharger.

This review primarily deals with the following:

- Section 2 deals with the short comings of the traditional fixed geometry turbine (FGT) turbochargers and highlights the correlation between throat area to radius ratio (A: R ratio) of the turbine housing and boost pressure.
- Section 3 is a discussion about the various types of variable geometry turbine (VGT) turbocharger and their application to gasoline engine downsizing.
- Section 4 deals with casing treatment for surge limits, choke flow improvement methods and other method to obtain a wide flow range in the turbocharger compressor.

2. Short comings of FGT turbochargers

The conventional or fixed geometry turbine turbo chargers have had some inherent short comings like turbo lag and inadequate low end torque [8]. This lack of transient response which is due to the inertia of the turbine rotor can be reduced to an extent by using lighter materials.[14] But as the turbine is constantly subjected to a highly oxidizing environment and extreme high temperatures of the exhaust gases, reducing the mass of the rotor material has proved to be rather difficult. However, they have become much lighter over the years, and the turbine rotors today are much lighter than what they used to be. One way to obtain a better transient response is by using fixed geometry turbine (FGT) turbo chargers in a two stage turbo charging system, using a low pressure turbo charger at low engine speeds and high pressure turbo charger at high engine speeds. This means using turbocharger turbine housing with a smaller throat area to radius ratio at low speeds and a turbocharger turbine housing with a larger throat area to radius ratio at high engine speeds [15, 16]. Lowering the A: R ratio increases the exhaust gas velocity in the turbine increasing the kinetic energy of the gas and thereby leading to a better throttle response.

Taking the concept of matching the A: R ratio of the turbocharger with the engine speed a step forward leads us to variable geometry turbochargers (VGT) [17, 18, and 19]. The VGTs unlike FGTs provide a whole range of A: R ratios as per the engine speed rather the just two sets of A:R ratios as in the case of two stage turbo charging. Moreover, in some cases, the usage of waste gates can be completely eliminated while using a variable geometry turbocharger [17]. However, it should be advisable to maintain an external waste gate in case of abnormal boost pressure over shooting. The various types of VGTs, their mechanisms, characteristics and usage to gasoline engine downsizing have been further described in the next section.

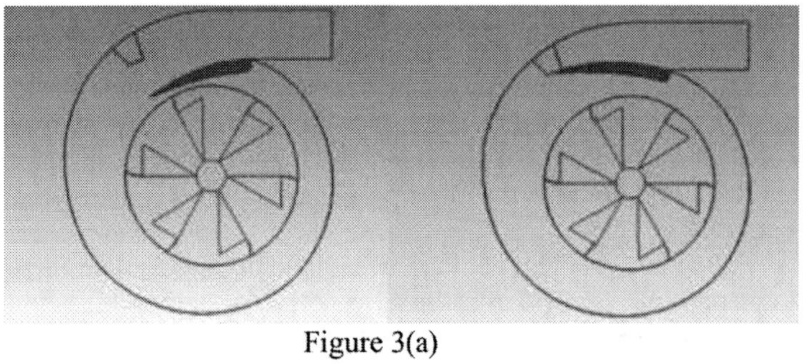

Figure 3(a)

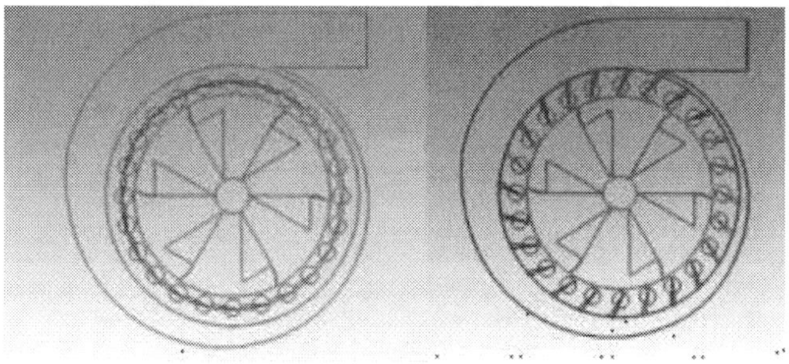

Figure 3(b)

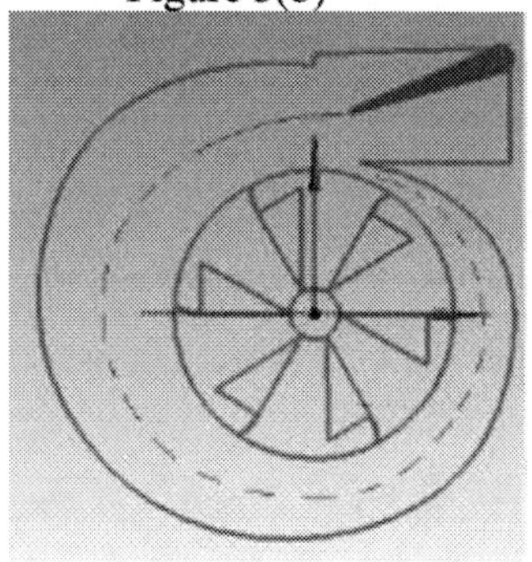

Figure 3(c)

Figure 3. Types of VGT: (a) variable turbine housing throat area (VAT), (b) variable geometry nozzle turbine (VNT) (c) variable flow turbine (VFT).

3. Variable Geometry Turbine (VGT) Turbochargers: scope, studies and gasoline engine applications

The variable geometry turbines have proved to be the best way of matching the A: R ratio of the turbine as per the engine speed. The various types of VGTs mainly include; variable throat area turbine (VAT) [20], variable flow turbine (VFT) [21] and variable nozzle turbine (VNT) [22].
Kawaguchi et al. reported a variable flow turbine (VFT), as novel technique to vary turbine inlet area [21]. In a VFT, the task of altering the flow dynamics as per the engine speed is achieved by the movement of a flow control vane on the radial plane as shown in figure 3(b). In addition to that, the flow within the turbine casing is confined within two scrolls, one primary and one secondary scroll in the axial direction. The wall separating the two scrolls within the turbine housing is an array of strategically well positioned stationary vanes [24]. The flow of the exhaust gases into the turbine is governed by the position of the flow control vane at the throat as mentioned above. At low engine speeds, when the exhaust flow is low, the vane directs the exhaust towards the primary (inner) scroll and as the flow increases with the engine speed, the vane gradually opens, thereby, increasing the throat area and directing more of the flow towards the secondary scroll. When the control vane comes to its fully opened position, the VFT starts behaving like a conventional FGT as the flow is now controlled by the waste gate.

Kawaguchi et al. [21] suggested the major positive point of the VFT to be the single moving component (the flow control valve) compared to the multiple nozzle vanes movement in a VNT. Kawaguchi et al. [21] also documented a 10 kPa boost pressure improvement in the VFT compared to an equivalent VGT under similar inlet exhaust pressure.

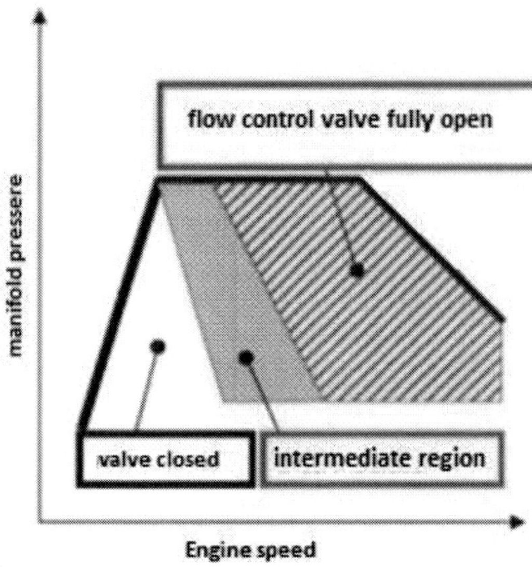

Figure 4. Flow control strategy of the VFT. [23]

The above mentioned flow control strategy of a VFT becomes more clear from figure 4 extracted from Ito et al. [23] (2007) which divides the flow into three regions, namely the closed, the intermediate and the fully open zone. Such VFT systems have been used in Honda Acura RDX 2.3 liter production engines along with variable valve timing (VVT) to obtain a further better engine aspiration [24]. Also a water cooled exhaust manifold (WCEM) to control the turbine inlet temperature is used. One problem with water cooled exhaust manifolds is that, cooling the exhaust also reduces the energy available at the turbine especially at low engine speeds. A solution to this problem could be – a slight delay in ignition timing. This could lead to a higher exhaust temperature at low engine speeds leading to better low end torques.
The VFT, which is kind of a synergy between a small (small A:R) and a large (large A:R) provide a better low end torque at low engine speeds and a better high end torque at higher engine speeds as compared to a large and a small FGT turbo charger respectively. The plot between engine speed and output break torque for a 4-cylinder gasoline adapted from Ito et al. [23] (figure 5) explains it better. A much wider torque plateau can be seen for the VFT.

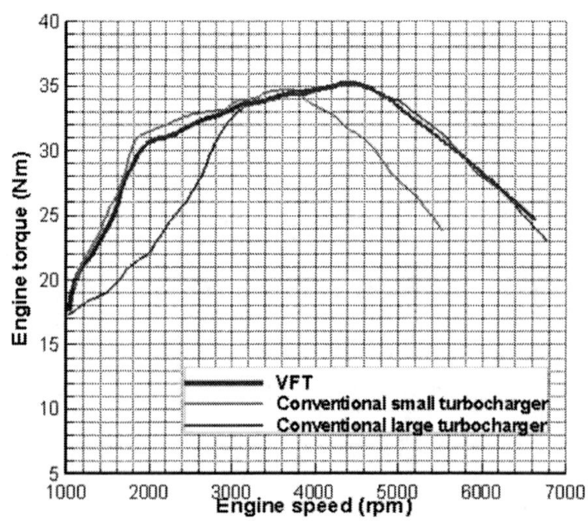

Figure 5. Comparison of VFT with conventional small and large FGT turbochargers. [23]

Moving towards variable throat area turbine (VAT). Flaxington and Szczupak [25] studied some methods of variable area devices for turbines. These include, in general, the variable area via regulation of the rotor exit area, volute exit area and volute tongue area (throat). They further investigated the latter two methods, which are shown in Figure 6 and 7 with its relevant engine performance results. They concluded, in general, an improved torque, wider speed range and improved transient performance. It was, however, concluded that no single method to area control was superior for all application, and each has its superiority in particular operating regions.

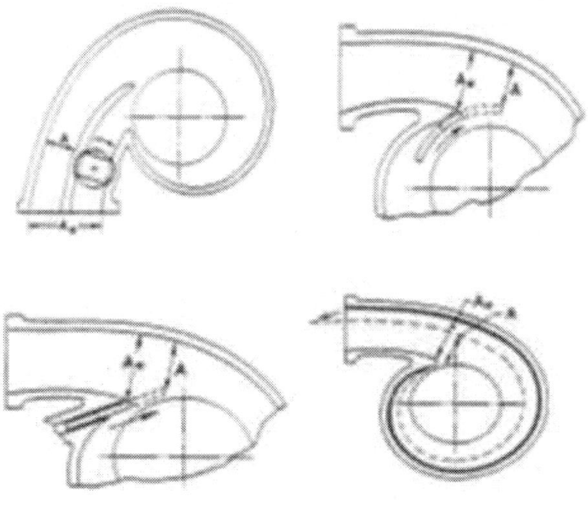

Figure 6. Methods of volute tongue area control studied by Flaxington and Szczupak.[25]

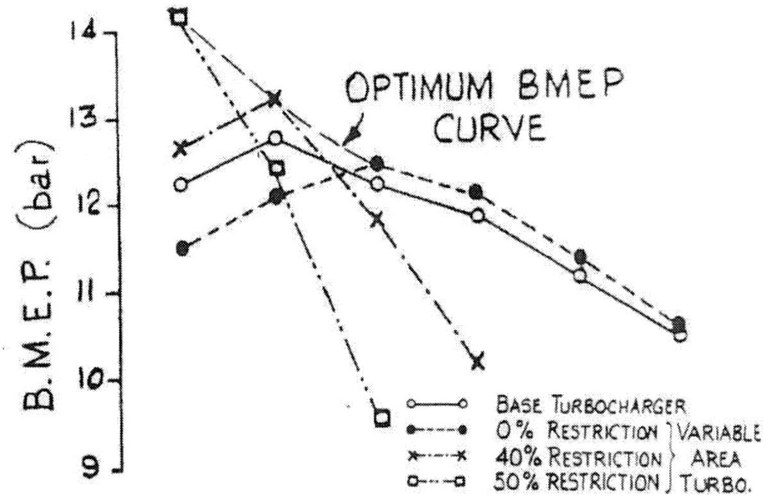

Figure 7. Engine performances under the various methods of volute tongue area control studied by Flaxington and Szczupak. [25]

VATs have been considered as a better and efficient approach to turbocharger- engine matching as compared to FGTs, and in some cases, better than FGT compounding. VATs can provide satisfactory boost over a wider range by varying the turbine housing throat area. At lower speeds, the reduction of the turbine inlet area increases the momentum impact on the rotor while at higher speeds the increase in throat area controls over boosting in the intake manifold. However, the scope of an internal or external waste gate cannot be eliminated.

Two of most widely used methods used to vary the throat area of VAT include the pivoting nozzle vane [20,26,27] and the sliding wall technique[27,28], the latter being a bit less popular due to causes of non-aerodynamic shape and subsequent efficiency losses[28]. In the first type, the flow rate is governed by the movement of the vane, and in the second, by movement of the sliding wall.

Pesiridis and Martinez- Botas [28] (2007) presented a technique to continuously regulate a turbine inlet area so as to adapt to the continuous exhaust gas pulsation. This system was called Active Control Turbocharger (A.C.T.). Here, a variable throat area turbine operation was extended to consider the pulsating nature of the exhaust flow. The throat inlet area opens progressively as the exhaust pressure increases and closes gradually at the end of the pulse. Some power improvements were seen, but efficiency losses contributed largely by the aerodynamically poor sliding nozzle employed was a drawback. Figure 8 further explains the mechanism.

Petitjean et al. [29] in 2004 also proposed a method of varying the turbine inlet area using a mechanism similar to Pesiridis and Martinez- Botas for application in gasoline engines, and claimed an advantage over variable nozzle turbine (VNT) in terms of durability. In a variable nozzle turbine (VNT), the flow dynamics of the exhaust gases is controlled by the movement of the vanes as shown in figure 3(a). The movement of the vanes increases or decreases the area between the vanes so as to alter the flow speed and angle of impingement on the rotor. At low engine speeds, reducing the area between the vanes increases the velocity at the expense of pressure, making the vanes function as sub sonic nozzles, and vice-versa. The VNTs therefore provide a better engine – turbocharger matching over a wider engine speed range. This may also lead to a complete elimination of a waste gate as well.[17,20]

Capobianco and Gambarotta analyzed a variable area turbine (VAT), a variable nozzle turbine (VNT) and a fixed geometry turbine (FGT) under steady and pulsating flow conditions and established relative aerodynamic performances. The VNT showed a better engine turbo matching with a wider operating range but with a loss in efficiency at the top end. The VNT unit, thus, proved to be a little better than its VAT counterpart. The FGT, VAT and VNT units used were Garrett TB025, Garrett VAT025 and Garrett VNT025, respectively. The aforementioned would become more clear from figure 9 adapted from Capobianco and Gambarotta [20].

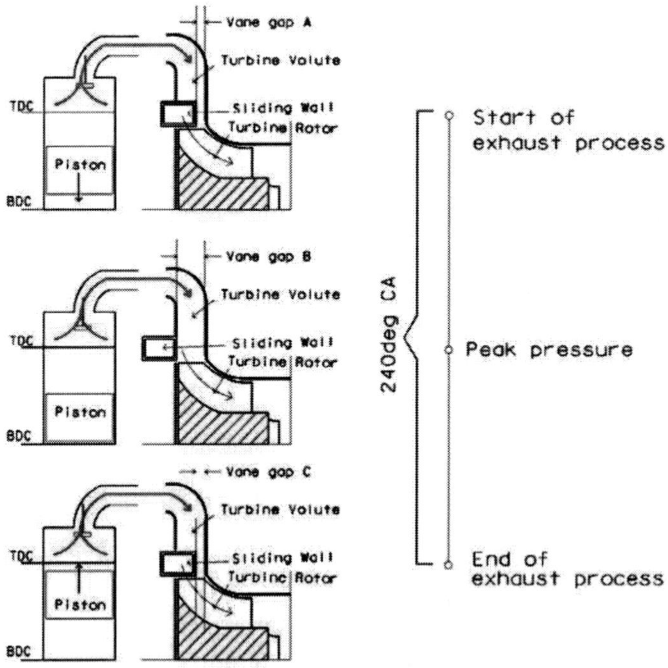

Figure 8. Active Control Turbocharger (A.C.T.); nozzle gap progressively opens towards the peak pressure and closes lower end of the pulse. [28]

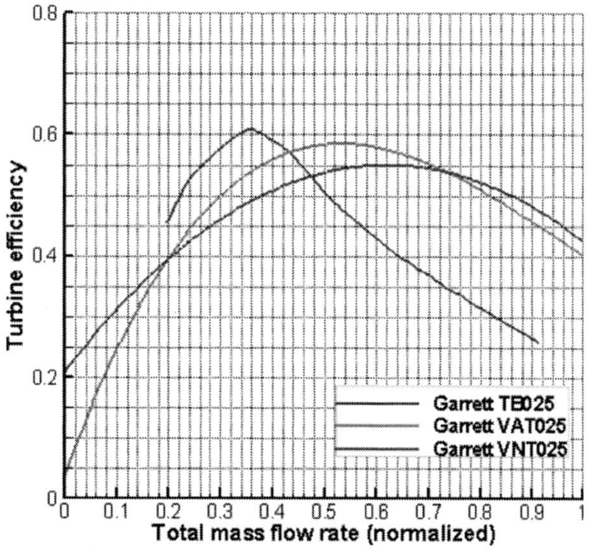

Figure 9. efficiency comparison of same sized VAT,VNT and FGT, the control vanes are at fully open and fully closed at one and zero respectively on the abscissa .[20]

However, in the study conducted by Capobianco and Gambarotta [20], only the static pressure at turbine inlet and outlet were measured and the rest were just time average values. Therefore, no conclusive correlation could be established for the entire flow range. However, a decreasing mass flow rate factor with increasing pulse amplitude was deduced.

Again, Gabrial et al. [17] in 2007, highlighted that exhaust gas pulsation damping at low vane position resulted in an increased turbine inlet temperatures leading to pumping losses. However, it was also claimed that exhaust gas pulsation damping gave way to transient compressor surge reduction, thereby having a positive end to it. It can also be seen from the plot in figure 9 that the peak efficiency of a FGT is slightly better than that of a VFT or even VNT. This is mainly due to the fact that a FGT has comparatively less hindrances in the flow passage. Even though a FGT (with waste gates) provides a better peak efficiency, it fails to provide sufficient boost pressure at higher engine speed levels as compared to its VGT counterparts [20].

Arnold et al. [30] highlighted the need for clearance between vane and housing of a VNT in order to prevent the sticking of vanes. Again in 2011, Hu Liangjun et al. [31] presented a research which showed dramatic decrease in turbine performances with increasing nozzle clearances. Various simulations were carried out with different nozzle clearances and also with no clearance. It was reported that an increase of 1% nozzle clearance led to a drop of 2.5% in turbine efficiency. This drastic drop in turbine efficiency can be further realized from the curve in figure 10 adopted from Liangjun et al. [31]. This dramatic drop in turbine performance is credited to two reasons. Firstly, the inclusion of clearance towards the hub and shroud side of the nozzle leads to back flow of the fluid (exhaust gases) from the pressure side to the suction side. This reverse flow tends to induce a component of flow velocity opposite to the direction of rotation of the rotor, leading to a fall in rotor efficiency. Secondly, due to the nozzle clearance and the above mentioned back flow, the fluid at impeller hub has comparatively low energy and is therefore unable to overcome the centrifugal forces of the rotor. The fluid hence has a tendency to migrate from the hub to the shroud side at the impeller inlet. As a result, low energy fluid occupies greater amount of flow passage as compared to a VNT without clearance.

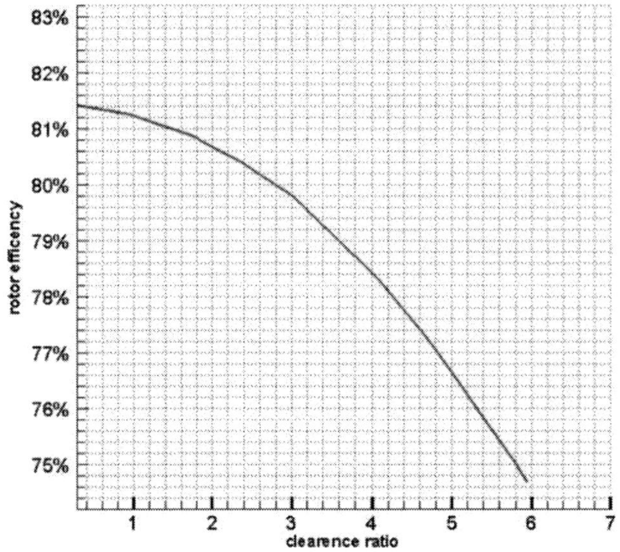

Figure 10. drastic fall of rotor efficiency with increasing clearance ratio in a VNT. [31]

As discussed previously the working mechanism of VNTs is much more complex than other VGTs (VFTs and VATs). This is greatly due to the greater number of moving parts. [28, 31]Also in case of high speed gasoline engine with a VNT complete elimination of a waste gate may not possible due a wide range of flow rates [32]. Again, in a simulation conducted by Yang and Wang, it was claimed that that using a VNT in a gasoline engine may lead to increased knocking at the beginning of combustion [33]. Moreover, in case of a gasoline engine, the exhaust temperature is higher than a diesel engine which makes the application of VNTs to gasoline engine much more difficult. These hindrances to gasoline VNT application have largely restricted the use of VNTs on gasoline production engines. Most of the VNT gasoline engines today are in prototype stage with massive future scope. In spite of all the hindrances, Porsche and Volkswagen have been quite successful in their gasoline VNT ventures.

One of the very few production gasoline engines using a VNT turbocharger can be seen in the Porsche 911 Turbo S [22] engine, which uses two VNTs in parallel. It is 3.8 liter engine which shows a enormous specific power of 112 kW/l. Other gasoline VNT application includes the new Volkswagen 1.5l EA211 TSI evo gasoline engine [34, 35]. The engine is expected to replace the 1.4l TSI and is expected to be 10% more efficient than its predecessor. Thanks to the VNT, the turbocharged four- cylinder engine that makes its maximum torque available really early, offering 200 Nm from just 1.300rpm in the 128 hp versions.

More recently in 2016, the 2.5 liter Porsche 718 Boxster S [36] and 718 Cayman S [37, 38] have been equipped with VNTs. This resulted in power output of 256 kW (235hp) at 6500 rpm and a maximum torque of 420 Nm available across a wide range from 1900 rpm to 4500 rpm. As compared to the 2 liter base model (718 Boxster and 718 Cayman) the 2.5 liter S model with the VGT provides an improvement of 25 kW (35hp) and 100 Nm but with a 14% decrease in fuel consumption and emission, thanks to the VGT. The turbochargers used in the 2.5 litre Porsche 718 Boxster S and 718 Cayman S have been derived from the majestic 3.8 liter Porsche 911 Turbo S and adapted for the 718 S to obtain excellent transient response at low engine speeds and maximum power output.

4. Increasing the flow range of the VGT turbocharger compressor

As discussed in the previous sections, variable geometry turbine turbochargers have helped largely in better turbocharger – engine matching and to overcome, to great extent, the problems of transient response. But, the compressor of the turbocharger has a surge limit related to the flow rate of the compressor on the low flow side and choke on the high flow side. Lowering the flow rate below the surge limit leads to pressure fluctuations characterized by loud noises. Solutions involving two or 3 turbochargers have been developed and put to production but issues like packaging, durability and high heat dissipation have been a major hindrance in its way. Therefore using a single turbocharger with a compressor that has a wide flow range seems to be a better alternative. One way to prevent the flow rate from going below the surge limit is increasing the rotational speed of the turbocharger, either by using a variable nozzle turbine (VNT) or by using motor assistance a lower engine speeds [39].

The surge limit can be improved by what is known as the casing treatment. Yamaguchi et al. [40] in 2002 developed a casing treatment for turbocharger compressors. There have also been many other works on improving the surge limit of a turbocharger compressor. Uchida et al. [13] made use of casing treatment and variable inlet guide vane (VIGV) to develop a turbo charger compressor with wide flow range and improved surge limits. Figure 11[13] shows the structure of compressor with VIGV and casing treatment. The VIGV located upstream to the casing treatment was able to control the pre-whirl of the inlet flow to the impeller. It was found from CFD analysis that vortexes were formed at the side of the shroud at the surge limits. This problem was catered by casing treatment i.e. by re-circulating the air from the impeller to point downstream of the variable inlet guide vane. The casing treatment implemented by Uchida et al.[13] had spoon type guide walls and was able to reduce the surge flow rate by 30% as compared to a conventional compressor at a pressure ratio of 2.5, not to mention the increment in compressor efficiency. Moreover the excellent synergy of the VIGV with the casing treatment yielded a reduction in surge flow rate by 59% as compared to a conventional compressor at a pressure ratio of 2.5. The structure of the compressor with the VIGT and the casing treatment is shown in figure 11(a) and the improvement in the surge limit has been shown in figure 11(b). It was also seen that the installation of a rib led to improvement in the surge limits at higher engine speeds.

Other studies on the effect VIGT on the flow range of a turbocharger compressor include, Fraser et al.[41], Mohtal et al.[42], Yamaguchi et al.[40]. Even though, the VIGT successfully widens the compressor's operating range, problem of throttling losses due to the variable inlet guide vanes are seen as low flow rates.

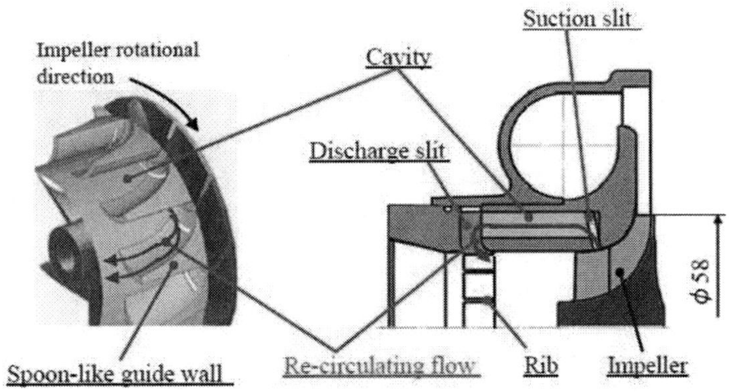

Figure 11(a). Structure of the casing treatment studied by Uchida et al.[13]

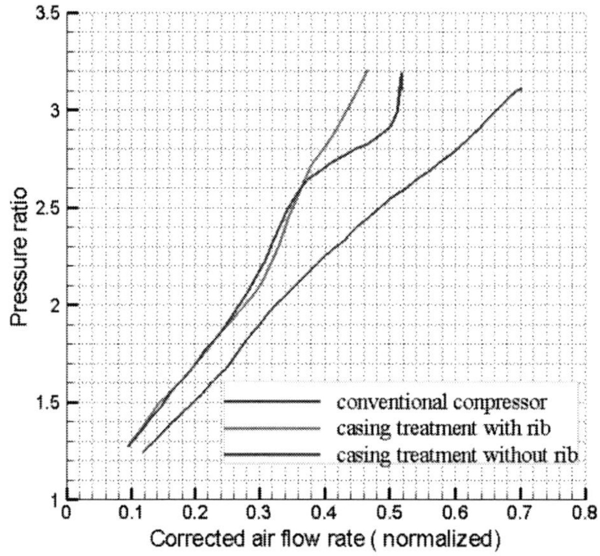

Figure 11(b). Effect of casing treatment (with and without rib installed in the discharge slit) on the surge limit of the compressor.[13]

Xinqian Zheng and Chaunjie Lan [43] improved the performance of a turbocharger compressor by a successful synergy of self-recirculation casing treatment (SRCT) and blade bowing. Six cases of blade bowing were studied which included both positive and negative bowing. It was reported that the negative bowing decreased the surge mass flow rate as well as the choke mass flow rate. Significant effects of blade bowing were seen at high rotational speeds. At near-coked conditions negative blade bowing enlarged the throat area, thereby changing the choke mass flow rate. With the synergy of SRCT and negative blade bowing improvements were seen on the surge flow rate of the compressor. An increase of 5.84% in the surge flow rate of the compressor was reported.

H Sun et al.[44] proposed a dual port casing treatment scheme (aka active casing treatment) to enhance the turbocharger compressor's operating range. It was an enhancement over the classical ported shroud recirculation casing treatment which successfully obtained a 12% additional flow capacity. The surge and choke were addressed separately with two different slots on the compressor shroud. Figure 12 further explains the design.

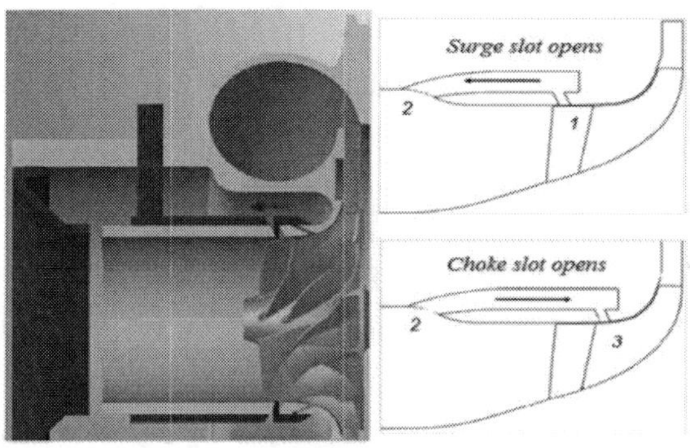

Figure 12. Dual-port casing treatment with bleeding and injection slot to extend the surge margin and the choke flow capacity of the compressor.

The slot between the full blades and the splinter blades serves to provide optimum surge control and the other slot downstream enhanced the choke flow capacity. As a shock wave involves a compression wave followed by expansion wave, the introduction of a second slot downstream could successfully take advantage of the following expansion and improve the choke flow capacity. An improvement in the flow capacity by 12% compared to aforementioned classical recirculation casing treatment without any compromise in low end efficiency and surge performance was reported.

Table 1 gives a brief description to recent gasoline production engines making use of VGT turbochargers. Apart from those mentioned in the table 1, Honda equipped the Legend with a VGT system starting in 1988, but it was only in production for two years. Chrysler equipped the Dodge Shelby SCX with a VGT in the late 1980s, but this was a very limited production run. With the recent trend in VGT application in gasoline production engines, it seems that VNT would be leading the way in the coming future. Though VGTs gasoline engine application seems to have been quite successful in few gasoline production engines, issues of production costs and durability still need some addressing. Also there seem to be scope for further development by addressing the issues of optimum vane clearances and vane sticking. Also some synergies of VGTs with Energy Recovery Systems may lead to a better control over the boost pressure. A brilliant application of the synergy Energy recovery system and turbocharger can be seen in the 2014 Mercedes Formula-1 engine.

Table 1. Recent gasoline production engines featuring VGTs

Engine	No of cylin ders	Displac ement	Boostin g system	Compre ssion ratio	Fuel consumptio n	CO2 emission	Specif ic power	torque	Power
Porsch e 911 turbo S Coupé	6	3.8 liter	VGT (VNT)	9.8:1	9.1 liter/100km	212 g/km	112.3 kW/liter	700 Nm (2,100–4,250 rpm) 750 Nm 2,250–4,000 rpm (with over boost	427 kW (580hp) 6,750 rpm
Porsch e 718 Boxter and 718 Cayma n	4	2.5 liter	VGT (VNT)	9.5:1	8.1 liter/100k m (manual) 7.3 liter/100k m (PKD)***	184 g/km (manual) 167 g/km (PKD)	102.8 kW/liter	420 Nm (1900 rpm to 4500 rpm)	257 kw (300hp) at 6500 rpm
911 Turbo S Cabrio let	6	3.8 liter	VGT (VNT)	9.8:1	9.3 liter/100km	216 g/km	112.3 kW/liter	700 Nm 2,100–4,250 rpm 750 Nm 2,250–4,000 rpm (with over boost****)	427 kW (580hp) 6,750 rpm
Volks wagen EA211 TSI evo	4	1.5	VGT (VNT)	12.5:1	Not released yet*	Not released yet*	73.33 kW/liter	Not released yet**	96 kW and 110 kW
Honda acura RXD 2.3L engine	4	2.3	VGT (VF T)	8.8:1	19/24/21 mpg (2WD) 17/22/19 mpg (SH-AWD	Not released yet	Not released yet	260 ft-lbs at 4500 rpm.	240hp 6000rpm

*Volkswagen claims the new EA211 TSI evo to have efficiency benefits of up to 10 per cent compared with the previous 1.4l TSI (92 kW).

**Volkswagen claims that compared with a 1.4l TSI (92 kW), the step change in load to the maximum torque takes place some 35 per cent faster.

*** Porsche Doppelkupplung (PDK) is 7- speed transmission which offers extremely fast gear changes. . The engine is decoupled from the transmission to avoid deceleration caused by engine braking

****over boost function gives the Porsche 911 Turbo an extra 37 lb.-ft. of torque for up to 10 seconds.

5. Conclusion

In the past VGT turbochargers have already proven their worth in diesel engine downsizing. Today, with the stringent emission norms and with the heavy reliance of gasoline engine downsizing on turbocharging, VGT application to gasoline engines seem inevitable. The VGT benefits of increased fuel economy, improved low end torques , better transient response, better steady state performances, increased power density and greater boost pressure over a wider range would be quite difficult to overlook in the next generation gasoline engine designs. VGTs have also proven to be a better solution over the more complex two or three stage turbocharging in many ways.

Though the high exhaust gas temperatures of gasoline engines and the perception of VGT (VNT) gasoline applications not being cost efficient have historically been taken as a hindrance to VGT (VNT) application to gasoline engines, Volkswagen 1.5 liter TSI evo engine seems to have successfully overthrown this perception by being a cost sensitive gasoline with a VNT turbocharger.

Despite all hindrances of high exhaust gas temperature and cost constraints, VGT turbochargers have successfully paved their way to gasoline engine application and have been wonderfully successful in providing a better transient response along with better steady state performance, not to mention the contribution of technologies like WCEM, variable valve timing etc. towards making this possible. At the same time, various improvements in the flow range of the turbocharger compressor have also been instrumental in the development of the VGTs so far. It therefore, seems that VGTs would possibly be a major driving force in the near future gasoline engine downsizing process.

6. References

[1] History, BorgWarner Turbo & Emission Systems, URL:http://www.turbodriven.com/en/turbofacts/default.aspx [accessed 18th April, 2016]

[2] Zacharof, N., Tietge, U. and Mock, P., 2014. CO2 emissions from new passenger cars in the EU: Car manufacturers' performance in 2014. The International Council of Clean Transportation, URL: http://www.theicct.org/co2-new-cars-eu-manufacturer-performance-2014

[3] Mathur, M.L. and Sharma, R.P., 2005. Internal combustion engines. 1st edition,Dhanpat Rai Publication.

[4] Leduc, P., Dubar, B., Ranini, A. and Monnier, G., 2003. Downsizing of gasoline engine: an efficient way to reduce CO2 emissions. Oil & gas science and technology, 58(1), pp.115-127.

[5] Turner, J.W.G., Pearson, R.J., Bassett, M.D. and Oscarsson, J., 2003. Performance and fuel economy enhancement of pressure charged SI engines through turboexpansion: An initial study. SAE transactions, 112(3), pp.593-604.

[6] Salamon, C., McAllister, M., Robinson, R., Richardson, S., Martinez-Botas, R., Romagnoli, A., Copeland, C. and Turner, J., 2012. Improving fuel economy by 35% through combined turbo and supercharging on a spark ignition engine. In 21st Aachen Colloquium Automobile and Engine Technology, 2012. University of Bath.

[7] IMechE., 2011. Innovations in Fuel Economy and Sustainable Road Transport. Elsevier Science.

[8] Fraser, N., Blaxill, H., Lumsden, G. and Bassett, M., 2009. Challenges for increased efficiency through gasoline engine downsizing. SAE International Journal of Engines, 2(1), pp.991-1008.

[9] Lundstrom, R.R. and Gall, J.M., 1986. A comparison of transient vehicle performance using a fixed geometry, wastegated turbocharger and a variable geometry turbocharger (No. CONF- 860263-). Vehicle Dynamics Lab., Lawrence Inst. of Technology.

[10] Uchida, H., 2006. Transient performance prediction for turbocharging systems incorporating variable-geometry turbochargers. R&D Review of Toyota CRDL, 41(3).

[11] Shahed, S.M., Allen, J., Lin, W., Barrios, E. and Birch, K., 2000. Turbocharging Systems with Electrical Assistance for Conventional, Hybrid and Fuel Cell Powertrains. FORTSCHRITT BERICHTE-VDI REIHE 12 VERKEHRSTECHNIK FAHRZEUGTECHNIK, 1, pp.310-324.

[12] Iwakiri, Y. and Uchida, H., 2006. Numerical fluid analysis of a variable geometry

compressor for use in a turbocharger. R&D Rev. Toyota CRDL, 41(3), pp.15-21.

[13] Uchida, H., Kashimoto, A. and Iwakiri, Y., 2006. Development of wide flow range compressor with variable inlet guide vane. R&D Review of Toyota CRDL, 41(3), pp.9-14.

[14] Rahnke, C.J., 1985, March. Axial Flow Automotive Turbocharger. In ASME 1985 International Gas Turbine Conference and Exhibit (pp. V001T04A001-V001T04A001). American Society of Mechanical Engineers.

[15] Sauerstein, R., Dabrowski, R., Becker, M. and Bullmer, W., 2010. Regulated two-stage turbocharging for gasoline engines. BorgWarner Turbo Systems.

[16] Galindo, J., Serrano, J.R., Climent, H. and Varnier, O., 2010. Impact of two-stage turbocharging architectures on pumping losses of automotive engines based on an analytical model. Energy Conversion and Management, 51(10), pp.1958-1969.

[17] Gabriel, H., Jacob, S., Münkel, U., Rodenhäuser, H. and Schmalzl, H.P., 2007. The turbocharger with variable turbine geometry for gasoline engines. MTZ worldwide, 68(2), pp.7- 10.

[18] Baines, N.C., 2005. Fundamentals of turbocharging. Development, 2004, pp.03-08.

[19] Andersen, J., Karlsson, E. and Gawell, A., 2006. Variable turbine geometry on SI engines (No. 2006-01-0020). SAE Technical Paper.

[20] Capobianco, M. and Gambarotta, A., 1992. Variable geometry and waste-gated automotive turbochargers: measurements and comparison of turbine performance. Journal of Engineering for gas Turbines and Power, 114(3), pp.553-560.

[21] Kawaguchi, J., Adachi, K., Kono, S. and Kawakami, T., 1999. Development of VFT (variable flow turbocharger) (No. 1999-01-1242). SAE Technical Paper.

[22] Aucock, R., 2010. Porsche 911 Turbo S. Automotive Engineer, 35(5), pp.11-12.

[23] Ito, N., Ohta, T., Kono, R., Arikawa, S. and Matsumoto, T., 2007. Development of a 4-Cylinder Gasoline Engine with a Variable Flow Turbo-charger (No. 2007-01-0263). SAE Technical Paper.

[24] Boyle, S. and Komnick, B., 2008. Late Model Hondas and High-Tech Diagnostics. Presentations, p.1.

[25] Flaxington, D. and Szczupak, D.T., 1982. Variable area radial-inflow turbine. In 2nd Int. Conf. on Turbocharging and Turbochargers.

[26] Kaupert, A.W., 1987. Diesel Engine Technology Update (No. 83-021-DET). AIR FORCE RESERVES DETROIT DETACHMENT ANN ARBOR MI.

[27] Rajoo, S. and Martinez-Botas, R., AUTOMOTIVE TURBOCHARGING.

[28] Pesiridis, A. and Martinez-Botas, R.F., 2007. Experimental evaluation of active flow control mixed-flow turbine for automotive turbocharger application. Journal of Turbomachinery, 129(1), pp.44-52.

[29] Petitjean, D., Bernardini, L., Middlemass, C. and Shahed, S.M., 2004. Advanced gasoline engine turbocharging technology for fuel economy improvements. SAE transactions, 113(4), pp.521-529.

[30] Arnold, S., Groskreutz, M., Shahed, S.M. and Slupski, K., 2002. Advanced variable geometry turbocharger for diesel engine applications (No. 2002-01-0161). SAE Technical Paper.

[31] Hu, L., Yang, C., Sun, H., Zhang, J. and Lai, M., 2011. Numerical analysis of nozzle clearance's effect on turbine performance. Chinese Journal of Mechanical Engineering, 24(4), p.618.

[32] Shimizu, K., Sato, W., Enomoto, H. and Yashiro, M., 2009. Torque control of a small gasoline engine with a variable nozzle turbine turbocharger (No. 2009-32-0169). SAE Technical Paper.

[33] Wang, L.S. and Yang, S., 2006. Turbo-cool turbocharging system for spark ignition engines. Proceedings of the Institution of Mechanical Engineers, Part D: Journal of Automobile Engineering, 220(8), pp.1163-1175.

[34] VW 1.5L TSI Evo petrol motor with VGT unveiled, URL:https://www.rushlane.com/vw-1-5l- tsi-evo-petrol-motor-vgt-unveiled-12195255.html (viewed on June 25, 2016)

[35] Volkswagen announces new 1.5 TSI 'EA211 evo' engine URL:http://performancedrive.com.au/volkswagen-announces-new-1-5-tsi-ea211-

evo-engine- 2910/ (viewed on May 2, 2016)

[36] porsche 718 catalog, URL:http://www.porsche.com/international/models/718/718-boxster-s/ (viewed on May 2, 2016)

[38] The new 718 Boxster and 718 Cayman For the sport of it, URL: http://www.porsche.com/international/models/718/718-cayman-s/)

[39] Uchida, H., 2006. Trend of turbocharging technologies. R&D Review of Toyota CRDL, 41(3), pp.1-8.

[40] Yamaguchi, S., Yamaguchi, H., Nakao, H. and Nakamura, F., 2002. The Development of Effective Casing Treatment for Turbochargers Compressors, IMechE Conference Transactions: 7th International Conference on Turbochargers and Turbocharging.

[41] Fraser, N., Fleischer, T., Thornton, J. and Rueckauf, J., 2007. Development of a fully variable compressor map enhancer for automotive application (No. 2007-01-1558). SAE Technical Paper.

[42] Mohtar, H., Chesse, P., Yammine, A. and Hetet, J.F., 2008. Variable inlet guide vanes in a turbocharger centrifugal compressor: local and global study (No. 2008-01-0301). SAE Technical Paper.

[43] Zheng, X. and Lan, C., 2014. Improvement in the performance of a high-pressure-ratio turbocharger centrifugal compressor by blade bowing and self-recirculation casing treatment. Proceedings of the Institution of Mechanical Engineers, Part D: Journal of Automobile Engineering, 228(1), pp.73-84.

[44] Sun, H., Hu, L., Zhang, J., Hanna, D., Krivitzky, E., Larosiliere, L., Lai, M.C. and Yang, C., 2014. Switchable dual-port casing treatment scheme for an enhanced turbocharger compressor operating range. Proceedings of the Institution of Mechanical Engineers, Part D: Journal of Automobile Engineering, 228(3), pp.235-244.

[45] http://www.acura.com/engine.aspx?model=rdx#~pKmwsdBZfcrw7a

[46] Volkswagen at the 37th Vienna Motor Symposium (press release), URL:https://www.volkswagen-media-services.com/en/detailpage/-/detail/Volkswagen-at-the- 37th-Vienna-Motor Symposium/view/3451577/7a5bbec13158edd433c6630f5ac445da?p_p_auth=9s2ksFjm

IOP Conf. Series: Materials Science and Engineering **943** (2020) 012011 doi:10.1088/1757-899X/943/1/012011

Investigation on the effect of ultrasonic-assisted transesterification for green synthesis of glycerol carbonate from crude glycerol

P K Lo[1], S Y Leong[1] and C Y Tan[1]

[1] Department of Petrochemical Engineering, Faculty of Engineering and Green Technology, Universiti Tunku Abdul Rahman, 31900 Kampar, Perak, Malaysia

Corresponding author: lopk@utar.edu.my

Abstract. The present work demonstrates the utilization of ultrasonic-irradiation for synthesis of glycerol carbonate (GC) with direct use of crude glycerol (C.Gly) and dimethyl carbonate (DMC). This transesterification reaction was catalysed by calcium oxide (CaO) and the effect of ultrasonic-assisted transesterification reaction was studied. In order to verify the contents of C.Gly, the C.Gly obtained from biodiesel production plant was characterized and the results showed that C.Gly consists of 71.21%w/w glycerol, 16.01%w/w of moisture, 7.10%w/w of methanol, 2.76%w/w of ash, 3.60 %w/w of soap and 10.02%w/w of matter organic non-glycerol (MONG). Subsequently, effects of reaction temperature, reaction time, molar ratios of reactants and catalyst loading on C.Gly conversion and GC yield have been investigated. The highest yield of GC (95.41%) was attained with 9 mol% of CaO catalyst loading, 3:1 molar ratio of (DMC:C.Gly) at 70 °C for 90 min. The yield of GC was observed to rise with all the reaction parameters till the optimum conditions obtained. Moreover, the yield of GC obtained from ultrasonic-assisted was compared with the conventional-heating method done in the previous study. It was noticeable that the yield of GC obtained via ultrasonic-assisted was found to be 51.44% better than conventional-heating transesterification. In conclusion, the production of GC via ultrasonic-assisted transesterification shows better feasibility than that of the conventional-heating method.

1. Introduction

Glycerol also known as 1,2,3-propanetriol and it is widely applied in pharmaceutical, cosmetics, food and chemical industries. Glycerol can be produced as a by-product from transesterification reaction in biodiesel industry as well as hydrolysis and saponification in oleochemical industry. The tremendous development and oversupply in biodiesel and oleochemical production has led to the rising of crude glycerol (C.Gly), a by-product from biodiesel and oleochemical industries.[1]. Therefore, to address these circumstances, transforming C.Gly into marketable and value-added chemical products is necessary. Glycerol can be converted into valuable glycerol derivatives by steam forming, dehydration, oxidation, carboxylation, hydrogenolysis and chlorination. Glyceric acid, acrolein, epichlorohydrine, propanediols and glycerol carbonate (GC) are some important chemicals that derived from glycerol [2]. Glycerol carbonate is one of the most notable derivatives of glycerol due to its physicochemical properties such as high boiling point, notable chemical reactivity, low toxicity, biodegradable, slight viscosity and low flammability [3,4]. Moreover, GC has been extensively used as protic solvent and additives in polymers and resins [4], a membrane component for gas-separation [5], surfactants [6], curing agent for concrete, electrolytes component in lithium-ion batteries [7], bio-lubricants and as solvent in cosmetics [8]. GC is a promising precursor for synthesis of valuable chemicals and a potential substitute for petroleum derivatives, such as ethylene carbonate or propylene carbonate.

The production of GC using glycerol can be obtained from several reactions pathway such as (i) carboxylation with CO_2 [9], (ii) glycerolysis with urea [10], (iii) phosgenation with phosgene [11] and (iv) transesterification with carbonation source (i.e. alkylene and dialkyl carbonate) [3]. Production of GC via catalytic transesterification of glycerol with dimethyl carbonate (DMC) is mostly preferred over other reactions. Since, the by-product (methanol) from transesterification between glycerol and DMC is recovered, recycled and reused as a reactant for other processes like biodiesel production. Meanwhile, the by-product (glycerol) obtained from biodiesel production could be reused as the feedstock for GC synthesis. In another words, this reaction not only offers a close ended process but providing more feasibility and sustainable platforms with economic merits for industrial production apart from its mild operating conditions, non-corrosive, having excellent solvent properties and produced higher GC yield [11–13]. The yield of GC is also governed by the synthesis techniques, choice of catalyst use and reaction parameters.

Heterogeneous, homogeneous and enzymatic catalysts have reportedly been used for catalytic transesterification of GC [14,15]. The typical transesterification for GC production were mainly catalyzed by heterogeneous base catalysts such as zeolites, alkaline earth metal oxide, mixed metal oxide derived from hydrotalcite [16]. Heterogeneous catalysts are more favorable than that of homogeneous catalysts due to its recovery, reusability, economical and high catalytic activity. Calcium oxide (CaO) has gained much attention owing to it has relatively high basic strength [17] and it is an environment-friendly catalyst. It is derived from low cost raw materials such as limestone, calcium hydroxide and calcium carbonate [18]. Numerous studies have been reported that CaO is one of the suitable catalysts to synthesize GC via catalytic transesterification [17–21].

Conventionally, transesterification of glycerol with DMC to synthesize GC using typical catalysts namely acid or base happens to be in immiscible phases of the reactants. It is important to maximize the interfacial surface area across the two or more immiscible phases in order to enhance the overall reaction rate. The conventional mechanical agitation technology is one of the processes that has the potential improve the mass transfer and interfacial surface area of the reactants in the reaction. Besides, advances in ultrasonication have been reported that it has the potential to enhance both chemical reactions and mass transfer in a shorter reaction time [22]. The fluid flows in a parallel direction similar with the propagation of ultrasound waves when energy of the ultrasound is applied in the liquid medium. Acoustic streaming referred to as a uniform flow induced by the absorption of acoustic energy during the passage of acoustic waves [23]. This acoustic streaming flow creates pressure-gradient difference resulting from the attenuated from the sound and this pressure gradient difference generates a circumstance including the sequential formation, growth and instantaneously implosive collapse of the cavitation of the bubbles in a liquid medium. This phenomenon results in generate local hot spot fields with enormous temperatures and pressures [24] with a lifespan of few microseconds [25]. These extreme conditions can lead to physical and chemical effects emerge at local positions of the solution in a reactor. In addition, the formation and/or collapse of the cavitation bubbles lead to the formation of physical effects, namely shockwaves, microstreaming and microjets [26]. These effects contribute in enhancing the effectiveness of mixing of solution [27], disrupting the immiscible liquid media and boosting mass transfer by disrupting the solid-liquid interface and liquid-liquid interface boundary layers.

Thus, ultrasound is a versatile, sustainable and promising environmentally friendly and sustainable technology [28] that has been extensively applied in engineering, material science, chemical, food and medicine industries [26]. Ultrasound has been applied in various divisions of a chemical process: (i) as separation of compounds [29], (ii) decompose chemical species, (iii) enhance chemical reaction rate [30], (iv) modify physical or/and chemical properties of a material [31], or (v) to identify and analyse the compounds [32]. Moreover, ultrasound is used in degassing, emulsification and filtration where these processes are required in the food industry.

To the best of our knowledge, C.Gly has not been studied in GC synthesis via ultrasound-assisted CaO catalytic transesterification. Therefore, this work is to study the effect of ultrasound on the synthesis of GC by transesterification of C.Gly and DMC using CaO catalyst. The aims of this study is to explore the effects of ultrasonic irradiation on synthesis of GC with various reaction parameters such as reaction time and temperature, molar ratio of reactants and CaO catalyst loading, analysis was mainly carried out by using gas chromatography.

2. Materials and methods

2.1. Materials

Crude glycerol (C.Gly) was obtained from a local biodiesel production plant located in Carey Island, Selangor, Malaysia and was used directly without further purification. The chemicals used in this study were glycerol carbonate standard (GC, Fisher Scientific), dimethyl carbonate (DMC, 99%, Merck), ethyl alcohol (Merck), calcium oxide (CaO, 95%, Systerm®.), glycerol (99%, R&M Chemicals), sodium hydroxide (NaOH, R&M Chemicals), ortho-phosphoric acid (H_3PO_4, 85%, Fisher Scientific). All purchased chemicals were analytical grade.

2.2. Synthesis of glycerol carbonate

A schematic diagram of the experimental setup was illustrated in figure 1. The transesterification experiments were performed using a 250 mL screw cap flask with a diameter and height of 60 mm and 135 mm, respectively. CaO was used as the catalyst and was calcined at 900 °C for 12 hours in a furnace. The screw cap flask was initially loaded with the corresponding amount of C.Gly, DMC and calcined CaO, depending on the case. The reactants were pre-mixed at a stirring speed of 1000 rpm with a magnetic stirrer for 15 minutes at room temperature. The screw cap flask was placed above a pre-heated (desired reaction temperature) ultrasonic water bath (Elmasonic S 180 (H), power rating at 1000 W and ultrasonic frequency of 37 kHz). The reaction temperature, reaction time, molar ratio of (DMC:C.Gly) and catalyst loading ranges employed is depicted in table 1. Upon completion of the reaction, the mixture was immediately cooled to room temperature in order to halt the reaction. Subsequently, the mixture was filtered using a syringe filter (0.45 μm) to remove the used catalyst. Later, the product mixture was subjected to evaporation at 90 °C to eliminate the unreacted DMC and methanol as by-product. Finally, the product was analysed using gas chromatography (Model: GC-2010 Plus; Brand: Shimadzu Japan) equipped with flame ionization detector, a capillary column (Forte GC BPX-5: 30 m × 0.25 mm × 0.25 μm) and helium gas was used as the carrier gas. Effects of the reaction parameters conversion C.Gly and yield of GC were analysed using external standard method. Ethyl alcohol was added into the product sample. Subsequently, the result obtained was used to determine the C.Gly conversion and GC yield by the following equation (1) and equation (2) [33,34]. A plausible CaO-catalysed transesterification mechanism is depicted in scheme 1.

$$Conversion\ (\%) = \frac{(C.Gly_{initial\ mole} - C.Gly_{final\ mole})}{C.Gly_{initial\ mole}} \times 100\% \quad (1)$$

$$Yield\ (\%) = \frac{GC_{mole}}{C.Gly_{initial\ mole}} \times 100\% \quad (2)$$

IOP Conf. Series: Materials Science and Engineering **943** (2020) 012011 doi:10.1088/1757-899X/943/1/012011

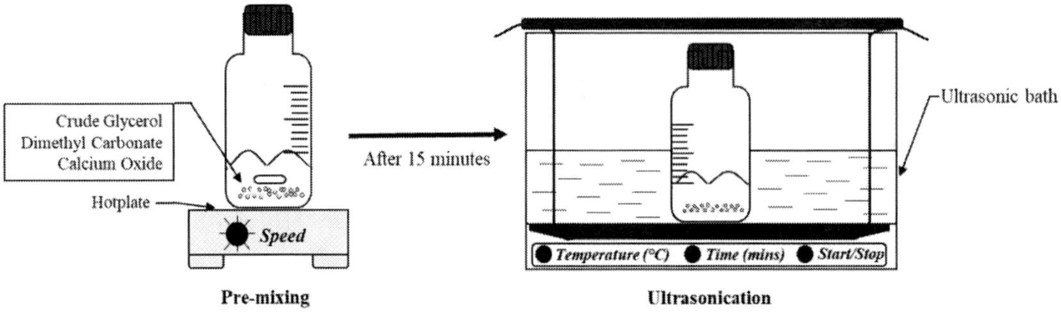

Figure 1. Experimental set-up for the ultrasonic assisted transesterification reaction.

Table 1. The experimental parameters and ranges in this work.

Operating Parameter	Operating Parameter Range
1. Reaction temperature, (°C)	50-85
2. Reaction time, (mins)	15-120
3. Molar ratio of (DMC: C.Gly)	1-5
4. Catalyst loading, (mol%)	3-15

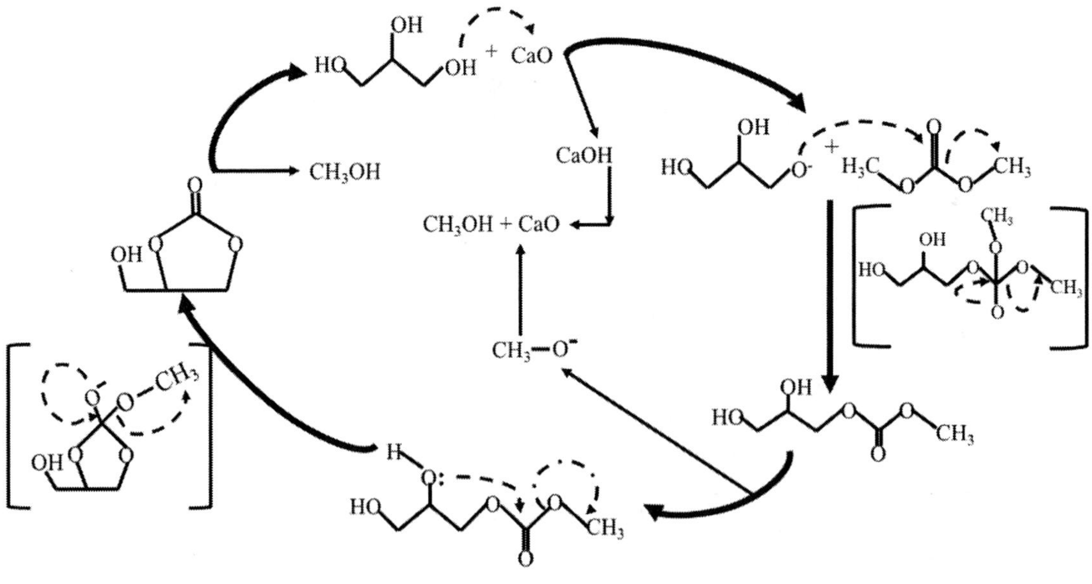

Scheme 1. A plausible CaO-catalysed transesterification mechanism.

2.3. Crude glycerol characterization methods

2.3.1. UV-Vis spectroscopy. The colour of C.Gly sample was determined by using a UV-Vis Spectrophotometer (HACH-DR6000™). Approximately 1 g of C.Gly sample was dissolved with 20 mL of deionized water. The absorbance value of C.Gly sample solution was recorded between the wavelength of 200 nm to 800 nm.

2.3.2. Density. Density of C.Gly sample was determined by using a density determining kit (Precisa). A beaker containing 450 mL of water and placed on the platform of the density determining kit. A cup together with a stopper for sample with density < 1 g/cm^3 was suspended on a holder. The C.Gly sample was placed on the cup and the mass of C.Gly sample was weighed in air and recorded. The weight of C.Gly sample was submerged in the filled beaker was recorded

2.3.3. pH. Prior to pH measurement, a pH meter (Eutech Instruments PC 2700) was calibrated using buffer solutions with of pH 4, 7 and 10. Subsequently, the pH value was recorded by dissolving 1 g of C.Gly sample in 50 mL of deionized water.

2.3.4. Viscosity. Viscosity of C.Gly was measured using a Rheometer (TA Instruments-Discovery Hybrid Rheometer). The accessory (plate SST ST 40mm SMART-SWAP) was placed on bearing of the Rheometer and the accessory was perpendicularly aligned with the bearing. The temperature of the Rheometer was set at 40 °C. Appropriate amount of sample was placed on bottom platform and the accessory was adjusted until it compresses the sample. Finally, the viscosity value of sample was obtained.

2.3.5. Alkalinity. Alkalinity number of C.Gly sample was calculated according to ASTM D 4662-08 as shown in equation (3).

$$Alkalinity\ number,\ (mgNaOH/g) = \frac{[(B-A)39.997N]}{W} \quad (3)$$

where:
B: NaOH solution required for titration of the blank (mL)
A: NaOH solution required for titration of the sample (mL)
N: normality of the NaOH solution (N)
W: weight of sample consumed (g)

2.3.6. Glycerol Content. Glycerol content from C.Gly was analysed using gas chromatography (Model: GC-2010 Plus; Brand: Shimadzu, Japan) equipped with a flame ionization detector (FID) and a capillary column (Model: Forte-BPX-5; 30 m x 0.25 mm x 0.25 μm). The operating conditions for sample analysis were reported in previous work [35].

2.3.7. Ash content. The ash content of C.Gly was determined according to ISO 2098-1972. Approximately 1 g of C.Gly sample was placed in a crucible. The ashing was done in a chamber furnace at 750 °C for 3 hours.

2.3.8. Methanol content. Methanol content from C.Gly was determined using a rotary evaporator (BUCHI). Approximately 1 g (W_1) of C.Gly was placed into a round bottle neck flask and clipped onto the rotary evaporator. The presence of methanol (W_2) in the C.Gly was evaporated under vacuum in a water bath at 60 °C. Methanol content in C.Gly was calculated based on equation (4):

$$Methanol\ content\ (\%\ w/w) = \left(\frac{W_2}{W_1}\right) \times 100\% \tag{4}$$

2.3.9. Water content. Approximately 1 g (W_3) of C.Gly was heated at 105 °C until a consistent weight of C.Gly (W_4) is obtained. The weight loss of the sample during the heating process was referred to the evaporation of methanol and water. The total of water and methanol content (W_5) was calculated using equation (5) and the water content was calculated based on equation (6):

$$W_5(\%\ w/w) = \left(\frac{W_3 - W_4}{W_3}\right) \times 100\% \tag{5}$$

$$Water\ content\ (\%\ w/w) = W_5 - Methanol\ content \tag{6}$$

2.3.10. Soap Content. Approximately 15 g of C.Gly was weighed; the initial pH value of the C.Gly was adjusted to pH 1 using 85% of ortho-phosphoric acid (H_3PO_4) at constant stirring speed for at least an hour. Upon reaching the desired pH, the solution was left at room temperature for 12 hours until two layers of solution was observed. The top layer which was free fatty acid (FFA) was collected and weighed. Soap content was determined according to the following equation (7) [36]:

$$Soap\ content = \left(\frac{304 \times mass\ of\ FFA_{(g)}}{282}\right) \times 100\% \tag{7}$$

Where average soap molar mass = 304; average FFA molar mass = 282

3. Results and discussion

3.1. Characterization of C.Gly

The C.Gly obtained from biodiesel production plant was characterized and the presence of impurities were presented in table 2. The physical appearance of C.Gly is in dark brown soft solid , which depicted in table 2. The C.Gly has a pH value of pH 12, it with density of 1.20 g/mL and viscosity of 48.45 Pa.s. It was observed that, C.Gly consists 71.21% of glycerol content, with high moisture content at 16.01%, ash content at 2.76%, soap content at 3.60% and matter organic non-glycerol (MONG) at 10.02%. As shown in figure 2, the UV-Vis spectroscopy result shows that C.Gly has greater absorbance of radiation and minimum transparency that led to a lower transmittance taking into the account the presence of impurities in C.Gly. Moreover, the dark brownish appearance of C.Gly can be observed through naked eye.

Table 2. Characteristics of crude glycerol.

Properties	Crude glycerol
Appearance	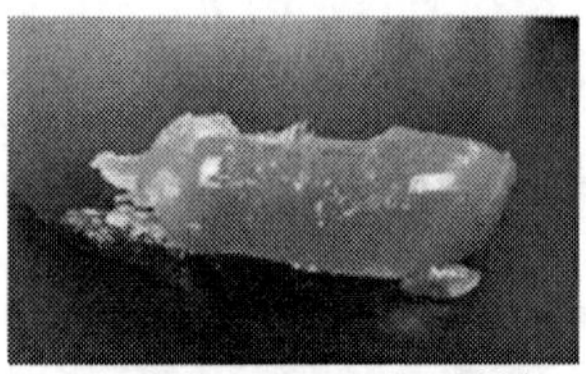 Dark brown soft solid
Density (at 21°C, g/mL)	1.20

pH	12.0
Viscosity (at 40 °C, Pa.s)	48.45
Glycerol Content (% w/w)	71.21±2.14
Water Content (% w/w)	16.01±2.57
Methanol Content (% w/w)	7.10±0.07
Ash Content (% w/w)	2.76±0.83
Soap Content (% w/w)	3.60±0.06
MONG Content[a] (%)	10.02±3.69
Alkalinity (mgNaOH/g)	8.28±0.39

[a]Matter organic non-glycerol (MONG) content = 100% - (glycerol + water + ash content) %

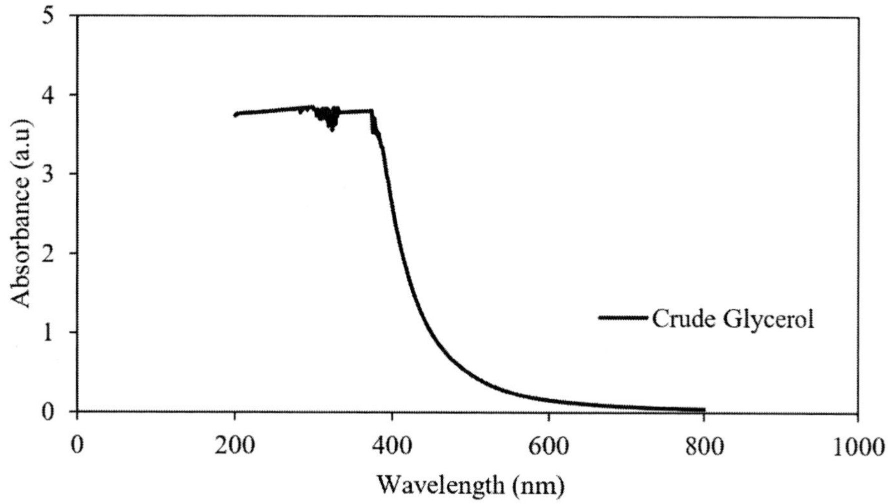

Figure 2. UV- Vis Spectrum of crude glycerol.

3.2. Effect of reaction temperature

Ultrasonic reaction temperature is one of the critical parameters for transesterification of glycerol with DMC as this reaction is reversible. This phenomenon explains that transesterification of glycerol is an endothermic reaction [37]. The equilibrium of the reaction follows the Le Chatelier's principle, a raise of reaction temperature in a system would gradually increase the substrate consumption, which promote the reaction equilibrium to shift towards forward direction and eventually increase the product formation. Hence, transesterification of C.Gly is thermodynamically favorable for GC synthesis [38]. In the transesterification of hydrophilic C.Gly and hydrophobic DMC, both of the reactants are immiscible. Solid base catalyst such as CaO added into the mixture of C.Gly and DMC leads to formation of three-phases. At low reaction temperature, the existence of three-phases leads to a slow diffusion and limited mass transfer rate of reaction. Therefore, a high reaction temperature of a system is required to increase the average kinetic energy of its constituent particles among reactant molecules. As kinetic energy increases, the effective collisions among particles increases and this encourage effective mixing in the bulk solution, eventually increase the reaction rate. The effect of reaction temperature for transesterification reaction was investigated with CaO as catalyst in the range of 50-85 °C. Figure 3 illustrates the effect of various temperature on C.Gly conversion and GC yield using molar ratio of (DMC:C.Gly) at 3:1, catalyst loading of 9 mol% at 90 min. At 50 °C, the conversion of C.Gly and yield of GC attained were 38.96 and 7.45%, respectively and this shows that low reaction rate at low reaction temperature. The conversion of C.Gly and yield of GC increases drastically from 50 °C to 70 °C and the yield of GC decreased slightly with further increased of the reaction temperature beyond 80 °C. The highest yield of 95.41% of GC was achieved at 70 °C. These results indicated that, as temperature increases, the number of oscillating bubbles increases. Thus, leads to more energy collision among the molecules and reduce the viscosity of mixture [12] in which overcome the immiscibility problem between C.Gly and DMC. As a result, high temperature promotes the interaction between the reactants molecules and catalyst consequently speed up the reaction rate.

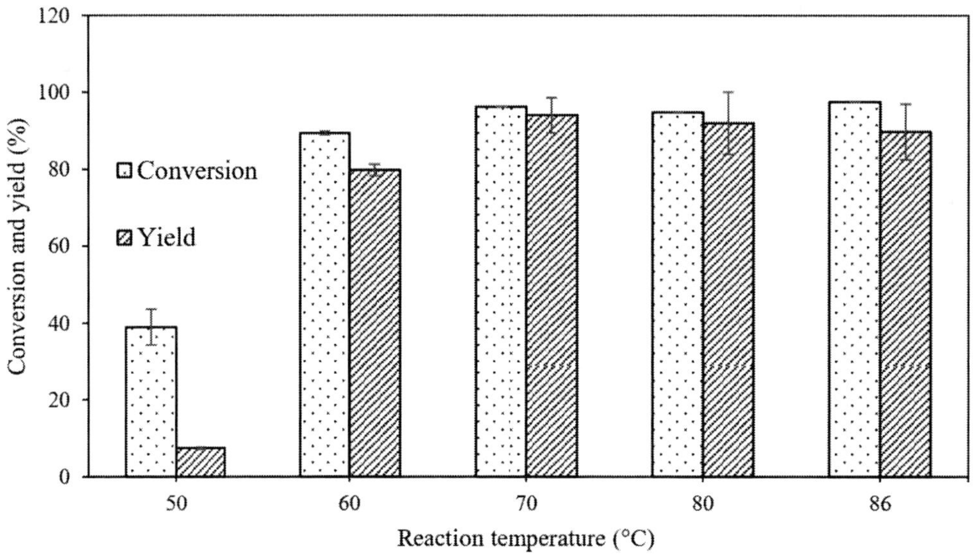

Figure 3. Effect of reaction temperature on GC synthesis via ultrasonic assisted transesterification reaction of C.Gly with DMC with the presence of CaO. Operating conditions: amount of C.Gly = 0.1 mol, time = 90 mins, molar ratio of (DMC:C.Gly) = 3:1 and catalyst loading = 9 mol%.

3.3. Effect of reaction time

The ultrasonic irradiation time in the reaction was evaluated for the production of GC from C.Gly and DMC with the presence of CaO. Both reactants are immiscible therefore a longer reaction time was required to overcome mass transfer resistance. By employing ultrasound technique, lesser reaction time is required to achieve higher yield of product since ultrasonic irradiation has the capability to overcome mass transfer barrier [39]. Figure 4 revealed that the effect of reaction time between from 15 min to 120 min for transesterification of C.Gly with DMC while maintaining all other operating parameters constant. Reaction time at 15 min, the conversion of C.Gly was 96.74%. However, low yield of GC was obtained at 20.87%. This could possibly due to insufficient of reaction time between C.Gly and DMC to shift forward and to achieve equilibrium in the reaction. The yield of GC was significantly increased from 15 min to 90 min and the highest yield of GC was obtained at 95.41% with reaction time of 90 min. However, prolonged reaction time beyond 90 min did not affect the conversion of C.Gly but had resulted in decreased of GC yield. This may possibly due to the effect of solubility of the by-product, methanol in the reaction mixture and the possibility of methanol-DMC azeotrope formation [40] in the reaction.

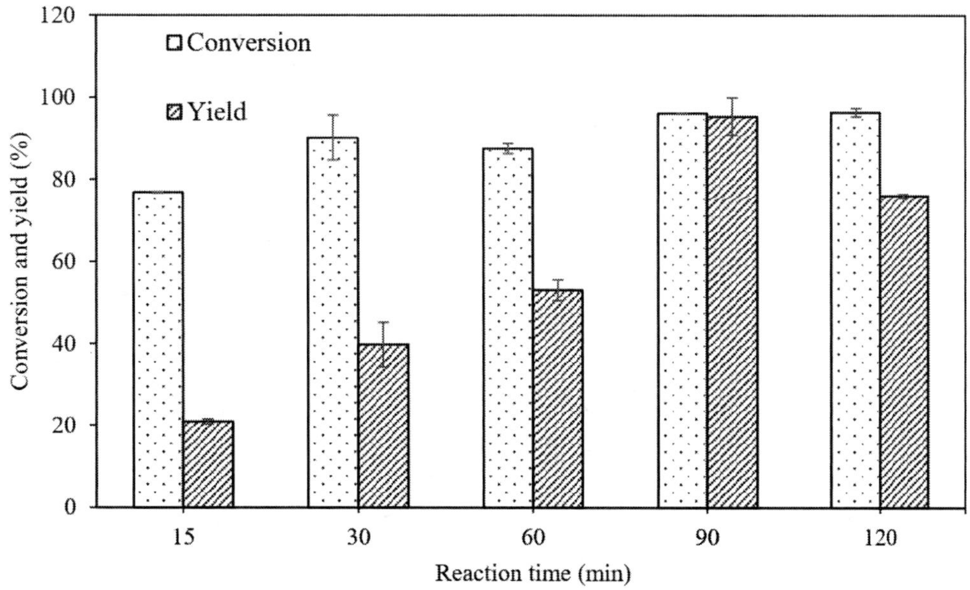

Figure 4. Effect of reaction time on GC synthesis via ultrasonic assisted transesterification reaction of C.Gly with DMC with the presence of CaO. Operating conditions: amount of C.Gly = 0.1 mol, temperature = 70 °C, molar ratio of (DMC:C.Gly) = 3:1 and catalyst loading = 9 mol%.

3.4. Effect of DMC to C.GLY molar ratio

The reaction rate of transesterification can be influenced by varying the molar ratio of reactants. Theoretically, an equal molar of glycerol and carbonate source is required to synthesize 1 mol of GC and 2 mol of by-product, methanol. The effect of molar ratio for DMC to C.Gly was investigated under ultrasonic irradiation with catalyst loading of 9 mol% at 70 °C for 90 min. As illustrated in figure 5, low C.Gly conversion of 33.88% and GC yield of 23.37% were obtained when equimolar of C.Gly and DMC was employed in the reaction. This observation is similar with the findings reported by Li and Wang [20], showing a low yield of GC was obtained when equimolar of reactants was employed in the reaction. This result implied that, using equimolar reactants is insufficient for transesterification reaction to shift the equilibrium to proceed forward by converting C.Gly to GC. Moreover, strong preferential adsorption of C.Gly towards active sites of CaO catalyst due to the polar protic structure in C.Gly material, which resulting in lesser conversion at equimolar of DMC:C.Gly ratio. The maximum yield of GC attained was 95.41% with (DMC:C.Gly) molar ratio of 3:1. As mentioned earlier, C.Gly and DMC reactants are immiscible in the reaction. Excessive molar ratio of alkyl carbonate is required to overcome the immiscibility problem during the reaction. Moreover, the transesterification of C.Gly with DMC is a reversible nature reaction, it is necessary to utilize excess DMC in the reaction to shift positively the chemical equilibrium towards the greater GC production [41,42]. Based on the Le Chatelier's principle, adding extra amount of reactant, in this case is DMC, to the transesterification reaction will shift the position of equilibrium toward formation of GC as showed in scheme 2, this phenomenon enhanced the C.Gly conversion to produce GC and methanol. However, further increase with molar ratio of (DMC:C.Gly) beyond 3:1 has an adverse effect towards the GC yield where the yield of GC decreased to 39.09% and 30.43% at molar ratio of 4:1 and 5:1, respectively. The rapid drop of GC yield probably due to C.Gly as the limiting reactant. The concentration of C.Gly decreases with the increasing molar ratio of (DMC:C.Gly) and subsequently led to low reaction rate [17]. There was no improvement in conversion of C.Gly with further addition of (DMC:C.Gly) molar ratio beyond 4:1. This result is attributed to the dilution of C.Gly with DMC. Dilution of C.Gly in the reaction gave rise to the lowering of collision frequency between the C.Gly molecules, DMC molecules and molecules of CaO and eventually led to decrease in reaction rate. Moreover, as molar ratio of (DMC:C.Gly) increases, the ultrasonic energy per unit volume of the mixture decreases and lead to lesser formation of local hot spot fields with high temperature. Based on figure 5, this study proved that the optimum molar ratio of (DMC:C.Gly) for GC synthesize was attained at 3:1.

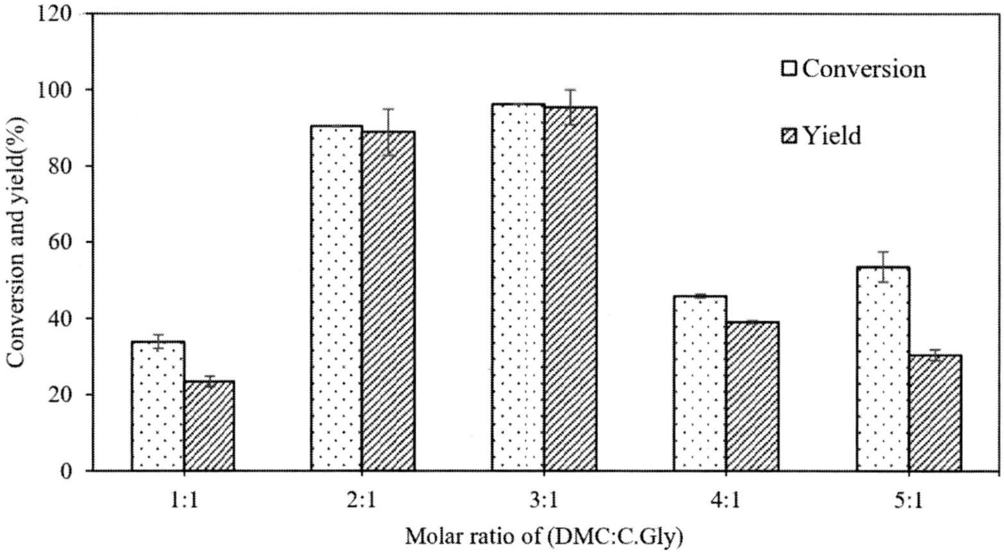

Figure 5. Effect of molar ratio of (DMC:C.Gly) on GC synthesis via ultrasonic assisted transesterification reaction of C.Gly with DMC with the presence of CaO. Operating conditions: amount of C.Gly = 0.1 mol, temperature = 70 °C, time = 90 mins and catalyst loading = 9 mol%.

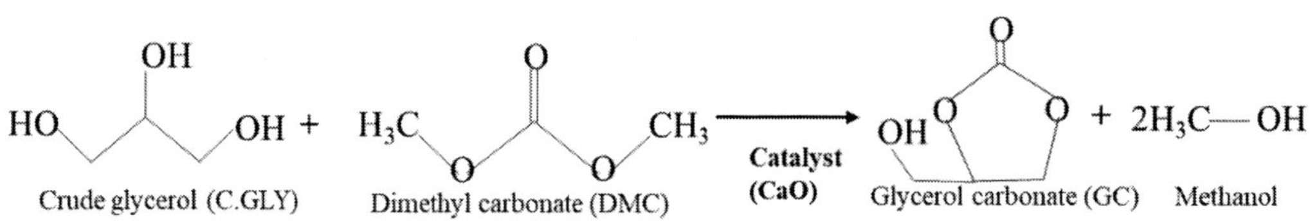

Scheme 2. Synthesis of GC via transesterification of crude glycerol and dimethyl carbonate [35].

3.5. Effect of catalyst loading

Catalyst of CaO is a controversial basic catalyst that has capability of abstracting a proton from the primary hydroxyl group from C.Gly. Thus, CaO was nominated as catalyst in this study. The catalyst of CaO used in this study was prior calcined to minimize the presence hydroxyl group's in CaO and to enhance the catalyst activity. The effect of catalyst loading ranged between 3 mol% to 15 mol% under ultrasonic irradiation was investigated with (DMC:C.Gly) molar ratio of 3:1 at 70 °C for 90 min (figure 6). The increased in catalyst loading, significantly increases the conversion of C.Gly and in GC yield . However, the conversion of C.Gly and yield of GC were remained almost constant when catalyst loading beyond 9 mol%. This is due to more vacant and available active site of CaO per glycerol molecule and the high amount active site of CaO would promote the effective collision between the reactants and catalyst to increase the reaction rate. The enhancement in conversion of C.Gly and yield of GC is closely related to the existence of catalyst particles at an optimum loading in the sonochemical reaction, where the optimum catalyst loading boosted the cavitation activity by ultrasonic irradiation. Conversely, further increased in catalyst loading beyond 9 mol%, a slight decline in the yield was noticed. This is attributed to the high loading of CaO particles which causes the limitation of mass transfer from the bulk to the active sites of catalyst and eventually induce particles agglomeration [43] and sedimentation [44] during the reaction. Furthermore, high loading of catalyst would cause deactivation of catalyst by collision with ground state molecules of catalyst [45]. The optimum CaO loading of 9 mol% gave the highest yield of GC which was attained at 95.41%.

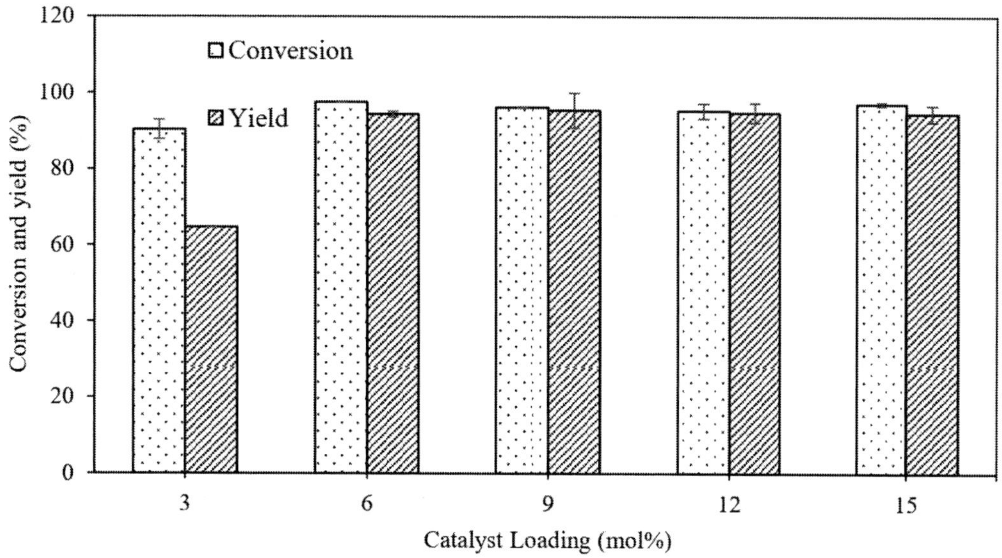

Figure 6. Effect of catalyst loading on GC synthesis via ultrasonic assisted transesterification reaction of C.Gly with DMC with the presence of CaO. Operating conditions: amount of C.Gly = 0.1 mol, temperature = 70 °C, time = 90 mins and molar ratio of (DMC:C.Gly) = 3:1.

3.6. Comparison of conventional-heating transesterification and ultrasonic-assisted transesterification
In previous study, conventional-heating assisted transesterification of C.Gly was conducted under conditions at 70 °C, molar ratio of (DMC:C.Gly) at 3:1 and 9 mol% of CaO for 90 min [35]. A comparison between conventional-heating and ultrasonic-assisted transesterification on conversion and yield were presented in table 3. Ultrasonic and conventional-heating methods were performed under the same reaction conditions and the results in table 3 shows that ultrasonic-assisted transesterification of C.Gly with DMC showed that higher yield of GC was attained at 95.41% than that of the conventional-heating method which the maximum yield of 63%. The yield of GC synthesized using ultrasonic-assisted method was found to be 51.44% better as compared to conventional-heating method. The transesterification reaction rate is closely associated with the value of activation energy. Theoretically, the lower the activation energy, the faster the reaction rate. Ultrasonic irradiation has the potential in enhancing the mass transfer characteristics for transesterification of C.Gly with reduction of activation energy, than that of mechanically stirring systems. Furthermore, ultrasonic technology has a beneficial impact, it produces fine emulsion between the non-miscible phases of the reactants. As a result, the interfacial area between the phases improves markedly, leading more kinetics reaction in the

heterogeneously catalyzed transesterification reaction. As shown in table 3, optimization study on conventional-heating method was also conducted in previous study [35], the yield of GC attained was 79% with (DMC:C.Gly) molar ratio at 3 and catalyst loading of 8.35 mol% at 80 °C for 105 min. In view of this study, the ultrasonic-assisted transesterification of C.Gly with DMC gave better performance than that of conventional heating. Higher GC yield was attained at 95.41% with a shorter reaction time and lower reaction temperature as compared to conventional-heating method at optimum condition.

Table 3. The comparison of conversion C.Gly, GC yield, selectivity in conventional transesterification and ultrasonic-assisted transesterification of industrial C.Gly.

Method	Grade of glycerol	Conversion of C.Gly (%)	Yield of GC (%)	Reference
Conventional- heating[a]	Crude	94.6	63.0	[35]
Conventional- heating[b]	Crude	97.3	79.2	[35]
Ultrasonic irradiation[a]	Crude	96.19	95.41	current work

[a] Operating conditions: reaction temperature = 70 °C, reaction time = 90 min, molar ratio of (DMC:C.Gly) = 3:1 and catalyst loading 9 mol%

[b] Operating conditions (Optimization): reaction temperature = 80 °C, reaction time = 105 min, molar ratio of (DMC:C.Gly) = 3:1 and catalyst loading 8.35 mol%

4. Conclusion

The present study rendered an environmentally friendly route for transesterification of C.Gly using DMC. Moreover, the study has shown that C.Gly can be used directly without further pre-treatment to synthesis GC via transesterification reaction. The influence of critical parameters such as reaction temperature, reaction time, molar ratio of (DMC:C.Gly) and catalyst loading to transesterification were studied. The results showed that ultrasonic irradiation transesterification of C.Gly shows positive effect towards conversion of C.Gly and yield of GC. The maximum C.Gly conversion and GC yield achieved were at 96.19% and 95.41%, respectively under operating conditions: reaction temperature at 70 °C, reaction time at 90 min, molar ratio of (DMC:C.Gly) at 3 and 9 mol% of CaO. Ultrasonic favors the transesterification reaction since the yield of GC was higher in ultrasonic-assisted method as compared to conventional-heating method. Ultrasonic irradiation is more energy efficient due to shorter reaction time and high GC yield attained. Hence, ultrasonic-assisted transesterification is suitable to valorize the surplus C.Gly to synthesize GC.

Acknowledgement

The authors acknowledged the research grant (IPSR/RMC/UTARRF/2019-C1/L09) provided by the Universiti Tunku Abdul Rahman, Postgraduate Research Grant Scheme. The authors wish to express their gratitude to the Sime Darby Biodiesel Sdn. Bhd. through their generosity by sponsoring industrial grade crude glycerol in this work.

References

[1] Rodrigues A, Bordado J C and Santos R G dos 2017 Upgrading the Glycerol from Biodiesel Production as a Source of Energy Carriers and Chemicals—A Technological Review for Three Chemical Pathways *Energies* **10** 18171–36

[2] Konwar L J, Mikkola J-P, Bordoloi N, Saikia R, Chutia R S and Kataki R 2018 Sidestreams From Bioenergy and Biorefinery Complexes as a Resource for Circular Bioeconomy *Waste*

Biorefinery (Elsevier) pp 85–125

[3] Wu Y, Song X, Cai F and Xiao G 2017 Synthesis of glycerol carbonate from glycerol and diethyl carbonate over Ce-NiO catalyst: The role of multiphase Ni *J. Alloys Compd.* **720** 360–8

[4] Wang D, Zhang X, Cong X, Liu S and Zhou D 2018 Influence of Zr on the performance of Mg-Al catalysts via hydrotalcite-like precursors for the synthesis of glycerol carbonate from urea and glycerol *Appl. Catal. A Gen.* **555** 36–46

[5] Wang S, Hao P, Li S, Zhang A, Guan Y and Zhang L 2017 Synthesis of glycerol carbonate from glycerol and dimethyl carbonate catalyzed by calcined silicates *Appl. Catal. A Gen.* **542** 174–81

[6] Granados-Reyes J, Salagre P and Cesteros Y 2017 Effect of the preparation conditions on the catalytic activity of calcined Ca/Al-layered double hydroxides for the synthesis of glycerol carbonate *Appl. Catal. A Gen.* **536** 9–17

[7] Wang J, Yong T, Yang J, Ouyang C and Zhang L 2015 Organosilicon functionalized glycerol carbonates as electrolytes for lithium-ion batteries *RSC Adv.* **5** 17660–6

[8] Shikhaliyev K, Okoye P U and Hameed B H 2018 Transesterification of biodiesel byproduct glycerol and dimethyl carbonate over porous biochar derived from pyrolysis of fishery waste *Energy Convers. Manag.* **165** 794–800

[9] Razali N A, Conte M and McGregor J 2019 The role of impurities in the La_2O_3 catalysed carboxylation of crude glycerol *Catal. Letters* **149** 1403–14

[10] Nguyen-Phu H, Park C and Shin E W 2018 Dual catalysis over ZnAl mixed oxides in the glycerolysis of urea: Homogeneous and heterogeneous reaction routes *Appl. Catal. A Gen.* **552** 1–10

[11] Algoufi Y T, Kabir G and Hameed B H 2017 Synthesis of glycerol carbonate from biodiesel by-product glycerol over calcined dolomite *J. Taiwan Inst. Chem. Eng.* **70** 179–87

[12] Lanjekar K and Rathod V K 2013 Utilization of glycerol for the production of glycerol carbonate through greener route *Biochem. Pharmacol.* **1** 1231–6

[13] Syamsuddin Y and Hameed B H 2016 Synthesis of glycerol free-fatty acid methyl esters from Jatropha oil over Ca–La mixed-oxide catalyst *J. Taiwan Inst. Chem. Eng.* **58** 181–8

[14] Parameswaram G, Rao P S N, Srivani A, Rao G N and Lingaiah N 2018 Magnesia-ceria mixed oxide catalysts for the selective transesterification of glycerol to glycerol carbonate *Mol. Catal.* **451** 135–42

[15] Liu Z, Wang J, Kang M, Yin N, Wang X, Tan Y and Zhu Y 2015 Structure-activity correlations of $LiNO_3/Mg4AlO_{5.5}$ catalysts for glycerol carbonate synthesis from glycerol and dimethyl carbonate *J. Ind. Eng. Chem.* **21** 394–9

[16] Silva L L, Alkimim I P, Costa J P V ., Meneghetti S M . and Cardoso D 2019 Catalytic evaluation of MCM-41 hybrid silicas in the transesterification reactions *Microporous Mesoporous Mater.* **284** 265–75

[17] Ochoa-Gómez J R, Gómez-Jiménez-Aberasturi O, Maestro-Madurga B, Pesquera-Rodríguez A, Ramírez-López C, Lorenzo-Ibarreta L, Torrecilla-Soria J and Villarán-Velasco M C 2009 Synthesis of glycerol carbonate from glycerol and dimethyl carbonate by transesterification: Catalyst screening and reaction optimization *Appl. Catal. A Gen.* **366** 315–24

[18] Roschat W, Phewphong S, Kaewpuang T and Promarak V 2018 Synthesis of glycerol carbonate from transesterification of glycerol with dimethyl carbonate catalyzed by CaO from natural sources as green and economical catalyst *Mater. Today Proc.* **5** 13909–15

[19] Lu P, Wang H and Hu K 2013 Synthesis of glycerol carbonate from glycerol and dimethyl carbonate over the extruded CaO-based catalyst *Chem. Eng. J.* **228** 147–54

[20] Li J and Wang T 2010 Coupling reaction and azeotropic distillation for the synthesis of glycerol carbonate from glycerol and dimethyl carbonate *Chem. Eng. Process. Process Intensif.* **49** 530–5

[21] de Caro P, Bandres M, Urrutigoïty M, Cecutti C and Thiebaud-Roux S 2019 Recent Progress in Synthesis of Glycerol Carbonate and Evaluation of Its Plasticizing Properties *Front. Chem.* **7**

1–13

[22] Gude V G 2015 Synergism of microwaves and ultrasound for advanced biorefineries *Resour. Technol.* **1** 116–25

[23] Sajjadi B, Asgharzadehahmadi S, Asaithambi P, Raman A A A and Parthasarathy R 2017 Investigation of mass transfer intensification under power ultrasound irradiation using 3D computational simulation: A comparative analysis *Ultrason. Sonochem.* **34** 504–18

[24] Kojima Y, Takayasu M, Toma M and Koda S 2019 Degradation of cellulose in NaOH and NaOH/urea aqueous solutions by ultrasonic irradiation *Ultrason. Sonochem.* **51** 419–23

[25] Safarifard V and Morsali A 2012 Sonochemical syntheses of a nanoparticles cadmium(II) supramolecule as a precursor for the synthesis of cadmium(II) oxide nanoparticles *Ultrason. Sonochem.* **19** 1227–33

[26] Chemat F, Rombaut N, Sicaire A G, Meullemiestre A, Fabiano-Tixier A S and Abert-Vian M 2017 Ultrasound assisted extraction of food and natural products. Mechanisms, techniques, combinations, protocols and applications. A review *Ultrason. Sonochem.* **34** 540–60

[27] Poosumas J, Ngaosuwan K, Quitain A T and Assabumrungrat S 2016 Role of ultrasonic irradiation on transesterification of palm oil using calcium oxide as a solid base catalyst *Energy Convers. Manag.* **120** 62–70

[28] Yüceer M and Caner C 2018 Ultrasound ; a Novel and Innovative Processing Method for Egg and Egg Products Preservation Keywords : Introduction *J. Chem. Biol. Pharm. Chem.* **1** 3–5

[29] Trojanowska A, Tsibranska I, Dzhonova D, Wroblewska M, Haponska M, Jovancic P, Marturano V and Tylkowski B 2019 Ultrasound-assisted extraction of biologically active compounds and their successive concentration by using membrane processes *Chem. Eng. Res. Des.* **147** 378–89

[30] Marimuthu E and Murugesan V 2017 Ultrasonic condition boosts up the rate of phase transfer catalyzed polymerization of acrylonitrile in two-phase system *Appl. Petrochemical Res.* **7** 85–96

[31] Kojima Y, Imazu H and Nishida K 2014 Physical and chemical characteristics of ultrasonically-prepared water-in-diesel fuel: Effects of ultrasonic horn position and water content *Ultrason. Sonochem.* **21** 722–8

[32] Figueiredo M K K, Costa-Felix R P B, Maggi L E, Alvarenga A V. and Romeiro G A 2012 Biofuel ethanol adulteration detection using an ultrasonic measurement method *Fuel* **91** 209–12

[33] Ishak Z I, Sairi N A, Alias Y, Aroua M K T and Yusoff R 2016 Production of glycerol carbonate from glycerol with aid of ionic liquid as catalyst *Chem. Eng. J.* **297** 128–38

[34] Teng W K, Ngoh G C, Yusoff R and Aroua M K 2016 Microwave-assisted transesterification of industrial grade crude glycerol for the production of glycerol carbonate *Chem. Eng. J.* **284** 469–77

[35] Tan C Y, Leong S Y and Lo P K 2018 Direct use of crude glycerol in synthesis of glycerol carbonate via transesterification *AIP Conference Proceedings* vol 2020 p 020063

[36] Chen J, Yan S, Zhang X, Tyagi R D, Surampalli R Y and Valéro J R 2018 Chemical and biological conversion of crude glycerol derived from waste cooking oil to biodiesel *Waste Manag.* **71** 164–75

[37] Praikaew W, Kiatkittipong W, Kiatkittipong K, Laosiripojana N, Viriya-empikul N, Boonyasuwat S, Aiouache F, Najdanovic V and Assabumrungrat S 2018 Synthesis of glycerol carbonate from dimethyl carbonate and glycerol using CaO derived from eggshells *MATEC Web Conf.* **192** 03045

[38] Li J and Wang T 2011 Chemical equilibrium of glycerol carbonate synthesis from glycerol *J. Chem. Thermodyn.* **43** 731–6

[39] Islam A, Taufiq-Yap Y H, Chan E, Moniruzzaman M, Islam S and Nabi M N 2014 Advances in solid-catalytic and non-catalytic technologies for biodiesel production *Energy Convers. Manag.* **88** 1200–18

[40] Okoye P U, Abdullah A Z and Hameed B H 2016 Glycerol carbonate synthesis from glycerol and dimethyl carbonate using trisodium phosphate *J. Taiwan Inst. Chem. Eng.* **68** 51–8
[41] Waghmare G V., Vetal M D and Rathod V K 2015 Ultrasound assisted enzyme catalyzed synthesis of glycerol carbonate from glycerol and dimethyl carbonate *Ultrason. Sonochem.* **22** 311–6
[42] Wang S, Wang J, Sun P, Xu L, Okoye P U, Li S, Zhang L, Guo A, Zhang J and Zhang A 2019 Disposable baby diapers waste derived catalyst for synthesizing glycerol carbonate by the transesterification of glycerol with dimethyl carbonate *J. Clean. Prod.* **211** 330–41
[43] Okoye P U, Abdullah A Z and Hameed B H 2017 Stabilized ladle furnace steel slag for glycerol carbonate synthesis via glycerol transesterification reaction with dimethyl carbonate *Energy Convers. Manag.* **133** 477–85
[44] Jamalluddin N A and Abdullah A Z 2011 Reactive dye degradation by combined Fe(III)/TiO_2 catalyst and ultrasonic irradiation: Effect of Fe(III) loading and calcination temperature *Ultrason. Sonochem.* **18** 669–78
[45] Zhou M, Yu J, Cheng B and Yu H 2005 Preparation and photocatalytic activity of Fe-doped mesoporous titanium dioxide nanocrystalline photocatalysts *Mater. Chem. Phys.* **93** 159–63

Direct Fermentation of Oil Palm (*Elaeis guineensis*) Trunk Sap to Bioethanol by *Saccharomyces cerevisiae*

Norhazimah A. H., Siti F.M.N., Aida M., Dilaeleyana A.B., Nur Shahirah M. A.

Department of Science and Mathematics, Centre for Diploma Studies, Universiti Tun Hussein Onn Malaysia, Johor, Malaysia

norhazimah@uthm.edu.my, fatimah@uthm.edu.my, aidamohd@uthm.edu.my, dila@uthm.edu.my, nshahirah@uthm.edu.my

Abstract. Oil palm agricultural biomass has a great potential for biofuel production due to its abundance and less pretreatment needed. In this study, bioethanol production from oil palm trunks sap was carried out by using yeast *Saccharomyces cerevisiae*. The suitability of the oil palm trunk sap for bioethanol production was tested by investigating the relationship between the oil palm trunk sap concentration and ethanol produced. The fermentation was conducted in static condition and without pH adjustment using 60, 80, and 100% oil palm trunk sap concentration. Results showed that the highest bioethanol production rate was achieved for 100% oil palm trunk sap (1.033 g/L.h), followed by 80% (0.744 g/L.h), and 60% (0.645 g/L.h). These results were influenced by the presence of sugars in the initial sap, suggesting that high substrate concentration does not inhibit bioethanol production. Therefore, the oil palm trunks sap showed significant potential as a renewable feedstock for the production of bioethanol without dilution.

Keywords: Bioethanol, *Elaeis guineensis*, palm oil, sap, trunk

1. Introduction

The recovery of natural energy resources such as petroleum and coal can result in global warming, which affects human lives and cause environmental problems due to the alleviation of earth's temperature [1, 2]. Rapid expansion of human population and industrial prosperity in developing countries increase the global demands for energy; making the use of bioethanol to replace the non-renewable fuels as attractive. Complete combustion of ethanol produces only carbon dioxide and water which are not harmful to the environment and this has become one of the advantages to accelerate the research momentum.

There is a growing interest among research communities to find alternative bioresource apart from sugar cane and starchy crop for bioethanol production such as biomass [3]. Production of ethanol from sugars in lignocellulosic has been showed to be promising [4] and despite the challenges, it comes with many opportunities for enhancement. The production costs are still the key impediment to the widely use of ethanol as fuel although the fermentative process for ethanol production is well known [4].
Utilization of agricultural wastes helps in reducing the impact of energy demands on food-derived-ethanol without having to compete with the food requirement. Besides that, the massive volume of such waste increase the production capacity while the lower price reduce the production costs [5]. In general, the locally grown agricultural crops are a good choice for reducing the transportation cost and guaranteed continuous supply of biomass. During replantation stage, the oil palm trunk sap is usually

chopped into small pieces, rearrange in rows to avoid road blockage, and left to be rotten naturally in the plantation area (Fig. 1).

Fig. 1: OPT was left to be rotten naturally during replantation

Leaving the trunk just like that is not good for the plantation because the high sugar and starch content in the oil palm trunk sap will attract microflora and microfauna thus increase the possibility of plant diseases [6], especially the younger plants. The presence of starch granules in the parenchyma lumen promotes the growth of the biodeterioration organism [7]. Although oil palm lumber has been successfully utilized as the main material in the production of blackboard, not all parts can be used for plywood manufacturing because only the outer part is relatively strong [2].

The oil palm trunk sap cannot be used as building structure too because it has low specific density, must be dried to remove moisture, and transportation problem. Hence, the inner part will be discarded as a waste due to its weak physical properties but can be considered as an important biomass in Malaysia. Oil palm trunk sap contains liquid (sap) with lower lignin percentage and readily fermentable sugars compared to the other parts of oil palm trees. Therefore, less or no pretreatment (chemical or biological) is needed to delignify or convert the lignocellulose to fermentative sugar, compared with other parts of oil palm tree.

Using oil palm trunk sap have advantages due to simplicity of the procedure, low energy consumption, and lesser time and production cost. Oil palm trunk sap exhibits desirable characteristics as an alternative raw material due to its abundancy and properties. Therefore, the finding of novel innovative method to utilize a cheap material for bioethanol production is important. In order to justify the potential of bioethanol production from oil palm trunks sap to be used as an energy source, it is necessary to analyze the amount of sugar consumption rate and bioethanol. This can provide useful data concerning the relationship between substrate utilization and bioethanol production.

2. Materials and Methods

2.1. Oil Palm Trunks Sap

Oil palm trunk (OPT) sap was used as fermentation medium. Five oil palm trees (26 years old) felled for replantation purposes were freshly obtained from a plantation in Negeri Sembilan, Malaysia. The upper part of the oil palm that contains fruit and frond was removed together with the root. Next, OPT

IOP Conf. Series: Materials Science and Engineering **943** (2020) 012012 doi:10.1088/1757-899X/943/1/012012

sap (10 m) was cut into several small pieces to undergo the squeezing process by using sugar cane press machine (Robin brand, 5 HP) to obtain the liquid sap. The OPT sap was collected in a big container, mixed well before divided into 5 L bottles, and stored under −20 °C. The OPT sap media was then filtered with 9.0 μm filter prior to use.

2.2. Microorganism and preculture development

For inoculation purpose, the microorganisms were aerobically incubated in OPT sap medium at 30 °C in static condition for 24 h. The microorganism used are Japan sake yeast *Saccharomyces cerevisiae* Kyokai no.7 (ATCC 26622 obtained from Japan Collection of Microorganisms (JCM)).

2.3. Fermentation

The fermentation was conducted at 30 °C in static condition without pH adjustment. About 10% v/v of inoculum was used as inoculum for fermentation medium. The shake flask was purged with nitrogen to remove oxygen to create anaerobic condition. Samples were withdrawn periodically at predetermined time intervals for analysis. Fermentation to study the effects of substrate inhibition were conducted using 60, 80 and 100% sap concentrations as in Table 1.

Table 1: Effect of Substrate Inhibition

Percentage OPT Sap	OPT sap	Distilled Water
60%	60%	40%
80%	80%	20%
100%	100%	0%

2.4. Method of Analysis

Sugar concentration was determined by High Performance Liquid Chromatography (HPLC) with refractive index (RI) detector (Agilent Carbohydrate Analysis Column, ratio acetonitrile to water 4:1, flow rate 1.4 mL/min). Column temperature was set at 60 °C. Injection volume was 10.0 μL. The solvent used for washing and dilution was ultrapure water. All individual sugars in OPT sap can be expressed as equivalent glucose concentration and summed up as total sugar using Equation 1.

$$\text{Total sugar} = \text{glucose} + \text{fructose} + (1.0526 \times \text{sucrose}) \quad (1)$$

All calculations are in gram per liter (g/L).

Gas chromatography (Agilent Technologies 6890 Series) equipped with a flame ionization detector (FID) was set up to determine bioethanol concentrations. The column used was HP-INNOWax Polyethylene Glycol (30 m × 250 μm × 0.25 μm nominal). Initial temperature, maximum temperature, and temperature rate in the oven were 50 °C, 170 °C and 20 °C/min, respectively. Temperature of injector and detector were set at 250 °C. Helium carrier gas at a flow rate of 45 mL/min was used. The measured value is "peak area" which was then converted into ethanol concentration (g/L) using a standard curve. Serial dilution was performed using pure ethanol. The solvent used for washing and dilution was n-propanol. One part of sample was mixed with 9 part of n-propanol and filtered through 0.2-micron nylon membrane (Fisher Scientific™). Ethanol productivity was calculated as the amount of ethanol concentration per time (h).

3. Results

Comparison for the effect of substrate concentration on the bioethanol production curve is presented in Table 2.

Table 2: Comparison between ethanol productions using different substrate concentrations at 30 °C.

	60% OPT sap		80% OPT sap		100% OPT sap	
Time	Ethanol (g/L)	Productivity (g/L.h)	Ethanol (g/L)	Productivity (g/L.h)	Ethanol (g/L)	Productivity (g/L.h)
0	2.20	-	1.94	-	2.10	-
2	2.52	1.259	2.92	1.460	2.68	1.341
4	3.60	0.900	3.73	0.932	4.10	1.026
6	5.16	0.861	5.40	0.901	5.88	0.980
8	5.71	0.714	7.47	0.934	8.00	1.000
10	9.10	0.910	8.50	0.850	11.37	1.137
12	8.80	0.734	10.86	0.905	14.29	1.191
18	12.52	0.698	12.05	0.669	20.45	1.136
24	12.06	0.502	17.03	0.710	18.33	0.764
30	13.91	0.464	18.02	0.601	21.88	0.729
36	14.95	0.415	17.42	0.484	22.72	0.631
42	13.55	0.323	18.69	0.445	21.89	0.521
48	12.96	0.270	18.19	0.379	20.95	0.436
54	14.67	0.272	18.51	0.343	22.39	0.415
60	14.16	0.236	18.59	0.310	21.78	0.363
66	14.52	0.220	18.33	0.278	20.90	0.317
72	14.86	0.206	16.96	0.236	18.17	0.252

Ethanol productivity reached the highest at initial stage of fermentation indicating that the *Saccharomyces cerevisiae* able to utilize sugar to produce bioethanol within a short time. However, the maximum bioethanol was only produced after 30 h of fermentation. The fermentation time was important since increasing time gave opportunity for the yeast to utilize more sugar. However, it was observed that increasing time to more than 36–48 h can cause declination of ethanol concentration due to the consumption of the accumulated ethanol by the organism. Similar phenomenon was reported by Kuhad et al. [8] in their work by using red sage as sugar sources.

At high concentration level, the production of bioethanol is limited by concentration of substrate and bioethanol itself. Sugar cannot be fully utilized and cause the production to be inefficient. The graph of glucose consumption rate at different substrate concentrations was plotted to calculate the rate of glucose consumption and cell growth. Fig. 2 and Fig. 3 show the glucose consumption and bioethanol production at different substrate concentrations.

IOP Conf. Series: Materials Science and Engineering **943** (2020) 012012 doi:10.1088/1757-899X/943/1/012012

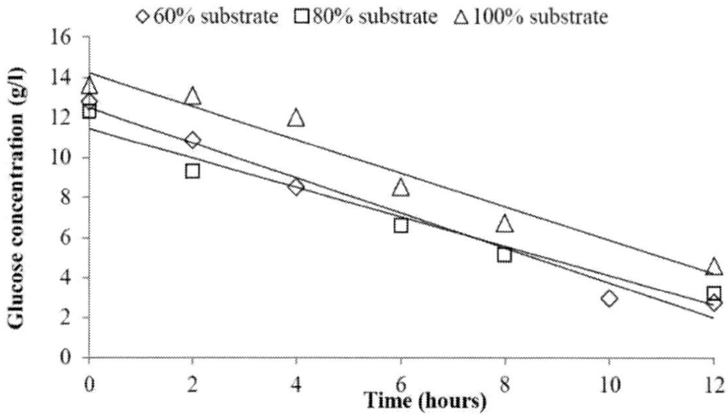

Fig. 2: Glucose consumption in different substrate concentrations (60, 80 and 100%) via fermentation process

At the lowest substrate concentration (60%), the glucose consumption rate was 0.8747 g/L.h (y = -0.8747x + 12.477). This is slightly higher than glucose consumption rate at 100% substrate (0.8353 g/L.h) (y = -0.8353x + 14.223) and at 80% substrate (0.7337 g/L.h) (y = -0.7337x + 11.435). This should be evaluated together with bioethanol production rate to select the best oil palm trunk sap concentration that should be used in further study.

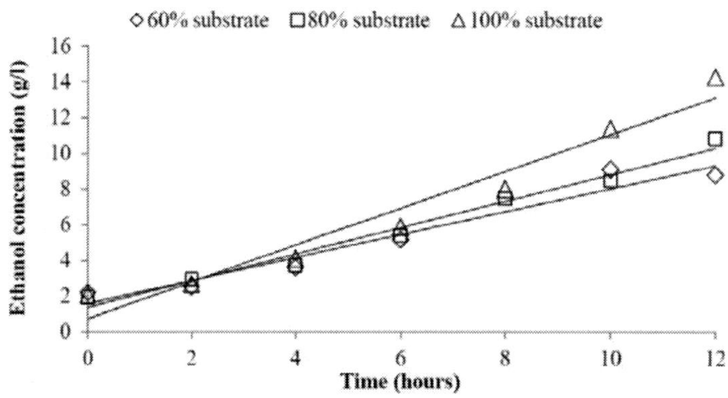

Fig. 3: Ethanol production in different substrate concentrations (60, 80 and 100%) via fermentation process.

Maximum bioethanol concentration was achieved by using 100% of OPT sap. This indicates that fermentation of OPT sap to bioethanol can be done without the need to be diluted. Low concentration of bioethanol was obtained at 60% and 80% since reduction of sugar amount occurred when OPT sap was diluted, which subsequently reduces the bioethanol production. The rate of bioethanol production is highest at 100% substrate concentration since its contain more sugar. High substrate concentration does not inhibit bioethanol production. At 60% substrate concentration, the rate of bioethanol of 0.645 g/L.h (y = 0.645x + 1.5747) was obtained while at 80%, 0.7441 g/L.h (y = 0.7441x + 1.3663) was obtained. At 100% substrate concentration, 1.033 g/L.h bioethanol productivity (y = 1.033x + 0.7195) was obtained.

4. Conclusion

In this work, the effect of different concentrations of OPT sap on bioethanol production has been investigated. OPT sap can be used effectively without dilution since no substrate inhibition was detected. Maximum bioethanol production was produced in fermentation of 100% OPT sap which is 1.033 g/L.h. This finding supported the claim made by other researchers which mentioned that oil palm trunk sap does not contain any inhibiting substances for bioethanol production [9, 10]. The use of waste, especially biomass can reduce the impact of energy demands on food-derived-ethanol. Further studies should be conducted to check the effect of chemical and physical factors which can affect bioethanol production from OPT sap.

5. Acknowledgement

This work was financially supported by research grant Tier 1 Vot U928 from Universiti Tun Hussein Onn Malaysia.

References

[1] C.S. Goh, K.T. Tan, K.T. Lee, S. Bhatia, Bio-ethanol from lignocelluloses: status, perspectives and challenges in Malaysia, Bioresource Technology. 101 (2010) 4834- 4841.

[2] S.H. Shuit, K.T. Tan, K.T. Lee, A.H. Kamaruddin, Oil palm biomass as a sustainable energy source: a Malaysian case study, Energy. 34 (2009) 1225-1235.

[3] M.R. Swain, S. Kar, A.K. Sahoo, R.C. Ray, Ethanol fermentation of mahula (Madhuca latifolia L.) flowers using free and immobilized yeast Saccharomyces cerevisiae, Microbiological Research. 162 (2007) 93-98.

[4] P.F. Siqueira, S.G. Karp, J.C. Carvalho, W. Sturm, J.A. Rodriguez-Leon, J. Tholozan, R.R. Singhania, A. Pandey, C.R. Soccol, Production of bio-ethanol from soybean molasses by Saccharomyces cerevisiae at laboratory, pilot and industrial scales, Bioresource Technology. 98 (2008) 8156-8163.

[5] J. Sumphanwanich, N. Leepipatpiboon, T. Srinorakutara, A. Akaracharanya, Evaluation of dilute-acid pretreated bagasse, corn cob and rice straw for ethanol fermentation by Saccharomyces cerevisiae, Annals of Microbiology. 58(2) (2008) 219-225.

[6] O. Sulaiman, N. Salim, K. Hashim, L.H.M. Yusof, W. Razak, Evaluation on the suitability of some adhesives for laminated veneer lumber from oil palm trunks, Materials and Design. 30(9) (2009) 3572-3580.

[7] O. Sulaiman, R. Hashim, R. Wahab, H.W. Samsi, A. Mohamed, Evaluation on some finishing properties of oil palm plywood, Holz Roh Werkst. 66(1) (2008) 5-10.

[8] R.C. Kuhad, R Gupta, Y.P Khasa, A. Singh, Bioethanol production from Lantana camara (red sage): pretreatment, saccharification and fermentation, Bioresource Technology. 101 (2010) 8348-8354

[9] K. Murai, R. Kondo, Extractable sugar contents of trunks from fruiting and nonfruiting oil palms of different ages, J Wood Sci. 57(2) (2010) 140-148.

[10] A. Kosugi, R. Tanaka, K. Magara, Y. Murata, T. Arai, O. Sulaiman, R. Hashim, Z.A.A. Hamid, M.K.A.Y. Yahya, M.N.M. Yusof, W.A. Ibrahim, Y. Mori, Ethanol and lactic acid production using sap squeezed from oil palm trunks felled for replanting, Journal of Bioscience and Bioengineering. 110(3) (2010) 322- 325.

IOP Conf. Series: Materials Science and Engineering **943** (2020) 012013 doi:10.1088/1757-899X/943/1/012013

Experimental Investigation of the Flexural Behavior of Steel Fiber Reinforced Concrete

A S Karzad, Z Al-Sadoon, M Leblouba, M Maalej

Department of Civil and Environmental Engineering, Sustainable Construction Materials and Structural System Research Group, University of Sharjah, Sharjah, UAE

E-mail: mmaalej@sharjah.ac.ae

Abstract. This research study is conducted to investigate the flexural behaviour of steel fiber reinforced concrete (SFRC) prisms incorporating different types and volume fractions of steel fibers, various grades of concrete, and different specimen depths, all tested under displacement control mode using four-point loading configuration. Two types of steel fibers (hooked end and straight) were incorporated into the concrete mixes at two different dosages (1% and 2%). Concrete mixes with compressive strengths of 40, 50, and 70 MPa were used to cast the samples. Specimen depth was also among the test variables, where prisms of depth 100, 150 and 200 mm were tested to study the size effect on the flexural behaviour of SFRC. Test results indicated that for the straight-end steel fibers, the increase in the fiber volume fraction from 1 to 2% had a noticeable effect on both flexural strength (or Modulus of Rupture, MOR) and associated prism mid-span deflection. However, for the hooked end steel fibers, there was only a limited increase in the same measured values due to reduced workability of the SFRC fresh mix. The test results also have shown that the increase in the concrete compressive strength had only a modest effect on the average MOR value, but a noticeable effect on average mid-span deflection at peak load. The latter was attributed to the relatively long length and the hooked end geometry of the steel fibers. Finally, test results revealed that specimen depth had an impact on the measured flexural strength, where increase in specimen depth was associated with reduced MOR values, a phenomenon that is commonly referred to in the literature as size effect.

1. Introduction

Concrete is a widely-used construction material due to its excellent overall mechanical properties and economical benefits. However, concrete is also known for its low tensile strength, quasi-brittleness, and low fracture toughness. These drawbacks lead to the cracking of concrete and subsequent deterioration of reinforced concrete structures due to corrosion of steel reinforcement when exposed to aggressive environments. To overcome these issues, several studies focused in the past decades on improving the tensile properties of concrete through the use of discontinuous metallic and synthetic fibers in the concrete mix, resulting in what is commonly known as Fiber Reinforced Concrete (FRC) [1]–[3].

Currently, there are different types of discontinuous fibers available in the market, such as steel, polyethylene, polypropylene, and carbon. These fibers come in different geometrical shapes or cross-sections (such as rectangular, circular, and irregular), with deformed or fully-straight ends, and can be either short or long length [2, 4]. Steel Fiber Reinforced Concrete (SFRC) is among the most commonly used FRC material and is produced by randomly distributing macro steel fibers into the fresh concrete mix during mixing. SFRC has been used in several applications such as tunnel lining, slabs on grade,

IOP Conf. Series: Materials Science and Engineering **943** (2020) 012013 doi:10.1088/1757-899X/943/1/012013

pavements, slope stabilization, composite metal decks, concrete pipes, dam construction, rehabilitation of marine structures, etc. [1].

The addition of steel fibers in concrete has shown the ability to improve the tensile/flexural properties, fracture energy, and cracking behaviour of an otherwise a quasi-brittle material. However, the compressive strength of the concrete material is not particularly affected by the addition of steel fibers to the mix. In tensile/flexural members, steel fibers act as crack arrestors by bridging the crack surfaces and restrict the propagation of cracks. This mechanism results in an improved energy absorption capacity and toughness of the FRC materials, ranging from subtle to substantial, depending on a number of factors, including fiber type, fiber properties, fiber dosage, and grade of concrete [2,5-7].

The present study aims to contribute further to the existing literature on the flexural behaviour of SFRC and how it is affected by the steel fiber type, the fiber volume fraction, the concrete compressive strength and the specimen depth. The research presented herein is part of a larger study that incorporates the development of a predictive analytical model of the flexural strength of SFRC using a fracture mechanics approach. The experimental program in the present research involves flexural testing of 27 SFRC prisms of varying depths (100, 150, and 200 mm), where different dosages (1% and 2 %) of steel macro fibers (hooked end and straight end) were used to reinforce concrete of varying compressive strengths (40, 50, and 70 MPa). The test results will be discussed in terms of the prisms load-mid-span deflection response, the measured flexural strength, and the cracking behavior and how these are affected by the fiber type, the fiber volume fraction, and the compressive strength and specimen depth.

2. Experimental Program

2.1. Materials and Specimens Preparation

The FRC prisms used in this study (Figure 1), were cast from two types of steel fibers (hooked end and straight end) of two different aspect ratios of fibers (L_f/ d_f= 55, and 65) and two dosages of steel fibers (1% and 2%). Three different concrete compressive strengths (40, 50, and 70 MPa) and three different specimen depths (100, 150, and 200mm) were used to cast these FRC prisms. Tables 1 and 2 show the specifications of the macro steel fibers and the concrete mix design used, respectively.

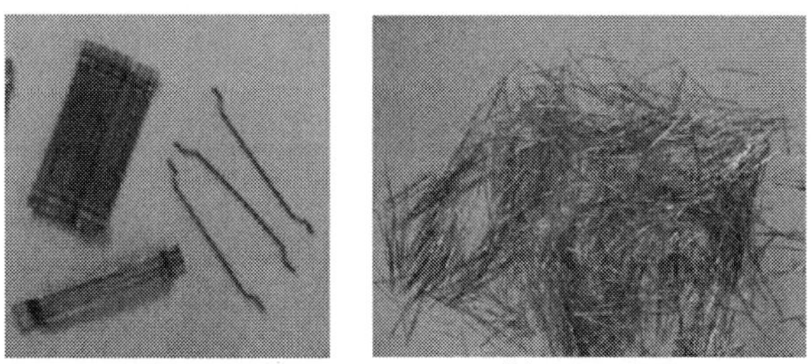

(a) ZP305 (b) OL13/0.2

Figure 1 – Photo of steel fibers dispersed in concrete mix

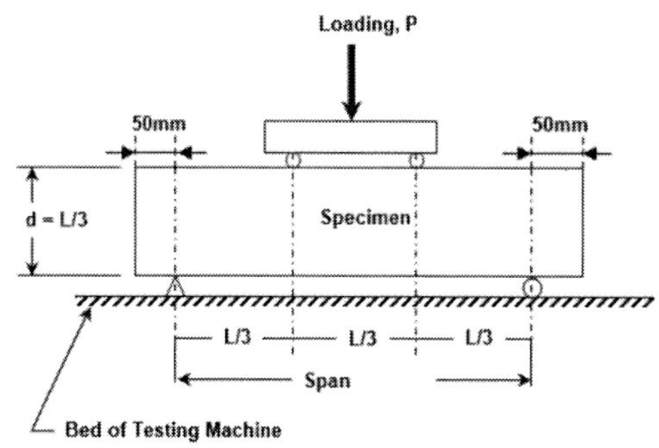

Figure 2 – Experimental setup for four-point flexural tests

Table 1 - Specification of the steel fibers

Fiber Type	Length L_f(mm)	Diameter d_f (mm)	Aspect ratio (L_f/d_f)	Fiber tensile Strength (MPa)	Fiber density Kg/m^3
ZP305	30	0.55	55	1100	7.850
OL13/0.2	13	0.2	65	2600	7.850

Table 2 - Mix design of concrete mixes

Compressive cube strength (MPa)	Cement Qty	Coarse aggregate-cement ratio	Fine aggregate-cement ratio	W/C ratio
40	1	1.74	2.36	0.58
50	1	1.46	1.90	0.49
70	1	0.925	1.09	0.37

2.2. Test Setup

Four-point deflection controlled flexural tests were carried out to determine the load–mid-span deflection response and the flexural strength of the various beams (prisms) according to ASTM C1018 [8] as shown in Figure 2. The SFRC prisms were tested under displacement control using an initial constant loading rate of 0.1mm/min, which was later increased to 0.3mm/min after the peak load is reached. The loading was terminated once the load on the specimen dropped to 2 kN.

3. Results and Discussions

A summary of the experimental results is presented in Table 3. For each test parameter, four identical test samples are tested and the corresponding load-deflection (P-Δ) curves are plotted in a single graph, which also shows a calculated average modulus of rupture (MOR). Each test sample is identified with a label showing the depth of the sample, type of fiber, the fiber dosage, and ends with the compressive strength of the sample. For example, test sample 100-ZP-1-40 represents a specimen of a depth of 100 mm, having fiber type ZP305, fiber volume fraction of 1% and cast with 40 MPa compressive strength concrete.

Table 3 – Summary of experimental results

Specimen index	Fiber Type	V_f(%)	Depth, d (mm)	Cube concrete strength, f_{cu} (MPa)	Average flexural strength, MOR (MPa)	Mid-span deflection at peak load Δ_u (mm)
100-0-40	-	0	100	40	4.70	0.25
100-OL-1-40	OL13/0.2	1	100	40	5.45	0.35
100-OL-2-40	OL13/0.2	2	100	40	7.85	0.45
100-ZP-1-40	ZP305	1	100	40	5.69	1.0
100-ZP-2-40	ZP305	2	100	40	5.82	1.2
150-ZP-1-40	ZP305	1	150	40	4.59	1.0
200-ZP-1-40	ZP305	1	200	40	3.56	1.4
100-ZP-1-50	ZP305	1	100	50	5.63	1.0
100-ZP-1-70	ZP305	1	100	70	6.62	1.8

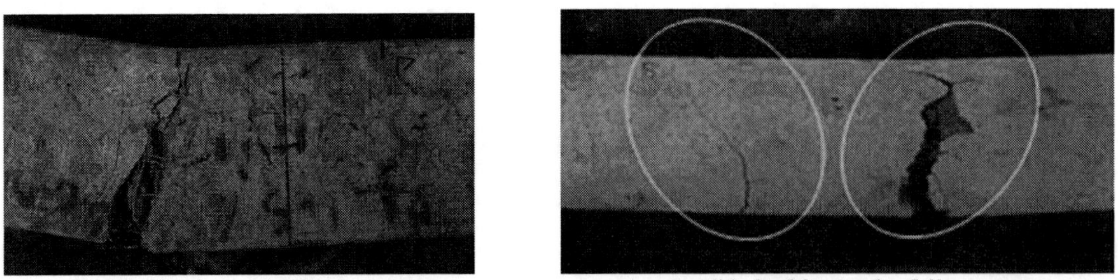

(a) single crack failure pattern (b) double cracks failure pattern

Figure 3. Observed crack patterns in sample test specimens

The tested SFRC prisms in this study failed after forming single or double cracks as shown in figure 3 with an observation, in some of the samples, of initial crack propagation not starting at the middle length of the prisms. i.e. cracks were initiated within the constant-moment region but away from the mid-span. Some of the tests prisms exhibited dual crack initiation and propagation as shown in Figure 3b. However, as the load increased, only one of the cracks propagated further until failure. This is in line with the assumption that single crack will propagate until failure.

3.1. Effect of Volume Fraction of Fibers (V_f) on the load-deflection Response

3.1.1. OL13/0.2 Fibers (Straight End):

Figures 4a and 4b show the load-deflection curves for two series of specimens (Series 1 and Series 2); each consists of three duplicate 100 mm-prisms cast using 40 MPa concrete; incorporated 1 and 2 % of OL13/0.2 straight end fibers, respectively. For both series, one or two significant cracks had formed on the prism tensile face within the constant moment region and continued to propagate towards the compression face with an increased mid-span deflection. This crack propagation was initially associated with an increase in applied load until a peak value was reached, and then started to drop as the crack continued to propagate and the mid-span deflection continued to increase. At the peak load, the measured flexural strengths (MOR) and associated mid-span deflection (Δ_u) were 5.45 MPa and 0.35 mm for Series 1, and 7.85 MPa and 0.45 mm for Series 2, respectively. It is evident in this case that by increasing the fiber volume fraction from 1 to 2%, the MOR and Δ_u had increased by 44% and 30%, respectively.

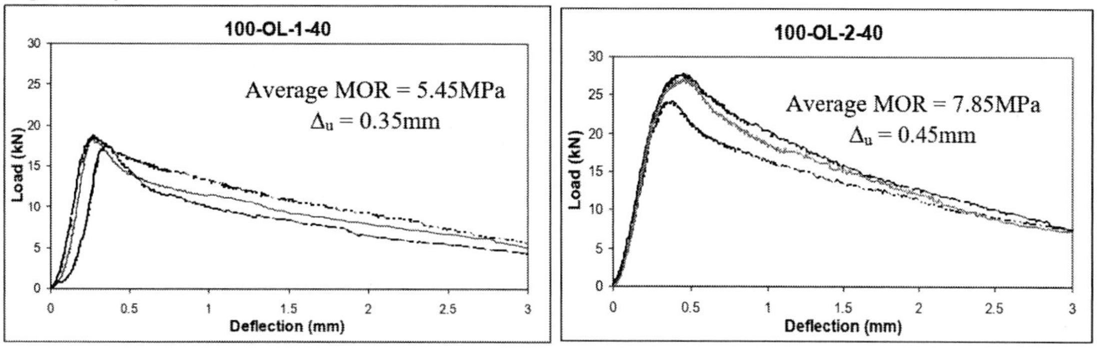

(a) 1% of OL13/0.2 Fibers (b) 2% of OL13/0.2

Figure 4. Effect of volume fraction of OL13 fibers on the load-deflection response of the specimens

3.1.2. ZP305 Fibers (Hooked End):

Figures 5a and 5b show similar curves for other two series of specimens (Series 3 and Series 4) that incorporated 1 and 2 % of a longer length (30 mm) ZP305 hooked-end steel fibers, respectively. For these series of tests, the observed crack initiation and propagation and its impact on the peak load and associated mid-span deflection were similar to those discussed earlier for Series 1 and 2. However, for the former Series (3 and 4), the increase in the fiber volume fraction had a minimal impact on the measured MOR, but a definite impact on the measured mid-span deflection Δ_u. This behavior was attributed to the relatively long length of the ZP305 fibers and the geometry of the end of the fibers (hooked end) which resulted in reduced workability of the SFRC fresh mix and hence reduced fiber-reinforcing efficiency in terms of limited increase in the measured MOR. The longer length of the ZP305 fibers has, on the other hand, a contribution to bridging major crack and thus allowing the specimen to sustain high load levels while increased mid-span deflections. This resulted in the apparent strain hardening behavior shown in Figure 5b for Series 4. The measured increase in Δ_u for Series 4 compared to Series 3 was 20%.

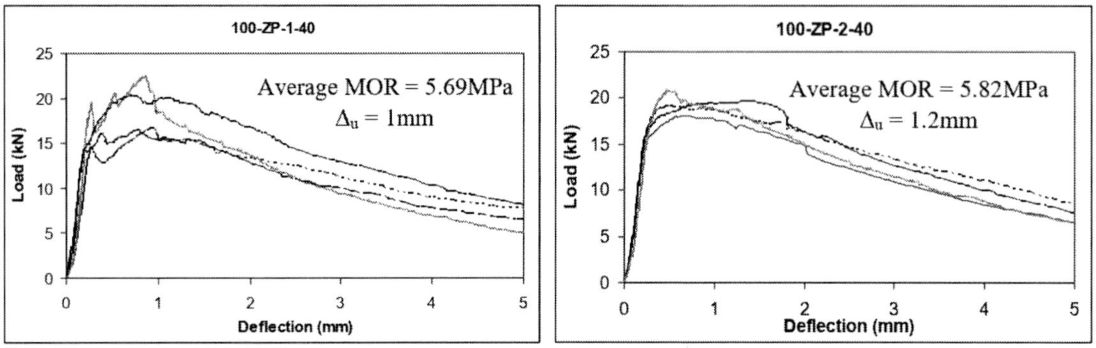

(a) 1% of ZP305 Fibers (b) 2% of ZP305 Fibers

Figure 5. Effect of volume fraction of ZP305 fibers on the load-deflection response of the specimens

3.2. Effect of Concrete Strength on the load-deflection Response

Figure 6 shows the load-deflection response of specimens of concrete strength of 50 MPa and 70 MPa. When comparing these figures with Figure 5a, it can be observed that with an increase in concrete strength, the specimens were able to sustain a higher load over an increased mid-span deflection. For specimens with concrete strength of 50MPa (Figure 6a), the load started to decrease when the deflection reached about 1 mm. However, for specimens with higher concrete strength (70 MPa), the specimens were able to sustain the applied load for a larger deflection of around 1.8 mm as shown in Figure 6b. This indicates that with the increase of concrete compressive strength (achieved with reduced w/c ratio), the bonding between the fibers and the concrete matrix has increased as well. Publsihed data in the literature [9] suggest that the reduction of w/c ratio does indeed lead to increased fiber/matrix bond strength. This increase had probably enhanced the crack arrest mechanism of the steel fibers, and thus, increased the capability of the specimens to sustain higher loads for larger deflections. Compared to the 40 MPa SFRC, the 70 MPa SFRC showed a corresponding increase of 16% and 80% in the average MOR and associated mid-span deflection values, respectively.

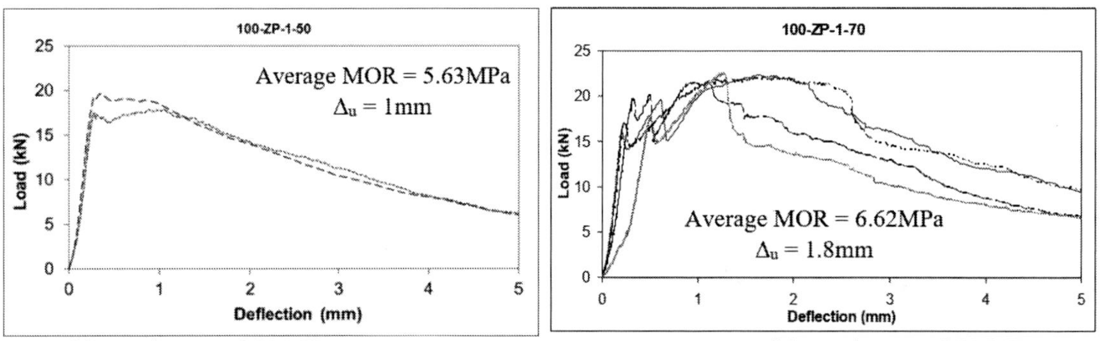

(a) specimens of 50 MPa (b) specimens of 70 MPa

Figure 6. Effect of concrete strengths on the load-deflection response of specimens

3.3 Effect of Depth on the load-deflection Response

Figure 7 shows the load-deflection response of ZP305 fibers specimen of (1%) but with increased depth from 100 mm (shown earlier in Figure 5a), to depths of 150 and 200 mm. Upon comparison, it can be observed that with the increase of the specimens depths from 100 to 200 mm, the specimens showed higher peak loads but lower MOR values over an increased deflection (around 1.4mm). This decrease in the measured MOR with increased specimen depth is indicative of the increased apparent brittleness of the SFRC prisms associated with the increased specimen depth. This is what is commonly referred to in the literature as the size effect [4].

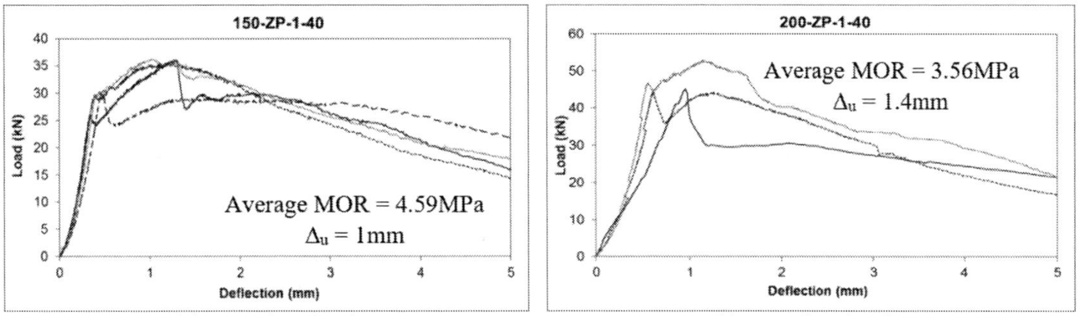

(a) samples depth 150 mm (b) samples depth 200 mm

Figure 7. Effect of depth on the load-deflection response of specimens

4. Conclusions

In this study, 27 SFRC prisms were cast and tested under a four-point flexural loading configuration. The parameters investigated were the type and volume fraction of steel fibers used, the compressive strength of the concrete, and the depth of specimens. Two types of steel fibers (hooked end and straight end) were used in two different dosages in the concrete mix.

Test results showed that the increase in the volume fraction of OL13/0.2 steel fibers (straight end) from 1% to 2% resulted in 44% and 30% increase in the average MOR and the associated mid-span deflection values, respectively. However, for the longer, hooked-end ZP305 steel fibers, there was very little increase in the average MOR value and only a 20% increase in the associated mid-span deflection was observed. This was attributed to the reduced workability of SFRC concrete mix related to the longer length of the fiber and due to the hooked end geometry. For the same fiber (ZP305), test results has also shown that that the concrete compressive strength had a modest effect on the average MOR value (only 16% increase), but a noticeable effect on the average mid-span deflection at peak load; with an increase of about 80% between the 40 MPa and 70 MPa concrete strength. As for the effect of specimen depth on the flexural behaviour of the SFRC, test results had shown that by increasing the SFRC specimen depth to 150 mm and 200 mm (original length 100 mm), the average MOR values has reduced by 19% and 37%, respectively.

5. References

[1] F. Altun, T. Haktanir, and K. Ari, "Effects of steel fiber addition on mechanical properties of concrete and RC beams," Constr. Build. Mater., vol. 21, no. 3, pp. 654–661, 2007.

[2] D. V. Soulioti et al. "Effects of fibre geometry and volume fraction on the flexural behaviour of steel-fibre reinforced concrete," Strain, vol. 47, no. 1, pp. 535–541, 2011.

[3] M. Maalej and V.C. Li, "Flexural Strength of Fiber Cementitious Composites," J. Mater. Civ. Eng. ASCE, vol. 6, no. 3, pp. 390–406, 1995.

[4] D. Y. Yoo, N. Banthia, J. M. Yang, and Y. S. Yoon, "Size effect in normal- and high-strength amorphous metallic and steel fiber reinforced concrete beams," Constr. Build. Mater., vol. 121, pp. 676–685, 2016.

[5] J. Lee et. al., "Flexural capacity of fiber reinforced concrete with a consideration of concrete strength and fiber content," Constr. Build. Mater., vol. 138, pp. 222–231, 2017.

[6] F. Bencardino et. al., "Composites : Part B Experimental evaluation of fiber reinforced concrete fracture properties," Compos. Part B, vol. 41, no. 1, pp. 17–24, 2010.

[7] H.P. Behbahani et al., Flexural behavior of steel fiber added RC beams with C30 and C50 classes of concrete, Inter. J. of Sust. Const. Eng. & Tech. , vol. 3, issue 1, 2012.

[8] ASTM C1018, "Standard Test Method for Flexural Toughness and First-Crack Strength of Fiber-Reinforced Concrete (Using Beam With," vol. 4, no. October, pp. 1–8, 1998.

[9] S. Abdallah, M. Fan, and D. W. A. Rees, Bonding Mechanism and Strength of Steel Fiber Reinforced Cementitious Composites:Overview, J. Mater. Civ. Eng (ASCE) vol.30, no.3, 2018

IOP Conf. Series: Materials Science and Engineering **943** (2020) 012014 doi:10.1088/1757-899X/943/1/012014

Photo-voltaic Powered Electric Vehicle Fast Charger

C P Y Lai[1], K H Law[2], K H Lim[3]

[1] Department of Electrical and Computer Engineering, Faculty of Engineering and Science, Curtin University Malaysia, CDT 250, 98009 Miri, Sarawak, Malaysia
[2] Department of Electrical and Computer Engineering, Faculty of Engineering and Science, Curtin University Malaysia, CDT 250, 98009 Miri, Sarawak, Malaysia
[3] Department of Electrical and Computer Engineering, Faculty of Engineering and Science, Curtin University Malaysia, CDT 250, 98009 Miri, Sarawak, Malaysia

E-mail: lawkahhaw@curtin.edu.my [2]

Abstract. In the last years, dominant usage of petroleum-based transportation and consumption of fossil fuels results in greenhouse gas (GHGs) emissions into the atmosphere which causes negative impacts across the globe. As the situation worsens, more people are motivated towards opting for electrified transportation. However, there are still a lot of challenges to be overcome with one of the main issues concerning the development and availability of fast charging technologies in electric vehicles (EVs). To decrease high energy demand from the grid, usage of renewable energy systems (i.e., photo-voltaic) to fast charge EVs are preferred. In this paper, high performance DC-DC boost converter controlled by PI controller is proposed with the objective to mitigate the current state of fast charging and environmental problems. The proposed converter is able to generate different output power level based on the desired preset reference voltage value. All simulation results are documented to verify the proposed photo-voltaic (PV) powered EV fast charger. For simplicity, EV batteries are being replaced with one resistor as the load whereas PV panels are used to generate input power for the converter.

1. Introduction

Vehicles have been shaping human civilization for centuries and in the past few decades. However, increasing usage of petroleum-based vehicles produce harmful exhaust components namely carbon dioxide (CO_2), oxides of nitrogen, and sulphur dioxide (SO_2). The United States Environmental Protection Agency analyzes that the transportation sector for more than a quarter of GHG emissions [1]. Fortunately, there has been a significant evolution from the usage of fossil fuel to electric based transportation in all sectors. The aforementioned transition is expected to accelerate exponentially, especially in EVs, through government support, technology advancement, increasing involvement of private sectors and reductions in prices due to large-scale production as it plays a huge role in aiding the market [2–4]. In addition, there will only be sufficient crude oil supply until about the year 2050 [5], it is clear that fossil fuels substitutes needs to be generated for all sectors, particularly in transportation. Furthermore, the global stock of EVs passed 5 million in 2018 which is a significant increase of 63% compared to the previous year [6]. EVs boosts energy saving (efficiency) through improved fuel economy, helps provide a wider range of fuel choices for transportation and decreases pollution/emissions especially with the integration of renewable energy sources (RES), such as wind and PV in electricity generation. These RES in the power systems have since matured. Although their

IOP Conf. Series: Materials Science and Engineering **943** (2020) 012014 doi:10.1088/1757-899X/943/1/012014

forecast is quite erratic, penetration of the RES into the power market has significantly increased to fulfill strict energy policies and energy security issues.

Despite the advancement in electrified transportation technologies, the charging times are still the chink in the armour of EVs as it has been quicker and handier to refuel from petroleum-based sources. As of 2018, publicly available fast chargers are still low at 0.14 million [6] globally. EV charger can be categories into AC and DC types. These are further categorised into three bodies; namely the IEC, Society of Automotive Engineers (SAE), and CHAdeMO, that are strive to elected as the standard for EV battery charging. Table 1 shows various charging levels and standards in Europe, Japan and North America.

Table 1. Charging levels and standards [7].

Charging Level	Voltage Level	Maximum Power (kW)	Charging Time (h)	Standards		
				Europe	Japan	North America
Level 1 (Slow)	120 VAC	3.7	10-15	Private Outlets		SAE J1772 (Type 1)
Level 2 (Slow)	220 VAC	3.7-22	3.5-7	IEC 62196 (Type 2)	SAE J1772 (Type 1)	SAE J1772 (Type 1)
Level 3 (Fast)	3-phase 480 VAC	22-43.5	0.17-0.5	IEC 62196 (Type 2)		SAE J3068
	200-600 VDC	<200		CCS Combo 2	CHAdeMO	CCS Combo 1
		<150			CHAdeMO	

The classified EV chargers from [6] are as: Level 1 with charge power less than 5 kW, Level 2 charges between 5 kW to 50 kW through three-phase outlets, and Level 3 is DC fast charge which is characterized using 480 V AC input with charging power greater than 50 kW. However, the power conversion and control stage becomes more expensive and bulky due to higher power capability. Due to size constraints, the required power electronics for Level 3 charging are not carried on board. Hence, to decrease the mass of on-board electronics, the power transferred to the vehicle for Level 3 charging is in DC while AC energy transfer is allowed in Level 1 and Level 2 chargers as they contain on-board electronic converters.

In recent literature, environmental consequences, integration issues, optimization, technological aspects, and the impact on the grid such as peak load demand and voltage malfunctions [8] are covered in existing PV-EV charging papers. In addition, combination of RES and the grid is considered appealing in regards of future smart grid framework . However, there are still limited amount of papers regarding the methodology of PV-EV fast charging. For example, in [9; 10], the design of a high efficient battery charger and control technique for EV fast charger is proposed but no RES is implemented. Not only that, in [11], a solar grid-connected charger for EV is designed but there is no fast charging.

This paper proposes a design and implements a PV-powered fast charger to produce different output power level from the source to the load. To achieve the aforementioned task, DC-DC boost converter topology is chosen. MATLAB/Simulink is utilized to carry out simulation to develop the model. For simplicity, the EV batteries are replaced by a resistive load during the modelling phase to incorporate the control of the charge current into the battery.

2. Battery

The main energy resource for EVs is rechargeable battery. There are several types of rechargeable battery integrated in EVs including lead-acid battery and lithium-ion battery. Lead-acid battery was the first battery technology being used in transportation. However, the downsides of utilizing it are heavy, low energy density, and not environmental friendly when compared with lithium-ion battery. There are several methods to charge an EV battery. For instance, constant current (CC), constant voltage (CV), constant power (CP), taper charging, and trickle charging [12]. The combination of these traditional charging techniques leads to the advance development of rapid battery charging methods such as CC-CV, pulse and negative pulse [13] charging [14].

2.1. Slow Charging Techniques

The following are the conventional methods used for charging and recharging of EV batteries.

2.1.1. Constant Voltage (CV). The charging voltage is maintained at the maximum voltage depending on the type of battery. As the battery is approaching full charge, the charging current slowly decreases. This method is suitable for lower voltages due to temperature control. However, long charging periods are of main concern [15].

2.1.2. Constant Current (CC). A CC is outputted to charge the battery. However, the reliability of the battery can still be affected due to overheating even if the charging current is not exceeding the rated current of the battery [15].

2.1.3. Constant Power (CP), Taper Charging, and Trickle Charging. For CP method, the battery is charged with constant regulated power. Taper charging method is performed with an unregulated voltage source, which increases the chance of damaging the battery due to overcharge [16]. Trickle charging charges the battery with small currents to atone for self-discharge in batteries.

2.2. Fast Charging Techniques

The following describes the rapid charging techniques in rechargeable batteries utilized in EVs.

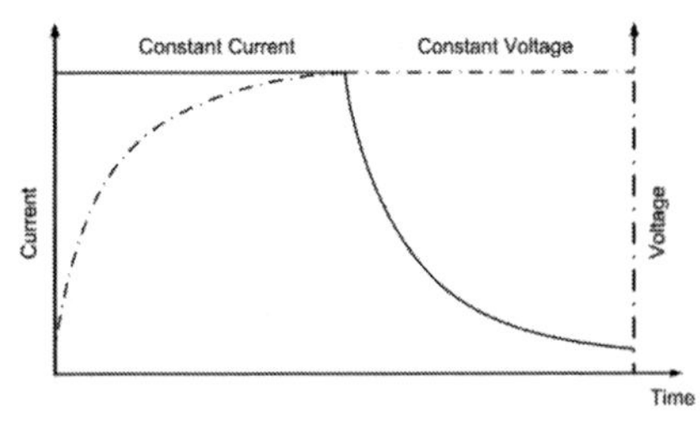

Figure 1. CC-CV charging.

2.2.1. Constant Current-Constant Voltage (CC-CV). CC-CV charging method as shown in Figure 1, is used in the majority of commercial chargers to charge lithium-ion batteries as they have higher output power and energy densities when compared to other battery types. Several benefits of CC-CV technique includes controlled charging voltage and current to prevent over-voltages and reduce thermal stress. With this method, the charger outputted CC until the predefined voltage potential of the battery has reached. Then, the battery voltage will be held constant while decreasing the charging current until reaching full charge [17]. The charging period is almost similar or more than the CC method due to its capability of reducing the charging current upon completed the charging process [18; 19].

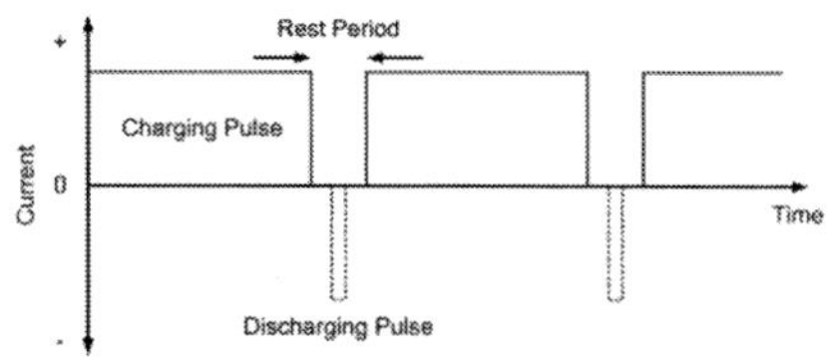

Figure 2. Pulse charging and negative pulse charging.

2.2.2. Pulse Charging and Negative Pulse Charging. Pulses from the charging current is applied through the battery. The charging rate can be approximately controlled by varying the pulse width. This method optimizes the charging time while taking into consideration of polarization, temperature, State-of-Charge (SoC) and variable battery impedance [20]. The charging impedance of the battery varies with frequency which in turn affects the charging current due to the battery's capacitive nature during charging. Hence, charging impedance can be minimized to maximize the charging current by selecting the suitable pulse frequency to achieve shorter charging times.

In [13], a duty-varied voltage pulse charging strategy was proposed to increase charge speed and efficiency to the battery. In [21], another pulse charging technique using optimal frequency and duty searching modes was proposed. The finding of this method proves that pulse charging speed is twice the rate when compared with CC-CV charging method.

Negative pulse charging, which initially being developed to increase the charging efficiency of converters for lead acid batteries, has been extended to lithium ion batteries [15]. This method helps to lower the stresses of the battery's cell through temperature monitoring. A faster charging period can be achieved through continuation of high currents pumped into the battery by depolarizing the cell periodically. Figure 2 illustrates the charging profile for pulse charging and negative pulse charging. Negative pulse charging is part and parcel of pulse charging which has been claimed to enhance overall charging process and prolong the battery life.

3. State-of-the-art DC Fast Chargers

Conventional state-of-the art DC fast charger consists of two power conversion stage from three-phase AC to DC which includes Power Factor Correction (PFC) circuit and a DC-DC stage with galvanic isolation [7]. AC voltage is rectified into immediate DC voltage during AC-DC rectification stage while the immediate DC voltage is controlled to the desired charging voltage and current levels, as defined by EV battery whilst providing galvanic isolation. PFC functions by ensuring the grid power quality adheres with the grid codes while the galvanic isolation enables the charger output stages to be connected in parallel plus separating EVs from the grid [22]. The DC-DC conversion stage implements a dynamic control method to reduce temperature and battery polarization for fast-charging applications [23]. A typical DC fast charger will also utilize an active pulse width-modulated (PWM) rectifier whereby the pulse generator would be implemented for pulse and negative pulse fast charging [20].

Commercially available DC fast chargers typically utilizes CC-CV technique. The battery voltage rises proportionally with the state-of-charge (SOC) when the battery charges in the CC region. Upon reaching CV phase, the power delivered to the battery decreases. Table 2 shows the technical specifications of commercially available DC fast charger.

Table 2. Commercially available DC fast chargers technical specifications [22].

Manufacturer Model	**ABB Terra 53**	**ABB Terra HP**	**Tritium Veefil-RT**	**Tesla Supercharger**	**EVTEC espresso& charge**
Rated Power (kW)	50	350	50	135	150
Supported standards	CCS Type 1 CHAdeMO 1.0	SAE Combo 1 CHAdeMO 1.2	CCS Type 1 & 2 CHAdeMO 1.0	Supercharger	SAE Combo 1 CHAdeMO 1.0
Input voltage	480 VAC	400 VAC ±10%	380-480 VAC 600-900 VDC	200-480 VAC	400 VAC ±10%
Output DC voltage (V)	200-500 50-500	150-920	200-500 50-500	50-410	170-500
Output DC current (A)	120	375	125	330	300
Peak efficiency	94%	95%	<92%	92%	93%
Volume (L)	758	1,894	495	1,047	1,581
Weight (kg)	400	1,340	165	600	400

4. Methodology

The future concept and perception of EV fast charging is shown in Figure 3 and Figure 4 below.

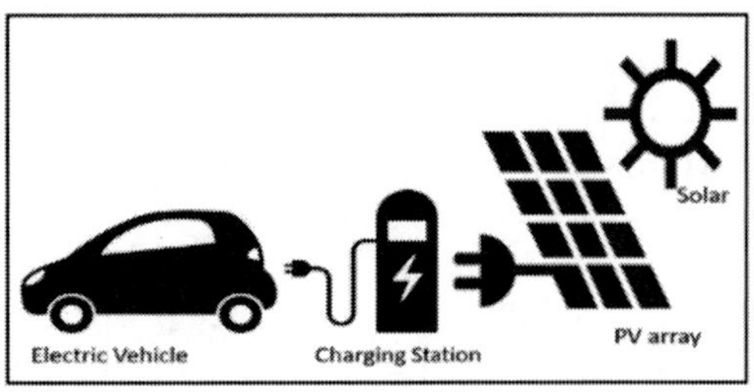

Figure 3. General Scheme of the power system for EV.

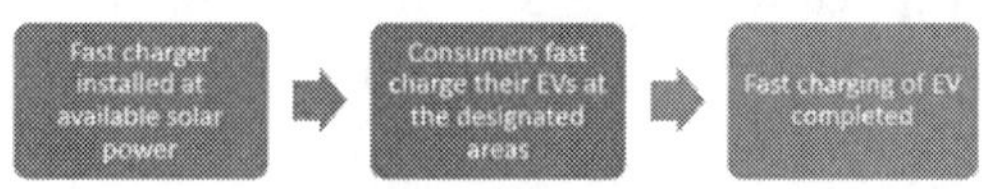

Figure 4. Pipeline of fast charging for EVs.

The proposed fast charger will be installed at designated areas nearby the available PV power allowing consumers to charge their EVs for a short period of time. Figure 5 below illustrates the pipeline to design the PV-powered fast charger. This project is carried out in four phases with different approach.

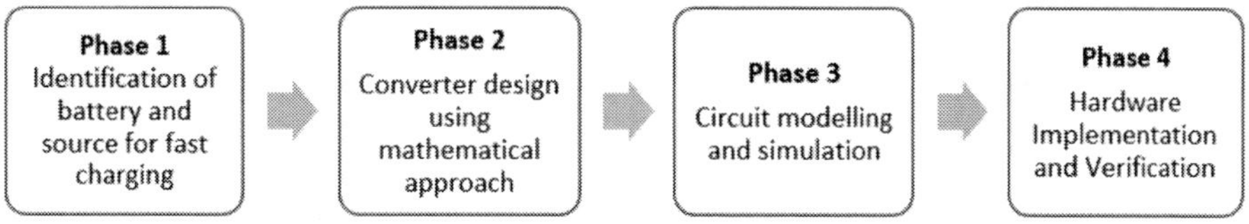

Figure 5. Phases of Project

In Phase 1, EV battery specification and its charging characteristic is identified and investigated. A review on the usage of PV energy to fast charge EV in terms of power level competency were also carried out. For Phase 2, the transfer function and state-space equations for the design of the DC-DC boost converter are derived [24–30]. The selection of the size of passive components are then carried out. After that, the development and selection of appropriate controller (Proportional Integral (PI) controller) is administered to conclude its gain values through root locus diagram. The conceptual design of the proposed PV-powered fast charger established for this work is shown in Figure 6.

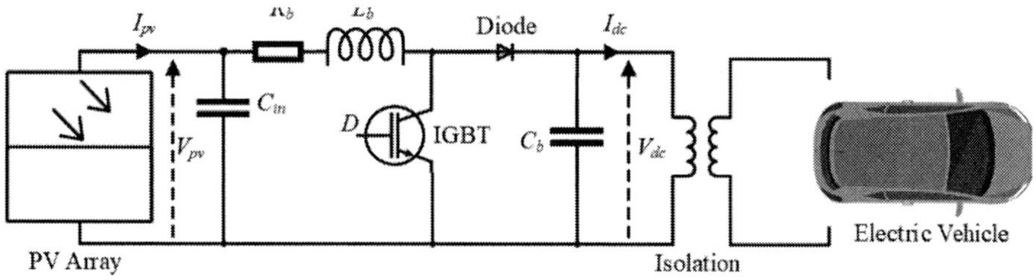

Figure 6. Boost converter with PV array for EV charging.

To make sure that power regulation and DC-link voltage stabilization can be achieved by the proposed PV-powered fast charger according to the desired predefined reference voltage value, MATLAB/Simulink is utilized. MATLAB/Simulink is also used to examine and measure he performance of the proposed topology based on the dynamic and transient response as well as the conversion effiency. Last but not least, to verify the optimization and real-time behaviour of the aforementioned works, a prototype of the proposed topology will be developed in the future works. The prototype will also be used to identify the discretization issues, modelling issues, and the accuracy of data supplied to the model.

5. Mathematical Modelling

Figure 7 shows the principle of operation of the proposed PV-powered fast charger based on DC-DC boost converter. From Figure 7(a), when the switch is on, the input source V_{SO} causes inductor L to charge and capacitor C to discharge to the load. However, as from Figure 7(b), when the switch is off, the energy stored in the inductor L discharges. A much higher DC output voltage V_{DC} when compared to V_{SO} is produced due to the summation of the input source V_{SO} with the energy. This simultaneously charge the capacitor C and drive the load.

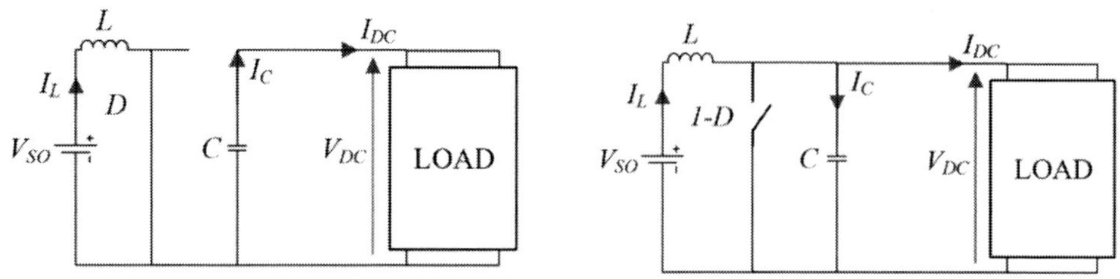

(a) Switch On (b) Switch Off

Figure 7. Proposed boost converter with different switch configurations.

The overall power rating of the converter together with the voltage and current ripple content produced by the converter can be determined by the size of the capacitor C and inductor L. The size of inductor L and capacitor C can be calculated using the equations defined in (1) and (2) as follows:

$$L = \frac{V_{SO}}{\Delta I_{Lmax}} DT_{SW} \tag{1}$$

$$C = \frac{V_{DC}}{\Delta V_{DCmax} R} DT_{SW} \tag{2}$$

where V_{SO} signifies the input voltage source, V_{DC} signifies the output voltage, ΔI_{Lmax} signifies the maximum ripple current flowing through the inductor L, ΔV_{DCmax} signifies the maximum voltage ripple measured across the capacitor C, R signifies the resistance of the load, D signifies the steady-state duty cycle generated from the PI controller and T_{SW} signifies the switching period.

The equations for ΔI_{Lmax} and ΔV_{DCmax} used in this work can be seen in 3 and 4 respectively, as follows:

$$\Delta I_{Lmax} = 1\% \times I_L \tag{3}$$

$$\Delta V_{DCmax} = 1\% \times V_{DC} \tag{4}$$

where I_L signifies the input current flowing through the inductor L.

Using the state-space averaging modeling technique, the voltage loop controller of the proposed DC-DC boost converter operating in continuous conduction mode (CCM) is designed. Two states of operations (i.e., when the switch is on (D)(see Figure 7(a)) and when the switch is off $(1\text{-}D)$(see Figure 7(b)) for the DC-DC boost converter are derived in 5 and 6 from figure 7 respectively as follows:

$$\begin{pmatrix} \frac{dI_L}{DT_{SW}} \\ \frac{dV_{DC}}{DT_{SW}} \end{pmatrix} = \begin{pmatrix} 0 & 0 \\ 0 & -\frac{1}{RC} \end{pmatrix} \begin{pmatrix} I_L \\ V_{DC} \end{pmatrix} + \begin{pmatrix} \frac{1}{L} \\ 0 \end{pmatrix} V_{SO} \tag{5}$$

$$\begin{pmatrix} \frac{dI_L}{(1-D)T_{SW}} \\ \frac{dV_{DC}}{(1-D)T_{SW}} \end{pmatrix} = \begin{pmatrix} 0 & -\frac{1}{L} \\ \frac{1}{C} & -\frac{1}{RC} \end{pmatrix} \begin{pmatrix} I_L \\ V_{DC} \end{pmatrix} + \begin{pmatrix} \frac{1}{L} \\ 0 \end{pmatrix} V_{SO} \tag{6}$$

The linear time invariant state-space average equations over the switching period T_{SW} are then derived by taking into consideration of the small deviation from the steady-state operating point of the boost converter for inductor current i_L and capacitor voltage v_C are derived in 7 and 8 respectively as follows [31]:

IOP Conf. Series: Materials Science and Engineering **943** (2020) 012014 doi:10.1088/1757-899X/943/1/012014

$$\frac{di_L}{dt} = -v_{DC}\left(\frac{1-D_O}{L}\right) + V_{DC}\left(\frac{d}{L}\right) + v_{SO}\left(\frac{1}{L}\right) \tag{7}$$

$$\frac{dv_{DC}}{dt} = i_L\left(\frac{1-D_O}{C}\right) - V_{DC}\left(\frac{1}{RC}\right) - I_L\left(\frac{d}{C}\right) \tag{8}$$

where the uppercase letters (i.e., D_O, V_{DC}, and I_L) represents the steady-state parameters whereas the lowercase letters (i.e., v_{DC}, d, v_{SO}, and i_L) represents the deviations or small perturbations from the steady-state parameters.

$$v_{DC}(s) = \left(\frac{v_{SO}(s)}{V_{SO}(s)^2(1-D_O)} + \frac{d(s)}{(1-D_O)^2)}\left(1 - \frac{sL}{(1-D_O)^2R}\right)\right) * \left(\frac{V_{SO}(s)}{(\frac{s^2LC}{(1-D_O)^2} + \frac{sL}{(1-D_O)^2R} + 1}\right) \tag{9}$$

where the first bracket of 9 represents the input voltage deviation during an operating point and the second bracket represents the transfer function of the boost converter $G_{conv}(s)$ (see Figure 8) which are required to calculate the gains of PI controller.

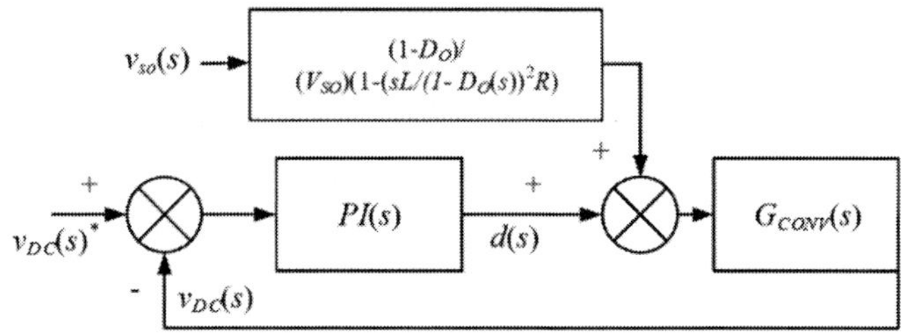

Figure 8. Controller modelling for the proposed boost converter.

As derived from 9, the instability effect caused by the right-half plane zero is due to the size of the load resistance R and the inductance L [31]. Hence, to eliminate the aforementioned effect, higher switching frequency f_{SW} must be chosen to reduce the size of inductor L. The proposed voltage loop feedback control for DC-DC boost converter is illustrated in Figure 8.

The DC voltage level that should be generated by the boost converter is determined by the DC output voltage reference $v_{DC}{}^*$ which is set by the user. To measure the source voltage deviation v_{SO}, the transfer function derived from 9 which can be seen in Figure 8 is utilized. The result from the transfer function is then summed with the duty cycle d to provide the resultant duty cycle D for the proposed boost converter to maintain its desired DC output voltage. Simply put, the proposed boost converter is able to regulate its DC output voltage in accordance to the DC output voltage references $v_{DC}{}^*$ despite change in input source voltage as well as loading conditions.

Integral feedback can be utilized to eliminate steady state error and reduce forward gain required. Integral action also mitigates offset issues. PI controller is used to remove delay and provide fast control over the converter. Using the tune feature in MATLAB/Simulink, the integral gain of the PI controller (i.e. 0.0280411089392773) was successfully identified. 10 shows the transfer function of the PI controller.

$$PI = K_P + \frac{K_I}{s} \tag{10}$$

where K_P signifies the proportional gain whereas K_I signifies the integral gain.

IOP Conf. Series: Materials Science and Engineering **943** (2020) 012014 doi:10.1088/1757-899X/943/1/012014

With the objective to increase the input voltage by 5 times, the desired output voltage V_{DC}, of 60 V is set in accordance to the output DC voltage, V_{DC}, from Table 2 (i.e. Tesla Supercharger and ABB Terra 53). Upon determining the intended output voltage, V_{DC}, the input voltage, V_{SO}, of 12 V can now be calculated using 11.

$$V_{SO} = \frac{1}{5}V_{DC} \tag{11}$$

Once the desired input and output voltage is determined, the duty cycle of 0.8 can now be calculated using 12

$$\frac{V_{DC}}{V_{SO}} = \frac{1}{1-D} \tag{12}$$

The value for inductor, L, and capacitor, C, is chosen based on the availability of the components. As for the resistance, R, the value of 60 Ω is selected according to the desired output voltage (i.e. 60 V) and current (i.e. 1 A) using 13

$$R = \frac{V}{I} \tag{13}$$

6. Results and Discussion

Using MATLAB/Simulink software package [32], the proposed PV-powered fast charger designed based on DC-DC boost converter is analysed to examine and study its dynamic and transient performances. For load, the EV batteries (load) are replaced by a resistor whereas the input source of the converter is retrieved from PV panels.

6.1. Simulation Results

The system parameters used to carry out simulation study of the proposed PV-powered fast charger designed based on DC-DC boost converter is listed in Table 3. The total simulation period is 15 s. Figure 9 shows the simulation result obtained from Figure 8 in regards to two different DC output voltage references v_{DC}^* (i.e, from 0 V to 60 V) which step changes at 5 s.

Table 3. System Parameters Used In Computer Simulation

Boost Converter Steady State Parameters	Value
Input voltage source, V_{SO}	12 V
Rated input current, I_L	5 A
Input power rating, P_{SO}	60 W
Rated output voltage, V_{DC}	60 V
Rated output current, I_{DC}	1 A
Switching frequency, f_{SW}	10 kHz
Filter inductance, L	100 μH
Filter capacitance, C	1000 μH
Load resistance, R	60 Ω
Steady-state duty cycle, D	0.8
Proportional gain, K_P	0
Integral gain, K_I	0.0280411089392773

IOP Conf. Series: Materials Science and Engineering **943** (2020) 012014 doi:10.1088/1757-899X/943/1/012014

Based on Figure 9, it can be observed that although the boost converter plant is unsteady, it is still able to remain stable and continue tracking with the preset DC output voltage references $v_{DC}{}^{*}$ until it reaches steady-state at 0.46 s. Besides, undershoot and overshoot behaviours are mitigated during the changes of DC output voltage V_{DC} due to care and experience when designing the DC-DC boost converter.

Figure 10 depicted the measurement result of four parameters obtained from the circuitry of boost converter (see Figure 7) after the schematic circuit is replaced by the boost converter transfer function G_{CONV}.

As seen from Figure 9, not only was the proposed voltage loop feedback control able to ensure zero tracking performance. The calculated values in 3 were also similar to the magnitude of the parameters obtained from Figure 10.

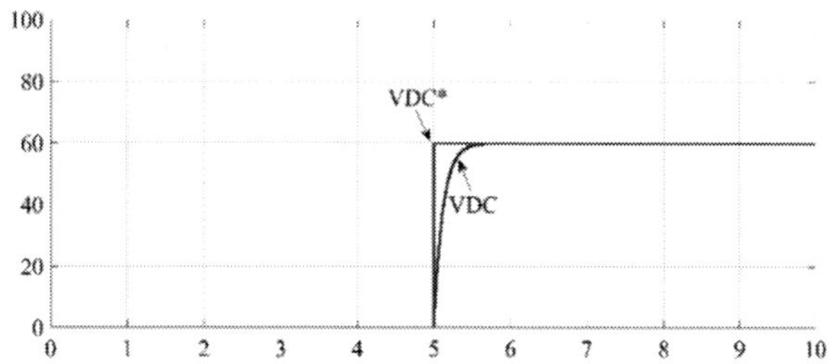

Figure 9. Proposed voltage loop feedback control for boost converter simulation result.

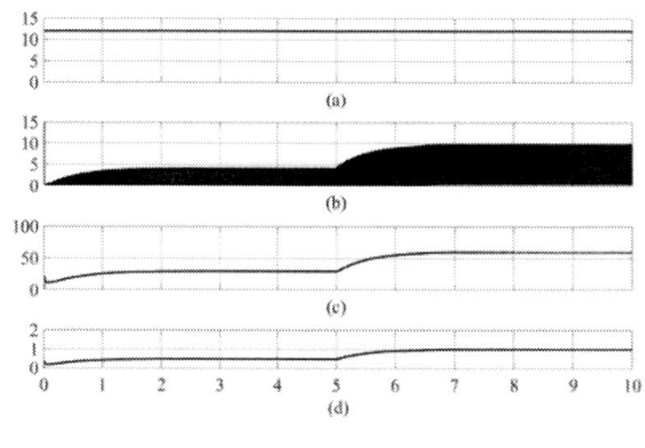

Figure 10. Simulation results of boost converter for (a) input voltage V_{SO}, (b) input current I_L, (c) output voltage V_{DC}, and (d) output current I_{DC} when nominal input source voltage-level is deviated.

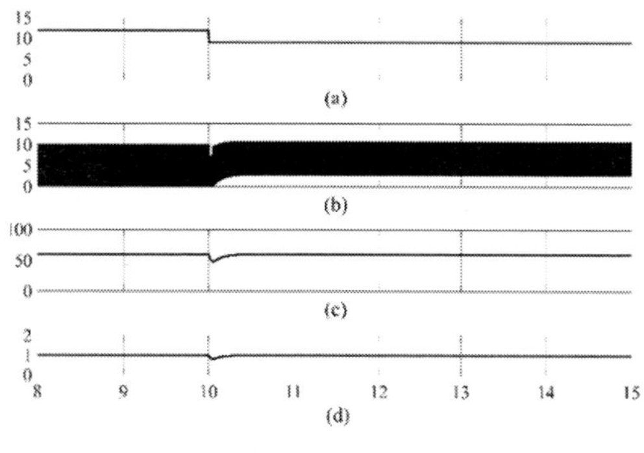

Figure 11. Simulation results of boost converter for (a) input voltage V_{SO}, (b) input current I_L, (c) output voltage V_{DC}, and (d) output current I_{DC} when nominal input source voltage-level is deviated.

Figure 11 illustrates the precedence of the proposed voltage loop feedback control as the nominal input source voltage level is varied from 12 V to 9 V under similar loading condition. It is observed from Figure 11 that the proposed controller has the ability to control and retain the desired DC output voltage-level of the boost converter to 60 V although the input source voltage V_{SO} is lower than 12 V.

Figure 12 shows the resultant duty cycle D obtained from the proposed voltage loop feedback controller throughout the 15 s simulation period. This results proves that the proposed boost controller is able to regulate its DC output voltage based on the DC output voltage references $v_{DC}{}^{*}$ despite changes in input source voltage as well as loading conditions.

IOP Conf. Series: Materials Science and Engineering **943** (2020) 012014 doi:10.1088/1757-899X/943/1/012014

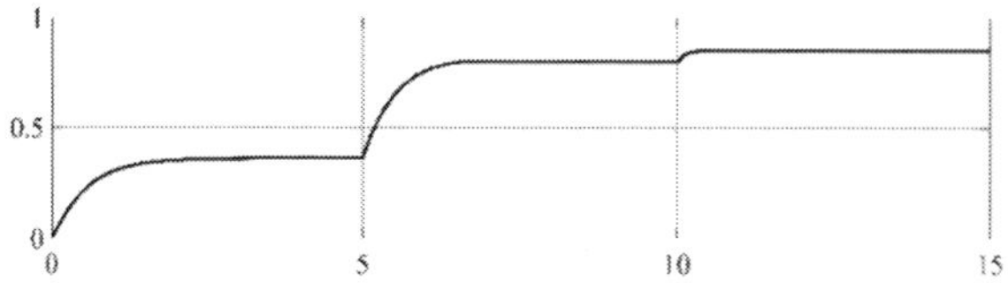

Figure 12. Resultant duty cycle from the proposed voltage loop feedback controller simulation result.

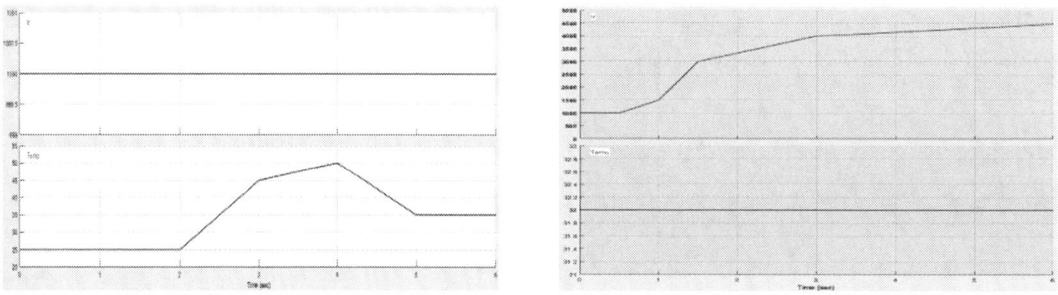

(a) (b)

Figure 13. Simulation results of PV panel with (a) constant irradiance variable temperature and (b) variable irradiance constant temperature.

Simulation for the proposed PV-powered fast charger designed based on DC-DC boost converter with PV panel (i.e., Suntech power STP180S-24-Ad+) is carried out. Figure 13 illustrates the results obtained when two set of irradiance and temperature characteristics are employed for a simulation period of 6 s. For Figure 13(a), the irradiance was maintained constant at 1000 W/m^2 whereas the temperature was increased from 25°C to 45°C at 2 s. Then 45°C to 50°C at 3 s after which it was decreased to 32°C at 5 s. For Figure 13(b), the temperature was maintained constant at at 32°C whereas the irradiance was increased from 1000 W/m^2 to 4500 W/m^2.

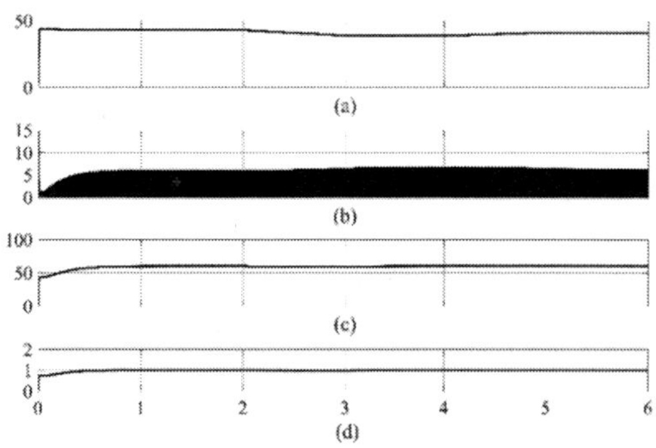

Figure 14. Simulation results of boost converter for (a) input voltage V_{SO}, (b) input current I_L, (c) output voltage V_{DC}, and (d) output current I_{DC} at constant irradiance variable temperature.

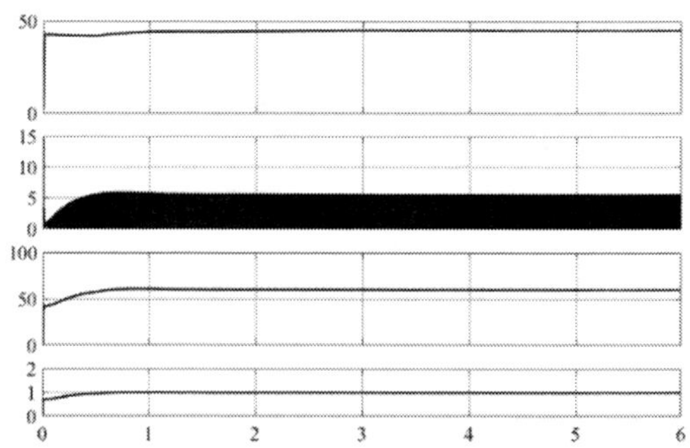

Figure 15. Simulation results of boost converter for (a) input voltage V_{SO}, (b) input current I_L, (c) output voltage V_{DC}, and (d) output current I_{DC} at constant temperature variable irradiance.

Figure 14 depicts the simulation results obtained from the proposed converter controlled by the proposed controller when subjected to the weather condition described in Figure 13(a).

Figure 15 illustrates the simulation results obtained from the proposed system when subjected to the weather condition described in Figure 13(b).

From Figure 14 and Figure 15, it is observed that the proposed system are still able to regulate its output voltage to 60 V although change in temperature of the PV panel results in a non-constant DC voltage output.

Figure 16(a) and Figure 16(b) illustrates the resultant duty cycle D of the proposed system when it is subjected to the weather conditions described in Figure 13(a) and Figure 13(b) respectively.

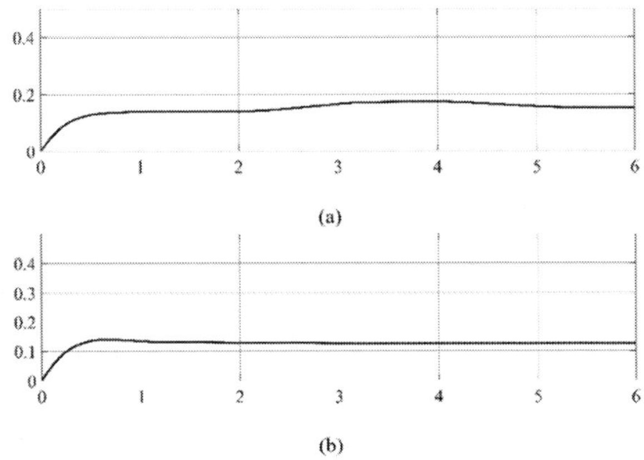

Figure 16. Resultant duty cycle from the proposed voltage loop feedback controller simulation result.

7. Conclusion

This paper presents the design of a PV-powered fast charger for electrical vehicles. PI controller is used to control provide automatic control of the system. To ensure that the design proposed meets the required specifications and needs, various values for various parameters were calculated, derived and chosen. Before proceeding to the next phase (i.e. hardware implementation), MATLAB/Simulink was utilised to carry out simulations for the proposed design to make sure that expected results are achieved. Simulation results obtained show the expected results which demonstrated the ability to generate different output power level based on the desired preset reference voltage value. Hardware implementations will be carried out in future works to ensure the usability of the designed fast charger and potential maximization of system efficiency and performance. In addition, analysis will be carried out to distinguish areas of improvement to reduce charging period through RES. This includes identifying better matching values and tool dynamics of capacitor used. Finally, implementation and integration of PV power for direct fast charging of EVs to increase efficiency during power conversion will be carried out in future works.

References

[1] Sources of greenhouse gas emissions URL `https://www.epa.gov/ghgemissions/`
[2] Blanning B 2013 *2013 World Electric Vehicle Symposium and Exhibition (EVS27)* pp 1–6
[3] Pillot C 2013 *2013 World Electric Vehicle Symposium and Exhibition (EVS27)* pp 1–6
[4] Hannisdahl O H, Malvik H V and Wensaas G B 2013 *2013 World Electric Vehicle Symposium and Exhibition (EVS27)* pp 1–13

[5] U.s. energy information administration - eia - international energy outlook 2017 URL `https://www.eia.gov/ieo`

[6] Global ev outlook 2019 URL `https://www.iea.org/gevo2019/`

[7] Ronanki D, Kelkar A and Williamson S S 2019 *Energies* **12** 3721

[8] Lai C P Y, Law K H and Lim K H 2019 *2019 7th International Conference on Smart Computing Communications (ICSCC)* pp 1–6

[9] Kim J M, Lee J, Eom T H, Bae K H, Shin M H and Won C Y 2018 *2018 21st International Conference on Electrical Machines and Systems (ICEMS)* (IEEE) pp 2603–2607

[10] Lee J H, Moon J S, Lee Y S, Kim Y R and Won C Y 2011 *IECON 2011-37th Annual Conference of the IEEE Industrial Electronics Society* (IEEE) pp 4577–4582

[11] Akmal M, Jawad A and Al Tarabsheh A 2018 *2018 UKSim-AMSS 20th International Conference on Computer Modelling and Simulation (UKSim)* (IEEE) pp 108–113

[12] Yong J Y, Ramachandaramurthy V K, Tan K M and Mithulananthan N 2015 *Renewable and Sustainable Energy Reviews* **49** 365–385

[13] Chen L R 2008 *IEEE Transactions on Industrial Electronics* **56** 480–487

[14] Shareef H, Islam M M and Mohamed A 2016 *Renewable and Sustainable Energy Reviews* **64** 403–420

[15] Trivedi N, Gujar N S, Sarkar S and Pundir S 2018 *2018 IEEMA Engineer Infinite Conference (eTechNxT)* (IEEE) pp 1–5

[16] Lynch W and Salameh Z 2007 *2007 IEEE Power Engineering Society General Meeting* (IEEE) pp 1–5

[17] Zhang S S 2006 *Journal of power sources* **161** 1385–1391

[18] Dharmakeerthi C, Mithulananthan N and Saha T 2012 *2012 IEEE Power and Energy Society General Meeting* (IEEE) pp 1–8

[19] Electropaedia Battery chargers and charging methods URL `http://www.mpoweruk.com/chargers.htm`

[20] Collin R, Miao Y, Yokochi A, Enjeti P and von Jouanne A 2019 *Energies* **12** 1839

[21] Di Yin M, Youn J, Park D and Cho J 2015 *2015 Ninth International Conference on Frontier of Computer Science and Technology* (IEEE) pp 40–45

[22] Srdic S and Lukic S 2019 *IEEE Electrification Magazine* **7** 22–31

[23] Law K H, Ng W P Q and Au P I 2019 *IOP Conference Series: Materials Science and Engineering* (IOPSCIENCE) pp 1–11

[24] Law K H, Dahidah M S, Konstantinou G S and Agelidis V G 2012 *Proceedings of The 7th International Power Electronics and Motion Control Conference* vol 1 (IEEE) pp 330–334

[25] Law K H and Dahidah M S 2014 *2014 IEEE 5th International Symposium on Power Electronics for Distributed Generation Systems (PEDG)* (IEEE) pp 1–7

[26] Law K H, Dahidah M S and Almurib H A 2014 *IEEE Transactions on Power Electronics* **29** 6433–6444

[27] Law K H, Ng W P Q and Wong W K 2017 *International Journal of Power Electronics and Drive Systems* **8** 100–108

[28] Law K H, Dahidah M S, Sim S Y, Ng W P Q, Masaoud A and Abu-Siada A 2018 *TENCON 2018-2018 IEEE Region 10 Conference* (IEEE) pp 0020–0025

[29] Law K H and Ng W P Q 2018 *2018 IEEE 7th International Conference on Power and Energy (PECon)* (IEEE) pp 56–61

[30] Yong S K, Law K H, Ng W P Q and Dahidah M S 2018 *Journal of Telecommunication, Electronic and Computer Engineering* **10** 39–44

[31] Law K H and Dahidah M S 2014 *2014 International Power Electronics Conference (IPEC-Hiroshima 2014-ECCE ASIA)* (IEEE) pp 1283–1290

[32] Mathworks URL `https://www.mathworks.com/`

IOP Conf. Series: Materials Science and Engineering **943** (2020) 012015 doi:10.1088/1757-899X/943/1/012015

A laboratory investigation of the compaction properties of road sub-base stabilised with cement and latex copolymer.

E H Tan[1], E M M Zahran[2]

[1] Civil Engineering Programme Area, Universiti Teknologi Brunei (UTB), Brunei Darussalam
[2] Centre for Transport Research, Universiti Teknologi Brunei (UTB), Brunei Darussalam

E-mail(s): angelteh86@gmail.com, elsaid.zahran@utb.edu.bn

Abstract. The quality of road pavements is an important factor to ensure an acceptable level of service and safety to the road users. When flexible road pavements are to be constructed on weak sub-grade soils, it is often necessary to provide a stabilised sub-base to improve their relevant engineering properties. This paper presents the results and observations from a laboratory investigation evaluating the compaction properties of road sub-base stabilised with Type I cement (ordinary Portland cement) and styrene-butadiene latex copolymer. The proposed polymer contents were 0.5%, 0.75%, 1% and 2% and the proposed cement contents were 2%, 3% and 5%. The 2.5kg rammer compaction test was conducted on the gravelly SAND and sandy GRAVEL samples. It was found that the maximum dry densities (MDD) for the untreated soil samples were higher than the cement-polymer-treated soil samples, indicating that the MDD values decrease when cement and polymer are added. The MDD values for gravelly SAND and sandy GRAVEL treated with cement-polymer lied between 2Mg/m^3±0.03Mg/m^3 and 2.09Mg/m^3±0.04Mg/m^3 respectively. The optimum moisture contents for gravelly SAND and sandy GRAVEL treated with cement-polymer were maintained at 9%±0.5%. The air void lines for gravelly SAND and sandy GRAVEL lied between 7%±1% and 3%±1% respectively.

1. Introduction

It is important to provide road pavements with adequate quality and performance, as they serve several significant functions within a transport system, such providing an acceptable level of service to the road users. Poor quality pavements could compromise the safety of road users. Previous studies, such as reported by Reference [1], had found possible correlations between the pavement quality and the rate of road traffic accidents (RTA). References [2] and [3] proposed how road sections with safety concerns, such as pavement defects, may be identified and ranked for remedial actions. However, rather than adopting a reactive approach, it is often beneficial to ensure that the pavement structural layers are adequately designed and constructed to avoid performance and safety issues during the service life of the road. This is especially crucial when flexible road pavements are to be constructed on weak sub-grade soils, such as those frequently encountered in Brunei Darussalam and its neighbouring countries.

In flexible pavements, the sub-base serves a significant role. It is only optional when the sub-grade is strong or contains high amount of granular soils, else the sub-base is required to act as a structural layer within the pavement system to further distribute the wheel loads from the roadbase to the weaker sub-grade [4]. The sub-base provides a stress-transmitting layer to spread the wheel loads to a larger area to reduce shear and consolidation deformations [5]. It also improves load carrying capacity by

IOP Conf. Series: Materials Science and Engineering **943** (2020) 012015 doi:10.1088/1757-899X/943/1/012015

providing higher stiffness and resistance to fatigue, and building up a relatively thick layer to distribute the wheel loads through a finite thickness [6]. A sub-base that is well-graded and dense prevents upward migration of fine-grained sub-grade soils and water into the roadbase [4]. As the quality of the low-cost sub-base granular soil is inferior to that of the high-cost roadbase granular soil [4, 5, 7], engineers and contractors are motivated to improve the engineering properties of the sub-base. Improving the properties of sub-base could also enable recycled soils, such as those reported by Reference [8], to be used in the sub-base layer for sustainability reasons. One such method is to stabilise unbound granular soils with soil/aggregate stabilisers to create bound granular soils. Bound granular soils provide additional strength and support without increasing the thickness [9]. Additionally, the thickness can be reduced due to the high bearing strength when compared to unbound granular soils [9]. Reference [10] found that, while the use of polymer emulsion for the stabilisation of granular bases has clear advantages and is becoming increasingly popular, there are relatively fewer studies that investigate the improvements to the granular properties using this method of stabilisation.

Compaction of granular soils, with or without soil/aggregate stabiliser added to water, is a crucial work package in road construction. The main objective of compaction is to increase the bulk density of the granular soils by reducing the volume of air voids. It also helps to reduce permeability and swelling. A well-compacted granular soil provides higher bulk density, bearing capacity, and durability to ensure long-term performance [11]. When compaction is insufficient, it can lead to settlement and inadequate stiffness can lead to structural distress [12]. On the other hand, excessive compaction is not desired as it can lead to unnecessarily high construction cost and time delay [12].

The objective of this paper is to present the results and observations of a laboratory investigation of the compaction properties of granular soils stabilised with cement and latex copolymer. In particular, this study focused on the effect of cement and latex copolymer contents variation on the compaction properties of the sub-base upper and lower gradation limits. The maximum dry density (MDD) and optimum water content (OMC) are expected to vary when chemical stabilisers are added, and when the ratio of gravel-to-sand changes. The compaction properties of granular soils essentially provide important indices for the construction of soil-aggregate structures. While the compaction parameters of fine-grained soils are influenced by their Atterberg's limits, the compaction parameters of coarse-grained soils are influenced by their gradations [13]. The results from this laboratory investigation will then be used to plan subsequent tests – California Bearing Ratio (CBR) test, unconfined compressive strength (UCS) test for both dry and wet conditions and (water) permeability test. This study is part of an ongoing research effort aimed to develop more sustainable and mechanically durable road as well as addressing the persisting flexible pavement distresses.

The stability of unbound soils depends on the particle size distribution, particle shape, relative density, internal friction and cohesion [6]. Unbound soil that is designed for maximum stability should possess high internal friction to resist deformation under load [6]. The internal friction and shearing resistance depend largely on the particle shape, particle size distribution and density [6]. The proportion of fine to coarse fraction is considered very important, and the presence of a wider range of particle size makes the soil mixture denser and more compacted, which gives the soil mixture greater strength under shearing [14]. When the cement and latex copolymer are added, they bind the granular soils together to form a soil-cement-polymer soil matrix when cured. While unbound granular soils may still have high tendency to slide, move and/or displace, bound granular soils have a much lower tendency to slide, move and/or displace as they are physically bound to each other.

2. Compaction test methods

The 2.5kg rammer compaction test outlined in BS 1377-4: 1990 [15] is widely used to determine the dry density and moisture content relationship of soils with particles up to medium-gravel size. Thus, no more than 30% by mass of soils should be retained on the 20mm and 37.5mm sieves (coarse gravel)

[15]. Reference [16] suggested that the ratio of mould diameter to the largest nominal particle size should not be less than 5 to 6. The ratio used in this experiment is 5.5 (105mm:19mm). Typically, a 2.5kg rammer compaction (equivalent to standard Proctor compaction) is used for normal traffic loading situations [17]. A 4.5kg rammer compaction (equivalent to modified Proctor compaction) is used to simulate field condition where a higher compactive effort is required for airfield pavements [17, 18].

3. Materials

3.1. Aggregate and gradation

In this laboratory investigation, crushed sandstone aggregate was used. The aggregate was sourced from a local quarry in the Temburong District – Brunei Darussalam which is mainly used for the sub-base layer. The physical and mechanical properties of the aggregate satisfied the local specification – GS1: 1998 Flexible Pavement [19]. A gradation envelope is established to provide quality control on the granular soils to be used in road construction, therefore to study the effects of the upper and lower bounds, it is necessary to select the appropriate envelope [18]. The upper and lower bounds for the sub-base specified by GS1: 1998 Flexible Pavement [19] are as shown in Figure. 1. Figure. 2 (a) and (b) illustrate the gravelly SAND and sandy GRAVEL samples before compaction.

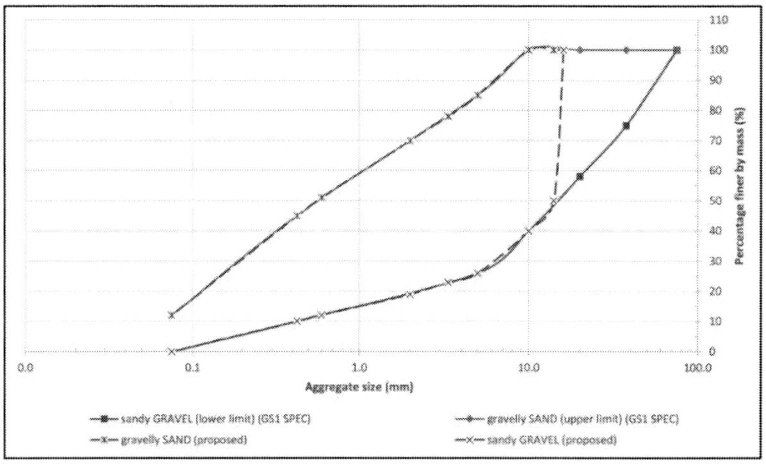

Figure. 1 Upper and lower gradation limits for the specified and proposed sub-base layer.

Figure. 2 (a) Uncompacted gravelly SAND and (b) uncompacted sandy GRAVEL samples.

Since the 2.5kg rammer compaction test was used, and the ratio of mould diameter to the largest nominal particle size was to be maintained, the gradation for the sandy GRAVEL was altered to satisfy the test and ratio requirements. BS 1377-4: 1990 [15] allows the same quantity of material of similar characteristics which was retained on the 20mm sieve, to be replaced with soil particles of smaller sizes. The standard stated that the substitution of large soil particles by smaller and similar soil particles generally gives dry densities (in the laboratory) which are reasonably comparable with those obtained in the field.

3.2. Cement

The cement used in the laboratory investigation was the locally manufactured Class 62.5N (62.5N/mm^2) Type I ordinary Portland cement (OPC). In Brunei Darussalam, road construction contractors typically add between 2% and 7% of cement by weight to the unbound soils to produce hydraulically-bound soils. Cement content from 6% to 10% onwards in normal condition will cause severe shrinkage cracking over a long period of time [20]. According to Reference [21], a well-graded soil mixture of stone fragments or gravel, coarse sand, and fine sand either with or without small amount of slightly plastic silt- and clay-sized soil particles will require 5% or less cement by weight for stabilisation. Based on the limitations stated above, the proposed cement contents were 2%, 3% and 5% by weight of sample.

3.3. Latex copolymer

The liquid additive used in the laboratory investigation was a white colour water-based dispersion of a latex copolymer, under the tradename Terratech T-Pro® 500. The main components of T-Pro® 500 are water (between 40% and 55% by weight) and styrene-butadiene (SB) copolymer (between 45% and 55% by weight) [22]. The latex copolymer has a specific gravity (S.G.) between 0.95 and 1.1, the dynamic viscosity is less than 500 mPa.s and the size of the SB particle is 125 nm [22]. The recommended polymer content and OMC based on the manufacturer-specified gradation is 0.75% and 9% respectively. Reference [23] used 0.5%, 0.75%, 1% and 2% of the polymer by weight to stabilise clay-soil mixture, and it was concluded that the optimum polymer content was 0.75% polymer content. The 3-day and 7-day compressive strengths for the 0.75% polymer content was 0.68MPa and 0.79MPa respectively [23]. The proposed polymer contents were 0.5%, 0.75%, 1% and 2% by weight of sample.

3.4. Latex copolymer and cement

It is common to have two additives added together to stabilise unbound soils. Unbound granular soil may exhibit poor engineering performances such as low bearing capacity, susceptibility to moisture damage and volume change, and susceptibility to environmental conditions; all of which would result in substantial pavement distress and shortening of pavement life [24]. The addition of one or two soil stabilisers to the unbound granular soils can improve its engineering properties. Previous research by References [24], [25] and [26] have concluded that 8% by weight of SB latex copolymer (tradename Mallard Creek Tylac® 4190) and 4% by weight of cement (Type II OPC) were effective to stabilise soil-aggregate roadbase. The results from References [24], [25] and [26] showed that the compressive strength for 8% polymer-4% cement increased by 94.43% when compared to 4% cement, the CBR for unsoaked condition for 8% polymer-4% cement was 412.2% while the CBR for 4% cement was 289.7% and the CBR for soaked condition for 8% polymer-4% cement was 452.8% while the CBR for 4% cement was 292.6%. It is worthwhile to note that the CBR for soaked soil samples were greater than unsoaked soil samples. The same authors found that there was an increase in 81.4% for 8% polymer-4% cement when compared to 4% cement, and 288.2% when compared to 8% polymer.

4. Results and discussion

The sizes of the granular soil particles affect the engineering behaviours of a given soil mixture, and the bulk density of a soil mixture depends mainly on the weight of the individual soil particle and the amount of water present [18]. The shape and texture of the soil particles also influence the bulk density of the

IOP Conf. Series: Materials Science and Engineering **943** (2020) 012015 doi:10.1088/1757-899X/943/1/012015

granular soil, though there is no refined geotechnical procedure to quantify it [18]. Reference [18] found that typical MDD values for granular soils are between 1.6Mg/m^3 to 2Mg/m^3, with limits between 1.3Mg/m^3 to 2.4Mg/m^3, and the typical OMC values are between 10% to 20%, with limits between 5% to 40%. The MDD and OMC obtained from the compaction test under a given compactive effort provide indications of the porosity of the soil mixture – lower MDD and higher OMC correspond to greater porosity [27].

Figure. 3 shows the compaction curves for the gravelly SAND and sandy GRAVEL treated with the latex copolymer only. It was observed that the MDD values for gravelly SAND were lesser than sandy GRAVEL. Generally, coarser soils can be compacted to a higher bulk density than finer soils [18].

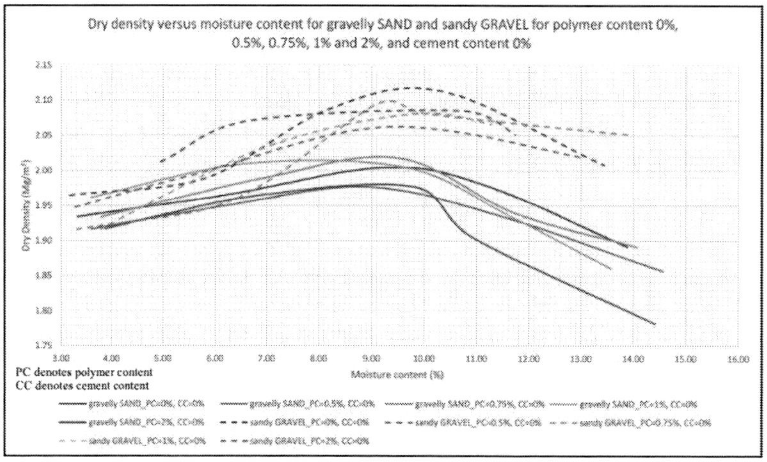

Figure. 3 Dry density versus moisture content of soil samples treated with polymer only.

Figure. 4 shows the compaction curves for gravelly SAND and sandy GRAVEL treated with 0.75% and 1% polymer content with 0%, 2%, 3% and 5% cement. It was also observed that the MDD values for gravelly SAND were lesser than sandy GRAVEL.

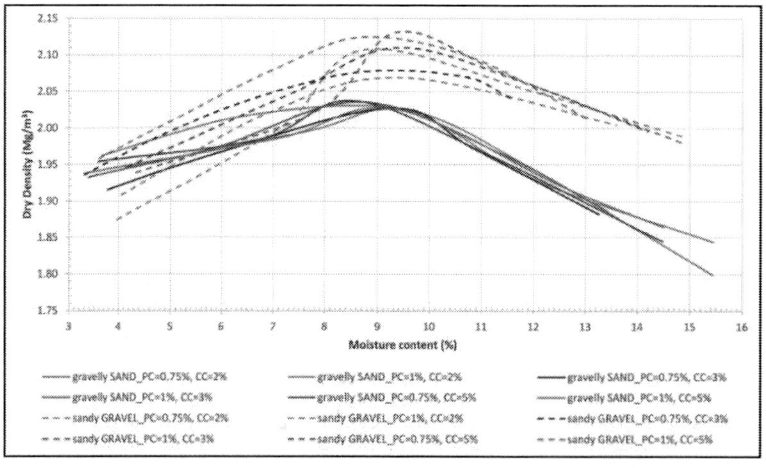

Figure. 4 Dry density versus moisture content of soil samples treated with polymer and cement.

While it was expected that the MDD values in this laboratory investigation should decrease with increasing polymer content as observed in Reference [23], as in both experiments, the same latex copolymer was used but with different gradations. However, the MDD values in this experiment did not show the expected decreasing trend. However, in the laboratory investigation by Reference [24], the MDD values increased to a peak and then decreased when 5% to 10% of SB polymer was added (peaked at 8%). This downward concave trend was as observed in this experiment when 0.5% to 2% of polymer was added. As shown in Figure. 5 (a), MDD peaked at 0.75% for both gravelly SAND and sandy GRAVEL. The MDD value for gravelly SAND was 2.02 Mg/m^3 when 0.75% polymer was added, and this value was slightly greater than when no polymer was used (2.01Mg/m^3). The drop in MDD values corresponding 0% and 0.5% polymer contents could be the result in the change of dynamic viscosity (i.e. a measure of a liquid's resistance to movement) between 100% water (< 1mPa.s) and (the addition of) polymer (< 500mPa.s). The kinematic viscosity (i.e. a measure of resistance to the fluid flowing) of water and polymer are less than 1×10^{-6} m^2/s and less than 456×10^{-6} m^2/s respectively. As the dynamic viscosity increases, the fluid impedes soil particle movement, thus preventing effective (soil particle) packing by compaction. Reference [28], in their experimental investigation, suggested that the decrease in MDD values (when soils were treated with increasing polymer content) to the differences in S.G. of water and polymer (1 and 1.1) and the soils (2.57 and 2.61), which is unlikely the case. The increase in MDD values until peak with increasing polymer content, as explained by Reference [24], was an indication of the consolidation of the rigid styrene chains and flexible butadiene chains of the SB molecular structure which improved the mechanical properties of the soil mixture. The nano-sized polymer particles penetrated and spread throughout the soil mixture to provide both toughness and flexibility [24]. The decrease in MDD values after peak was due to the presence of excessive water that inhibited effective compaction, which in turn reduced toughness and flexibility [24].

It was expected that for the polymer contents 0.75% and 1%, when the cement content increases, the MDD values should increase. However, this expectation was only observed in the gravelly SAND of polymer content 0.75% and sandy GRAVEL of polymer content 1% as shown in Figure. 5 (b). The cement acted as fine filler to fill the air void thus increasing the bulk density. The sandy GRAVEL of polymer content 0.75% observed decreasing MDD values until cement content of 3% and increased when cement content is 5%. One highly suspected contributory factor to this contradictory trend is the gravel shapes and sizes passing the 20mm sieve and retained on the 14mm sieve. It is sometimes difficult to get a consistent soil mixture of similar shapes and sizes for all the soil samples. In such a situation, it is recommended to repeat the experiment for the outliers to ascertain whether or not the gravel shape and size were the main reason for this error.

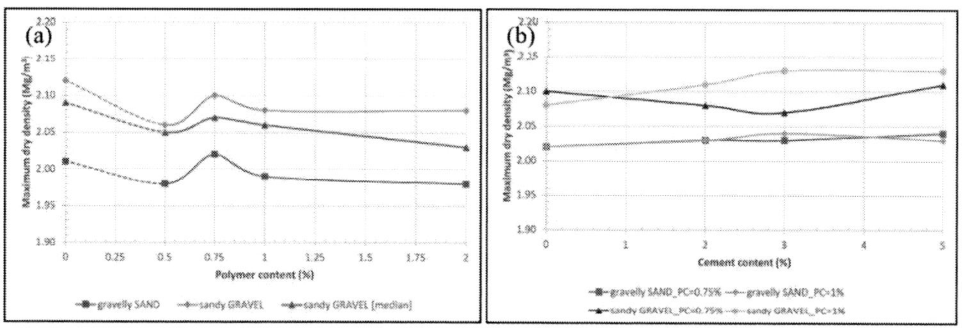

Figure. 5 (a) MDD against polymer content for 0% cement content, (b) MDD against cement content for 0.75% and 1% polymer content.

Figure. 6 (a) and (b) show the soil packing features of sandy GRAVEL and gravelly SAND close to the peak MDD values when 1% of polymer and 3% of polymer were added. Figure. 7 (a) and (b) show the packing features when 2% of polymer was added to the soil samples.

Figure. 6 (a) Compacted gravelly SAND and (b) compacted sandy GRAVEL.

Figure. 7 (a) Compacted gravelly SAND and (b) compacted sandy GRAVEL.

Since the latex copolymer contains between 40% and 55% by weight of water, the increase in polymer content will result in the increase in the soil sample's water content. However, as shown in Figure. 8 (a), there was no consistent trend for the relationship between OMC and polymer contents from 0% to 0.75% for the gravelly SAND and sandy GRAVEL, but from 1% to 2% polymer content, the OMC values increased for the two investigated gradations. Nonetheless, the OMC values for 0% cement content for gravelly SAND and sandy GRAVEL maintained between 9%±0.5% and 10%±0.5% respectively as shown in see Figure. 8 (a).

For the polymer content 0.75% and 1%, an increase in cement content should observe slight decrease in OMC as the tricalcium silicate (C_3S) and dicalcium silicate (C_2S) react with water to produce calcium silicate hydrate (CSH) and crystallised calcium hydroxide (CH). Type I cement contains 59% C_3S and 15% C_2S [29]. The rate of reaction of C_3S is much faster than C_2S (slowest) with the former producing

CSH (Type E) and CH within the hour [30]. An initial water-to-cement ratio of 0.40 to 0.42 is required for the complete hydration of cement [30-32]. The initial setting time of Type I cement is between 70mins to 120mins, and the final setting time is between 140mins to 240mins [33]. However, these setting time ranges can change, as (anionic) SB polymer retard/suppress the hydration reaction [34]. Nonetheless, the OMC values for gravelly SAND and sandy GRAVEL treated with cement were maintained between 9%±0.5%, as shown in Figure. 8 (b).

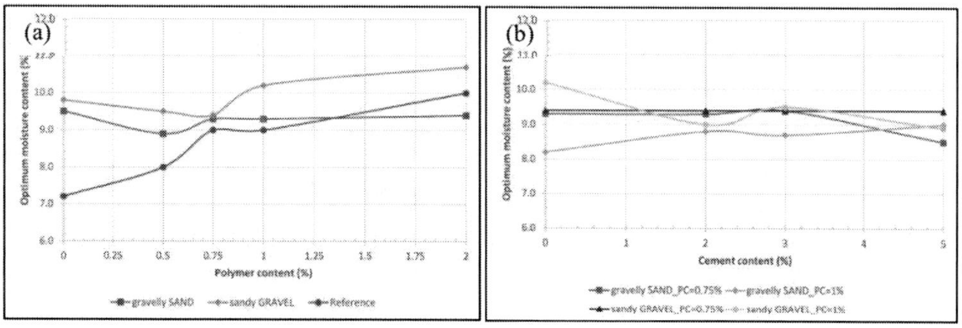

Figure. 8 (a) OMC against polymer content for 0% cement content, (b) OMC against cement content for 0.75% and 1% polymer content.

Figure. 9 shows a typical compaction curve illustrating the relationship between moisture content and dry density. The 0% air void line (AVL) corresponds to the 100% saturation line (SL), and theoretically, the compaction line should not intersect with the 0% AVL. An intersection observed in the test results may be due to incorrect test measurement and/or an incorrect specific gravity value. This was found to be a problem for the sandy GRAVEL samples at high water content (> 10%) as the water distribution throughout the soil samples was inconsistent as illustrated in Figure. 11 (c), where water was concentrated at the top of the sample and gradually decreased with increasing depth of the soil sample. In the laboratory investigation, most of the compaction curves for the sandy GRAVEL samples after the third addition of water intersected with the 100% SL, and the soil samples were all observed to be highly saturated with water.

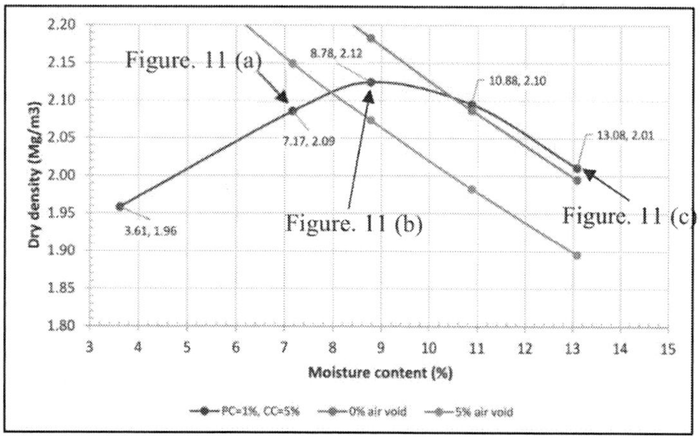

Figure. 9 Dry density against moisture content curve for sandy GRAVEL (PC = 1%, CC = 5%).

Figure. 10 (a), (b) and (c) and Figure. 11 (a), (b) and (c) illustrate the sandy GRAVEL at 0% and 5% cement content respectively. The sample with 5% cement content was able to hold the shape and was less slurry.

Figure. 10 (a) and (b) Saturated sandy GRAVEL and (c) highly saturated sandy GRAVEL for polymer content 1% and cement content 0% (W denotes amount of water added).

Figure. 11 (a) and (b) Saturated sandy GRAVEL and (c) highly saturated sandy GRAVEL for polymer content 1% and cement content 5%.

The decrease in saturation when 0.5% of polymer was added, as shown in Figure. 12 (a), could be explained by the change in the kinematic viscosities of the water-polymer mixture. The increase in the kinematic viscosity meant that the movement of soil particles and water-polymer was impeded for effective packing when compacted. The increase in the saturation with increasing polymer content could be due to the additional (free) water from the polymer. When 0.5% of polymer was added, an additional of 6.25mL of water from the polymer was added, and when 2% of polymer was added, an additional of 25mL of water from the polymer was added, to the same weight of soil sample. The increase in water content can displace the finer soil particles, thus reducing the MDD values. When 1% and 2% of polymer was added, while the saturation degree increased, the MDD values decreased. Furthermore, sandy GRAVEL samples inherently have larger air void, and when water cannot escape the compaction mould,

IOP Conf. Series: Materials Science and Engineering **943** (2020) 012015 doi:10.1088/1757-899X/943/1/012015

upon compaction, water navigated through the large voids and was subsequently trapped, increasing the degree of saturation. Theoretically, gravelly SAND samples have larger surface area to volume, therefore, more water was absorbed into the soil particles and some was adsorbed on the surface, leaving the voids filled with air than water. Unlike sandy GRAVEL, the compaction curves of the gravelly SAND did not intersect with the 100% SL. The AVL for the gravelly SAND and sandy GRAVEL samples were between 7%±1% and 2%±1.5% respectively, as shown in Figure. 12 (a). Figure. 12 (b) illustrates the SL for the gravelly SAND and sandy GRAVEL samples when cement was added to 0.75% and 1% polymer. The SL and AVL are inverse of each other.

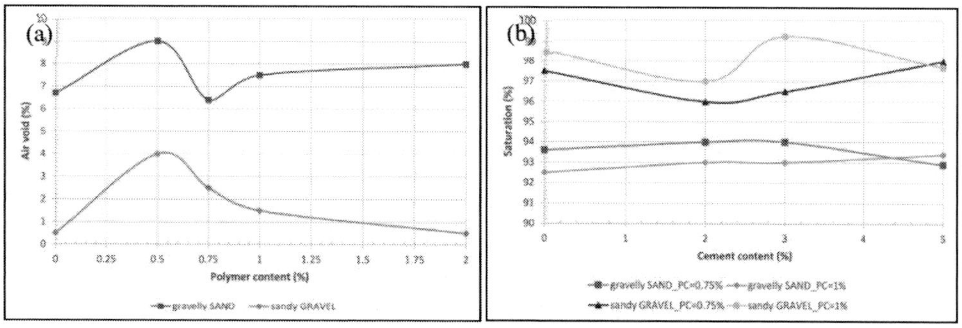

Figure. 12 (a) Air void against polymer content for 0% cement content, (b) Saturation against cement content for 0.75% and 1% polymer content.

5. Conclusion

The aim of the laboratory investigation to evaluate the compaction properties of the granular soil samples stabilised with cement and latex copolymer using 2.5kg rammer compaction test has been accomplished. The following are the conclusions drawn from the study:

- The peak MDD values of the sandy GRAVEL and gravelly SAND were when 0.75% polymer-0% cement was added to stabilise the soil samples. When 2%, 3% and 5% of cement were added, there was no consistent trend, but the MDD values for gravelly SAND and sandy GRAVEL lied between $2Mg/m^3$±$0.03Mg/m^3$ and $2.09Mg/m^3$±$0.04Mg/m^3$ respectively.
- While it was expected that the OMC values would gradually increase with increasing polymer content, the OMC values determined in the laboratory investigation did not follow the expected trend, but the OMC values for 0% cement content for gravelly SAND and sandy GRAVEL lied between 9%±0.5% and 10%±0.5% respectively. When 2%, 3% and 5% of cement was added, it was expected that the OMC would decrease as water was used during the hydration reaction. Contrary to the expectation, there was no definite increasing or decreasing trend for gravelly SAND and sandy GRAVEL, but the OMC values maintained between 9%±0.5%.
- For 0% cement content, the gravelly SAND has lower range of saturation degree – between 92%±2% (corresponded to 8%±2% air void) than sandy GRAVEL – between 98%±1.5% (corresponded to 2%±1.5% air void). When 2%, 3% and 5% of cement was added, there was no definite trend for gravelly SAND and sandy GRAVEL, but the air void lines lied between 7%±1% and 3%±1% respectively.

Subsequent laboratory tests – California Bearing Ratio (CBR) test, unconfined compressive strength (UCS) test for both dry and wet conditions and (water) permeability test – must to be conducted for the gravelly SAND and sandy GRAVEL to determine quantitatively the structural stability, strength and permeability properties/characteristics. While the compaction test gave initial results on the MDD values, they cannot be used as indication of strength gained. The soil samples have to be cured – allowing the hydration products to form and bind the soil particles, and water to evaporate to form the polymer film to bind the soil particles, CSH and CH – to determine the final strength gained.

Acknowledgement

The authors are grateful to Universiti Teknologi Brunei for their funding support, and would also express appreciation to Pg. Dr. Hj. Ismail, Mr. Chua Min Chu and Mr. Tang Kum Chuen (SWEE Sdn. Bhd.), Mr. Dhammika Hewage (Meraih Jaya Sdn. Bhd.) and Mr. Mahri Matahir (Terratech Earth Solution Ltd.) for their contribution towards technical and resources supports.

Reference

[1] Hikmah N, Tan S J, Zahran E-S M M, Yap Y H and Taib H 2018 *Statistical correlation between road surface roughness and traffic accidents Brunei Int. Conf. Eng. Tech.*, (Brunei Darussalam: Institution of Engineering and Technology)

[2] Zahran E-S M M, Tan S J, Yap Y H, Rahman E K A and Husaini N H 2017 A novel approach for identification and ranking of road traffic accident hotspots *MATEC Web of Conferences* **124**

[3] Zahran E-S M M, Tan S J, Tan E H A, Mohamad 'Asri Putra N A A B, Yap Y H and Abdul Rahman E K 2019 Spatial analysis of road traffic accident hotspots: evaluation and validation of recent approaches using road safety audit *J. Trans, Safety & Security* 1-30

[4] O'Flaherty C A 2002 *Highways the location, design, construction and maintenance of road pavements* (Maine, USA: Butterworth-Heinemann)

[5] Manjunatha H 2014 Laboratory studies on granular sub base *Int. J. Eng. Res. Tech.* **3** 2535-2538

[6] Yoder E J and Witczak M W 1975 *Principles of pavement design* (USA: John Wiley & Sons, Inc)

[7] Mallick R B and El-Korchi T 2013 *Pavement engineering - principles and practice* (Boca Raton, Florida: CRC Press)

[8] Khan S A, Tan S J, Rahman E K A and Muneerah D N 2014 *Effect of hydrocarbon contamination and subsequent remedial treatment on the engineering properties of soil Brunei Int. Conf. Eng. Tech.*, (Brunei Darussalam: Institution of Engineering and Technology)

[9] Aher D D, Sangale Y B, Pager S R, Yadnesh P and Rahul R 2018 *Cement treated sub-base for bituminous pavement Int. Conf. Recent Trends in Eng. Tech.*, (Maharashtra, India:) p 875-880

[10] Tan E H A, Zahran E-S M M and Tan S J 2019 *A review of chemical stabilisation in road construction International Conference on Materials Technology and Energy*, (Miri, Malaysia: IOP Publishing)

[11] Wu P and Houben L J M 2013 Investigation of the effects of different laboratory compacting methods on the properties of cement stabilized materials *Int. Conf. on the Bearing Capacity of Roads, Railways and Airfields*, (Trondheim, Norway:)

[12] Fragaszy R J 1991 *Compaction control of granular soils*. (Washington, USA: Washington State Department of Transportation)

[13] Verma G and Kumar B 2019 Prediction of compaction parameters for fine-grained and coarse-grained soils: a review *Int. J. Geotech. Eng.*

[14] Sharmila K 2017 Efficacy of cement on stabilization of gravel soils as road construction material *Int. Res. J. Eng. Tech.* **4** 1211-1215

[15] British Standards Institutions 1990 *BS 1377-4: 1990*. In: Methods of test for soils for civil engineering purposes - part 4: compaction-related tests, (UK: British Standards Institutions)

[16] Donaghe R T and Torrey V H 1994 A compaction test method for soil-rock mixture in which equipment size effect are minimized *Geotech. Testing J.* **17** 363-370

[17] Viji V K, Lissy K F, Sobha C and Benny M A 2013 Predictions on compaction characteristics of fly ash using regression analysis and artificial neural network analysis *Int. J. Geotech. Eng.* **7** 282-291

[18] Worku A and Shiferaw D 2004 Prediction of maximum dry density of local granular fills *Journal of EEA* **21** 59-70

[19] Construction Planning and Research Unit 1998 *GS 1: General specifications for flexible pavement.* (BSB, Brunei Darussalam: Construction Planning and Research Unit)

[20] Gafar H S H and Qing W D 2002 *Review of chemical stabilisation technologies and applications for public roads in Brunei Darussalam The Regional Seminar on Quality Roads - The Way Forward,* (BSB, Brunei Darussalam:)

[21] Portland Cement Association 1992 *Soil-cement laboratory handbook* (Illinois, USA: Portland Cement Association)

[22] Terratech Inc. 2012 *T-PRO®500 technical data sheet.* (U.S.A.: Terratech Inc.)

[23] Hadi N H A, Jiann T S, Zahran E-S M M, Jeludin M and Hie T E 2019 *Laboratory investigation into the compaction and strength properties of road subbase infused with a latex copolymer Int. Conf. Geotech. Res. Eng.*, (Rome, Italy:)

[24] Baghini S M, Ismail A, Karim M R, Shokri F and Firoozi A A 2014 Effect of styrene–butadiene copolymer latex on properties and durability of road base stabilized with Portland cement additive *Const. Build. Mater.* **68** 740-749

[25] Baghini M S, Ismail A, Naseralavi S S and Firoozi A A 2016 Performance evaluation of road base stabilized with styrene–butadiene copolymer latex and Portland cement *Int. J. Pav. Res. Tech.* **9** 321-336

[26] Baghini M S, Ismail A, Asghar M P, Fendereski G and Sadeghi M 2018 Measuring the effects of styrene butadiene copolymer latex-Portland cement additives on properties of stabilized soil-aggregate base *Int. J. Pav. Res. Tech.* **11** 458-469

[27] Xiao Y, Tutumluer E, Qian Y and Siekmeier J A 2012 Gradation effects influencing mechanical properties of aggregate base–granular subbase materials in Minnesota *J. Trans. Res. Board* **2267** 14-26

[28] Kolay P K, Dhakal B, Kumar S and Puri V K 2016 Effect of liquid acrylic polymer on geotechnical properties of fine-grained soils *Int. J. Geosyn. Ground Eng.* **2**

[29] Neville A M and Brooks J J 2010 *Concrete technology,* (UK: Pearson Education Limited)

[30] Gartner E M, Young J F, Damidot D A and Jawed I 2002 *Structure and performance of cements,* ed J Bensted and P Barnes (London, UK: Spon Press) pp 57-113

[31] Alesiani M, Pirazzoli I and Maraviglia B 2007 Factors affecting OPC early-age hydration studied by NMR: fineness, water-to-cement ratio and curing temperature *Appl. Magn. Reson.* **32** 1-10

[32] Wang R, Lackner R and Wang P M 2011 Effect of styrene-butadiene rubber latex on mechanical properties of cementitious materials highlighted by means of nanoidentation *Strain* **47** 117-126

[33] Hodhod H and Abdeen M A M 2010 Experimental comparative and numerical predictive studies on strength evaluation of cement types: effect of specimen shape and type of sand *Eng.* **2** 559-572

[34] Baueregger S, Perello M and Plank J 2015 Influence of carboxylated styrene–butadiene latex copolymer on Portland cement hydration *Cem. Concr. Compos.* **63** 42-50

Surface Engineering of Titanium Using Anodization and Plasma Treatment

Manjaiah Mallaiah[1] and Ravi Kumar Gupta[2,*]

[1]Department of Mechanical and Manufacturing Engineering, Manipal Institute of Technology, Manipal Academy of Higher Education, Manipal-576104, Karnataka, India; manjaiah.m@nitw.ac.in
[2]Department of Mechanical Engineering, Manipal University Jaipur, Dehmi Kalan, Near GVKToll Plaza, Jaipur-303007, India; ravikumar.gupta@jaipur.manipal.edu

[*]Corresponding author's e-mail: rkgmuj@gmail.com

Abstract. Ti64 alloy is a well-known material for biomedical applications due to high corrosion resistance and biocompatibility properties. Surface properties of implants plays a vital role in bone and cell growth in the human body. With the anodization process, we can increase the surface porosity, which will be adequate for surface fascination of the implant screw to the bone and appropriate mechanical properties. Hence, the present study attempted to improve the surface properties of Ti64 by anodization and plasma treatment that may be promising method to increase the biocompatibility of Ti materials. Anodization process is the cheapest one to improve the surface properties of Ti alloy and riskless process. To intensify the open pores on the Ti64 surface plasma treatment was performed. Also, the aim of this study was to improve the aesthetic appearance of the dental implants and reproduce interference of colours. With the help of UV-VIS spectrophotometer the colour and spectral reflectance were investigated. The oxide layer thickness, chemical composition and nanosurface roughness was measured. These results suggests the surface modification of Ti64 alloy by anodization can produce interference of colours and are dependent on the applied voltage, oxide layer thickness. The surface oxidation consisting of anatase and rutile phase and change in nanosurface roughness, may improve the biocompatibility of Ti64.

Keywords. Ti64; Nanosurface Roughness; Anodization; Oxide Film and XRD.

1. Introduction

Metallic biomaterials often require surface modifications in order to improve corrosion and wear resistance, biocompatibility, and surface wettability. Mechanical holding between two surfaces can be improved by modifying one or both surfaces to increase effective surface area by a mechanical modification methods such as sandblasting, shot peening, or laser peening. *Osteointegration* is defined as a "direct structural and functional connection between ordered living bone and the surface of a load-carrying implant." Using current materials and techniques, a titanium implant requires few months to osteointegrate together with surrounding tissue and bone. Although for successful implants that had continued for up to 17 years, proposing that the osteointegration has been imperfect for a long period of time [1]. Surface structure plays a significant role in the operational performance of dental implants. Average surface roughness (S_a) is a surface integrity descriptor that is most often used to quantify the topography. As typically used, it does however have limitations and may not contain sufficient information to adequately describe and quantify the surface topography as relevant to dental implants. The physiological responses of the cells, bone which are in contact of the implant depend on the surface properties of the implant [2].

For dental and orthopedic implant applications, Titanium is a well-proven implantable material [3] as it is biocompatibility and has anti-corrosion surface properties. Furthermore, extremely porous titanium is a smart material for biomedical implantations because of its particular properties such as strength, weightlessness and high resistance to corrosion. Modulus of elasticity of porous titanium is alike to that of human bone and desired pore size helps in bone ingrowth.

Anodic oxidation was performed to create porous titanium oxide film on commercial Ti by Dikici et al. [4]. It was found that, the film with non-porous structured TiO_2 can be created by the anodic oxidation method which is transferred into anatase phase later annealing in air at 480°C for 2hrs. The TiO_2 film is good candidate for the photocatalytic degradation of harmful organic impurities in wastewater.

The bioactive spongy porous coating with nano surface structure and exceptional apatite formation capability on titanium were effectively attained by a hybrid process combining sand blasting, acid etching and anodization process, heat treatment and hydro-thermal treatment [5]. The processed samples have very good apatite creation capability due to the formation of nano anatase and brushite, which is having outstanding bioactivity.

Recently, selective laser metal (SLM) deposition is considered as a hopeful technology for manufacturing of scaffold Titanium implants with definite porosities. Wally et al. (2019) [6] found that it is possible to produce porous Ti64 structures with identical and graded porosity for dental implants by using SLM technique. Ti 64 with graded lattice structure of implants were fabricated by SLM technique, their biocompatibility results were showed that better osteointegration than the existing implants due to porous structure. This might be the efficient process for the fabrication of porous dental implants but the cost of the implants will be very high because of the powder cost and process is not commercialized still.

One go with lowest manufacturing cost by the anodization and plasma treatment of titanium implants which also have better results in terms of biocompatibility and bone tissue growth.

The plasma treatment has been confidently did on polymeric materials like polyethylene and polyethylene terephthalate to enhance biocompatible properties such as [7,8] wettability, roughness, cell adhesion, spreading and proliferation [9]. The plasma treatment is also done on metals to improve surface properties i.e. titanium implants to progress mechanical properties and biocompatibility [10].

Therefore, combining anodization and plasma treatment applied to titanium material to enhance the surface roughness and biocompatible properties is considered in the current study. The consequence of plasma treatment on the Ti samples was examined and particular properties of the samples were analyzed therefore.

2. Experimental Setup

2.1. Plasma Treatment of Ti Plates

Ti64 (grade 4) 2mm thick and 25mmx25mm samples were prepared. The Ti64 samples were polished mechanically with different SiC grit papers. Afterwards the samples were sand blasted with 120µm SiO_2 from a distance of 10mm perpendicular to the specimen surface at a pressure of 0.35MPa for 5sec. The samples were then cleaned ultrasonically in acetone and distilled water for 15min, respectively. The specimens were then subjected to argon oxygen plasma treatment under atmospheric pressure to clean and activates the surface of the samples as shown in Fig. 1. The plasma treatment was performed at a discharge power of 200W with argon and oxygen for 3min with a standoff distance as 5mm. Anodization were conducted using DC voltage source. The Aluminium plate of 50mmx50mm and 5mm thick was used as cathode, Ti samples were used as anode electrode. Diluted sulfuric acid was used as electrolyte and maintained a proper distance between the anode and cathode. The anodic oxidation were carried out for a constant time of 2min for different voltages between 10V to 90V.

Figure 1. Atmospheric plasma spray setup.

2.2. Reflectance Characteristics and TiO2 Layer Thickness

The colors of anodized samples were measured using UV-VIS spectrophotometer at an observation angle 20°-70°. The reflectance data was measured for a wavelength of 200-900nm. The oxide layer thickness of the anodized samples was calculated by the wavelength and the maximum reflectance with the following equations (1) and (2) [11]. The maximum reflectance for particular wavelength was chosen for measuring the refractive index.

$$t = \frac{\lambda_{max}}{4n_o} \quad (1)$$

$$n_o^2 = 5.193 + \frac{2.441\times10^7}{\lambda_{max}^2 - 0.803\times10^7} \quad (2)$$

Where; t - thickness of the oxide film in nm, λ_{max} -wavelength of highest film intensity absorption in nm, and n_o - film refractive index. The oxide layer refractive index varied with wavelength, which is obtained by the Eq (2).

2.3. Surface Characterization

The surface topography of anodized samples was measured by using atomic force microscope (AFM; Model: Nanosurf Easyscan 2) to understand the nanosurface roughness and porous structure. The X-ray diffraction (XRD; Model: RIGAKUD/MAX2500) was used to examine the surface phase compositions and their percentages of the anodized samples using CuKα radiations with scanning speed at a rate of 2°/min.

3. Results and Discussion

The interference of colors are attained on titanium during anodizing at various voltages in diluted sulfuric acid electrolyte. The oxygen is formed on the titanium plates and reacts with titanium to form TiO_2. The thickness of the oxide layer increases with increase in quantity of voltage applied as a function of time. For variation in the voltage, oxide layer thickness increases/ grows but halts when the resistance has reached at a point where the current decomposed to a value at which only few OH- ions are existing to support to continue film growth. Generally, it is known that titanium anodization produces an inert transparent oxide film capable to produce interference of colors. This process is very simple and economical to produce stable oxide with different colors at room temperature within short time (2-3min) [12]. The oxide film has a capability to refract, reflect, and observe the light. The light from the light source falling on to the TiO_2 oxide film is partly reflected and partly transmitted. The portion of transmitted rays attaining the titanium surface is again partially fascinated but mostly reflected to oxide film. A phase shift occurred through this method composed with many reflections. The amount of absorption is reliant on the thickness of oxide layer.

The different thickness of oxide film causes the disparities of luminous flux, reflective index and refractive index, harvests various precise colors of the anodized titanium. Golden yellow color is obtained at 40V-45V; seemed that the color oxide films are appropriate in dental clinic to colorize the titanium dental outlines of dentures. This will enhance the aesthetic of implants by anodization due to golden yellow color symphonic with the color of the teeth. The other colors produced by anodizing may be used for color coding of the abutments to recognize implanted size and type. The obtained results are further discussed in the following Sub-sections.

3.1. Refractive Index

Figure 2 illustrates the spectral reflectance curves of anodized samples at diverse voltages from 10V to 90V, which demonstrates that the reflectance index of titanium anodized at different voltage characterized tiny yellow to dark gray colors. The spectral reflectance of individually anodized samples were measure by using spectrophotometer. The oxide film thickness and the refractive index was calculated from λ_{max} using the Eq (1) and (2).

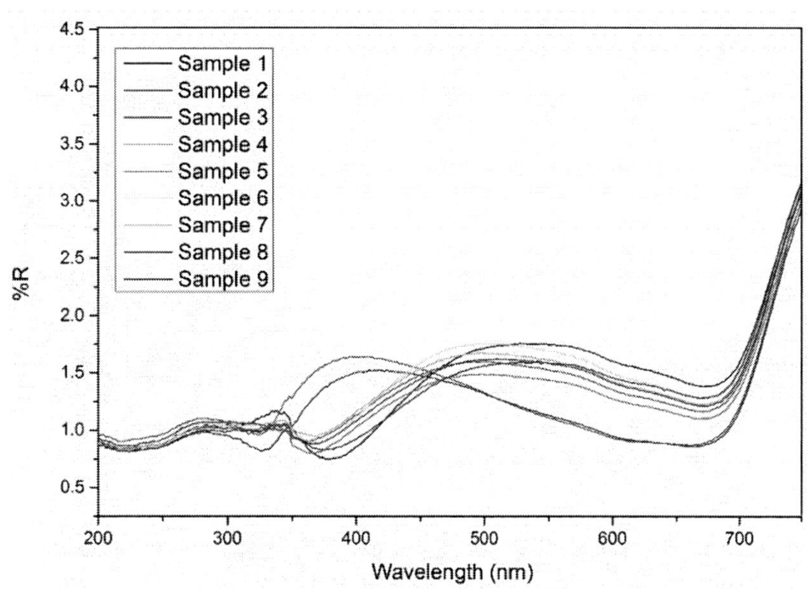

Figure 2. Spectral reflectance versus wavelength for different applied voltage 10V-90V.

Figure 3 illustrates the variations in the refractive index versus the applied anozining voltage. It's observed that, the layer thickness increases with increasing applied voltage, which implies decrease in the refractive index of the oxide film. This may be due to the increasing of anodizing voltage which is responsible for decreasing density and compression of the anodic film [7]. The highest refractive index was obtained for the lowest applied voltage (10V) sample. In principle, the oxide layer thickness will be increases constantly with increased applied voltage for a fixed time. When the Ti start to react with oxygen, the color will be altered, the current density decreases intensely and the film is basically stable. Few researchers have examined the relationship of atmospheric plasma treatment on titanium sample to enhance the biocompatibility. The plasma treatment of titanium improves the cell adhesion and osteogenic differentiation. Atmospheric plasma treatment imparts hydrophilicity to the surface of pure titanium and eliminates impurities from it. The plasma treatment induces adhesion bone marrow cells and protein to the material and trigger the surface tissue formation [13,14]. Based on this, plasma treated plus anodized samples may have high adhesion of cell growth and higher biocompatibility of the titanium material. It may improve the higher degree of tendency of bone contact with the implants and tissue growth.

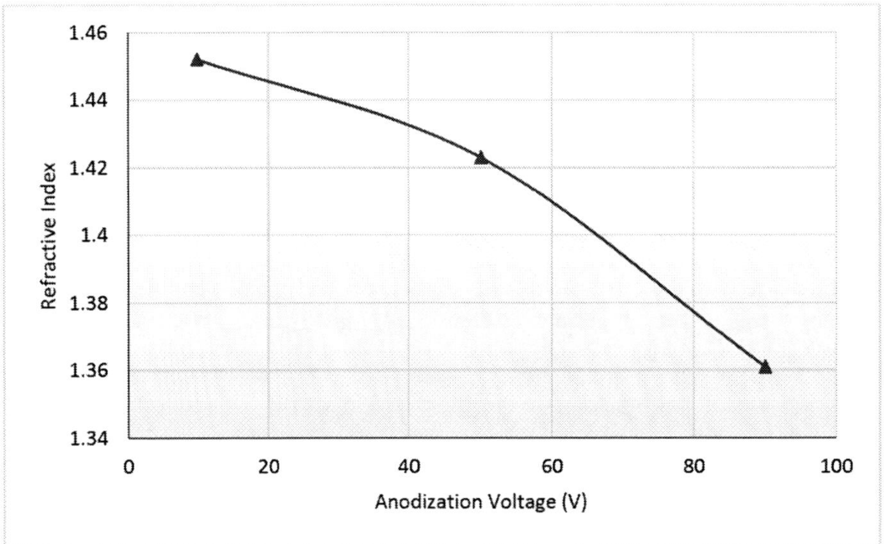

Figure 3. Refractive index of anodized samples.

3.2. Surface Topography
Figure 4 describes the increase in nanosurface roughness with applied anodization voltage. The TiO_2 layer with 3D structures consisting of numerous open pores was made on the surface of the titanium plate during anodization. The rate of oxide layer growth is slow up to 50V, hence less peak and valley were observed in 3D topographical images (Fig. 5). Further, the valleys and peaks are more caused by the formation of volcano shaped oxide layer completely covering surface area of the titanium specimen. It has increased size and number of open pores on the surface of specimen with higher applied voltage, caused for increase in the nanosurface roughness. Atomic force microscope images clearly notices differences among the oxidized samples for varying voltage. It can observed that, sharp peaks and valleys in a range of width, with a roughness Sa of 9nm to 33nm. Due to anodic reaction, the oxide layer becomes thick. This is because of oxygen ion diffusion moved into the titanium interface, a migration of Ti^{4++} ions from the titanium plate to electrolyte interface at the anode. The anodic oxidation is capable to change the surface roughness from smooth to rough by creating pores on the surface of the plate. The sulfuric acid is more effective to produce pores on the titanium plate compared to other acids. The formation of apatite is more in anodized specimen in sulfuric acid [15]. The increase in the apatite formation will enhance the bonding between the implant and bone when inserted into human body [16].

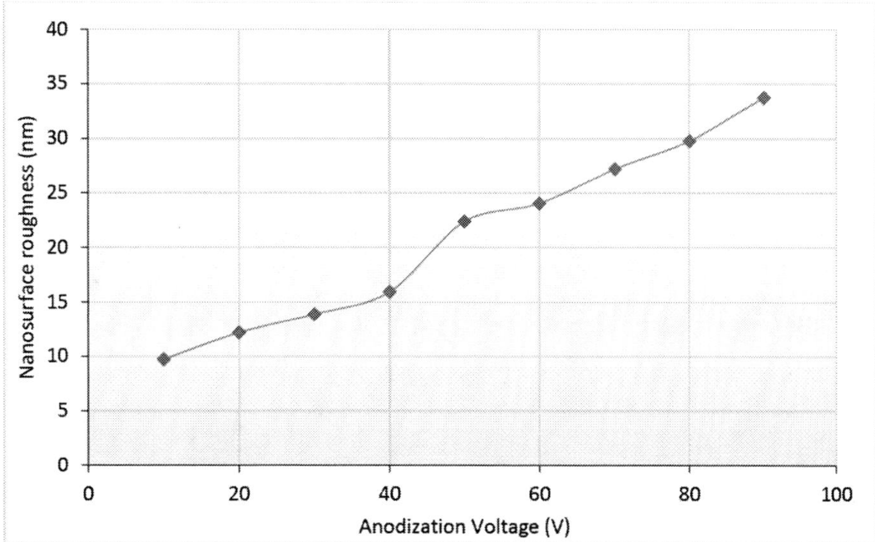

Figure 4. Nanosurface roughness for the applied voltage.

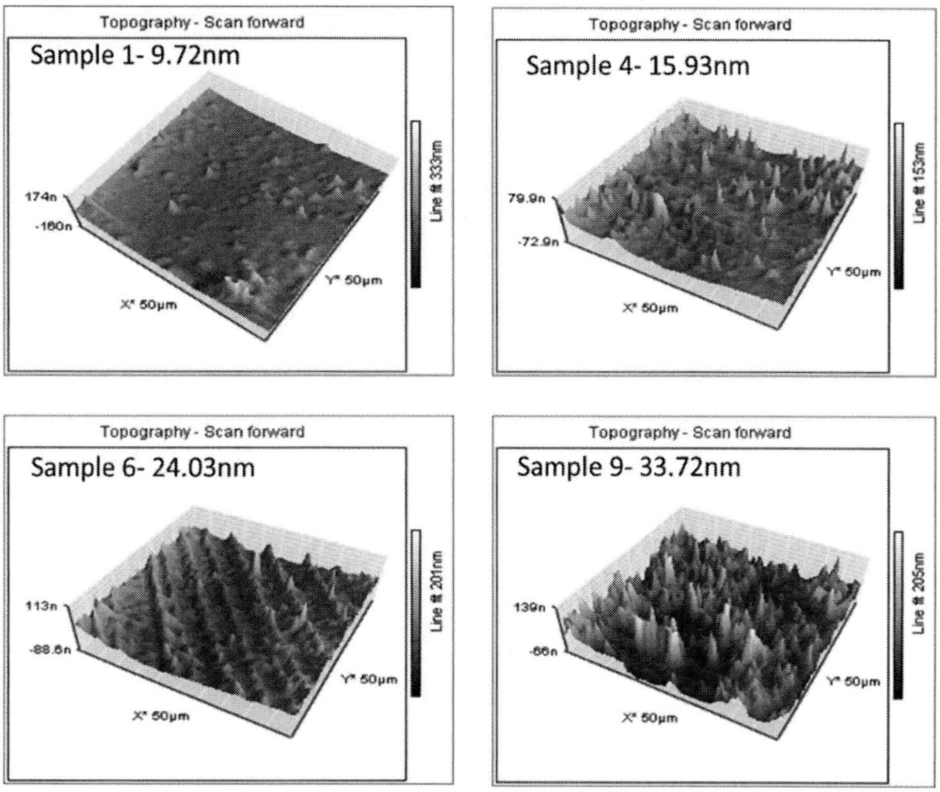

Figure 5. 3D topography examination of the anodized TiO_2 surface at different voltages.

3.3 Phase Analysis

Figure 6 describes the phase pattern of the anodized sample at 80 voltage. The XRD pattern endorses the presence of TiO_2 on the anodized surface. The XRD pattern clearly shows us the presence of anatase and rutile structure on the anodized samples. The rutile peak intensity were relatively higher than that anatase peak. Based on the published results, the structure of the oxide layer formed on titanium plate has characteristically stated to be amorphous at low voltage (<20V) and crystallization takings place at greater voltage (80V) [17] which are dominant phases with biocompatibility. This crystalline structure may be beneficial for bone growth and less healing time. Cells and bacterial growth will be high when compared to the amorphous structure.

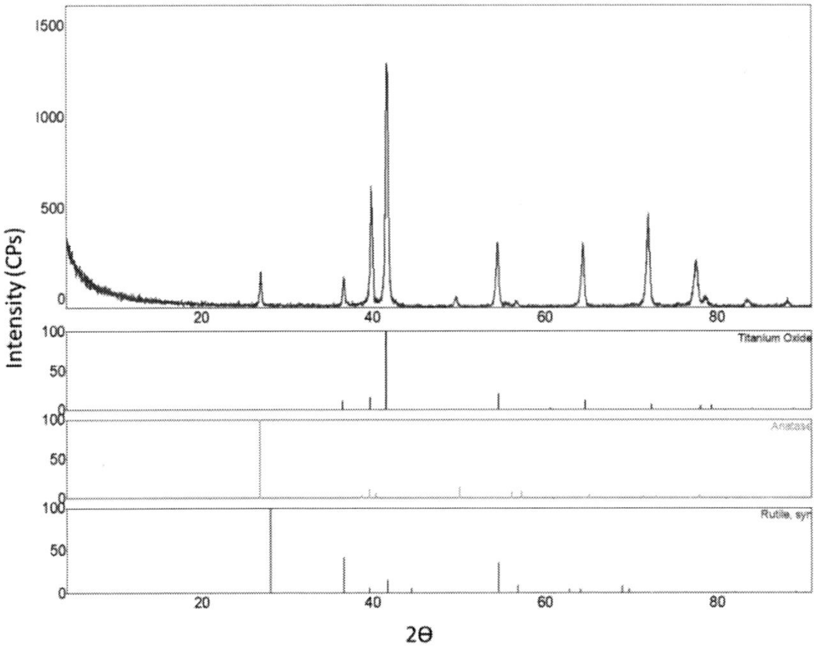

Figure 6. Phase pattern of Anodized Titanium sample at 80V.

4. Conclusions

Anodization, an electrochemical process, was employed for oxidation of a titanium alloy. It concludes that colors of the samples and the anodization voltage can be related to one another. It is found that uniform interference of colors obtained by anodization process. These data may employed for dental implant aesthetic appearance and coding the size of abutments, screw diameter etc. Significant improvement in surface roughness was obtained for an increased voltage. The amount of surface roughness increased for 90V by 71% compared to 10V. Nanosurface roughness is on par with increased oxide layer thickness. The atmospheric plasma treatment and anodization may have significant influence on porous structure on titanium plates. The phase analysis evidently reveals the existence of two different anatase and rutile crystalline phases on the anodized titanium surface.

5. Future scope

The work needs to be continued to evaluate the biological properties of anodized samples for different applied voltages. In addition, the effect of atmospheric plasma treatment on the biocompatibility of titanium will be further investigated.

References

[1] Elias C N, Lima J H C, Valiev R and Meyers M a 2008 Biomedical Applications of Titanium and its Alloys *J. Miner. Met. Mater. Soc.* 46–9

[2] Sista S, Wen C, Hodgson P D and Pande G 2013 Expression of cell adhesion and differentiation related genes in MC3T3 osteoblasts plated on titanium alloys: Role of surface properties *Mater.*

Sci. Eng. C **33** 1573–82
[3] Kent D, Wang G, Yu Z, Ma X and Dargusch M 2011 Strength enhancement of a biomedical titanium alloy through a modified accumulative roll bonding technique *J. Mech. Behav. Biomed. Mater.* **4** 405–16
[4] Dikici T, Erol M, Toparli M and Celik E 2014 Characterization and photocatalytic properties of nanoporous titanium dioxide layer fabricated on pure titanium substrates by the anodic oxidation process *Ceram. Int.* **40** 1587–91
[5] Wei D, Feng W, Du Q, Zhou R, Li B, Wang Y, Zhou Y and Jia D 2015 Titania nanotube/nano-brushite composited bioactive coating with micro/nanotopography on titanium formed by anodic oxidation and hydrothermal treatment *Ceram. Int.* **41** 13115–25
[6] Wally Z, van Grunsven W, Claeyssens F, Goodall R and Reilly G 2015 Porous Titanium for Dental Implant Applications *Metals (Basel).* **5** 1902–20
[7] Kieswetter K, Antonio S, Schwartz Z, Hospital H, Dean D D and Boyan B D 2015 the Role of Implant Surface Characteristics **7** 329–45
[8] Placko, H. E., Mishra, S., Weimer, J. J., & Lucas L C 2000 Surface characterization of titanium-based implant materials *" Int. J. Oral Maxillofac. Implant.* **15.3**
[9] Sato M and Webster T J 2004 Nanobiotechnology: implications for the future of nanotechnology in orthopedic applications. *Expert Rev. Med. Devices* **1** 105–14
[10] Sunny M C and Sharma C P 1991 Titanium-Protein Interaction: Changes with Oxide Layer Thickness *J. Biomater. Appl.* **6** 89–98
[11] Sharma A K 1992 Anodizing titanium for space applications *Thin Solid Films* **208** 48–54
[12] Oliveira N C M, Moura C C G, Zanetta-Barbosa D, Mendonça D B S, Cooper L, Mendonça G and Dechichi P 2013 Effects of titanium surface anodization with CaP incorporation on human osteoblastic response *Mater. Sci. Eng. C* **33** 1958–62
[13] Ujino D, Nishizaki H, Higuchi S, Komasa S and Okazaki J 2019 Effect of plasma treatment of titanium surface on biocompatibility *Appl. Sci.* **9**
[14] Kim I-H, Son J S, Kwon T-Y and Kim K-H 2015 Effect of Atmospheric Plasma Treatment to Titanium Surface on Initial Osteoblast-Like Cell Spreading *J. Nanosci. Nanotechnol.* **15** 134–7
[15] Vermesse E, Mabru C and Arurault L 2013 Surface integrity after pickling and anodization of Ti-6Al-4V titanium alloy *Appl. Surf. Sci.* **285** 629–37
[16] Simka W, Sadkowski A, Warczak M, Iwaniak A, Dercz G, Michalska J and Maclej A 2011 Characterization of passive films formed on titanium during anodic oxidation *Electrochim. Acta* **56** 8962–8
[17] Kalucrossed D Signerović M R, Schreckenbach J P and Graf H L 2014 First titanium dental implants with white surfaces: Preparation and in vitro tests *Dent. Mater.* **30** 759–68

Temperature on the Mechanical, Thermal and Barrier Property of Polyurethane

P I Yeap[1], N Y Yuhana[2], S Fariz[3], and M Z Otoh[3]
[1]Department of Chemical and Process Engineering, Universiti Kebangsaan Malaysia, Bangi, Selangor, Malaysia
[2] Research Centre for Sustainable Process Technology (CESPRO), UniversitiKebangsaan Malaysia, 43000 Bangi, Selangor, Malaysia
[3]Eureka Efektif Sdn Bhd, No. 30, Jalan Perindustrian Suntrack Hub Perindustrian Suntrack Off Jalan P/1A, Seksyen 13, Bandar Baru Bangi, 43650 Bangi, Selangor, Malaysia

Email: yuliana@ukm.edu.my

Abstract. Polyurethane is a type of segmented polymer which is synthesized from polyisocyanate, polyol and chain extender. The reaction between isocyanate and hydroxyl bond is known as polymerization. Two-step synthesis is used in synthesizing polyurethane, where polyisocyanate reacts with polyol to form pre-polymer, follows by the reaction between pre-polymer and diol. In this research, different weight percentages of triethylenediamine catalyst (0.0%, 0.2%, 0.4%, 0.6% and 0.8%) are used. Moreover, curing time and temperature that examined in this research is at a temperature of 70°C for 24 hours (Set A), which is practised in the industry. Curing at shorter time and higher temperature is proposed, which is 70°C for 2 hours followed by 100°C for 3 hours (Set B). Mechanical property and absorption property of polyurethane is studied by Shore A hardness test and water absorption test respectively. Morphology of polyurethane is studied by Scanning Electron Microscopy (SEM). Moreover, density test, Thermogravimetric Analysis (TGA) and Fourier Transform Infrared Spectroscopy (FTIR) analysis are carried out.

1. Introduction

Polyurethane is a type of segmented polymer, where it consists of soft and hard segment alternately [7]. The soft segment consists of polyol that gives chemical resistance whereas hard segment consists of isocyanate that gives hardness and tensile strength to polyurethane [8]. Polyurethane is produced by the chemical reaction between polyol and polyisocyanate in the presence of catalyst, chain extender, filler, colorant and additive. Additional polymerization occurs between the hydroxyl (-OH) and isocyanate group (-NCO) to produce urethane group (-NHCOO-), which is the functional group of polyurethane.

Polyurethane has high tensile strength and high melting point. These make polyurethane even more durable. Polyurethane is a good replacement for plastic as it has degradation resistance towards water, oil and solvent [3]. However, polyurethane is weak in terms of thermal stability and barrier property. Polyurethane is difficult to be melted or recycled as most polyurethane is thermoset polymer [10].

Triethylenediamine, is a tertiary amine that widely used in polyurethane industry. According to Sharmin & Zafar, the catalytic activity of polyurethane is determined by the structure and basicity of amine, in which the reaction increases with increasing of basicity and decreasing of steric hindrance [9]. Triethylenediamine based catalyst is often used to avoid fogging and odour problems [6].

Curing of polyurethane involves the formation of a three-dimensional network through reactions among polyisocyanates and polyols. The curing process starts with linear growth of chain, follows by branching and lastly cross linking [4]. High curing temperature will increase the mobility of molecules and thus promotes the cross linking to occur at higher rate [5].

Density is an important parameter in influencing the properties and performances of polyurethane. Density affects the water absorption property, mechanical property and morphology as well as the thermal stability of polyurethane. Research shows that increasing in density will cause the increase in hardness and decrease in the water absorption property [12]. As reported by Dahlia Z. et al., polyurethane synthesized by using waste paper as filler has higher hardness in the presence of triethylenediamine catalyst. This shows that catalyst will produce a more compact structure. Thus increases the hardness of polyurethane [2]. High curing temperature also increases the hardness of polyurethane [1]. Network of cross links will prevent the diffusion of water molecules and thus prevent the water molecules from residing in available volume [11]. Therefore, high curing temperature will decrease the water absorption property of polyurethane. As reported by Anna Andersson, high curing temperature with short curing time is insufficient for isocyanate to react completely. The high percentage of absorption peaks at 2260cm^{-1} indicating that there are lots of isocyanate present in polyurethane [1].

Therefore, the objective of this study is to investigate the mechanical property, absorption property and morphology of polyurethane by varying the weight percentage of triethylenediamine catalyst. Moreover, the properties of polyurethane are also investigated at different curing time and temperature.

2. Methodology

2.1 Materials

In this study, we use methylene diphenyl diisocyanate (MDI), polyester polyol (Polyol), 1,4-butanediol (BDO), triethylenediamine (TEDA) and polydimethylsiloxane-polyol (AAA). All of the materials were supplied by Covestro, France.

2.2 Sample Preparation

The raw materials were prepared according to the mass as shown in Table 1 below by using weighing device PB4002-S. MDI was prepared and heated to 40°C by heating plate MS-H280-Pro. Then polyol, AAA and TEDA were added accordingly into the container. Mixture was stirred at 195rpm for 15minutes (off heating). Lastly, BDO was added and stirred at 195rpm for 3minutes. Polyurethane is then cured at 70°C for 24hours in oven. Different weight percentages of catalyst are repeated. The samples prepared are denoted as set A. Another set of polyurethane is prepared by using different weight percentage of catalyst and being cured at 70°C for 2hours followed by 100°C for 3hours. These samples are denoted as set B. The sample codes used in this study are shown in Table 2.

Table 1. Composition of Raw Materials

Raw Material	Weight Percentage (%) per 100g
Methylene Diphenyl Diisocyanate (MDI)	38.34
Polyester Polyol (Polyol)	57.52
1,4-butanediol (BDO)	4.14
Polydimethylsiloxane-Polyol (AAA)	0.30
Triethylenediamine (TEDA)	0.00

Table 2. Sample Coding and Catalyst Amount in Polyurethane

Sample	Curing Condition	Weight Percentage (%)
PU0.0A	A	0.0
PU0.2A	A	0.2
PU0.4A	A	0.4
PU0.6A	A	0.6
PU0.8A	A	0.8
PU0.0B	B	0.0
PU0.2B	B	0.2
PU0.4B	B	0.4
PU0.6B	B	0.6
PU0.8B	B	0.8

Curing Condition of A = 70°C (24hours), B = 70°C (2hours) 100°C (3hours)

2.3 Characterization

All the samples were prepared then sent to the following characterization:-

- Hardness

Testing was done by using Durometer with accordance to standard ASTM D2240, where the sample size is 50mm x 75mm x 6mm. The sample was placed on a flat hard surface. The indenter was pressed into the sample and Shore A hardness value was recorded within one second.

- Water Absorption

Testing was done with accordance to standard ASTM D570, where the sample size is 20mm x 20mm x 6mm. Firstly, sample was dried in oven at 50°C for 24hours. The sample was weighted by using weighing scale HR-250AZ to obtain original weight before immersed in water. Then, the sample was immersed in distilled water at 24°C for 24hours. The sample was removed and dried by using cloth. Finally, the sample was weighted. Placed back the sample and repeated the process of weight measurement for one week. Percentage of water absorption was calculated according to equation (1).

$$\text{Percentage of Water Absorption (\%)} = \frac{W_t - W_o}{W_o} \times 100\% \quad (1)$$

where W_t and W_0 are the weight at time t and original weight respectively.

- Scanning Electron Microscope (SEM)

Leo 1450VP SEM was used in SEM analysis. Samples were prepared at low temperature by immerging in liquefied nitrogen. All samples were coated with a layer of gold before characterization.

- Density Meter

Density of sample is tested by using density meter MH-300A, where the sample size is 50mm x 25mm x 6mm. Make sure the sample and density meter were free from dust. Distilled water was filled into the container of density meter. The sample density was calculated by using the equation (2).

$$\text{Density} \left(\frac{kg}{m^3}\right) = \text{Specific Gravity x Distilled Water Density} \left(\frac{kg}{m^3}\right) \quad (2)$$

- Fourier Transform Infrared Spectroscopy (FTIR)

Measurements were performed at room temperature. Single beam spectra of the samples were obtained after averaging 40 scans between 4000cm^{-1} and 500cm^{-1}.

- Thermogravimetric Analysis (TGA)

Thermogravimetric analysis was carried out at temperature range of 25°C to 600°C, with the heating rate of 10°C/min.

3. Results and Discussion

3.1 Hardness

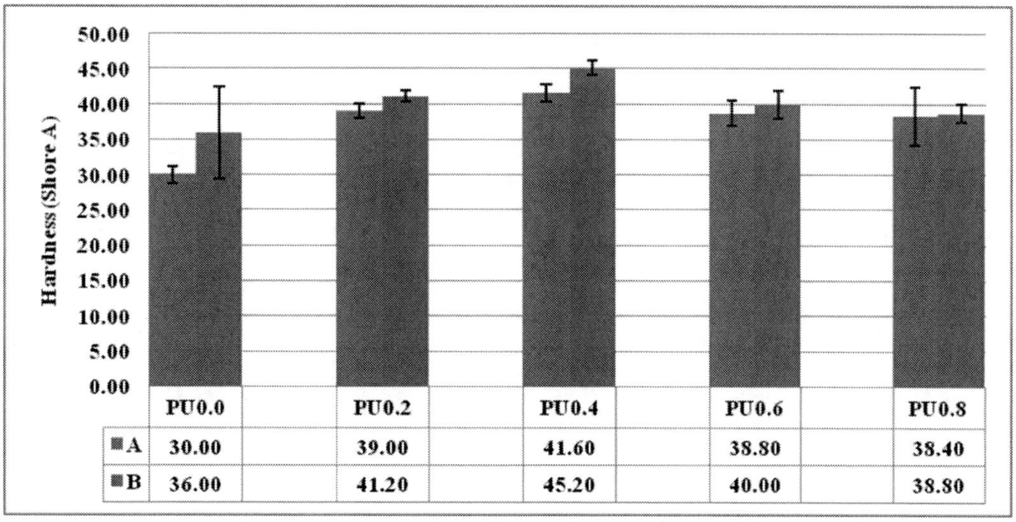

Figure 1. Hardness of Polyurethane Set A and Set B

Polyurethanes with the presence of catalyst have hardness greater than that of polyurethanes without catalyst. Polymerization reaction catalyzed by triethylenediamine catalyst has greater reaction rate, which is then further increased the cross linking level. Higher cross linking creates closer packed structure and increases the hardness. Polyurethanes with 0.4% of catalyst have the highest hardness, where hardness of PU0.4A and PU0.4B are 41.60 Shore A and 45.20 Shore A respectively. Hardness of polyurethane for set A and set B decrease in the order of: PU0.4, PU0.2, PU0.6, PU0.8 and PU0.0. When the catalyst weight percentage achieves optimum value, 0.4%; hardness of polyurethane decreases. High amount of catalyst triggers fast gelling reaction between polyisocyanate and polyol. Therefore, diffusional restriction of curing agent, 1,4-butanediol reduces the extent of conversion. Decrement of cross linking level reduces the polyurethane hardness. Hardness of polyurethane set B is observed to be greater than that of set A. At higher curing temperature, mobility of reactants is increased. This allows densification of urethane segment and follows by increasing in hardness.

3.2 Density

Increasing of density leads to the increasing of cross linking and thus increases the hardness of polyurethane. Polyurethanes PU0.2, PU0.4, PU0.6 and PU0.8 for set A and set B have high density and hardness. However, polyurethane PU0.0A and PU0.0B have the highest density with lowest hardness. This is due to the porous structure that retains water molecules on the surface during density test. High curing temperature increases mobility of reactants and thus increases the cross link and density. Therefore, polyurethane set B has higher density than set A.

IOP Conf. Series: Materials Science and Engineering **943** (2020) 012017 doi:10.1088/1757-899X/943/1/012017

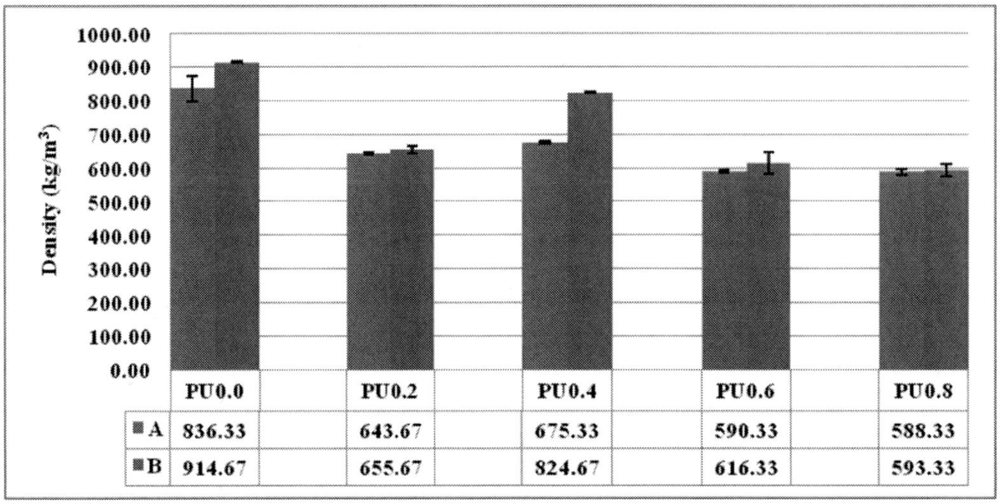

Figure 2. Density of Polyurethane Set A and Set B

3.3 Water Absorption Property

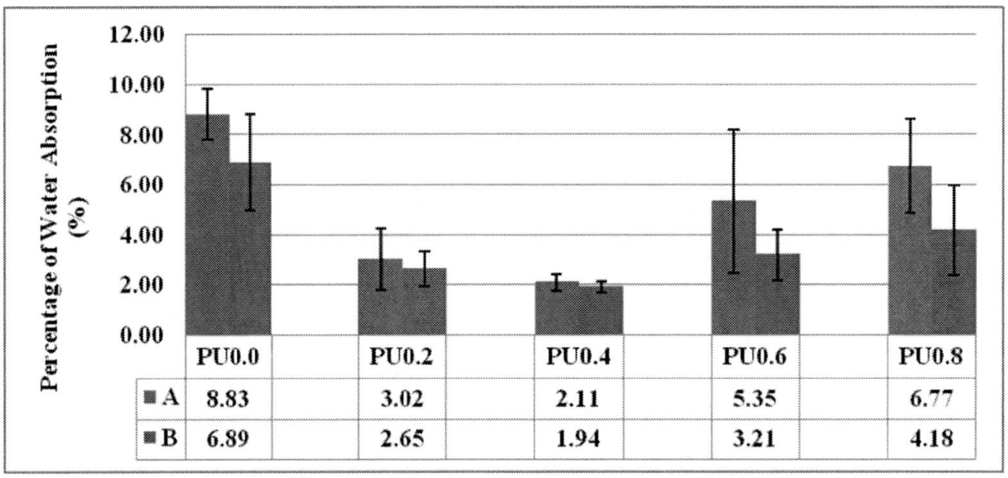

Figure 3. Water Absorption Percentage of Polyurethane Set A and Set B for 24 hours Immersion

The decline in polyurethane density shows the decline in cross linking, where the diffusion of water molecules into polyurethane network will be increased. This is because the cross link will prevent water molecules to penetrate and reside in the polyurethane structure. Therefore, water absorption properties increase with decreasing density. However, polyurethane PU0.0A and PU0.0B have high density with high water absorption property. The high density of PU0.0A and PU0.0B is due to water molecules that remain on porous structure during density test. Polyurethane for set B gives lower water absorption percentage. High curing temperature increases mobility of reactants and thus increases the cross link and density. High cross link level can resist the diffusion of water molecules. Therefore, water absorption property decreases with increasing of density.

3.4 Fourier Transform Infrared Spectroscopy (FTIR) Studies

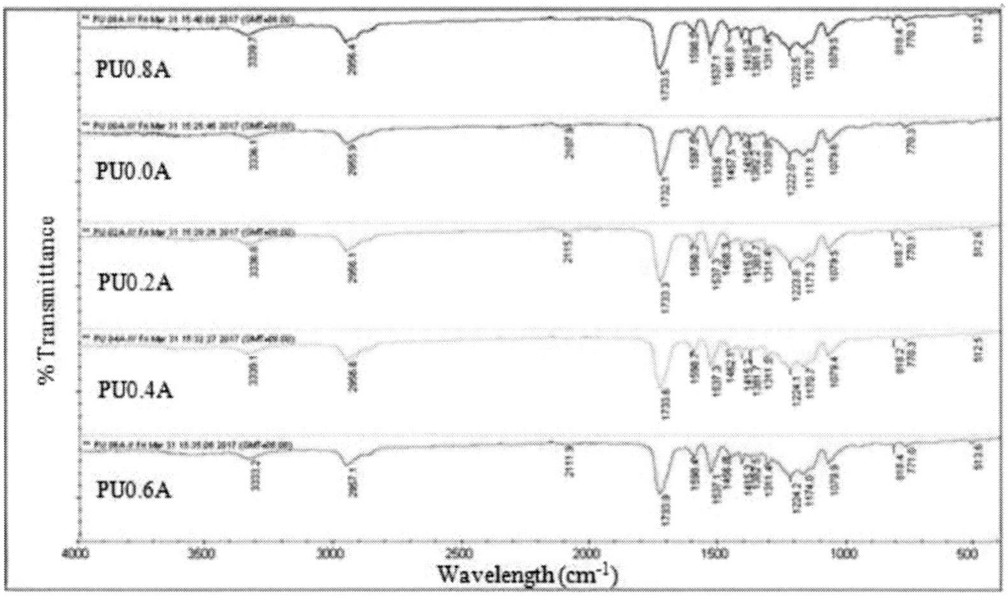

Figure 4 Comparison of FTIR Spectrum for Polyurethane Set A

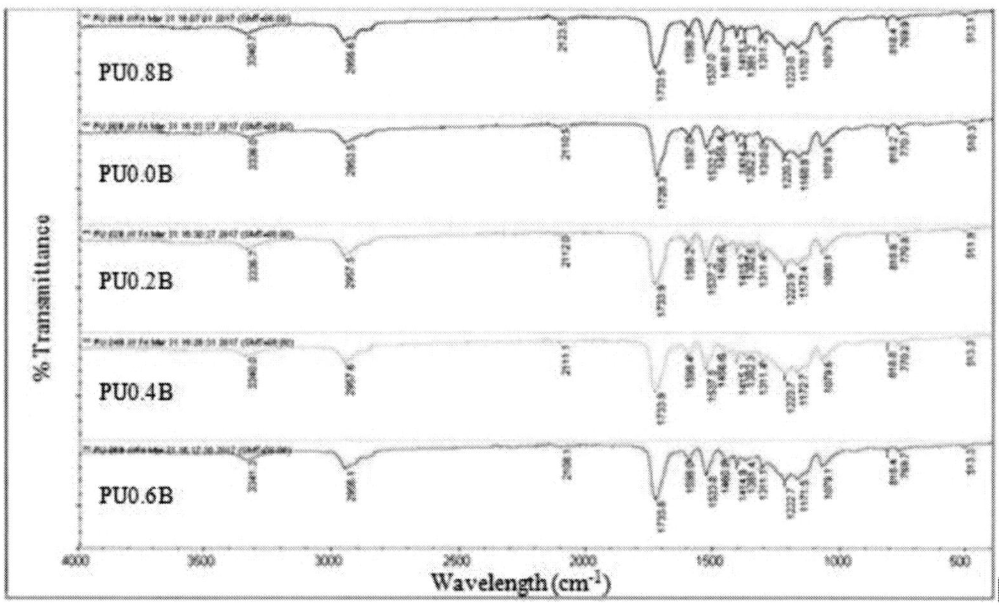

Figure 5 Comparison of FTIR Spectrum for Polyurethane Set B

No isocyanate peak at 2270cm^{-1} was observed in IR spectra for polyurethane set A and set B. This indicates that the isocyante is converted completely into polyurethane and no residues of isocyanate. Peak at 3300cm^{-1} indicates that all N-H group polyurethanes involve in hydrogen bond formation. Peaks at around 1220cm^{-1} and 1079cm^{-1} represent O-C=O bond at soft segment and C-O-C bond at urethane respectively. The strong and sharp absorption peaks at 1220cm^{-1} and 1079cm^{-1} are shown at IR spectra of polyurethane without catalyst (PU0.0A and PU0.0B). This indicates that the urethane bond formed is lesser than that of polyurethane with catalyst.

3.5 Thermal Analysis

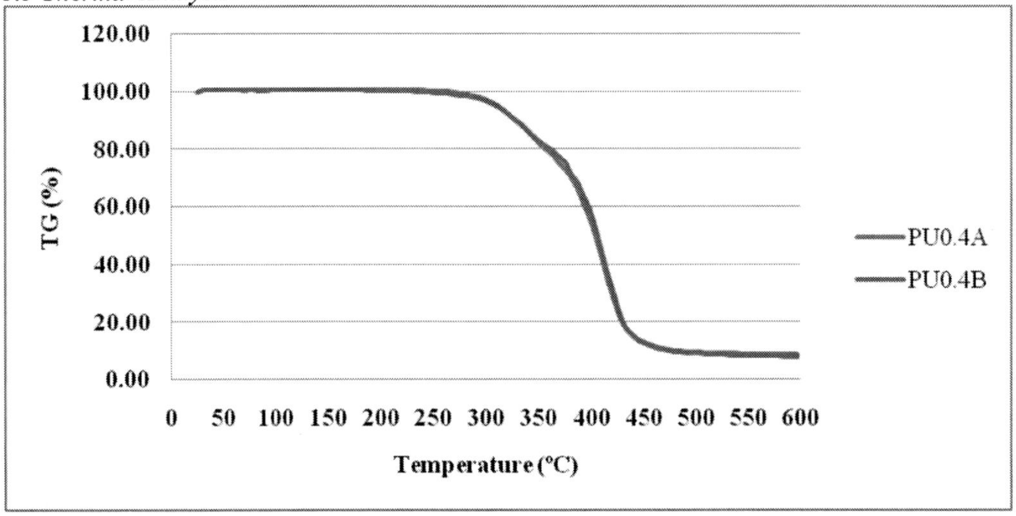

Figure 6. TG Curve for Polyurethane PU0.4A and PU0.4B

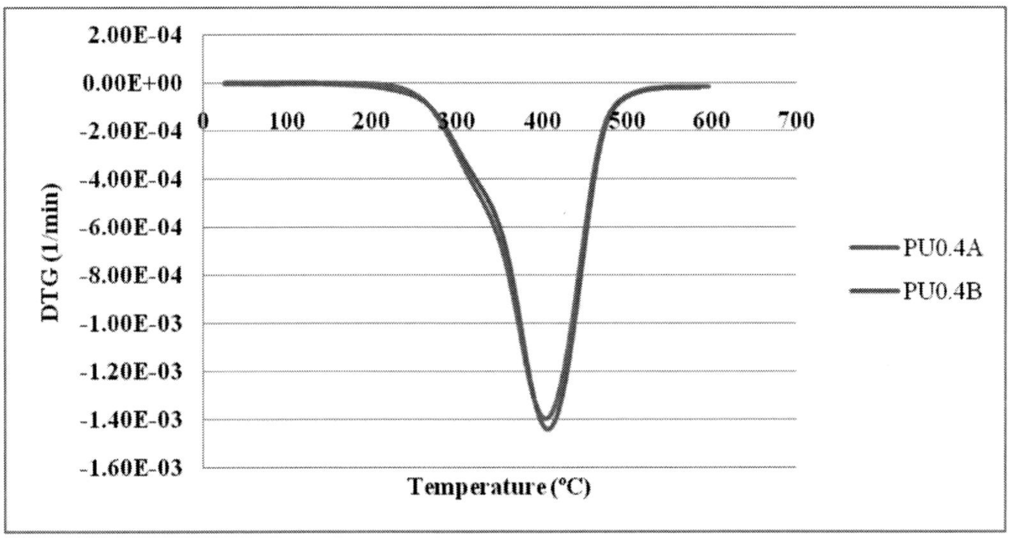

Figure 7. DTG Curve for Polyurethane PU0.4A and PU0.4B

TG curve shows that there are two stages of degradation, at which the weight of polyurethane is decreased. For both polyurethanes, decrement of weight at first stage happened at the range of 210°C and 310°C. This indicates that the degradation of polyurethane starts at urethane bond where C-O bond is broken. At second stage, degradation of urethane bonds lead to the formation of primary amine and olefin or formation of secondary amine and carbon dioxide. C-H and C-N bonds are broken at second stage of degradation [13]. Moreover, TG curve of polyurethane PU0.4A is almost similar with polyurethane PU0.4B. This shows that curing time and temperature does not have significant effect on thermal stability. Polyurethane PU0.4B has greater maximum degradation temperature and degradation onset temperature shows that it has better thermal stability compared to polyurethane PU0.4A. However, only slight differences. DTG curve with only one peak indicates that mixture of pre-polymer reacted completely.

Table 3 TGA Data for Polyurethane PU0.4A and PU0.4B

Sample	Maximum Degradation Temperature (°C)	Degradation Onset Temperature (°C)	Weight at 600°C (%)
PU0.4A	404.16	226.64	8.5826
PU0.4B	404.66	228.95	8.0523

3.6 Morphology

Polyurethane PU0.4A has porous structure which consists of cell. The surface of polyurethane is rough and filled with spherical cell. In comparison, structure of polyurethane PU0.4B is packed structure with little cell. High curing temperature of polyurethane PU0.4B has higher cross linking level due to high mobility of reactants. Therefore, structure of polyurethane is much closer and packed.

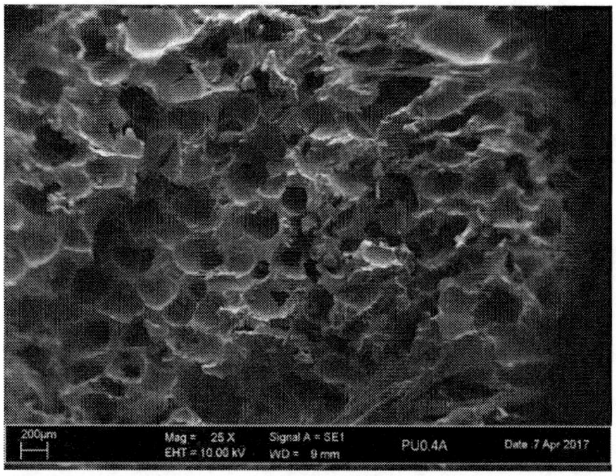

Figure 8. SEM Cross Sectional View of Polyurethane PU0.4A

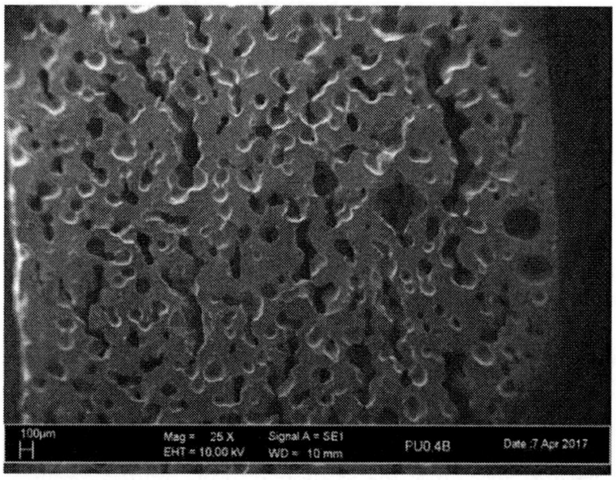

Figure 9. SEM Cross Sectional View of Polyurethane PU0.4B

Figure 10. SEM Surface of Polyurethane PU0.4A

Figure 11. SEM Surface of Polyurethane PU0.4B

Figure 10 and Figure 11 shows that polyurethane PU0.4A and PU0.4B have a uniform matrix with small whitish speckled. Moreover, there are also spherulitic-like structures protruded on the surface of both polyurethanes. These represent the content of hard segment. However, there are no significant differences between polyurethane PU0.4A and PU0.4B.

4. Conclusion

Polyurethane with 0.4% of triethylenediamine catalyst gives the best mechanical property; highest Shore A hardness. It has also the weakest water absorption property. Polyurethanes that are being cured at 70°C for 2 hours followed by 100°C for 3 hours have higher hardness, higher density and poorer water absorption property compared to polyurethanes that are being cured at 70°C for 24 hours. The thermal stability of polyurethane PU0.4B is observed to be slightly better than that of polyurethane PU0.4A. Polyurethane PU0.4B has more closed and packed structure compared to polyurethane PU0.4A. All of the polyurethanes reacted completely to form urethane bond with no residues of isocyanate as shown in IR spectra.

References

[1] Andersson A. 2011. Characterisation of the influence of curing temperature on the properties of 2K waterborne topcoat waterborne topcoat. Sweden.
[2] Dahlia Z., Salmah H. & Azlin O. 2009. The Effect of Triethylene Diamine on the Properties

of Waste Paper Foam Composites. *Journal of Physical Science*, *20*(1), 49–57. Retrieved from http://www.usm.my/jps/20-1-09/JPS 20(1) ART 5(49-57) NEW.pdf

[3] Howard G. T. 2011. *7* . Microbial biodegradation of polyurethane. University Hammond Louisiana.

[4] Li S., Vatanparast R. & Lemmetyinen H. 2000. Cross-linking kinetics and swelling behaviour of aliphatic polyurethane. *Polymer*, *41*(June 1999), 5571– 5576.

[5] Lucio B., Luis J. & Fuente D. 2014. Thermochimica Acta Rheological cure characterization of an advanced functional polyurethane. *Thermochimica Acta*, *596*, 6–13. doi:10.1016/j.tca.2014.09.012

[6] Parasar B., Jing G. W., Yuan D., Kun W., Wang P. & Dasgupta A. (n.d.). Copper Replaces Tin : A Copper based Gelling Catalyst for from Discarded Motherboard 1–24.

[7] Rehman F.-. 2010. Synthesis and Characterization of Speciality Polyurethane Elastomers.

[8] S. Gopalakrishnan and T. Linda Fernando. 2011. Studies of Aging Performance of Some Novel Polyurehanes. *Journal of Chemical and Pharmaceutical Research*, *2*, 848–862.

[9] Sharmin E. & Zafar F. 2012. Polyurethane: An Introduction. *Polyurethane*, 3–16. doi:10.5772/51663

[10] Sivertsen K. 2007. Polymer foams 1–17.

[11] Tan S. G. & Chow W. S. 2011. Thermal properties , curing characteristics and water absorption of soybean oil-based thermoset. *eXPRESS Polymer Letters*, *5*(6), 480–492. doi:10.3144/expresspolymlett.2011.47

[12] Thirumal M., Khastgir D., Singha N. K., Manjunath B. S. & Naik Y. P. 2008. Effect of foam density on the properties of water blown rigid polyurethane foam. Retrieved from http://onlinelibrary.wiley.com/doi/10.1002/app.27712/abstract

[13] Zamzam, M. M. 2005. A Study on the Thermal Degradation Resistance of Thermoplastic Polyurethane Coatings.

IOP Conf. Series: Materials Science and Engineering **943** (2020) 012018 doi:10.1088/1757-899X/943/1/012018

The Mechanical and Thermal Properties of Polyurethanes/Precipitated Calcium Carbonate Composites

Q Y Lai[1], N Y Yuhana[2], S Fariz[3], and M Z Otoh[3]
[1]Department of Chemical and Process Engineering, Universiti Kebangsaan Malaysia, Bangi, Selangor, Malaysia
[2] Research Centre for Sustainable Process Technology (CESPRO), UniversitiKebangsaan Malaysia, 43000 Bangi, Selangor, Malaysia
[3]Eureka Efektif Sdn Bhd, No. 30, Jalan Perindustrian Suntrack Hub Perindustrian Suntrack Off Jalan P/1A, Seksyen 13, Bandar Baru Bangi, 43650 Bangi, Selangor, Malaysia

Email: yuliana@ukm.edu.my

Abstract. The properties of polyurethane containing precipitated calcium carbonate (PCC) were studied. The PCC were produced from acetylene gas waste, and its usage as filler material for polymer were explored in this study. Present findings suggested the addition of 5 per hundred gram of PU(phr) significantly increases the shore A hardness to 39.2, in addition to improve the degradation mperature and specific gravity with decrease in water and oil absorptions capacity. The addition of 5 phr PCC to polyurethane improves the material structure and the morphology as calcium carbonate were dispersed when observed using microscopes. This research extends our knowledge of the addition of PCC to strengthen the mechanical and thermal properties of polyurethanes.

1. Introduction

Polyurethane (PU) presents in large range of mechanical properties [1] and commonly used as pigging in oil and gas industries with deep water flow line coatings and seals protection [2]. PU can be produced through polymerization reaction, where isocyanates react with polyols. The bond between the isocyanate group (-NCO) and hydroxyl group (OH) bond yield urethane (-NHCOO-) [3]. Urethane bond is the functional groups for polyurethane and it usually made either from polyester or polyether. Previous study conducted by Rutkowska. et. al. suggested polyether PU contains higher retention strength when compared to polyester PU [4].

Polyurethane has three dimension (3D) structure which is cross-linked and does not degrade easily by the application of heat. The reaction between two monomers such as formation of cross-linked networks or a huge number of linear prepolymers cross-linking routes can form thermoset polymers. The thermoset polymers have high properties compared to thermoplastic such as thermal stability, hardness and strength [5, 6].

Filler is added to a polymer so that a conjugated biphasic material can form [7]. The forces that applied to the matrix will be changed to another form and showed on the disperse filler phase [8]. Researchers have investigated few types of organic and inorganic fillers that can improve the properties of polyurethane. Among the inorganic materials include: aluminium hydroxide, calcium carbonate, silica, dolomite, talc [9] and titanium dioxide; while most of the organic materials used are natural fibers and carbon black [10]. The increase in density and compression resistance is promoted by filler in PU flexible foam. However, the resilience is reduced and it contributes to a permanent deformation. Plus, fillers influenced the properties of PU such as water absorption, tear strength, oil absorption, the mechanical properties and morphology of PU [11]. Therefore, it is crucial to determine the amount of filler added so results of reliable quality can be obtained [12].

In order to reduce cluster that caused by interactions of particles and high surface energy, the surface modifiers of $CaCO_3$ with stearic acid is used. The adhesion properties wetting which involved two phases are affected [13]. In [14, 15] $CaCO_3$ nanofillers are been used to improve the PU tensile properties. $CaCO_3$ filler particles with micro-sized improve the properties of rheological, cost of composition and thermal stability of thermoplastic PU matrix [16] is reduced.

The adaptability of calcium carbonate as filler in flexible PU foam composite has been observed. Sabina et. al. [17] studied the properties of physico-chemical of flexible PU foam which containing the excess of commercial $CaCO_3$ and commercial $CaCO_3$ used in industry. The results show that the rise of hysteresis can cause deformations permanently and it can also harm the final product quality. Latinwo et. al. [18] investigated the distributions of particle size on the mechanical properties of flexible PU foam and the effect of $CaCO_3$ of different compositions. A study showed that filler which is finely divided has increased the hardness characteristics of the foam to compositions which is up to 35 wt%. Meanwhile, there is no significant improvement in the property of coarse filler. The elongation at break and tensile strength decreased when filler content is increased for all particle sizes.

Sepulcre-Guilabert et al. [19] showed that the thermal, mechanical, rheological, adhesion properties and surface of a TPU are not affected by the addition of natural micronized calcium carbonate filler. Without the net PU and filler interactions, the scarcity of filler influence was formed. It caused by the absence of surface functional groups in the filler and the large size of particles. Jiang et al. [20] investigated the effect fatty acid coating of PCC on the properties of mechanical and rheological of TPU. They demonstated that a fatty acid coating a particular rheological behaviour produced decreased at high frequencies with PCC loading but storage modulus(G_0) increased at low frequencies. The viscosity of the solvent-based TPU solutions is decreased with the addition of PCC. It is caused by the fatty acid coating partially removed from the particle surface of PCC which is being transferred to the TPU matrix.

2. Methodology

2.1 Materials
The prepolymer (MDQ 24163), chain extenders (BAYTEC D24), curative (1- butanediol) and additives (AAA) were obtained from Covestro Elastomers SAS, Romans cedex France. The catalyst that used was triethylenediamine (SD 2.4) and was obtained from Covestro AG, Leverkusen, Germany. The filler that added to the poyurethane is precipiteted calcium carbonate ($CaCO_3$) and can be obtained from Mineral Research Centre, Ipoh, Malaysia.

2.2 Preparation of polyurethane
Prepolymer MDQ 24163 and calcium carbonate are stirred for three minutes. The calcium carbonate weight is varied from 0, 5 , 10, 15 g, 20 and 25 phr . After they are mix, the mixture is heated to 40°C. All the ingredients are added except the curative (BDO) when the temperature reached 40°C. Stir the mixture for 15 minutes at 196 rpm in without heating. When 15 minutes have elapsed, the BDO is added to the mixture and stirred for 3 minutes. The mixture is poured immediately to the mold and then be heated in the oven at 70°C for 24 hours. After 24 hours, the sample is removed from the oven and cut into different size for different test purposes.

Table 1: The composition to produce polyurethane

Raw materials	Percentage in mass (%) For every 100 g of PU
Prepolymer (MDQ 24163)	38.34
Chain Extender (Baytec)	57.52
Curative (BDO)	4.14
Additive (AAA)	0.3
Catalyst (SD 2,4)	0.4

2.3 Testing Method

Fourier Transform Infrared Spectroscopy (FTIR)
FTIR analysis was carried out in FTIR Thermo Fisher Nicolet C700 (Thermo Scientific, USA) to investigate the density of cross linking of the polyurethane. Each of the spectra was recorded as the average of 32 scans taken with a resolution of 4 cm^{-1} and was collected at an operating wave number ranging from 4000 to 425 cm^{-1}. The FTIR analysis was performed to identify the N=C=O, O-H and H-N=C=O bonds.

Optical Microscope
The optical microscope uses lens and light system to magnify small samples' images. The image of the optical microscope can be captured by normal light camera to generate a micrograph. Phenix brand optical microscope with Phenix Micro Image software were used.

Scanning electron microscope (SEM)

SEM analysis was performed to observe the morphology and dispersion of calcium carbonate in the polyurethanes. The scanning electron microscope that used is Leo 1450 VP SEM. Before examination, all the samples were covered with a layer of gold to get rid of the charge of electrons in the sample.

Thermogravimetri analysis (TGA)
TGA is a technique that uses heat to observe the reaction and the physical changes of a material that provides a quantitative measure of any change in the mass of the polyurethane. A Netztech STA 449 F3 Jupiter model was used to study the range of temperature between 25 °C to 600 °C with a heating rate of 10 ° C/min.

Specific gravity (SG)
The type MH 300A density meter with a static support for an immersion vessel above or below the pan was used to determine the SG of polyurethane. The sampel in air was weighed and then immersed in distilled water with the help of sinker. The weight of the specimen was calculated.

$$\text{Specific gravity of specimen} = \frac{a}{a+w} - b \quad (1)$$

where a = sample in air weight, b = sample with sinker in water weight, w = totally immersed sinker weight

Shore A Hardness
The shore A hardness was performed in Eureka Efektif. Durometer type A was used when conducting the experiments. The samples were tested and readings were taken at four different points on the sample.

Water absorption
Water absorption test conducted in accordance with ASTM D570 and the sample size used in the experiment is 10 mm x 10 mm x 5 mm. The equation to calculate the mass of the sample after immersed in water and the results are recorded after 24 hours.

$$M_w = \frac{W_t - W_o}{W_o} \times 100\% \quad (2)$$

Where,
Mt = Percent increase in mass,%, Wt = Mass of sample at time t, g, Wo = original mass of sample, g

Oil absorption
Oil absorption test is conducted and the sample size used in the experiment is 25 mm x 25 mm x 3 mm. The equation to calculate the mass of the sample after 4 days.

$$M_o = \frac{W_t - W_o}{W_o} \times 100\% \quad (3)$$

where Mt = Percent increase in mass,%, Wt = Mass of sample at time t, g, Wo = original mass of sample, g.

3. Results and Discussion

3.1 FTIR

FTIR measurements are performed to prove the bonds in the polyurethane.There is no drastic changes when calcium carbonate is added. Disappearance of peak at 2200-2300 cm^{-1} proved that there is no unreacted isocyanates left.For polyurethane without calcium carbonate, peaks at 3336.6 cm^{-1} corresponds to the NH stretching. Sharp peaks at 2956.1 and 1415.2 cm^{-1} indicate the presence of OH stretching variation. The CH_2 bend are identified at the height of 1456.1, 1382.4 and 1311.4 cm^{-1}. In addition, the peak at 1734.1 cm^{-1} is associated with the group C=O in polyurethane. NH bend is identified at the peak of 1598.1 and 1536.9 cm^{-1}. The peak 1224.0 and 1174.2 cm^{-1} indicate C-O-C stretch while the peak 1080.1 cm^{-1} indicates C-N. In addition, the peak of 818.5 and 770.8 cm^{-1} are related to the CH bend.

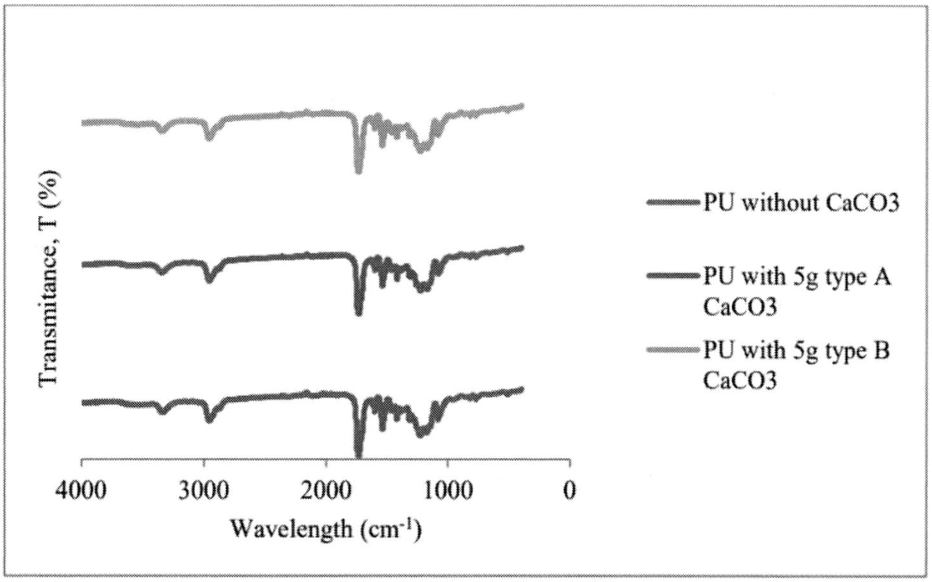

Figure 1. FTIR spectra of pure polyurethane and polyurethane with 5g of type A and type B calcium carbonate

3.2 Morphological Analysis

The SEM images in Figures 2(a-b) showed the image of two different types of the calcium carbonate used at magnifications of 500 ×. Figure 3a illustrated the type A calcium carbonate existed in round and circular forms whereas Figure 2(b) type B calcium carbonate structure looked like needle structure. It was proved that calcium carbonate did not evenly distribute in the liquid PU mixture, and the existence of calcium carbonate agglomeration were observed. The optical microscope images of liquid PU were showed in Figure 3(a). The image showed granules of solid PU which indicated curing process took place and clear liquid phase indicated the unreacted mixtures. The mixture of viscous mixtures of PU and PCC morphology were shown in Figures 3(b-c).

The SEM images of cured neat PU and composite samples were shown in Figure 4. The cross section images indicated the presence of micropores in the neat and composite samples.

Figure 2. SEM of (a) type A calcium carbonate (b) type B calcium carbonate under magnification 500x.

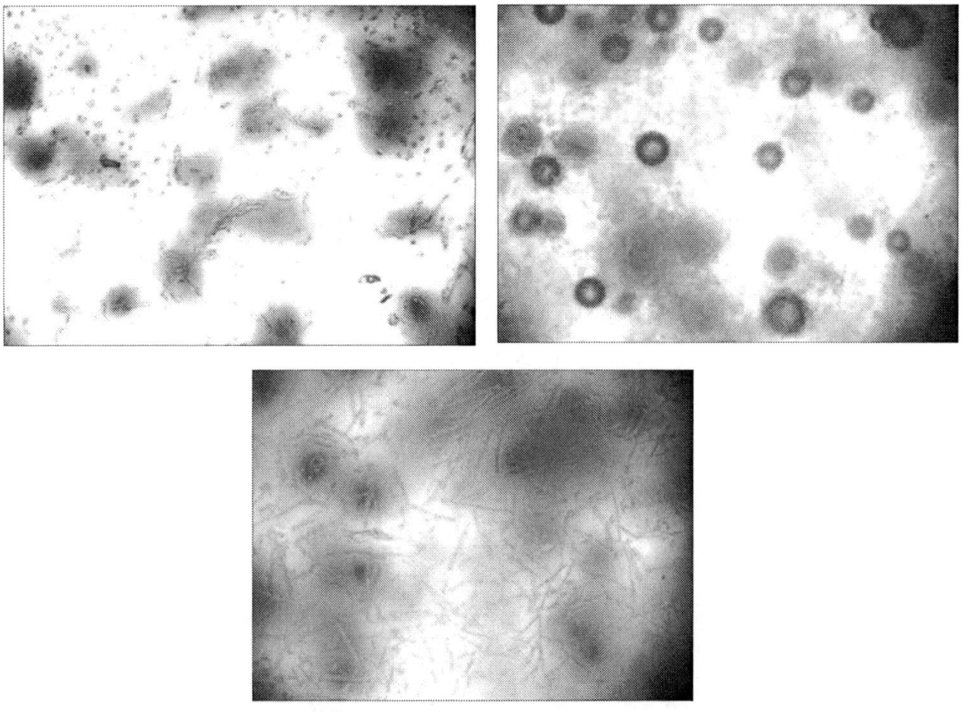

Figure 3. Images of optical microscope of (top, left) PU without $CaCO_3$ (top, right) PU with 5g type A $CaCO_3$ (bottom) PU with 5g type B $CaCO_3$ under magnifications 10x.

IOP Conf. Series: Materials Science and Engineering **943** (2020) 012018 doi:10.1088/1757-899X/943/1/012018

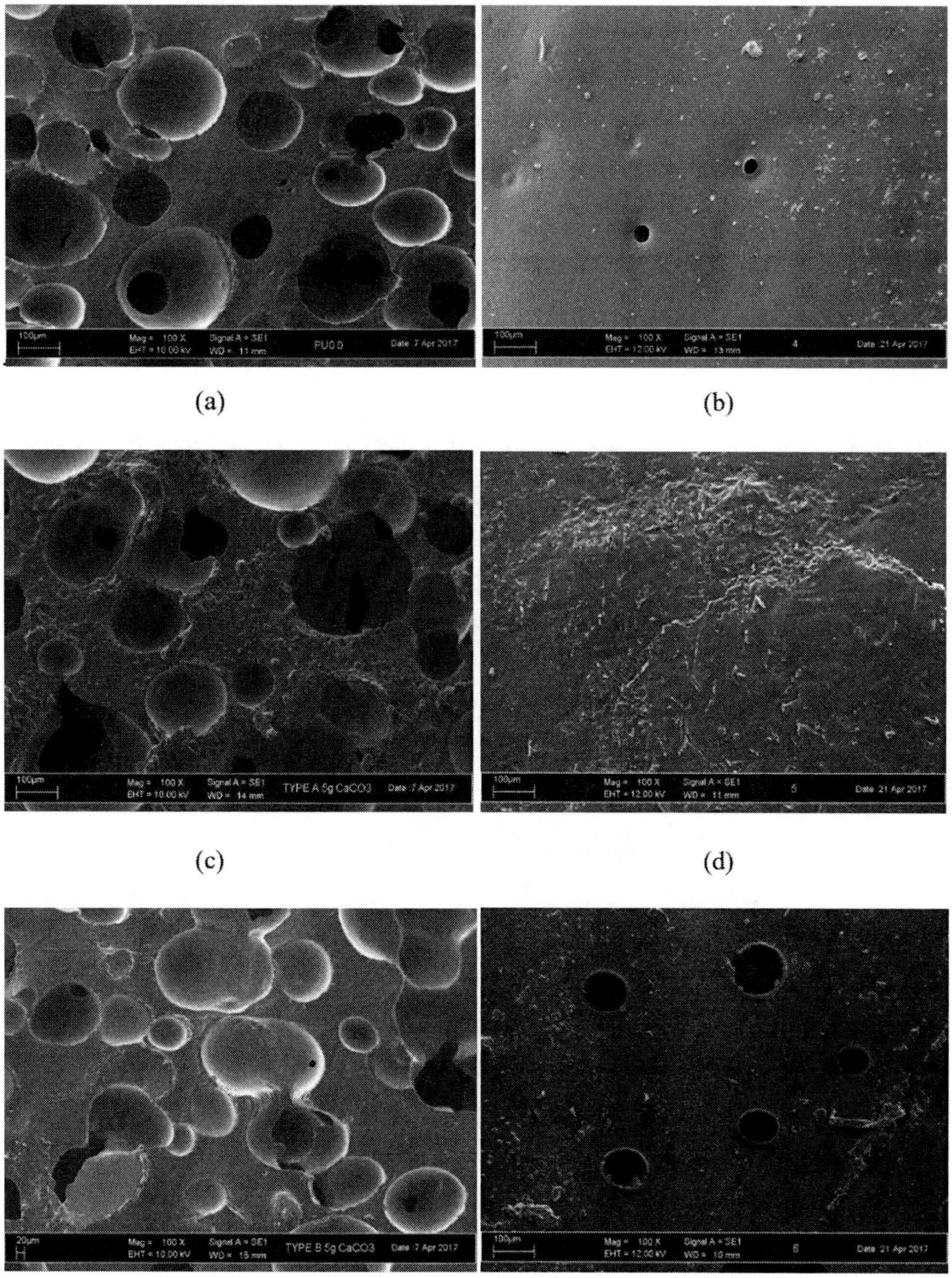

(a) (b)

(c) (d)

(e) (f)

Figure 4. SEM images of cross section and surface area (a-b) PU without $CaCO_3$ (c-d) PU with 5 phr type A $CaCO_3$ (e-f) PU with 5 phr type B $CaCO_3$

3.3 Thermal Analysis

Based on the TGA curve, it showed the percentage of mass left for polyurethane without calcium carbonate and polyurethane with 5 g type A and type B calcium carbonate. The polyurethane showed the decomposition temperature between 190° C and 600° C and produced 9:51 % residue for polyurethane without calcium carbonate. The decomposition temperature of

IOP Conf. Series: Materials Science and Engineering **943** (2020) 012018 doi:10.1088/1757-899X/943/1/012018

calcium carbonate was between 680° C and 875° C where carbon dioxide will be lost and calcium oxide was generated. Polyurethane with calcium carbonate had a higher residual mass. Higher values of this sample showed calcium carbonate containing inert compounds which are in correlation with EDS analysis.

The polyurethane showed one stage of decomposition and the maximum temperature for the decomposition was recorded. It was proved that the calcium carbonate was homogenously distributed throughout the polyurethanes samples as the percentage of mass left corresponds to the mass of calcium carbonate added.

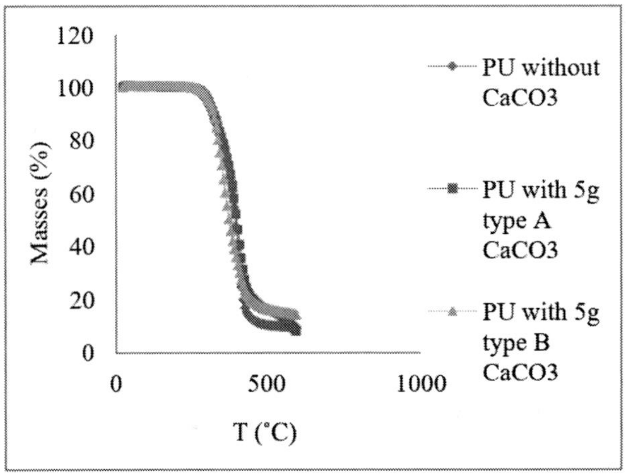

Figure 5. Thermal Gravimetric Analysis of polyurethane without calcium carbonate and polyurethane with 5g type A and 5g type B calcium carbonate

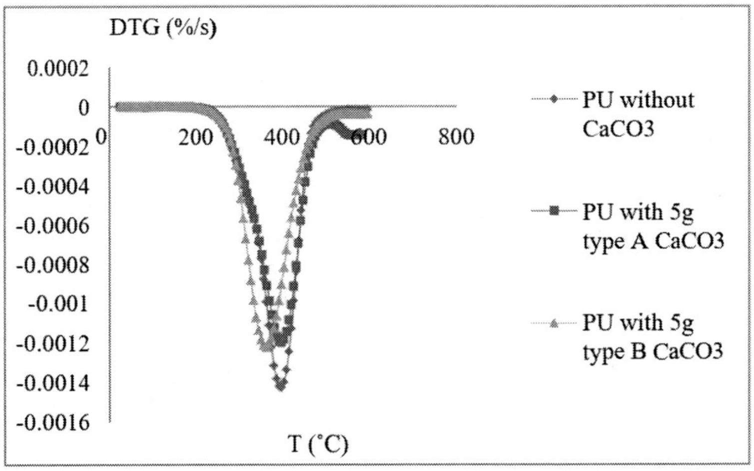

Figure 6. Thermal Gravimetric Analysis of polyurethane without and polyurethane with calcium carbonate

3.4 Specific Gravity

There is an increment for the specific gravity when the mass of calcium carbonate increases. By comparing the density between the two graphs, it was found that the density of the PU with type A calcium carbonate higher than the PU with type B calcium carbonate. It has been verified that the size of type A calcium carbonate bigger compared to type B in SEM analysis. Specific gravity is proportional to the size of calcium carbonate. The higher the specific gravity was due to the formation of bigger number of closed size and the bigger facilitation of calcium carbonate in the polyurethane. The bigger the mass of calcium carbonate, the particles move more efficiently which caused the relative density increases.

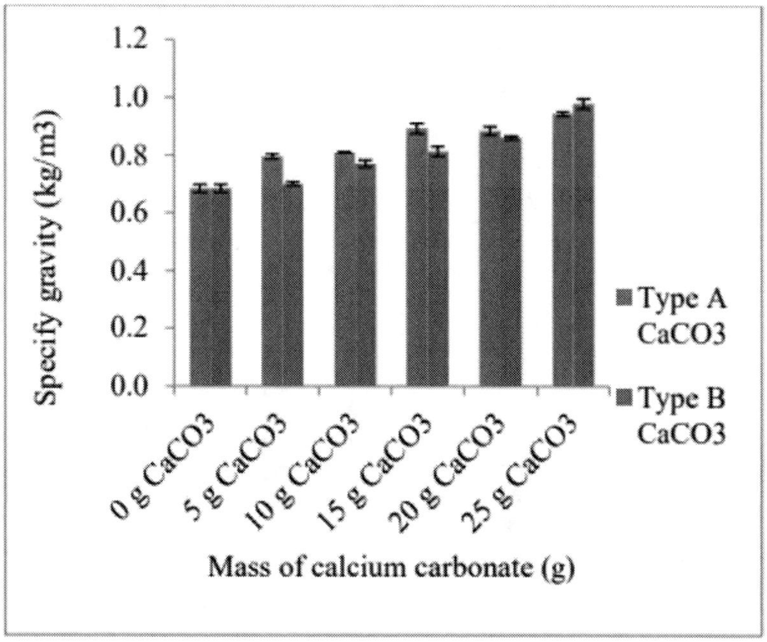

Figure 7. Specific gravity of polyurethane with increasing mass of calcium carbonate

3.5 Hardness

The effect of mass and size of filler on hardness of the polyurethanes were investigated. The polyurethane density played an important role which affects the mechanical properties. When the isocyanate index increases, the hardness increases. The matrix experienced plasticizing when the fillers are treated organically. When the temperature is high for long period of time, the hardness of the polyurethanes increases because over curing happened. At higher filler content, the impact strength and flexural modulus increases. Furthermore, incomplete curing and high closed cell caused the hardness decreases [22].

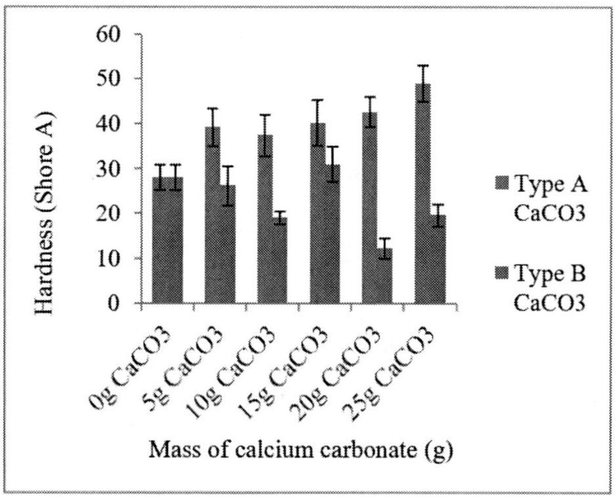

Figure 8. Hardness of polyurethane with increasing mass of calcium carbonate

3.6 Water absorption

Water absorption increases when the mass of calcium carbonate increases but decreases when achieve the optimum mass. The percentage of water volume were obtained after 24 hours. Less increment in water absorption volume is due to the number of closed cell. When the mass of calcium carbonate increases, the cell contents decreases which cause the water absorption decreases.

Saijun *et al.* conducted a study and showed that polymeric systems that high crosslink density will have lower water absorption [23]. When equilibrium stage achieved, the water absorption of polyurethane reached the saturation where all the 3- dimensional crosslinked network structure are filled with water molecules [24].

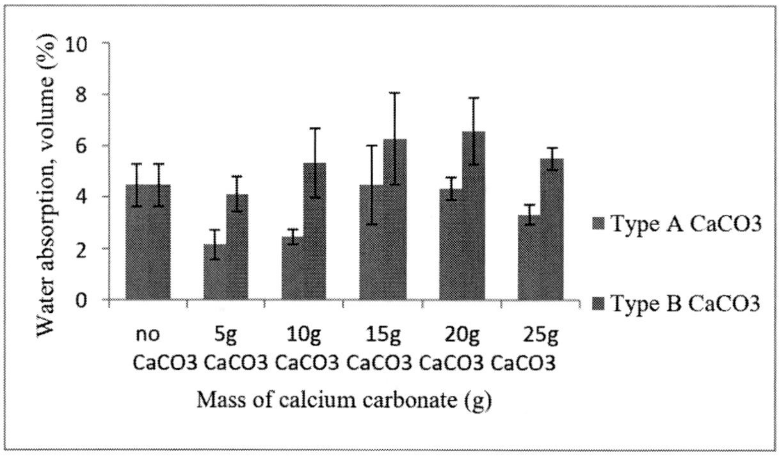

Figure 9. Water absorption of polyurethane with increasing mass of calcium carbonate

3.7 Oil absorption

Oil absorption increases when the mass of calcium carbonate increases but decreases when achieve the optimum mass. The percentage of oil absorption was measured for four consecutive day. Less increment in water absorption volume is due to the number of closed cell. When the mass of calcium carbonate increases, the cell contents decreases which cause the oil absorption decreases.

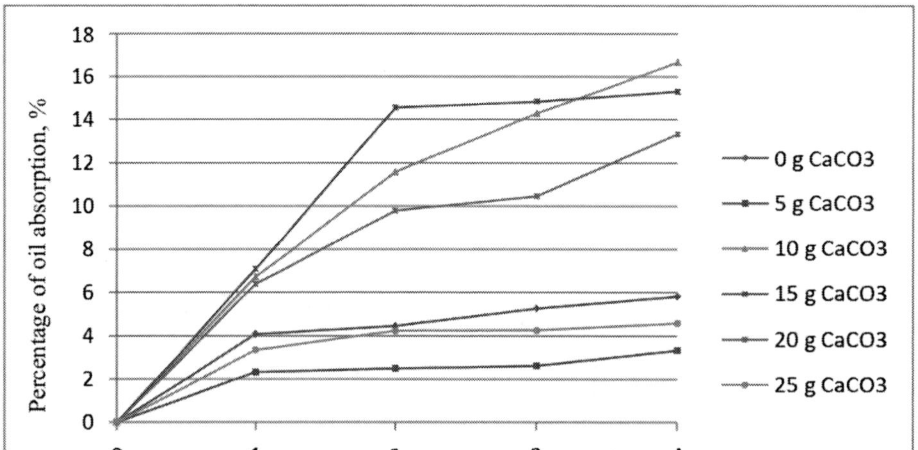

Figure 10. Oil absorption of polyurethane with type A calcium carbonate

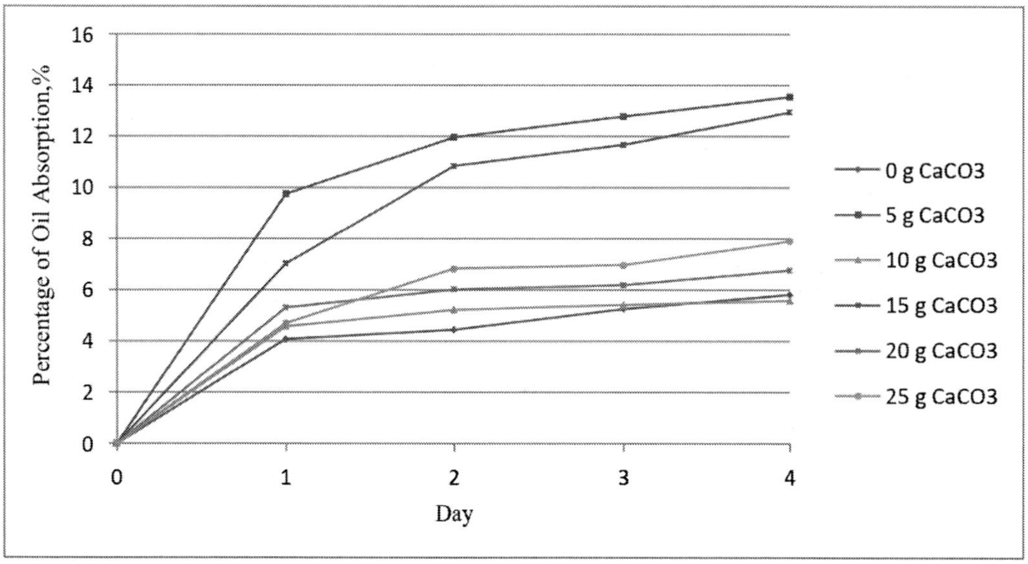

Figure 11. Oil absorption of polyurethane with type B calcium carbonate

4. Conclusion

The FTIR spectra did not show significant changes in the density of cross linking. Polyurethane with different size and mass of calcium carbonate have been produced. The morphology of calcium carbonate and polyurethane with and without calcium carbonate were obtained by using optical microscope and SEM. The result of TGA experiments showed that addition of filler increases the thermal stability of the polyurethanes. The results showed that specific gravity and hardness of the polyurethanes increases when the filler contents increases. The water and oil absorption showed increasing trend but decreased when achieved a saturated state.

Acknowledgements

The authors would like to acknowledge Universiti Kebangsaan Malaysia for the GUP- 2017-041 grant for conducting related research.

References

[1] Hepburn C. 1992. Polyurethane elastomers. 2nd ed. *Elsevier Applied Science.*
[2] J.-P. Pascault , H. Sauterea , J. Verdu , 2002. Thermosetting Polymers, *Basel New York* .
[3] Sharmin, E. & Zafar, F. 2012. Polyurethane: An Introduction. Polyurethane, 3–16.
[4] Rutkowska, M., Krasowska, K., Heimowska, A., Steinka, I. & Janik, H. 2002. Degradation of polyurethanes in sea water. *Polymer Degradation and Stability*, 76(2), 233–239.
[5] Akshay Chavan. 2016. The Plastic Wars: Difference Between Thermoplastics and Thermosets. Buzzle. http://www.buzzle.com/articles/difference-between- thermoplastics-and-thermosets.html. [25 November 2016]
[6] Randall, David; Lee, Steve. 2002.. The Polyurethanes Book. New York: *Wiley*. ISBN 0-470-85041-8.
[7] M.A. Usman, S.O. Adeosun, G.O. Osifeso. 2012. Optimum Calcium Carbonate Filler Concentration for Flexible Polyurethane Foam Composite. *Journal of Engineering & Materials Characterisation & Engineering*. Vol. 11, No. 3, pp.311- 390.

[8] Callister WD. 2000. *Materials Science and Engineering*: An Introduction. 5 ed. New York: John Wiley & Sons

[9] M. Rabello, Aditivação de Polímeros, Artliber, São Paulo, 2000.

[10] N. Rattanasom, T. Saowapark, C. Deeprasertkul, Polym. Test. 26(3) (2007) 369.

[11] Vilar W. Química e Tecnologia de Poliuretanos. 2 ed. Rio de Janeiro: Vilar Consultoria; 1998.

[12] International Organization for Standardization. 1986. ISO document 3386-1. Polymeric materials, cellular flexible: Determination of stress-strain characteristics in compression. Part 1, Low-density materials. Geneva.

[13] Armstead, J. C. 1997. Precipitated Calcium Carbonate's Particle Size, Surface Treatment Affect Sealant Rheology Adhesives Age 40 (8): pp. 18 – 22.

[14] Gao, X., Zhou, B, Guo, Y., Zhu, Y., Chen, X., Zheng, Y., Gao, W., Ma, X., Wang, Z. 2010. Synthesis and Characterization of Well-dispersed Polyurethane/$CaCO_3$ Nanocomposites Colloids and Surfaces A: *Physicochemical and Engineering Aspects*. 371 (1 – 3): pp. 1 – 7.

[15] Gao, X., Zhu, Y., Zhou, S., Gao, W., Wang, Z., Zhou, B. 2011. Preparation and Characterization of Well-dispersed Waterborne Polyurethane/$CaCO_3$ Nanocomposites Colloids and Surfaces A: *Physicochemical and Engineering Aspects* 377 (1 – 3): pp. 312 – 317.

[16] Donate-Robles, J., Martín-Martínez, J. M. 2011. Addition of Precipitated Calcium Carbonate Filler to Thermoplastic Polyurethane Adhesives. *International Journal of Adhesion & Adhesives*. 45 (8) 2011: pp. 795 – 804.

[17] Sabrina Sá e Sant'Annaa, Denilson Arlindo de Souzaa, Danielle Marques de Araujoa, Cornélio de Freitas Carvalhob, Maria Irene Yoshida, 2008. Physico- chemical Analysis of Flexible Polyurethane Foams Containing Commercial Calcium Carbonate. *Material Research*. 11(4). 433 – 438.

[18] Latinwo, G.K., Aribike, D.S., Oyekunle, L.O., Susu, A. A., Kareem, S.A. 2010. Effects of Calcium Carbonate of Different Compositions and Particle Size Distribution on the Mechanical Properties of Flexible Polyurethane Foam, *Nature and Science*. 8(9). 92-101.

[19] Sepulcre-Guilabertj, Ferra´ndiz-Go´mez TP, Martı´n-Martı´nez JM. 2001. Properties of polyurethane adhesives containing natural calcium carbonate þfumed silica mixtures. *JAdhesSciTechnol*. 15(2):187–203.

[20] Jiang L,Lam Y C,Tam K C,Li D T,Zhang J. 2006. The influence of fatty acid coating on the rheological and mechanical properties of thermoplastic polyurethane (TPU) / nano-sized precipitated calcium carbonate (NPCC) composites. *Polym Bull*. 57:575–86.

[21] S.S.Sant Anna, D.A. Souza, C. F. Carvalho, M.I. Yoshida. 2008. Morphological and thermal analyses of flexible polurethane foams containing commercial calcium carbonate, Ecl. Quim 33 (2): 55-60.

[22] S Mishra, Arindam Mukherji. 2006. *Journal of Scientific & Industrial Research*. Vol. 65 pp. 582-589.

[23] Saijun D., Nakason C., Kaesaman A., Klinpituksa P. 2009. Water absorption and mechanical properties of water swell able natural rubber. *Songklanakarin Journal of Science and Technology*. 31. 561–565.

[24] Chow W. S. 2007. Water absorption of epoxy/glass fiber/organo-montmorillonite nanocomposites. *Express Polymer Letters*. 1. 104–108.

Finite Element analysis: Predicting cutting force in turning of Inconel 625 using ceramic tools

M M Reddy* and L C S William

Department of Mechanical Engineering, Faculty of Engineering and Science, Curtin University Sarawak Malaysia.

Email: mohan.m@curtin.edu.my

Abstract. Inconel 625 is a unique material due to its characteristic properties of high strength at high temperatures, high corrosion and oxidation resistance. These unique properties thus pose a challenge to machine Inconel 625. Characteristics of work hardening in Inconel 625 generates high cutting force which results in poor surface finish and reduce the performance of cutting tool. Two-dimensional turning operation simulation is considered to evaluate cutting force in this research study. Box- Behnken design of experiment under Response Surface Methodology is used to identify the cutting force with 15 different combinations of cutting speed, feed rate and depth of cut parameters. Predictive mathematical model for cutting force is developed and obtained. From Analysis of Variance (ANOVA), depth of cut had significant influence on cutting force followed by feed rate and cutting speed. Optimization of cutting parameters was carried out and it is recommended that a low combination of depth of cut and feed rate should be selected with a high cutting speed to achieve minimum cutting force. The effect of rake angle on cutting force was investigated for ceramic cutting tools.

1. Introduction

Nickel-based superalloys are a class of material with exceptional combination of high temperature strength, high hardness, high corrosion/oxidation resistance and low thermal conductivity. These characteristics are excellent in the application of aerospace, nuclear power plant, offshore oil & gas, and chemical processing plants. One of the well-known families of nickel-based superalloy is Inconel 625. Inconel 625 is known as a difficult to cut material due to its characteristics that tends to work hardened during machining and high hardness. Machining usually generates high cutting force and high temperature that leads to rapid tool wear, tool failure, and poor surface finish. With suitable selection of cutting parameters (cutting speed, feed rate, depth of cut) may improve Inconel 625 machinability. The current trend of the manufacturing sectors is looking towards advanced ceramic tool in machining superalloys as these ceramics are cheaper and have high temperature resistance. Studies have shown that whisker reinforced ceramic is excellent in high speed machining of nickel- based superalloys due to improved properties such as fracture toughness, thermal shock resistance, and wear resistance [1-3].

Venkatesan et al [4] and Marimuthu & Baskaran [5] optimized cutting parameters in dry turning of Inconel 625 using coated carbide by employing Taguchi's technique design of experiment to achieve optimal cutting force and excellent surface finish. Amini, Fatemi & Atefi [6] investigated the influence of cutting parameters on cutting force and surface roughness in high speed turning of Inconel 718 using ceramic and coated carbide tools. From the results, it was observed that cutting speed had no effect on cutting force for ceramic tools within the range of 150-300 m/min. Depth of cut was found to be most significant on cutting force followed by feed rate. Sonawane, Patil & Pawade [7] conducted experimental investigation of cutting force in high speed dry turning of Inconel 718 using whisker reinforced ceramic tools where all the cutting parameters are significant on cutting force. The authors concluded that cutting speed, feed rate, and depth of cut had significant influence on cutting force components.

In recent years, finite element analysis numerical technique had been used widely in machining process to reduce time and cost for experimental studies as well as to accurately predict machinability factors. Lofti, Jahanbakhsh & Farid [8] developed a 3D finite element modeling using DEFORM 3 software to predict tool wear of whisker reinforced ceramic and coated carbide in turning of Inconel 625. The workpiece Inconel 625 was based on Johnson-Cook constitutive material model. Usui's wear rate model was used to predict the tool wear rate and the simulation results were validated with experimental results which showed good agreement. Samanta, Teli & Singh [9] performed laser- assisted mechanical micro turning of Inconel 625 by finite element modeling simulation. The material model used for Inconel 625 was Johnson-Cook model. Experimental and simulation results were compared and predicted cutting force showed good agreement with experimental values. Sulaiman et al [10] performed finite element analysis simulation by studying the influence of tool rake angle on cutting force and tool temperature in high speed turning of AISI 4340 steel using carbide tools. From their findings, cutting force decreases with the increase of positive rake angle.

In this study, statistical method is utilized to analyze the cutting force results obtained from finite element method simulation. The simulation of turning process is based on Thirdwave AdvantEdge Software. This research aim is to obtain predictive mathematical model for cutting force at 95% confidence. The mathematical model for cutting force is based on cutting speed, feed rate, and depth of cut parameters. Analysis of variance (ANOVA) is conducted to determine the cutting parameters that influence the most to cutting force. Optimization of cutting parameters are also carried out.

2. Methodology

The 2D simulation of turning process is conducted by using Thirdwave AdvantEdge software. Thirdwave AdvantEdge offers the ability to choose constitutive model for workpiece. Constitutive model of workpiece is important to describe its machinability. The workpiece is defined as custom material under user-defined yield stress (UDYS) constitutive model. UDYS is based on Johnson-Cook material constitutive model. Johnson-Cook is described as the following:

$$\sigma = [A + B\epsilon^n]\left[1 + C\ln\frac{\epsilon}{\epsilon_0}\right]\left[1 - \left(\frac{T - T_{room}}{T_{melt} - T_{room}}\right)^m\right] \quad [1]$$

Where σ is the equivalent stress, A is the yield strength, B is the hardening modulus, C is the strain rate sensitivity coefficient, ϵ is the plastic strain equivalent, is the plastic strain rate, ϵ_0 is the reference plastic strain rate, m is the thermal softening coefficient, n is the hardening coefficient, T is the temperature of workpiece, T_{melt} is the workpiece melting temperature, and T_{room} is the room temperature. Inconel 625 Johnson-Cook flow stress model parameters are presented by the following:

The rake angle and relief angle for cutting tool is set fixed at $+6^{\circ}$ and $+6^{\circ}$ and cutting edge radius of 0.04 mm. Rake length of 2 mm and relief length 2 mm. AdvantEdge utilizes Coulomb friction model for friction coefficient between tool-workpiece meshing. The Coulomb friction equation is defined as:

$$F_f = \mu . F_n \quad (2)$$

Where F_f is the frictional force, μ is the friction coefficient, F_n is the normal force. The friction coefficient was set at 0.5.

Table 1. Johnson-Cook Parameters for Inconel 625 [9]

A	B	C	n	m	ϵ_0
400 MPa	1798 MPa	0.031	0.91	1.00	1

Alumina Silicon Carbide also known as Silicon Carbide Whisker reinforced ceramic is chosen as the cutting tool. With the addition of silicon carbide whiskers to alumina, strength and fracture toughness of this tool increase therefore it is widely use in the application as a cutting tool for nickel based superalloys [11]. The cutting tool is set as custom material in the AdvantEdge and defined as a rigid body where it cannot undergo plastic deformation. The cutting tool is based on CC670 whisker reinforced ceramic by Sandvik Coromont. The properties of Inconel 625 and whisker reinforced ceramic are shown in Table 2.

Table 2. Mechanical Properties of Inconel 625 and Whisker Reinforced Ceramic [12, 13]

Material	**Inconel 625**	**Ceramic Tool**
Young's Modulus	205 GPa	390 GPa
Thermal Conductivity	9.8 W/mK	18 W/mK
Heat Capacity	430 J/Kg.°C	750 J/Kg.°C
Density	8440 Kg/m^3	3740 Kg/m^3
Poisson's Ratio	0.308	0.25
Thermal Expansion	1.28 E-5	-
Melting Temperature,	1350 °C	-

2.1 Design of Experiment:

The cutting parameters are in the control level range of minimum and maximum based on the Box-Behnken 3 level 3 factorial design of experiment and a total of 15 combination simulation run was performed. The selection of cutting parameters are based on Sandvik Coromont Handbook and previous studies. The cutting parameters are listed below in Table 3.

Table 3. Cutting parameters at various levels

	CONTROL LEVEL		
CUTTING PARAMETERS	-1	0	+1
Cutting speed, v (m/min)	100	175	225
Feed, f (mm/rev)	0.1	0.15	0.2
Depth of cut, d (mm)	**0.2**	**0.4**	0.6

3. Results and Discussions

The aim of this research is to study the effect of cutting parameters on cutting force and obtain its respective predictive mathematical model. Box-Behnken Design (BBD) design of experiment is used for this statistical parametric study under response surface methodology. BBD three level factorial of selected cutting parameter are inputted into Design-Expert 11.0 software and total of 15 combinations of run are obtained. The cutting parameters consist of cutting speed, depth of cut, and feed rate.

The predictive mathematical model for resultant cutting force is obtained to predict the values of response and graphical correlation of predicted and simulated values. The predictive mathematical model for resultant force is two-factor interaction (2FI) as it is found to be significant. The predictive mathematical regression model derived for resultant force from Design Expert 11.0 is denoted by:

$$\text{Resultant Force} = +440.83 - 6.80\,A + 134.42\,B + 219.32\,C - 3.43\,AB + 0.6950\,AC + 67.58\,BC$$

Where A is cutting speed (m/min), B is feed rate (mm/rev), and C is depth of cut (mm).

Table 4. Simulation Results

Run	Cutting speed, v (m/min)	Feed rate, f (mm/rev)	Depth of cut, d (mm)	Resultant force, Fr (n)
1	225	0.2	0.4	559.92
2	175	0.2	0.6	862.45
3	225	0.1	0.4	299.40
4	175	0.15	0.4	441.59
5	225	0.15	0.6	658.30
6	125	0.15	0.2	227.45
7	175	0.15	0.4	441.59
8	125	0.15	0.6	663.54
9	175	0.2	0.2	287.48
10	175	0.15	0.4	441.59
11	125	0.1	0.4	313.13
12	175	0.1	0.2	152.32
13	125	0.2	0.4	587.36
14	225	0.15	0.2	219.43
15	175	0.1	0.6	456.96

3.1 Analysis of Variance:

ANOVA study is performed to analyze the significance of cutting parameter and check the validity of developed predictive mathematical model equation. The significance of cutting parameter on resultant force can be identified by Fisher's statistical test (F-test) value in ANOVA. It should be noted that for whichever factor parameter to be significant, F-value should be higher than its P-value where P value must be less than 0.05 (or 95% confidence). Table 5 shows the calculated variable for resultant force using Design Expert 11.0 ANOVA.

Table 5. Analysis of Variance

Source	Sum of Squares	DF	Mean Square	F-Value	P-Value
Model	5.481E+05	6	91343.99	6051.74	< 0.0001
A-Cutting Speed	370.33	1	370.33	24.54	0.0011
B-Feed Rate	1.446E+05	1	1.446E+05	9577.46	< 0.0001
C-Depth of Cut	3.848E+05	1	3.848E+05	25494.81	< 0.0001
AB	46.99	1	46.99	3.11	0.1157
AC	1.93	1	1.93	0.1280	0.7298
BC	18269.58	1	18269.58	1210.40	< 0.0001
Residual	120.75	8	15.09		
Lack of Fit	120.75	6	20.13		
Pure Error	0.0000	2	0.0000		
Cor Total	5.482E+05	14			

From ANOVA study in Table 5, the variables sum of square and F-value for the two factor interaction (2FI) model obtained for resultant force is high, the high value indicates that the model is significant. The high F-value (6051.74) of the model implicates that only 0.01% chance it could occur due to noise. The maximized Adjusted R-Squared and Predicted R-Squared value in Table 6 for resultant cutting force model is 0.9996 and 0.9988 respectively which further validates the model. Due to the high degrees of freedom (DF) number compared to the number of simulation run as shown in Table 5, the values of Adjusted R-Squared and

Predicted R-Squared are close 1 (Unity). Both Adjusted R- Squared and Predicted R-Squared is in good agreement with each other at a difference of less than 0.2. Adequate precision is the measurement of signal to noise ratio and the ratio is desirable if it is greater than 4. The adequate precision is 266.5745, meaning that the predictive model is adequate. Based on the individual factor and its interaction parameters in ANOVA, A, B, C, and BC are the significant terms as their F-values are higher and P-values are less than 0.05 respectively. The interaction term of AB and AC has P-value of larger than 0.05, both the model terms are not significant. Depth of cut (C) has the highest value for Sum of squares (384800) and F-value (25494.81). These high values indicate that depth of cut has significant impact on resultant cutting force. Feed rate (B) has the second highest value of Sum of squares (144600) and F-value (9577.46). In other words, depth of cut is the most influential factor affecting resultant cutting force followed by feed rate.

Table 6. Statistical Parameters for resultant force model

Parameters	Resultant Force
Std. Dev.	3.89
Mean	440.83
C.V. %	0.8813
R^2	0.9998
Adjusted R^2	0.9996
Predicted R^2	0.9988
Adeq Precision	266.5745

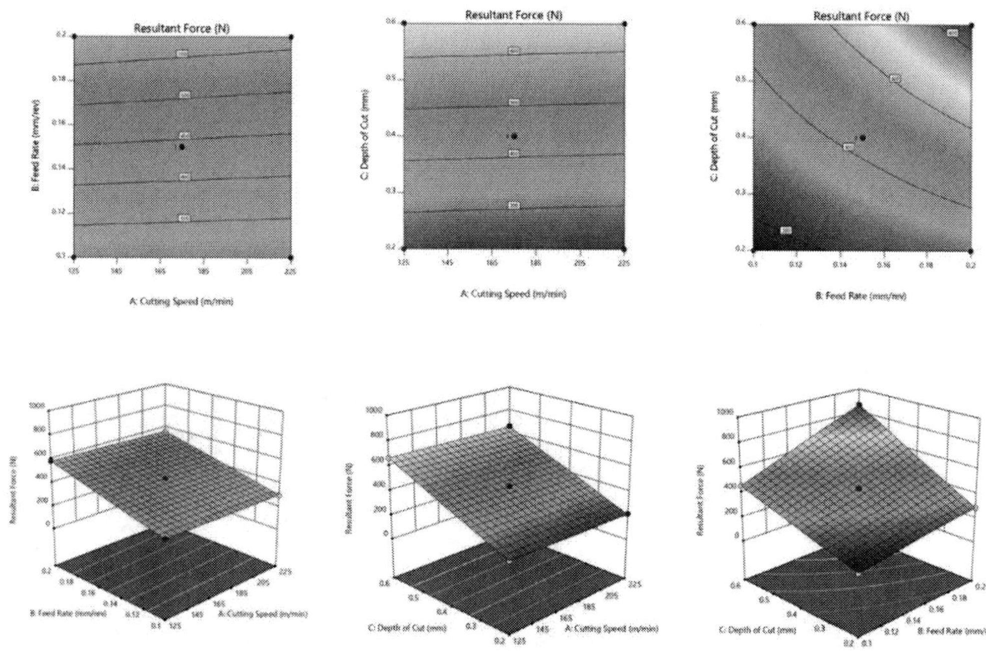

Figure 1. Contour and 3D Surface Response Interaction Plot for Resultant Force

Contour and 3D surface interaction plots model of the combination of different cutting parameters on resultant cutting force response are shown in Figure 1. From the figure, it shows that cutting speed has little influence on resultant cutting force even by increasing the cutting speed. However, resultant cutting force decreases slightly with increase of cutting speed from 125 – 225 m/min. Similar results were obtained by *Amini, Fatemi and Atefi* [6] using ceramic tool in high speed machining of nickel- based super alloy where cutting speed had small impact on the cutting force in the range of 150 – 300 m/min and can be neglected.

Resultant cutting force is increased by increasing the feed rate. Feed rate has significant influence on resultant cutting force and this finding is similar to *Sonawane, Patil and Pawade (2014)* [7]. By increasing feed rate, cutting force increases due to the increment of friction coefficient at tool tip where sheared chip thickness increases in the shear zone and thus requires higher force for chip removal. In other words, increasing feed rate increases the amount of chip being removed and this produces higher cutting force.

The impact of depth of cut and feed rate on resultant cutting force is significant. Increasing depth of cut and feed rate simultaneously generates high resultant cutting force. This is similar to the findings of *Amini, Fatemi and Atefi (2014)* where depth of cut and feed rate has significant influence on cutting force and cutting speed had little effect [6]. Predominantly depth of cut has a higher effect on cutting force as was indicated in ANOVA. This is due of the large volume of material deformation required to shear between workpiece contact length and tool, higher cutting force are exerted to cut the workpiece specific volume.

3.2 Rake Angle effect on Cutting Force:

The investigation of the influence of rake angle on cutting force is studied. The cutting parameters are set fixed at cutting speed of 225 m/min, 0.1 mm/rev, and depth of cut of 0.4 mm.

Table 7. Comparison on the effect of rake angle of ceramic tool on cutting force

Rake Angle	Cutting Force, Fx (N)	Thrust Force, Fy (N)	Resultant Force, Fr (N)
$+6^{\circ}$	271.72	125.74	299.40
0°	286.69	153.93	325.40
-6°	309.39	190.52	363.35

Based on the result data of different rake angle in Table 7, positive rake angle gave a lower resultant force as compared to negative rake angle. The cutting forces of negative rake angle are higher than using positive rake angle in turning. Positive rake angle generates lower cutting force because of easier tool dive into workpiece material as the rake angle produce higher shear angle. With positive rake angle for cutting tool, lower plastic material deformation is required and thus easier flow chip shearing. This result concurred with the investigation conducted by *Sulaiman et al* [10] where positive tool rake angle had positive influence on cutting force.

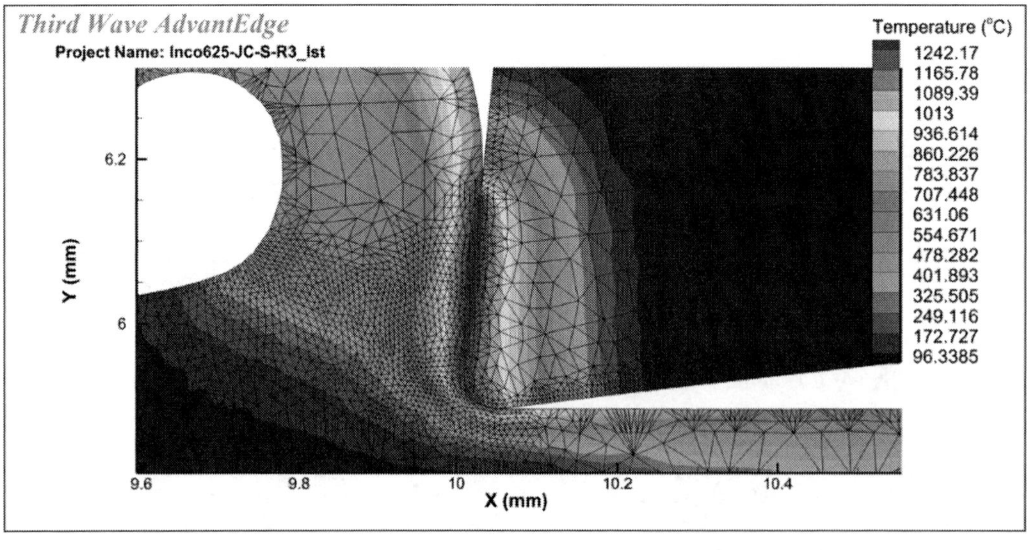

Figure 2. Temperature distribution at tool-chip interface for cutting speed at 225 m/min

The temperature contour plot in Figure 2 is based on simulation number 3 where the cutting speed is 225 m/min with feed rate of 0.1 mm/rev and depth of cut of 0.4 mm. It is observed that the highest temperature is concentrated at the rake face of the cutting tool which is the secondary deformation zone. At this zone, friction slide energy is converted to heat energy during plastic deformation. The simulation was conducted in a dry cutting environment and at high cutting speed of 225 m/min which explains the high cutting temperature between tool-chip interfaces.

3.3 Optimization of Cutting Parameter on Resultant Force:

From the developed predictive mathematical model of resultant force, the model can be used to optimize cutting parameters.

Table 8. Solution with highest desirability

Cutting Speed (m/min)	Feed Rate (mm/rev)	Depth of Cut (mm)	Resultant Force (N)	Desirability	
220.35	0.10	0.20	150.98	0.999	Selected

The optimal cutting force is suggested with cutting speed of 220.35 m/min, feed rate of 0.1 mm/rev, and depth of cut of 0.2 mm. This would help maximize ceramic tool performance in machining Inconel 625.

4 Conclusion

From the finite element analysis simulation performed in Thirdwave AdvantEdge software, cutting force results have been acquired and analyzed in turning of Inconel 625 using ceramic cutting tools. The specific findings from the result are as the following:

1. The predictive mathematical model for resultant cutting force in terms of cutting speed, feed rate and depth of cut has been developed.
2. From Analysis of Variance (ANOVA), depth of cut is found to be the most significant influence on resultant cutting force followed by feed rate, whereas cutting speed shows little influence. The results are only valid in the range of cutting speed 125 – 225 m/min, feed rate 0.1 – 0.2 mm/rev, and depth of cut 0.2 – 0.6 mm.
3. From the investigation of rake angle effect on resultant cutting force, the investigation shows that positive rake angle generates lower cutting force compared to negative rake angle. Positive rake angle had positive influence on resultant cutting force.
4. Based on temperature distribution at tool-chip interface, peak tool temperature is concentrated on the rake face at the secondary shear zone. Highest temperature is located at this zone.
5. From the predictive model, optimization of cutting parameters combination for resultant cutting force is obtained. The recommended cutting parameters to achieve lower cutting force is with high cutting speed of 220.35 m/min, lower feed rate of 0.1 mm/rev and lower depth of cut of 0.2 mm. Tool performance and surface finish of Inconel 625 would be improved.

References

[1] M. A. Xavior, M. Manohar, P. Jeyapandiarajan and P. M. Madhukar, "Tool Wear Assessment During Machining of Inconel 718," *Procedia Engineering,* vol. 174, pp. 1000-1008, 2017.

[2] R. P. Zeilmann, F. Fontanive and R. M. Soares, "Wear mechanisms during dry and wet turning of Inconel 718 with ceramic tools," *The International Journal of Advanced Manufacturing Technology,* vol. 92, no. 5-8, p. 2705–2714, 2017.

[3] A. Altin, M. Nalbant and A. Taskesen, "The effects of cutting speed on tool wear and tool life when machining Inconel 718 with ceramic tools," *Materials & Design,* vol. 28, no. 9, pp. 2518- 2522, 2007.

[4] K. Venkatesan, R. Ramanujam, V. Saxena, N. Chawdhury and V. Choudhary, "Influence of cutting

parameters on dry machining of inconel 625 alloy with coated carbide insert - a statistical approach," *ARPN Journal of Engineering and Applied Sciences 9,* pp. 250-258, 2014.

[5] P. Marimuthu and R. Baskaran, "Optimal Setting of Machining Parameters for Turning Inconel 625 Using Coated Tool," *Applied Mechanics and Materials,* vol. 573, pp. 632-637, 2014.

[6] S. Amini, M. H. Fatemi and R. Atefi, "High Speed Turning of Inconel 718 Using Ceramic and Carbide Cutting Tools," *Arabian Journal for Science and Engineering,* vol. 39, p. 2323–2330, 2014.

[7] G. S. Sonawane, C. V. Patil and R. S. Pawade, "Analysis of Cutting Forces in High Speed Turning of inconel 718 by using Ceramic Tools," *International Conference on Multidisciplinary Research & Practice,* vol. 1, no. VII, pp. 107 - 112, 2014.

[8] M. Lotfi, M. Jahanbakhsh and A. A. Farid, "Wear estimation of ceramic and coated carbide tools in turning of Inconel 625: 3D FE analysis," *Tribology International,* vol. 99, pp. 107-116, 2016.

[9] A. Samanta, M. Teli and R. K. Singh, "Experimental characterization and finite element modeling of the residual stresses in laser-assisted mechanical micromachining of Inconel 625," *Proceedings of the Institution of Mechanical Engineers, Part B: Journal of Engineering Manufacture,* vol. 231, no. 10, pp. 1735-1751, 2015.

[10] S. Sulaiman, M. Ariffin, M. K. Anuar and R. A, "Finite Element Modeling of the effect of tool rake angle on tool temperature and cutting force during high speed machining of AISI 4340 steel," *IOP Conference Series: Materials Science and Engineering* , vol. 50, 2013.

[11] T. Tiegs, "SiC Whisker Reinforced Alumina," in *Handbook of Ceramic Composites*, Springer, Boston, MA, 2005, pp. 307-323.

[12] Precision Castparts Corp , "INCONEL alloy 625," [Online]. Available: http://www.specialmetals.com/assets/smc/documents/alloys/inconel/inconel-alloy-625.pdf. [Accessed 25 August 2018].

[13] M. A. Khan and A. S. Kumar, "Machinability of glass fibre reinforced plastic (GFRP) composite using alumina-based ceramic cutting tools," *Journal of Manufacturing Processes,* vol. 13, no. 1, pp. 67-73, 2011.

IOP Conf. Series: Materials Science and Engineering **943** (2020) 012020 doi:10.1088/1757-899X/943/1/012020

Telerobot: Failure Detection Using Imaging Technique for Piping Internal Surfaces

H.M.A. Hussein[1], Aya A. A. Ramadan[2] and Ravi Kumar Gupta[3,*]

[1]Mechanical Engineering Department, Faculty of Engineering, Helwan University, Cairo, Egypt; Email: hmahuss@hotmail.com
[2] Mechanical Engineering Department, Faculty of Engineering, Helwan University, Cairo, Egypt; Email: ayaabdalla@h-eng.helwan.edu
[3]Mechanical Engineering Department, Manipal University Jaipur, Dehmi Kalan, Near GVKToll Plaza, Jaipur-303007, India; ravikumar.gupta@jaipur.manipal.edu

[*]Corresponding author's e-mail: rkgmuj@gmail.com

Abstract. Pipelines are the energy lifelines of almost every activity of everyday life. It is a safe way of transporting water, gas and oil over long distances. It is also the most practical way to deliver oil and gas across countries as being an extremely safe and economically viable solution to transport energy. Cracking and corrosion on the external and/or internal surfaces of in-service pipes, reduces the integrity of the material and potentially reduces the service life of the transportation system and equipment. Defects in pipelines may have various forms and may be initiated by one or more mechanisms potentially resulting in corrosion and/or cracking. In this paper we propose a system that detects the internal cracks in pipelines through a telerobot which can be inserted into the pipelines and detects the internal Failures (cracks) through a special modern high resolution digital camera and image processing techniques. The telerobot is a programmable and can be controlled as per the requirements such as speed and direction of the telerobot and sensors using mobile application and can be connected to a computer and smart phone. The telerobot captures a high definition images and also captures the precise location. These images are transferred to a computing facility for the analyses and to detect cracks.

Keywords. Fault Detection; Inspection; Pipelines; Telerobot.

1. Introduction

Failure detection using robot and imaging technique for piping internal surfaces were the robot is inserted into pipes to check for failure or damage has many advantages. These robots differ from totally machine-controlled robots in the sense that the advantage of presence of man is taken in situations where it is not possible to expect all the motion and handling requirements in such details as to render them programmable for machine control. Telerobot are human controlled robots that take routine, dangerous or long-duration tasks out of human hands. Most telerobots are multi-functional for example a medical robot has arms to grab, cut, and move skin. Telerobotics is the area of robotics concerned with the control of semi-autonomous robots from a distance, using Wireless network (like Wi-Fi, Bluetooth, and similar).

The technique is designed to remove the human factor from labor intensive or dangerous work and also to act in inaccessible environment. The use of such devices is more common today than ever before and it is no longer exclusively used by the heavy production industries. The inspection of pipes may be relevant for improving security and efficiency in industrial plants. These specific operations as inspection, maintenance, cleaning etc. are expensive, thus the application of the robots appears to be one of the most attractive solutions. Inspection devices are used in many fields of industry. One application is monitoring the inside of the pipes and channels, recognizing and solving problems through the interior of pipes or channels. In October 2013, it has been discussed to presents an autonomous robotic system to perform pipeline inspection for early detection and prevention of leakages in thermal power plants, based on the work developed within the MAINBOT [1] European project. The formally reported work on design and fabrication of in pipe inspection robots is presented by Prasad er al., [2], Yousef et al., [3] and Barane and Mujbaile [4].

In this paper we propose a system that detects the internal cracks of pipelines through a Telerobot which can be inserted into the pipe and detects the internal failures (cracks) through a special modern high resolution digital camera. The Telerobot captures a very clear and high definition images which are analyzed as per the requirements. The remaining paper is organized as, the proposed Telerobot: failure detection using imaging technique for piping internal surfaces long with its design and fabrication is presented in Sec 2; discussion on usages and limitations are presented in Sec 3; final conclusion and further enhances are described in Sec 4.

2. Telerobot: Failure detection Using Imaging Technique for Piping Internal Surfaces

The main objective for developing Telerobot is to provide non-destructive testing inspection solution by inserting the robot into pipes to check for obstructions and damages and distant controlling the same. Controlling mechanisms for in pipe inspection robots are presented in papers [5,6]. A specially desinged rotating probe for the inspection is suggested by Sun et al., [7]. The proposed solution is to inspect the interior of pipelines through a Telerobot which can be inserted into the pipe and detects the internal failures (cracks) through a special modern high resolution digital camera. The Telerobot captures a very clear and high definition images which are analyzed to find the interior information of the pipelines. One application is monitoring the inside of pipes and channels, recognizing and solving problems through the interior of the pipes or channels. In this work, Failure detection using imaging technique for piping internal surfaces with ability to move inside horizontal pipes has been designed and fabricated.

2.1. Design of Telerobot

We identified the need, dissatisfaction, improved existing products, determined specifications and requirements & collected relevant design information feasibility study. Then we designed mechanical parts by using SolidWorks applying stress analysis and mode of failures of each part of the mechanism. After that we designed control system by determining the required applications of the Telerobot which contains DC motors, ultrasonic sensor, temperature sensors & LED lights and we can control it by using specially designed mobile application.

The Telerobot is collapsible in-order to fit in different sizes of pipe lines. Sensors and controlling mechanism used are concealed in the central pipe (body). All legs (Links) dimensions are equal for execution of uniform motion. The angle at the maximum diameter must not exceed 80 degree and not go below (15 -20) degree for minimum diameter for proper functioning as shown in Fig. 1. The wheels of the Telerobot is chosen such that they should be capable of moving without slipping in the vertical direction and should not wear out easily with the use. Stress analysis is performed on all the components in the Telerobot to validate and optimize the material used and dimensions. The feasibility analysis is also performed. The overall dimensions of the Telerobot can be increased or decreased based on the requirements. The challenge is to make this robot adaptable to vary its diameter by adding one sliding mechanisms.

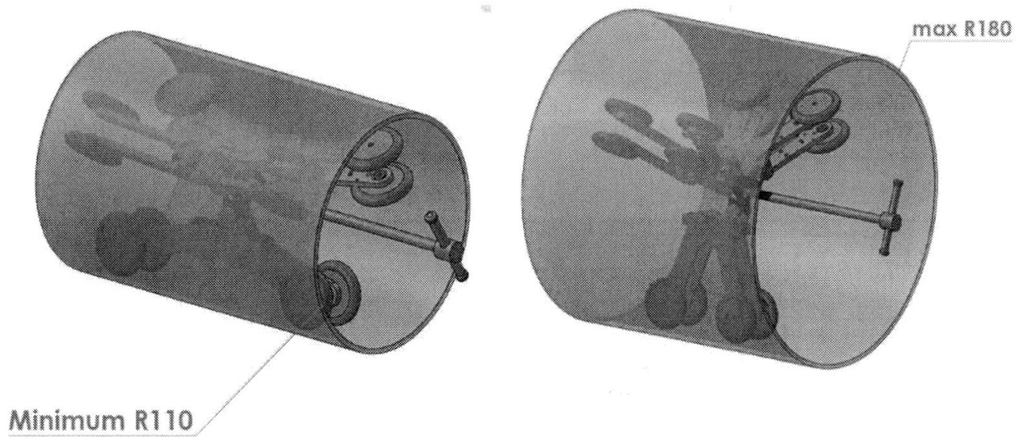

(a) (b)

Figure 1. Pipe line internal radius for Telerobot insoection (a) Minimum radius as 100mm and (b) Maximum radius as 180mm.

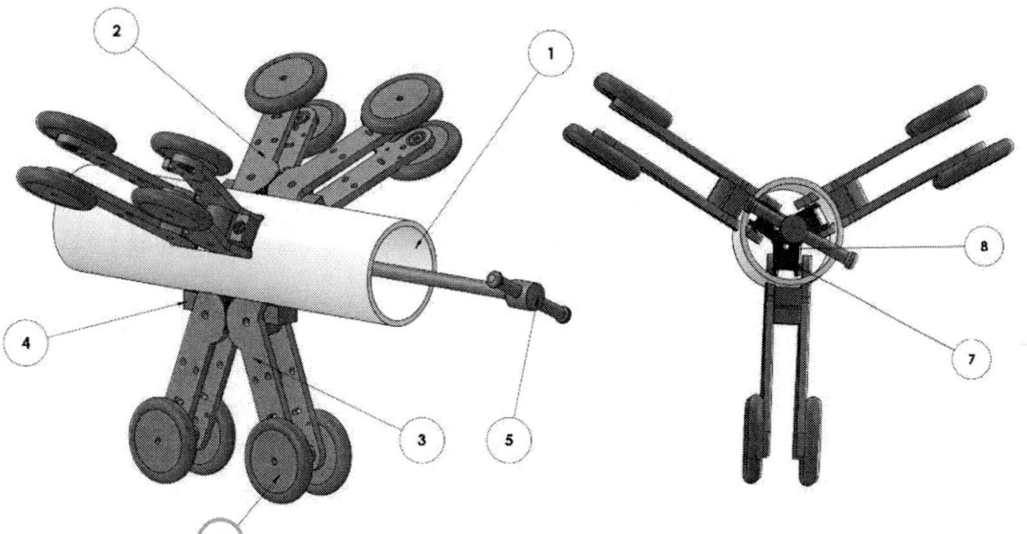

Figure 2. Telerobot design.

The developed Telerobot is shown in Fig. 2. The Telerobot consists of one sliding mechanisms fitted onto a PVC central tube, referred as body. Sliding mechanism has six legs work as a driver and another six legs work as a driven, each two legs fixed together by a part (Collector, component 4 in Fig. 2) between them. To slide, a nut connected to power screw (component 5 in Fig. 2) makes the linear movement possible, if the nut move outside legs are moved away from each other, the telerobot's diameter decreases. Once the nut return to its position or reverse its direction, the telerobot extends automatically. Inside the PVC tube, there is a holder of the electronic components. Each two legs are controlled by two geared DC motors that make them powerful enough to drive the whole mechanism. The components of the Telerobot are presented in Table 1.

Table 1. Components of the Telerobot

Component No. as in Fig. 2	**Name**	**Material**	**Number of components**
1	Body	PVC	1
2	Arm1	Acrylic	6
3	Arm2	Acrylic	6
4	Arm Collector	Acrylic	3
5	Power Screw	Steel 37	1
7	The Core	Acrylic	2
8	Washer	Acrylic	1

The Telerobot is fitted with Mini Panoramic 360 Degree Fisheye WiFi Wireless Camera for capturing images and see through the pipe which is inspected. This camera facilitates discovering the insides obstacles and inner cracks inside the inspected pipe so that these can be rectified.

2.2. Control system of Telerobot

A programmable microcontroller is used in the Telerobot as a self-contained system with peripherals, memory and a processor that is used as an embedded system. 10 steps procedure (starting from hardware requirements and interfaces to actual experiments through software architecture, cost and power constraints) is followed to ensure that the right choice is made.

The selected Arduino (an open-source prototyping platform based on easy-to-use hardware and software) board is able to read inputs - light on a sensor or a finger on a button, and turn it into an output - activating a motor, turning on an LED, publishing something online or send required information to mobile application. The mobile application interacts with the Arduino board what to do by sending a set of instructions to the microcontroller on the board. To do so the Arduino programming language and the Arduino Software integrated development environment (IDE) are used. The important hardware components used in the Telerobot are described in the following paragraphs.

Mini DC gearbox motors, selected based on the requirements (size, torque, inertia and expected motion profile), are used as actuator of the Telerobot and connected directly to the tires without any couplings. This elements the assembly complications and ensures the compactness. Motor driver kit, dual H-bridge DC motor drive controller board module for Arduino L298 is used to control motion of the Telerobot.

Temperature sensors (LM35) are fitted in the Telerobot to measuring temperature of a particular environment, providing thermal shut down for a circuit/component and monitoring battery temperature. Ultrasonic sensor (HC-SR04) is used to avoid and detect obstacles, measure the distance and depth of pits, walls, obstacles. Mini Panoramic 360 Degree Fisheye WiFi Wireless Camera is fitted to capture images and see through the pipe and also to discover the obstacles and inner cracks inside the inspected pipe.

Li-ion Battery 18650 3.7v 1100mAh as in-built power supply is fitted. HC-05 Bluetooth module is used for communication between the mobile application and Arduino board.

The programing is done to control the motion, speed, camera, sensors and lights inside the inspected pipe using a mobile application which permit us to control it completely with no external wires (remote control).

2.3. Telerobot Fabrication and Testing

The fabricated model for inspection of pipes is shown in Fig. 4. Materials used for the Telerobot are light and rigid. Different materials can be used for different parts of the robot. Table 1 shows the materials used for different components. The fabricated components are shown in Fig. 3. It has be tested in empty horizontal pipelines. The Telerobot is working fine and controlled remotely. Pictures are taken the transferred to the personal computer. Two images taken by the Telerobot are represented in Fig. 5.

Limitations are identified as restriction in turning in T-shaped pipe and moving in a plug valve. It did not work in water-filled pipes. In order to work in water-filled pipes and T-shaped pipe and plug valve, the Telerobot design and controlling mechanism need to be modified.

Fig. 3. Components in Telerobot.

Fig. 4. Fabricated Telerobot for piping inspection.

(a) (b)

Fig. 5. Images from inside pipes taken by Telerobot.

3. Discussion

The inspection of pipes may be relevant for improving security and efficiency in industrial plants. These specific operations as inspection, maintenance, cleaning etc. are expensive, thus the application of the robots appears to be one of the most attractive solutions. It reduces the frequency of entering human into the testing environment. Operating cost related to other method is low. Save comprehensive investment, improve work efficiency, more accurate detection. The pipe inspection robot inspects situation inside the pipe which will be recorded and displayed on the monitor screen, it also facilitates working personnel for effective observation, detection, quick analysis and diagnosis operating cost related to other method is low. The robot has great application in accessing the regions of pipe in which human doesn't have reach. It is mounted with a camera to take pictures of inside and help in complete inspection.

By taking advantage of the non-destructive testing inspection methods, the proposed robot offers to identify faults in-order to avoid increasing the 'up and running' operational time of all kinds of pipe systems. It has wide application including power plants, chemical and petrochemical plants, gas pipelines, communal waste water pipe systems. Leakages in long distance heat conduits, cause through external corrosion, cause energy and water losses resulting in damage to, among others, subterranean constructions. Minimizing energy loss during the transport of heat from source to end user is one of the most important requirements in order to exclude danger to people and the environment. The proposed Telerobot can help in distant inspection in such kind of conditions. The hygiene standard in the food and drinks industries can be maintained by inspection of the individual pipe networks using the proposed Telerobotic solution.

4. Conclusion

Telerobot for complete inspection, detecting internal cracks of pipelines is developed. The Telerobot is inserted into a pipe, takes pictures and detects internal failures (cracks). The components of the Telerobot are designed in Solidworks by applying stress analysis and mode of failures of each component of the mechanism. The mechanism consists of DC motors to move forward and backward, Ultrasonic sensor, Temperature sensor, and LED lights. The movement, speed, direction and all the sensors of the Telerobot are controlled by mobile application which can be connected to other computing devices (i.g. computer or other mobile device) for further analysis and visualization. It offers many advantages over conventional methods.

In order to work in water-filled pipes and T-shaped pipe and plug valve, the Telerobot design and controlling mechanism need to be modified which is consider as future work.

References

[1] Ibarguren A, Molina J, Susperregi and Maurtua I, 2013. Thermal Tracking in Mobile Robots for

Leak Inspection Activities. Sensors, vol. 13, pp. 13560-13574, Oct. 2013. (http://www.mainbot.eu)

[2] Prasad EN, Kannan M, Azarudeen A and Karuppasamy N, 2012. Defect identification in pipe lines using pipe inspection robot. IJMERR, vol. 1, pp. 1-14, July 2012.

[3] Yousef BF and Bastaki N, 2014. Worm Robot with Dynamic Adaptation to Pipe Diameter for In-Pipe Inspection. IJEIT, vol. 3, pp. 286-292, April 2014.

[4] Bakane PP and Mujbaile P. 2015. Pipe inspection system: a review. IRJET, vol. 02, pp. 1865-1868, May 2015.

[5] Kuber KH, Kumar P, Patil A, Yadav U, Nair J, 2016. Review of Design and Fabrication of in Pipe Inspection Robots. IJRRCME, vol. 2, pp. 81-84, Oct. 2015 – Mar. 2016.

[6] Cuong NC, Phuong NT and Van DV. 2013. Design and Control of In pipe Inspection Robot. International Journal of Engineering and Science, vol. 3, pp. 38-46, Nov. 2013.

[7] Sun KNY, Srivastava RK, Ogai H and Bishakh. 2010. Advanced pipe inspection robot using rotating probe. ISAROBAROB, pp. 573-576, Oita, Japan, Feb. 2010.

IOP Conf. Series: Materials Science and Engineering **943** (2020) 012021 doi:10.1088/1757-899X/943/1/012021

Extraction of reinforced epoxy nanocomposite using agricultural waste biomass

Manpreet Kaur[1], N. M. Mubarak[1]*, Bridgid Lai Fui Chin[1]*, Mohammad Khalid[2], Rama Rao Karri[3], Rashmi Walvekar [4], E.C. Abdullah[5], Faisal Amri Tanjung[6]

[1]Department of Chemical Engineering, Faculty of Engineering and Science, Curtin University, 98009 Sarawak, Malaysia

[2]Graphene & Advanced 2D Materials Research Group (GAMRG), School of Science and Technology, Sunway University, No. 5, Jalan Universiti, Bandar Sunway, 47500 Subang Jaya, Selangor, Malaysia

[3]Petroleum and Chemical Engineering, Universiti Teknologi Brunei, Brunei Darussalam

[4]School of Engineering, Taylor's University Lakeside Campus, Jalan Taylor's, Subang Jaya, 47500, Selangor, Malaysia

[5]Department of Chemical Process Engineering, Malaysia-Japan International Institute of Technology (MJIIT) Universiti Teknologi Malaysia (UTM), Jalan Sultan Yahya Petra, 54100 Kuala Lumpur, Malaysia. Universiti Teknologi Malaysia (UTM)

[6]Department of Mechanical Engineering, Universitas Medan Area, Jalan Kolam No. 1, Medan Estate, 20223, Medan, North Sumatera, Indonesia.

E-mail: mubarak.mujawar@curtin.edu.my; mubarak.yaseen@gmail.com

Abstract. The growing awareness and concern of having greener and biodegradable materials has promoted researchers over the world to focus their attention on studies related to natural fibre reinforced polymer composites. In this study, the mechanical and thermal properties of the coconut shell powder and lignin obtained from different extraction methods such as Alkaline and Organosolv, on the epoxy composites were studied. The composite was fabricated using hand lay-up method, where the coconut shell and lignin fibers of 1, 2, 3 and 4 wt.% were mixed with the matrix consisting of epoxy resin and hardener in the ratio of 10:3. The obtained result showed that the usage of coconut shell powder and lignin as reinforcing fillers enhances the tensile strength and Young's modulus of composite when compared to neat epoxy composite by 16.4%, 14.6%, 9.70%, and 31.4%, 25.3%, 17.2% respectively at filler content of 3 wt. % but reduced the elongation at break up to 49.9% with respect to higher content of filler. Similarly, the flexural strength and Young's modulus was enhanced by 13.5%, 10.0%, 6.57% and 49.6%, 39.1%, 19.6% respectively at filler content of 3 wt.% but reduced the elongation at fracture up to 19.0% with respect to higher filler content. Structural characterization was also carried out where fillers were proven to enhance the purity and strength of the lignin.

1. Introduction

Polymers utilization has reliably developed globally in the past decade by playing an important role in many aspects of our daily lives, observing an increase of 45 million tonnes to 202 million tonnes in between 2005 till 2013, with an expectation of a further 40 million tonnes by 2017 [1]. Polymer composites are typically utilized due to advantages such as being able to be processed easily, and has a high rate of production with a reduction in cost they offer in comparison to conventional materials [2]. One drawback concerning the usage of polymer composites is that the unsustainable consumption of fossil fuels and raw materials, such as petroleum, coal and natural gas. The huge consumption of fossil fuels causes greenhouse gas emission that harms the environment, affecting changes in climate and global warming such as the rising of sea levels, rising of average global temperatures, decrease of the polar ice cap, and the rapid depletion of petroleum resources [3-6]. In the recent decade, industries have been using renewable energy that is obtained from potential sources of materials; including residues obtained from forestry, biomass, and agriculture, and are used as natural filler raw materials for polymer composite industries. Typically, in today's market, conventional fibres such as glass and carbon have been the reinforcement of choice due to its abundant presence and low cost. However, many industries are now using natural fibres as reinforcement materials in composite, mainly due to the increase in the thought of obtaining a green environment.

Besides that, the usage of natural fibres in polymer composites is also due to their advantageous assets for instance, being eco-friendly, biodegradable, renewable, cost effective, economically feasible, having low density with a specific strength, and a greater recovery of energy [2]. Some of the common natural fibres includes hemp [7], bamboo [8], jute [9], kenaf [10], and coconut [11]. In the context of this study, coconut shell is chosen as the biomass due to its absence in commercial uses.

Coconut (Cocos nucifera) tree is primarily a plantation corp in some of its origin tropical countries like India, Malaysia, Thailand and Sri Lanka [12]. Usually, the coconut shell is discarded after the fruit and water is consumed. These discarded coconut shells are known as bio waste, which today contributes, to more than 60% of the national waste volume [13]. Coconut shell comprises of organic substances such as cellulose (a glucose polymer), hemicellulose (a sugar polymer mainly containing pentoses) and lignin (a polymer of phenols) [14]. Lignin is an abundant polymer and important organic substance that is present in lignocellulosic biomasses. Lignin is deposited in the walls of certain specialized plant cells, which undergoes secondary wall biosynthesis, making them rigid and water impermeable [15]. It has potential to be converted into desired high quality phenolic products that can be used as a replacement of products that are derived from petroleum and fossil fuels [16]. Soda process [17], Kraft process [18], Sulfite process [19] Organosolv process [20], and steam explosion [21] are methods that can be used to extract lignin from a lignocellulosic biomass.

The field of polymer nanocomposites is presently one of the utmost popular scopes of research and development among researchers from different backgrounds. The term nano specifically refers to the smallest length lower than 100 nm [22], whereas the combination of two or more materials with certain properties that forms a new material that has higher properties is known as a composite. Generally, composites have three basic physical phases, where one phase in known as matrix, the other is known as reinforcement, and the final one is known as composites interphase [23] (Durowaye et al., 2014). A nanocomposite is formed when a filler is reinforced into a matrix in a nanometre (nm) scale (10-9 m) [24].

Natural fibres used as fillers are often the desired choice for reinforcements. Natural fillers are advantageous to the environment in comparison to conventional fillers as they decrease the dependency on non-renewable sources, which usually causes pollution and green house emission [11]. As natural fillers are of natural fibres, they too have similar properties of natural fibres such as they are cheaper, tougher, lower density, has a good specific strength properties, does not corrode easily, and a reduced tool wear [25]. However, they also do have disadvantageous in a sense that, they have got low tensile strength, low melting point, unfit for high temperature uses, poor surface adhesion to hydrophobic polymers, uneven filler sizes, and they tend to degrade by moisture [26]. Therefore, to amend these disadvantageous, chemical treatments are carried out to alter the properties of the fibre surface [26].

The aim of this study shows the different extraction methods of obtaining lignin from coconut shell fibre, which is the Alkaline and Organosolv process. The lignin that is obtained is used as reinforcement in a nanometre scale, while, epoxy resin is used as the thermosetting polymer matrix, where the epoxy resin matrix acts in bonding the reinforcement particles together.

2. Materials and Methods

2.1. Materials

Coconut shells purchased from a market in Miri Sarawak. USA brand epoxy resin and hardener purchased from MS Diamond and Cut Rock Trading. Mirror Glaze maximum mold release wax purchased from Poly Glass Aluminium and Construction Works. Chemicals such as ethyl alcohol (C2H6O), sulphuric acid (H2SO4), and sodium hydroxide (NaOH) purchased from Merck Chemicals.
2.2Material preparation

2.1.1 Collection and Processing of Coconut Shell

The preparation of the coconut shell powder was carried out by cleaning the coconut shells thoroughly using normal tap water to remove the dirt from the outer and inner surface of the shells. The coconut shells were then manually hammered using a sledgehammer and dried in the oven at 110°C for 24 hours. This was followed by the grinding of the coconut shells using a ball mill grinder for 12 hours to obtain them in powdery form. Then, the coconut shells powder was sieved using a sieve shaker in the range of 45 to 300 micrometre to fractionate the powder to obtain fine coconut shell powder.

2.1.2 Extraction of Lignin using Alkaline method

The extraction of lignin using alkaline method was first carried out by measuring 20g of Sodium Hydroxide (NaOH). The measured NaOH was mixed into 500mL of distilled water and was mixed thoroughly using a stirring rod. Next, 10g of coconut shell powder (CSP) was added into the solution, and the solution was again mixed thoroughly through constant stirring. When all the NaOH and CSP have dissolved into the distilled water, the mixture was covered and heated for 2 hours using a magnetic stirrer heater at a temperature of 80°C. A magnetic stirrer bar was placed in the mixture and a stir of 200 revolutions per minute (rpm) was set to allow a uniform mixing in the solution. The solution was then allowed to cool. Afterward, 130mL of sulphuric acid (H2SO4) was added into the mixture gradually. The mixture was allowed to stay overnight to allow separation to occur. When separation occurs, two layers can be seen in the solution. The bottom layer contains of the lignin. The solution was then filtered out using a Buchner funnel and a vacuum flask. The washing of the filtered mixture using distilled water until a clear filtrate was obtained followed this. The residue that was obtained is the lignin. The lignin obtained was then placed on a petri dish and put into the oven at a temperature of 80°C, for 24 hours to dry the lignin. When the lignin has dried, a mortar and pestle is used to crush the lignin into powder form. Equation (1) and (2) was used to determine the mass of solid NaOH needed in the solution, and equation (3) was used to determine the volume of H2SO4 needed in the solution.

$$C = \frac{n}{V} \tag{1}$$

Where

C represents the concentration of the solution (mol/L),
n represents the no. of moles of NaOH (mol),
V represents the total volume of the solution (L).

$$m = n \times MW \tag{2}$$

Where

IOP Conf. Series: Materials Science and Engineering **943** (2020) 012021 doi:10.1088/1757-899X/943/1/012021

m represents the mass of NaOH (g),
n represents the no. of moles of NaOH (mol),
MW represents the molecular weight of NaOH (g/mol).

$$M_1V_1 = M_2V_2 \tag{3}$$

Where

$\mathbf{M_1}$ represents concentration in molarity of the concentrated solution (mol/L),
$\mathbf{V_1}$ represents the volume of the concentrated solution (mL),
$\mathbf{M_2}$ represents the concentration in molarity of the diluted solution (mol/L),
$\mathbf{V_2}$ represents the volume of the concentrated solution (mL).

2.1.3 Extraction of Lignin using Organosolv method

The extraction of lignin using Organosolv method was carried out by mixing ethyl alcohol (C2H6O) and distilled water in the ratio of 60:40, respectively. The ethyl alcohol of 300mL was measured out into a 500mL beaker. This was then followed by the addition of 100mL of distilled water, 10g of coconut shell powder (CSP) and 8mL of sulphuric acid (H2SO4). The H2SO4 here acts as a catalyst. Next, the solution was mixed thoroughly using a stirring rod. When all the C2H6O and CSP have dissolved into the distilled water, the mixture was covered and heated for 2 hours using a magnetic stirrer heater at a temperature of 75°C. A magnetic stirrer bar was placed in the mixture and a stir of 200rpm was set to allow a uniform mixing in the solution. The solution was then allowed to cool and to allow separation to occur. When separation occurs, two layers can be seen in the solution. The bottom layer, also known as black liquor was discarded. Here, a centrifuge was used to separate the layers accurately. The centrifuge was run for 10 minutes, and the speed was set at 3500rpm. Afterwards, the filtrate which was obtained from the top layer was precipitated with distilled water until precipitation occurred. The precipitation process was carried out using the filtrate and distilled water in the ratio of 1:3, respectively. The precipitate contains of the lignin. The precipitate was then filtered out using a Buchner funnel and a vacuum flask. The washing of the filtered mixture using distilled water until a clear filtrate was obtained followed this. The residue that was obtained is the lignin. The lignin obtained was then placed on a petri dish and put into the oven at a temperature of 80°C, for 7 hours to dry the lignin. When the lignin has dried, a mortar and pestle is used to crush the lignin into powder form.

2.1.4 Composite fabrication

Hand lay-up technique, also known as open molding, is one of the most common technique used to fabricate the polymer matrix composites. The fabrication was prepared using two different molds, each for their own testing. Three batches of fabrication were carried out with different filler. The first batch was carried out using pure coconut shell powder (1, 2, 3 and 4 wt. %) and was mixed with the matrix consisting of epoxy resin and hardener in the ratio of 10:3. Similarly, the second and third batch was carried out using Alkaline-lignin and Organosolv lignin with the same weight percentage as the CSP and was mixed with the matrix consisting of the same ratio as of the first batch. Table 1 describes the mixing formulation of filler, epoxy and hardener. The fabrication process was started by polishing the molds with an anti-adhesive agent, in this case, wax to prevent the composite from sticking onto the mold upon removal. Then, the required epoxy resin, hardener and filler were measured out in three different paper cups, on an electronic weighing balance. The measured materials were poured into a new paper cup, with the sequence of epoxy resin, hardener, and filler, followed by a constant moderate stirring for 5 minutes using a popsicle stick. It must be ensured not to stir the mixture rapidly as it will trap more bubbles in the solution. Next, the mixture was immersed into an ultrasonic bath filled with 1500mL of water for sonication. The sonication took place for 3 minutes at a temperature of 0°C, with maximum power of 5 and high frequency. This was to enhance the dispersion of filler at the surface of the matrix. The mixture was then immediately placed in a vacuum chamber for 5 minutes to allow degassing process to occur. Degassing process is carried out to remove all bubbles that have been trapped in the mixture during the fabrication of the composites. After that, the mixture was poured into the molds and left for cool mounting at room temperature for 24 hours until the mixture was hardened. When the composite has hardened, they were cured in an oven at a temperature of 100°C for 3 hours. After the curing, the composites were allowed to cool to room temperature, and then removed from the molds by bringing the mold onto a 30 to 45-degree bend.

Table 1 Mixing formulation of filler, epoxy and hardener

No.	Filler loading (wt. %)	Filler (g)	Epoxy resin (g)	Hardener (g)
1	0	0	70	30
2	1	1	76.2	22.8
3	2	2	75.4	22.6
4	3	3	74.6	22.4
5	4	4	73.8	22.2

2.2 Material Testing

2.2.1 Tensile testing

Tensile testing was carried out on the specimens to determine the tensile properties of the neat epoxy, as well as the pure coconut shell powder (CSP), alkaline-lignin (AL), and organosolv-lignin (OL) reinforced epoxy composites following the ASTM D638-91 test method. The force required to break the sample specimen and the extent to which the specimen elongates to break point were measured. The mold used for this specimen is based on the Type 1 tensile bar dimension and thickness outlined by ASTM international, which was a dog-bone shape mold. The tensile testing was conducted using LLYOD LR 10K Plus Advanced Universal Testing Machine (UTM), with cross-head speed of 20mm/min at the temperature of 25±3°C. The testing was carried out by placing the specimens at a specified grip separation in the grips of the tester and pulled until the specimen fractures.

2.2.2 Flexural testing

Flexural testing was carried out on the specimens to determine the bending strength of the composite materials under three-point loading condition, which indicates the materials stiffness when being flexed, according to the ASTM D790 test method. The mold used for this specimen is based on the flexural bar dimension and thickness outlined by ASTM international. The flexural testing was conducted using LLYOD LR 10K Plus Advanced Universal Testing Machine (UTM), with cross-head speed of 1.3mm/min. The testing was carried out by placing the specimens at a specified grip separation in the grips of the tester and pulled until the specimens reach 5% deflection or fractures before 5% deflection.

2.3 Characterization

2.3.1 Fourier Transform Infrared Spectroscopy

The surface chemistry of the pure and extracted specimens in terms of functional groups attached to the surface was examined using the Perkin Elmer FT-IR Spectrometer Frontier model. The specimens were exposed to infrared radiation, where the range of infrared region was set to 4500 to 500 cm-1. The scanning was regulated at 4 cm-1 resolution performing 50 scans for every specimen. The specimens absorb the radiation of the wavelengths and cause changes of the dipole moment of the sample molecules. The transmittance spectrum was observed in the region ranging of 4500 to 500 cm-1.

2.3.2 Thermogravimetric Analysis

Thermogravimetric analysis (TGA) of the pure and extracted specimens was conducted to examine the thermal degradation and stability by using a thermogravimetric analyzer (Perkin Elmer STA 6000). The testing was conducted by placing a specimen weight of about 10 to 15 mg an open alumina pan under an air flow of 20mL per minute. The specimens were then heated from ambient temperature of 25

IOP Conf. Series: Materials Science and Engineering **943** (2020) 012021 doi:10.1088/1757-899X/943/1/012021

to 700°C, at a heating rate of 10°C per minute. The temperature was recorded for further analysis and discussion.

3. Result and Discussion

3.1 Tensile properties

Fig. 1 (a) shows the effect of different filler content on tensile strength of coconut shell powder (CSP)/epoxy composite, alkaline lignin (AL)/epoxy composite and organosolv lignin (OL)/epoxy composite. The tensile strength graph illustrates that, as filler content increase, tensile strength shows a linear increase for all composites from 1 wt. % until 3wt%, and then decreases at 4wt%. A similar trend to the present study can be observed by studies carried out by Abdul Khalil et al. [27] and Tangimul Islam et al. [28]. In both studies, CSP fillers were reinforced to different composites. The tensile strength increases with the filler content up to a certain content, and thereafter decreased at higher filler concentration. In the case of where lignin was used as reinforcing fillers, Behin et al. [29] and Yu et al. [30] both respectively show the same behaviour of tensile strength that increases as filler content increases. Here, we can perceive that despite the different matrixes were used in the studies a similar trend can be observed when CSP or lignin is used as reinforcing fillers. We can deduce that the increase of the tensile strength as filler content increases may be due to the increased packing fraction and the lower void content in the composite [31]. For the lignin fillers, the reason of the increase in tensile strength is due to the effective transfer of stress from the molecular chains to the lignin networks, which shows a good reinforcement by the rigid lignin network [30].

As for the decrease of the strength at weight of 4%, it could be due to the poor surface adhesion of the reinforcing fillers with the epoxy matrix, as there was no coupling or dispersing agents used in the composites [32]. Also, aggregations and agglomerations of the reinforcing fillers due to the increase in interfacial area that may have formed within the epoxy matrix that caused interruption to the filler-matrix bonding [33]. This interruption lead to the reduction of an efficient load transfer from the matrix to the filler phase, which then easily causes the formation and propagation of crack within the matrix of the composite, and thus leads to a reduction of tensile strength of both the CSP and lignin epoxy composites [34]. Moreover, the reduced tensile strength for the lignin indicates the existence of a plasticizing effect of the lignin at 4 wt. %, filler content. The plasticizing effect occurred due to the poor interfacial adhesion between the lignin filler and the epoxy matrix, which caused lack of infusion of the composite [35].

Besides that, when compared to the tensile strength of neat epoxy composite that shows strength of 49.465 MPa, it can be observed that all the reinforced epoxy composites show a greater tensile strength in the range of lowest 52.483 MPa at 1 wt. % filler content, and highest of 57.575 MPa at 3 wt. % filler contents. Pure CSP, AL and OL composites show the maximum strength of 54.264, 57.575 and 56.674 MPa at 3 wt. % filler content, respectively. AL composite demonstrates the highest tensile strength of all composite with a significant increase of 16.4% compared to the neat epoxy composite. This is followed by OL and CSP, with a significant increase of 14.6% and 9.70% when compared to neat epoxy composite, respectively. From the highest tensile strength at 3 wt. % filler content, we can deduce that all reinforced composites could withstand a stress before it breaks. However, lignin reinforced composites have the best tendency to withstand a stress. The ability of the CSP and lignin to reinforce the epoxy matrix and withstand stress that produces high tensile strength could be due to the intermolecular bonding between the functional groups of the both the filler and epoxy matrix, which results a uniform distribution of the filler on the surface of the epoxy matrix. Moreover, CSP and lignin filler can easily disperse on the surface of matrix at lower filler content, which enables it to improve the filler-matrix interfacial bonding significantly. The good interfacial interaction between the epoxy matrix and filler allow the composite to withstand a great load, which results in a better stress transfer from the matrix to the filler.

Fig. 1 (b) shows the effect of different filler content on tensile Young's modulus of coconut shell powder/epoxy composite, alkaline lignin/epoxy composite and organosolv lignin/epoxy composite. The tensile Young's modulus graph exhibits the same behaviour of the tensile strength graph, where it can be observed that as filler content increase, tensile modulus shows a linear increase for all composites from 1 wt. % until 3wt%, and then decreases at 4wt%. Sarki et al. [25], Wood et al. [35] and Oliveira et al. [36] shows a similar trend of tensile young's modulus in their studies. Sarki et al. [25] reinforced coconut shell particles into the epoxy composite, Wood et al. [35] reinforced lignin into the hemp/epoxy composite, and Oliveira et al. [36] reinforced eucalyptus fibre into the polyester epoxy composites. In all three studies conducted by the respective authors, they observe that the tensile modulus increases with the filler content, but after that decrease at a higher filler concentration. Hence, similarly to the tensile strength, we can perceive that despite the different matrixes were used in the studies a similar trend can be observed when CSP or lignin is used as reinforcing fillers. The increase of the tensile modulus up to 3 wt. % filler contents could be due to the stiffness of the reinforcement which strengthens as the filler content increases. Also, the increase from 1 to 3 wt. % filler contents is due to the fact as the filler increase, it causes a more uniform distribution of the epoxy matrix, resulting in the improved bonding between the matrix and filler interphase. As for the decrease at 4 wt.% filler content of tensile modulus, the reason could be due to the agglomeration of the filler molecules around the matrix, which causes an improper curing of the composite [37]. The decrease could also be due to the increased micropores between the filler and the matrix that weakened the adhesion between the epoxy matrix and the filler interface [38]. It is also understood that fillers at a higher filler content act as flaws due to the lack of resin which wets the fibre surface, and results a weak stress transfer.

Besides that, when compared to the tensile modulus of neat epoxy composite that shows a modulus of 1484.39 MPa, it can be observed that all the reinforced epoxy composites show a greater tensile modulus in the range of lowest 1617.19 MPa at 1 wt. % filler content, and highest of 1950.67 MPa at 3 wt. % filler content. This clearly shows that the reinforcing fillers demonstrate a potential of increasing the young's modulus. Pure CSP, AL and OL composites show the maximum modulus of 1740.27, 1950.67 and 1860.17 MPa at 3 wt. % filler content, respectively. AL composite demonstrates the highest tensile modulus of all composite with a significant increase of 31.4% compared to the neat epoxy composite. This is followed by OL and CSP, with a significant increase of 25.3% and 17.2% when compared to neat epoxy composite, respectively. Here, it can be deduced that AL composite have got the highest resistance to deformation. Furthermore, the increase of the tensile modulus as filler content increases from 1 to 3 wt. % may be due to the improved rigidity of the composite. The rigidity could be improved by the good interfacial bonding between the CSP and lignin fillers and the epoxy matrix. The incorporation of the CSP and lignin filler into the epoxy matrix limited the molecular motion of the epoxy chain. The limitation enables the reduction of the composite deformation.

Fig. 1 (c) shows the effect of different filler content on elongation percentage of break of coconut shell powder/epoxy composite, alkaline lignin/epoxy composite and organosolv lignin/epoxy composite. The tensile elongation illustrates that, as filler content increase, tensile elongation shows a linear decrease for all composites from 1 wt. % until 4wt%. Also, when compared to the tensile elongation at break of neat epoxy composite that shows a 7.217% break, it can be observed that all the reinforced epoxy composites show a lower tensile elongation in the range of lowest 1.675% at 4 wt.% filler content, and highest of 4.654% at 1 wt.% filler content. Pure CSP, AL and OL composites show their maximum elongation at break of 4.654, 4.135 and 3.615% at 1 wt. % filler content, respectively. CSP composite demonstrates the highest tensile elongation of all composite with a significant decrease of 35.5% compared to the neat epoxy composite. This is followed by AL and OL composite with a significant decrease of 42.7% and 49.9% when compared to neat epoxy composite, respectively.

Based on the observation of the percentage of elongation at break, the elongation attributes are inversely proportional to the Young's modulus from 1 until 3 wt. % filler contents. This is because the stiffness of the sample affects the percentage of the elongation, and thus we can make a hypothesis that the stiffer the specimen, the lower the percentage of elongation at break of the tested specimen. Besides that, we can deduce that the decrease of the tensile elongation as the filler content increases may be due to the formation of agglomeration that lead to low adhesion between the epoxy matrix and the filler reinforcement, and consequently caused the composite to be brittle [39]. In other words, as the filler content increases, the composite becomes tougher, and therefore, the chain mobilization in the matrix reduces. The reduction of the chain mobility, where the forming of the cross-linked network in the polymer chain is restricted by the high attractive force, causes the brittleness of the composites to increase [28].

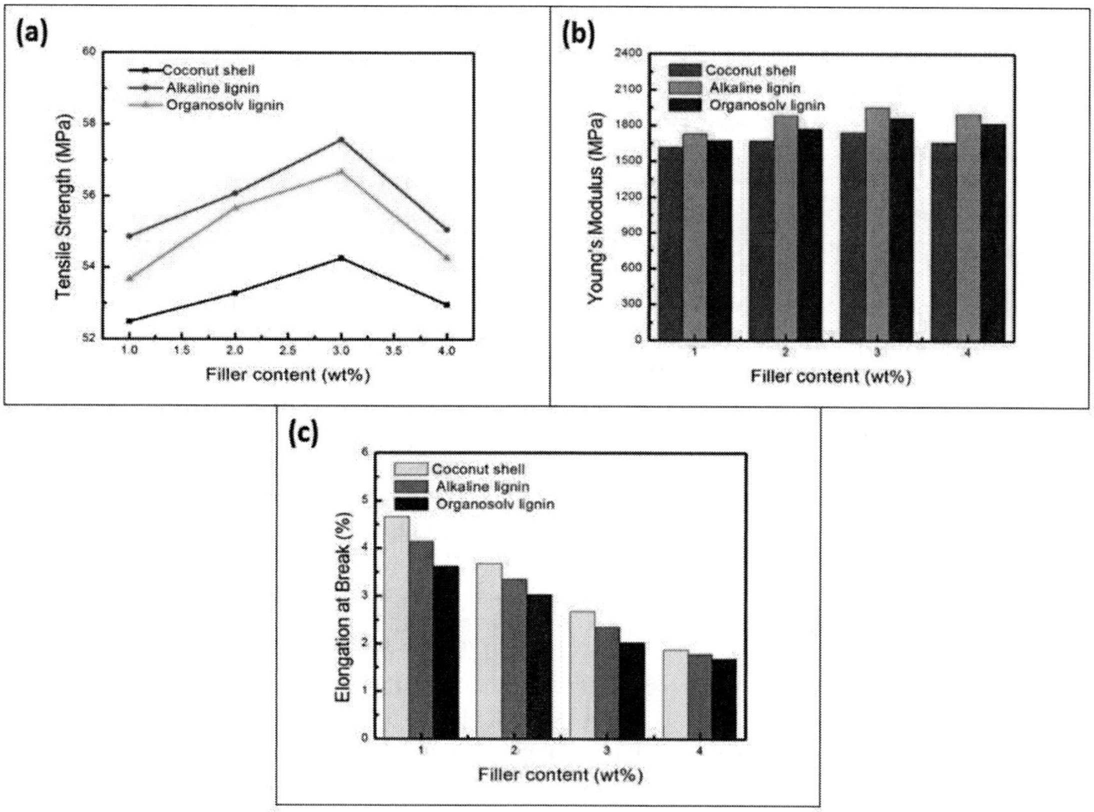

Fig. 1: (a) Tensile strength versus filler content, (b) Tensile Young's modulus versus filler content, and (c) Elongation at break versus filler content

3.2 Flexural properties

Fig. 2 (a) shows the effect of different filler content on flexural strength of coconut shell powder (CSP)/epoxy composite, alkaline lignin (AL)/epoxy composite and organosolv lignin (OL)/epoxy composite. The flexural strength graph illustrates that as filler content increase, flexural strength shows a linear increase for all composites from 1 wt. % until 3wt%, and then decreases at 4wt%. The same was reported by Tangimul Islam et al. [28] that studied the effect of filler from 0 to 100% and revealed that as filler content increases, flexural strength was observed increasing from 0 to 30%, and decreasing afterward. The increase of flexural strength from 1 to 3 wt. % filler contents could be due to strong interfacial bonding caused by the compatibility of the fillers and the polymer matrix. The high strength could also indicate good mixing and infusion of the reinforcing fillers. As for the decrease at 4wt%, the reason could be due to the weak interfacial adhesion to transfer the stress, or the presence of the pores at the surface interface of the filler and the matrix. Moreover, voids content and dispersion problem, and other factors like the size, shape and type of reinforcing filler may cause the strength to decrease Tangimul Islam et al. [28].

Besides that, when compared to the flexural strength of neat epoxy composite that shows strength of 110.58 MPa in Table 1, it can be observed that all the reinforced epoxy composites show a greater flexural strength in the range of lowest 112.89 MPa at 1 wt. % filler content, and highest of 125.47 MPa at 3 wt. % filler contents. This clearly shows that the reinforcing fillers demonstrate a role of increasing the flexural strength. Pure CSP, AL and OL composites show the maximum strength of 117.85, 125.47 and 121.64 MPa, at 3 wt. % filler content, respectively. AL composite demonstrates the highest flexural strength of all composite with a significant increase of 13.5% compared to the neat epoxy composite. This is followed by OL and CSP, with a significant increase of 10.0% and 6.57% when compared to neat epoxy composite, respectively. Here, it can be deduced that AL composite have got the highest ability to resist bending. Based on the observation of the flexural strength graph, the composites with filler content of weight 3% displayed greater load carrying ability and deflection. This indicates that the composites are less brittle and undergoes a slow rate of crack propagation [40].

Moreover, when compared to the flexural strength of neat epoxy composite that shows strength of 110.58 MPa, it can be observed that all the reinforced epoxy composites show a greater flexural strength in the range of lowest 112.89 MPa at 1 wt. % filler content, and highest of 125.47 MPa at 3 wt. % filler content. This clearly shows that the reinforcing fillers demonstrate a role of increasing the flexural strength. Pure CSP, AL and OL composites show the maximum strength of 117.85, 125.47 and 121.64 MPa, at 3 wt. % filler content, respectively. AL composite demonstrates the highest flexural strength of all composite with a significant increase of 13.5% compared to the neat epoxy composite. This is followed by OL and CSP, with a significant increase of 10.0% and 6.57% when compared to neat epoxy composite, respectively. Here, it can be deduced that AL composite have got the highest ability to resist bending. The enhancement of the surface roughness enables the lignin to have a better fibre-matrix adhesion. Besides that, based on the observation of the flexural strength graph, the composites with filler content of weight 3% displayed greater load carrying ability and deflection. This indicates that the composites are less brittle and undergoes a slow rate of crack propagation.

Fig. 2 (b) shows the effect of different filler content flexural Young's modulus of coconut shell powder/epoxy composite, alkaline lignin/epoxy composite and organosolv lignin/epoxy composite. The flexural Young's modulus graph exhibits the same behaviour of the flexural strength graph, where it can be observed that as filler content increase, flexural modulus shows a linear increase for all composites from 1 wt. % until 3wt%, and then decreases at 4wt%. Gopinath et al. [41] and Vasu et al. [42] shows a similar trend of flexural young's modulus in their studies. Gopinath et al. [41] reinforced jute fibre particles into the polyester and epoxy matrix, and Vasu et al. [42] reinforced coconut shell powder into the propylene composites. Both authors, each in their own respective study showed the results of the flexural modulus that increases with the filler content, however decreases at a higher filler concentration. Here, we can observe that despite the different matrixes and reinforcing fillers were used in the studies, a similar trend can be observed when CSP or lignin is used as reinforcing fillers. The increase of the flexural modulus is due to the stiffness of the reinforcement which strengthens as the filler content increases. As for the decrease of the modulus at 4 wt. %, the reason may be due to the poor mixing and infusion of the filler and matrix, or in other words, improper wetting of the matrix and filler which caused the epoxy matrix deficiency, and thus agglomeration.

Besides that, when compared to the flexural modulus of neat epoxy composite that shows a modulus of 3324.05 MPa, it can be observed that all the reinforced epoxy composites show a greater flexural modulus in the range of lowest 3724.85 MPa at 1 wt. % filler content, and highest of 4973.64 MPa at 3 wt. % filler content. This clearly shows that the reinforcing fillers demonstrate a potential of increasing the young's modulus. Pure CSP, AL and OL composites show the maximum modulus of 3973.76, 4973.64 and 4624.75 MPa at 3 wt. % filler content, respectively. AL composite demonstrates the highest flexural modulus of all composite with a significant increase of 49.6% compared to the neat epoxy composite. This is followed by OL and CSP, with a significant increase of 39.1% and 19.6% when compared to neat epoxy composite, respectively. Here, it can be deduced that AL composite indicates the highest value of stiffness, and thus have got the highest resistance to deformation which is closely followed by organosolv-lignin and CSP. Both the lignin's show higher modulus compared to CSP, which could be due to the fact that during the extraction of the lignin, the lignin fibres may have undergone swelling which was caused by the water absorption. The water absorption crowds the space between the fibre and the epoxy matrix, which led to an increase in the mechanical properties of the composites [43].

Fig. 2 (c) shows the effect of different filler content on elongation percentage of fracture of coconut shell powder/epoxy composite, alkaline lignin/epoxy composite and organosolv lignin/epoxy composite. The flexural elongation illustrates that, as filler content increase, flexural strength shows a linear decrease for all composites from 1 wt. % until 4wt%. Also, when compared to the flexural elongation at break of neat epoxy composite that shows a 18.713% fracture, it can be observed that all the reinforced epoxy composites show a lower flexural elongation in the range of lowest 13.218% at 4 wt.% filler content, and highest of 16.157% at 1 wt.% filler content. Pure CSP, AL and OL composites show their maximum elongation at break of 16.157, 15.635 and 15.158% at 1 wt. % filler content, respectively. CSP composite demonstrates the highest flexural elongation of all composite with a significant decrease of 13.7% compared to the neat epoxy composite. This is followed by AL and OL composite with a significant decrease of 16.5% and 19.0% when compared to neat epoxy composite, respectively. This clearly shows that the reinforcing fillers decrease the elongation percentage at fracture of the composites, which may be due to the reduction of polymer component inside the composite.

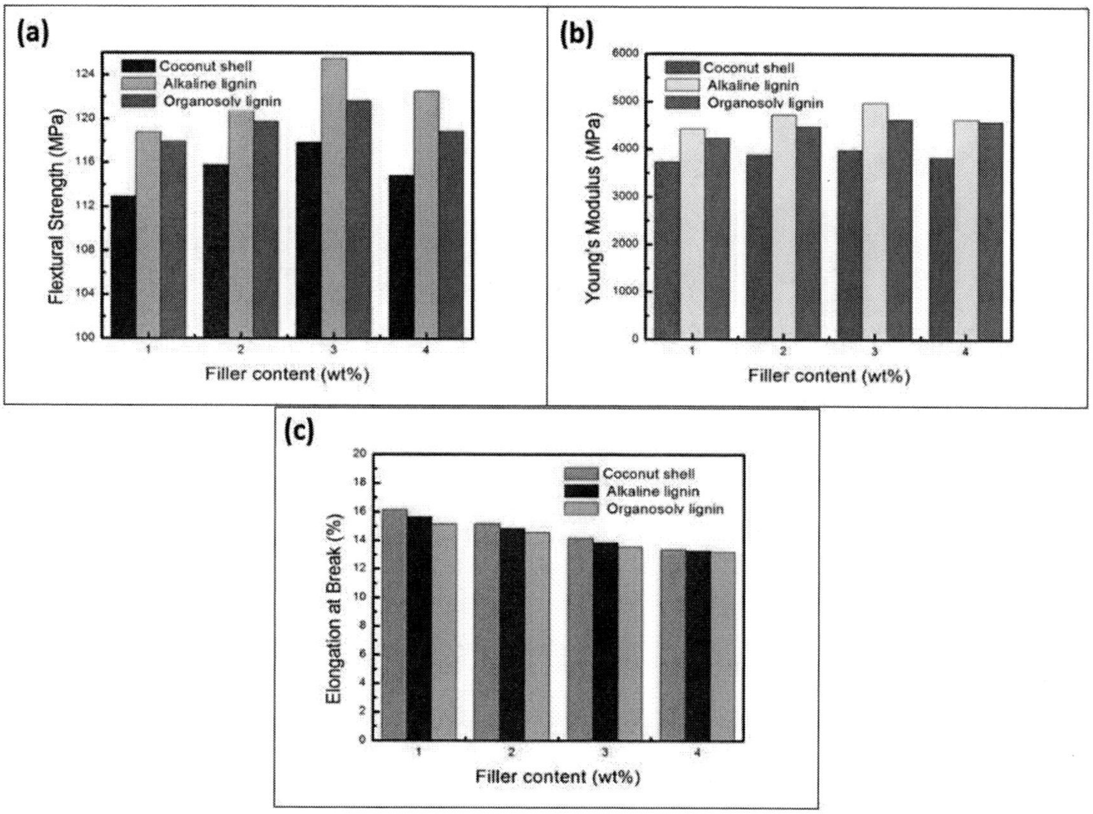

Fig. 2: (a) Flexural strength versus filler content, (b) Flexural Young's modulus versus filler content, and (c) Elongation at fracture versus filler content

3.3 Fourier Transform Infrared Spectroscopy

The Fourier Transform Infrared Spectroscopy (FTIR) spectrum for the coconut shell powder, alkaline lignin and organosolv lignin in this study is shown in Fig. 3. From the figure, it can be observed that different absorption bands ranges in the spectrum from 4000 to 300 cm-1. The absorption band around 3350 cm-1 is due to the hydroxyl (O-H) stretching vibration. According to Ibrahim et al. [44] that have quoted Khan et al. [45] in his study, the O-H stretching may be due to the absorbed alcohols that are found in cellulose, hemicellulose, lignin, extractives and carboxylic acids. Also, the relatively small O-H stretching band may be due to the limited reactive efficiency of the hydroxyl at the coconut shell and lignin fiber surface. This signifies the capability of easier surface water desorption, as well as hemicellulose/lignin decomposition for the CSP. The absorption band around 2950 cm-1 is due to the carbon-hydrogen (C-H) stretching, which is a characteristic of any natural fiber. The presence of the band which is only an indistinct peak in the spectrum is observed at around 2650 cm-1. However, this band could not be identified with any molecular origin. It is also observed that the bands in the range of 2350 cm-1 and around 2020 cm-1 may be related to carbonyl (C=O) stretching. As for the bands in the range of 1750 to 1300 cm-1, Ibrahim [46] suggested that this range is due to the deformation of the carbon-hydrogen (C-H) stretching. Ibrahim [46] also suggested that the deformation of the C-H and carbon-oxygen (C-O) are due to the ether linkage. In this spectrum, the deformation of C-H and C-O is noticeably observed to be at 1080 cm-1. This accentuated band could represent the difficulty for the ether radicals to decompose.

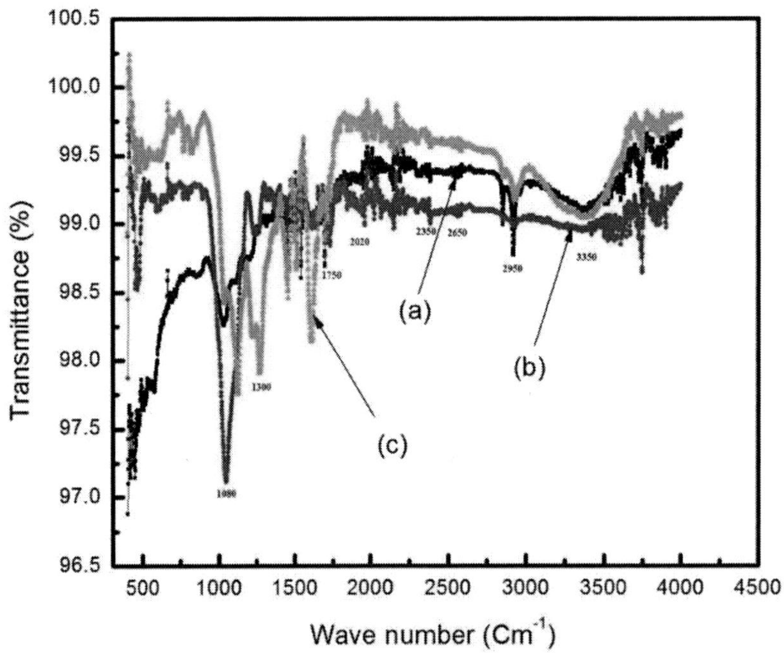

Fig. 3: FTIR spectrum of (a) alkaline-lignin, (b) pure coconut shell powder, and (c) organosolv-lignin

3.4 Thermogravimetric Analysis

Coconut shell powder (CSP), alkaline-lignin (AL) and organosolv-lignin (OL) were selected for testing to identify thermal stability and the purity of the composite materials. Fig. 4 shows the comparison for the thermogravimetry analysis (TGA) curves of the particulates for pure CSP, alkaline-lignin and organosolv-lignin. From the figure, it can be observed that there is a thermal degradation at much lower temperature and significant weight with respect to temperature. Fig. 8 depicts the TGA curve of pure

CSP and lignin fibres as a function of temperature. It shows the degradation of all three fibres occurred at three stages. A similar degradation of three stages was observed by Monteiro et al. [47] in his study. The first stage was at a heating range of 25 to 250°C. At this stage, all the fibre films show a relative weight loss, despite the thermal stability improvement of the fibre. The cause of the relative weight loss may be due to the loss of moisture content in the fibres. Here, the loss of moisture is because of the water loss and structural rearrangements with the release of small molecular weight compounds. According to Monteiro et al. [47], temperatures below 250°C usually corresponds to the water loss of the natural fibre, and it also indicates the condensation and splitting of the side chains of the lignin. Second stage occurred at the heating range of 250 to 400°C. At this stage, all the fibre films show a major weight loss. The cause of the major weight loss may be due to the degradation of components of carbohydrates in the CSP and lignin samples, which are converted to volatile gasses such as carbon monoxide (CO), carbon dioxide (CO2) and methane (CH4). Third stage occurred at the heating range of 400 to 660°C. Monteiro et al. [47] stated that the decomposition above 400°C of the lignin is associated with a series of degradation and condensation reactions with accumulation of aromatic products. Kabir et al. [48] stated that a greater thermal stability is obtained due to higher decomposition temperatures. Therefore, at this third stage, all the composite films show a significant weight loss. Here, the cause may be due to the degradation of the fibres due to burning and oxidation. The carbon-hydrogen (C-H) and carbon-oxygen (C-O) bonds in the organic coconut shell were cleaved to generate non-condensable volatile components. As for the lignin, the degraded volatile products include phenols, alcohols and aldehyde acids along with the formation of gaseous products that are getting removed. The mass change in this stage was extremely small. Besides that, it can be observed from the TGA curve that the thermal degradation for all the samples did not commence until the materials had adsorbed a certain amount of heat energy. Therefore, the heat initiated the degradation processes, whereas the breaking down of the structure caused the molecular chain to be broken. It can fairly be suggested that the CSP and lignin's thermal properties are dependent on their source.

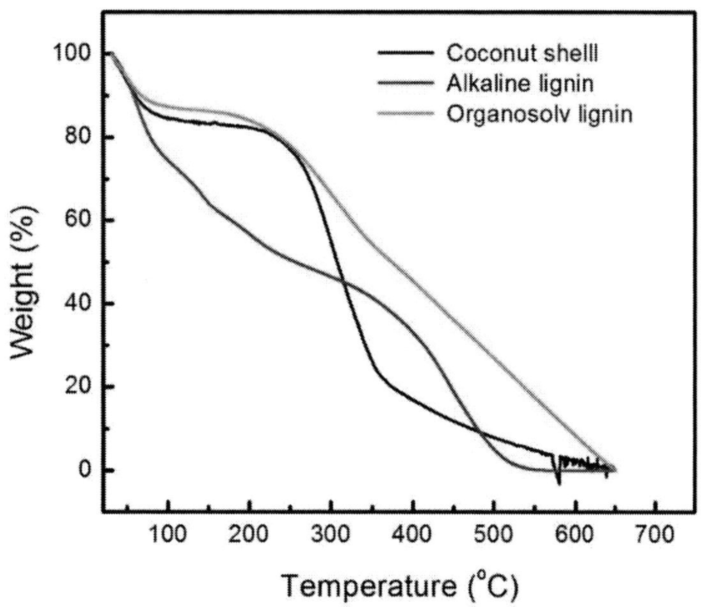

Fig. 4: Weight versus temperature of coconut shell powder

4. Conclusion

The present study investigated the usage of natural fibre polymer composites filler with addition of natural fine fillers into matrix to enhance the properties of the epoxy composite. The incorporation of nano-size reinforcement of functional filler is advisable to compensate with the limitation of the epoxy resin. Functional nanofillers like nanofibers are great filler to modify and improvise the mechanical and thermal properties of polymer matrix resin, like epoxy. The tensile strength, tensile Young's modulus, flexural strength and flexural Young's modulus was observed to increase until 3 wt.% of filler content, however degraded at a higher addition of 4 wt.%. The obtained results showed that the tensile strength of CSP, alkaline-lignin and organosolv-lignin shows a significant increase of 9.70%, 16.40%, 14.6% and the tensile modulus observes an increase of 17.2%, 31.4%, 25.3% when compared to neat epoxy composite. The flexural strength of CSP, alkaline-lignin and organosolv-lignin shows a significant increase of 6.57%, 13.5%, 10.0% and the flexural modulus observes an increase of 19.6%, 49.6%, 39.1% when compared to neat epoxy composite. In addition, the tensile elongation of break of CSP, alkaline-lignin, organosolv-lignin shows a decrease of 35.5%, 42.7%, 49.9% and the flexural elongation of fracture was observed a decrease of 13.7%, 16.5%, 19.0% when compared to neat epoxy composite. The findings of the present study draw the main conclusion that lignin/epoxy composites improve the mechanical and thermal properties of epoxy composite compared to coconut shell/epoxy composite and neat composite.

Acknowledgement
This work is carried out under the financial support of Fundamental Research Grant Scheme (FRGS) (MOHE Ref. No: FRGS/1/2016/TK05/CURTIN/03/1).

References

[1] [1] F. Galiè, C. Trabucchi, ICIS World Polymer Outlook, Reed Business Information Ltd (2014).
[2] D.N. Saheb, J.P. Jog, Natural fiber polymer composites: a review. Polymer Engineering Group, Chemical Engineering Division, National Chemical Laboratory 18 (1999) 351–363.
[3] B. Chanda, R. Kumar, K. Kumar, S. Bhowmik, Optimisation of Mechanical Properties of Wood Dust-reinforced Epoxy Composite Using Grey Relational Analysis, Springer, India, 2015.
[4] F. Ferdosian, Synthesis, Characterization and Applications of Lignin-Based Epoxy Resins. Electronic Thesis and Dissertation Repository, 2015.
[5] J. Jayaramudu, S.C. Agwuncha, S.S. Ray, E.R. Sadiku, A.V. Rajulu, Studies on the chemical resistance and mechanical properties of natural polyalthia cerasoides woven fabric/glass hybridized epoxy composites, Adv. Mat. Letters 6 (2015) 114-119.
[6] S.M. Mohomane, T.E. Motaung, N. Revaprasadu, Effect of alkali and acid treatment on thermal degradation kinetics of sugar cane bagasse. Ind. Crops Prod. 74 (2015) 472-477, https://doi.org/10.1016/j.indcrop.2015.05.062.
[7] H. Peltola, B. Madsen, R. Joffe, K. Nättinen, Experimental study of fiber length and orientation in injection molded natural fiber/starch acetate composites. Advances Mat. Sci. Eng. 2011 (2011) 1-7, http://dx.doi.org/10.1155/2011/891940.
[8] A. Gupta, A. Kumar, A. Patnaik, S. Biswas, Effect of different parameters on mechanical and erosion wear behavior of bamboo fiber reinforced epoxy composites. Int. J. Polym. Sci. 2011 (2011) 1-10, ttp://dx.doi.org/10.1155/2011/592906.
[9] M.K. Guptaa, R.K. Srivastavaa, H. Bisaria, Potential of jute fibre reinforced polymer composites: a review, Int. J. Fiber Textile Res. 5 (2015) 30-38.
ssssssssssssssssssssssssssssss

[10] T. Williams, M. Hosur, M. Theodore, A. Netravali, V. Rangari, S. Jeelani1, Time effects on morphology and bonding ability in mercerized natural fibers for composite reinforcement, Int. J. Polym. Sci. 2011 (2011)1- 9, http://dx.doi.org/10.1155/2011/192865.
[11] R. Udhayasankar, B. Karthikeyan, A review on coconut shell reinforced composites, Int. J. ChemTech Res. 11 (2015) 624-637.
[12] W.N.B.W. Othman, Characterization of oil palm shell nano filler in hybrid kenaf/coconut fibres reinforced polyester composites, PhD Thesis, National Institute of Rourkela, 2015.
[13] K. Gunasekaran, P.S. Kumar, M. Lakshmipathy, Mechanical and bond properties of coconut shell concrete, Constr. Build. Mat. 25 (2011) 92-98, https:// dx.doi.org/10.1016/j.conbuildmat.2010.06.053.
[14] S. Rodrigues, G. Pinto, Ultrasound extraction of phenolic compounds from coconut (Cocos nucifera) shell powder, J. Food Eng. 80 (2007) 869-872, https://doi.org/10.1016/j.jfoodeng.2006.08.009.
[15] R. Vanholme, B. Demedts, K. Morreel, J. Ralph, W. Boerjan, Lignin biosynthesis and structure, American Soc. Plant Biologists 153 (2010) 895-905, https://doi.org/10.1104/pp.110.155119.
[16] W. Ji, Z. Ding, J. Liu, Q. Song Q, X. Xia, H. Gao, H. Wang, W. Gu, Mechanism of lignin dissolution and regeneration in ionic liquid, Energy & Fuels 26 (2012) 6393-6403. https://doi.org/10.1021/ef301231a.
[17] C. Watt C, H. Burgess, Improvement in the manufacture of paper from wood, US11343A. St..Albans Terrace, Kensington. John Watt, 1 Ampton Street, Grays Inn Road, 1854.
[18] O.F. Dahl, Process of Manufacturing Cellulose from Wood, 1884:US 296935A.
[19] W. Glasser, V. Davé, C. Frazier, Molecular weight distribution of (semi-) commercial lignin derivatives, J. Wood Chem. Tech. 13 (1993) 545-559, https://doi.org/10.1080/02773819308020533.
[20] K. Theodor, T. Kurt T, Process of decomposing vegetable fibrous matter for the purpose of simultaneous recovery both of the cellulose and of the incrusting ingredients, US1856567A, 1930.
[21] W.H. Mason, Process and apparatus for disintegration of wood and the like, US1578609A, 1926.
[22] J.H. Koo, Polymer nanocomposites: processing, characterization, and applications. New York: The McGraw-Hill Companies, Inc; 2006.
[23] S.I. Durowaye, G.I. Lawal, M.A. Akande, V.O. Durowaye, Mechanical properties of particulate coconut shell and palm fruit polyester composites, Int. J. Mat. Eng. 4 (2014)141-147, https://doi.org/10.5923/j.ijme.20140404.04.
[24] D.G. Raghavendra, Mechanical And tribological behavior of nanofiller reinforced polymer nanocomposite, PhD Thesis, Department of Mechanical Engineering, National Institute of Technology Rourkela, 2015.
[25] J. Sarki, S.B. Hassan, V.S. Aigbodion, J.E. Oghenevweta, Potential of using coconut shell particle fillers in eco-composite materials, J Alloys Comp. 509 (2011) 2381-2385, https://doi.org/10.1016/j.jallcom.2010.11.025.
[26] S.M. Sapuan, M. Harimi, M.A. Maleque, Mechanical properties of epoxy/coconut shell filler particle composites, Arabian J. Sci. Eng. Section B: Eng. 28 (2003) 171-181.
[27] H.P.S. Abdul Khalil, M. Masri, K.S. Chaturbhuj, M.R.N. Fazita, A.A. Azniwati, N.A.S. Aprilia, E. Rosamah, D. Rudi, Incorporation of coconut shell based nanoparticles in kenaf/coconut fibres reinforced vinyl ester composites, Mat. Res. Express 4 (2017): 035020.
[28] M. Tangimul Islam, S. Das, J. Saha, D. Paul, M. Islam, M. Rahman, M. Khan, 2017. Effect of coconut shell powder as filler on the mechanical properties of coir-polyester composites, Chem. Mat. Eng. 5 (2017) 75-82, https://doi.org/10.13189/cme.2017.050401.
[29] J. Behin, L. Rajabi, H. Etesami, S. Nikafshar, Enhancing mechanical properties of epoxy resin using waste lignin and salicylate alumoxane nanoparticles, Korean J. Chem. Eng. 35 (2018) 602-612, https://doi.org/10.1007/s11814-017-0301-0.
[30] P. Yu, H. He, C. Jiang, D. Wang, Y. Jia, L. Zhou, D.M. Jia, Reinforcing styrene butadiene rubber with lignin-novolac epoxy resin networks, Express Polym. Letters 9 (2015) 36-48, https://doi.org/10.3144/expresspolymlett.2015.5.

[31] G.U. Raju, S. Kumarappa, Experimental study on mechanical properties of groundnut shell particle-reinforced epoxy composites, J. Reinforced Plastics Compos. 30 (2011) 1029-1037.
[32] V. Singh, J. Bhaskar, Physical and mechanical properties of coconut shell particles reinforced-epoxy composite, J Mat. Environ. Sci. 4 (2013) 227-232.
[33] Z. Mosadeghzada, I. Ahmad, R. Daika, A. Ramli, Z. Jalaludin, Preparation and properties of acacia sawdust/UPR composite based on recycled PET, Malaysian Polym. J. 4 (2009) 30-41.
[34] J. Ervina, M. Mariatti, S. Hamdan, Effect of filler loading on the tensile properties of multi-walled carbon nanotube and graphene nanopowder filled epoxy composites, Procedia Chem. 2016;19:897-905, https://doi.org/10.1016/j.proche.2016.03.132.
[35] B.M. Wood, S.R. Coles, S. Maggs, J. Meredith, K. Kirwan, Use of lignin as a compatibiliser in hemp/epoxy composites. Compos. Sci. Technol. 71 (2011) 1804-1810, https://doi.org/10.1016/j.compscitech.2011.06.005.
[36] C.G.D. Oliveira, F.M. Margem, S.N. Monteiro, F.P.D. Lopes, Comparison between tensile behavior of epoxy and polyester matrix composites reinforced with eucalyptus fibers. J Mat. Res. Tech. 6 (2017) 406-410, https://doi.org/10.1016/j.jmrt.2017.08.002.
[37] K. Rahul, K. Kausik,S. Prasanta, S. Bhowmik, Study of mechanical properties of wood dust reinforced epoxy composite, Procedia Mat. Sci. 6 (2014) 551-556, https://doi.org/10.1016/j.mspro.2014.07.070.
[38] K. Hardinnawirda, A. Sitirbiatull, Effect of rice husks as filler in polymer matrix composites, J. Mech. Eng. Sci. 2 (2012) 181-186, http://dx.doi.org/10.15282/jmes.2.2012.5.0016.
[39] S. Shuhadah, S.A. Ghani, LDPE-isophthalic acid-modified egg shell powder composites (LDPE/ESPI), J. Phys. Sci. 20 (2009) 87-98.
[40] C.G.H. Manjunatha, A.K.S. Sabeel, Experimental characterization of coconut shell particle reinforced epoxy composites, J. Mat. Environ. Sci. 8 (2017) 1661-1667.
[41] A. Gopinath A, M.S. Kumar, A, Elayaperumal, Experimental investigations on mechanical properties of jute fiber reinforced composites with polyester and epoxy resin matrices, Procedia Eng. 97 (2014) 2052-2063, https://doi.org/10.1016/j.proeng.2014.12.448.
[42] A.T. Vasu, C. Reddy, S. Danaboyina, G.K. Manchala, M. Chavali, The improvement in mechanical properties of coconut shell powder as filter in HDPE composites, J. Polym. Sci. Appl. 1 (2017) 1-6,
[43] M. Mohammed, A.R. Rozyanty, B.O. Beta, T. Adam, A.F. Osman, I.A.S. Salem, O.S. Dahham, M.N. Al-Samarrai, A.M. Mohammed, Influence of weathering effect in natural environment on thermal properties hybrid kenaf blast/glass fibre and unsaturated polyester composite, J Physics: Conf. Series 908 (2017); 012004, https://doi.org/10.1088/1742-6596/908/1/012004.
[44] M.M. Ibrahim, A. Dufresne, W.K. El-Zawawy, F.A. Agblevor, Banana fibers and microfibrils as lignocellulosic reinforcements in polymer composites. Carbohydrate Polym. 81 (2010) 811-819, https://doi.org/10.1016/j.carbpol.2010.03.057.
[45] M.A. Khan, K.M. Idriss Ali, S.C. Basu, IR studies of wood plastic composites, J. Appl. Polym. Sci. 49 (1993) 1547-1551, https://doi.org/10.1002/app.1993.070490905.
[46] M. Ibrahim, Preparation of cellulose and cellulose derivative azo compound, Cellulose 9 (2000) 337-349.
[47] S.N. Monteiro, V. Calado, R.J.S. Rodriguez, F.M. Margem, Thermogravimetric stability of polymer composites reinforced with less common lignocellulosic fibers – an overview, J. Mat. Res. Tech. 1 (2012) 117-126, https://doi.org/10.1016/S2238-7854(12)70021-2.
[48] M.M. Kabir, M.M. Islam, H. Wang, Mechanical and thermal properties of jute fibre reinforced composites, J. Multifunctional Compos. 1 (2013) 71-77.

IOP Conf. Series: Materials Science and Engineering **943** (2020) 012022 doi:10.1088/1757-899X/943/1/012022

A study on the challenges of implementing green building concept in Sarawak, Malaysia

Yee Yong Lee[1], Muhd Syaznie Ikqmal Azmi[1] and Yeong Huei Lee[2*]

[1]Department of Civil Engineering, Faculty of Engineering, Universiti Malaysia Sarawak, 94300 Kota Samarahan, Sarawak, Malaysia.
[2]Department of Civil and Construction Engineering, Faculty of Engineering and Science, Curtin University Malaysia, CDT250, 98009 Miri, Sarawak, Malaysia.

Email: yeong.huei@curtin.edu.my

Abstract. Malaysia is looking forward in achieving developed country status which drives several strategic plans in construction, as it plays a vital role in country"s economic growth and development. Construction Industry Transformation Programme (CITP) has been introduced to boost its performance with sustainable construction. However, most of the construction industry players are not involved in sustainable construction especially on green building concept and thus decelerating the development. Therefore, this study is conducted to identify the drivers and challenges faced by the construction industry players through questionnaire survey where the market demand from stakeholders is placed primarily in implementing green building. The findings showed that awareness and understanding of the stakeholders about green building concept are lacked due to unforeseen circumstances. Detail explanations or descriptions on the benefits and contributions from green building application are needed to be delivered to stakeholders. A framework is proposed for future implementation by improving the education and delivered information, as well as providing the mechanism of promoting the green building concept among the stakeholders.

1. Introduction

During the last 20 years, the average contribution of services, manufacturing, agriculture and construction industries in gross domestic product (GDP) were 48.3%, 28.2%, 9.3% and 4.1% respectively [1]. Being the top 4 GDP contributors, construction industry provides significant developments to the country. As one of the strategic planning, Construction Industry Transformation Programme (CITP) has been introduced to boost its performance with sustainable features. However, most construction industry players are not involved in this transformation and thus decelerating the development. There are four strategic thrust that has been highlighted in the programme. One of the four strategic thrusts is Environmental Sustainability. In order for the environmental sustainability policy to be achieved, green building is one of the current practice approached by Malaysia"s government to implement in the construction industry.

Green building, also known as sustainable building, is the method of building structures using environmental- friendly and resource- efficient processes throughout the life cycle of a building, starting with design, construction, operation, maintenance, renovation and deconstruction [2]. The "green movement" started around 1960"s where people have the impression that Earth"s resources were being salvage rather than smartly exploited. Moreover, the increases of oil prices in 1970"s contributed the factor which a lot of research and activity has been conducted to improve energy efficiency and finding renewable energy sources.

Different organizations from all over the world have set up their own codes and rating systems to evaluate a building towards sustainability. These are some examples of green building rating guidelines: BREEAM (United Kingdom), CASBEE (Japan), LEED (United States), Green Star (Australia) and Green Mark (Singapore). In Malaysia, Green Building Index is set up as a rating tool for green building. Green Building Index (GBI) is developed in 2009 by Pertubuhan Arkitek Malaysia (PAM) and the Association of Consulting Engineers Malaysia (ACEM). Recently in 2015, Malaysian Carbon Reduction and Environmental Sustainability Tool (MYCREST) has been developed to manoeuvre, support, quantify, and therefore decreasing the impact of the built environment in terms of carbon emissions and environmental impact, and taking consideration for a comprehensive perspective of the built environment through the life cycle.

There is a rapid development in the capital of Sarawak, Kuching. It was found that there is a low application of green building in Kuching. Therefore, this paper addresses the perceptions and opinions from the industry players on the implementation of green building concept in their construction practices. The obtained information should be useful to assist in adapting the green building practices in their construction work. A framework of solution is proposed to the relevant organization such as the state government and public sector in finding the alternatives for construction players to implement the green building concept and thus, promoting the Malaysia national agenda in achieving sustainable construction.

2. Questionnaire Survey

Quantitative method was applied for data collection. According to [3], quantitative data collection concentrates on surveys, statistics and measurement, pre-existing data and questionnaires. When analysing quantitative research, validity and reliability are necessary in order to evaluate theories, developed facts and state the relationship between the variables [4]. The data for this research use both primary and secondary data to achieve the objective. Primary data was obtained by from questionnaire survey. The questionnaires have been distributed electronically and manually to the developers in Kuching area. Secondary data was collected based on the literature review of the past studies, journals and articles. Data analysis was performed by data collection from the questionnaire survey.

The questionnaire were categorized into 3 parts, namely

- Part A: Respondent Background (Demography)
- Part B: Drivers Contributing In Implementation of Green Building
- Part C: The Challenges in Implementing Green Building Concept.

Part A provided the general background information about the respondents and their organization. Part B and C were focusing on the research questions in this study. For part B, it was to identify the drivers contributing to implementation of green building concept among the developers. For part C, the question aimed to identify the challenges by the developer in order to implement green building concept in their construction project.

As recommended by [5], the Likert scale of five points (i.e. 1= strong disagreement, 2= disagreement, 3= neutral, 4= agreed and 5= strongly agreed) was used in Part B and C. Likert scale used to determine the respondents" level of agreement on the variables. The population of this research study were the developers certified under the Sarawak Housing and Real Estate Developers" Association, SHEDA. There were 154 of developer companies located in Kuching, Sarawak which registered under SHEDA. Based on [6], the population was 160, and needed 113 responses.

3. Data Analysis

3.1. Demographic analysis

The respondents were developers companies specifically located in Kuching, Sarawak. A total of 160 forms has been distributed to the developers. The number of returned questionnaire was 114 and making the response rate of 71.2%. For the role/position of respondents, 32 or 28% of the respondents were civil engineers, 19 or 17% were quantity surveyors, 14 or 12% were general manager, 12 or 11% were sales manager and 10% for architects, 10 or 9% were admin, 9 or 8 % were project manager and the 6 or 5% were secretary.

For the respondent"s level of education, 68% owned a Bachelor Degree, 30% possessed a Diploma and 2% had a Master Degree. Their knowledge and education on Green Building also have been assessed in this questionnaire where 51% answered no while 49% answered yes. For the respondent"s experience aspect in construction industry, 34% have 7 to 9 years of experience, 28% have 10 years and above, while 21% have 4 to 6 years of experience, and 17% have 1 to 3 years of experience in the industry.

In the aspects of the company"s experience, 42% companies have been established 11 to 15 years, 31% which established between 16 years and above, 19% operated between 6 to 10 years and 8% have been run for 0 to 5 years. The respondents also have been asked on their involvement in green building project where 89% not involved in, while 11% involved in green building project.

3.2. Drivers

Based on the descriptive analysis, the result in Table 1 shows that the market demands have the highest mean value (4.66) among the other drivers. The lowest mean value was identified for the educational programs (4.25).

Table 1 Drivers in implementing green building concept

Variable	**Mean**	**Rank**	**Means of Responses***
Market Demand	4.66	1	Very Imp.
Economic Incentives from Government	4.50	2	Very Imp.
Establishment of Green Organization	4.30	3	Imp
Enforcement of Environmental and Sustainable Legislation	4.27	4	Imp
Educational Programs	4.25	5	Imp

(Note: *mean 1.0 to 1.49 = unimportant, 1.5 to 2.49 = less important, 2.5 to 3.49 = moderately important, 3.5 to 4.49 = important, 4.5 to 5.0 = very important)

3.2.1. Market demands

There was 71.1 % of respondents strongly agree that the market demand is one of the drivers for construction industry player to implement the green building which is quite a big number compared to the others drivers. There is also 0.9 % of the respondent felt disagree on the statement yet the number not affecting the mean value much as the drivers of market demands scored the highest, 4.66. The market demand mostly contributed by the investment of stakeholders in implementing the green building project. Anyone interested in the process or outcome of a project is a stakeholder. Respondent believed that the strong marketability of green building which contributed by the stakeholders investment in construction market can attracted the involvement of construction industry players in green development project.

3.2.2. Economic incentives from government

Based on the analysis, 59.6 % strongly agree the economic incentive from government is a driver for implementing green building. Moreover, 9.6% remains neutral as they were not sure on it. The final mean value of this variable was 4.50. High cost is needed in implementing green building project due to the limited resources of green technology and materials. With the sufficient amount of incentives and financial support such as tax exemption as a driver, the risk of loss can be lowered to the developers, hence boosting the confidence of developers in implementing the green construction as they will not bear the high cost of green development.

3.2.3. Establishment of green organisation

There is no respondent chose the strongly disagree in this option of driver. Respondents mostly chose strongly agree with contribution of 47.4 % from 114 respondents. Even though the number of strongly agree is high, there is also some respondent which was 1.8 % disagree for the establishment of green organization to be one of the drivers that help in green building implementation.

3.2.4. Enforcement of environmental and sustainable legislation

For the Enforcement of Environmental and Sustainable Legislation variable as the drivers of green building implementation, the result chosen by respondents showed 2.6% for level 2 agreement and 44.7% for level 5 agreement. The majority chose strongly agree while minority chose to disagree on the statement. The mean value still considered high with mean value of 4.27. The legislation established by government are considered to be the first step to trigger construction industry players and sending a clear illustration that government is moving towards sustainable development. Hence, most of the respondents believed that this variable is one of the important drivers for green development in Kuching.

3.2.5. Educational program

By referring to the result analysis, the mean value is 4.25. Majority of the respondents which is 45.6 % chose agreed that educational programs as the drivers of green building implementation. The lowest percentage chooses by the respondent is neutral with 14.9 %. This driver tends to educate and introduce people about the green building concept which is important to boost the market demands.

3.3. Challenges

Based on the descriptive analysis, the results in Table 2 showed that the lack of market demands had the highest mean value of 4.40. Meanwhile, the lowest mean value, 3.20, is the challenge of time consuming variable. Nine out of the ten variables obtained mean value range between from 3.5 to 4.49 which can be denoted as an „important" variable. Only one variable scored mean value in the range of 2.5 to 3.49 which is time consuming variable. Hence, the variable is denoted as „moderately important" in the challenges aspect.

Table 2 Challenges in implementing green building concept

Variable	Mean	Rank	Means of Responses*
Lack of Demand from Stakeholders	4.40	1	Imp.
Lack of Public Awareness	4.35	2	Imp.
Lack of Incentives	4.31	3	Imp
Lack of Building Codes and Regulations	4.27	4	Imp
Lack of Technology	4.27	5	Imp
Lack of Expertise and Professional Knowledge	4.26	6	Imp
Higher Final Product Price	4.08	7	Imp
Higher Investment Cost	4.01	8	Imp
Risk of Investment	3.65	9	Imp
Time Consuming	3.20	10	Moderately Imp.

(Note: *mean 1.0 to 1.49 = unimportant, 1.5 to 2.49 = less important, 2.5 to 3.49 = moderately important, 3.5 to 4.49 = important, 4.5 to 5.0 = very important)

3.3.1. Lack of demand from stakeholders

There was 49.1 % strongly agree that the lack of market demand from stakeholder is one of the challenges for construction industry player to implement the green building. There was also 1.8 % of the respondent disagreed on this challenge statement but the highest value of 4.40 agreed with this challenge. The results also were corresponding to the top most driver which is market demand for implementing green building concept by the respondents. In other words, the demand from the market is considered as top most important criteria to implement green building concept in Kuching Sarawak. Sharif et al. [7] stated that the practices of sustainable development in Asia is caused by the lack of demand and it is proven the validity of the obtained data.

3.3.2. Lack of public awareness

Based on the analysis, 49.1 % was obtained from „agree" for the variable of lack of public awareness. Moreover, 1.8% remained disagree that the buyer awareness about green building is still lacking. The final mean value of this variable is 4.35. A lot of researchers agreed that public awareness related closely to demand. Said et al. [8]; Sharif et al. [7]; Zainordin et al. [9] found that public awareness about green development can drive and create demand for the construction industry. Sharif et al., [7] stated that public should be educated and informed on the importance of green building to the environment in order to create a surge of demands of green building development in Malaysia. From developers" view, without these demands from the public, the implementation may decelerate.

3.3.3. Lack of incentives

Majority of the respondents agreed that the lack of incentives barricade them from implementing green building with a percentage of 44.7%. The lowest value of 12.3% chose neutral. The mean value scored was 4.31. Respondents felt that the amount of incentives and support provided by the government are less as they are unable to cope the high price to construct green building. The results have been proven with [10] where they found that the current incentives are still inadequate to attract construction players to involve in green building project.

3.3.4. Lack of building codes and regulations

Based on the analysis, 53.5 % agreed towards the majority choice of lack of building codes and regulations. Moreover, 9.6% remained neutral as they were not sure about the challenge of implementation. The overall mean value of this variable was 4.27. Majority of the respondents agreed that current frameworks of regulations, standards or guidelines for green development established by authority are complicated to be complied. Construction players are still confused which codes and regulations needed to be applied as the guidelines are not being standardized. Current rules and regulations should be revised in order to support green construction development [10].

3.3.5. Lack of technology

For the lack of technology variable as the challenges of green building implementation, the results that have been chosen by respondents showed the range of 5.3 % and 48.2 %. The majority chose strongly agree while minority chose to disagree on this challenge. The mean value was 4.27. Most of the technology needed to be imported from the more developed and experienced countries. The construction industry players believed that the need of the technology to be imported will potentially contribute to the increment of the logistic cost.

3.3.6. Lack of expertise and professional knowledge

By referring to the result analysis, the mean value for the total respondents was 4.26 for this variable. Majority of the respondents which was 50.9 % chose agree that construction industry has low number of experts in green building that can give them guidance throughout the green development project. The lowest percentage of 1.8% chose by the respondent was disagreed with this challenge. Green building project considered as complex which needed a proper planning and monitoring. Mistake made during the process will increase their cost. Hence, developers are afraid to implement the green development with the lacking of expertise and professional knowledge.

3.3.7. Higher final product price

The mean value for the total respondents answered was 4.08. Majority of the respondents which was 39.5 % chose agree that high selling price made the green building to be less popular among the public where buyers prefer a low and affordable price. The lowest percentage of 5.3% by the respondent was the option of disagree. The challenge is high selling price towards end user and reduces the interest of developers in the participation of green development project.

3.3.8. Higher investment cost

From the analysis, the mean value for the total respondents answered was 4.01. Majority of the respondents, 41.2 %, chose „agree" that higher investment cost as the challenges of green building implementation. The lowest percentage chose by the respondent was neutral with 21.1 %. To implement green practices in the construction project, it required huge amount of cost compared to the uses of conventional project [7]. Research of [11] found that the high financial of risk may cause a lot of construction industry players decided not to involve in green building project. To support the statement, study of [12] also proved that the construction industry players are not interested in sustainable construction as they are concern about the upfront cost.

3.3.9. Risk of investment

For the risk of investment variable as the challenges of green building implementation, the results chose by respondents showed a range of 0.9 % to 40.4 %. The majority chose agree while minority chose strongly disagree on this challenge. As the level of agreement is varies between strongly disagree to strongly agree, the mean value scored for this variable was 3.65. It is estimated that the initial capital costs are generally in the range of 1 to 25% more than the conventional building. Construction industry players refuse to bear the high initial investment costs [13].

3.3.10. Time consuming

The results chose by respondents showed from 7.0 % to 38.6 %. The majority chose neutral while minority chose to strongly disagree on this statement. As the level of agreement is various between strongly disagree to strongly agree, the mean value scored for this variable is low among others variables, which was 3.20. Green building construction in Malaysia required a lot of time as the industry is shortage of expertise and resources. In order to have an expert, it require the construction player to send their staff for training which are time consuming as the company will have a shortage of hands to handle the project which lengthen the project time for completion.

4. Conclusions and recommendations

The drivers and challenges in implementing green building and faced by construction industry has been determined. The market demands and lack of market demand were identified as the number one drivers and challenges that needed to be overcome in order to promote sustainable construction in Kuching, Sarawak.

Some actions need to be initiated by solving the market demand problem. In order to solve this, level of awareness and understanding of the stakeholders must be boosted by improving the education, information and the mechanism of promoting the green building concept among the stakeholders.

Acknowledgements

The authors would like to acknowledge the financial support to carry out the research from the Universiti Malaysia Sarawak by research grant Small Grant Scheme (SGS) for vote number of F02/SGS/1797/2019. Special gratitude also extended to Curtin University Malaysia for conference related supports.

References

[1] Khan, R. A., Liew, M. S., & Ghazali, Z. Bin. (2014). Malaysian Construction Sector and Malaysia Vision 2020: Developed Nation Status. Procedia - Social and Behavioral Sciences, 109, 507–513. https://doi.org/10.1016/j.sbspro.2013.12.498

[2] U.S. Environmental Protection. (2016). Basic Information of Green Building.

[3] Anupama, P. (2017). Quantitative Data Analysis and Representation, 7(3), 4853–4856.

[4] Chua, Y.P. (2012). Mastering Research Methods. McGraw-Hill Education (Malaysia) Sdn. Bhd.

[5] Losby, J. & Wetmore, A. (2012). CDC coffee break: using Likert scale in evaluation survey work. http://www.cdc.gov/dhdsp/pubs/docs/cb_february_14_2012.pdf

[6] Krejcie R.V. and Morgan D.W. (1970). Determining sample size for research activities. Educational and Psychological Measurement, 30:607-610.

[7] Sharif, S., Nizam, S., & Pitt, M. (2012). Implementation of Green Building Projects in Malaysia : Issues and Challenges, 1–13.

[8] Said, I., Osman, O., Shafiei, M. W. M., Razak, A. A., & Kooi, T. K. (2009). Sustainability in the Housing Development among Construction Industry Players in Malaysia. The Journal of Global Business Management, Vol. 5,15.

[9] Zainordin, N., Nur, S., & Mohd, A. (2017). Green Building Concept Implementation for Residential Project: An Insight Among Construction, (July), 1–5.

[10] Samari, M., Godrati, N., Esmaeilifar, R., Olfat, P., & Shafiei, M. W. M. (2013). The investigation of the barriers in developing green building in Malaysia. Modern Applied Science, 7(2), 1–10. https://doi.org/10.5539/mas.v7n2p1

[11] Wilson, J. L., & Tagaza, E. (2012). [GB] Wilson & Tagaza, 2004.

[12] Powmya, A., & Abidin, N. Z. (2014). The Challenges of Green Construction in Oman. International Journal of Sustainable Construction Engineering & Technology, 5(1), 2180–3242. https://doi.org/10.5829/idosi.mejsr.2014.21.06.21202

[13] Niig, L. (2016). The Challenges in Implementing Green Building Concept – A Study Among Contractors in Sibu, Sarawak (Malaysia).

IOP Conf. Series: Materials Science and Engineering **943** (2020) 012023 doi:10.1088/1757-899X/943/1/012023

Monolayer grafting of APTES modified graphene oxide on silica coated magnetite: Synthesis, characterization and application in heavy oil removal

Chung Ye Wong[1], Inn Shi Tan[2,*], Henry Chee Yew Foo[2], Daniel Kuok Ho Tang[3], Yie Hua Tan[2], Yin Fong Yeong[3], Man Kee Lam[3, 4]

[1]Department of Petroleum Engineering, Curtin University Malaysia, 98009 Miri, Sarawak, Malaysia

[2]Department of Chemical Engineering, Curtin University Malaysia, 98009 Miri, Sarawak, Malaysia

[3]Department of Environmental Engineering, Curtin University Malaysia, 98009 Miri, Sarawak, Malaysia

[4]Department of Chemical Engineering, Universiti Teknologi PETRONAS, 32610, Bandar Seri Iskandar, Perak

[5]Centre for Biofuel and Biochemical Research, Institute of Self-Sustainable Building, Universiti Teknologi PETRONAS, 32610 Seri Iskandar, Perak, Malaysia

*Corresponding author: innshitan@outlook.com or tan.s@curtin.edu.my

Abstract. Petroleum industries have been creating negative environmental and ecological impacts due to the formation of oilfield produced water (OPW) during refining processes. Graphene oxide (GO) as a chemical demulsifier was found non-recyclable; hence, it could cause environmental pollution. Thus, magnetic graphene oxide (M-GO) was proposed to prevent any possible contamination. In this work, GO, and M-GO were successfully synthesized and the produced M-GO composites were found coated with silica-coated magnetite (Fe3O4@SiO2) nanoparticles. As such, they possess excellent superparamagnetic properties. The demulsification tests demonstrated that 0.25 wt% of M-GO aqueous suspension achieved 98.53% of demulsification efficiency on the as-prepared crude oil-in-water emulsion, which was more superior than GO as the suspending oil floccules attached to M-GO can be separated effectively using magnetic separation. The optimizations of the demulsification process were conducted, and significant findings were included as the major novelty of this work. The recycling tests have proven that M-GO can be reused to perform demulsification effectively for ten times. Despite being able to reduce the demulsification costs in the industry, this research suggests that the excellent properties of M-GO can minimize the risk of polluting the environment due to leakage of demulsifier through the application of magnetic separation.

1. Introduction

In the petroleum industry, large amounts of sewages such as sludge, oil, and water are generated from different activities and processes at different stages. For instance, oilfield produced water (OPW) can mainly be found from the refinery sector of the petroleum industry, as vast volumes of water are required in the oil refineries [1]. OPW is known as an oil-in-water emulsion (O/W emulsion), and it is the largest waste stream that contains several organic and inorganic compounds, such as suspended solids, oil, and grease (O&G), water-soluble metals, and salts in the petroleum industry. Oils and greases are the primary contaminants in the wastewater produced from the petroleum industry. In offshore oil recovery operations, the produced water can affect the ecological resources and functions, cause environmental pollutions, destruct the natural landscape, and threaten aquatic resources and human health if OPW is discharged to the underground or surface water environment without proper treatment [2, 3]. OPW can even increase the lethal or sublethal effect of cellular functions, and it is also the root of high toxicity to the catalysts in the refining process and the corrosion of production facilities, pipe surface and distillation equipment. Oil-in-water emulsions with oil droplet size larger than 20 μm can be treated easily using physical treatments such as gravity separation, centrifugation, and ultrasonic application. In contrast, O/W emulsions with emulsified oil droplets (oil droplet size smaller than 20 μm) are difficult to be treated by common physical treatments due to its high stability of the oil-water interface protective film [4]. Thus, O/W emulsions in the wastewater (OPW) are immensely unsolicited in the petroleum industry, and they have to be treated by an effective demulsification process. Demulsification techniques such as membrane, chemical, biological, and electrical demulsification are reported to have some shortcomings such as long separation time, low separation efficiency, high cost, the formation of pollutants, fouling issues and poor recyclability [5]. Consequently, new sustainable methods focusing on economic savings and environmental preservation are essential to be invented for oily wastewater (OPW) treatment [6, 7].

Graphene oxide (GO) possesses small size (nano), large specific surface area, and intrinsic optical properties. As a result, many industry experts and researchers have been inventing useful applications of GO [8]. For instance, nano-size GO was proposed for biomedical applications [8]. GO is also an excellent amphipathic surfactant which can reduce the interfacial energy between the interfaces of oil and water [9, 10]. In emulsion systems, GO can be applied in both emulsion stabilization [11] and demulsification processes [12]. It was discovered that GO could demulsify O/W emulsions effectively by causing repulsion force (nonbonding π-π interactions) between oxygenated functional groups of GO and the aromatic rings of asphaltene in the interlayer space [13]. GO as a chemical demulsifier can demulsify oil-in-water emulsion efficiently at low costs, and it can be synthesized efficiently using Hummers method. However, the Hummers method may lead to the liberation of toxic gases such as dinitrogen tetroxide (N_2O_4) and nitrogen dioxide (NO_2) [14]. Then, sodium nitrate ($NaNO_3$) was found to be replaceable by a mixture of sulphuric acid (H_2SO_4) and phosphoric acid (H_3PO_4) in the Hummers method to produce exceptional quality of GO without production of toxic gases [15].

Consequently, the modified Hummers method (Tour method) became more widely accepted due to its strengths, such as equivalent conductivity, able to produce GO on a large scale and emit zero toxic gases. However, nano-size GO could not be recycled effortlessly when they are well dispersed in a solvent, leading to the causes of negative ecological and environmental impacts and the threats of human health if they are released to the surface along with the treated water [13]. Various kinds of synthesis methods of magnetic-functionalized GO using γ-Fe_2O_3/Fe_3O_4 compounds were invented to further develop the recyclability and reusability of GO for industrial applications, and they were reported to be successful [16-18]. Therefore, it is proposed to decorate GO nanosheets with magnetic nanoparticles (MNPs) in order to produce recyclable magnetic GO nanosheets for the demulsification of OPW, and it is expected to be feasible, high efficiency, and cost-effective. Magnetic functional particles with interfacial activities can accumulate at the interlayer space between oil and water, and impart their magnetic properties on the dispersed droplets. As a result, they can separate droplets from multiple phases with the aid of an external magnetic field.

Additionally, some influential parameters in chemical demulsification process that can be altered to increase or decrease the stability of the emulsion. It was stated that the hydrophilic-lipophilic deviation (HLD) value could be shifted from negative (O/W emulsion) to approximately zero through increasing the salinity of emulsions, hence improving the demulsification effectiveness [19]. The impacts of pH of water on the stability of emulsion usually depend on the brine compositions and oil phase in the emulsion [20]. Generally, the surfactant hydrophilicity increases as the pH value of emulsion increases. It was reported that O/W emulsions are more likely to be generated in basic conditions (high pH), whereas W/O emulsions are prone to be generated in acidic conditions (low pH) [21]. Asphaltenic interfacial films are stronger in acidic environments, whereas resinous films are more rigid in the alkaline environment. Nevertheless, extremely low pH (too acidic) environment may cause corrosion of equipment in the treatment plants which results in unnecessary extra cost; hence, precautions have to be taken to achieve optimal pH for demulsification.

Hence, this research aims to study the synthesis of recyclable magnetic graphene oxide (M-GO) by utilizing co-precipitation and a series of cross-linking reactions, including Stöber process. The effect of influential parameters such as salinity and pH of emulsion on the demulsification performance of M-GO has been investigated. Moreover, the recyclability of M-GO has also been tested and proved to reduce the demulsification cost and contamination through the application of magnetic separation after performing the demulsification process.

2. Experimental

2.1 Materials and chemicals

The chemicals H_2SO_4, H_3PO_4, $KMnO_4$, H_2O_2, NaOH, n-Hexane, and NH_4OH were purchased from Merck and used without further treatment. Graphite (<20 μm, synthetic powder), ethanol, 1-(3-dimethylaminopropyl)-3-ethylcarbodiimide hydrochloride (EDC-HCl), N-hydroxy succinimide (NHS), $FeCl_2 \cdot 4H_2O$, $FeCl_3 \cdot 6H_2O$, sodium citrate ($Na_3C_6H_5O_7$), tetraethyl orthosilicate (TEOS), toluene and (3-aminopropyl) triethoxysilane (3-APTES) were purchased from Sigma-Aldrich and used as received. The crude oil sample used to prepare synthetic O/W emulsion was provided by Miri Crude Oil Terminal (MCOT, Malaysia).

2.2 Synthesis of Graphene Oxide (GO)

Graphene oxide (GO) nanosheets were synthesized from graphite powder according to the modified Hummers method (Tour method) [15] with some modifications. Graphene oxide (GO) nanosheets were synthesized from graphite powder according to the modified Hummers method (Tour method) (Marcano, D. C. et al., 2010). Briefly, 0.75 g graphite powder was added into a mixture of acid solutions consisted of concentrated H_2SO_4/H_3PO_4 (90:10 mL) in a 500 mL beaker and stirred at room temperature for 0.5 h. 4.5 g of $KMnO_4$ was then added gradually into the mixture underwater bath stirred for another 0.5 h. Then, the mixture was stirred at 50 °C for another 2 h. The resultant mixture was then ultra-sonicated for 1 h for further exfoliation. The solution was allowed to cool down to room temperature, and 200 mL of deionized ice water was added into the mixture. 3 mL of 30% H_2O_2 was then added into the mixture to remove excessive $KMnO_4$, and the colour of the mixture became brilliant yellow. The reaction was conducted underwater bath conditions, and 0.5 h of stirring was used to achieve a completed reaction. The mixture was centrifuged (4000 rpm, 20 min) and the supernatant was removed, 10% HCl aqueous solution was used to wash the remained solids (GO), followed by the subsequent centrifugation. This cycle was repeated for 4 times until the pH value of the suspension was nearly 6. Then, the final product (GO solid) was dried in a vacuum oven (TVAC-92) at 60 °C under 0.05 MPa for 16 h.

2.3 Synthesis of magnetite nanoparticles (MNPs)

As shown in Figure 1a, the synthesis of magnetic particles (Fe_3O_4) was achieved based on a co-precipitation method [22] with some modifications. Briefly, 0.02 mol (5.41 g) of $FeCl_3 \cdot 6H_2O$ and 0.01 mol (1.99 g) of $FeCl_2 \cdot 4H_2O$ were added into 100 mL of degassed deionized water in a 250 mL three-neck flask under a nitrogen atmosphere with constant agitation at room temperature for 1 h. While keeping the constant flow of nitrogen gas into the three-neck flask, the 5M NaOH solution was added into the reaction solution gradually, until the resultant dispersion pH reached 11, according to Equation (1):

$$Fe^{2+}(aq) + 2Fe^{3+}(aq) + 8OH^{-}(aq) \rightarrow Fe_3O_4(s) + 4H_2O(l) \quad (1)$$

After the system was stirred for 4 h, the produced insoluble precipitates (Fe_3O_4) were separated from the solution using a strong magnet. Lastly, the precipitates were dried in a vacuum oven (TVAC-92) for 12 h at 60 °C and pressurized under 0.05 MPa after they were repeatedly washed with degassed deionized water four times.

2.4 Synthesis of mesoporous Fe_3O_4@SiO_2 core-shell microspheres

The preparation of silica-coated Fe_3O_4 was completed based on the Fe_3O_4 Stöber process [23]. 2 g of the magnetic Fe_3O_4 nanoparticles were added and dispersed into 200 mL of 0.5 M sodium citrate in order to reduce aggregation. The dispersed Fe_3O_4 nanoparticles were isolated using a magnet and then added into a 400 mL mixture of ethanol (75 vol. %), water (23.5 vol. %), and ammonia solution (1.5 vol. %). Then, a 16 mL mixture of TEOS and ethanol solution (4:12 mL) was gradually added to the dispersion under mild agitation at 40 °C in a water bath for 24 h. The as-prepared particles were repeatedly washed with ethanol and water, and subsequently dried at 80 °C in the vacuum oven for 24 h under 0.05 MPa.

2.5 Fe_3O_4@SiO_2 surface functionalization

The core-shell microspheres were functionalized with 3-APTES. 1 g of produced silica-coated Fe_3O_4 nanoparticles was added into 20 mL of toluene and mixed under vigorous stirring for 0.5 h. 20 mL of toluene was used to dilute 10 mL of 3-APTES, and they were gradually added into the mixture under constant agitation in the water bath and N_2 atmosphere at 120 °C for 5 h. Then, the produced amino-functionalized magnetite (Fe_3O_4@SiO_2—NH_2) nanoparticles were collected using a magnet after cooling down to room temperature. Lastly, the nanoparticles were washed by ethanol and toluene and dried in a vacuum oven (60 °C, 0.05 MPa, 12 h).

2.6 Fabrication of GO onto Fe_3O_4@SiO_2 core-shell microspheres

The preparation progressions of M-GO were demonstrated in Figure 1a. Firstly, 0.1 g of produced GO was dispersed in 100 mL of deionized water. The resulting dispersion was sonicated using Probe type Ultrasonic Liquid Processor (SONOMECHANICS LSP-500, 500 W, 20 kHz) at a 40% amplitude for 1 h in order to achieve a GO suspension with a concentration of 1 mg/mL. Then, 40 mg of NHS and 50 mg of EDC.HCl was added to the GO suspension. By using the NaOH solution, the pH of the suspension was adjusted to ~5.5. In order to obtain a homogeneous suspension, the mixture was mechanically stirred for 0.5 h, followed by another 0.5 h of sonication. Then, 0.3 g of prepared amino-functionalized Fe_3O_4 nanoparticles were added into the GO suspension and ultra-sonicated for 0.5 h, followed by mechanical stirring for 1 h at 80 °C for the grafting reaction [17]. The synthesized M-GO was washed with deionized water and separated using a magnet. Lastly, the M-GO solids dried in a vacuum oven (60 °C, 0.05 MPa, 12 h). The mechanism of synthesis of M-GO through cross-linking reaction as illustrated in Figure 1b.

IOP Conf. Series: Materials Science and Engineering **943** (2020) 012023 doi:10.1088/1757-899X/943/1/012023

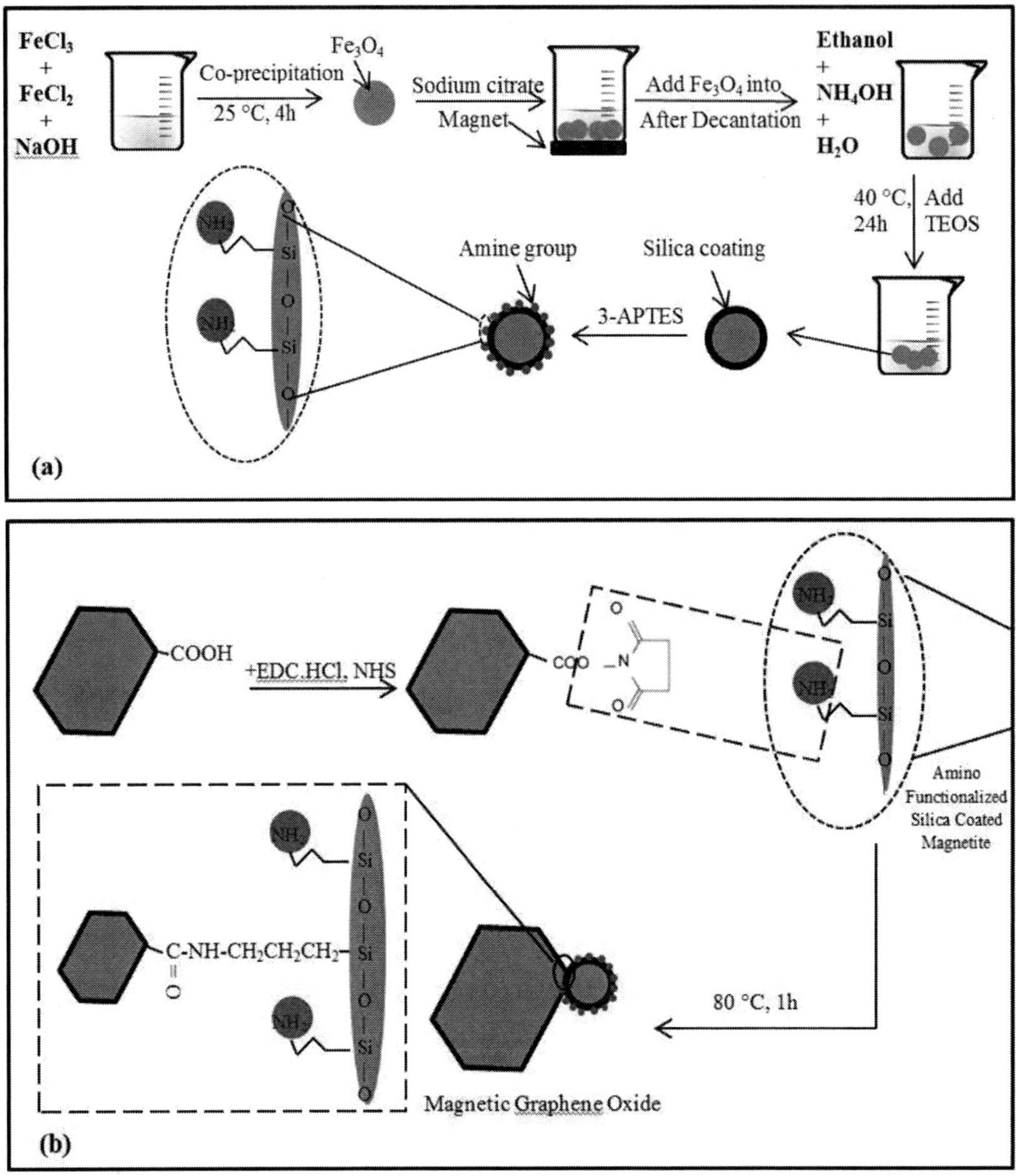

Figure 1. Mechanism of synthesis of desired nanoparticles and magnetic graphene oxide (M-GO). (a) Amino functionalized $Fe_3O_4@SiO_2$ Nanoparticles, (b) M-GO.

2.7 Preparation of crude oil-in-water emulsion

10 g of crude oil sample obtained from MCOT was directly added into 190 g of deionized water in a glass bottle. The mixture was then homogenized by a high-speed homogenizer (Silverson L5M) at 6000 rpm for 30 min; and eventually, a crude oil-in-water emulsion with 5.0 wt% oil content was formed. The prepared emulsion was further used for the demulsification test, recycling test of M-GO, and to investigate the effect of influential parameters on the demulsification efficiency of M-GO.

2.8 Characterization methods

An attenuated total reflection Fourier transform infrared (ATR-FTIR) spectrometer (Agilent Technologies, Cary 630 FTIR) was used to identify the presence of functional groups in the produced GO, various magnetic nanoparticles obtained from each stage and M-GO. The FTIR spectra and characteristic peaks of the samples were obtained and recorded in a computer at a wavelength range between 550 and 4000 cm^{-1} at ambient temperature and pressure.

Zetasizer Nano-ZS (Malvern Instruments, UK) was used to conduct DLS analysis and zeta potential measurement on the produced GO, various nanoparticles, and M-GO. 30 mg of produced GO solid was added and homogenized in 300 mL of deionized water by Silverson L5M Homogenizer at 2000 rpm for 10 min. For DLS analysis, three samples (GO suspension without sonication (GO1), GO suspension with 15 min sonication (GO2), and GO suspension with 60 min sonication (GO3)) were measured in this analysis to investigate the effect of the ultra-sonication technique on the oxidation and exfoliation degree of GO. Besides, the measurements of zeta potential for produced GO, various magnetic nanoparticles obtained from each stage, and M-GO will be performed in aqueous solutions at pH of ~7.

The morphology and physical structure of various magnetic nanoparticles and M-GO were observed by a scanning electron microscope (SEM, Hitachi S-3000N) at an essential magnification at ambient temperature. Dried solid samples were observed on an aluminium stub, and the images were obtained subsequently. An X-ray diffractometer (XRD, STOE Stadi-P diffractometer) was used to investigate the phase and crystal structure of the as-prepared magnetic nanoparticles.

The thermal characteristics of GO and M-GO were analyzed by the thermogravimetric analyzer (TGA) and differential scanning calorimetry (DSC) using a simultaneous thermal analyzer (METTLER TOLEDO, Switzerland). Briefly, 10 mg of samples were weighed into the ceramic crucible with a constant purging of pure nitrogen at a rate of 100 mL/min and heated from 25 °C to 800 °C at a heating rate of 10 °C/min.

The magnetic hysteresis loops of various magnetic nanoparticles and M-GO were investigated using the vibrating sample magnetometer (Lake Shore 7400 Series VSM system) at room temperature to determine their magnetization.

2.9 Demulsification test

The mechanism of demulsification processes, as illustrated in Figure S1. 0.1 wt% of M-GO aqueous suspension was shaken and added into the as-prepared oil-in-water emulsion. The M-GO dosage can be determined based on the concentration of mass. Thorough agitation was required to achieve the demulsification, and the mixture was allowed to settle at room temperature for the oil and water separation. An as-prepared emulsion without any addition of M-GO was employed as a reference for comparison afterwards. A magnet was then used to quicken the residual oil floccules separation from the water phase. Subsequently, a separatory funnel was used to separate the separated water samples from the newly formed mixture.

The demulsification efficiency can be calculated by measuring the remaining oil content in the sample of separated water. The treated separated water was collected from a separatory funnel. Hexane was used to extract the residual oil from the separated water. The residual oil content possesses a linear relation with the absorbance of hexane-diluted-oil, and it was analyzed by UV-visible Spectrophotometer, Lambda 25 UV/Vis Double Beam (Perkin Elmer, USA) at a wavelength of 570 nm (1 cm path length). Each of the tests with the exact dosage of M-GO was conducted three times to get the average values for higher accuracy, and the demulsification efficiency was calculated using Equation (2) and Equation (3):

$$c = \frac{m_o}{V_w} \tag{2}$$

$$DE(\%) = \frac{c_o - c_i}{c_o} \times 100\% \tag{3}$$

Where m_o is mass of the oil, V_w is the water samples volume (in mL); c, c_o, and c_i (mg/mL) are the oil concentration, oil concentration of emulsion and oil concentration of separated water respectively. DE is the calculated demulsification efficiency of M-GO by determining the oil removal percentage from the original emulsion.

2.10 Effect of influential parameters on demulsification efficiency

0.01 wt%, 0.05 wt%, 0.1 wt%, and 0.25 wt % M-GO aqueous suspensions were used to perform demulsification on the as-prepared emulsion respectively to determine the effect of dosage of M-GO on the demulsification performance. O/W emulsions with different pH values (2, 4, 6, 8, and 10) were prepared, and the demulsification efficiencies achieved by 0.25 wt% M-GO aqueous suspensions were compared to investigate on the effect of emulsion pH on the demulsification efficiency. Also, 0.1 wt% of M-GO aqueous suspensions were added into O/W emulsions with different salinity (0 - 0.5 M NaCl) at neutral pH to determine the impact of salinity of O/W emulsion on the demulsification efficiency.

2.11 Recycling and reuse of M-GO

The produced M-GO was expected to be recycled and reused for demulsification. The recyclability of M-GO was investigated in the optimal condition of pH value at 4, with a 0.25 wt% dosage of M-GO added on the O/W emulsion with initial oil concentration at 50 mg/mL. The M-GO was separated and collected using a magnet after performing demulsification. Successively, the collected M-GO was washed by toluene 3 times repeatedly in order to remove the asphaltene molecules which were previously adsorbed onto the M-GO surface. M-GO was then collected and dried in a vacuum dryer at 60 °C for 6 h. Finally, the M-GO was used for the same demulsification test, and the first cycle of the recycling tests was completed. Numerous cycles of recycling tests were performed until the oil content in the separated water was above 10 mg/mL after the demulsification test.

3.0 Results and discussion

3.1 Characterization of graphene oxide, various nanoparticles, and magnetic graphene oxid

3.1 (a) Fourier transform infrared (FTIR) spectra. The FTIR spectra of graphite and the produced GO are shown in Figure 2. From the FTIR spectrum of graphite, a broad absorption band is exhibited at 3559.6 cm^{-1}, which is corresponded to O-H stretching vibration. A broad peak between 3200 cm^{-1} and 3400 cm^{-1} was also observed, which indicated the presence of the C-OH stretch of the hydroxyl functional group and –OH stretching vibrational mode from the carboxylic acid group in the GO sample [24]. The apparent broad peak at 3213 cm^{-1} is corresponded to the water molecules adsorbed onto the GO surface. The narrow peak at 1718.3 cm^{-1} is attributed to the C=O stretch of the carboxyl group of GO, whereas the peak at 1628.8 cm^{-1} indicates the presence of C=C stretches from the incompletely oxidized (sp^2 character) graphitic domain in the GO sample. Furthermore, there are a total of 3 characteristic peaks shown in the fingerprint region of the IR spectra, which are 1155.5 cm^{-1}, 1028.7 $cm^{-1,}$ and 872.2 cm^{-1}. Both of the peaks at 1155.5 cm^{-1} and 1028.7 cm^{-1} correspond to the C-O stretch of an epoxy group and C-O stretching vibration of the carboxylic acid group in the GO samples, respectively [25]. The peak at 872.2 cm^{-1} was corresponded to the out of plane wagging vibration of C-H stretches of the GO [26]. The presence of epoxy and carboxyl groups in the FTIR spectra of GO specifies that the initial extended conjugated π-orbital system of the graphite (precursor) was destroyed, followed by the insertion of oxygen-containing groups into the carbon skeleton during the oxidation process, hence proving the completion of the oxidation process of graphite into GO [27].

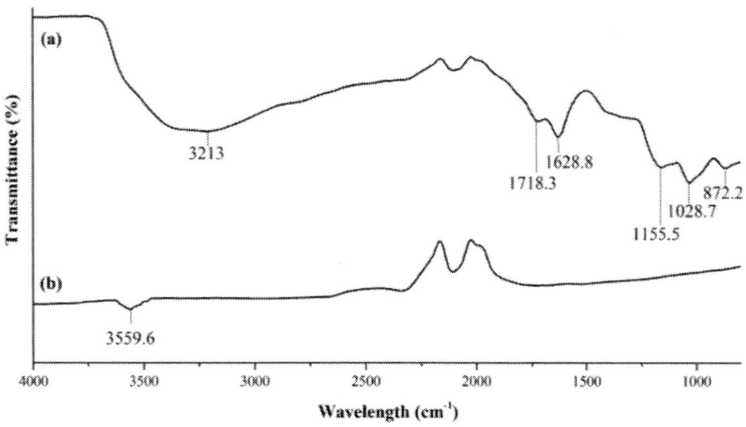

Figure 2. FTIR spectra of (a) graphite and (b) GO sample.

As shown in Figure 3, the IR spectrum of all nanomaterials showed a predominant peak at ~600 cm^{-1}, which indicated the characteristic of Fe-O vibrational stretch from the magnetite (Fe_3O_4) phase. The IR spectrum of Fe_3O_4@SiO_2, Fe_3O_4@SiO_2-NH_2, and Fe_3O_4@SiO_2-GO nanoparticles showed stable broadband at ~1058 cm^{-1}, which was attributed to the Si-O-Si asymmetric stretching vibration or the Fe-O-Si stretching vibration at the silica coating layer, proving that the layer of silica was coated onto the surface of Fe_3O_4. A minor peak at 2907 cm$^{-1,}$ which is attributed to the vibrational stretching of –CH_2 groups of aminopropyl from the 3-APTES, was observed after the silane treatment (addition of 3-APTES) on the Fe_3O_4@SiO_2 nanoparticles. The characteristics peaks showed the vibrational stretching and bending of amino groups at 1528 cm^{-1} and 650 cm^{-1}. This has indicated that the 3-APTES was successfully grated onto the surface of silica-coated Fe_3O_4 nanoparticles. The IR spectrum of M-GO in Figure 4a has shown apparent peaks at 1625-1655 cm^{-1}, 1401 cm^{-1} and 1066 cm^{-1}, which were attributed to the characteristic bending of –OH from hydroxyl groups, a vibration of -C=O from carbonyl groups and stretching of C-O-C from alkoxy groups of GO respectively. A strong broad peak at 3354-3332 cm$^{-1,}$ which is likely attributed to the vibrational stretching of N-H in the amide group (CO-NH-) of MGO and the stretching of –OH from hydroxyl groups have determined the existence of amino-functionalized magnetic particles on the GO surface.

3.1 (b) Dynamic light scattering (DLS) analysis. A full DLS analysis was applied to analyze the polydispersity index (PDI), average crystallite size, and size distribution of the GO nanosheets dispersed in deionized water. With the application of the ultra-sonication technique, the optimal average particle diameter size of the GO3 sample was determined to be 215.7 nm with a PDI value of 0.3 as shown in Table 1. Hence, the application of sonication plays a significant role in controlling the size and the dispersity of GO in its solvent (water). The size measurement of Fe_3O_4 and produced M-GO was conducted using DLS analysis with the application of ultra-sonication technique, and their size was determined to be 37 nm and 304 nm, respectively, each having PDI values of 0.3 and 0.294.

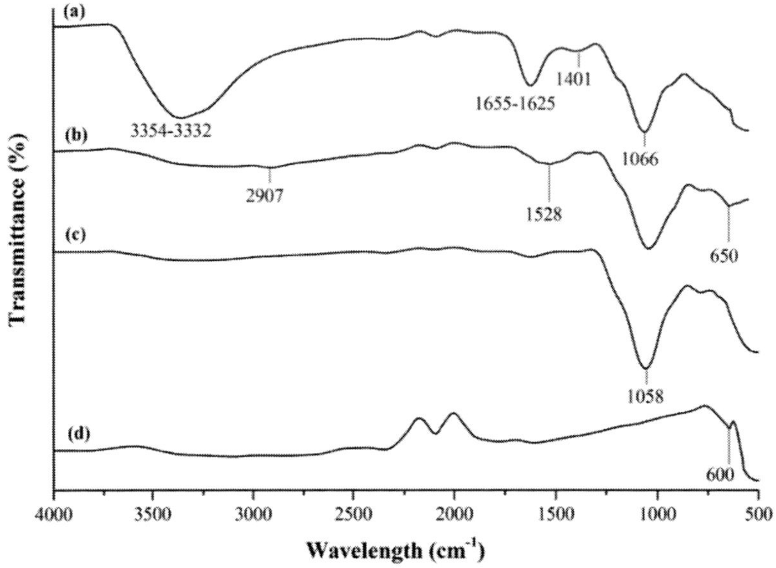

Figure 3. FTIR spectra of (a) M-GO, (b) amino functionalized $Fe_3O_4@SiO_2$ NPs, (c)$Fe_3O_4@SiO_2$ NPs, and (d) Fe_3O_4 NPs.

Table 1. Results of DLS analysis on different GO samples.

Sonication Duration (min)	Sample	Average Particle Diameter (d.nm)	PDI value	Average Size Distribution		
				Size (d.nm)	Intensity (%)	Volume (%)
0	GO1	703.3	0.546	404.2	100	100
15	GO2	353.6	0.371	356.8	84.9	16.5
				4635	11.8	82.1
				77.65	3.3	1.4
60	GO3	215.7	0.301	222.9	95.4	28.6
				5072	4.6	71.4

3.1 (c) Zeta potential measurement. Zeta potential measurements of all products were also completed. As shown in Figure 4, the average surface zeta potential of GO and Fe_3O_4 were determined to be -47 mV and +5.17 mV correspondingly. The coating of the silica layer on the surface of the magnetite was confirmed because the zeta potential value of Fe_3O_4@SiO_2 became -26.8 mV due to the negatively charged silica surface layer [28]. Then, the Fe_3O_4@SiO_2-NH_2 nanoparticles became positively charged with its zeta potential value of +1.53 mV due to the amino groups' protonation on the silica surface. It was determined that the molecules of 3-APTES were successfully grafted onto the silica layer of magnetite. After the cross-linking reaction between GO and produced Fe_3O_4@SiO_2-NH_2, the surface charge of the produced M-GO was measured as -18.2 mV. As compared to GO (-47 mV), the zeta potential value of M-GO is smaller due to the influence of unreached amino groups and acylamino groups on the silica layer [29]. Thus, it has convincingly evidenced that the synthesis of M-GO was successful.

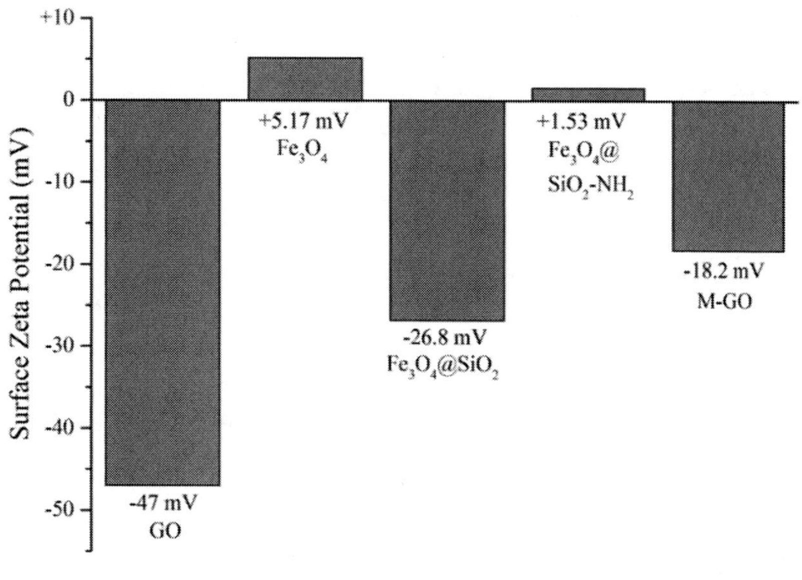

Figure 4. Zeta potential measurements.

3.1 (d) Scanning electron microscope (SEM). Scanning electron microscope (SEM) analysis was performed to investigate the morphology and structure of the Fe_3O_4, Fe_3O_4@SiO_2 nanoparticles, and M-GO, as shown in Figure S2, and a factor of 5000 magnified all the images. The SEM image of Fe_3O_4 shown in Figure S2a indicates that the synthesized spherical Fe_3O_4 nanoparticles were highly aggregated into clusters due to their magnetism [30]. In Figure S2b, it can be observed that the spherical Fe_3O_4 nanoparticles were coated by nonporous silica layers, hence forming near-spherical core-shell Fe_3O_4@SiO_2 nanoparticles. The decreased of the degree of aggregation was evidenced by the bunches of Fe_3O_4 nanoparticles were separated and became more dispersed after wrapped by the nonporous silica layer. The SEM image of M-GO shown in Figure S2c indicates that the magnetic nanoparticles were decorated in uniform bunches and well dispersed on the GO sheets. Therefore, it is suggested that M-GO was successfully formed.

3.1 (e) X-ray diffraction (XRD) analysis. X-ray diffraction analysis was conducted to determine further the phase and structure of GO, Fe_3O_4 nanoparticles and M-GO, and the summary of the results is shown in Table 2. A characteristic diffraction peak of GO was observed at 2θ value of 10.3° (assigned to (001) plane), corresponding to the 0.75 nm interlayer spacing of GO, which is comparatively larger than the typical interlayer spacing of graphite as the GO layers are loosely stacked [31]. From the XRD measurement of synthesized magnetic nanoparticles, numerous well-defined significant peaks at 2θ value of 30.3°, 35.6°, 42.8°, 53.4°, 57.3°, 62.7° which can be assigned to the (220), (311), (400), (422), (511) and (440) plane reflections correspondingly were found to be consistent with the XRD patterns of Fe_3O_4 nanoparticles (JCPDS card No. 19-0629) [32]. The fitted lattice constant of unit cell dimension (*a*) was found to be 8.37 Å, suggesting the good agreement a the cubic spinel crystal structure of Fe_3O_4 nanoparticles [30]. Besides, the diffraction peaks observed from the XRD measurement of M-GO were found to match well with the characteristic peaks of GO and Fe_3O_4 nanoparticles, which indicated the existence of Fe_3O_4 particles in the M-GO composites [33]. Nevertheless, a small amount of maghemite (γ-Fe_2O_3) may exist amongst the Fe_3O_4 nanoparticles due to defection. Similar results were also obtained by other research [34].

Table 2. Summary of results of XRD measurements of GO, Fe_3O_4 and M-GO.

Sample	Diffraction Peak (2θ)	Reflection (Plane)
Graphene oxide	10.3°	(001)
Magnetite (Fe_3O_4)	30.3°	(220)
	35.7°	(311)
	42.8°	(400)
	53.4°	(422)
	57.3°	(511)
	62.7°	(440)
Magnetic graphene oxide	30.1°	(220)
(M-GO)	35.3°	(311)
	43.1°	(400)
	52.9°	(422)
	57.1°	(511)
	62.7°	(440)

3.1 (f) Simultaneous thermal analyzer (STA). Thermogravimetric Analysis (TGA) is the analytical technique that was used to characterize the reaction progress of GO, Fe_3O_4, and M-GO during thermal reduction. Also, derivative weight loss curves were generated to reveal the most significant weight loss point, and the Differential Scanning Calorimetric (DSC) were analyzes for the samples. Based on the TGA curve of GO shown in Figure 5a, the GO sample became thermally unstable and resulted in losing weight when the heating temperature was increased at a rate of 10 °C/min. This has proven that GO has low thermal stability due to the weak Van der Waals forces interaction between its structure [35]. Three main stages of the GO weight loss can be observed. The first apparent gradual weight loss occurred below 100 °C, which is corresponded to the evaporation of absorbed water molecules from the GO surface [36]. The main weight loss was observed between 100 °C and 350 °C, which was attributed to the decomposition of the oxygen functional groups from the GO surface, resulting in the yield of steam, carbon monoxide (CO) and carbon dioxide (CO_2) [37]. It was shown that the pyrolysis of the oxygen-containing functional groups (hydroxyl, epoxy, carboxyl, and carbonyl groups) had caused the increase in a weight loss of GO from 43% to 76% in the temperature range from 150 °C to 350 °C. The oxygen to carbon ratio of the GO was significantly decreased as most of the oxygen functional groups were removed. The third gradual weight loss was observed between 350 °C and 800 °C, which is ascribed to the pyrolysis of the GO carbon skeleton and oxygen functionalities that were more stable [36]. Figure 6a has shown the DSC curve of GO samples, and a few distinct endothermic and exothermic peaks were observed. The first endothermic peak was observed between 110 °C

IOP Conf. Series: Materials Science and Engineering **943** (2020) 012023 doi:10.1088/1757-899X/943/1/012023

and 130 °C, which is corresponded to the vaporization of water molecules. A sharp exothermic peak was observed at 150 °C, which can be attributed to the significant weight loss due to the decomposition of oxygen functionalities such as epoxide and carboxyl groups of GO. The DSC curve also demonstrated the exothermic nature of the pyrolysis of the carbon skeleton in the temperature at 400 - 500 °C.

Figure 5b has shown the TGA curve of Fe_3O_4 nanoparticles in a temperature range between 25 °C and 800 °C under the nitrogen atmosphere. 10% weight loss of Fe_3O_4 was observed at 25 – 110 °C, which has indicated the removal of absorbed water molecules from the Fe_3O_4 surface. Another weight loss (~3%) occurred at 110 – 310 °C was observed, and it corresponded to the removal of water molecules bonded with Fe_3O_4 surface and chemical absorbed water molecules [38]. A minor gradual weight loss (~1%) occurred at 310 - 800 °C was observed, which can be ascribed to the phase transition of magnetite to maghemite due to the higher stability of FeO when the temperature is 600 °C and above [39]. No major exothermic peaks can be observed from the DSC curve of Fe_3O_4 in Figure 6b, and its small weight loss (~15%) has determined that magnetites are highly thermally stable materials.

As shown in Figure 5c, the TGA curve of M-GO demonstrated a weight-loss trend similar to GO. Nevertheless, the amount of weight loss of M-GO at any given temperature is much lesser than that of GO. At 150 °C, the total weight loss of GO is 44%; however, the total weight loss of M-GO is only 10.4%. Additionally, only 18.17% weight of GO residue existed while 69.15% of the original weight of M-GO was left at 800 °C, indicating that M-GO has distinctively higher thermal stability as compared to GO due to the coating of Fe_3O_4 nanoparticles on the GO sheet [40]. As shown in Figure 6c, no abrupt exothermic or endothermic peaks can be observed from the DSC curve of M-GO. Hence, it was determined that M-GO are thermally stable materials [41].

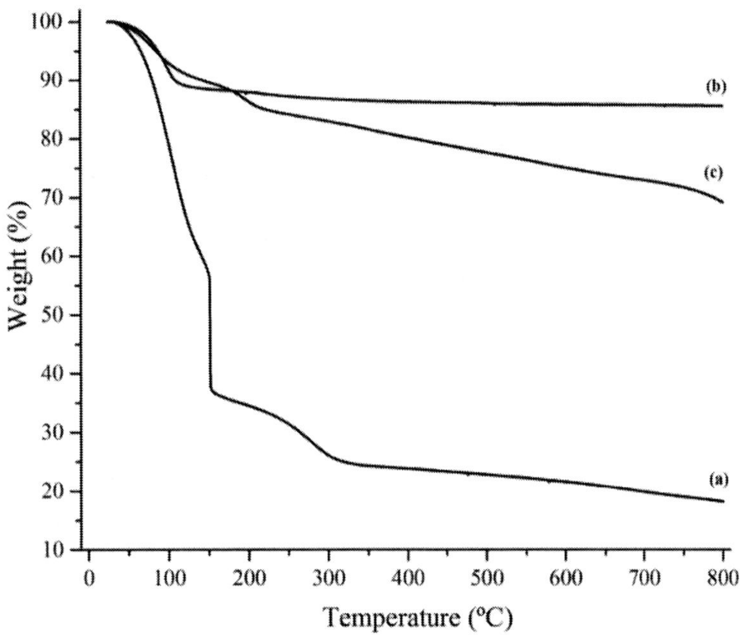

Figure 5. Thermogravimetric weight loss (TGA) curves of (a) GO, (b) Fe_3O_4 and (c) M-GO.

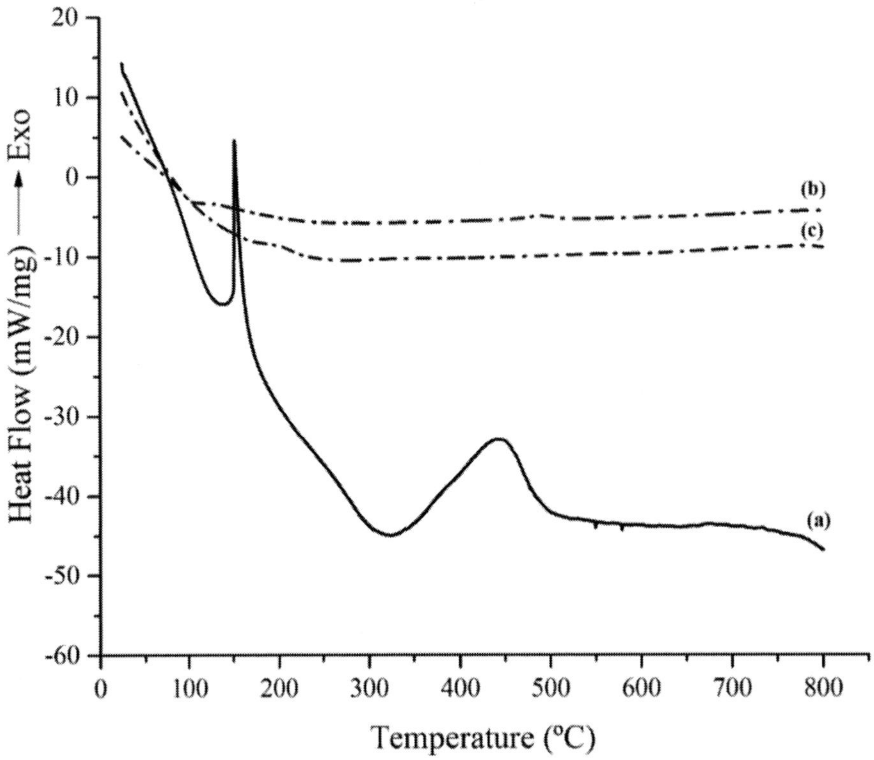

Figure 6. DSC curves of (a) GO, (b) Fe_3O_4 and (c) M-GO.

3.1 (g) Vibrating sample magnetometer (VSM) test. The magnetic properties of various magnetic nanoparticles and M-GO were investigated by Lake Shore 7400 Series VSM system. The magnetic hysteresis loops of Fe_3O_4, Fe_3O_4@SiO_2, amino-functionalized Fe_3O_4@SiO_2 and M-GO has shown in Figure 7 confirmed that they possessed excellent superparamagnetism as the magnetic remanence values of all samples were determined to be 0 emu/g, which indicated that there was no residual magnetization after the removal of external magnetic field [30]. The saturation magnetization of amino-functionalized Fe_3O_4@SiO_2 particles was reduced to 46.67 emu/g, while that for the original Fe_3O_4 nanoparticles was 51.73 emu/g initially. This was ascribed to the coating of nonporous silica layer on Fe_3O_4 surface, and the decoration of amino groups on the silica layer [40]. The saturation magnetization of M-GO was then decreased to 32.67 emu/g due to the presence of non-magnetic GO in the M-GO composite. In other words, the considerably high saturation magnetization of M-GO particles has confirmed the attachment of Fe_3O_4 nanoparticles on the GO sheets [42]. Figure S3 has also shown the magnetic separation of M-GO particles, and it was observed that all M-GO particles were rapidly attracted to the applied magnetic field, followed by the quick re-dispersion of M-GO particles with a slight agitation after the removal of the magnetic field. Thus, it was indicated that M-GO particles which possess excellent magnetic properties are suitable for the practical demulsification. Similar results were reported by other researchers as well [41].

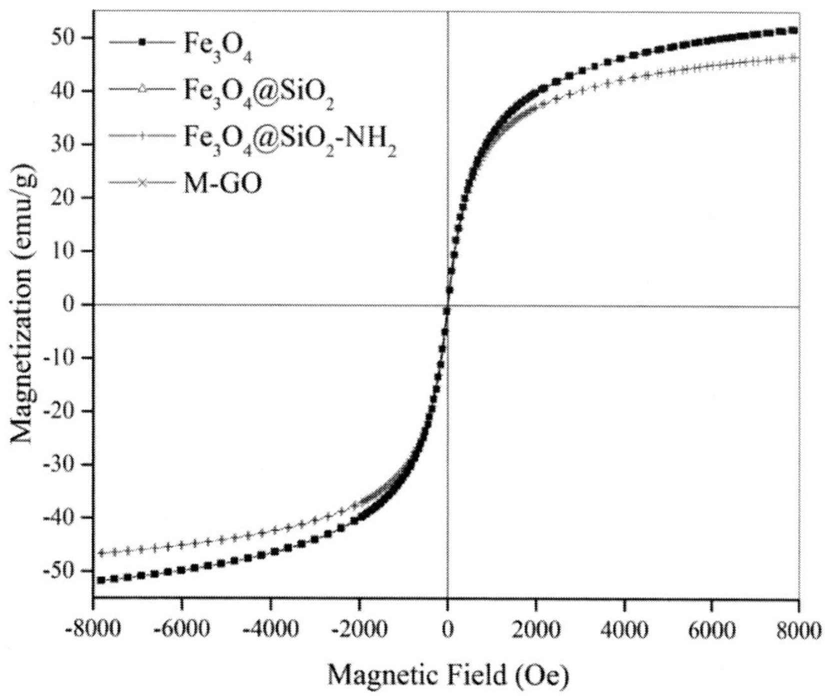

Figure 7. Magnetic hysteresis loops of various magnetic materials and M-GO.

3.2 Demulsification test

The bottle test method was applied to investigate on the demulsification performance of GO and M-GO in destroying the interfacial film of O/W emulsion. Two bottles containing the same oil content of O/W emulsion were prepared and stored under ambient conditions for 6 h. The emulsions remained stable in each bottle, and there was no distinct phase separation. Subsequently, 0.1 wt% of the produced GO and M-GO aqueous solution was added into the prepared O/W emulsion in bottle 1 and bottle 2 respectively, with mechanical vibration. An immediate color change of the emulsion was observed in the first ten seconds after the vibration, indicating the breaking of the emulsion stability in both bottles. Small oil floccules from the emulsion were coalesced and formed large oil droplets, resulting in floating onto the top of the water phase to form the oil phase. As shown in Figure S4, very obvious phase separations were accomplished within 5 min after the addition of GO and M-GO into the emulsion correspondingly.

On the other hand, a very light yellow color was exhibited by the newly formed water phase in bottle 1 as shown in Figure S4a. This could be attributed to the suspension of tiny oil floccules and GO in the water phase. In contrast, magnetic separation was conducted on bottle 2, and the M-GO nanoparticles were isolated along with absorbed oil droplets at the bottom of the bottle for further recycling purpose as shown in Figure S4b.

The collected water was then separated into different bottles, with each bottle will then be investigated to determine the demulsification efficiency of GO and M-GO. The residual oil content of the separated water was obtained by correlating its absorbance to the generated calibration curve by using a UV-visible spectrophotometer absorbance test at a wavelength of 570 nm. The oil content of the emulsion in each bottle was determined to reduce from 50 mg/mL to ~4 mg/mL (bottle 1) and from 50 mg/mL to ~2.2 mg/mL (bottle 2) in the treated water phase. GO is an efficient and low-cost chemical demulsifier as it has achieved a 91.97% of demulsification efficiency with only 0.1 wt% addition. Additionally, M-GO has been proved to be more efficient than GO in performing demulsification as it has achieved a higher demulsification efficiency of 95.59% with the same dosage of addition into the O/W emulsion with same initial oil content.

3.3 The demulsification mechanism

GO is consisted of a flexible planar sheet which has a network of hydrophobic polyaromatic islands of benzene rings on the basal plane and hydrophilic polar oxygen functionalities decorated on the periphery [43]. Therefore, GO is an amphipathic absorbent/surfactant that can be utilized in an application in emulsion systems such as demulsification processes and emulsion stabilization [12, 44]. In the emulsion system, the crude oil-in-water emulsion stability is dependent on the electrical double layer repulsive force between the protective film and oil droplets (Figures 12a1-b1) [45]. Thus, reducing the repulsive force or extinguishing the viscoelastic film is crucial in demulsifying the stable emulsion at the oil/water interfaces. Amphipathic GO nanosheets with a hydrophobic basal plane and hydrophilic edges can well disperse in the water phase [46]. As GO is added into an oil-in-water emulsion, it can be uniformly dispersed in the water phase caused by strong mechanical agitation (Figures 12b2-b3). Then, the GO in the emulsion can contact to the asphaltenes/resins molecules when it reaches the oil/water interfaces [13]. Due to the strong interaction (attraction) between asphaltenes/resins and GO (Figure S5a2), the cortical protective films of the oil droplets in emulsion are partially destroyed with the aid of collision between water and oil, which ultimately resulted in the formation of a non-continuous protective film at the O/W interface (Figure S5b4). Subsequently, the partially protected oil droplets can 'escape' from the partial destroyed protective film by coalescing with other oil droplets (Figure S5b5). The small oil droplets start to coalesce and aggregate to become big oil droplets and eventually an oil phase is formed and floated up to the liquid surface (Figure S5a3 and Figure S5b6). With the mechanisms mentioned above, the dispersed oil droplets are successfully separated from the water phase.

Asphaltenes and resins are structurally similar to GO as they both have massive delocalized π systems, which are attributed to their conjugated aromatic rings. π-π interactions are commonly known as non-covalent interactions, and they can be utilized to interpret either self-assembly or strong adsorptions of molecules with delocalized π systems [47]. Generally, there is a π-donor and a π-acceptor among the molecules in stacking interactions. Resins and asphaltenes possess more negative electron affinity than GO, which explains that resins and asphaltenes can easily acquire electrons from GO. Thus, it is very likely that GO acts as a π-donor, whereas resins and asphatenes are the π-acceptors. When polarized π orbitals of GO interact with the π-conjugated system of resins/asphaltenes, strong π-π interactions will be formed between them. Furthermore, the adsorption strength is also further enhanced due to the σ-π interaction between the lowest unoccupied molecular orbital (antibonding) of resins/asphaltenes and the electron-rich functional groups in GO. The analysis of non-covalent interaction has determined that the strong affinity between asphaltenes/resins and M-GO are caused by the strong π-π interactions and σ-π interactions (between the hexagonal rings of GO and the asphaltenes/resins aliphatic side chain).

With the aid of magnetic NPs attached to the amino-functionalized GO, the sedimentation of the small oil floccules in the water phase can be improved through using a magnet. As the M-GO has strongly interacted with the oil droplets, most of them can be brought into the oil phase after the demulsification process. Hence, it is evident that both the demulsification efficiency and kinetics can be enhanced with the aid of the external magnetic field. As illustrated in Figure S4, a magnet was used to enhance the sedimentation of oil floccules and M-GO and form a colorless water phase within 5 min. With the aid of an external magnetic field, it was proved that the fine oil floccules were also separated efficiently from the water phase along with M-GO. GO only achieved 91.97% while M-GO has achieved 95.59% of demulsification efficiency with 0.1 wt% dosage added into the O/W emulsions with the same initial oil content. Therefore, M-GO is strongly recommended for the application of demulsification based on its excellent properties and the obtained experimental results. Similar findings were also reported by the other research [28].

3.4 Effect of influential parameters on demulsification efficiency

3.4 (a) Effect of emulsion pH. The interfacial activity GO is highly sensitive to the pH value of the solvent, and it was reported that the pH-dependent behaviour of GO is mainly attributed to the deprotonation degree of its carboxyl groups at the edges [48]. Thus, the effect of pH value of emulsion on the demulsification performance of M-GO was investigated. Two central regions can be observed from the results shown in Figure S6a. In the first region (pH 2-4), the lowest separated water oil concentration was reduced from 50 mg/mL to ~0.7 mg/mL (98.6% demulsification efficiency) in acidic condition with pH value at 4 while the oil concentration was decreased to ~1.85 mg/mL (96.3% demulsification efficiency) at a pH value of 2 with 0.25 wt% of M-GO addition. It was observed that M-GO sheets are less hydrophilic at lower pH as they tended to rise to the oil phase along with the absorbed oil droplets after demulsification. Some M-GO sheets that were initially isolated at the bottom of the bottle (with an application of magnetic separation) started to rise to the oil-water interface or float in the separated water phase when the pH value was lowered below 4, which results in higher remaining oil concentration [13]. In the second region (pH 4-10), it was shown that the surface activity level of M-GO increases at acidic conditions [21]. 0.25 wt% M-GO was only able to reduce the oil concentration from 50 mg/mL to 4.29 mg/mL at pH ~10, with a demulsification efficiency of 91.42% while the demulsification efficiency increased obviously when the pH value was lowered down gradually. In other words, M-GO performed better in demulsification in acidic conditions than alkaline conditions. Therefore, the pH value at 4 was suggested to be the optimal condition for demulsification performed by M-GO.

3.4 (b) Effect of salinity of emulsion. The salinity of emulsion was reported to have considerable impacts on the demulsification performance [19]. The improvement in the effectiveness of the demulsification performance of M-GO was then investigated based on the increasing salinity of the produced O/W emulsion at neutral pH. The results shown in Figure S6b has depicted that the demulsification efficiency was increased as the salinity of O/W emulsion increased. The demulsification efficiency of 0.1 wt% M-GO was increased from 95.59% to 98.12% when the salinity of O/W emulsion was increased to 0.5 M NaCl. It was mainly due to the "salting-out" effect which reduced the stabilizing surfactants' hydrophilicity hence resulting in lower O/W emulsion interfacial tension [49].

3.4 (C) Effect of dosage of M-GO addition on the demulsification performance. The demulsification efficiency can be impacted significantly by the dosage of added M-GO. The results showed that ~88% of oil removal was achieved with 0.01 wt% M-GO addition as the oil concentration was reduced from 50 mg/mL to 6.1 mg/mL. Furthermore, 98.53% of demulsification efficiency (0.74 mg/mL in oil concentration) was achieved with a 0.25 wt% M-GO addition. As shown in Figure S6c, it can be observed that M-GO possesses better demulsification ability as compared to GO. A higher and optimal dosage of M-GO exhibits higher demulsification efficiency as more possible surface areas are exposed to the emulsion. Also, it was concluded that an external magnetic field could effectively eliminate the residual oil by combining extremely small oil floccules in the emulsion with M-GO distributed in the separated water hence resulting in better demulsification.

3.5 Recyclability of magnetic graphene oxide

Since M-GO was determined to possess superparamagnetic properties, it was expected to be recyclable and reusable after performing demulsification with the application of magnetic separation. During the separation process between the oil and water phase, it was observed that the M-GO absorbed the asphaltene, and was isolated to the bottom of the bottle due to the applied magnetic field. Toluene was used to remove the adsorbed oil for the reactivation of M-GO as the adsorbed oil (asphaltenes/resins) may remain on the M-GO surface after demulsification. As shown in Figure S6d, the recycling tests were successfully repeated for 10 times. In the first 5 cycles, M-GO was able to reduce the oil concentration in the water phase from 50 mg/mL (high concentration O/W emulsion) to <1 mg/mL, achieving ~98% of demulsification efficiency. The demulsification efficiency of M-GO started to decrease from the sixth cycle onwards. In the last cycle, the oil concentration was only reduced to 9.4 mg/mL, and the demulsification efficiency was dropped to 81.19%. It was suggested that the toluene was not able to remove the adsorbed oil molecules that were firmly attached to the M-GO surfaces after multiple cycles of demulsification. As a result, magnetic demulsifier M-GO was proved to be recyclable and reusable for demulsification due to its outstanding magnetic properties, leading to the advantages of low chemical cost and reduction in adverse environmental impacts in industrial petroleum applications as compared to the other conventional non-recyclable chemical demulsifier.

4. Conclusions and recommendations

GO, and M-GO were successfully synthesized by conducting modified Hummers method, surface functionalizations of GO and magnetite, and the cross-linking reactions of GO with magnetic nanoparticles correspondingly. Subsequently, the properties of obtained GO, various magnetic nanoparticles, and M-GO were investigated by several characterizations. In summary, it was determined that the produced M-GO composites which contain important oxygen functionalities such as carboxyl groups were successfully coated with silica-coated magnetite ($Fe_3O_4@SiO_2$) nanoparticles and they possess excellent superparamagnetic properties. A comprehensive understanding of the mechanisms of formation of GO and M-GO and the demulsification of crude oil-in-water emulsion were also presented. The demulsification tests demonstrated that 0.25 wt% of M-GO aqueous suspension could achieve a 98.53% of demulsification efficiency within 5 min on the high concentration (5 wt%) as-prepared crude oil-in-water emulsion. The demulsification performance of M-GO was found to be superior, in comparison to GO as the suspending tiny oil floccules that were attached to M-GO can be separated from the water phase more effectively using magnetic separation. The optimizations of the demulsification process were also conducted to investigate the effect of dosage, pH and salinity and significant findings were presented. Also, the recycling tests have proven that the M-GO can be reused to perform demulsification effectively for 10 times. In addition to being able to reduce the demulsification costs in the industry, the excellent properties of M-GO can also minimize the risk of polluting the environment due to leakage of demulsifier through the application of magnetic separation for recycling purposes.

A water bath is suggested for the synthesis of GO instead of an ice bath, as a more constant reaction temperature could be achieved. Also, the non-covalent interaction (NCI) analysis is recommended to further investigate the inherent interaction between M-GO and asphaltene molecules during the demulsification process in future work.

Acknowledgements

The authors would like to acknowledge Curtin University Malaysia for supporting this research.

Disclosure statement

This article was not published and is not under consideration for publication anywhere else. We also do not have any conflicts of interest to disclose.

References

[1] S. Jafarinejad. Petroleum waste treatment and pollution control. Butterworth-Heinemann2016.
[2] L. Yu, M. Han, F. He. A review of treating oily wastewater. Arabian journal of chemistry. 10 (2017) S1913-S22.

[3] C. An, G. Huang, Y. Yao, S. Zhao. Emerging usage of electrocoagulation technology for oil removal from wastewater: A review. Science of the Total Environment. 579 (2017) 537-56.
[4] S. Wong, J. Lim, S. Dol. Crude oil emulsion: a review on formation, classification and stability of water-in-oil emulsions. Journal of Petroleum Science and Engineering. 135 (2015) 498-504.
[5] R. Zolfaghari, A. Fakhru'l-Razi, L.C. Abdullah, S.S. Elnashaie, A. Pendashteh. Demulsification techniques of water-in-oil and oil-in-water emulsions in petroleum industry. Separation and Purification Technology. 170 (2016) 377-407.
[6] J. Pérez-Calderón, M.V. Santos, N. Zaritzky. Optimal clarification of emulsified oily wastewater using a surfactant/chitosan biopolymer. Journal of Environmental Chemical Engineering. 6 (2018) 3808-18.
[7] A. Karkooti, A.Z. Yazdi, P. Chen, M. McGregor, N. Nazemifard, M. Sadrzadeh. Development of advanced nanocomposite membranes using graphene nanoribbons and nanosheets for water treatment. Journal of Membrane Science. 560 (2018) 97-107.
[8] G. Gonçalves, M. Vila, M.T. Portolés, M. Vallet-Regi, J. Gracio, P.A.A. Marques. Nano-Graphene Oxide: A Potential Multifunctional Platform for Cancer Therapy. Advanced healthcare materials. 2 (2013) 1072-90.
[9] A.J. Mendoza, E. Guzmán, F. Martínez-Pedrero, H. Ritacco, R.G. Rubio, F. Ortega, et al. Particle laden fluid interfaces: dynamics and interfacial rheology. Advances in colloid and interface science. 206 (2014) 303-19.
[10] O. Jankovský, A. Jiříčková, J. Luxa, D. Sedmidubský, M. Pumera, Z. Sofer. Fast Synthesis of Highly Oxidized Graphene Oxide. ChemistrySelect. 2 (2017) 9000-6.
[11] T.M. McCoy, M.J. Pottage, R.F. Tabor. Graphene oxide-stabilized oil-in-water emulsions: pH-controlled dispersion and flocculation. The Journal of Physical Chemistry C. 118 (2014) 4529-35.
[12] V. Rajak, I. Singh, A. Kumar, A. Mandal. Optimization of separation of oil from oil-in-water emulsion by demulsification using different demulsifiers. Petroleum Science and Technology. 34 (2016) 1026-32.
[13] J. Liu, X. Li, W. Jia, Z. Li, Y. Zhao, S. Ren. Demulsification of Crude Oil-in-Water Emulsions Driven by Graphene Oxide Nanosheets. Energy & Fuels. 29 (2015) 4644-53.
[14] H. Yu, B. Zhang, C. Bulin, R. Li, R. Xing. High-efficient synthesis of graphene oxide based on improved hummers method. Scientific reports. 6 (2016) 36143.
[15] D.C. Marcano, D.V. Kosynkin, J.M. Berlin, A. Sinitskii, Z. Sun, A. Slesarev, et al. Improved synthesis of graphene oxide. ACS nano. 4 (2010) 4806-14.
[16] H. Ravishankar, J. Wang, L. Shu, V. Jegatheesan. Removal of Pb (II) ions using polymer based graphene oxide magnetic nano-sorbent. Process Safety and Environmental Protection. 104 (2016) 472-80.
[17] Z. Hu, S. Qin, Z. Huang, Y. Zhu, L. Xi, Z. Li. Recyclable graphene oxide-covalently encapsulated magnetic composite for highly efficient Pb (II) removal. Journal of environmental chemical engineering. 5 (2017) 4630-8.
[18] H. Li, X. Zhu, H. Sitinamaluwa, K. Wasalathilake, L. Xu, S. Zhang, et al. Graphene oxide wrapped Fe2O3 as a durable anode material for high-performance lithium-ion batteries. Journal of Alloys and Compounds. 714 (2017) 425-32.
[19] A. Bera, A. Mandal, B. Guha. Synergistic effect of surfactant and salt mixture on interfacial tension reduction between crude oil and water in enhanced oil recovery. Journal of Chemical & Engineering Data. 59 (2013) 89-96.
[20] N.N.A. Ling, A. Haber, B.F. Graham, Z.M. Aman, E.F. May, E.O. Fridjonsson, et al. Quantifying the Effect of Salinity on Oilfield Water-in-Oil Emulsion Stability. Energy & Fuels. 32 (2018) 10042-9.
[21] I. Silva, B. Borges, R. Blanco, M. Rondón, J.-L. Salager, J.C. Pereira. Breaking of water-in-crude oil emulsions. 5. Effect of acid-alkaline additives on the performance of chemical demulsifiers. Energy & Fuels. 28 (2014) 3587-93.
[22] M.C. Mascolo, Y. Pei, T.A. Ring. Room temperature co-precipitation synthesis of magnetite nanoparticles in a large pH window with different bases. Materials. 6 (2013) 5549-67.

[23] X.-B. Zhang, H.-W. Tong, S.-M. Liu, G.-P. Yong, Y.-F. Guan. An improved Stöber method towards uniform and monodisperse Fe 3 O 4@ C nanospheres. Journal of Materials Chemistry A. 1 (2013) 7488-93.
[24] L. Shahriary, A.A. Athawale. Graphene oxide synthesized by using modified hummers approach. Int J Renew Energy Environ Eng. 2 (2014) 58-63.
[25] O. Jankovský, P. Marvan, M. Nováček, J. Luxa, V. Mazánek, K. Klímová, et al. Synthesis procedure and type of graphite oxide strongly influence resulting graphene properties. Applied Materials Today. 4 (2016) 45-53.
[26] P. Dash, T. Dash, T.K. Rout, A.K. Sahu, S.K. Biswal, B.K. Mishra. Preparation of graphene oxide by dry planetary ball milling process from natural graphite. RSC Advances. 6 (2016) 12657-68.
[27] T. Soltani, B.-K. Lee. Low intensity-ultrasonic irradiation for highly efficient, eco-friendly and fast synthesis of graphene oxide. Ultrasonics sonochemistry. 38 (2017) 693-703.
[28] J. Peng, Q. Liu, Z. Xu, J. Masliyah. Synthesis of interfacially active and magnetically responsive nanoparticles for multiphase separation applications. Advanced Functional Materials. 22 (2012) 1732-40.
[29] J.-H. Deng, X.-R. Zhang, G.-M. Zeng, J.-L. Gong, Q.-Y. Niu, J. Liang. Simultaneous removal of Cd (II) and ionic dyes from aqueous solution using magnetic graphene oxide nanocomposite as an adsorbent. Chemical Engineering Journal. 226 (2013) 189-200.
[30] S. Balasubramanyan, S. Arayathody, S. Sugunan, B.N. Narayanan. Selective liquid phase oxidation of cyclohexene over magnetic Fe3O4/graphene oxide nanocomposite. Materials Chemistry and Physics. 211 (2018) 23-33.
[31] S. Zhu, J. Guo, J. Dong, Z. Cui, T. Lu, C. Zhu, et al. Sonochemical fabrication of Fe3O4 nanoparticles on reduced graphene oxide for biosensors. Ultrasonics sonochemistry. 20 (2013) 872-80.
[32] I. Tyagi, V. Gupta, H. Sadegh, R. Ghoshekandi, A.S.H. Makhlouf. Nanoparticles as adsorbent; a positive approach for removal of noxious metal ions: a review. Science Technology and Development. 34 (2017) 195-214.
[33] S.M. Sharif, N.F.A. Bakar, M.N. Naim. Deposition of fine iron oxide particles in tap water using electrophoretic deposition (EPD) technique. Journal of Water Process Engineering. 7 (2015) 123-30.
[34] N.H. Othman, N.H. Alias, M.Z. Shahruddin, N.F.A. Bakar, N.R.N. Him, W.J. Lau. Adsorption kinetics of methylene blue dyes onto magnetic graphene oxide. Journal of Environmental Chemical Engineering. 6 (2018) 2803-11.
[35] J. Shen, M. Shi, B. Yan, H. Ma, N. Li, Y. Hu, et al. Covalent attaching protein to graphene oxide via diimide-activated amidation. Colloids and Surfaces B: Biointerfaces. 81 (2010) 434-8.
[36] M. Mirza-Aghayan, M.M. Tavana, R. Boukherroub. Palladium nanoparticles supported on reduced graphene oxide as an efficient catalyst for the reduction of benzyl alcohol compounds. Catalysis Communications. 69 (2015) 97-103.
[37] S. Mallakpour, A. Abdolmaleki, S. Borandeh. Surface functionalization of GO, preparation and characterization of PVA/TRIS-GO nanocomposites. Polymer. 81 (2015) 140-50.
[38] B. Bhanvase, M. Patel, S. Sonawane. Kinetic properties of layer-by-layer assembled cerium zinc molybdate nanocontainers during corrosion inhibition. Corrosion Science. 88 (2014) 170-7.
[39] M. Izadi, T. Shahrabi, B. Ramezanzadeh. Synthesis and characterization of an advanced layer-by-layer assembled Fe3O4/polyaniline nanoreservoir filled with Nettle extract as a green corrosion protective system. Journal of Industrial and Engineering Chemistry. 57 (2018) 263-74.
[40] N.A. Hussien, N. Işıklan, M. Türk. Aptamer-functionalized magnetic graphene oxide nanocarrier for targeted drug delivery of paclitaxel. Materials Chemistry and Physics. 211 (2018) 479-88.
[41] X. Shen, J. Wu, S. Bai, H. Zhou. One-pot solvothermal syntheses and magnetic properties of graphene-based magnetic nanocomposites. Journal of Alloys and Compounds. 506 (2010) 136-40.

[42] B. Yu, Y. Bai, Z. Ming, H. Yang, L. Chen, X. Hu, et al. Adsorption behaviors of tetracycline on magnetic graphene oxide sponge. Materials Chemistry and Physics. 198 (2017) 283-90.
[43] L. Imperiali, K.-H. Liao, C. Clasen, J. Fransaer, C.W. Macosko, J. Vermant. Interfacial rheology and structure of tiled graphene oxide sheets. Langmuir. 28 (2012) 7990-8000.
[44] Z. Sun, T. Feng, T.P. Russell. Assembly of graphene oxide at water/oil interfaces: tessellated nanotiles. Langmuir. 29 (2013) 13407-13.
[45] F. Goodarzi, S. Zendehboudi. A Comprehensive Review on Emulsions and Emulsion Stability in Chemical and Energy Industries. The Canadian Journal of Chemical Engineering. 97 (2019) 281-309.
[46] D.W. Johnson, B.P. Dobson, K.S. Coleman. A manufacturing perspective on graphene dispersions. Current Opinion in Colloid & Interface Science. 20 (2015) 367-82.
[47] M. Mousavi, T. Abdollahi, F. Pahlavan, E. H. Fini. The influence of asphaltene-resin molecular interactions on the colloidal stability of crude oil2016.
[48] C.-J. Shih, S. Lin, R. Sharma, M.S. Strano, D. Blankschtein. Understanding the pH-dependent behavior of graphene oxide aqueous solutions: a comparative experimental and molecular dynamics simulation study. Langmuir. 28 (2011) 235-41.
[49] M.V.G. Paixão, R. de Carvalho Balaban. Application of guar gum in brine clarification and oily water treatment. International journal of biological macromolecules. 108 (2018) 119-26.

IOP Conf. Series: Materials Science and Engineering **943** (2020) 012024 doi:10.1088/1757-899X/943/1/012024

Influence of flow limiter and grain shape factor of sand particles on erosion activity in pipeline

Genevieve A T K Tang, Henry C Y Foo, I S Tan*, Y F Yeong[3], M K Lam[3, 4]
[1]Department of Petroleum Engineering, Curtin University Malaysia, 98009 Miri, Sarawak, Malaysia

[2]Department of Chemical Engineering, Curtin University Malaysia, 98009 Miri, Sarawak, Malaysia

[3]Department of Chemical Engineering, Universiti Teknologi PETRONAS, 32610, Bandar Seri Iskandar, Perak

[4]Centre for Biofuel and Biochemical Research, Institute of Self-Sustainable Building, Universiti Teknologi PETRONAS, 32610 Seri Iskandar, Perak, Malaysia

Email: tan.s@curtin.edu.my

Abstract. Erosion has been evaluated as the main cause of failure in pipework. The focus of this work is the effect of erosion activity on two-phase flow (solid-liquid) condition. Pressure distribution of liquid flow, velocity profile of liquid flow and trajectories of sand particles are scrutinized. Based on the experimental data, different limiter pipeline was simulated to validate the data and an overall grain shape factor of 0.265 was obtained. From the qualitative and quantitative analysis of grain shape factor (GSF), it was identified that the 75 % of sand are composed of high sphericity particles (GSF=0.2), 25 % of sand composed of slightly lower sphericity particles (GSF=0.45). For the simulation work, the fluid flow rate, grain shape factor and sand concentration is altered ranging between 25 litres/s to 50 litres/s, 0.2 to 1.0 and 0.5 % to 1.0 % respectively. Through ANOVA analysis and it demonstrated the significant influence on erosion activity in descending order (fluid flow rate > shape factor > sand concentration). After numerical optimization, the minimum erosion activity on limiter was obtained by obeying the settings of fluid flow rate, shape factor and sand concentration as 50 litres/s, 0.2 and 1.0 % respectively.

1. Introduction

In the oil and gas industry, transportation of fluids during production is normally entrained with solid particles. The sand production in pipelines is one of the main contribution to erosion, thus it has been scrutinized as the cause of pipe work failures [1]. The solid particle found in pipeline is originated from sandstone where oil and gas is extracted. The presence of this sand particle has been a common issue in the oil and gas industry [2]. There are some areas inside the pipeline which are more susceptible to erosion, such as integral parts, elbows, plugged tees, choke and choke valves [3]. There are many preventive measures taken to minimize the erosion problem. For instances, the use of gravel packs and sand screen to remove the sand particles in the fluids [3]. However, these two equipment could not prevent smaller sand particles from flowing into the pipeline. In order to resolve this problem, researchers have been using an assortment of models and approaches to study the erosion process. Among empirical, mechanistic and computational fluid dynamics (CFD) approaches, this work is focusing on CFD approach to investigate the controlling parameters as well as easy visualization of erosion process. With the help of this theoretical experiment, this paper aims to find the solution to minimize the erosion problem found in the transportation pipeline.

In this present work, STAR CCM+ will be used to study the changes in erosion rate based on different studied parameters. The impact location of the sand particles and its velocity are strongly related to the fluid flow phenomena [3]. CFD's approach is widely used as it enables researcher to study the areas of interest that is prone to erosion, in particular of a complicated geometry that is difficult for researcher to collect the erosion's data from or to set-up the experiment [4]. Generally, CFD-based erosion modeling involves three main steps, namely (i) flow modeling, (ii) particle tracking, and (iii) the information of particles that are going to impact on the wall pipeline [5]. Due to the fact that CFD can simulate a real fluid flow and also able to complement with experimental approaches, it is considered as a cost-effective means of tool [6]. Researchers are attempting to tackle this erosion problem from both experimental and theoretical approaches [7,8]. Due to difficulties in building industrial-scaled transportation piping, theoretical simulation has gained much attention recently

However, based on the state-of-art technologies, commercial computational fluid dynamics (CFD) software such as STAR CCM+, COMSOL and Fluent also have their own draw backs in simulating erosion activity in pipeline. For instance, on the erosion equation side, despite of utilizing a large number of physical properties in existing model and yet there is some erosion phenomenon still remained unknown [9]. One of the most important challenges is the effect of particle's morphology. Sizes, shapes and composition of sand particles are the common studied subjects in literature. Effects of particle size and shape on erosion of materials as well as transition between brittle to ductile properties of the target material require further investigation. Lastly, mechanism of how erosion rate being different for multi-phase flow regimes and the effect of fluid viscosity on erosion rate are not fully understood as well. Therefore, there are still plenty of room of improvement on erosion modeling. Currently available erosion models are too conservative and mostly based on correlation method for aforementioned physical properties which affects the efficiency of oil and gas production. Developing more accurate erosion models will definitely be helpful for optimizing production and reducing risk [4]. Thus, the objective of this study is to analyze the grain shape factor of the sand particles, to study effects of flow rate, grain shape factor and sand concentration on erosion rate in pipeline. Hence, validate and evaluate developed erosion model with experimental data.

2. Simulation and numerical method

In this paper, the numerical simulation is performed for different sand particle sizes in multiphase flows (liquid and solid) to study the effect of particle size on erosion activity. The grain shape factor coefficient used in this study ranging between 0.2 and 1.0. Details of the numerical approaches are exemplified below.

2.1 Mathematical Model

In this study, the approaches taken was Lagrangian approach and k-omega turbulent model to simulate the multiphase flow (water and sand particle). In trajectory models, the Lagrangian model is usually assumed to be discrete. In this approach, the fluid phase is solved using Newtonian equations of motion to determine trajectories of individual particles. The Lagrangian method is essential to identify appropriate model to describe particle motion, using calculation particle loading and Stokes number. The particle mass loading equation expressed as follows:

$$\beta = \frac{r_p \rho_p}{r_f \rho_f} \tag{1}$$

where r is the fraction of volume, ρ refers to density (kg/m^3), and the subscripts p and f is the particle and fluid phases respectively. Based on Newton's law of motion, the governing particle motion equation is given as:

$$m_p \frac{dV_p}{dt} = F_d + F_p + F_b + F_a \tag{2}$$

where m_p is mass of particle(kg), $V_p$is local particle velocity(m/s) and t denotes time(s). The right hand side terms is the drag force acting on surface due to viscous effect of fluid medium, taking accounts on the cohesion between particle and fluid streamline.

The continuous liquid phase is investigated by Navier-Stokes solver to develop the Reynolds Averaged Navier Stokes equations, where the liquid and solid particles are spherical properties added into continuous phase flow field known as discrete phase, which are captured by discrete phase model (DPM). The water is assumed as incompressible flow, no heat exchange between water and sand particle, and the sand particles are known as diluted phase in water. The governing equations of water phase is described as:
Turbulent kinetic energy (k) equation:

$$\frac{\partial k}{\partial t}+\bar{u}_j\frac{\partial k}{\partial x_j}=\frac{\partial}{\partial x_j}\left[\left(v+\sigma^*\frac{k}{\omega}\right)\frac{\partial k}{\partial x_j}\right]-\beta*k\omega+\tau_{ij}\frac{\partial \bar{u}_i}{\partial x_j} \quad (3)$$

The specific dissipation rate $(\omega)$is given by:

$$\frac{\partial \omega}{\partial t}+\bar{u}_j\frac{\partial \omega}{\partial x_j}=\frac{\partial}{\partial x_j}\left[\left(v+\sigma\ \frac{k}{\omega}\right)\frac{\partial \omega}{\partial x_j}\right]-\beta\omega^2+\frac{\sigma_d}{\omega}\frac{\partial k}{\partial x_j}\frac{\partial \omega}{\partial x_j}+a\frac{\omega}{k}\tau_{ij}\frac{\partial \bar{u}_i}{\partial x_j} \quad (4)$$

Where turbulence kinetic energy$k(m^2/s^2)$, dynamic viscosity μ (Pas), dynamic viscosity for$\mu_t(Pas)$, velocities vector$\vec{u}$ (m/s), specific dissipation rate$\omega(m^2/s^3)$, and t denotes time. The $k-\omega$ model is preferred for several reasons. For instance, it is sensitive to near-wall interactions and more accurate for free shear flows and separated flows.

2.2 Erosion Mechanics and Modeling Strategies

The solid particle studies are done by many researches due to the significant of the industrial requirement. Hence, the erosion equation development is advancing empirically or numerically. In the studies aforementioned, the characteristic of solid particle concentration is described via suggesting different geometries and mechanisms. The earliest proposed erosion equations is derived via solving equations of motion for particle as it cut the target surface [10]. In this research, the erosion equations and modeling approaches are reviewed to select the suitable erosion equation [4]. The erosion equations are exemplified in Table 1.

Table 1: Evaluation of erosion equations.

Model	Equation	References
Ahlert	$ER=A(BH)^{-0.59}F_s u_p^n f(\alpha)$ $f(\alpha)=a\alpha^2+b\alpha \quad \alpha\le\alpha_0$ $f(\alpha)=x\cos^2\alpha\sin(w\alpha)+y\sin^2\alpha+z \quad \alpha>\alpha_0$	[11]
DNV	$ER=Cf(\alpha)u_p^n$ $f(\alpha)=\sum_{i=1}^{8}(-1)^{i+1}A_i\alpha^i$	[12]
Neilson- Gilchrist	$ER=\frac{u_p^2\cos^2\alpha\sin\frac{\pi\alpha}{2\alpha_0}}{2\varepsilon_C}+\frac{u_p^2\sin^2\alpha}{2\varepsilon_D} \quad \alpha<\alpha_0$ $ER=\frac{u_p^2\cos^2\alpha}{2\varepsilon_C}+\frac{u_p^2\sin^2\alpha}{2\varepsilon_D} \quad \alpha>\alpha_0$	[13]
Oka	$E_{90}=81.714(Hv)^{-0.79}(\frac{u_p}{u_{ref}})^{k2}(\frac{D_p}{D_{ref}})^{k3}$ $g(\alpha)=(\sin\alpha)^{n1}(1+Hv(1-\sin\alpha))^{n2}$	[14, 15]
Zhang E/CRC	$\boldsymbol{ER=C(BH)^{-0.59}F_sV_p^nF(\Theta)}$ $F(\theta)=5.40\theta-10.11\theta^2+10.93\theta^3-6.33\theta^4+1.42\theta^5$	[16]

From all the models stated above, the erosion equations are normally based on various testing conditions influenced by the effect of erosion parameters. Thus, it can be deduced that the erosion equations required experimental data to validate the given constants to generate erosion equation. Currently, there is no specified erosion equation suitable for all erosion conditions. Furthermore, the values of parameters in erosion models are defined as follows:

Table 2: Parameters values of erosion models [12, 15, 17].

	Parameters							
1	A	B		w	x	y	z	
	38.4	22.7		1	3.147	0.3609	2.532	
				$A4$				
2	$A1$	$A242.$	$A3$	175.80	$A5$	$A6$	$A7$	$A8$
	9.370	295	110.864	4	170.137	98.398	31.211	4.170
3	$k0$	$k1$	$k2$	$k3$	$n1$	$n2$	$V'/(mls)$	$d'/\mu m$
	65	-0.12	$2.3(Hv)^{0.038}$	0.19	$0.71(Hv)^{0.14}$	$2.4(Hv)^{-0.94}$	104	326

2.3 Modified erosion model

The erosion rate depends mainly on the alteration of particle shape coefficient in addition to impact velocity, impact angle and target material properties. The Zhang E/CRC erosion model improved the particle near-wall velocity and erosion prediction through two refinement via standard wall function application at near-wall particle tracking and rebounding the particle at a radius from the wall, thus avoiding nonphysical impacts [18]. These correlations have remarkable quality of results which taken into account of the complexity of the phenomenon [19]. In this study, the empirical model developed by Zhang et al. [16] is modified by varying the particle shape coefficient. The empirical equation defined is given by:

$$\boldsymbol{ER = C(BH)^{-0.59} F_S V_p^n F(\theta)} \quad (5)$$

$$F(\theta) = 5.40\theta - 10.11\theta^2 + 10.93\theta^3 - 6.33\theta^4 + 1.42\theta^5 \quad (6)$$

where ER is erosion ratio, C and n are empirical constants given as $2.17\text{x}10^{-7}$ and 2.41 respectively, BH is Brinell hardness of wall material, F_s is particle shape coefficient while 1.0 for sharp (angular), 0.53 for semi-rounded, or 0.2 for fully rounded sand particles, V_p is particle impact velocity and $\theta$is the particle impact angle in radian. Similarly, to Oka correlation, this is a robust model that relies exclusively on the flow data and the eroded material properties.

2.4 Computational domain and boundary condition

The geometry prepared in this work is 0.033 m diameter pipeline attached with the flow limiter. The flow limiter used throughout this thesis is a valley limiter, fluted limiter and upstand limiter with 0.0265m. The geometry generated in STAR CCM+ in a quarter of the geometry from a perspective view via boolean operation on faces. According to Lanasa [20], the upstream and downstream pipe diameter is larger than the pipe size to ensure a well advanced flow in pipelines.The restriction in the pipeline component is the limiter identifying the erosion. The computational domains are discretized using polyhedral mesh grid type. The meshing strategy adopted is the parts-based meshing (PBM) where the initial surface was re-tessellated to obtain a finer triangulation for generation of high quality volume mesh. In the current study, the surface remesher was employer to re-triangulate and optimize the surface quality for volume meshing. The polyhedral mesh type was adopted considering factors such as the desired solution accuracy and convergence rate, the computational memory available, the quality of the starting surface mesh and quality of solution. The three different limiters studied is expressed in Fig. 1.

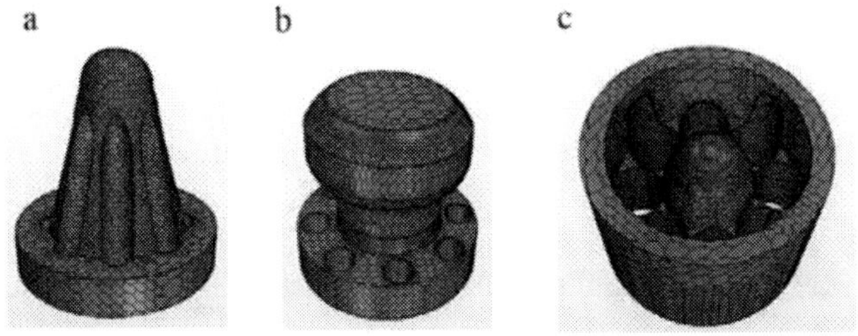

Figure 1: Limiter with polyhedral mesh: (a) Fluted (b) Upstand (c) Valley.

The effect of erosion activity is investigated on the three limiter namely, upstand limiter, fluted limiter and valley limiter with the following parameters fixed at diameter of 0.033 m, fluid density of 1000 kg/m^3 and injector mass flow rate of 0.027787 kg/s.

Different limiter is considered in this study. The boundary conditions (which are inlet, outlet, symmetry and walls) is assigned to region per part and boundary per part surface. The flow profiles specified at inlet sections with distributions of fluid velocity with dynamic viscosity of $8.8871x10^{-4}Pa.s$, constant density $997.561kg/m^3$, turbulence intensity of 0.029 and turbulence length scale of 0.0037 m. At the outlet boundaries the mean pressure is specified and remain constant to zero value. The particles exit the domain. The walls is assumed under no slip condition for fluid phase. The solid properties is specified with a density of $2650 \ kg/m^3$ and a diameter of $2.75e^{-4}\ m$. The physics continuum for multiphase (liquid-solid) used for present study consisted of the models $k-\varepsilon$ standard and Lagrangian phase. Lagrangian multiphase model is the solid material while the continuum is defined as liquid material. In this model, the liquid phase properties are defined and the parameters that can be altered is inlet velocity. Moreover, the segregated model is selected to ensure the equations of fluid flow is solved separately via finite volume method. The quasi steady state is preferable as it has digitize into small element. The proposed workflow for STAR CCM+ is described in Fig. 2.

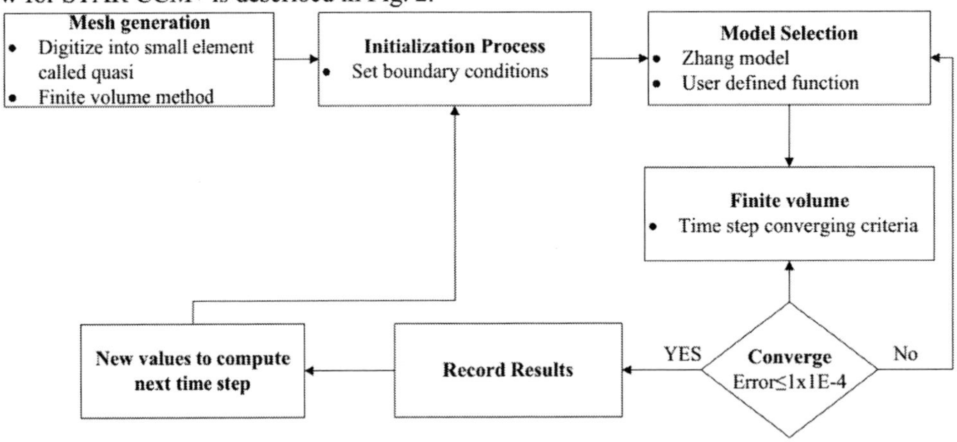

Figure 2: STAR CCM+ Flowchart.

2.5 Grain Shape Factor of sand particle

The photogrammetry method adopted have to ability to produce photo real texturing, lower cost when compared to the equipment required when using other methodologies, and its portability.

The research was completed using Autodesk's Recap to create three dimensional point clouds to visualize the model with high accuracy and efficiency. It is also compatible with AutoCAD, Revit, Infraworks, Civil 3D, Navisworks, Remake and more [21]. The ReCap also known as Reality Capture is used extract the

IOP Conf. Series: Materials Science and Engineering **943** (2020) 012024 doi:10.1088/1757-899X/943/1/012024

generated three dimensional meshes from ReCap for further optimization of the three dimensional models. Remake major workflows typically fall into two main categories, preparation and editing. The preparation is to capture at least forty pictures for each set of sample and uploaded to the ReCap dashboard to be processed into three dimensional model. The editing part is when the ReMake is used to clean, fix, sculpt, edit, retopo, optimize, compare, export or publish [22]. The completed three dimensional models is then sliced via Slic3r for each image to evaluate grain shape factor shown in Fig. 3.

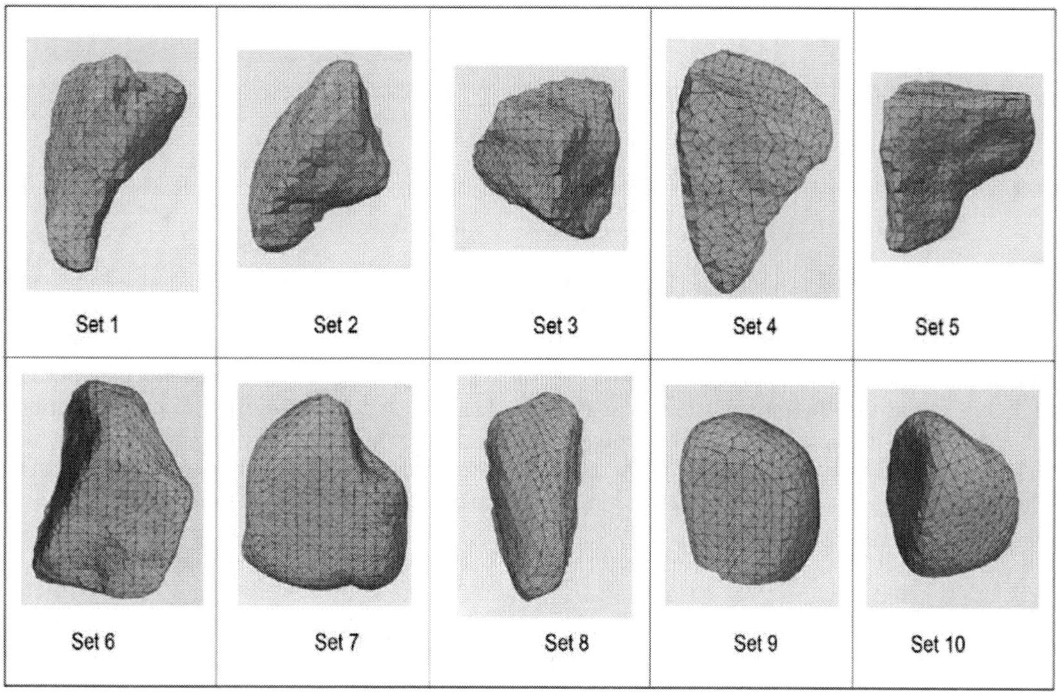

Figure 3: Three-dimensional (3D) of solid particles.

Slic3r free software is used to convert the digital 3D model into different slice of layers to evaluate the shape transformation of the selected grain particle [23]. The proper model file generated in Autodesk ReCap is transferred into Slic3r in .stl format to slice to Scalable Vector Graphics (SVG) format. The SVG format is preferred as it can be scaled in vector format. In addition, each layer created is represented as image.

2.6 Microstructure analysis on sand particle

The sand particles grain properties is evaluated via ImageJ digital image processing program followed by MATLAB. Ten reference particles with various shapes is investigated to study the qualitative and quantitative evaluation. The ImageJ software is utilized to analyze the particle whereas the conversion of captured image into a binary is recommended [24]. With the integration of Image Processing Toolbox in MATLAB, PolyLX, which undertake the review of reference-standard algorithm, functions, and applications for image analysis, processing, visualization and algorithm development, thus the shapes of the particles is measurable [25].

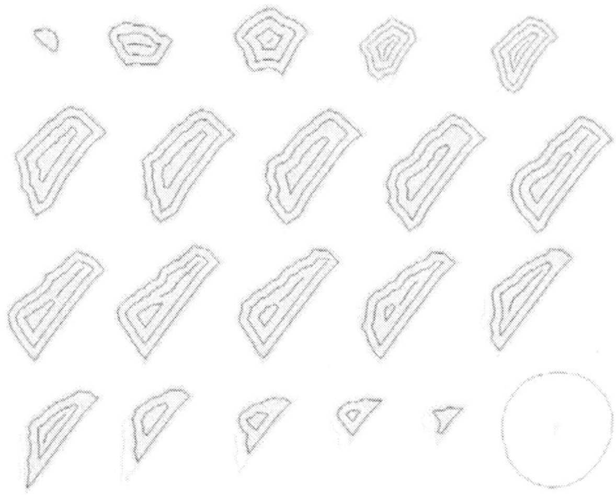

Figure 4: Sliced selected particle in Slic3r.

In the current study, Fig. 4 were calibrated and converted to 8-bit grayscale images. The following approach executed in ImageJ software to obtain confident results. In ImageJ, the set measurements function are area, minimum and maximum gray value and mean gray value. The grayscale threshold is used to optimize for discrimination of particle background. The basis of cut-off value in thresholding is pixel values which will be affected by any alteration in image acquisition, thus the final binarization result depends on the enhancing on the areas of interest and fostering the binarization process to ensure the reproducibility. After thresholding, further measurements on area of interest is corrected by closing the internal holes via analyze particles [26]. The particle grains results is then key in into PolyLX MATLAB toolbox to analyze the grain particle quantitatively.

After generating the result in qualitative evaluation, the grain shape particle was further analyzed through the quantitative evaluation via MATLAB PolyLX toolbox. The PolyLX toolbox is utilized to solve the image files from ImageJ digital image processing program to analyze the textural analysis of each object in pixels [27]. The quantitative textural analysis depend on detailed and precise description of grain sizes, grain shapes, grain boundaries as well as preferred orientations of grain and grain boundaries. The selected particle shape parameters used in MATLAB toolbox are perimeter, area, length of the longest axis and length of the shortest axis which is stated in Table 3.

Table 3: Parameters averaged values results of the selected particles.

Particles (Fig. 3)	Length (Pixel)	Width (Pixel)	Area (Pixel)	Perimeter (Pixel)
Set 1	1564.837	774.32666	876117.79	5309.1429
Set 2	923.40156	530.21394	327322.76	3054.2927
Set 3	845.0879	680.97733	376549.74	3206.5807
Set 4	1462.9529	563.70606	495456.38	4406.6667
Set 5	1310.0961	844.46465	624014.41	4398
Set 6	967.30292	496.43234	370839.24	3360.7059
Set 7	1095.9541	825.12551	655308.09	3934.875

Set 8	1164.4261	573.90971	516451.24	3573.6191
Set 9	1149.7786	1031.8025	865971.2	4234.5125
Set 10	1192.1593	808.18467	705026.74	4319.551

The mathematical formulation of grain shape factor properties was examined via the PolyLX toolbox grain shape factor parameters defined as Eq. (7):

$$GSF = \left(\frac{L}{W}\right)^{0.318} \cdot \frac{P}{2\sqrt{A}} \quad (7)$$

whereas W represents the width in pixel, P is the perimeter in pixel, A is the grain area in pixel and L is length in pixel. Nevertheless, the grain shape factor values obtained is hard to quantify the grain shape factor of sand as the variance between the values is large. In order propose the equation, the calculated values should be normalized to the range of 0.2 to 1.0 by using Eq. (8). The min-max normalization method is adopted to scrutinize the normalized grain shape factor in the range desired [28].

$$Normalize\ Shape\ Factor = \mathbf{1} + \frac{(x-2.46279)(1-0.2)}{4.4072743-2.46279} \quad (8)$$

Table 4: Normalization of grain shape factor.

Particles (Fig. 3)	Average Grain Shape Factor	Normalized GSF
Set 9	2.441528788	0.2
Set 3	2.954148387	0.402626939
Set 10	3.020467347	0.429975628
Set 8	3.7595	0.499070596
Set 5	3.399502381	0.586282534
Set 7	2.89341875	0.700929404
Set 6	3.688241176	0.705352972
Set 1	3.7892	0.746986488
Set 2	3.907587805	0.795807388
Set 4	4.402742857	1

Through the application of regression evaluation in the data analysis built in Microsoft Excel (Appendix A), a grain shape factor equation is generated according to the normalized shape factor in Table 4.

$$\text{New Shape Factor} =- 0.39657086 \left(\frac{L}{W}\right) -0.1225171 \left(\frac{P}{\sqrt{A}}\right) + 1.93211497 \quad (9)$$

The following particle shape distribution are tabulated into respective modified grain shape factor value shown in Fig. 5. Although in Zhang E/CRC erosion model, the assumption made for particle shape coefficient was a constant of 0.2 for fully rounded sand particles, the existence of various particle shape was found. Thus, the shapes of sand particles is analyzed via implementing the grain shape factor analysis. The grain shape factor analysis is completed through the MATLAB software via the iteration of the particle grain shape factor illustrated in Appendix B. The distribution of particle shape is examined and categorized in Table 5. Therefore, there is 75 % of high sphericity of the sand particles in set 9 whereas 25 % in set 10 were identified to be slightly higher in terms of roundness and angularity that contribute to the erosion rate of 0.58 g/hr.

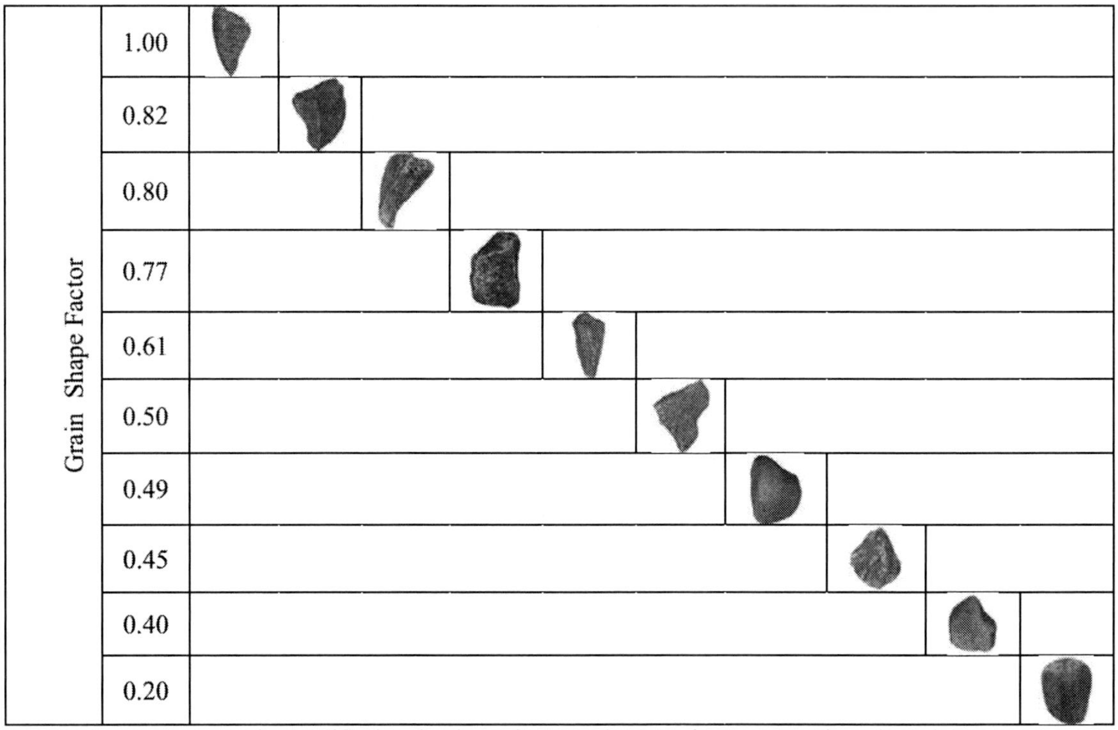

Figure 5: The modified grain shape factor values on the ten sets of sand particles.

Table 5: Particle shape distribution for shape factor of 0.265.

	Grain Shape Factor	Particles fluid distribution (%)
Set 4	1.00	0
Set 2	0.82	0
Set 1	0.80	0
Set 6	0.77	0
Set 7	0.61	0
Set 5	0.50	0
Set 8	0.49	0
Set 10	0.45	25
Set 3	0.40	0
Set 9	0.20	75

3. Results and discussion

3.1 Minimization of erosion rate based on response surface methodology

The application of the statistical analysis tool utilized is the design of experiment (DOE) and response surface methodology (RSM). The basic fundamentals of DOE composed of one or a series of data where the input parameters are modified to monitor the difference in the result output. The models is classified as theory and empirical models, which the models are following the statistical and machine learning method. Response surface methodology combines experimental methods and mathematical statistics, continuously testing the specified points until the relationship between parameters is solved. Also, response surface methodology explains the response tool and future values via the statistical evaluation of regression coefficient from the result [29]. Moreover, the second-order polynomials is examined, to evaluate the response value and defined effect size. The general function of RSM is expressed as:

$$y^{\lambda} = \beta_0 + \sum_{i=1}^{k} \beta_i x_i + \sum\sum_{i<j} \beta_{ij} x_i x_j + \sum_{i=1}^{k} \beta_{ii} x_i^2 + \varepsilon \quad (10)$$

where $y$is response purpose, lamda value$\lambda$is Box-Cox power transformation value, $\beta_0$is constant of real regression, β_i, β_j, $\beta_{ij}$and β_{ii} are main regression coefficients values, interaction and quadratic terms,$x_i$are the ith function independent variables and ε is the random error.

3.2 Effect of particle properties and contact parameters on erosion process

In this study, the parameters influencing erosion in pipeline with different geometry limiter is studied. The comparison between the fluted limiter, valley limiter and upstand limiter under different conditions is simulated for 15 runs and evaluated to ensure consistency analysis of result. The effect of erosion activity is evaluated via the statistical analysis of fluid flow rate ranging from 25 $litres/s$ till 50 $litres/s$. The effect of mass flow rate is illustrated in Fig. 6.

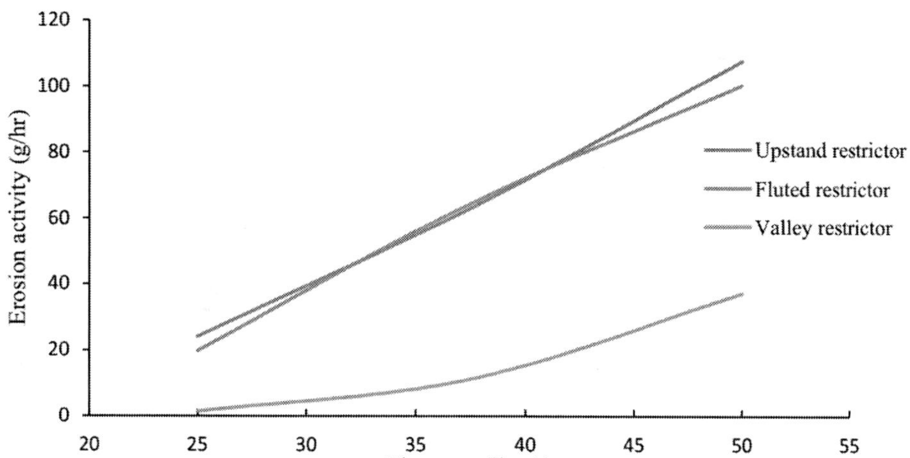

Figure 6: Effect of flow rate influencing erosion activity.

According to Fig. 6, it is shown that the flow rate is directly proportional to the erosion activity. The flow rate is influenced by the particle kinetic energy, hence the kinetic energy will increase as the eroded target surface area increases. This indicated that the material removal of the target surface is caused by the higher momentum of particles. Moreover, the relationship between flow rate and erosion activity can be proven using the exponential function where the power exponent is known to be equal to 2 [30].

The Fig. 7 shown the effect of grain shape factor influencing the erosion activity. The linear correlation between erosion activity and shape factor is applied to the three limiter simulated. Hence, the higher the angularity of the grain, the greater the erosion activity. The shape factor 0.2 pointed out that it is a rounded grain whereas 0.9 shown that it is an angular grain. This is due to the higher angularity grain have a more shaper edge compared to the rounded grain. Bukhaiti et al. [31] stated that the higher the angularity of grains, the sharper the edge of the angular grain, as a result the greater the eroded target surface. As the point pressure caused by the sharpness of the angular grain has reached a certain point, it will create scars, thus it contributes in the removing of surface material.

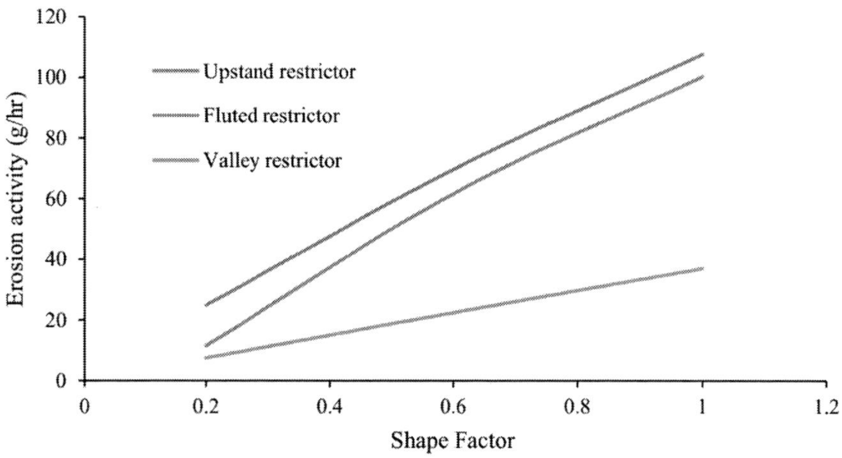

Figure 7: Effect of shape factor influencing erosion activity.

From the Fig. 8, it evaluated effect of sand concentration influencing the erosion activity. The sand concentration was set from 0.5 % to 1.0 % by weight. The erosion activity is linearly proportional to sand concentration. This relationship of the effect of sand concentration and erosion activity can be interpreted as the amount of sand particles in fluid. When the amount of sand concentration is significant, the greater the amount of solid particles collides at the target surface, hence greater rebounding action and wearing of the target surface. The results are qualitatively similar to those research published by Peng and Cao [11].

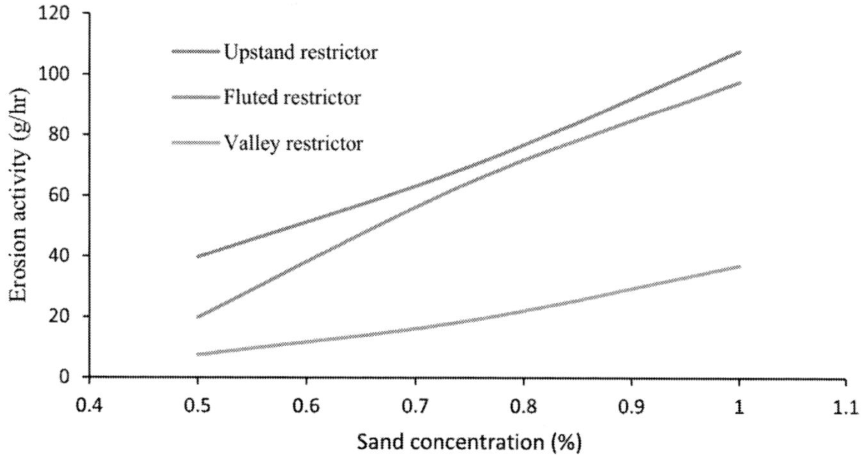

Figure 8: Effect of sand concentration influencing erosion activity.

3.3 Validation of STAR CCM+ model

To demonstrate the CFD methodology, the validation case by Wallace [32] is presented in Table 6 to validate the simulation results obtained. The three different limiter in STAR CCM+ is simulated to evaluate the erosion activity of each case under the same parameters to provide a good validation case.

Table 6: Erosion experiment results [32]

No	Description	Time	Mass Before	Mass After	Sand Conc.	Mass loss	Erosion rate	Flow rate	Cv Start	Cv End
		(hrs)	(g)	(g)	(% mass)	(g)	(g/hr)	(litres/s)	SI	SI
1	US, 4130	14	688.98	669.09	0.53	19.89	1.42	28.14	11.8	13.5
2	US, Duplex	15	688.37	673.66	0.39	14.71	0.98	28.49	11.7	13.5
3	Fluted,4130	14	690.30	681.15	0.44	9.15	0.65	29.34	12.6	14.4
4	Valley,4130	14	838.93	830.77	0.39	8.16	0.58	28.57	12.9	14.7

The simulated data obtain is evaluated through the calculation percentage of error shown in Table 7. It can be seen that fluted and upstand has an error percentage greater than 5%, due to the different in geometry dimension constructed as insufficient information provided on the dimensions of limiter. In addition, the material of limiter studied is AISI 4130, which composed of hardness 240.9 Hv, density of 7790 kg/m^3 and surface roughness 0.28μm [32].

Furthermore, the shape factor is evaluated where Zhang et al. [33] set the grain shape factor as a constant value of 0.2. The difference between simulated results and experiment data was overcome via changing the shape factor in erosion equation until it is an acceptable range close to the experimental data. Therefore, shape factor required is 0.265 and thus, the shape of sand particles varies in angularity and roundness. Consequently, the effect of grain shape factor is significant in influencing erosion activity and should be taken into consideration for further investigation.

Table 7: Comparison between experiment and simulation erosion activity.

Limiter		Flow Rate (l/s)	Shape Factor	Sand Conc. (%)	Erosion rate (g/hr)
Upstand	Experiment	28.14	-	0.53	1.42
	Simulation	28.14	0.265	0.53	1.573
			Percentage of Error		9.724
Fluted	Experiment	29.34	-	0.44	0.65
	Simulation	29.34	0.265	0.44	0.769
			Percentage of Error		15.437
Valley	Experiment	28.57	-	0.39	0.580
	Simulation	28.57	0.265	0.39	0.580
			Percentage of Error		0.000

3.4 Formulation of new erosion

The significant of the effects of mass flow rate, shape factor and sand concentration was evaluated via analysis of variance (ANOVA) with the combination of Fisher's statistical test. The ANOVA method utilized is to interpret the importance of the respective variable to the erosion rate, and it examine based on sum of squares and Percentage of contributions (PC) for individual values. By assuming the coupling effects of the input parameters are infinitesimal, the output response can be accurately predicted via low-dimensional functions, namely ANOVA components. Table 8 are the simulated results of a number of 20 simulation run completed. In this research, the simulation runs done is to evaluate the effects of the variables on the erosion activity. The results acquired is then analyzed via the regression analysis function in the Design Expert Software.

Table 9 reported the results of ANOVA analysis to investigate the effect of erosion activity. The model F value is monitored, where a higher F-value implies that the model is significant. Moreover, another value to identify the importance of model is through analyzing the p-value which have to be less than 0.005. Therefore, the selected model is based on the criteria mentioned. The model selected is reflected in the equation for each limiter respectively.

In order to identify the greatest effect on erosion activity, the F-value was evaluated. The greater the F value, the greater the effect on the respective model. Accordingly, the greatest to smallest F value for upstand limiter is flow rate (927.54), shape factor (865.10) and sand concentration (1.14) consecutively. The following concept applies to fluted limiter and valley limiter. This means that the influence of flow rate on erosion activity is remarkably larger compared to shape factor and sand concentration, considering that flow rate has a great impact of energy colliding onto the target surface when flow rate increases. Hence, as the flow rate increases, the larger the F value, as F value depends on the effect of flow rate whereas the effect of shape factor and sand concentration is infinitesimal. The proposed model for the erosion model is defined as shown in Eq. (11), Eq. (12) and Eq. (13):

$$Erosionrate_{upstand}(g/hr) = 33.76 + 27.14A + 26.21B + 17.29AB \quad (11)$$

$$Erosionrate_{Fluted}(g/hr) = 31.07 + 24.69A + 23.8B + 15.27AB + 0.49AC \quad (12)$$

$$Erosionrate_{Valley}(g/hr) = 11.33 + 9.04A + 8.56B + 6.0014AB + 2.00A^2 \quad (13)$$

The model validation is performed to recognize the good agreement of data acquired by predicting the response value and compared with the simulation result. The correlation coefficient, R^2value was 0.9953 which is an acceptable value to proceed to predict the model response and parity plot. The graph of predicted against actual is illustrated in Fig. 9 to shown its good agreement of data between erosion activity from the equations and the actual erosion activity. Fig.10 addressed the three-dimensional plot and Fig. 11 demonstrated the desirability function established through the consideration of the key parameters of flow rate, shape factor and sand concentration.

Table 8: The simulated results and the response of each limiter.

Run	Variables			Response		
				Upstand	Fluted	Valley
	A: Flow rate (litres/s)	B: Shape Factor	C: Sand concentration (%)	Erosion rate (g/hr)	Erosion rate (g/hr)	Erosion rate (g/hr)
1	25	0.6	0.75	12.52	10.63	4.29
2	37.5	1	0.75	63.06	64.43	18.86
3	50	0.2	1	21.75	25.48	7.49
4	37.5	0.6	0.75	31.55	28.41	11.29
5	37.5	0.6	0.75	31.98	31.47	11.29
6	50	1	1	107.76	97.63	36.99
7	50	0.6	0.75	69.69	61.71	22.41
8	25	1	1	24.12	19.28	7.05
9	37.5	0.6	0.5	39.70	30.00	11.23
10	37.5	0.2	0.75	10.86	11.49	3.77
11	37.5	0.6	0.75	30.35	29.68	11.36
12	25	0.2	0.5	4.08	3.64	1.41
13	50	1	0.5	113.03	100.40	37.27
14	37.5	0.6	0.75	32.03	30.84	11.36
15	37.5	0.6	0.75	32.05	30.22	11.48
16	37.5	0.6	1	34.27	29.91	11.39
17	25	0.2	1	4.61	3.96	1.41
18	25	1	0.5	20.34	19.77	7.04
19	50	0.2	0.5	24.87	18.97	7.48
20	37.5	0.6	0.75	35.50	28.70	11.21

Table 9: Analysis of variance (ANOVA) (a)Upstand (b) Fluted (c) Valley.

(a) Upstand limiter

Source	Sum of Squares	df	Mean Square	F Value	p-value Prob > F
Model	16857.91	9	1873.10	235.82	< 0.0001
A	7367.23	1	7367.23	927.54	< 0.0001
B	6871.31	1	6871.31	865.10	< 0.0001
C	9.04	1	9.04	1.14	0.3112
AB	2394.03	1	2394.03	301.41	< 0.0001
AC	20.19	1	20.19	2.54	0.1419
BC	1.51E-01	1	0.15	0.02	0.8932
A^2	70.07	1	70.07	8.82	0.0140
B^2	2.22	1	2.22	2.80E-01	0.6083
C^2	2.35	1	2.35	2.96E-01	0.5982

Residual	79.43	10	7.94		
Lack of Fit	64.57	5	12.91	4.35	0.0663
Pure Error	14.86	5	2.97		
Cor Total	16937.34	19			

(b) Fluted limiter

Source	Sum of Squares	df	Mean Square	F Value	p-value Prob > F
Model	13821.42	9	1535.71	234.93	< 0.0001
A	6096.08	1	6096.08	932.57	< 0.0001
B	5662.51	1	5662.51	866.24	< 0.0001
C	1.22	1	1.22	0.19	0.68
AB	1864.74	1	1864.74	285.26	< 0.0001
AC	1.91	1	1.91	0.29	0.60
BC	12.74	1	12.74	1.95	0.19
A^2	30.21	1	30.21	4.62	0.06
B^2	71.68	1	71.68	10.97	0.01
C^2	23.21	1	23.21	3.55	0.09
Residual	65.37	10	6.54		
Lack of Fit	58.23	5	11.65	8.16	0.02
Pure Error	7.14	5	1.43		
Cor Total	13886.79	19			

(c)Valley limiter

Source	Sum of Squares	df	Mean Square	F Value	p-value Prob > F
Model	1858.4	9	206.5	772.4	< 0.0001
A	817.7	1	817.7	3058.9	< 0.0001
B	733.3	1	733.3	2743.1	< 0.0001
C	9.02E-04	1	9.02E-04	3.38E-03	0.9548
AB	288.1	1	288.1	1077.9	< 0.0001
AC	9.79E-03	1	9.79E-03	3.66E-02	0.8521
BC	9.04E-03	1	9.04E-03	3.38E-02	0.8578
A^2	11.1	1	11.1	41.4	< 0.0001
B^2	2.71E-03	1	2.71E-03	1.02E-02	0.9217
C^2	4.89E-03	1	4.89E-03	1.83E-02	0.8951
Residual	2.7	10	2.67E-01		
Lack of Fit	2.6	5	5.26E-01	62.1	0.0002
Pure Error	4.24E-02	5	8.48E-03		
Cor Total	1861.1	19			

IOP Conf. Series: Materials Science and Engineering **943** (2020) 012024 doi:10.1088/1757-899X/943/1/012024

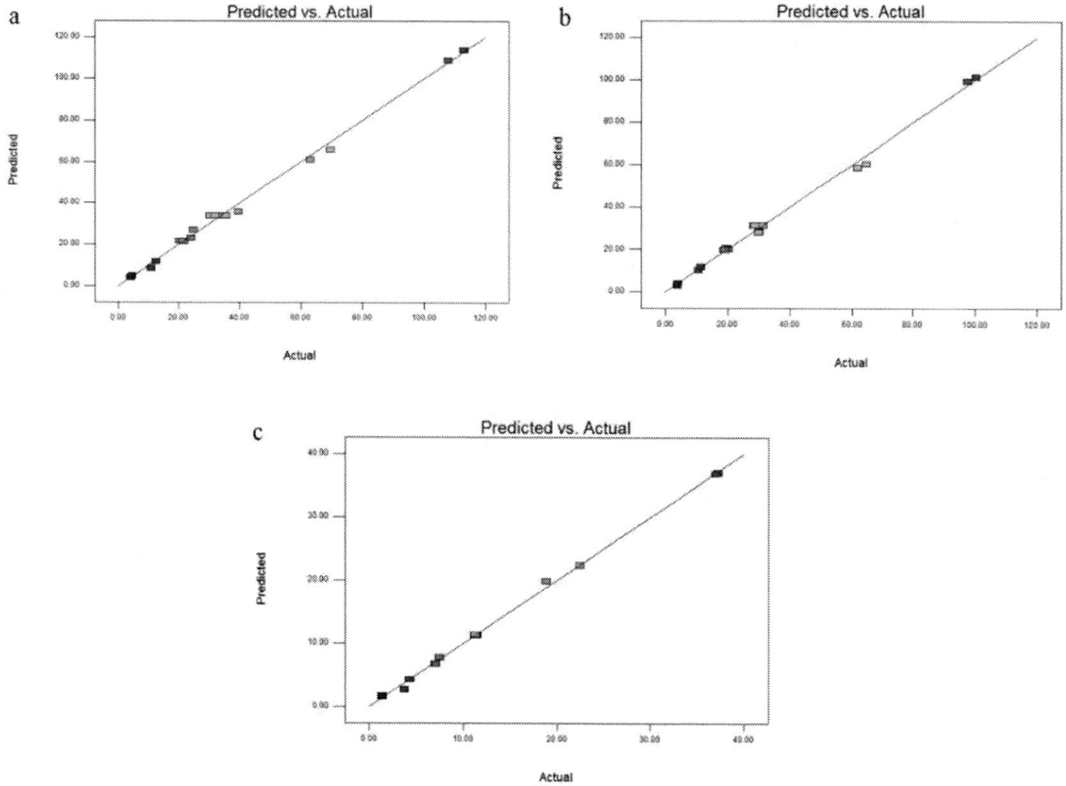

Figure 9: Parity plot of the predicted against actual simulation erosion activity (a) Upstand (b) Fluted (c) Valley.

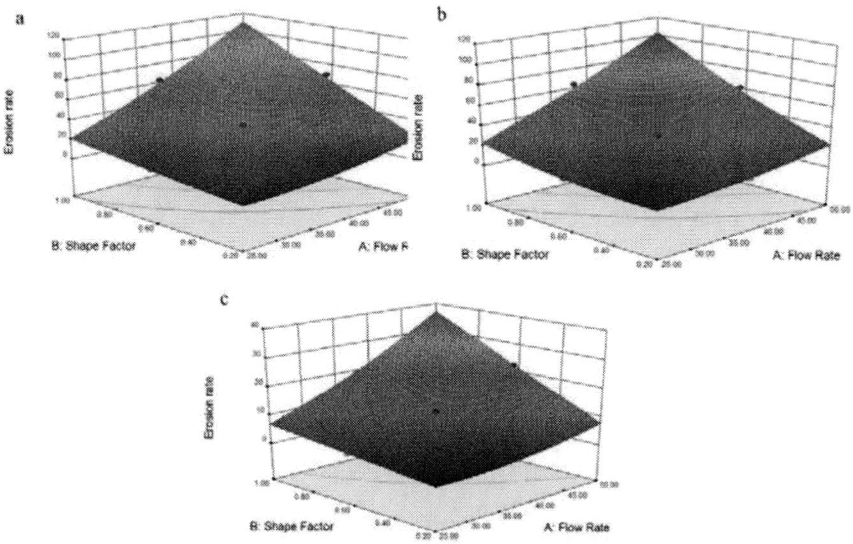

Figure 10: Three dimensional plot erosion activity (a) Upstand; (b) Fluted; (c) Valley.

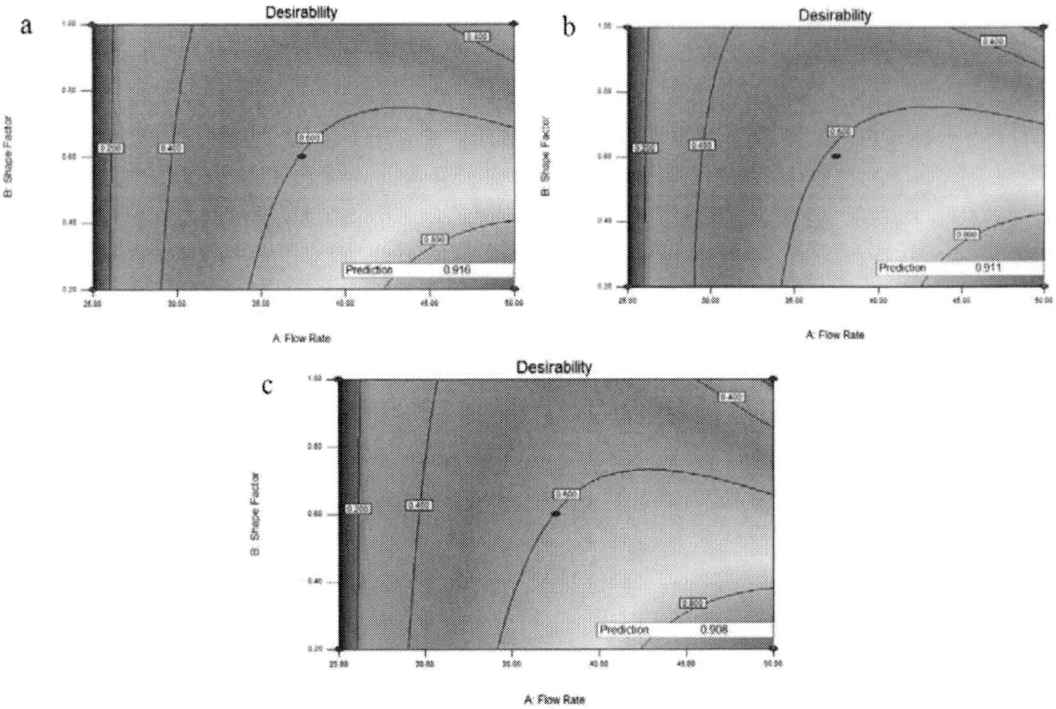

Figure 11: Contour graph of erosion activity (a) Upstand (b) Fluted (c) Valley.

Numerical optimization introduced remarkable insights through the evaluation of the optimum case to minimize the solid particles erosion activity in pipeline. The desirability function utilized the regression analysis approach in the Design Expert Software to analyze the ideal case through filtering the variables which comply with the constraints fixed for the parameters chosen. The desirability function is commonly used in the numerical optimization routine where it range from 0 (not acceptable) to 1 (ideal) based on the desired goal of each response and the priorities set by the user. This function is expressed in the Myers and Montgomery response surface methodology (RSM) text [34].

In the current work, the ideal case was to maximize the flow rate and minimize the erosion activity. Table 10 expressed the solutions provided by the Design Expert numerical optimization results following the criteria mentioned.

The Table 10 parameters are FR is flow rate in l/s, SF is shape factor, SC is sand concentration in %, ER is erosion rate g/hr and D is desirability. The highest desirability for each limiter for each case is selected, where it is shown in Table 15 below. Moreover, to validate the results obtained from Design Expert the following variables set is simulated to compare both results between numerical optimization from Design Expert and the simulation STAR-CCM+. The simulated result for each limiter is tabulated in Table 11 respectively.

Table 10: Optimum case by Design Expert of different limiter with ideal conditions.

Upstand limiter						Fluted limiter						Valley limiter					
No.	**FR (l/s)**	**SF**	**SC. (%)**	**ER (g/hr)**	**D**	**No.**	**FR (l/s)**	**SF**	**SC. (%)**	**ER (g/hr)**	**D**	**No.**	**FR (l/s)**	**SF**	**SC. (%)**	**ER (g/hr)**	**D**
1	50	0.2	1	21.6	0.916	1	50	0.2	0.5	20.1	0.911	1	50	0.2	1	7.74	0.908
2	50	0.2	0.99	21.6	0.916	2	49.88	0.2	0.5	20.0	0.91	2	50	0.2	0.96	7.75	0.907
3	50	0.2	0.99	21.6	0.916	3	50.0	0.21	0.5	20.6	0.908	3	50	0.2	0.95	7.75	0.907
4	50	0.2	0.98	21.6	0.916	4	50	0.21	0.5	20.8	0.907	4	50	0.2	0.95	7.76	0.907
5	50	0.2	0.98	21.7	0.916	5	49.95	0.22	0.5	21.4	0.903	5	50	0.2	0.5	7.76	0.907
6	50	0.2	0.97	21.7	0.916	6	49.99	0.22	0.5	22.0	0.9	6	50	0.2	0.51	7.76	0.907
7	50	0.2	0.96	21.8	0.915	7	50	0.2	0.59	22.5	0.897	7	50	0.2	0.92	7.76	0.907
8	50	0.2	0.94	21.8	0.915	8	50	0.2	0.59	22.6	0.897	8	50	0.2	0.53	7.77	0.907
9	50	0.2	0.94	21.9	0.915	9	50	0.2	0.59	22.7	0.896	9	50	0.2	0.91	7.77	0.907
10	50	0.2	0.93	21.9	0.915	10	50	0.2	0.6	22.7	0.896	10	50	0.2	0.53	7.77	0.907
11	50	0.2	0.92	21.9	0.914	11	50	0.2	0.6	22.9	0.895	11	50	0.2	0.89	7.77	0.907
12	50	0.2	0.92	21.9	0.914	12	50	0.2	0.63	23.5	0.892	12	50	0.2	0.89	7.77	0.907
13	50	0.2	0.92	22.0	0.914	13	50.0	0.2	1	24.3	0.887	13	50	0.2	0.55	7.77	0.907
14	50	0.2	0.91	22.0	0.914	14	50	0.2	1	24.6	0.885	14	50	0.2	0.57	7.78	0.907
15	49.88	0.2	1	21.6	0.914	15	50	0.2	0.97	24.7	0.884	15	50	0.2	0.58	7.78	0.907
16	50	0.21	1	22.4	0.912	16	50	0.2	0.71	24.7	0.884	16	50	0.2	0.85	7.78	0.907
17	50	0.2	0.84	22.5	0.912	17	50	0.2	0.95	25.0	0.883	17	50	0.2	0.59	7.78	0.907
18	50	0.2	0.84	22.5	0.911	18	50	0.2	0.94	25.0	0.883	18	50	0.2	0.59	7.78	0.907
19	50	0.2	0.82	22.7	0.911	19	50	0.2	0.92	25.2	0.882	19	50	0.2	0.83	7.78	0.907
20	50	0.2	0.82	22.7	0.911	20	50	0.2	0.92	25.2	0.881	20	50	0.2	0.61	7.78	0.907

Table 11: Model validation simulation data and Design Expert optimum condition.

Limiter		Flow Rate (l/s)	Shape Factor	Sand Conc. (%)	Erosion rate (g/hr)
Upstand	Simulation	50	0.2	1	21.595
	DOE	50	0.2	1	21.748
			Percentage of Error		0.703 %
Fluted	Simulation	50	0.2	0.5	19.633
	DOE	50	0.2	0.5	19.768
			Percentage of Error		0.681 %
Valley	Simulation	50	0.2	1	7.483
	DOE	50	0.2	1	7.493
			Percentage of Error		0.128 %

The calculation percentage of error is examined to validate the optimized model. Therefore, the error is less than less than 0.8% thus it is reliable for optimization of erosion rate. This proven the statistical analysis is reliable for determination of the optimized erosion activity.

4. Conclusions

The outcome of this project is a numerical approach for the simulation on effect of two-phase flow in pipelines, for fluted limiter, upstand limiter and valley limiter. The proposed E/CRC Modified erosion model is utilized to simulate the erosion activity for 11mm diameter for the limiter with length of 0.033 m pipeline. STAR CCM+ software was used to simulate the specified conditions via different limiter's constructed through 3D-CAD. Based on the model, the erosion activity was simulated according to the experiment data in terms of flow rate, grain shape factor and sand concentration, and the percentage error is lower than 16% and this could be explained as the simulated limiter dimension is not constructed based on the actual experiment limiter dimension, thus discrepancy of values of erosion activity occurs Wallace [32] and further investigation is suggested. The shape factor from Zhang E/CRC erosion model was altered until it matches with the experimental erosion rate from Wallace's research, thus the shape factor 0.265 was reported.

Apart from that, the shape distribution of solid particle in fluid flow was investigated through the demonstration of microstructure analysis of ten selected sand particles using ImageJ and MATLAB. The formulation of new grain shape factor equation is generated for the selection of the suitable shape factor of solid particles in evaluating the erosion rate. The newly generated grain shape factor equation demonstrated that shape factor particles are defined as 75 % of 0.20 shape factor and the remaining 25 % of 0.45 shape factor present in fluid. These particles are tabulated through consideration of roundness and sphericity to conclude an overall grain shape factor of 0.265 which corresponds to the erosion activity of the limiter's respectively.

One of the main scope of this project is to develop, analyze and validate the different limiters affecting the erosion activity on the targeted surface. Results from CFD simulations were generally used to compare with the experimental results to study the erosion activity, where the velocity parameter is considered. Moreover, the numerical optimization approach is taken to analyze the parameters are through the engineering-statistical tools, DOE and RSM. The ANOVA technique is implemented to evaluate the importance of each factor and the response rate. The parameters are scrutinized to investigate the effects of erosion activity. In order to study the effect of mass flow rate, shape factor and sand concentration on erosion activity, the numerical method was applied. The statistical analysis conducted on each factor to analyze the effect of erosion activity, hence it can be concluded that the relationship between the parameters is presented (fluid flow rate > shape factor > sand concentration). The optimum condition to minimize erosion activity in the limiter is under the settings of the following sequence flow rate, shape factor and sand concentration presented as 50 litres/s, 0.20 and 1.0 % respectively.

The numerical results are promising and encouraging for future research areas. Therefore, some areas for further study are the prediction of piping maintenance for coat surface or sample collection of changing pipe to evaluate sand particle erosion rate can be done at a certain period. Furthermore, additional storage tank to settle down the solid particle to produce low grain shape factor and low sand concentration for pipe transportation.

References

[1] N. R. Kesana, R. Vieira, B. S. McLaury, and S. Shirazi, "Experimental study of sand particle concentration profiles in straight and pipe elbow for horizontal multiphase flows," *Journal of Energy Resources Technology,* vol. 136, p. 033001, 2014.

[2] G. V. Messa, G. Ferrarese, and S. Malavasi, "A mixed Euler–Euler/Euler–Lagrange approach to erosion prediction," *Wear,* vol. 342–343, pp. 138-153, 11/15/ 2015.

[3] P. G. Kumar, B. R. Smith, D. Vedapuri, H. J. Subramani, and L. D. Rhyne, "Sand fines erosion in gas pipelines–Experiments and CFD modeling," in *CORROSION 2014*, 2014.

[4] M. Parsi, K. Najmi, F. Najafifard, S. Hassani, B. S. McLaury, and S. A. Shirazi, "A comprehensive review of solid particle erosion modeling for oil and gas wells and pipelines applications," *Journal of Natural Gas Science and Engineering,* vol. 21, pp. 850-873, 2014.

[5] M. Parsi, M. Agrawal, V. Srinivasan, R. E. Vieira, C. F. Torres, B. S. McLaury, *et al.*, "CFD simulation of sand particle erosion in gas-dominant multiphase flow," *Journal of Natural Gas Science and Engineering,* vol. 27, pp. 706-718, 2015.

[6] J. Tu, G. H. Yeoh, and C. Liu, *Computational fluid dynamics: a practical approach*: Butterworth-Heinemann, 2012.

[7] M. Liu, H. Liu, and R. Zhang, "Numerical analyses of the solid particle erosion in elbows for annular flow," *Ocean Engineering,* vol. 105, pp. 186-195, 2015.

[8] H. Pouraria, J. K. Seo, and J. K. Paik, "Numerical study of erosion in critical components of subsea pipeline: tees vs bends," *Ships and Offshore Structures,* vol. 12, pp. 233-243, 2017.

[9] N. Barton, M. Lewis, and P. Emmerson, "CFD Erosion Prediction in Gas-Liquid-Sand Flow," in *SPE International Oilfield Corrosion Conference and Exhibition*, 2016.

[10] I. Finnie, "The mechanism of erosion of ductile metals," in *3rd US national congress of applied mechanics*, 1958.

[11] W. Peng and X. Cao, "Numerical simulation of solid particle erosion in pipe bends for liquid–solid flow," *Powder Technology,* vol. 294, pp. 266-279, 6// 2016.

[12] D. N. Veritas, "Recommended Practice, Erosion Wear of Piping Systems," *DNV RP O,* vol. 501, 1996.

[13] J. Neilson and A. Gilchrist, "Erosion by a stream of solid particles," *Wear,* vol. 11, pp. 111-122, 1968.

[14] Oka and T. Yoshida, "Practical estimation of erosion damage caused by solid particle impact: Part 2: Mechanical properties of materials directly associated with erosion damage," *Wear,* vol. 259, pp. 102-109, 2005.

[15] Oka, K. Okamura, and T. Yoshida, "Practical estimation of erosion damage caused by solid particle impact: Part 1: Effects of impact parameters on a predictive equation," *Wear,* vol. 259, pp. 95-101, 2005.

[16] Y. Zhang, E. P. Reuterfors, B. S. McLaury, S. A. Shirazi, and E. F. Rybicki, "Comparison of computed and measured particle velocities and erosion in water and air flows," *Wear,* vol. 263, pp. 330-338, 9/10/ 2007.

[17] K. R. Ahlert, *Effects of particle impingement angle and surface wetting on solid particle erosion of AISI 1018 steel*: BUniversity of Tulsa, 1994.

[18] K. Wang, X. Li, Y. Wang, and R. He, "Numerical investigation of the erosion behavior in elbows of petroleum pipelines," *Powder Technology,* vol. 314, pp. 490-499, 6/1/ 2017.

[19] G. C. Pereira, F. J. de Souza, and D. A. de Moro Martins, "Numerical prediction of the erosion due to particles in elbows," *Powder Technology,* vol. 261, pp. 105-117, 2014/07/01/ 2014.

[20] P. J. Lanasa and E. L. Upp, *Fluid Flow Measurement: A Practical Guide to Accurate Flow Measurement*: Butterworth-Heinemann, 2014.

[21] A. ReCap, "Autodesk ReCap," *Accessed March,* vol. 30, 2014.

[22] V. Bevilacqua, G. F. Trotta, A. Brunetti, G. Buonamassa, M. Bruni, G. Delfine, *et al.*, "Photogrammetric Meshes and 3D Points Cloud Reconstruction: A Genetic Algorithm Optimization Procedure," in *Italian Workshop on Artificial Life and Evolutionary Computation,* 2016, pp. 65-76.

[23] A. Ranellucci, "Reprap, Slic3r and the future of 3D printing," *Canessa, E., Fonda, C., and Zennaro, ed., "Low-cost 3D Printing for Science, Education, and Sustainable Development,* pp. 75-82, 2013.

[24] M. Duris, Z. Arsenijević, D. Jaćimovski, and T. Kaluđerović Radoičić, "Optimal pixel resolution for sand particles size and shape analysis," *Powder Technology,* vol. 302, pp. 177-186, 2016/11/01/ 2016.

[25] T. Morris, "Image Processing with MATLAB," *Supporting Material for COMP27112,* 2016.

[26] W. Burger and M. J. Burge, *Digital image processing: an algorithmic introduction using Java*: Springer, 2016.

[27] G. Desbois, J. L. Urai, S. Hemes, S. Brassinnes, M. De Craen, and X. Sillen, "Nanometer-scale pore fluid distribution and drying damage in preserved clay cores from Belgian clay formations inferred by BIB-cryo-SEM," *Engineering Geology,* vol. 179, pp. 117-131, 2014/09/04/ 2014.

[28] C. Saranya and G. Manikandan, "A study on normalization techniques for privacy preserving data mining," *International Journal of Engineering and Technology (IJET),* vol. 5, pp. 2701-2704, 2013.

[29] H. R. Karami, M. Keyhani, and D. Mowla, "Experimental analysis of drag reduction in the pipelines with response surface methodology," *Journal of Petroleum Science and Engineering,* vol. 138, pp. 104-112, 2// 2016.

[30] M. A. Islam, T. Alam, and Z. Farhat, "Construction of erosion mechanism maps for pipeline steels," *Tribology International,* vol. 102, pp. 161-173, 10// 2016.

[31] M. Al-Bukhaiti, A. Abouel-Kasem, K. Emara, and S. Ahmed, "Particle Shape and Size Effects on Slurry Erosion of AISI 5117 Steels," *Journal of Tribology,* vol. 138, p. 024503, 2016.

[32] M. S. Wallace, "CFD-based erosion modelling of simple and complex geometries," University of Strathclyde, 2001.

[33] R. Zhang and H. Liu, "Numerical simulation of solid particle erosion in a 90 degree bend for gas flow," in *ASME 2014 33rd International Conference on Ocean, Offshore and Arctic Engineering,* 2014, pp. V06AT04A044-V06AT04A044.

[34] R. M. Khan, "Design of Experiment," *Problem Solving and Data Analysis using Minitab: A clear and easy guide to Six Sigma methodology,* pp. 407-455, 2013.

IOP Conf. Series: Materials Science and Engineering **943** (2020) 012025 doi:10.1088/1757-899X/943/1/012025

Comparison of Soil Erosion Control by Mulch and Cultivated Grass

A.C. Tay and P. Ratnam

Department of Civil and Construction Engineering, Curtin University Malaysia, Miri, Sarawak, Malaysia

tayaichen@curtin.edu.my

Abstract. Soil erosion has deteriorated water quality and caused land degradation. As human industrialization quickens, the effects of soil erosion are getting worse. Therefore, it is essential to have an effective mitigation measure at hand to solve this problem. In this research, two types of erosion control methods were compared to determine their effectiveness in controlling soil erosion. The two methods used were mulch made from dried grass clippings and grass cultivation grown with *Axonopus compressus* or more commonly known as cow grass. The plots were placed under five rainfall events together with a control plot. The runoff volume, turbidity and total suspended solids (TSS) from the plot discharges were tested and compared. The effectiveness of each erosion control method under different rainfall conditions were also evaluated. Results indicated that the mulch plot produced a lower runoff volume but the grass plot did manage to improve performance over time. The turbidity of both plots were similar. The TSS produced by the grass plot was lower than the mulch plot in 4 out of the 5 testing periods. The grass plot also performed better in longer duration and high intensity rainfall events. The mulch was able to produce similar results with the grass in short duration, low or moderate intensity rainfall events. As a conclusion, the grass cultivation is a more effective erosion control method when compared to the mulch.

1. Introduction

Soil erosion is a removal of topsoil due to human, natural or animal activities in our environment. Generally, the topsoil is removed faster than natural soil forming processes that can replace it [1].

Globally, there are numerous environmental, social and economic problems caused by soil erosion. The main environmental problem associated with soil erosion is the decrease in water quality as a result of runoff. Turbidity and total suspended solids (TSS) can prevent sunlight to penetrate water. In addition, the eroded soil can change the chemical composition of water as they might contain pesticides from agricultural land, heavy metals from industrial activities and nutrients from the soil itself. Excess nutrients in the water can cause eutrophication, which can lead to an increase in Biological Oxygen Demand (BOD) [2].

In recent times, Malaysia as a developing country has experienced significant environmental changes due to higher population, urbanization and faster economic growth. This has brought upon tremendous land use changes which has made the land in Malaysia more susceptible to soil erosion [3].

A slope is prone to soil erosion, but there are challenges to conserve the slope due to lacking of knowledge, technical support and cost constraint. Engineers introduce slope protection by designing gabion, retaining wall, riprap and other engineering methods [4, 5]. It has been much research done previously on the topic of soil erosion control using a plot of land [6]. The effectiveness of soil erosion control using mulch and grass has not been well identified. Mulch and grass cover represent the most cost effective and economic ways to control soil erosion at a slope among the common methods. The main objectives of this study were as follows:

1. To examine the quality of the surface runoff produced by measuring the physical properties of the water namely the runoff volume, turbidity and total suspended solids (TSS).
2. To evaluate the effectiveness of two methods of soil erosion control, namely mulch and cultivated grass under different rainfall conditions.

Furthermore, this research is able to contribute to water quality improvement as the effective soil erosion method at low cost with less turbidity, TSS and runoff volume at the soil erosion affected areas.

2. Materials and Methods

Different methods, designs, parameters and procedures were performed in previous studies [7, 8]. In this experiment, a small-scale plot of land will be constructed and placed on a slope to induce erosion. It was then placed outside during a natural rainfall and the runoff was collected. The runoff was checked for its turbidity, TSS and runoff volume and was measured to evaluate which erosion control method is more effective in reducing sedimentation.

Plot was designed according to other similar experiments done by other researchers [9, 10]. The plot designed into a width of 0.6m, length of 1m and a depth of 0.2m. The plot was made using thin waterproof plywood. The box was nailed together using nails and any gaps were closed using a hot glue gun. Three holes were drilled at the front of the box with a diameter of one and a half inches. PVC pipes were joined at each of the holes to direct the runoff out. The top of the PVC pipes was covered with a 0.1m long wooden cover. The collection buckets were placed below the PVC pipes.

2.1. Control plot

Bare soil was created as a control plot. This plot was filled to a depth of 6 inches. It was placed on an incline in the horizontal: vertical ratio of 3:1.

2.2. Mulch plot

The plot was filled with soil to a depth of 6 inches. Mulch was made using dried grass clippings and dried grass. It was laid onto the soil and spread evenly throughout the soil. The thickness of the mulch used was 1 – 1.5 cm.

2.3. Grass cultivation plot

The plot was filled with soil to a depth of 6 inches. Patches of cow grass (*Axonopus compressus*) was planted along the soil. The plot was watered with 1 litre of water every day for a month to ensure that it grows properly. The grass coverage had more than 90% observable coverage on the soil before beginning to record the data.

2.4. Soil analysis

A soil analysis test was conducted to determine the type of soil. The test conducted is a Standard Operating Procedure (SOP), which was based on ASTM D422-63 standard Test Method for Particle-Size Analysis of Soils through sieving [11].

2.5. Rainfall data

The rainfall data was collected by the weather station located in the Curtin Malaysia campus. The rainfall intensity was categorized as low, moderate or high using the mean rainfall intensity over the specific time period. It was assumed that a low intensity rainfall event has a mean intensity of lower than 10 mm/hr, a moderate intensity rainfall event as having a mean intensity of between 10 mm/hr and 20 mm/hr and a high intensity rainfall is as having a mean intensity of more than 20 mm/hr. For the duration of rainfall, a short rainfall duration was assumed as occurring for less than one hour while a long duration was quantified as occurring for more than 1 hour. The rainfall data for different test periods is shown in **Table 1**.

Table 1. Rainfall data for different test periods.

Period	Duration of rain exposure (min)	Average rainfall intensity (mm/hr)	Description
1 (02/09/2018)	35	32.15	The rain started intensely before dropping to a lower intensity and then rising back up to a higher intensity. It was a short duration and high intensity.
2 (05/09/2018)	60	9.09	During this rainfall event, the intensity was quite low throughout the rainfall period although it was for a longer duration. It was a long duration and low intensity.
3 (06/09/2018)	80	15.71	The rain started intensely for only 15 minutes before reducing to a very low intensity. It was a long duration and moderate.
4 (19/09/2018)	105	26.26	The rainfall stopped briefly after 55 minutes before resuming again. It was a long duration and high intensity.
5 (01/10/2018)	55	13.17	The rainfall indicated a high intensity when it started before dropping in intensity. It later increased back up after 30 minutes before subsiding. It was a short duration and medium intensity.

3. Results and Discussion

3.1. Soil analysis

Soil used for this experiment was put through a sieve analysis test and the data was recorded in **Table 2**. The soil comprised of 4% of silt, 52% of sand and 44% of gravel.

Table 2. Particle size distribution.

Silt (%)	Sand (%)			Gravel (%)		
	Fine	Medium	Course	Fine	Medium	Course
4	12	12	28	29	15	20

3.2. Runoff volume

The control plot consistently produced more runoff when compared to the other two plots. This is expected as the control plot has no erosion control method and is solely to provide comparison for the other two plots. Comparison of runoff volume between mulch and grass is shown in **Figure 1**.

Both plots produced the least amount of runoff volume during the first test period which was the period with the shortest rainfall duration. The rainfall intensity during this testing period started out high, but later subsided. The mulch plot only produced 80 mL of surface runoff while the grass produced 184 mL. The mulch was probably able to produce such a low runoff due to the decreased amount of rainfall as it was the shortest duration. This was mainly due to the function of the mulch whereby it prevented the surface water from accumulating and forming micro rills. The grass plot was able to produce a low amount as well due to the short rainfall duration, the water had been absorbed into the soil instead of accumulating on the surface. It produced more runoff than the mulch as it could not obstruct the flow of the surface runoff [12].

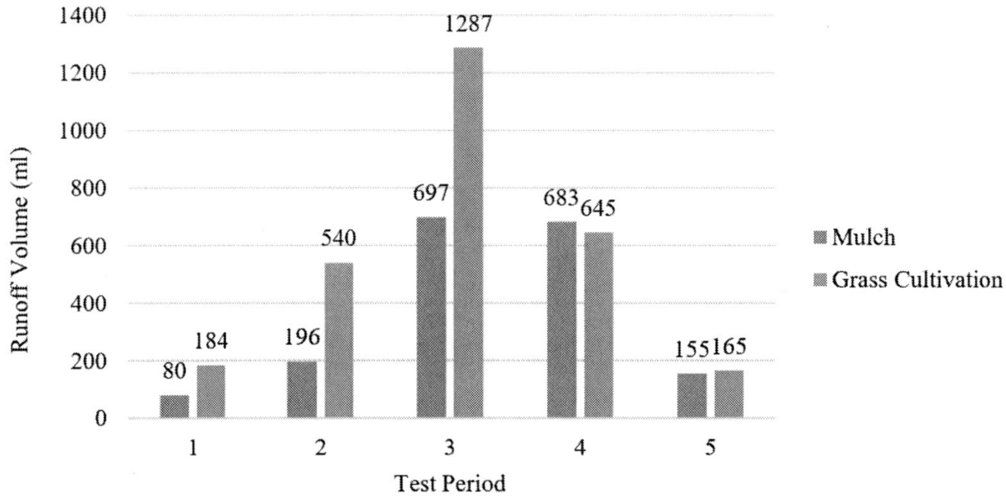

Figure 1. Runoff volume for mulch and grass.

In the second test period, the rainfall period was longer but was a low intensity rain throughout the rainfall period. Nevertheless, both plots produced more runoff with the mulch producing 196mL and the grass producing 540mL. This is a stark difference when compared to test period 1 where the difference between the two was only 104mL. In this case, it is more than double the amount. The longer duration could have caused the water to accumulate on the grass plot thereby producing more surface runoff. The low intensity rainfall did not seem to have much effect on the mulch plot as there was minimal difference between the test period 1 and test period 2. The low intensity rain caused the rain drops to be absorbed by the mulch material.

In the third test period, the rainfall was even longer and started out with a high intensity. But this high intensity only lasted for 10-15 minutes before reducing to a low intensity rain. It was a moderate intensity rainfall event. However, it must be noted that test period 3 took place only a few hours after test period 2. Hence, this is a chance to investigate the erosion control performance of the two methods after successive rainfall events. Both the test plots produced significantly higher amounts of runoff with the mulch producing 687 and the grass plot double that with 1287 mL. This could be attributed to the soil sealing which occurs in the grass plot reducing the infiltration rate of the soil [6]. Water accumulated on the soil surface resulting in more runoff produced. This soil sealing had a lesser effect on the mulch as the raindrops were absorbed by the mulch material as well as the soil.

In the fourth test period, the rainfall was the longest among the testing periods. The mulch produced more or less the same amount of runoff when compared to period 3. This could indicate that although the rainfall was longer and with a high intensity, the mulch plot was unaffected by extended periods of rainfall. The grass plot on the other hand, showed a significant reduction by producing 645 mL of runoff which was lower than the volume produced by mulch plot. The roots from the grass had grown more and increased infiltration rates in the soil thereby decreasing the occurrence of soil sealing [13].

In the last testing period, the rainfall was a short rainfall event with a medium intensity. Both plots produced similar results with the runoff volume produced being 155 mL and 165 mL for the mulch and grass plots respectively. The grass plot had once again showed significant improvement in the reduction in runoff produced.

The percentage reduction obtained shows that the mulch had more or less the same performance throughout the testing period as its percentage reduction ranged between 83% – 95.04%. The grass plot showed more variance as its percentage reduction ranged from 53.17% - 94.72%. When compared to results obtained from previous work done [10], the engineered fiber matrix (EFM) with seed showed the highest percentage reduction of runoff volume at 68.0 %. These results from the previous work are based on average runoff volume. Nevertheless, the difference between the results obtained from this experiment and the results from the previous research are too large. Hence, both of the erosion control methods used in this experiment is more effective in controlling runoff.

The mulch consistently produced a lower runoff volume when compared with the grass plot except in the last two testing periods where the runoff volume produced was very similar. The grass required sufficient time to grow which caused the roots not deep enough into the soil, therefore, the porosity of the soil was low and the water accumulated on the surface of the soil. The grass plot was a better erosion control method in terms of runoff volume over time. The mulch on the other hand, did not show any significant improvements in its performance in reducing runoff over time.

3.3. Turbidity

The control plot showed a higher turbidity (ranged from 8,550 to 23,760 NTU) when compared to the mulch plot and grass plot. Comparison of turbidity for mulch plot and grass plot is shown in **Figure 2**.

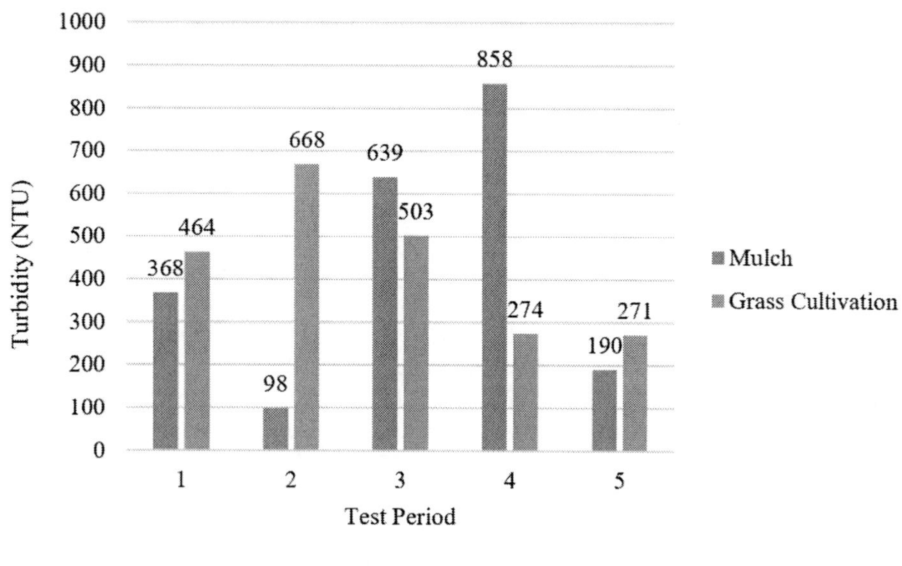

Figure 2. Turbidity of mulch and grass.

In period 1, the mulch plot had a turbidity of 368 NTU and the grass plot a turbidity of 464 NTU. Both control measures produced similar results with the mulch performing slightly better. This could be due to period 1, having a very short rainfall duration. In a short rainfall duration, the mulch could absorb the raindrops before it hit the soil surface and dislodged the soil particles thus, resulting in a lower turbidity [6].

In period 2, the mulch plot had turbidity of only 98 NTU and the grass plot a turbidity of 668 NTU. This could be due to the surface cover provided by the mulch. As the rainfall had a low intensity, there was not enough kinetic energy in the rain drop particles to dislodge the soil. The grass cover had not grown fully to cover the soil causing the raindrops to directly impact the soil and dislodge the particles.

During period 3 which occurred a few hours after test period 2, the mulch produced 639 NTU and the grass plot 503 NTU. This grass plot showed very similar readings when compared to test period 2 with slightly decreased. However, a drastic rise in turbidity for the mulch plot. After successive rainfall events, the mulch was still full of water absorbed from the previous rainfall event. So when it rained again a few hours after, there was more water produced as runoff instead of being absorbed by mulch. If there was time in between the rainfall events, the water absorbed by the mulch would have evaporated and it would have been able to absorb more water by the time the next rainfall event occurred.

In the fourth test period, the rainfall had the longest duration a high intensity. The mulch plot showed the highest turbidity at 858 NTU while the grass plot was 274 NTU. This could be due to the grass growing longer and thicker thereby providing more surface cover for the soil dissipating the high kinetic energy in the raindrops. The mulch plot on the other hand, was drastically affected by the high intensity and longer duration rainfall. By having these conditions, the raindrops had more kinetic energy to dislodge the soil particles, and by continuing at this high intensity for an extended period of time, the mulch cover was unable to fully dissipate the kinetic energy of the raindrops [14].

In the last test period, the mulch cover had a better performance by producing 190 NTU and the grass cover a turbidity of 271 NTU. The last test period had a short rainfall duration with a medium intensity. The grass cover produced a similar amount of turbidity as test period 4. The mulch on the other hand produced a lower amount, indicating that the mulch plot performed better when the rainfall duration is short.

The mulch plot had a reduction in turbidity with the range of 97.31% - 99.1%. The grass plot once again had a bigger variance as its range varied from 92.19% - 99.59%. These results are comparable to the previous study [10], whereby engineered fiber matrix (EFM) with seed indicated a percentage reduction of 98.7%. The research results are the average turbidity, hence it cannot be concluded that either of the methods used in this experiment are better as the percentage reduction from the previous research are still in the range obtained in this experiment.

The mulch showed the highest reduction most of the time when compared to the grass cultivation plot. Although the results are almost similar, the mulch showed a slightly higher reduction. This could be attributed to the mulch covering the surface of the top soil whereby reducing the kinetic energy of the raindrop particles which impact the surface thereby reducing the amount of soil being dislodged. They might even prevent the raindrops from even reaching the surface altogether. The surface cover on the grass cultivation plot was less due to the gaps. The gaps between grass cover caused the raindrops to fully reach the surface and dislodge the soil particles. However, the grass plot had a better performance compared to the mulch plot when it was a high intensity and prolonged rainfall event. The mulch lost the ability to dissipate the kinetic energy of the raindrops after a sustained rainfall.

The turbidity produced from both control measures was not up to the water quality standards in Malaysia as all the turbidity values are above 50 NTU. The water produced did not meet the Class III Standards as listed in the Malaysian national guidelines [15]. Hence, using these methods alone might not be enough to achieve the desired erosion control performance at construction sites.

3.4. Total suspended solids (TSS)

The control plot showed a higher TSS (ranged from 6.5 to 82 g/l). Comparison of TSS for mulch plot and grass plot is shown in **Figure 3**.

During the first test period. The grass plot produced 0.34 g/l as compared to the mulch which produced 0.7 g/l. During period two, the mulch plot had a better performance as it recorded 0.26 g/l as compared to the grass plot which recorded 0.42 g/l. For the following test periods, the mulch recorded 0.58 g/l, 0.64 g/l and 0.22 g/l for test periods 3, 4 and 5 respectively. The grass plot recorded 0.3 g/l, 0.16 g/l and 0.18 g/l for test periods 3, 4 and 5 respectively.

The grass plot indicated a better performance in 4 out of the 5 test periods. The reduced performance of the mulch could be due to the mulch biodegrading over time as it is an organic material. This could have contributed to tiny organic matter to be eroded together with the soil causing the TSS value to be higher. However, the rate of mulch decomposition is unclear. Nevertheless, when the mulch decomposes it could create a more conducive environment for the microorganisms in the soil underneath by stabilizing the temperature and moisture [16]. This advantage is nullified, however, the increase in TSS in the surface runoff causing increased pollution to water bodies.

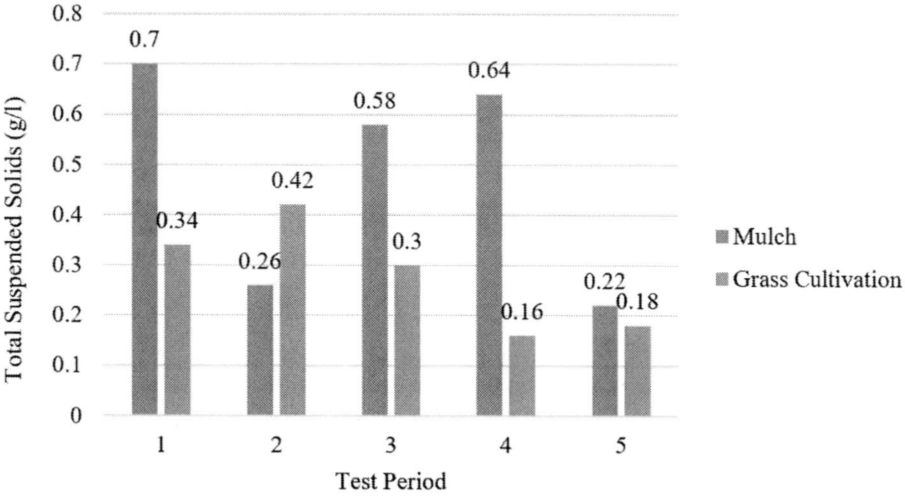

Figure 3. TSS of mulch and grass.

The higher amount of TSS detected in the grass plot during period 2 could be attributed to the soil not settling fully. Also, the roots of the grass had not grown fully at that time hence after the rainfall event, the soil on the surface was easily washed away. In addition, the rainfall conditions did not affect the TSS values obtained [8]. During period 4 when there was a high intensity and long duration rainfall event, the TSS value obtained by the grass plot was the lowest for all the test periods. The mulch plot produced a high amount of TSS during test period 4 but this amount is similar with the TSS amount produced during test period 3. Hence, the rainfall conditions did not affect the amount of TSS produced.

The percentage reduction of the mulch plot ranged from 96% - 99.2% while the grass plot ranged from 93.54% - 99.8%. During test period 4, when the rainfall event was the longest and had a high intensity. The grass plot showed the lowest TSS among all periods tested during period 4. When compared to previous work [10], the best performing control method which was the engineered fiber matrix (EFM) with seed achieved a 99.8% reduction. The engineered fiber matrix (EFM) with seed performs better than the two erosion control methods employed here in terms of TSS reduction.

The grass plot and mulch plot was able to occasionally achieve TSS values of below 300 mg/l which is the Class IV standard listed in the Malaysian national guidelines [18]. However, the grass cultivation plot is considered a more viable option as the biodegradation of the mulch over time was not fully quantified.

3.5. *Cover factor*

The cover factor is calculated by using the ratio of sediment yield from a surface with a specific control measure to a surface with no control measures. The cover factor is one of the most commonly used parameters to test the effectiveness of soil erosion control measures. The cover factor is scaled from 0 to 1 where zero means a 100% reduction of erosion and one meaning a 0% reduction in erosion or in other words it is the same as the bare soil. The average cover factor for mulch plot and grass plot is shown in **Table 3.**

Table 3. Average cover factor.

Plot	Average Cover Factor
Mulch	0.024
Grass Cultivation	0.022

The cover factor is one of the parameters used in the Universal Soil Loss Equation (USLE). It is used to represent the control method used when compared to a bare unprotected soil. Among the many parameters in use for the USLE equation, the cover factor is the most easily influenced. It is the factor most considered by policy makers and farmers [17].

The grass cultivation showed the better cover factor as compared to the mulch. The grass was better to be used as a cover than the mulch. However, these results are not entirely accurate as the TSS from different test periods with varying rainfall conditions were averaged out. Mulch with a 100% cover which is used as an erosion control at construction sites have a C factor value of 0.02 based on the Malaysia Department of Irrigation and Drainage (DID) [18]. The grass with more than 90% cover in this experiment indicated the C factor value of 0.02. Hence, the cover factor is this experiment is similar to the guideline listed in DID.

4. Conclusion

The mulch plot was able to consistently produce a lower runoff volume when compared to the grass plot. The turbidity produced by both sets of erosion control measures are very similar although the mulch produced a slightly lower value. The mulch acted as a better surface cover, which is effective in reducing the kinetic energy of the raindrops. However, the mulch was biodegrading over time that causing small organic particles to be eroded as well. In terms of runoff volume and turbidity, the grass plot was able to perform better in high intensity and longer duration rainfall events when compared to the mulch plot. During low and moderate intensity rainfall events, the mulch plot produced slightly better results when compared with the grass plot. It proved that the rainfall conditions had no effect on the TSS produced. Although there were certain instances when the results were very similar, after taking into account the increase in performance of the grass over time as well as its ability to perform better in harsher conditions, it is concluded that the grass cultivation is the better option to be used as an erosion control method. However, mulch can still be used as a temporary erosion control method on construction sites while it would be better to apply grass as a more permanent erosion control method.

References

[1] Fournier A J 2011 *Soil Erosion* (New York: Nova Science Publishers)

[2] Rickson R J 2014 Can control of soil erosion mitigate water pollution by sediments? *Sicence of the Total Environment* **468** 1187

[3] Obaid H A and Shahid S 2017 Soil Erosion Susceptibility Of Johor River Basin *Water And Environment Journal* **31** 367

[4] Mentani *et al* 2016 A New Approach to Evaluate the Effectiveness of Rockfall Barriers *Procedia Engineering* **158** 398

[5] Theisen M M 1992 The role of geosynthetics in erosion and sediment control: An overview *Geotextile and Geomembranes* **11** 535

[6] Khan *et al* 2016 Effect of slope, rainfall intensity and mulch on erosion and infiltration under simulated rain on purple soil of south-western Sichuan Province, China *Water* **8** 528

[7] Durán *et al* 2008 Soil-erosion and runoff prevention by plant covers. A review *Agronomy For Sustainable Development* **28** 65

[8] Schiff *et al* 2016 Effects of rainfall intensity and duration on the first flush from parking lots *Water* **8** 320

[9] Shoemaker A 2009 Evaluation of Anionic Polyacrylamide as an erosion control measure using intermediate-scale experimental procedures *Master Thesis* (Alabama: Auburn University)

[10] Sidhu *et al* 2015 Cost effectiveness of erosion control covers during vegetation establishment under simulated rainfall *Journal of Water Resource and Protection* **7** 119

[11] Geoinfo.nmt.edu. 2008 *Standard operating procedure no. 33* https://geoinfo.nmt.edu/staff/mclemore/projects/environment/documents/SOP_33v8_Particle SizeAnalyses_FINAL.pdf

[12] Merritt E 1984 The identification of four stages during micro-rill development *Earth Surface Processes and Landforms* **9** 493

[13] Zadeh K and Sepaskhah A R 2016 Effect of tree roots on water infiltration rate into the soil *Iran Agricultural Research* **35** 13

[14] Moussouni *et al* 2018. The effect of raindrop kinetic ernergy on soil erodibility *International Journal Of Environmental And Ecological Engineering* **8** 878

[15] National Water Quality Standards For Malaysia 2017 *WEPA* http://wepa-db.net/3rd/en/topic/waterstandard/index.html.

[16] Maggard *et al* 2018 Tree-based mulches influence soil properties and plant growth Horttechnology **22** 353

[17] Panagos et al 2015 Estimating the soil erosion cover-management factor at the European scale. *Land Use Policy* **48** 38

[18] Department of Irrigation and Drainage 2010 *Guideline for Erosion And Sediment Control in Malaysia* (Kuala Lumpur: Department of Irrigation and Drainage)

IOP Conf. Series: Materials Science and Engineering **943** (2020) 012026 doi:10.1088/1757-899X/943/1/012026

A Review on Biodiesel Synthesis using Iron Doped Catalyst

Yu Herng Foo[1], Yie Hua Tan[1*], Nabisab Mujawar Mubarak[1], Jibrail Kansedo[1], Inn Shi Tan[1], Henry Chee Yew Foo[1], Mohd Lokman Ibrahim[2,3,4]

[1]Department of Chemical Engineering, Faculty of Engineering and Science, Curtin University Malaysia, Miri, Malaysia.
[2]School of Chemistry and Environment, Faculty of Applied Sciences, Universiti Teknologi MARA, 40450 Shah Alam, Selangor, Malaysia
[3]Centre of Nanomaterials Science, Institute of Science, Universiti Teknologi MARA, 40450 Shah Alam, Selangor, Malaysia
[4]Industrial Waste Conversion Technology Research Group, Green Technology and Sustainable Development, Universiti Teknologi MARA, 40450 Shah Alam, Selangor, Malaysia

Email: tanyiehua@curtin.edu.my

Abstract. This paper summarized on the history of biodiesel synthesis, biodiesel synthesis using iron doped catalyst. Biodiesel are gaining enormous attention from researches and manufacturers considering its advantages of non-toxic, biodegradable, renewable, and environment-friendly in order to improve the air quality over the atmosphere followed by reducing the global warming problems effectively. Regarding the conventional biodiesel production method, manufacturers generally utilize vegetable oil and alcohols to pass through the transesterification process with the presence of homogeneous base catalyst. The utilization of this types of catalyst lead to another environmental issues that most of the researches are concerning about since the homogeneous catalysts are not reusable and causing separation problems between oil products and the catalyst itself. Therefore, it is highly necessary for the researches and manufacturers to further explore and investigate some catalyst types that are truly feasible for high-scale or industrial usage. Heterogeneous catalysts are known to consists of an identical phase with the reacting medium where the liquid phase contains the feedstock and reaction medium whereas the catalyst retain within the solid phase, hence resolving the separation problems that the homogeneous catalyst would lead to. Recently, researches have introduced the novel method of magnetic properties impregnation of catalytic active component onto the carbonaceous compound. This proposed action tends to provide the advantages of high surface area, separation ease, and preventing the leaching of catalytic active components from the catalyst. Both the potential iron doped carbon catalyst have been extensively researched and studied within this report. Besides, the catalyst properties have discussed further for analyzing the effects of the biodiesel yield that introduced by the chemical and physical properties of various types of catalysts.

1. Introduction

Biodiesel, is also known as fatty acid methyl ester (FAME) is being acceptable by the populations due to its reusability, biodegradability, eco-friendly properties, hence reducing the probability of global warming while having the similar features with the comparison of petrodiesel or fossil fuels [1].

Biodiesel is commonly established via transesterification process between the triglycerides (mainly fats or oils) and alcohols (mainly methanol or ethanol) with the presence of catalysts. Technically, biodiesel is yet to be considered as the complete alternative energy source for the conventional petrodiesel throughout the global due to its high production cost which take into account of the raw materials cost and labour cost [2]. In addition, insufficient stock of the raw materials and ineffective catalytic system are also the considerable challenges faced while manufacturing the biodiesel to replace the conventional fossil fuels as this issue might cause difficulties for meeting the population's demand across the global and/or other environment policy related problems could introduce at the same time as well [3]. Despite all of that, biodiesel is still not fully commercialized and practicable globally due to the low biodiesel demand but an enormous capital production cost, hence resulting the manufacturer to sell the biodiesel in a higher price. Regarding the biodiesel production, there are generally 2 kinds of catalyst which are acid catalyst and base catalyst. According to the researches, the biodiesel production via vegetable oil mainly utilized the homogeneous base catalyst with the consideration of reduced mass transfer resistance effect, hence faster reaction efficiency compared to the heterogeneous catalysts [4]. However, the homogeneous base catalyst like NaOH introduces the drawback of soap foaming and emulsions within the solutions due to presence of high free fatty acid (FFA) content > 1% inside the feedstock whereas the homogeneous acid catalyst is able to encounter this issue [5]. The liquid form conventional catalyst that commonly used for the biodiesel production is not reusable and causing corrosion towards the reactor and storage tank. Besides that, it requires several pre-treatments before the catalyst get discharged, otherwise it might bring along the negative effects towards the aquatic lives. Therefore, heterogeneous catalyst is preferable to be utilized for catalysing the biodiesel production since the solid catalyst is reusable, hence reducing the cost of purchasing the new catalyst. In order to improve the separation efficiency and recoverability for the catalyst during the final stages of transesterification process, the heterogeneous catalyst which complied with magnetic properties are proposed by impregnating it with ferrite ion. Therefore, the costly mechanical separation methods such like filtration and centrifugation are not necessary as the magnetic decantation method has provided effective and desirable separation performance for the catalyst from the biodiesel [6].

This research paper aims to inspire and motivate more researches to develop in-depth investigation and exploration regarding the biodiesel synthesis using iron doped carbon catalyst that have modified with magnetic properties. Besides that, this study brings out the idea of synthesizing the high-yield green fuel (biodiesel) via transesterification instead of the non-renewable fossil fuel through conventional method. Hence, this paper is capable to prove the sustainability and eco-friendliness of the biodiesel production via the iron doped carbon catalyst.

2. Latest Technologies for Biodiesel Production

2.1. Microwave Assisted Process

The biodiesel production via microwave assisted process basically describing the transesterification process that is proceeding under the microwave irradiation condition, hence shorten the reaction time and improving the production yield. The utilization of microwave for the biodiesel production based on the concept of emerging the electromagnetic radiations with the frequency ranges from 300 MHz to 30 GHz for affecting the molecular movement such as ion migration or dipole rotations but remained the structure of the molecule [7]. Followed by that, the microwaves effects slightly triggered the variance of the involved polar molecules and ions hence resulting molecular friction, and finally initiating the chemical reactions between the raw materials (oil, methanol, and base catalyst that contains both polar and ionic components). Since the energy provided by the microwave interacts with the molecules with a rapid rate, the transesterification process is able to accelerate within a short reaction time. Therefore, the biodiesel production via microwave assisted process produced high purity product, relatively low amount of by-product and a shorter separation time could be obtained [8]. However, the scale-up process for the laboratory-scale microwave assisted process is still containing much uncertainties and remain a challenge for the researches to improve it in order to upgrade this technique to become feasibly applicable for industrial biodiesel production.

2.2. Supercritical Process

Supercritical fluid, also known as the fluid consists of a temperature and pressure above its critical point, which means the liquid and gas phases do not exist at the same time, hence it is able to diffuse through any solids like a gas or dissolve any materials like a liquid. Researches have wisely utilized this unique property of the supercritical liquid and apply it as a part of the biodiesel production process. During the biodiesel production via supercritical process, researches have generally treated the water, carbon dioxide, and alcohol as the supercritical fluids [9]. As the liquid methanol is known to be a polar solvent and has a pair of hydrogen bonding in order to form the methanol clusters between the OH oxygen and OH hydrogen from another two methanol molecules. However, a supercritical methanol has a hydrophobic nature with a lower dielectric constant in order to allow the non-polar triglycerides can well-isolated with the supercritical methanol to form a single-phase oil-methanol mixture (methyl ester). With this reason, the conversion rate for the oil into methyl ester was found to largely increase compared to the one that produced from triglycerides and liquid phase methanol via the conventional method, transesterification process [8].

On the other hand, there were several factors are affecting the performance of the transesterification process via the supercritical process such like temperature effect, pressure effect, and the effect alcohol-oil molar ratio. Firstly, temperature is the prior factor among all factors that affects the transesterification process under the supercritical condition. According to the study of Kusdiana & Saka [10], the conversion rate for the triglycerides to methyl ester was 70 wt % which is relatively low when the supercritical fluid temperature of 200-230°C was applied as the process conditions whereas 95 wt % of triglycerides to methyl ester conversion rate was obtained at 350°C for only 4 minutes of the reaction time [11]. Additionally, the increment of temperature effects leads to the rising of pressure effects which could increase the triglycerides' solubility, hence close contact between the alcohol and triglyceride has established. Finally, a high alcohol-oil molar ratio could result a favourable and desire transesterification process due to the increment of contacted area between the oil and alcohol under supercritical conditions. In overall, biodiesel production via supercritical process has the advantage of high product recovery rate, but also with the drawbacks of high energy consumption and excess alcohol usage in the supercritical conditions [8].

2.3. Ultrasonic Irradiation Process

The ultrasonic irradiation supplied a large negative pressure gradient to the liquid followed by causing the liquid to break down into cavitation bubbles. These small cavities grow rapidly under high ultrasonic intensities and collapse immediately after seconds, hence rising the mass transfer condition by interrupting the interfacial boundary layers within the mixture of involved oil and alcohols [12]. Followed by that, the emulsification between the immiscible mixture (oil and alcohol) is able to improve effectively due to the mixture boundary layer enlargement which caused by the ultrasonic cavitation. For transesterification process, ultrasonic mixing provides a better mixing with the improved mass transfer rates compared to the one that using conventional batch reactor for mixing since a well-mixing is the key for a high production of biodiesel. In overall, ultrasonic biodiesel production could be beneficial for small-scale production in terms of short reaction time and low energy consumption, but it would be a huge challenge to utilize this method if large scale industrial production is required due to the necessity of enormous increment of ultrasound probes and energy supplier.

2.4. Membrane Biodiesel Production

The membrane system that designed for the biodiesel production generally possessed with the potential characteristics of high selectivity, high surface area per unit volume, and the ability for controlling the level and concentration of the involved components during the mixing between the two phases. In the biodiesel production process via membrane system, the membrane reactor system is the first phase of the two phases to trans-esterify the triglycerides and alcohols to form biodiesel. Followed by that, the second phase is the separative membrane system which responsible for isolating the impurities such like catalysts, soap, organic or inorganic solvents, and absorbents from the crude biodiesel. Furthermore, the organic membrane is commonly suitable for the process which excluded high acidic and basic conditions whereas inorganic membranes such as metallic and ceramic are preferable while dealing with harsh conditions such as high temperature, high pressure, and high vibration [13]. However, there are several prospects and challenges that might encounter during the biodiesel production through membrane system which has shown as Table 1. below, therefore, this technique can only be utilized feasibly after the listed challenges have been overcame.

Table 1. Prospects and challenges for biodiesel production via membrane system [14], [15].

Prospects	Challenges
1. Can be operated continuously	1. High expectation for membrane systems
2. Can combine both transesterification reaction and separation simultaneously	2. Membrane systems are limited by pore size and shape of materials to be filtered
3. Can provide both technical and economic advantages over alkaline catalyst when process low quality feed	3. Soap formation when homogeneous alkaline catalyst is used
4. Easy product removal during transesterification	4. Leaching of solid alkaline catalyst
5. Can reduce foaming of by-products, glycerol	5. Yield are lower compared to those obtained heterogeneous catalyst in batch reactors
6. Improving the efficiency of hot water washing step	

2.5. *In-situ Transesterification*

In-situ transesterification, also known as the reactive extraction which processed based on the concept of simplifying the biodiesel production process by allowing both the oil extraction and transesterification process to perform within a single step at the same time. In in-situ transesterification process, the lipids in the oil-bearing seeds directly contacts with the selected chemical solvent with the presence of a catalyst. The differences between the conventional method and in-situ transesterification for biodiesel production has shown as Figure 1. below. According to the research in recent years, the feedstock commonly consists of Jatropha oil, soybeans oil, and sunflower seed oil followed by mixing with the alcohol directly for the triglycerides transesterification process instead of the extracted oil mixed with the alcohols and catalysts, hence skipping the oil extraction process and resulting the reduction in operating cost [16]. Additionally, the use of co-solvent in in-situ transesterification process could stimulate the methanolysis process, hence allowing the maximum amount of alcohol to consume during the transesterification process followed by obtaining a considerably high yield of biodiesel. According to researches, an approximately yield of 95% Chlorella pyrenoidosa biodiesel was obtained with the utilization of hexane as the co-solvent with the molar ratio of 76:1 (hexane to lipid) at a reaction temperature of 95°C and catalyst loading of 0.5 M H_2SO_4 [17]. In conclusion, the Figure 2. has shown all the schematic diagrams of the latest technology in biodiesel production whereas Table 2. has summarized all the latest technologies in biodiesel production that have been carried out in experimental basis.

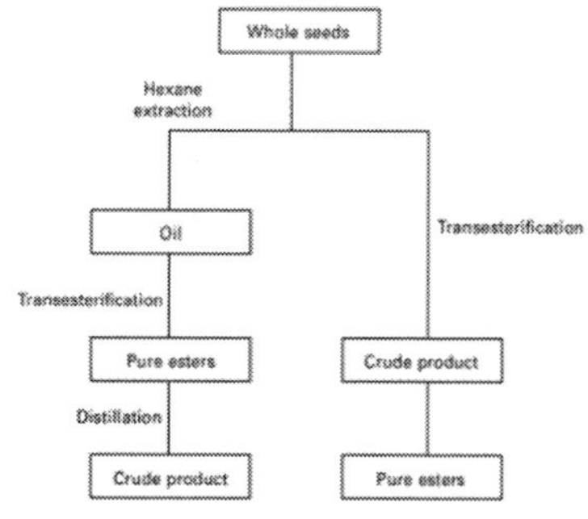

Figure 1. Differences between conventional method and in-situ method for biodiesel production

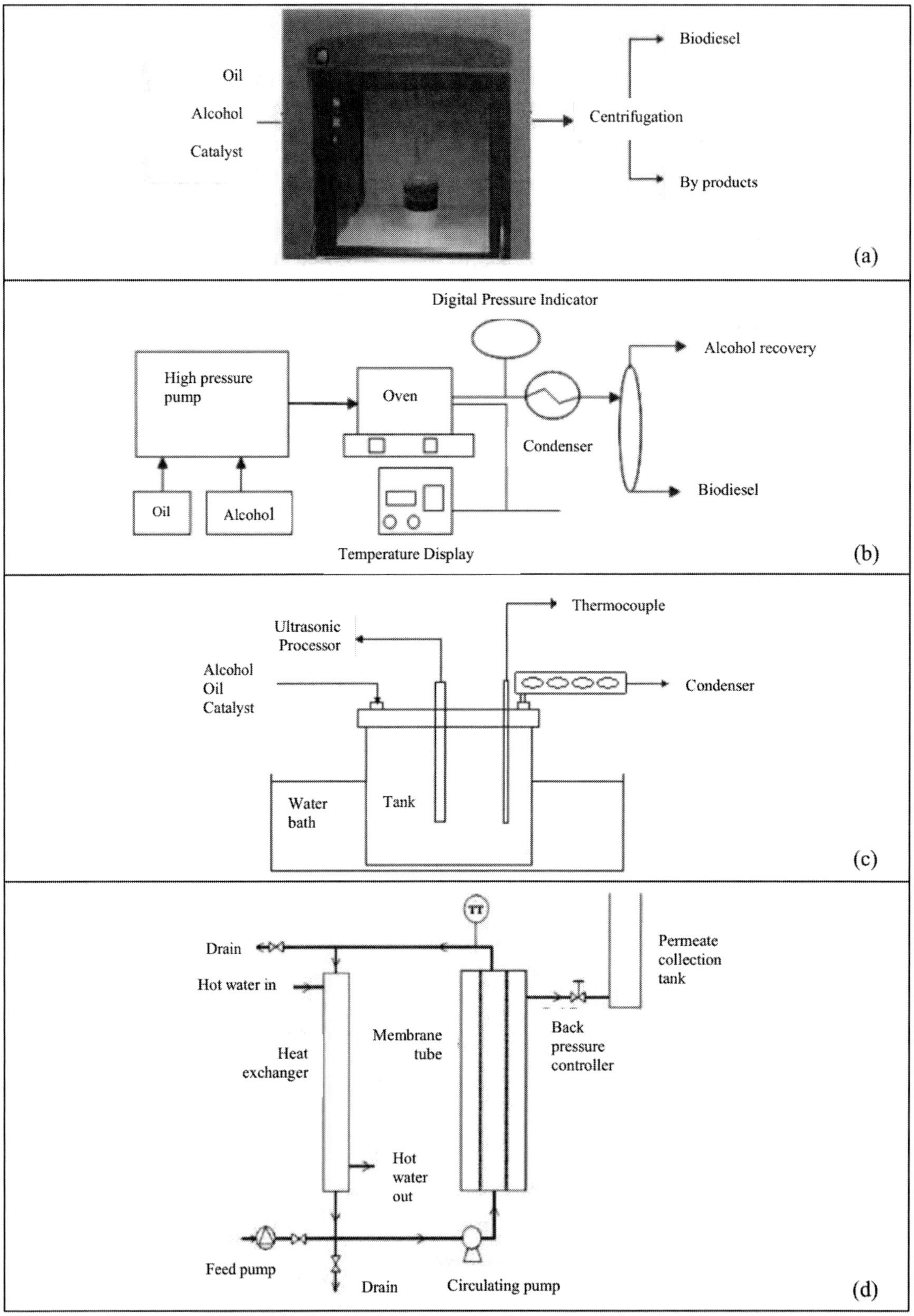

Figure 2. Latest technology in biodiesel production, (a) Microwave assisted, (b) Supercritical, (c) Ultrasonic irradiation, (d) Membrane

IOP Conf. Series: Materials Science and Engineering **943** (2020) 012026 doi:10.1088/1757-899X/943/1/012026

Table 2. Latest technologies for biodiesel production.

Method	Advantages	Limitations	Feedstock	Optimal operating conditions	Yield	Ref.
Supercritical Process	• Short reaction time • High purity product • Well-mixing reaction • Catalyst is not required • Allow large-scale continuous production	• High equipment capital cost • High energy usage • Excess alcohol usage	Soybean oil	**Temperature**: 280°C **Pressure**: 12.8 MPa **M/O ratio**: 24:1 **Reactor type**: Batch-type vessel	98%	[18]
Microwave Assisted Process	• High conversion rate • Short reaction time • Less energy consumption	• Difficult to scale-up from laboratory scale to industrial scale • Solid feedstock is not preferable	Safflower seed oil	**Temperature**: 333 K **Catalyst loading**: 1 wt% **Reaction time**: 6 mins **M/O ratio**: 10:1	98.4%	[19]
Ultrasound Assisted Process	• Well-mixing between oil and alcohol • Less energy usage • Short reaction time	• Large catalyst usage • Not fully-commercialized due to its infeasibility	Canola oil	**Temperature**: 55°C **Reactor conditions**: 450 W **M/O ratio**: 6:1 **Catalyst loading**: 1 wt%	95%	[20]
In-situ Transesterification Process	• Less alcohol required • Most of the oil types are applicable • Eliminate the selection of costly oil feedstocks	• Difficult to scale-up from laboratory scale to industrial scale • Not-fully commercialized industrially due to its low profitability	Linseed oil	**Temperature**: 40°C **Pressure**: 60 MPa **M/O ratio**: 10:1 **Reaction time**: 90 mins **Catalyst loading**: 6.8 wt%	94%	[21]
Microwave Assisted Process	• High selectivity • Efficient separation • Environmental-friendly	• Not easy to scale into industrial scale • High membrane cost • Fouling issues	Soybean oil	**Temperature:** 65°C **Catalyst loading:** 40.5 wt% **Reaction time:** 150 mins **M/O ratio:** 6:1	94.5%	[22]

3.0 Catalyst Heterogeneous Catalyst

3.1. Heterogeneous Catalysts Iron Doped Carbon Catalyst

Catalyst plays an extremely significant role within the modern era of scientific and technologies as it is able to improve the reaction rate, reduce the chemical process temperature, and provide specific selectivity during the asymmetric synthesis. Both the homogeneous and heterogeneous catalysts have their benefits. For example, heterogeneous catalysts are readily to be separated from the product mixture solution but the reaction rate is limited due to the catalyst's insufficient surface area, whereas the homogeneous catalyst provides a fast reaction rate and high product conversion, however, separation problem between the catalyst and the product will occur since the catalyst is miscible within the reaction medium, hence causing a series of pollution problems after it being discharged into the environment [23]. Therefore, the iron doped carbon catalyst has highly sought the attentions from the researches due to its high reusability and self-isolation ability. Based on the information above, iron doped nanoparticles are potentially utilized for attaching with the catalyst support material which is usually a solid with high surface area in order to maximize the magnetic decantation strength for the catalyst to isolate from the mixture solution [24].

The attached nanoparticles onto the solid catalyst does not only provide the advantages of increased catalyst surface area which could lead to high reaction rate, but also forms a stable suspension within the reaction medium when the nanoparticles are well dispersed into the catalyst, so that allowing the elevation of reaction rate [25]. The mechanism of iron doped catalyst is shown as Figure 3. below. These nanoparticles generally constituted of a series of metals such like iron, cobalt, or nickel, and alloys like iron oxides or ferrites which consist of high-strength magnetic moments when the external magnetic field has applied [26]. For example, the iron doped nanoparticles might disperse throughout the mixture solution with the absence of applied magnetic field, however, the iron doped nanoparticles will be deposited selectively towards a specific direction with the presence of applied external magnetic field. This kind of operations enable the iron doped nanoparticles catalyst to be recovered and reused again and again after every cycles of the clean and convenient magnetic decantation process. According to the study of Guo et al. 2006 [27], the recovery rate of an iron doped catalyst is about 2 times faster compared to the non-iron doped catalyst [24].

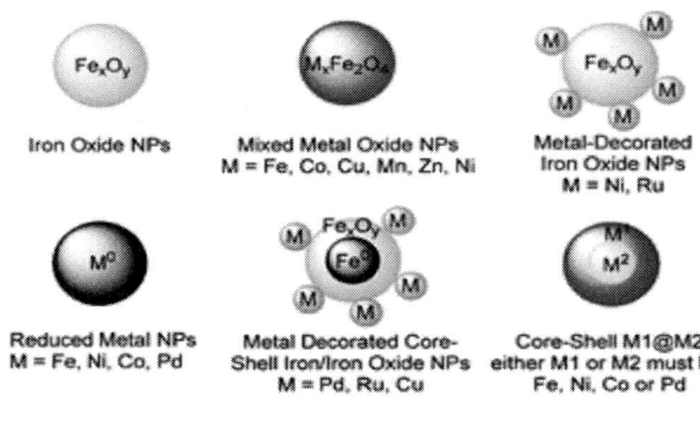

Figure 3. General mechanism of iron doped catalyst [26].

According to the study of Kang Liu and Rui Wang, 2017 [28], their conducted experiment was successfully synthesized a novel bifunctional bamboo charcoal-based iron doped solid base catalyst (K/BC-Fe_2O_3) via the in-situ impregnation-calcination method. The schematic diagram of the formation mechanism within the alkaline active sites on the iron doped catalyst (K/BC-Fe_2O) surface has shown as Figure 4. as below. At the beginning of the activation process, the gasification reaction undergone continuously within the active sites around the vicinity of the potassium salts compound followed by forming the process intermediates, finally the K_2O species formed. Thus, the transesterification process has occurred with high efficiency within the active sites of the K_2O species [29].

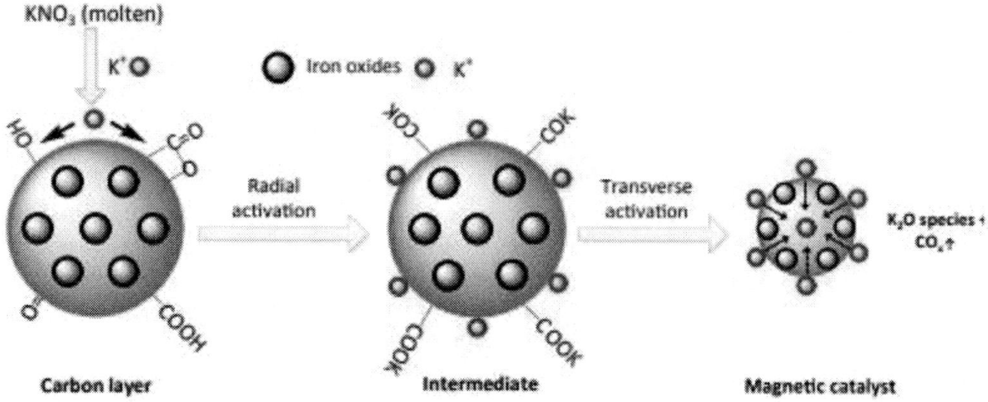

Figure 4. Formation mechanism of the alkaline active sites onto the iron doped catalyst surface [29].

The maximum production yield was achieved 98% conversion rate with the optimal operating conditions: 2.5 wt% of catalyst amount, a methanol to oil molar ratio of 8:1, 1-hour reaction time, and a temperature of 60°C. Nevertheless, the conducted regeneration experiments in this research highlighted the high catalytic activity of the iron doped catalyst that could still maintained after the 4 times catalyst reused. Therefore, this research has presented a truly potential novel bifunctional heterogeneous iron doped catalyst in the biodiesel production field [29].

According to the study of Yi-Tong Wong et al, 2019 [30], the sulfonated iron doped solid acid catalyst, ZrFe-SA-SO_3H and ZrFe-CMC-SO_3H were synthesized via the 4-step process. Firstly, the Fe^{3+} ion undergoes the chelation process to produce the $-(COO)_3Fe$ structure, followed by reducing the $-(COO)_3Fe$ structure into Fe_3O_4 via the subsequent calcination process at a temperature of 400°C since the stable iron doped core of the Fe_3O_4 does not allow the reduction reaction to occur below 400°C. Then, the chelation and embedding process with Zr^{4+} tend to construct a dense carbon shell with the $-(COO)_3Zr$ structure which act as the protection for the magnetic core from the sulfuric acid dissolution process during the sulfonation reaction. Finally, the external part of the carbon shell was partially carbonized during the sulfonation, hence creating a strong Bronsted acid sites for the carbon skeleton structure in order to proceed the biodiesel production process via the esterification of oleic acid [30]. The preparation process for this catalyst has shown as Figure 5. schematically as below.

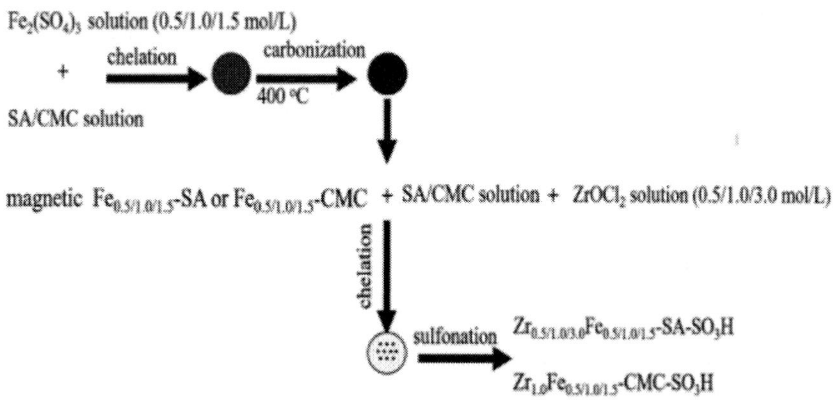

Figure 5. A 4-step process for synthesizing the sulfonated iron doped solid acid catalyst [30].

This sulfonated iron doped acid catalyst is able to supply both high surface-acidity density and considerable magnetization. The synthesized catalyst was tested to provide 92.5-99.5% of biodiesel yield for the first catalytic cycle at temperature of 90°C with 4-hours of reaction time. However, the biodiesel yield has shown to be decreased 9% after five cycles of catalytic activity were proceeded by using the ZrFe-SA-SO_3H catalyst. Therefore, this catalyst indicated a relatively low recoverability compared to other industrialized catalyst but still feasible and practicable for industrial usage considering the acceptable yield amount of the biodiesel production [30].

According to the study of Indu Ambat (2019) [31], the potassium impregnated nano-magnetic ceria catalyst was synthesized from the rapeseed oil for the biodiesel production, water treatment, and bio-catalysis. This catalyst is basically prepared based on the concept of impregnating the cerium oxide, CeO_2 iron doped nanoparticles into the potassium ions [32]. Moreover, a 25 wt% of potassium impregnated Fe_3O_4-CeO_2 nano-magnetic catalyst performed the best yield of biodiesel production. In addition, a 96.13% yield of biodiesel production could be obtained with the utilization of the following reaction parameters: a 4.5 wt% amount catalyst added, temperature of 65°C, 1:7 molar ratio of methanol to oil, and 120 minutes of reaction time [33]. The resulted biodiesel properties were determined using EN 14214 method, hence all these results are potentially indicated that the Fe_3O_4-CeO_2 nano-magnetic catalyst is a feasible catalyst for the desired quality biodiesel production by using the rapeseed oil as the oil feedstock [34].

Pingbo Zhang (2014) [35] had reported the novel iron doped solid base catalyst CaO/$CoFe_2O_4$ for the biodiesel production. The iron doped catalyst in this study was synthesized by first preparing a $CoFe_2O_4$ iron doped cores from the reaction between $CoSO_4 \cdot 7H_2O$ and $FeCl_3 \cdot 6H_2O$. The $CoFe_2O_4$ iron doped core was then mix with the anhydrous $CaCl_2$ with a Ca^{2+} to $CoFe_2O_4$ molar ratio of 5:1. After the NaOH solution was added into the mixture, the mixture solution was stirred for 30 minutes at room temperature and aged for 18 hours at 65°C. Finally, the iron doped solid base catalyst, CaO.$CoFe_2O_4$ was obtained after the magnetic precursor was dewatered at 80°C for 12 hours and undergone the calcination process at 800°C for 3 hours [36]. The results from the characterization demonstrated that the stronger magnetic strength which provided from the CaO/$CoFe_2O_4$ catalyst compared to the CaO/$ZnFe_2O_4$ and CaO/$MnFe_2O_4$ catalysts due to its higher wettability and basicity as better wettability has the advantage to the enhanced contact between the catalyst and reactants, as well as better water resistance in order to protect the active sites of the catalyst support, CaO [35].

Another novel catalyst is the heteropolyacid-supported cotton-regenerated magnetic cellulose microsphere (MCM) catalyst which presented by Han et al, 2016 [36]. The catalyst was synthesized by first regenerating the cellulose microsphere (CM) by using the cotton. Then, the cellulose microsphere(s) were modified by mixing the triethylenetetramine (TETA) with the presence of $NaCO_3$ at 50°C for 8 hours followed by the co-precipitation process using the mixture solution of Fe (II) and Fe (III) salts in order to obtain the magnetic cellulose microsphere (MCM). Next, the mixture of heteroplyacid (HPW) and MCM was the heated in an oil-bath at a temperature of 60°C for an 8-hours continuous stirring, finally, the MCM was filtered and dried in order to obtain the MCM-HPW catalyst [38].

This catalyst was specifically utilized for the transesterification of the highly acidic Pistacia chinensis seed oil in order to produce biodiesel as the end-product. This study indicated that 93.1% of biodiesel yield has been converted from the FAME under the following optimal conditions: 15 wt% catalyst amounts, methanol to oil ratio of 10:1, temperature of 60°C and 80-minutes reaction time [39]. Nevertheless, the synthesized catalyst was observed to be separated efficiently from the mixture with the magnetic field application, as well as potentially reused for at least 4 cycles of catalytic activity and maintain the catalytic stability at the same time. Thus, the MCM-HPW catalyst fulfilled the requirements of a "green" biodiesel due to its reusability and environmental-friendly characteristics [38].

Fan Zhang et al, 2017 [40] study has reported another two type of novel catalysts which were the iron doped carbonaceous solid acid (C-SO_3H@Fe/JHC) and base (Na_2SiO_3@Ni/JRC) catalysts. These catalysts were synthesized by loading active groups into the carbonaceous catalyst support which derived from the Jatropha-hull hydrolysate and hydrolysis residue which were responsible for the esterification and transesterification process of Jatropha oil respectively to synthesize the biodiesel [41]. The Jatropha-hull carbon coated- iron doped catalyst was sulfonated by concentrated sulfuric acid to form the final catalyst product, iron doped carbonaceous solid acid catalyst C-SO_3H@Fe/JHC [42]. On the other hand, the iron doped solid base catalyst was synthesized by first incubating both the Jatropha hydrolysis residue and nickel nitrate mixture within the oil-bath to obtain Ni/JRC followed by calcinating the product under an extremely high temperature to give the Ni/JRC supporter [41]. Then, the $Na_2SiO_3 \cdot 9H_2O$ was set to load with the Ni/JRC supporter followed by the calcination process at 400°C for 2 hours, lastly the final product, iron doped carbonaceous solid base catalyst, Na_2SiO_3@Ni/JRC was obtained. The preparation route for these two catalysts was presented as Figure 6. below [40].

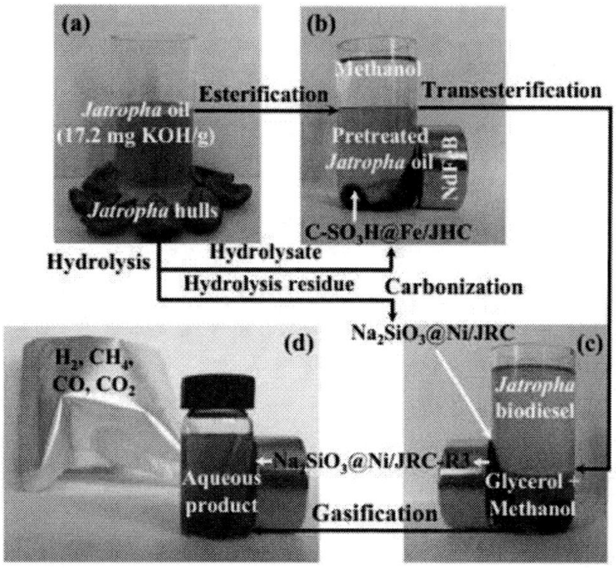

Figure 6. Preparation route of the iron doped carbonaceous solid acid and base catalyst [40].

Both these acid and base catalysts were observed to be effectively recovered with the average recovery yield of 90.3 wt% and 86.7 wt% respectively after at least 5 cycles of catalytic activity were proceeded. Besides that, a highest 96.7 % of biodiesel yield production could be obtained via the two-step biodiesel production (first esterification using the C-SO_3H@Fe/JHC catalyst, then transesterification using the Na_2SiO_3@Ni/JRC catalyst) [40]. Table 3. has summarized the different types of iron doped catalyst for biodiesel production in recent years. In conclusion, the iron doped bamboo charcoal solid base catalyst possessed the highest biodiesel yield among the other iron doped catalysts that listed in the Table 3.

IOP Conf. Series: Materials Science and Engineering **943** (2020) 012026 doi:10.1088/1757-899X/943/1/012026

Table 3. Different types of iron doped catalyst for biodiesel production.

Catalyst	Catalyst feedstock	Synthesis process	Catalyst characterization	Oil feedstock	Operating conditions	Catalyst reusability	Yield	Ref.
Iron doped bamboo charcoal solid base catalyst (K/BC-Fe_2O_3)	Support: bamboo charcoal Magnetic: γ – Fe_2O_3 Chemicals: KNO_3	1. In-situ synthesis 2. Impregnation 3. Calcination	Magnetic saturation: 35.4 emu/g Basic density: N/A	Soybean oil	**Reaction temperature**: 60°C **Catalyst loading**: 2.5 wt% **M/O ratio**: 8:1 **Reaction time**: 60 mins	Cycles number: 4 Biodiesel yield after the multiple cycles: 94%	98%	[29]
Sulfonated iron doped solid acid catalyst (ZrFe-SA-SO_3H), (ZrFe-CMC-SO_3H)	Support: carbon Magnetic: Fe_3O_4 Chemicals Oleic acid	1. Chelation 2. Calcination 3. Embedding 4. Sulfonation	Magnetization: 3.7-8.1 Am^2/kg Acid density: 4.4-8.4 mmol/g	Jatropha curcas oil	**Reaction temperature**: 90°C **Catalyst loading**: 2.5 wt% **Reaction time**: 4 hrs	Cycles number: 5 Biodiesel yield after the multiple cycles: 90.3%	92.7-99.5%	[30]
Nano-iron doped potassium impregnated ceria catalyst (Fe_3O_4-CeO_2)	Support: ceria Magnetic: $FeCl_2 \cdot 4H_2O$, $FeCl_3 \cdot 6H_2O$ Chemicals: KOH	1. Co-precipitation 2. Impregnation 3. Calcination	Magnetic saturation: 0.75 emu/g Basic density: 0.308 mg KOH/g	Rapeseed oil	**Reaction temperature**: 65°C **Catalyst loading**: 4.5 wt% **Reaction time**: 120 mins **M/O ratio**: 7:1	Cycles number: 5 Biodiesel yield after the multiple cycles: 80.94%	96.13 %	[31]

IOP Conf. Series: Materials Science and Engineering **943** (2020) 012026 doi:10.1088/1757-899X/943/1/012026

Table 3. Different types of iron doped catalyst for biodiesel production (cont'd).

Catalyst	Catalyst feedstock	Synthesis process	Catalyst characterization	Oil feedstock	Operating conditions	Catalyst reusability	Yield	Ref.
Iron doped solid base catalyst ($CaO/CoFe_2O_4$)	Support: CaO Magnetic: $CoFe_2O_4$ Chemicals: NaOH	Magnetic cores: 1. Titration 2. Mixing 3. Drying Catalyst: 1. Impregnation 2. Calcination	Magnetic saturation: 45 emu/g Basic density: N/A	Soybean oil	**Reaction temperature**: 80°C **Catalyst loading**: 1.0 wt% **M/O ratio**: 5:1 **Reaction time**: 3 hrs	Cycles number: 5 Biodiesel yield after the multiple cycles: 80%	87.4%	[35]
Magnetic cellulose microsphere supported heteropolyacid catalyst (MCM-HPW)	Support: cotton Magnetic: Fe_3O_4 Chemicals: heteropolyacid	1. Regeneration 2. Co-precipitation 3. Immobilization	Magnetization: N/A Acid density: N/A	Pistachia chinensis seed oil	**Reaction temperature**: 60°C **Catalyst loading**: 15 wt% **Reaction time**: 80 mins **M/O ratio**: 10:1	Cycles number: 4 Biodiesel yield after the multiple cycles: 80.7%	93.1%	[37]

IOP Conf. Series: Materials Science and Engineering **943** (2020) 012026 doi:10.1088/1757-899X/943/1/012026

Table 3. Different types of iron doped catalyst for biodiesel production (cont'd).

Catalyst	**Catalyst feedstock**	**Synthesis process**	**Catalyst characterization**	**Oil feedstock**	**Operating conditions**	**Catalyst reusability**	**Yield**	**Ref.**
Iron doped carbonaceous solid acid and base catalyst (C-SO_3H@Fe/JHC), (Na_2SiO_3@Ni/JRC)	Support: Jatropha hull Magnetic: Acid: Fe_3O_4 Base: $NiOH_2$ Chemicals: Acid: H_2SO_4 Base: $Na_2SiO_3 \cdot 9H_2O$	Acid: 1. Hydrothermal 2. Pyrolysis 3. Sulfonation Base: 1. Incubation 2. Calcination	Magnetic saturation: Acid: 11.2 emu/g Base: 15.0 emu/g Acid density: 2.96 mmol/g Basic density: 3.24 mmol/g	Jatropha curcas oil	Acid: **Reaction temperature**: 90°C **Catalyst loading**: 10 wt% **Reaction time**: 120 mins **M/O ratio**: 12:1 Base: **Reaction temperature**: 65°C **Catalyst loading**: 7 wt% **Reaction time**:120 mins **M/O ratio**: 9:1	Cycles number: 5 Biodiesel yield after the multiple cycles: Acid: 90.3% Base: 86.7%	96.7%	[40]

4. Conclusions

In conclusion, the use of the iron doped nanoparticles bounded catalyst is to create such high recoverability of the catalyst and reuse it in order to save the capital cost of the biodiesel production process. Besides that, the iron doped catalyst usage allows the elimination of additional purification treatments that required for the products when the homogeneous catalysts are utilized for catalysing the process, hence making the iron doped catalyst to be more commercialized and recognized by the public due to its environmental-friendliness and economical benign.

Acknowledgments

This project was sponsored by the Ministry of Higher Education, Malaysia under Fundamental Research Grant Scheme (FRGS), contract number: FRGS/1/2019/TK10/CURTIN/03/1.

References

[1] Tan, Y. H., Abdullah, M. O., Kansedo, J., Mubarak, N. M., Chan, Y. S., Hipolito, C. N. 2019. "Biodiesel production from used cooking oil using green solid catalyst derived from calcined fusion waste chicken and fish bones." Renewable energy 139: 696-706.

[2] Chung, Z. L., Tan, Y. H., Chan, Y. S., Kansedo, J., Mubarak, N. M., Ghasemi, M., Abdullah, M. O. 2019. "Life cycle assessment of waste cooking oil for biodiesel production using waste chicken eggshell derived CaO as catalyst via transesterification." Biocatalysis and Agricultural Biotechnology, 101317.

[3] Wong, J. J., Abdullah, M. O., Baini, R., Tan Y. H. 2017. "Performance monitoring: A study on ISO 14001 certified power plant in Malaysia." Journal of cleaner production 147: 165-174.

[4] Quah, R.V., Tan, Y. H., Mubarak, N. M., Khalid, M., Abdullah, E. C., Hipolito, C. N. 2019. "An overview of biodiesel production using recyclable biomass and non-biomass derived magnetic catalysts." Journal of Environmental Chemical Engineering, 103219.

[5] Hassan, Muhammad, Ghulam Abbas Ashraf, Bo Zhang, Yiliang He, Genxiang Shen, and Shuangqing Hu. 2019. "Energy-efficient degradation of antibiotics in microbial electro-Fenton system catalysed by M-type strontium hexaferrite nanoparticles." Chemical engineering journal 380:122483. doi: 10.1016/j.cej.2019.122483.

[6] Maniam, Gaanty Pragas. 2017. "Regeneration of calcium oxide (CaO) from waste gypsum via two-step reaction."

[7] Badoei, Arastoo, Saeid Malekabadi, Zahra Karami, and Ghasem Sargazi. 2019. "Magnetic cross-linked enzyme aggregates of Km12 lipase: A stable nanobiocatalyst for biodiesel synthesis from waste cooking oil." Renewable energy 141:874-882. doi: 10.1016/j.renene.2019.04.061.

[8] Yuchel, Sevil. 2017. "Novel Methods in Biodiesel Production."

[9] Abed, K. A., M. S. Gad, A. K. El Morsi, M. M. Sayed, and S. Abu Elyazeed. 2019. "Effect of biodiesel fuels on diesel engine emissions." Egyptian Journal of Petroleum 28 (2):183-188. doi: https://doi.org/10.1016/j.ejpe.2019.03.001.

[10] Kusdiana, D. & Saka, S. (2001). Kinetics of transesterification in rapeseed oil to biodiesel fuel as treated in supercritical methanol. Fuel, Vol. 80, pp. 693-698.

[11] Ahmad, Farah B., Zhanying Zhang, William O. S. Doherty, and Ian M. O'Hara. 2016. "Evaluation of oil production from oil palm empty fruit bunch by oleaginous micro-organisms." Biofuels, bioproducts & biorefining: Biofpr. 10 (4):378-392. doi: 10.1002/bbb.1645.

[12] Chavalparit, Orathai, Maneerat Ongwandee, and Krittiporn Trangkaprasith. 2013. "Production of Pelletized Fuel from Biodiesel-Production Wastes: Oil Palm Fronds and Crude Glycerin." Engineering Journal 17 (4):61-70. doi: 10.4186/ej.2013.17.4.61.

[13] Atadashi, I. M., M. K. Aroua, A. R. Abdul Aziz, and N. M. N. Sulaiman. 2011. "Membrane biodiesel production and refining technology: A critical review." Renewable & sustainable

energy reviews. 15 (9):5051-5062. doi: 10.1016/j.rser.2011.07.051.

[14] Dimian, Alexandre C., and Anton A. Kiss. 2019. "Eco-efficient processes for biodiesel production from waste lipids." Journal of cleaner production. 239:118073. doi: 10.1016/j.jclepro.2019.118073.

[15] Gardy, Jabbar, Mohammad Rehan, Ali Hassanpour, Xiaojun Lai, and Abdul-Sattar Nizami. 2019. "Advances in nano-catalysts based biodiesel production from non-food feedstocks." Journal of environmental management 249:109316. doi: 10.1016/j.jenvman.2019.109316.

[16] Technology, Biodiesel Science and. 2018. "Transesterification processes for biodiesel production from oils and fats."

[17] Bayramoglu, Gulay, Hacı Mehmet Kayili, Merve Oztekin, Bekir Salih, and M. Yakup Arica. 2018. "Hydrophilic spacer-arm containing magnetic nanoparticles for immobilization of proteinase K: Employment for speciation of proteins for mass spectrometry-based analysis." Talanta 206:120218. doi: 10.1016/j.talanta.2019.120218.

[18] Cao, W., Han, H. & Zhang, J. (2005). Preparation of biodiesel from soybean oil using supercritical methanol and co-solvent. Fuel,Vol. 84, pp. 347–351.

[19] Düz, M.Z., Saydut, A. & Öztürk, G. (2011). Alkali catalyzed transesterification of safflower seed oil assisted by microwave irradiation. Fuel Processing Technology, Vol. 92, pp. 308-313.

[20] Lee, S.B., Lee, J.D. & Hong, I.K. (2011). Ultrasonic energy effect on vegetable oil-based biodiesel synthetic process. Journal of Industrial and Engineering Chemistry, Vol. 510, pp. 1-6.

[21] Taherkhani, Mohammad. 2017. "Optimization studies of biodiesel production from linseed via in-situ transesterification

[22] Mootabadi, H., Salamatinia, B., Bhatia, S. & Abdullah, A.Z. (2010). Ultrasonic-assisted biodiesel production process from palm oil using alkaline earth metal oxides as the heterogeneous catalysts. Fuel, Vol.89, pp. 1818-1825.

[23] Ferreira, Elano Nery, Tathilene Bezerra Mota Gomes Arruda, Francisco Eduardo Arruda Rodrigues, David Thomas Duarte Arruda, Joadir Humberto da Silva Júnior, Dayanne Lopes Porto, and Nágila Maria Pontes Silva Ricardo. 2019. "Investigation of the thermal degradation of the biolubricant through TG-FTIR and characterization of the biodiesel – Pequi (Caryocar brasiliensis) as energetic raw material." Fuel 245:398-405. doi: 10.1016/j.fuel.2019.02.006.

[24] Wang, Anping, Hu Li, Hu Pan, Heng Zhang, Fusheng Xu, Zhaozhuo Yu, and Song Yang. 2018. "Efficient and green production of biodiesel catalyzed by recyclable biomass-derived magnetic acids." Fuel processing technology 181:259-267. doi: 10.1016/j.fuproc.2018.10.003.

[25] Gutiérrez, Norma, Esthela Ramos-Ramírez, Alma Serafín-Muñoz, Adrián Zamorategui-Molina, and Jesús Monjaraz-Vallejo. 2019. "Use of Co/Fe-Mixed Oxides as Heterogeneous Catalysts in Obtaining Biodiesel." Catalysts. 9 (5):403. doi: 10.3390/catal9050403.

[26] Gholami, Ali, Fathollah Pourfayaz, Ahmad Hajinezhad, and Majid Mohadesi. 2019. "Biodiesel production from Norouzak (Salvia leriifolia) oil using choline hydroxide catalyst in a microchannel reactor." Renewable energy 136:993-1001. doi: 10.1016/j.renene.2019.01.057.

[27] Guo, Yang, and Dennis Leung. 2006. "Transesterification of neat and used frying oil: Optimization for biodiesel production."

[28] Liu, Kang, Rui Wang, and Meiqing Yu. 2017. "Biodiesel production from soybean oils by a novel nano-magnetic solid base catalyst (K/ZrO2 / /γ-Fe2O3)."

[29] Liu, Kang, Rui Wang, and Meiqing Yu. 2018. "An efficient, recoverable solid base catalyst of magnetic bamboo charcoal: Preparation, characterization, and performance in biodiesel production." Renewable energy 127:531-538. doi: 10.1016/j.renene.2018.04.092.

[30] Wang, Yi-Tong, Xing-Xia Yang, Jie Xu, Hong-Li Wang, Zi-Bing Wang, Lei Zhang, Shao-Long Wang, and Jing-Long Liang. 2019. "Biodiesel production from esterification of oleic acid by a sulfonated magnetic solid acid catalyst." Renewable energy 139:688-695. doi:

10.1016/j.renene.2019.02.111.

[31] Ambat, Indu, Varsha Srivastava, Esa Haapaniemi, and Mika Sillanpää. 2019. "Nano-magnetic potassium impregnated ceria as catalyst for the biodiesel production." Renewable energy 139:1428-1436. doi: 10.1016/j.renene.2019.03.042.

[32] Rocha, Pablo D., Leandro S. Oliveira, and Adriana S. Franca. 2019. "Sulfonated activated carbon from corn cobs as heterogeneous catalysts for biodiesel production using microwave-assisted transesterification." Renewable energy 143:1710-1716. doi: 10.1016/j.renene.2019.05.070.

[33] Santosa, Sri Juari. 2018. "Palm Oil Boom in Indonesia: From Plantation to Downstream Products and Biodiesel." Clean: soil, air, water. 36 (5-6):453-465. doi: 10.1002/clen.200800039.

[34] Sani, Yahaya Muhammad, Aisha Olatope Raji-Yahya, Peter Adeniyi Alaba, Abdul Raman Abdul Aziz, and Wan Mohd Ashri Wan Daud. 2015. "Palm Frond and Spikelet as Environmentally Benign Alternative Solid Acid Catalysts for Biodiesel Production." Bioresources. 10 (2):3393-3408. doi: 10.15376/biores.10.2.3393-3408.

[35] Zhang, Pingbo, Qiuju Han, Mingming Fan, and Pingping Jiang. 2014. "Magnetic solid base catalyst CaO/CoFe2O4 for biodiesel production: Influence of basicity and wettability of the catalyst in catalytic performance." Applied surface science 317:1125-1130. doi: 10.1016/j.apsusc.2014.09.043.

[36] Bayo, Alejandra, Daniel López-Chicharro, Victoria Morales, Juan José Espada, Daniel Puyol, Fernando Martínez, Sergi Astals, Gemma Vicente, Luis Fernando Bautista, and Rosalía Rodríguez. 2019. "Biodiesel and biogas production from Isochrysis galbana using dry and wet lipid extraction: A biorefinery approach." Renewable energy 146:188-195. doi: 10.1016/j.renene.2019.06.148.

[37] Han, Ying Zhi, Li Hong, and Xi Qing Wang. 2016. "Biodiesel production from Pistacia chinensis seed oil via transesterification using recyclable magnetic cellulose-based catalyst."

[38] Liu, Chang, Pengmei Lv, Zhenhong Yuan, Fang Yan, and Wen Luo. 2010. "The nanometer magnetic solid base catalyst for production of biodiesel." Renewable energy 35 (7):1531-1536. doi: 10.1016/j.renene.2009.10.009.

[39] Sirrajudin, M. S. 2016. "Enhancing the energy properties of fuel pellets from oil palm fronds of agricultural residues by mixing with glycerin." ARPN journal of engineering and applied sciences. 11 (9):6122-6127.

[40] Zhang, Fan, Xiao-Fei Tian, Zhen Fang, Mazloom Shah, Yi-Tong Wang, Wen Jiang, and Min Yao. 2017. "Catalytic production of Jatropha biodiesel and hydrogen with magnetic carbonaceous acid and base synthesized from Jatropha hulls." Energy conversion and management 142:107-116. doi: 10.1016/j.enconman.2017.03.026.

[41] Zhang, Heng, Hu Li, Yulin Hu, Kasanneni Tirumala Venkateswara Rao, Chunbao Xu, and Song Yang. 2019. "Advances in production of bio-based ester fuels with heterogeneous bifunctional catalysts." Renewable & sustainable energy reviews. 114:109296. doi: 10.1016/j.rser.2019.109296.

[42] Yao, Chuangye, Muhammad Ismail, Aize Hao, Santhosh Kumar Thatikonda, Wenhua Huang, Ni Qin, and Dinghua Bao. 2019. "Au nanoparticles introduced to spinel Co3O4 thin films: Switching enhancement and magnetization modulation." Journal of magnetism and magnetic materials 493:165702. doi: 10.1016/j.jmmm.2019.165702.

IOP Conf. Series: Materials Science and Engineering **943** (2020) 012027 doi:10.1088/1757-899X/943/1/012027

Performance evaluation of horizontal flow constructed wetlands as primary and secondary treatment for university campus wastewater

Yee Yong Tan[1], Fu Ee Tang[1], Carrie Lee Ing Ho[1], Madeline Shu Zhen Wong[1]

[1]Department of Civil and Construction Engineering, Faculty of Engineering and Science, Curtin University Malaysia, CDT 250, Miri 98009, Sarawak, Malaysia

E-mail: tan.yee.yong@curtin.edu.my

Abstract. Horizontal flow constructed wetland (HFCW) is an attractive green technology for wastewater treatment. In the present study, four laboratory-scale HFCWs were constructed to serve as the primary and secondary treatment of university campus wastewater. The effect of hydraulic residence time (HRT) and pre-aeration of influent were investigated. The experimental results demonstrated that the wetland vegetation played an important role in the oxygen restoration and the influence of pre-aeration was minor. In the primary treatment, effective removals of TSS and NH_4-N were observed. However, the treatment of BOD and NO_3-N were insignificant due to the low influent concentrations. As for secondary treatment, the HFCWs effectively reduced the concentrations of BOD, NH_4-N, and NO_3-N in the effluent from the Intermittent Decanted Extended Aeration (IDEA) plant. The effluent quality of all HFCWs complied with the Standard A sewage discharge. Longer HRTs brought a positive effect to the removals of organic matter and nitrogen.

1. Introduction

Constructed wetlands have been well regarded as an attractive alternative for wastewater treatment due to its simple operation and maintenance, excellent energy efficiency, and reasonable construction cost [1]. Horizontal flow constructed wetland (HFCW) is a type of subsurface flow constructed wetland that the influent is fed at one side, and the treatment takes place when it flows horizontally towards the outlet zone. HFCWs have been widely used to treat municipal wastewater, stormwater runoff, agricultural wastewater, and industrial wastewater [2]. However, the application of such treatment technology is uncommon in Sarawak, where only a HFCW serves as a secondary treatment in a 600 population equivalent (PE) wastewater treatment system [3]. There was a pilot-scale HFCW treating greywater discharged from biofilters, but the system has been demolished due to some reasons [4]. The effluent from both HFCWs complied with the Standard A of the *Environmental Quality Act (Sewage and Industrial Effluents) 1979*, which revealed that HFCW is a feasible wastewater treatment option in Sarawak. Reference [5] indicated that the lack of interest in constructed wetlands in Sarawak was attributed to the insufficient existing system to demonstrate its treatment capability.

HFCW consists of a wetland bed planted with wetland vegetation. The wetland bed, which is also known as substrate, is typically constructed by sand or gravel [2]. Reference [6] emphasized that the selection of substrate materials is based on the permeability and retention capacity for the contaminants. The role of wetland vegetation in HFCW is to support the microbial growth and to release oxygen to the substrate via the roots-rhizome [7]. Besides, the presence of wetland vegetation brought a positive effect in removing organic matter and nitrogen [8]. As the HFCW is continuously fed, the wetland bed is saturated and the concentration of dissolved oxygen is limited. The suspended solids are removed via filtration and sedimentation and the efficiency is always satisfactory [2,8]. On the other hand, the degradation of organic matter and nitrate is dependent on the anoxic and anaerobic processes [2,9]. In HFCW, the removal of ammonium is less efficient as the occurrence of nitrification is limited by the anoxic and anaerobic condition [8]. As a result, HFCW was typically used as a secondary or tertiary treatment option [1,10]. Artificial aeration was proposed as an operational alternative to enhance the treatment efficiency of organic matter and nitrogen in HFCWs [11]. Reference [8] also emphasized the significance of hydraulic residence time (HRT) in HFCW, in which a longer HRT contributed to better treatment efficiencies.

The main aim of this study was to determine the performance of HFCWs as a primary and secondary treatment system for university campus wastewater. Four laboratory-scale HFCWs were constructed in Curtin University Malaysia to treat raw wastewater and treated wastewater from the existing Intermittent Decanted Extended Aeration (IDEA) treatment plant. The effluents were sampled at regular intervals to assess the influence of HRTs in HFCWs. The effect of pre-aeration of influent on the treatment performance was also investigated. The effluent concentrations from HFCWs were compared to national effluent standard and the effluent from the IDEA plant.

2. Materials and Methods

2.1. Laboratory-scale HFCWs

Four laboratory-scale HFCWs were constructed in plastic containers that were 0.70 m long, 0.40 m wide, and 0.33 m high. The surface area and depth of the laboratory-scale HFCWs were 0.28 m^2 and 0.23 m, respectively. The sand was used as the main substrate materials in the HFCWs. The effective size and uniformity coefficient of the sand were determined using sieve analysis, which was 0.16 mm and 1.74 respectively. Constant head permeability test was adopted to measure the hydraulic conductivity of the sand, where an average value of 0.09 mm/s was obtained. The average porosity of the sand was 46%. 4.75 mm-gravel was used in the inlet and outlet zone of the HFCWs, where the main function was to uniformly distribute the influent wastewater and collect the treated wastewater. The substrate profile of the HFCW was 0.10 m inlet zone (4.75 mm-gravel), 0.50 m treatment zone (sand), and 0.10 m outlet zone (4.75 mm-gravel). The top surface of the substrate was covered with a 20 mm thick gravel layer to prevent the sand erosion, and thus the total depth of the substrate was increased to 0.25 m. Six common reeds (*Phragmites karka*) were planted in each HFCW. The bottom slope of the HFCW was set as 1°. Prior to the commencement of experiments, each HFCW were acclimatized with wastewater up to eight weeks. Figure 1 illustrates the schematic representation of the laboratory-scale HFCW.

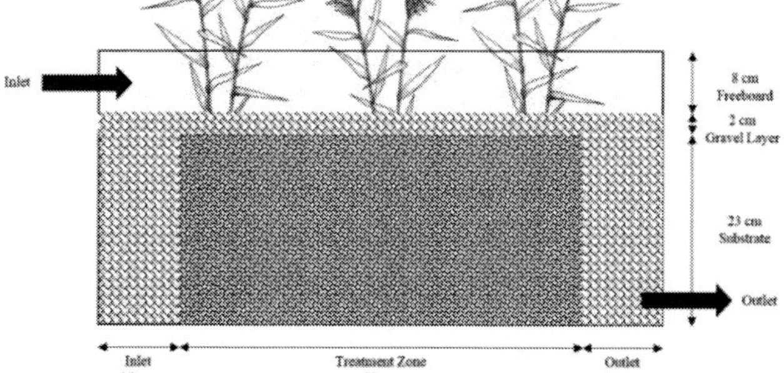

Figure 1: Laboratory Scale HFCW

2.2. Operation and analytical analysis of HFCWs

The laboratory-scale HFCWs were operated from May 2018 to August 2018. Two HFCWs were fed with raw wastewater, which was identified as primary treatment. Another two HFCWs were fed with treated wastewater from the IDEA plant located in Curtin University Malaysia, which were known as secondary treatment. To study the effect of pre-aeration on the treatment performance, one of each primary and secondary HFCW were fed with influent that was aerated until the dissolved oxygen (DO) reached saturation. Table 1 summarises the descriptions of each HFCW in this study. Three litters of influent were fed in each system, and the effluents were sampled every 12-hour until 60 hours. The experiments were repeated for four times in each HFCW. The parameters used to evaluate the performance included dissolved oxygen (DO), total suspended solids (TSS), 5-day biochemical oxygen demand (BOD_5), ammonium (NH_4-N) and nitrate (NO_3-N). DO, BOD_5, NH_4-N, NO_3-N were measured using Hach HQ40D Portable Multi Meter with IntelliCAL™ probes, while Hach DR2800 Spectrophotometer was used to determine the concentration of TSS.

Table 1: Descriptions of HFCWs

HFCW	Influent	Aeration of Influent
PN	Raw Wastewater (Primary)	No
PA	Raw Wastewater (Primary)	Yes
SN	Treated Wastewater (Secondary)	No
SA	Treated Wastewater (Secondary)	Yes

3. Results and Discussion

The results of the influent and effluent of each HFCW are presented in Table 2 to Table 5. According to Table 2 and Table 3, it was observed that the quality of raw wastewater was highly variable. The concentrations of DO fluctuated between 0.55 mg/L and 2.6 mg/L and TSS concentrations were within a range of 89 mg/L to 112 mg/L. The BOD_5 was unexpectedly low in raw wastewater, in which only two samples showed a concentration above 1 mg/L. The concentrations of NH_4-N were typically high, which ranged from 22.70 mg/L to 117 mg/L. In contrast, the concentrations of NO_3-N were mostly below 0.7 mg/L. According to Table 4 and Table 5, the wastewater was aerated in the IDEA plant, where the DO concentrations were above 4.3 mg/L in all cases. The concentrations of TSS and NH_4-N were significantly reduced after the treatment, where the TSS concentrations were within a range of 30 mg/L to 33 mg/L and NH_4-N concentrations were below 51 mg/L. However, it was observed that the concentrations of BOD_5 and NO_3-N increased in the IDEA plant. The BOD_5 concentrations fluctuated between 4.49 mg/L and 7.33 mg/L, while the NO_3-N concentrations were mostly higher than 1 mg/L.

3.1. Dissolved oxygen (DO)

Figure 2 and Figure 3 illustrate the influent and effluent DO concentrations from HFCWs as primary and secondary treatment. It was observed that the pre-aeration of influent had an insignificant effect on the effluent DO concentrations from HFCWs. On the other hand, the effluent DO concentrations were greatly influenced by the wetland vegetation. As the HFCWs were fed at 6.30 pm, the wetland vegetation underwent respiration only during night time and it resulted in lower effluent DO concentrations in the first 12-hour of treatment. The only exception was HFCW PN. During the daytime, the oxygen was released from the photosynthesis of wetland vegetation, which substantially increased the DO concentrations in all HFCWs at the 24-hour HRT. A similar pattern was recorded in the following 36 hours. This observation demonstrated that the well-developed root-rhizome of the common reeds (as shown in Figure 4) effectively supplement oxygen in the primary and secondary HFCWs. This finding agreed with reference [7] and reference [12]. It also explained that the rise in DO concentrations in HFCW PN in the first 12-hour was due to the high initial oxygen content in the wetland bed.

Table 2: Results of influent and effluent of HFCW PN

Samples	Experiments 1	2	3	4
		DO (mg/L)		
Influent	0.55	1.44	2.59	1.37
12 hours	2.38	2.34	1.96	2.4
24 hours	6.01	6.06	6.97	7.69
36 hours	2.59	2.32	3.38	4.75
48 hours	7.46	7.89	8.09	9.09
60 hours	3.7	3.32	3.84	4.78
		TSS (mg/L)		
Influent	112	100	- [(a)]	89
12 hours	34	31	- [(a)]	32
24 hours	32	31	- [(a)]	30
36 hours	33	28	- [(a)]	29
48 hours	33	32	- [(a)]	29
60 hours	31	34	- [(a)]	30
		BOD_5 (mg/L)		
Influent	0.82	6.71	0.51	0.29
12 hours	2.57	1.05	1.38	1.28
24 hours	1.37	0.82	1.74	1.89
36 hours	1.33	1.18	1.27	1.35
48 hours	0.73	0.69	0.74	0.88
60 hours	2.2	1.43	0.98	1.06
		NH_4-N (mg/L)		
Influent	117	95.5	24.8	30.3
12 hours	6.81	2.14	1.63	1.2
24 hours	5.1	1.36	0.611	1.08
36 hours	1.2	0.618	1.26	0.672
48 hours	2.31	2.89	1.32	1.32
60 hours	1.61	1.12	0.683	0.875
		NO_3-N (mg/L)		
Influent	0.312	0.25	12.5	0.632
12 hours	0.375	0.369	0.398	0.802
24 hours	0.406	0.484	0.409	0.402
36 hours	0.478	0.381	0.447	0.366
48 hours	0.352	0.716	0.446	0.439
60 hours	0.424	0.561	0.414	0.417

(a) Equipment was unavailable

Table 3: Results of influent and effluent of HFCW PA

Samples	Experiments 1	2	3	4
		DO (mg/L)		
Influent	7.44	7.29	7.50	7.04
12 hours	2.22	1.26	1.33	1.64
24 hours	5.84	5.47	5.74	5.34
36 hours	2.77	2.23	3.34	3.06
48 hours	7.28	7.18	8.58	7.25
60 hours	4.37	3.66	4.89	3.89
		TSS (mg/L)		
Influent	108	112	- (a)	116
12 hours	36	34	- (a)	37
24 hours	31	34	- (a)	30
36 hours	31	31	- (a)	31
48 hours	32	30	- (a)	29
60 hours	37	34	- (a)	32
		BOD_5 (mg/L)		
Influent	0.39	3.68	0.54	0.12
12 hours	2.30	0.99	1.37	1.14
24 hours	1.66	0.71	0.35	0.70
36 hours	1.12	0.66	0.16	0.17
48 hours	0.94	0.35	0.51	0.44
60 hours	0.62	0.46	0.61	0.53
		NH_4-N (mg/L)		
Influent	115	75.90	22.70	32.60
12 hours	3.41	2.53	0.893	0.838
24 hours	2.75	1.45	0.113	0.51
36 hours	0.418	0.607	0.303	0.334
48 hours	1.13	0.868	0.283	0.298
60 hours	0.716	0.441	0.194	0.184
		NO_3-N (mg/L)		
Influent	0.394	0.476	17.10	0.624
12 hours	0.427	0.218	0.313	0.838
24 hours	0.453	0.256	0.404	0.510
36 hours	0.522	0.283	0.435	0.334
48 hours	0.444	0.875	0.454	0.298
60 hours	0.479	0.706	0.446	0.184

(a) Equipment was unavailable

Table 4: Results of influent and effluent of HFCW SN

Samples	Experiments 1	2	3	4
		DO (mg/L)		
Influent	5.6	5.33	5.16	4.34
12 hours	2.96	2.29	1.92	2.01
24 hours	7.58	7.24	6.59	6.76
36 hours	3.94	3.96	3.21	3.37
48 hours	7.91	7.94	8.42	7.54
60 hours	4.53	4.26	3.89	3.28
		TSS (mg/L)		
Influent	30	33	- (a)	31
12 hours	33	30	- (a)	39
24 hours	31	32	- (a)	30
36 hours	30	32	- (a)	33
48 hours	30	30	- (a)	31
60 hours	30	33	- (a)	30
		BOD_5 (mg/L)		
Influent	5.91	6.14	5.10	4.49
12 hours	0.6	0.05	1.11	1.37
24 hours	1.6	0.23	1.15	1.4
36 hours	0.43	0.31	1.03	0.94
48 hours	0.7	0.19	0.36	0.85
60 hours	0.1	0.35	0.41	0.45
		NH_4-N (mg/L)		
Influent	50.6	32.3	8.92	15.80
12 hours	1.12	0.388	0.306	0.325
24 hours	0.906	0.204	0.268	0.248
36 hours	0.239	0.0599	0.0822	0.177
48 hours	0.57	0.636	0.116	0.251
60 hours	0.299	0.146	0.0147	0.047
		NO_3-N (mg/L)		
Influent	0.787	1.28	15.3	1.00
12 hours	0.436	0.202	0.286	0.504
24 hours	0.243	0.232	0.28	0.261
36 hours	0.244	0.136	0.287	0.236
48 hours	0.213	0.372	0.335	0.269
60 hours	0.58	0.33	0.346	0.318

(a) Equipment was unavailable

Table 5: Results of influent and effluent of HFCW SA

Samples	Experiments 1	2	3	4
		DO (mg/L)		
Influent	7.47	7.32	7.33	7.05
12 hours	3.25	2.37	2.17	2.08
24 hours	6.70	6.83	6.04	4.67
36 hours	4.09	4.16	3.51	2.25
48 hours	7.53	7.65	7.71	5.45
60 hours	4.93	4.17	4.09	3.05
		TSS (mg/L)		
Influent	33	32	- [(a)]	32
12 hours	31	31	- [(a)]	33
24 hours	29	34	- [(a)]	32
36 hours	32	34	- [(a)]	33
48 hours	31	35	- [(a)]	32
60 hours	31	34	- [(a)]	34
		BOD_5 (mg/L)		
Influent	6.34	7.33	6.54	6.03
12 hours	1.22	0.91	1.44	1.26
24 hours	0.67	0.51	0.78	0.64
36 hours	0.45	0.36	0.88	0.48
48 hours	0.65	0.49	0.75	0.62
60 hours	0.38	0.82	1.3	1.17
		NH_4-N (mg/L)		
Influent	43.7	21.4	8.68	16.8
12 hours	0.859	0.392	0.291	1.444
24 hours	0.744	0.267	0.219	0.456
36 hours	0.204	0.119	0.111	0.351
48 hours	0.652	0.315	0.0886	0.312
60 hours	0.376	0.18	0.0604	0.172
		NO_3-N (mg/L)		
Influent	0.881	1.21	17.8	1.11
12 hours	0.836	0.354	0.331	0.597
24 hours	0.396	0.197	0.295	0.329
36 hours	0.313	0.212	0.31	0.251
48 hours	0.294	0.401	0.36	0.319
60 hours	0.347	0.387	0.365	0.312

[(a)] Equipment was unavailable

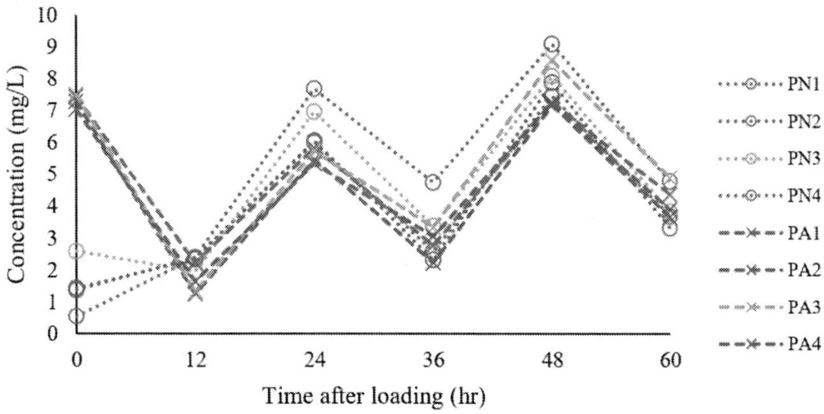

Figure 2: Influent and effluent DO concentration of HFCW PN and PA

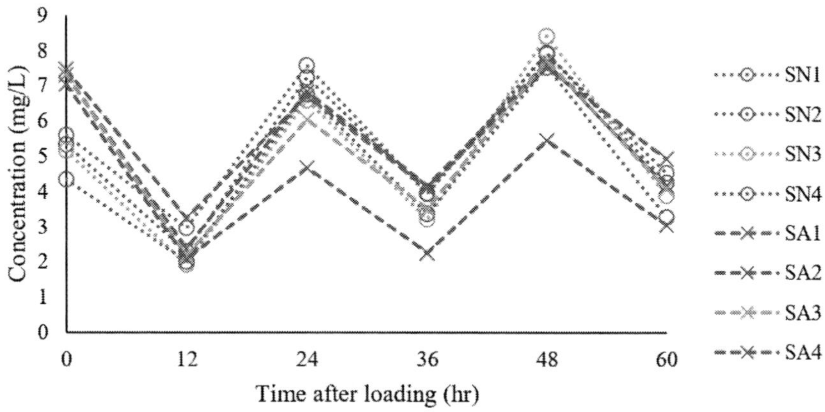

Figure 3: Influent and effluent DO concentration of HFCW SN and SA

Figure 4: Roots-rhizome of common reeds in HFCWs

3.2. Total suspended solids (TSS)

Figure 5 and Figure 6 illustrate the influent and effluent TSS concentrations from HFCWs as primary and secondary treatment. As the removal of TSS was dependent on filtration and sedimentation process [10], the effect of pre-aeration was insignificant to the treatment performance. According to Figure 5, the effluent TSS concentrations were within a range of 30 to 40 mg/Land were independent of influent concentrations and HRTs. The overall removal efficiencies of TSS in HFCW PN and PA were between 68 and 73%. As for-secondary treatment, no TSS removal was observed in SN and SA as the influent TSS concentrations were typically low. These observations showed that no further TSS removal was achieved in the HFCWs when the TSS concentration was around 30 mg/L. The removal efficiencies and effluent TSS concentrations were similar to reference [8] but were lower than reference [1] and reference [4]. Since the accumulation of solids and debris, as well as the continuous growth of microorganism and root-rhizomes, were beneficial to the TSS removals, the treatment capacity of HFCWs could be higher in the process of time [8].

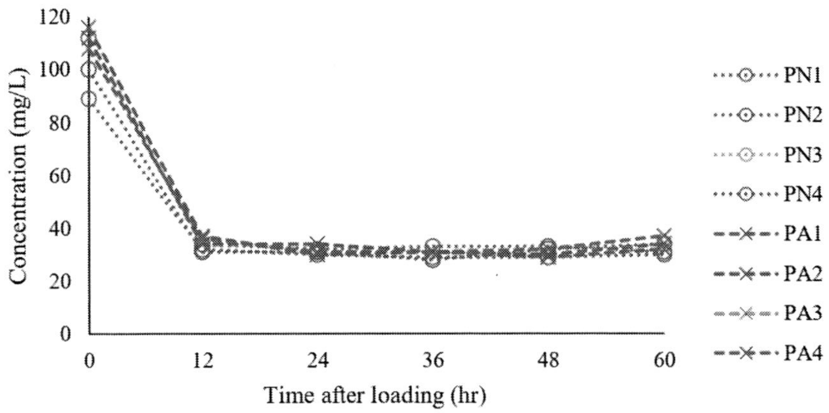

Figure 5: Influent and effluent TSS concentration of HFCW PN and PA

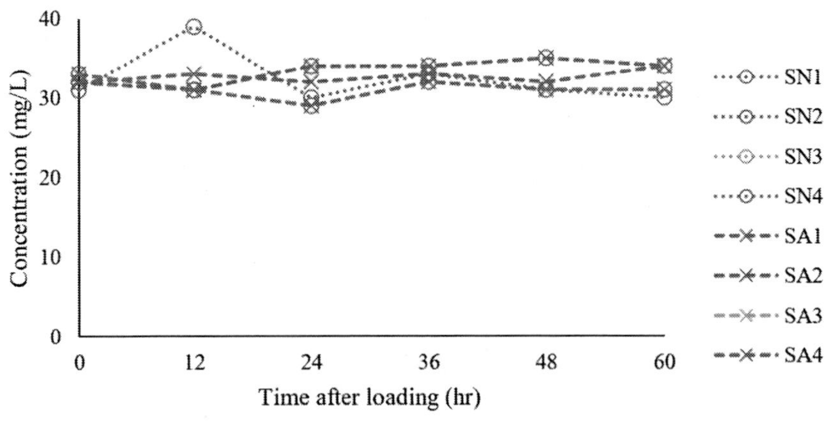

Figure 6: Influent and effluent TSS concentration of HFCW SN and SA

3.3. Biological Oxygen Demand (BOD)

Figure 7 and Figure 8 illustrate the influent and effluent BOD_5 concentrations from HFCWs as primary and secondary treatment. In HFCW PA and PN, only the third run of experiments demonstrated BOD removals in the first 12-hour. This was due to the low BOD_5 concentrations in raw wastewater, and thus the removal efficiencies of BOD in HFCWs as primary treatment were unable to be identified. The organic matter released from the vegetal debris and microorganism in HFCWs contributed to higher BOD_5 concentrations in effluent [10].

As for secondary treatment, the BOD removals were observed in both HFCW SN and SA. The mean removal efficiencies of BOD in HFCW SN and SA in the first 12-hour of treatment were 84.19% and 81.36%, respectively. This finding indicated that the effect of pre-aeration of influent was insignificant compared to the influence of wetland vegetation. It is also observed that a longer HRT resulted in a lower BOD_5 concentration. In HFCW SA, the removal efficiency increased to 90% after 24-hour HRT, while HFCW SN achieved 93% at 60th hour. Similar observations were reported in reference [8]. The overall removal efficiencies of BOD_5 in HFCW SN and SA were higher than the data reported in the literature [1,8]. The well-developed roots-rhizome of common reeds maintained high oxygen content and supported the biofilm growth in HFCWs, which stimulated the aerobic degradation of organic matter and guaranteed a desirable treatment performance [8,11,12].

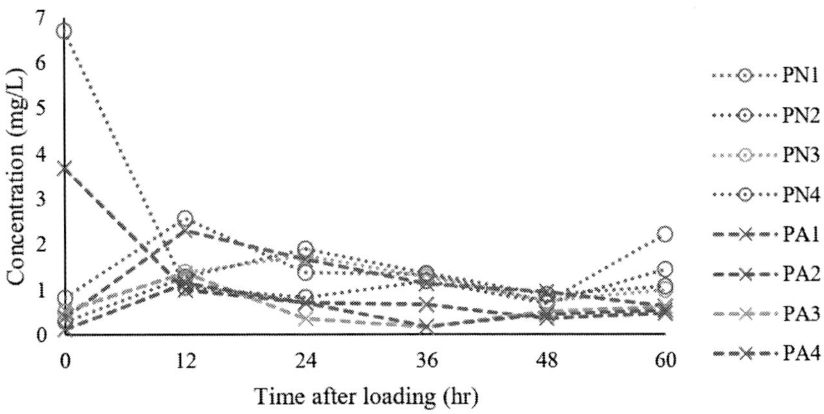

Figure 7: Influent and effluent BOD_5 concentration of HFCW PN and PA

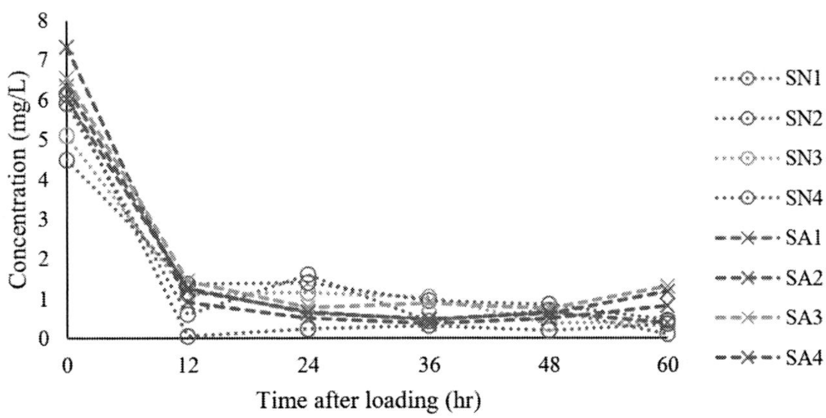

Figure 8: Influent and effluent BOD_5 concentration of HFCW SN and SA

3.4. Ammonium (NH_4-N)

Figure 9 and Figure 10 illustrate the influent and effluent NH_4-N concentrations from HFCWs as primary and secondary treatment. According to Figure 9, the mean removal efficiencies of NH_4-N in HFCW PN and PA in the first 12-hour were 95.35% and 96.80%, respectively. The removal efficiencies and effluent concentrations were much lower compared to the literature [1,2,8,11]. In HFCW, the treatment capacity of NH_4-N was limited by the DO concentrations, in which nitrification was less efficient under anoxic and anaerobic condition [9]. However, the DO concentrations were found to be high in the proposed HFCWs due to the presence of wetland vegetation, and thus the nitrification process acted an important role in removing NH_4-N. In addition, the influent concentrations of BOD were low, and thus the oxygen content did not pose as a limiting factor in the nitrification process. The adsorption and vegetation uptake also played important roles in NH_4-N removal. Reference [13] reported that the adsorption capacity of fine gravel for NH_4-N was more than 750 mg/kg. Reference [6] also indicated that the nitrogen uptake capability of common reeds was up to 2500 kg/ha/yr. Accordingly, the proposed HFCWs demonstrated excellent treatment capacity for NH_4-N. Similar to BOD, a longer HRT also contributed to a lower NH_4-N concentration. The removal efficiencies of NH_4-N achieved 97.96% and 99.34% in HFCW PN and PA, respectively. The removal efficiencies and effluent concentrations of HFCW SN and SA were highly similar to HFCW PN and PA.

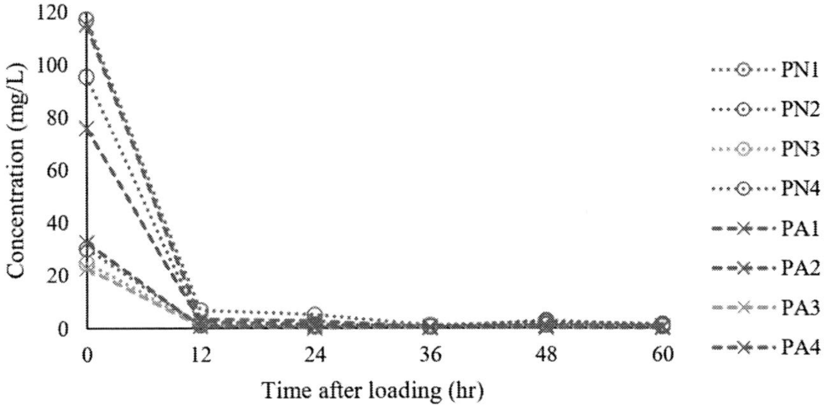

Figure 9: Influent and effluent NH_4-N concentration of HFCW PN and PA

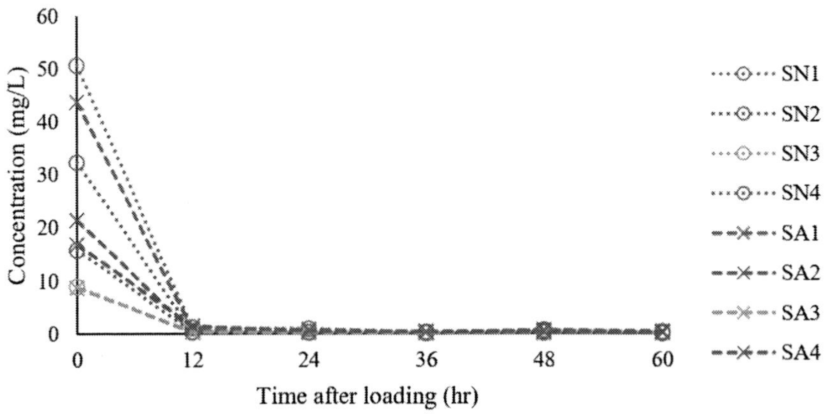

Figure 10: Influent and effluent NH_4-N concentration of HFCW SN and SA

3.5. Nitrate (NO_3-N)

Figure 11 and Figure 12 illustrate the influent and effluent NO_3-N concentrations from HFCWs as primary and secondary treatment. HFCW was well recognized as a treatment option that is favourable for denitrification and has excellent treatment capacity for NO_3-N [6]. However, the removals of NO_3-

N were insignificant in HFCW PN and PA due to the low influent concentrations. Only the third run of experiments demonstrated remarkable NO_3-N removal. The slight increase in the effluent concentrations of NO_3-N in the first 12-hour was attributed to the occurrence of nitrification in HFCWs, where NO_3-N was released from the degradation of NH_4-N. Based on the removal efficiencies of NH_4-N in HFCW PN and PA, the effluent concentrations of NO_3-N were expected to be much higher. Accordingly, the low NO_3-N concentrations in the effluent implied that denitrification and vegetation uptake removed a certain amount of NO_3-N in HFCWs. On the other hand, the HFCW SN and SA showed reasonable treatment for NO_3-N, which achieved mean removal efficiencies 69.13% and 55.05%, respectively, in the first 12-hour HRT. A longer HRT was advantageous to the denitrification, where the removal efficiencies reached the maximum at 36th hour in both HFCWs. The removal efficiencies and effluent concentrations of NO_3-N in HFCW SA and SN were comparable to reference [1] and reference [4]. The effect of pre-aeration of influent was not observed in the removal of NO_3-N in all HFCWs.

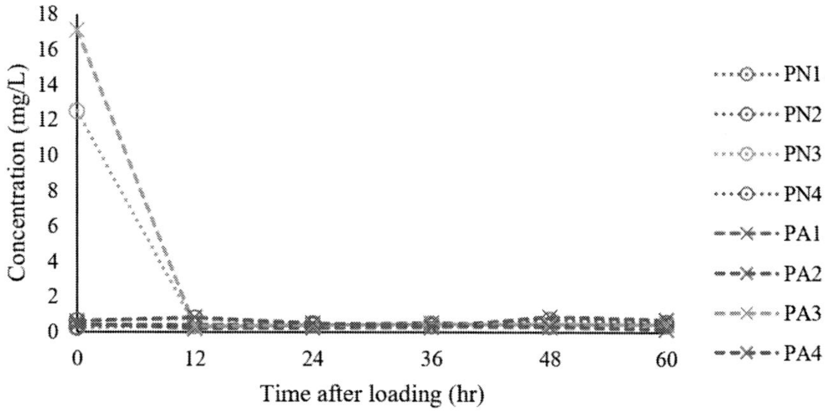

Figure 11: Influent and effluent NO_3-N concentration of HFCW PN and PA

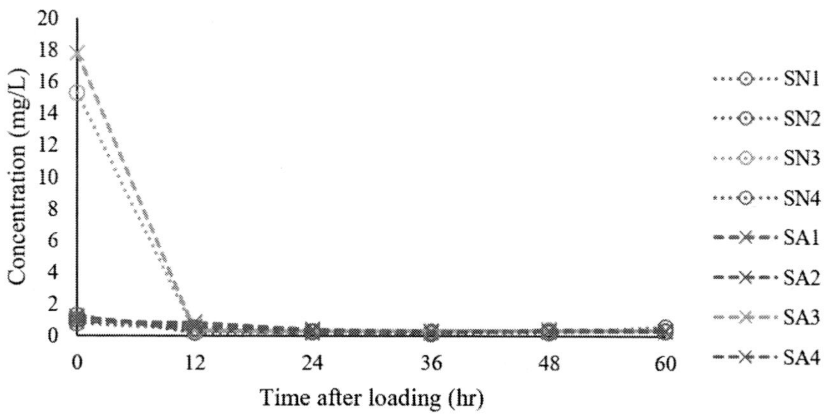

Figure 12: Influent and effluent NO_3-N concentration of HFCW SN and SA

3.6. Primary treatment and secondary treatment

In order to assess the feasibility of HFCW as a primary treatment option for university campus wastewater, the effluent quality of HFCW PN and PA were compared to the effluents from the existing IDEA plant. In addition to this, the effluent quality of HFCW SN and SA were compared to the performance of IDEA plant to evaluate the treatment efficiencies as a secondary treatment. The present study revealed that the influence of pre-aeration of influent was minor to the treatment performance of HFCWs, and thus the effluent quality of HFCW PN and PA were highly similar. Therefore, the effluent concentrations of HFCW PN and PA at each HRT (i.e. 12-hour, 24-hour, 36-hour, 48-hour, and 60-hour) were averaged to determine the mean effluent concentrations of each quality parameters. As significant treatment efficiencies were obtained in the first 12-hour, the 12-hour HRT was adopted in the analysis. A similar approach was carried out for HFCW SN and SA. Figure 13 presents the comparisons of the effluent quality from HFCWs, IDEA plant, and the acceptable conditions of sewage discharge of standard A in *Environmental Quality (Sewage) Regulations 2009*. The mean effluent concentration of DO from IDEA plant was higher than HFCWs, but the oxygen released from the photosynthesis of wetland vegetation substantially increased the DO content in the next 12 hours. The TSS concentrations from HFCW were slightly higher than the effluent from IDEA plant, but both systems fulfilled the requirement of Standard A sewage discharge (50 mg/L). The effluent BOD_5 and NO_3-N concentrations from both HFCWs and IDEA plant complied with the standard (20 mg/L and 10 mg/L, respectively), but the treatment performance of HFCWs were found to be better than the IDEA plant. The mean effluent concentration of NH_4-N from the IDEA plant was two times higher than the standard (5 mg/L), while the NH_4-N concentrations discharged from the HFCWs were below the limit. As for secondary treatment, Figure 14 demonstrated that the HFCWs improved the quality of the effluent from the IDEA plant, where the concentrations of BOD_5, NH_4-N, and NO_3-N were significantly reduced. However, no further removal of TSS was observed in HFCWs. The experimental results revealed that HFCW is a feasible option for the primary and secondary treatment of university campus wastewater.

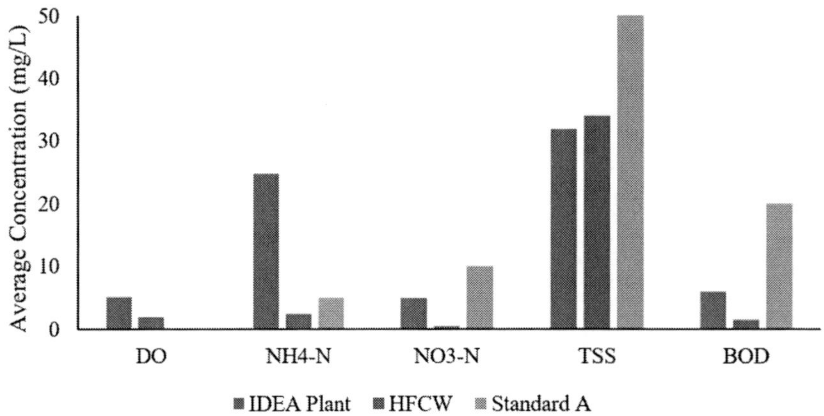

Figure 13: Performance of HFCWs as primary treatment

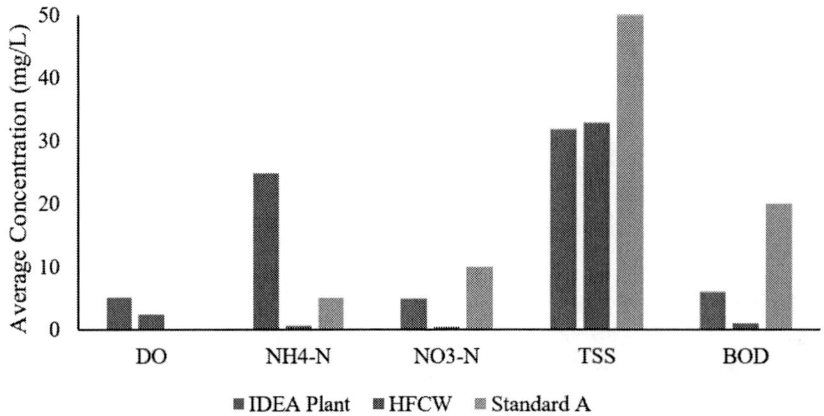

Figure 14: Performance of HFCWs as secondary treatment

4. Conclusion

This study demonstrated that the proposed HFCW is a feasible option for the primary and secondary treatment of university campus wastewater. The oxygen dynamics in HFCWs were governed by the wetland vegetation, and the pre-aeration of influent had a minor effect on the treatment performance. The removals of TSS and NH_4-N were excellent in the primary treatment, but the treatment capacity of BOD and NO_3-N were unable to be identified due to the low influent concentrations. As for secondary treatment, the concentrations of BOD, NH_4-N, and NO_3-N were significantly reduced. However, no further removal of TSS was observed. Additionally, longer HRTs contributed to higher removal efficiencies of BOD, NH_4-N, and NO_3-N.

References

[1] Thalla A K, Devatha C P, Anagh K and Sony E 2019 Performance evaluation of horizontal and vertical flow constructed wetlands a tertiary treatment option for secondary effluents *Appl Water Sci* **9** pp 147

[2] Vymazal J 2010 Constructed Wetlands for Wastewater Treatment *Water* **2(3)**, pp 530-549

[3] Sewerage Services Department Sarawak 2016 *Sewage Treatment Plant at Taman Boulder Built Kuching*

[4] Ling T Y, Apun K and Zainuddin S R 2009 Performance of a pilot-scale biofilters and constructed wetland with Ornamental plants in greywater treatment *World Applied Science Journal* **6(11)** pp 1555-1562

[5] Tan Y Y, Tang F E and Ho C L I 2019 Constructed Wetlands: Sustainable Solution to Managing Domestic Wastewater in the Rural Areas of Sarawak *IOP Conf. Series: Materials Science and Engineering* **495** pp 012063

[6] Kumar S and Dutta V 2019 Constructed wetland microcosms as sustainable technology for domestic wastewater treatment: an overview *Environmental Science and Pollution Research* ***26*(12)** pp 11662-11673

[7] Vymazal J 2011 Plants used in constructed wetlands with horizontal subsurface flow: a review *Hydrobiologia* ***674*(1)** pp 133-156

[8] Papaevangelou V, Gikas G D and Tsihrintzis V A 2016 Effect of operational and design parameters on performance of pilot-scale horizontal subsurface flow constructed wetlands treating university campus wastewater *Environmental Science and Pollution Research* ***23*(19)** pp 19504-19519

[9] Vymazal J 2007 Removal of nutrients in various types of constructed wetlands *Science of the total environment* ***380*(1-3)** pp 48-65

[10] Dotro G, Langergraber G, Molle P, Nivala J, Puigagut J, Stein O and Von Sperling M 2017 *Treatment wetlands* (Vol. 7) London UK: IWA publishing

[11] Zhang L Y, Zhang L, Liu Y D, Shen Y W, Liu H, and Xiong Y 2010 Effect of limited artificial aeration on constructed wetland treatment of domestic wastewater *Desalination* ***250*(3)** pp 915-920

[12] Rehman F, Pervez A, Khattak B N and Ahmad R 2017 Constructed wetlands: perspectives of the oxygen released in the rhizosphere of macrophytes *CLEAN–Soil, Air, Water* ***45*(1)** pp 1600054

[13] Zhu W L, Cui L H, Ouyang Y, Long C F and Tang X D 2011 Kinetic Adsorption of Ammonium Nitrogen by Substrate Materials for Constructed Wetlands *Pedosphere* **21(4)** pp 454-463

Acknowledgement

The authors would like to acknowledge Faculty of Engineering and Science, Curtin University Malaysia for supporting this study.

External time-influencing risk factors in infrastructure soft soils projects

W W Low[1], K S Wong[1] and J L Lee[1]

[1] Department of Civil and Construction Engineering, Faculty of Engineering and Science, Curtin University Malaysia, CDT 250, Miri, Sarawak, Malaysia

E-mail: low.wai.wah@curtin.edu.my

Abstract. Malaysian government has putting a lot of budget and resources in developing and upgrading infrastructure of the country. The covered area of an infrastructure project especially in a transportation project is quite significant compared to residential and commercial projects. As such, construction on soft soils is unavoidable and it increases uncertainties and risks in an infrastructure project. The current literature review showed that existing knowledge on risk assessment of infrastructure projects on soft soils is scarce. Hence, this research was focused on determining the critical time-influencing external risk factors during the construction phase. Analytical Hierarchy Process (AHP) was used to determine the most critical external risks in infrastructure projects on soft soils. Research method was questionnaire survey. It is found that the five most critical risk factors were thick soft soil layer, high water content, high organic content, price changes (material & energy) and karstic cavity formation. While the five least critical risk factors were archaeological/habitant issue, lack of political support, protectionism, corruption and change in government/legislative/regulatory. In conclusion, soil condition and economic are critical determining the time performance in infrastructure projects on soft soils, while political risk contributes little impact on this aspect.

1. Introduction

Malaysian government has putting a lot of resources and investment in respect of the infrastructure development. In the 11th Malaysian Plan (11MP) (2016-2020), one of the strategic thrusts is to strengthen infrastructure for economic expansion [1]. The examples strategies involved in 11MP are "transforming rural areas to uplift wellbeing of rural communities" (p. 16), "improving coverage, quality, and affordability of digital infrastructure", "continuing the transition to a new water services industry framework", "building an integrated need-based transport system" (p. 32) and "transforming construction" (p. 35) [1]. Some of the major investment in infrastructure projects include high-speed railway, LRT3 and Pan Borneo Highway [1]. In the Twelfth Malaysia Plan (12MP) (2021-2025), Malaysian government is planning to maintain their priority on the infrastructure development in which one of the main focuses is integrated regional development which include infrastructure connectivity [2]. According to the Economic Planning Unit, construction sector has contributed 11.1% of Gross Domestic Product (GDP) in the 10th Malaysian Plan (2011-2015) and the targeted contribution of this sector in the 11MP is 10.3% (Economic Planning Unit, n. d., p.11).

As the area that covered in infrastructure projects is large, it is unavoidable that part of the structure or the entire structure to be constructed on soft soils. Soft soils are problematic [3] in the fields of construction and geotechnical engineering. Examples of soft soils are organic soil and peat soil [4]. One of the significant issue caused by soft soils is low bearing capacity and this is because of its natural characteristics of "rich-water content, high natural void level, low shear strength, high compressibility" [5]. The low bearing capacity can affect project performance especially in terms of time, cost and quality. The traditional way of replacing soft soils is merely incurred high project cost [4], due to the complication in the process, it requires additional time. Any wrong or inaccurate decision will further increase project time and thus provide pressure on project time performance. Besides that, poor risk management in respect of risk identification and assessment could further aggravate project time performance. Risks and uncertainties are unique in every construction project. In view of the higher and unique risks and uncertainties of construction projects on soft soils, risk management plays a critical role in project management. In addition, the current literature review showed that existing knowledge on risk assessment of infrastructure projects on soft soils is scarce.

In view of the significance of infrastructure projects in developing countries as well as the issues and risks of construction projects on soft soils, this study intends to determine critical time-influencing external risk factors of infrastructure projects on soft soils. The focus of this study was on the project time performance as it is one of the key elements in determining the success of project management. As the types, probability and likelihood of risks are different at each project life cycle, this research focused on the construction phase. According to Edwards and Bowen [6], external risks such as political, economic and financial risks are critical and should be given priority in the construction research.

2. Literature Review

Review of the past studies showed that studies on risk assessment of infrastructure projects has been gaining considerable attention by researchers in the past. Some focused on general infrastructure projects [7], others on specific types of infrastructure projects such as bridge, roads, highway [8, 9], seaport [10] and railway [11]. As indicated by Low, Wong and Lee [7], managerial (soft) aspect is paying very little attention by past literature compared to the technical (hard) aspect of infrastructure projects on soft soils. Soft knowledge is essential in assuring the success of a construction project especially if the ground condition is problematic like soft soils. With reference to the work by Low, Wong and Lee [7], 4 types of external risks at level 1 (criteria) were used in this study. A total of 21 risk factors at level 2 (subcriteria) were extracted and used in studying the risk assessment of infrastructure projects on soft soils.

3. Research Methodology

The risk assessment of this study was conducted based on the Analytic Hierarchy Process (AHP) method. According to Saaty [12], AHP was introduced by Thomas Saaty and it is a multicriteria decision making approach that determining critical factors in a decision [12] based on pairwise comparison method. The main advantages of this approach is the input of experts' knowledge [13] and thus the reliability of findings is quite high in a small sample size [14]. This study adopted the relative importance scale of 1 (equal) to 9 (extreme importance) introduced by Saaty [12]. The research method of this study was questionnaire and the questionnaires were sent to the respondents by few techniques, namely, by post, email and face-face. This study was conducted based on cross-sectional approach. The experts of this study were professionals from the contractor organizations with civil engineering or other construction-related background, at least 15 years and above working experience in the construction sector and have been involved in at least one infrastructure project on soft soils.

4. Results and discussions

A total of 13 valid responses were used in this study. Five (5) industry experts have 16-20 years working experience in the construction and the remaining eight respondents have more than 20 years working experience in the construction. Three respondents have involved in one infrastructure project on soft soils and ten respondents have involved in at least 2 or more infrastructure project on soft soils. The consistency ratios at level 1 & 2 were less than 0.10 which is within the acceptable threshold. Table 1 indicates the results of the local and global weightage of the external risk factors. The significance level of external risks in descending order was as follows: "Soil condition" risk (0.47), "Economic" and "Political" risks (0.19) and "Environmental" risk (0.15). In respect of the global weightage, the five most critical external risk factors on infrastructure project time performance on soft soils were "Thick soft soil layer" (0.14), "High water content" (0.11), "High organic content" (0.08), "Price changes (material & energy)" (0.07) and "Karstic cavity formation" (0.06). Results of this study were consistent with past studies in which "Thick soft soil layer", "High organic content" and "Price changes (material & energy)" are not merely play a critical role on project cost performance as indicated by Low, Wong and Lee [7] but they play an equal important role in determining time performance of infrastructure projects on soft soils. Although the annual inflation rate in Malaysia marked an average +1.5% in August 2019, "Price changes (material & energy)" is still considered as an important risk factor in infrastructure projects on soft soils.

While the five least critical external risk factors on project time performance were "Archaeological/habitant issue" (0.02), "Lack of political support" (0.02), "Protectionism" (0.02), "Corruption" (0.02) and "Change in government/legislative/regulatory" (0.03). The results implied that Malaysian government provides support in infrastructure development which is in line with the 11th Malaysian plan and the change in government provides least significant impact on project time performance. In other words, it can be suggested that political risk plays a little role in determining the success of project time performance.

Table 1. Local and global weightage of external risks.

Risks Level 1		Risk Factors Level 2	Local weightage	Global weightage
Economic	0.19	Wages inflation	0.19	0.036
		Local inflation change	0.26	0.049
		Price changes (material & energy)	0.35	***0.065***
		Economic recession	0.20	0.038
Environment	0.15	Unfavourable weather condition	0.40	0.059
		Adverse environmental impact due to construction works	0.25	0.037
		Hard to access site and mobilize resources	0.26	0.038
		Archaeological/habitant issue	0.10	0.015
Soil condition	0.47	Thick soft soil layer	0.30	***0.144***
		High water content	0.23	***0.110***
		High organic content	0.17	***0.083***
		Karstic cavity formation	0.13	***0.064***
		Sand lenses/laminations/different soils mixture (complex soil layers)	0.08	0.036
		Mineralogy of soil (high activity/sensitivity of soils etc.)	0.08	0.040
Political	0.19	Change in government/legislative/regulatory	0.15	0.029
		Corruption	0.11	0.021
		Political opposition/interferences	0.16	0.031
		Lack of political support	0.08	0.015
		Protectionism	0.08	0.015
		Land acquisition issue	0.24	0.045
		Delay in obtaining consent/approval (access right etc.)	0.18	0.033

5. Conclusions

It can be concluded that soil condition risk and economic risk play a vital role in assuring the success of an infrastructure project on soft soils. While, political risk has little contribution in determining the success of an infrastructure project on software. As such, a further study is proposed to delve into the impact level of different soil condition on project time performance of infrastructure project on soft soils.

References

[1] Economic Planning Unit n.d Eleventh Malaysia Plan 2016-202: Anchoring growth on people.

[2] Ministry of Economic Affairs 2019 Twelfth Malaysia Plan.

[3] Jafer H M, Atherton W, Sadique M, Ruddock F and Loffill E 2018 Development of a new ternary blended cementitious binder produced from waste materials for use in soft soil stabilization *Journal of Cleaner Production* **172** 516-28

[4] Mujah D, Rahman M E and Zain N H M 2015 Performance evaluation of the soft soil reinforced ground palm oil fuel ash layer composite *Journal of Cleaner Production* **95** 89-100

[5] Yang G, Xue R and Li H 2011 *Rock mechanics: Achievements and ambitions,* pp 683-6

[6] Edwards P J and Bowen P A 1998 Risk and risk management in construction: A review and future directions for research. *Engineering Construction and Architectural Management* **5** 339-49

[7] Low W W, Wong K S and Lee J L 2019 Cost-influencing risk factors in infrastructure projects on soft soils *International Journal of Construction Management* 1-13

[8] Kumar L, Jindal A and Velaga N R 2018 Financial risk assessment and modelling of PPP based Indian highway infrastructure projects *Transport Policy* **62** 2-11

[9] Mousavi S M, Tavakkoli-Moghaddam R, Azaron A, Mojtahedi S M H and Hashemi H 2011 Risk assessment for highway projects using jackknife technique *Expert Systems with Applications* **38** 5514-24

[10] John A, Paraskevadakis D, Bury A, Yang Z, Riahi R and Wang J 2014 An integrated fuzzy risk assessment for seaport operations *Safety Science* **68** 180-94

[11] Hwang B-G, Zhao X and Yu G S 2016 Risk identification and allocation in underground rail construction joint ventures: contractors' perspective *Journal of Civil Engineering and Management* **22** 758-67

[12] Saaty T L 1990 How to make a decision: the analytic hierarchy process *European journal of operational research* **48** 9-26

[13] Saaty T L 2008 Decision making with the analytic hierarchy process *International Journal of Services Sciences* **1** 83-98

[14] Darko A, Chan A P C, Ameyaw E E, Owusu E K, Pärn E and Edwards D J 2018 Review of application of analytic hierarchy process (AHP) in construction *International Journal of Construction Management* 1-17

IOP Conf. Series: Materials Science and Engineering **943** (2020) 012029 doi:10.1088/1757-899X/943/1/012029

Dynamic Simulation of Hydrogen Iodide Decomposition in Catalytic Multi-Tubular Reactor

Jobrun Nandong[1,*], Akmal Sirajul[1] and Noraini Mohd[2]

[1] Department of Chemical Engineering, Curtin University Malaysia, 98009 Miri, Malaysia

[2] School of Energy and Chemical Engineering, Xiamen University Malaysia, 43900 Sepang, Selangor, Malaysia.

Email: jobrun.n@curtin.edu.my

Abstract. Sulfur-Iodine thermochemical cycle process has been considered a promising method to generate hydrogen from water. One of the main limitations of this process lies in the hydrogen iodide decomposer section, which requires substantial amount of heat at high temperature. The present study analyses the heat transfer limitation effect on the performance of hydrogen iodide decomposition in a multi-tubular catalytic reactor by means of dynamic simulation. The extensive numerical simulation shows that the inlet temperature heating medium (Helium gas) should be at least 700°C so to enable at least 90% conversion when the inlet molar fraction of hydrogen iodide is 0.6. Lowering the inlet heating medium will require the reduction in the inlet molar fraction of hydrogen iodide in order to keep the conversion at 90% and above.

1. Introduction

The water-splitting Sulfur-Iodine Thermochemical Cycle (SITC) process is one of the hydrogen production methods that has received intensive research attentions over the last two decades. This process has some advantages: (a) using abundant water supply, (b) easy separation and purification of hydrogen, (c) potential integration with nuclear reactor and other plants producing high-temperature waste heat, and (d) zero greenhouse gas emission. The SITC process has three major sections where each section corresponds to a major chemical reaction as follows [1]:

Section I: Bunsen reaction

$$I_2 + SO_2 + 2H_2O \rightleftharpoons H_2SO_4 + 2HI, \quad \Delta H_1 = -75 \pm 15 \text{ kJ/mol}$$

Section II: Sulfuric acid decomposition

$$H_2SO_4 \rightleftharpoons \frac{1}{2}O_2 + SO_2 + H_2O, \quad \Delta H_2 = 186 \pm 3 \text{ kJ/mol}$$

Section III: Hydrogen iodide decomposition

$$2HI \rightleftharpoons H_2 + I_2, \quad \Delta H_3 = 12 \text{ kJ/mol}$$

Figure 1 shows the interactions of three major sections in the SITC process. The inputs to the process is water and heat while the outputs are hydrogen (product) and oxygen (by-product). The reactants (sulfur dioxide and iodine) are fully recycled – thus no harmful waste is released to the environment. The major challenges in this process, however, arise from the need of a very high-temperature thermal energy (in Section II) and materials of construction that can cope corrosive and high-temperature environment. The closed-cycle operation of the SITC was successfully achieved at a production rate of 32 L/hr lasting for 20 hours – the automation of the process in a stable state is considered one of the challenges [2]. The hydrogen iodide decomposition section is very important because it is where hydrogen is produced. One of the main concerns in the SITC process is the hydrogen iodide decomposition reactor, which is thermodynamically limited due to the reversible reaction nature. The hydrogen iodide decomposition occurs in the presence of suitable catalysts (e.g., platinum catalyst) at the temperature range of 300-500°C. Due to the chromodynamics limitation, the conversion of hydrogen iodide into hydrogen is quite low.

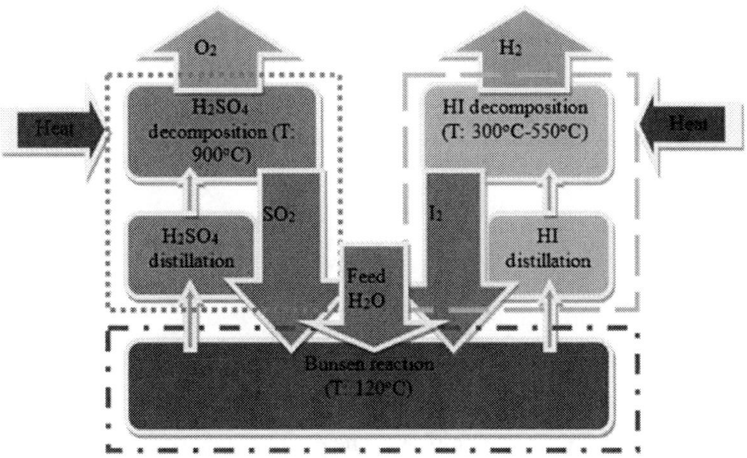

Figure 1. Sulfur-Iodine Thermochemical Process: Section I (dashed mix dotted area), Section II (dotted area), and Section III (dashed area) [3].

2. HI decomposition reactor modelling

2.1. Process description

Figure 2 illustrate the reactor configuration for HI decomposition. The catalyst pellets are packed inside the tubes in which the reaction occurs. Since the reaction is endothermic, heating medium is required to supply for the heat of reaction. In this study, the helium (He) gas is used as the heating medium which flows in the shell side. We choose He as the heating medium because of the high-temperature condition used in this study, i.e., the heating medium should be above 400°C. In this study, we choose co-current flow configuration because the reactor is thermodynamically limited especially at lower temperature. As the reaction proceeds along the reactor length, the rate of reaction becomes severely limited by the thermodynamics equilibrium at a higher temperature. Thus, reducing the temperature along the reactor helps reduce the equilibrium limitation. Consequently, with this co-current flow configuration it will be expected that the overall conversion of HI will increase but at the expense of larger reactor volume, i.e., compared to the counter-current flow configuration.

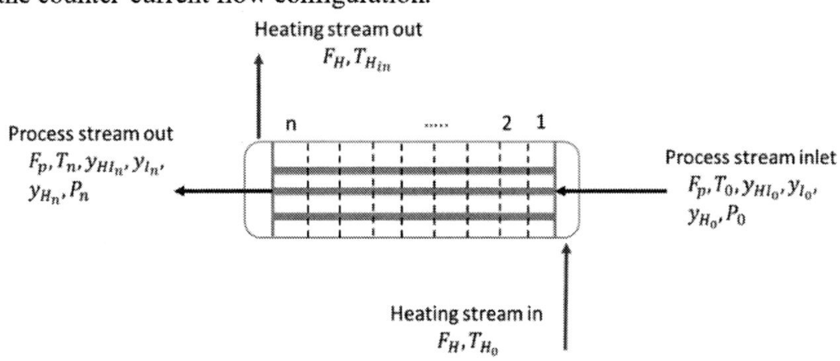

Figure 2. Schematic diagram of HI decomposition reactor configuration.

2.2. Mass and energy balance equations

In this study, the reactor is modelled as a set of ordinary differential equations (ODEs). Since the reactor is long and the spatial temperature and concentration gradients are unlikely to be constant along the reactor, it is necessary to divide the reactor into several zones. Each of the zones is then modelled using the ODEs, i.e., assumption of well mixing in each of the zones. In this case, the reactor is divided into 22 zones and each zone is 0.5 m in length. Given a zone j for $j = 1, 2, \dots n$, the pseudo homogeneous mass (mole) balance equations for HI, I_2 and H_2 denoted with the subscripts A, B and C respectively are given as below:

$$\frac{dn_{A_j}}{dt} = F_p\left(y_{A_{j-1}} - y_{A_j}\right) - r_{A_j}W_{cat} \quad (1)$$

$$\frac{dn_{B_j}}{dt} = F_p\left(y_{B_{j-1}} - y_{B_j}\right) + \frac{1}{2}\left(r_{A_j}W_{cat}\right) \quad (2)$$

$$\frac{dn_{C_j}}{dt} = F_p\left(y_{C_{j-1}} - y_{C_j}\right) + \frac{1}{2}\left(r_{A_j}W_{cat}\right) \quad (3)$$

where n_{k_j} denotes the number of moles of species k in the zone j, y_{k_j} the molar fraction of the species k in the zone j, r_{A_j} the rate of consumption of HI (kmol/kg-cat.s), W_{cat} the weight of catalyst per zone (kg) and F_p the process stream flow rate (kmol/s). The energy balance in the tube-side is as follows

$$\frac{d\left(\bar{C}_{P_j}N_{t_j}T_{P_j}\right)}{dt} = F_p\bar{C}_{P_j}\left(T_{P_{j-1}} - T_{P_j}\right) - \Delta H_r r_{A_j}W_{cat} + Q_{s_j} \quad (4)$$

Meanwhile, the energy balance in the shell-side is

$$V_s\left[\frac{d\left(C_{h_j}\rho_{h_j}T_{h_j}\right)}{dt}\right] = F_h\left(C_{h_{j-1}}T_{h_{j-1}} - C_{h_j}T_{h_j}\right) - Q_{s_j} \quad (5)$$

$$Q_{s_j} = UA_s\left(T_{h_j} - T_{P_j}\right) \quad (6)$$

In the equations (4) – (6), $\bar{C}_{P_j}$ denotes the average specific heat capacity of the process fluid ($kJ/kmol.K$), C_{h_j} the specific heat capacity of He gas (kJ/kg.K), T_{P_j} the process fluid temperature (°C), T_{h_j} the heating medium temperature (°C), ΔH_r the heat of reaction (kJ/kmol-HI), ρ_{h_j} the heating medium density (kg/m^3), N_{t_j} the total number of moles (kmol), U the overall heat transfer coefficient (W/m^2.K), V_s the volume of shell per zone (m^3), A_s the heat transfer area per zone (m^2) and Q_{s_j} is the heat supply from the heating medium (kJ/s). The average specific heat capacity is calculated as follows

$$\bar{C}_{P_j} = \sum_{k=1}^{3} MW_k C_{k_j} \quad (7)$$

where MW_k and C_{k_j} denote the molecular weight (kg/kmol) and specific heat capacity (kJ/kmol.K) of the species k respectively. The specific heat capacity (J/mol.K) of each of the species (gaseous phase) depends on the temperature and represented using empirical polynomial equation:

$$C_k = \beta_0 + \beta_1\tilde{T} + \beta_2\tilde{T}^2 + \beta_3\tilde{T}^3 + \beta_4\tilde{T}^{-2} \quad (8)$$

Here, $\tilde{T}=T/1000$ and T is the absolute temperature (K). The values of parameters of the equation (8) are given in the Table 1, from the NIST Chemistry WebBook [4].

Table 1. Parameters of specific heat capacity correlations [4].

Parameters	HI	I_2	H_2	He
β_0	26.0454	37.79763	33.066178	20.78603
β_1	4.689678	0.225453	-11.363417	4.8506E-10
β_2	4.911765	-0.912556	11.432816	-1.5829E-10
β_3	-2.654397	1.034913	-2.772874	1.5251E-11
β_4	0.121419	-0.083826	-0.158558	3.1963E-11

2.3. Reaction kinetics

Nguyen et al. [1] studied the hydrogen iodide reaction kinetics over the Pt/γ-alumina 1 wt% in a sulfur-iodine thermochemical cycle process. The researchers involved have established new kinetic parameters of the reaction estimated from the measured data. In this work, the kinetics information from Nguyen et al. [1] is adopted. Therefore, interested readers may refer to the paper itself for the details about the reaction kinetics of hydrogen iodide over Pt-based catalyst. Table 2 shows the values of parameters used for the simulation in this study. The rate of HI decomposition is expressed as follows

$$r_A = \frac{k(1-X) - k'\left(\frac{X}{2}\right)}{\left[1 + \left(\sqrt{K_I}\right)\left(\sqrt{P_{B_0} + \frac{P_{A_0}X}{2}}\right)\right]^2} \tag{9}$$

where $k' = k/K_p$ and the other parameters above are defined as below:

$$K_p = exp\left(-\frac{\Delta G_T^0}{RT}\right) \tag{10}$$

$$k = 31.3exp\left(-\frac{37.1}{RT}\right) \tag{11}$$

$$K_I = 1.8\times 10^{-9} exp\left(\frac{75.1}{RT}\right) \tag{12}$$

Table 2. Parameters and values used in the current simulation.

Parameters and units	**Values**
Number of tubes, No_t	1000
Tube internal diameter, d_t (m)	0.05
Tube length, L_t (m)	11.0
Ratio of total tube volume to reactor shell volume, rv	2.5
Reactor shell diameter, D_r (m)	2.5
Voidage factor in tube, ε	0.4
Catalyst pellet (spherical shape) dimeter, d_c (m)	0.005
Catalyst density, ρ_c (kg/m^3)	480.0
Nominal feed molar faction HI	0.6
Nominal feed molar fraction I_2	0.4
Nominal feed flow rate, F_p (kmol/s)	0.02
Nominal He flow rate, F_h (kg/s)	30.0
Nominal inlet process temperature, T_{p_0} (^{o}C)	350.0
Nominal He inlet temperature, T_{h_0} (^{o}C)	500.0
Nominal inlet pressure of process fluid, P_{p_0} (kPa)	500.0
Idea gas constant, R (kJ/mol.K)	8.3141E-3
Overall heat transfer coefficient, U (kW/m^2.K)	0.20

3. Results and discussion

One of the major challenges in the operation of HI decomposition reactor is due to the limitation imposed by the thermodynamics equilibrium. A large reactor volume is required to achieve significant conversion of the hydrogen iodide. There are a number of simulation studies on HI decomposition reactor reported in the literature. Hwang and Onuki reported that 90% conversion of HI is achieved in a membrane reactor [5]. Goswami and co-workers [6] conducted simulation study on HI decomposition in a packed bed membrane reactor (PBMR) and compared the results with that in a conventional packed bed reactor. Their study showed that the parameters having high impact on conversion are the wall temperature, feed temperature, reactor diameter and packed bed porosity. They suggested that the reactor wall temperature is recommended to be in the range of 690 –700 K, and the bed porosity is recommended to be in the range of 0.2 – 0.4. According to the simulation study of [7], the proper operating temperature of the HI thermal decomposer to satisfy the hydrogen production rate of 50 NL-H2/h and sustain the closed loop operation of the Sulfur-Iodine (SI) process is 587 °C. The simulation of SI thermochemical plant by [8], showed that that the hydriodic acid (HI) decomposition step is the rate-limiting step of the entire SI cycle. In the present simulation study, the focus will be on the effects of heating medium temperature and process fluid temperature on the HI conversion.

3.1. Effect of inlet temperature of heating medium

First, an open-loop simulation is conducted where the nominal operating conditions and parameters are shown in the Table 2. Here, the effects of inlet temperature of the heating medium (He gas) on the HI percentage conversion and H_2 production rate are examined. Figure 3 shows the effect of varying the inlet temperature of the heating medium from 500°C to 600°C and then to 700°C on the conversion at different inlet molar fraction of the hydrogen iodide. As the inlet temperature (Th0) increases, for a given inlet molar fraction of HI, the conversion also increases. This is accompanied by the increase in the hydrogen production rate (normal L/s). However, for a given inlet temperature of the heating medium, it can be seen that the rise of the hydrogen production rate shows rapid increase as the inlet molar fraction of HI increases up to a certain threshold value, thereafter the increase is much reduced. From the Figure 3, it can be seen that the threshold values are 0.25, 0.37 and 0.48 corresponding to the inlet temperature of heating medium at 500°C, 600°C and 700°C respectively. Increasing the inlet temperature of heating medium from 500°C to 700°C leads to about 90% increase in hydrogen production rate from 51 NL/s. This suggests that an adequate heating effect is crucial to achieve better conversion of HI. This is because the heat transfer from the heating medium to the process fluid is limited by the low overall heat transfer coefficient. To increase the heat transfer rate, it is necessary to increase the surface area for the heat transfer. This may be done by reducing the tube diameter and increasing the total number of tubes, but this has to be done with caution as not to have an excessively large pressure drop across the reactor – the smaller the tube diameter, the larger the pressure drop across the reactor. Another strategy is to increase the temperature difference between the heating medium and process fluid so to compensate for the low overall heat transfer coefficient. If molten salt is used as the heating medium, the maximum temperature for the heating medium is about 500°C. Therefore, to have a larger temperature of heating medium, helium (He) gas could be used. Note that, a higher heating medium temperature will result in a higher process fluid temperature, which will increase the reaction rate; the equilibrium constant increases with the increase in temperature because the reaction is endothermic. Thus, the decomposition reaction is favoured at higher temperature. The result of this simulation is in agreement with the previous simulation studies reported in literature, e.g., [6].

Figure 4 demonstrates the temperature profiles in the tube and shell sides of the reactor. The temperature inside the tube drops very drastically from 350°C to 275°C and 300°C when the inlet temperatures of the heating medium are 500°C and 700°C respectively. Along most of length of the reactor, the temperature inside the tube is nearly constant although a significant temperature gradient is observed in the shell side (heating medium). From this figure, it can be deduced that increasing the operating temperature of the process fluid from 275°C to 300°C will lead to about 90% increase in the hydrogen production rate, when the inlet molar fraction of HI is 0.6; equivalently, the conversion rises from41%

to about 74%. The question is, can the conversion be raised further to above 90% for the case of inlet temperature of heating medium at 700°C, and inlet molar fraction of HI at 0.6? To answer this question, it is necessary to develop a temperature control of the reactor. Here, the temperature control maintains the exit temperature of the process fluid at a setpoint by manipulating the heating medium flow rate.

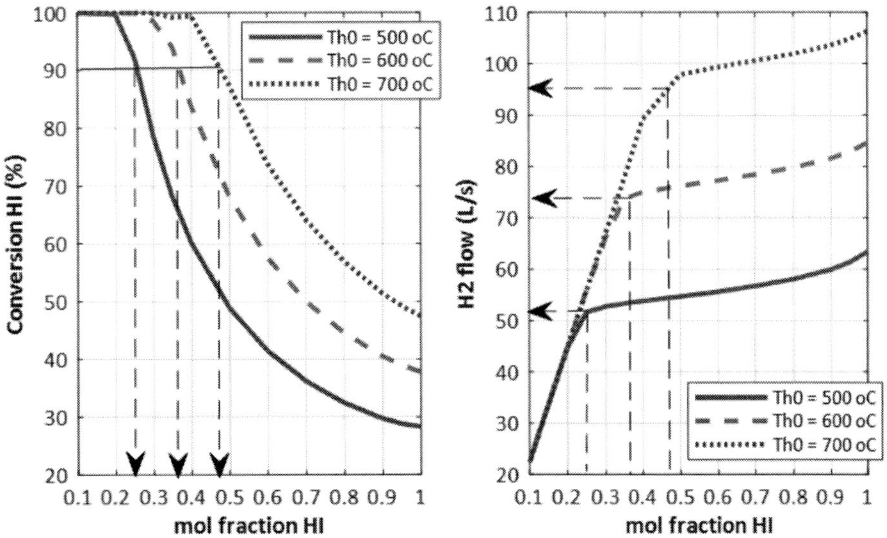

Figure 3. Conversion of HI and H_2 product flow rate against inlet molar fraction of HI in the process stream under different values of inlet temperature of the heating (He gas) stream.

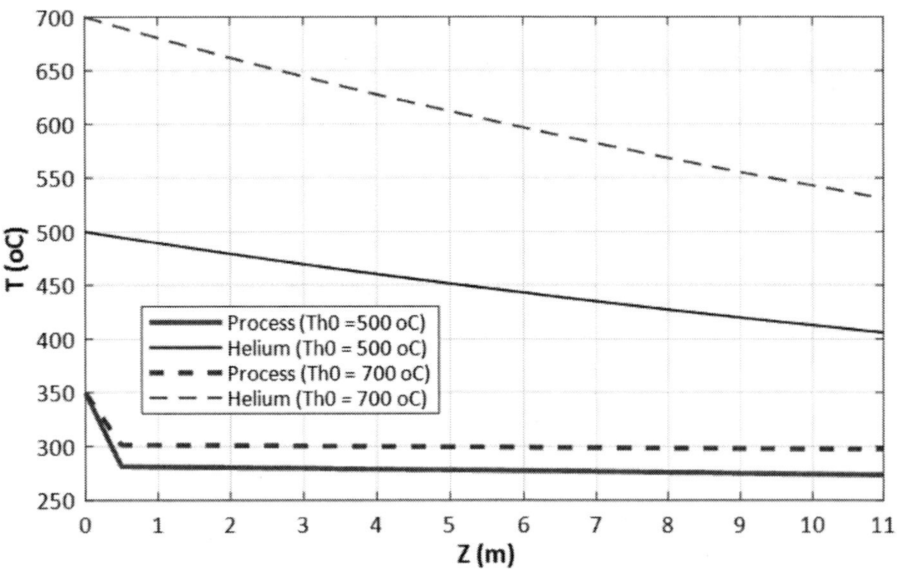

Figure 4. Temperature profile along the reactor under different values of inlet temperature of the heating medium for inlet HI molar fraction at 0.6.

3.2. Closed-loop responses

The exit temperature of process stream is controlled by manipulating the heating medium flow rate. A Proportional Integral (PI) controller is adopted where the tuning is done by using the linearized model:

$$G_p(s) = \frac{0.4\exp(-0.119s)}{1.65s + 1} \tag{13}$$

IOP Conf. Series: Materials Science and Engineering **943** (2020) 012029 doi:10.1088/1757-899X/943/1/012029

Based on the linearized model (13), the PI controller is tuned using the IMC tuning method (available in the Control System Designer, MATLAB R2017b). The PI controller is given below:

$$G_c(s) = 1.1\left(1 + \frac{1}{1.65s}\right) \quad (14)$$

A series of step changes to the temperature setpoint (5°C at each step) is applied to the reactor. The impact of the setpoint changes under different inlet temperatures of the heating medium, on the conversion and heating medium flow rate are shown in Figures 5 and 6.

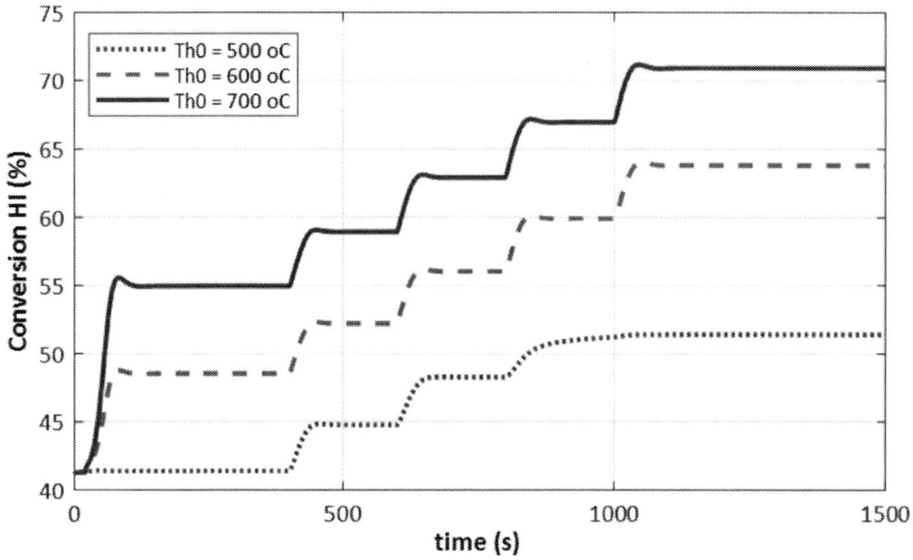

Figure 5. Conversion of HI in response to a series of temperature setpoint changes (5°C at a time) under different inlet temperatures of heating medium (inlet molar fraction of HI is 0.6).

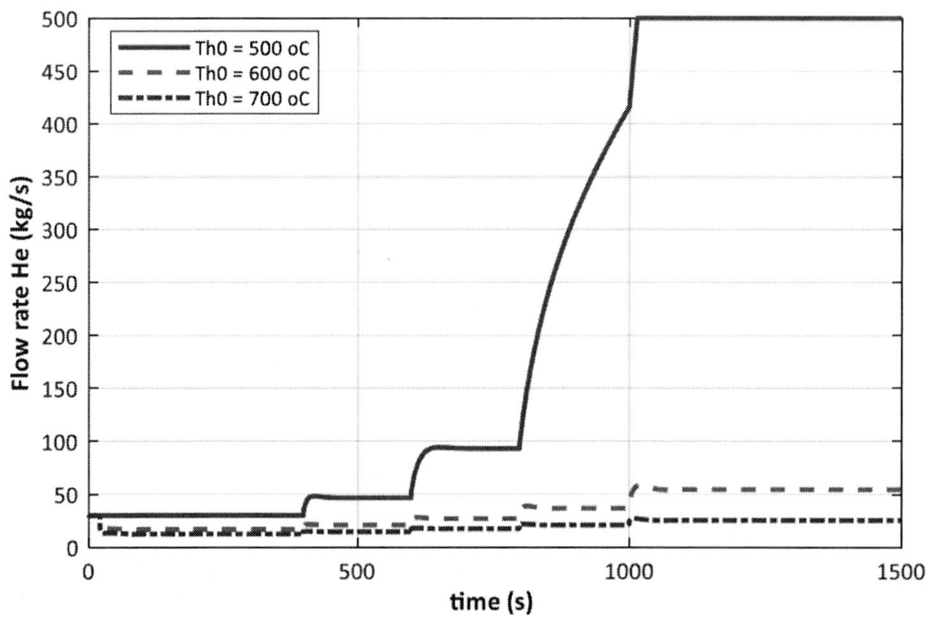

Figure 6. The profiles of heating medium flow rate in response to a series of temperature setpoint changes (5°C at each step change) where the inlet molar fraction of HI at 0.6.

IOP Conf. Series: Materials Science and Engineering **943** (2020) 012029 doi:10.1088/1757-899X/943/1/012029

From Figure 5, it can be seen that the conversion increases as the exit temperature setpoint increases. The higher the temperature of the heating medium the larger is the conversion. Figure 6 shows how the heating medium flow rate varies in response to the temperature setpoint changes. When the inlet temperature of the heating medium is low, the amount of the flow rate raises dramatically with the exit temperature. Eventually, the maximum heating medium flow rate is reached (due to the physical limitation of valve that allows only up to certain range of flow) – the valve becomes saturated and it is no longer possible to raise the exit temperature any further. As a result, the maximum conversion achievable depends on the inlet temperature of the heating medium. In this case, when the inlet temperature of heating medium is 500°C, the maximum conversion achievable is only about 52%, when the inlet molar fraction of HI is 0.6. Obviously, to achieve a higher conversion the inlet temperature of the heating medium has to be increased, for examples, to 600°C and 700°C where the heating medium valve will not saturate. For the case of inlet temperature of heating medium at 700°C, the maximum achievable conversion is about 73% and the corresponding exit temperature of process fluid is 294°C. Can the conversion be raised up to 90% and above by further increasing the exit temperature? Of course, this would depend on the heating medium flow rate required, i.e., whether the valve will encounter saturation or not in order to raise the temperature to the desired value leading to 90% conversion.

A simulation is further conducted in which a series of setpoint changes is applied with a magnitude of 15°C per step change. Figure 7 shows the profile of conversion in response to the setpoint changes of exit process temperature. Obviously, an increase of the 60°C in the exit temperature from 274°C can result in about 91% conversion. Figure 8 shows the exit temperature profile in response the sequential setpoint changes. In this case, the valve does not encounter saturation as shown in Figure 9. In Figure 9, it is noticed that the percentage increase of the heating medium per step change in setpoint increases as the setpoint increases. This implies that the process gain decreases with the setpoint changes, which further implies that the process controllability decreases with the temperature setpoint increase. This decrease in process controllability can be explained by the decrease in the temperature difference between the shell and tube sides. With the decrease in this temperature difference, it is becoming more and more difficult to meet the required heat by the reaction in the face of low overall heat transfer coefficient and limited heat transfer area. As a result of that, the effect of increasing the heating medium flow rate becomes less influential at a higher setpoint of exit temperature of process fluid – leading to the reduced process controllability as process fluid temperature increases.

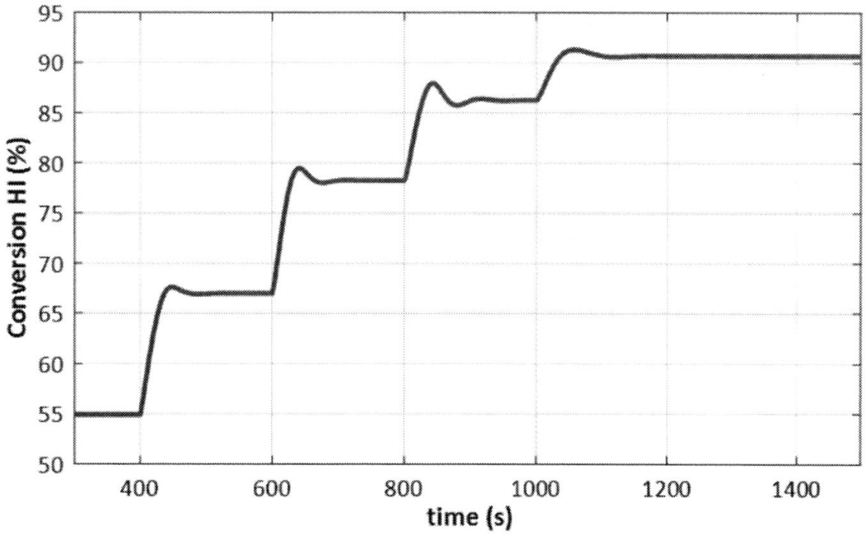

Figure 7. Conversion of HI in response to a series of temperature setpoint changes (15°C at a time) when the inlet temperatures of heating medium is at 700°C and inlet molar fraction of HI is 0.6.

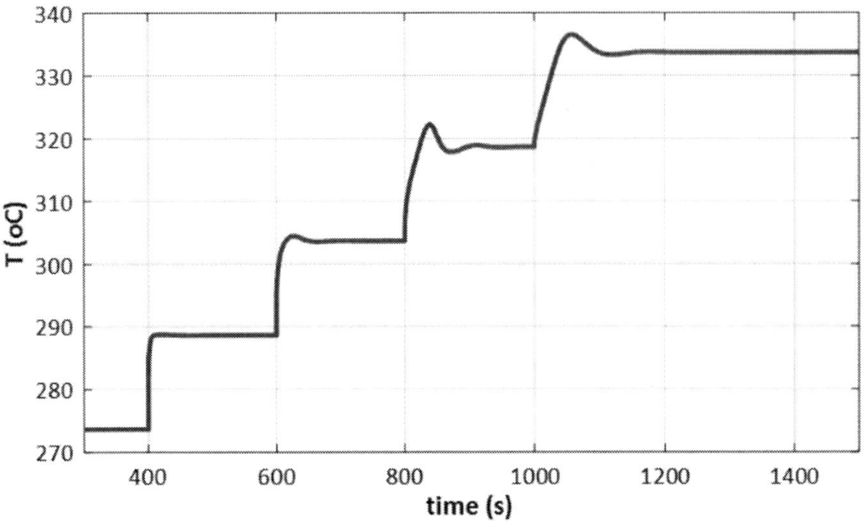

Figure 8. Tracking profile of the exit process stream temperature in responses to setpoint changes.

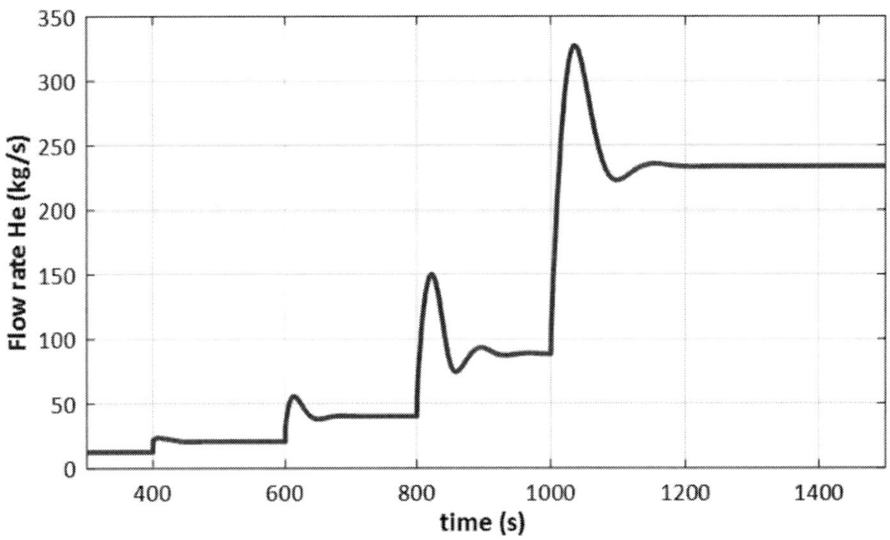

Figure 9. Profile of the heating medium flow rate in response to exit temperature setpoint changes.

4. Conclusion

In this study, the dynamic simulation of an industrial-scale multi-tubular catalytic reactor for HI decomposition to produce hydrogen shows that, a sufficiently high heating medium temperature is needed to achieve satisfactorily high conversion ($\geq$ 90%). This study suggests that the inlet temperature of the heating medium (He gas) should be at least 700°C to achieve over 90% conversion when the inlet molar fraction of HI is 0.6. If the inlet heating medium is reduced to 600°C, it is still possible to achieve above 90% conversion but the inlet molar fraction of HI should be less than 0.48. Please note that, at lower inlet temperature of heating medium the process controllability is low due to the near valve saturation. The main limitation for the HI reactor used as observed in this simulation is the limited heat transfer area. If the heat transfer area can be increased by 50% or more, then the process controllability can be greatly improved at a lower inlet temperature of the heating medium.

IOP Conf. Series: Materials Science and Engineering **943** (2020) 012029 doi:10.1088/1757-899X/943/1/012029

References

[1] Nguyen T D, Gho Y, Cho W C, Kang K S, Jeong S U, Kim C H, Park C and Bae K 2014 *Applied energy* 115 531-539.

[2] Kubo S, Nakajima H, Kasahara S, Higashi S, Masaki T, Abe H and Onuki K 2004 *Nucl. Eng. Des.* 233 347-354.

[3] Sakaba N, Kasahara S, Ohashi H, Sato H, Kubo S, Terada A, Nishihara T, Onuki K and Kunitomi K 2006 *Proc. of WHEC16.*

[4] NIST Chemistry WebBook 2019 *National Institute of Standards and Technology.*

[5] Hwang G J and Onuki K 2001 *J. Membrane Sci.* 194 207-215.

[6] Goswani N, Singh K K, Kar S, Bindal R C and Tewari P K, *Int. J. Hydrogen Energ.* 39 18182-18193.

[7] Shin Y, Lee T, Lee K and Kim M 2016 *Applied energy* 173 460-469.

[8] Brown N R, Oh S, Revankar S T, Vierow K, Rodriguez S, Cole Jr R and Gauntt R 2009 *Nucl. Technol.* 167 95-106.

Biogenic synthesis of magnesium oxide nanoparticles using *Aloe barbadensis* leaf latex extract

Jaison Jeevanandam, Yen San Chan*, Yee Jing Wong, Yiik Siang Hii
Department of Chemical Engineering, Faculty of Engineering and Science, Curtin University, CDT250, Miri, Sarawak 98009, Malaysia

Email: chanyensan@curtin.edu.my

Abstract. Biological methods are employed to yield less or non-toxic MgO nanoparticles to utilize them in biological applications. Among various biosynthesis approaches, plant extracts with phytochemicals, especially from leaves, are widely used to fabricate MgO nanoparticles, due to their high availability, rapid synthesis and ability to yield smaller stable nanoparticles. *Aloe barbadensis* is a succulent xerophytic plant with unique characteristics to withhold water in its leaf named parenchymal gel, which is protected by a chloroplast containing thick latex, to avoid transpiration in high temperature condition of the desert. These latex contains phytochemicals such as flavanol, quercetin, Kaempeferol, myricetin and fisetin, along with other common phytochemicals such as phenols and terpenoids, that are essential for nanoparticle formation. Further, these compounds also possess enhanced biological properties. Thus, the aim of the present study is to obtain crude phytochemical extracts from *Aloe barbadensis* latex and utilize them as reducing and stabilizing agent for the smaller MgO nanoparticle formation. In addition, the parameters which affect the formation of nanoparticles are identified and optimized to yield smaller MgO nanoparticles with phytochemicals as surface functional groups, to be beneficial in biomedical applications.

Keywords: *Aloe barbadensis*; Biogenic synthesis; MgO nanoparticles; Leaf latex; Phytochemicals; Functional groups

1. Introduction

Magnesium oxide (MgO) nanoparticles are currently popular among researchers, due to their exclusive biological, electronical and thermal properties [1]. This has led to the search of various synthesis approaches to yield smaller sized MgO nanoparticles with distinct morphologies [2]. Chemical approaches such as sol-gel, hydrothermal, co-precipitation and solvothermal methods are under extensive research to yield smaller sized nanoparticles [3-6]. However, the toxicity of these chemicals derived MgO nanoparticles are the major limitation, while utilizing these nanoparticles for biological applications [7]. Thus, biological methods are employed to yield less or non-toxic MgO nanoparticles to utilize them in biological applications [8]. Algae [8] and fungi [9] are used for the biosynthesis of MgO nanoparticles, however the time required for synthesis and limited availability of the organisms are the drawbacks to utilize them for large scale biological applications. Thus, plant extracts with phytochemicals, especially from leaves, are widely used to fabricate MgO nanoparticles, due to their high availability, rapid synthesis and ability to yield smaller stable nanoparticles [10-12].

Aloe barbadensis is an evergreen perennial succulent xerophytic plant that grows widely in places, where ground water is low [12]. The evolution to sustain the climatic conditions in the desert (xerophytic) ecosystem, these plants possess unique characteristics to withhold water in its leaf named parenchymal gel to avoid transpiration in high temperature condition of the desert [13]. Thus, the normal leaves in *Aloe barbadensis* are modified into pseudo-leaves with high water content (gel) [14], which is protected by a chloroplast containing thick latex [15]. These latex contains phytochemicals such as flavanol, quercetin, Kaempeferol, myricetin and fisetin [16], along with other common phytochemicals such as phenols [17] and terpenoids [18]. Both the parenchymal gel as well as latex is highly beneficial in biomedical applications, due to the presence of unique phytochemical combinations [19, 20]. Several metal oxide nanoparticles have been synthesized previously by using phytochemical extracts as reducing and stabilizing agents [21-23]. Thus, the aim of the present study is to obtain crude phytochemical extracts from *Aloe barbadensis* latex and utilize them as reducing and stabilizing agent for the smaller MgO nanoparticle formation. In addition, the parameters which affect the formation of nanoparticles are identified and optimized to yield smaller MgO nanoparticles with phytochemicals as surface functional groups, to be beneficial in biomedical applications.

2. Methodology

2.1. Materials

The magnesium nitrate hexahydrate ($Mg(NO_3)_2 \cdot 6H_2O$) with molecular weight of 256.41 g/mol and purity of 99.8% was used as magnesium precursor and are purchased from Alfa Acer® (Singapore). The *Aloe barbadensis* latexes were collected from the local farm in Miri, Sarawak, Malaysia. The ethanol for the phytochemical extraction was purchased from Sigma-Aldrich® (Singapore) with molecular weight of 46.069 g/mol and 70% purity.

2.2. Preparation of Aloe latex powder

The *Aloe barbadensis* leaf latex was collected from local farm and cleaned with distilled water. The latex was chopped into small pieces and homogenized in a bio-dryer for a day. Later, the chopped latexes were blended using a blender to obtain fine powders as shown in Figure 1. However, the latex contains water molecules along with leaf layer, which makes the powder to be rich in water content. The water molecules can be removed by freeze-drying the latex powder at -40°C and 0.133 mBar vacuum using Fisher brand Labconco* 1.5 L freeze dry system as shown in **Figure 1**. The resultant latex powders were stored in a desiccator for further experiment.

Figure 1. Processes involved in Aloe latex powder preparation

2.3. Preparation of Aloe solid and liquid extract

20 grams of latex powder were added to 100 ml of distilled water and 100 ml of ethanol in a 250 ml conical flask [24]. The mixture in the conical flask was allowed to constantly stir (450 rpm) for 24 h at room temperature for complete elucidation of active materials and to dissolve them in the solvent as shown in **Figure 2**. Later, the latex powder extract is filtered using vacuum filtration method, where both the residues and filtrate was collected as in **Figure 2**. The collected residues (solid state) and the filtrate (aqueous state) are subjected to liquid-liquid extract purification process via the rotary vacuum evaporator for 3 h to eliminate solvent (ethanol) solvent from the aqueous phytochemical extract using a water bath at 50°C of temperature as displayed in **Figure 2**. The resultant solid residue and the aqueous filtrate containing phytochemicals, that are free from ethanol, was collected and used for MgO nanoparticle synthesis.

IOP Conf. Series: Materials Science and Engineering **943** (2020) 012030 doi:10.1088/1757-899X/943/1/012030

Figure 2. Processes involved in phytochemical extraction from Aloe latex powder

2.4. Preparation of MgO nanoparticles

Both solid and liquid extracts of Aloe leaf latex powder were used for the biosynthesis of smaller MgO nanoparticles. The parameters that affect the synthesis such as reaction time, extract ratio, temperature and precursor concentration was optimized using one-factor-at-a-time (OFAT) approach. **Table 1** shows the design of experiment used for the optimization of parameters. In each step of the OFAT analysis, the parameters were optimized, and the optimized parameter was included in the next step. The flowchart for the synthesis of MgO with both extracts are shown in **Figure 3**.

Table 1. Design of experiment for the optimization of parameters

Solid or liquid extract

Sample name	Fixed parameters		Parameter to be optimized
			Extract ratio
SE1/LE1	60°C	0.01M precursor	1:10
SE2/LE2	temperature	(fixed)	1:20
SE3/LE3	(fixed)		1:30
SE4/LE4			1:40
SE5/LE5			1:50
Sample			**Temperature (°C)**
ST1	1:10 extract ratio	0.01M precursor	40
ST2	(selected)	(fixed)	50
ST3			60
ST4			70
Sample			**Precursor concentration (mol/L)**
SP1	60°C	1:10 extract ratio	0.01
SP2	temperature	(fixed)	0.055
SP3	(selected)		0.10
SP4			0.15
Sample			**Temperature (°C)**
LT1	1:50 extract ratio	0.01M precursor	40
LT2	(selected)	(fixed)	50
LT3			60
LT4			70
Sample			Precursor concentration (mol/L)
LP1	60°C temperature	1:50 extract ratio	0.01
LP2	(selected)	(fixed)	0.055
LP3			0.10
LP4			0.15

Note: S – Solid; L – Liquid; T – Temperature; P – Precursor; E - extract

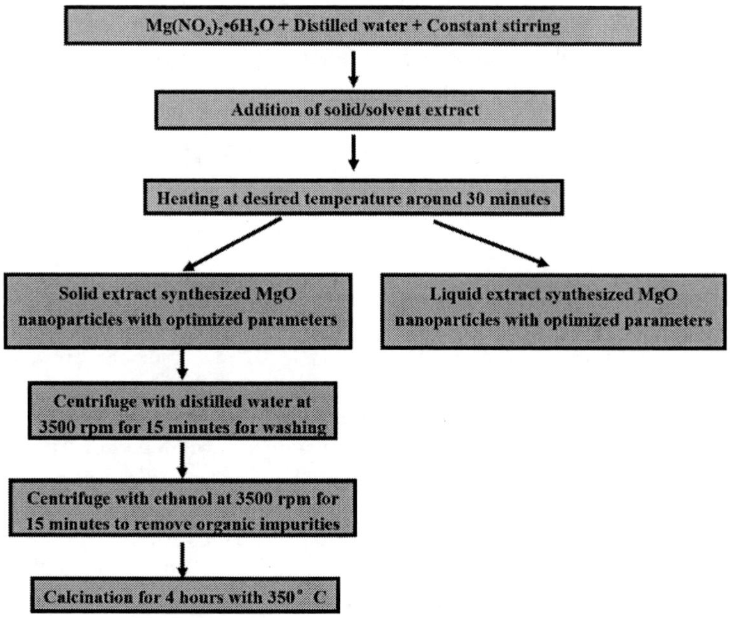

Figure 3. Synthesis of MgO nanoparticles using Aloe latex solid and liquid phytochemical

2.5. Characterization of MgO nanoparticles
The parameters were optimized based on the average particle size (particle size distribution) of the samples. Nano-ZS Zetasizer (Malvern Instruments, Worcestershire, UK) was used to analyze the average particle size and poly dispersity index of the biosynthesized MgO nanoparticles via dynamic light scattering (DLS) approach [25]. The formation of MgO nanoparticles via both solid and liquid Aloe extracts was confirmed by their surface plasmon resonance property via Perkin Elmer® Lambda 25 Ultraviolet-visible (UV-Vis) spectroscopy. The calcination temperature to obtain pure MgO nanoparticles from the solid extracts were identified via thermal degradation and heat flow patterns from Thermal gravimetric analyzer – Differential Scanning Calorimetry (TG-DSC), Mettler Toledo [26]. The functional groups present in the Aloe extracts and the MgO nanoparticles were identified using Thermo scientific NICOLET iS10 Fourier-Transform Infrared (FTIR) spectroscopy. Transmission electron microscope (TEM) from Tecnai^{TM} G^2 Spirit BioTWIN, FEI company, is used to determine the size, shape and morphology of MgO nanoparticles and the analysis was performed at an accelerating voltage of 200 kV.

3. Results and discussions
Both solid and liquid extracts of Aloe latex powder were green in color, which upon adding to the precursor dissolved in distilled water, does not change in color. The heating process to form nanosized MgO particles has led to a pale green color of the powder (solid extract) and the filtrate (liquid extract) which denotes the formation of MgO nanoparticles and are confirmed via DLS analysis.

3.1. Optimization of reaction time
Initially, the reaction time to form the MgO nanoparticles was optimized with 60°C as temperature, 1:50 as extract (extract: distilled water) ratio and 0.01 M as precursor concentration. It is noteworthy that 23°C and 85°C of temperatures may lead to precipitation of particles, which may increase the size of nanoparticles by irregular disintegration of phytochemicals in the extract [27]. Further, it has been demonstrated in our previous study that temperatures below 60°C is not sufficient to provide energy for the formation and stabilization of nanoparticles. Thus, 60°C was selected as the optimum

temperature to optimize reaction time. **Figure 4 (A)** shows the average size of MgO nanoparticles formed via reaction time of 10 to 60 min. The smallest nanoparticle size of 78.82 nm was obtained at 30 min of reaction time, which was selected as the optimum reaction time for the formation of MgO nanoparticles using both solid and liquid Aloe latex powder extract. The selected reaction time was similar to the experiment performed by Gopalakrishnan et al. (2014), where chemical precipitation approach was used to obtain molybdenum sulfate quantum dots [28]. Further, MgO nanoparticles synthesized via other approaches such as sol-gel, microemulsion, co-precipitation, hydrothermal/solvothermal methods require a reaction time of minimum 2 hours [29]. Hence, 30 min was selected as the optimum reaction time for the fabrication of MgO nanoparticles and to optimize other parameters that affect nanoparticle formation capability of Aloe latex powder extracts.

3.2. Optimization of extract ratio

The addition of water with aqueous phytochemical extract (both solid and liquid) is a significant factor to be optimized, as it determines the concentration of plant extract (dilutions) in the reaction. A reaction time of 30 min, temperature of 60°C and 0.01 M of precursor concentrations were used as constant. **Figure 4 (B)** shows the average MgO nanoparticle size that is formed, while using water in the range of 10 to 50 parts along with 1 part of the extract. It was observed that liquid extract: an aqueous ratio of 1:50 yields smaller MgO nanoparticles with size of 85.05 nm. This may be due to the ability of phytochemical extract to reduce the particle size upon increasing reduction rate by decreasing concentrations [30]. The result of experiments performed by Ali et al. (2016) is exactly similar, where synthesis of ZnO nanoparticles was reported with variations in the concentrations of *Aloe vera* extract such as 10, 20, 30, 40 and 50% to be added in 100 mL of 0.25M $ZnSO_4$ solution [31]. Similarly, a smaller MgO nanoparticle of 91.28 nm size was formed, when the extract ratio was 1:10 for the solid Aloe extract. The higher extract ratios yielded smaller nanoparticles, contrary to the liquid Aloe extract. However, it was reported that the increase in the concentration of solid extract may lead to decrement in the particle size of nanoparticles [32]. Thus, it is evident that both liquid and solid exhibits different mechanisms in the formation of nanoparticles, as there is a variation in the presence and quantity of certain phytochemicals.

3.3. Optimization of temperature

Figure 4 (C) shows the graph of average MgO nanoparticle size yielded by both solid and liquid extract at different temperatures ranging from 20 to 80°C. The precursor concentration of 0.01 M, a liquid extract ratio of 1:50 and reaction time of 30 min was used as constant. It is noteworthy that a smaller sized MgO nanoparticle was obtained between 50-60°C of reaction temperature for both solid and liquid extracts. MgO nanoparticle with an average size of 85.05 nm was obtained at 60°C, when liquid Aloe extract was used, whereas a further increase in the temperature has led to an increment in particle size. It is well known from previous literature that an increase in temperature will result in elevating diffusion coefficient, that reduces the reaction time required for the formation of stable nuclei and hence, reduces induction time. Typically, increasing temperature will increase the solubility and thus reduces the supersaturation. When the low supersaturation growth is more favour than nucleation, the growth rate increases and larger sized nanoparticles will be yielded [33]. Likewise, the effect of temperature in the average size of MgO nanoparticle was studied for solid Aloe extracts, which revealed that the nanoparticle size increases with an increase in temperature as explained in literatures [34]. A smaller nanoparticle size of 71.02 nm was obtained at 50°C, however, the PDI (supplementary information) revealed that the particles are polydispersed in nature. Thus, 60°C was selected as an optimum temperature to yield smaller MgO nanoparticles via both solid and liquid Aloe phytochemical extracts. These results can be attributed to the nucleation and growth of nanoparticles in a solution, where the particle morphology is affected by supersaturation, nucleation, and growth rates, colloidal stability, recrystallization, and aging process. Supersaturation is determined by the solution temperature, which has a predominant influence on the particle morphology and possesses high Gibbs free energy. The tendency of a system to lower its Gibbs free energy is the driving force in the processes of nucleation and growth of particles. At low temperature, a higher saturation result in a greater reduction of Gibbs free energy and this energy reduction will become the increased surface energy facilitating the continued nucleation to form smaller sized particles. The high temperature prefers a rapid hydrolysis and leads to a high supersaturation, which in turn lead to the formation of several nuclei. Thus, a rapid growth of nanoparticles size was observed at high temperatures [35], which has led to the selection of 60°C as the optimum reaction temperature for smaller sized MgO nanoparticle formation.

3.4. Optimization of precursor concentration

Figure 4 (D) shows the effect of different precursor concentration on the average size of MgO nanoparticles. It can be noted that an increment in the precursor concentration, increases the size of the nanoparticle. Further, the higher concentration of metal precursors enhances the size of the nanoparticle, independent of its nature. When liquid and solid Aloe latex powder extract were used with lower precursor concentration of 0.01 M, smaller average particle size of 85.05 nm and 72.3 nm was obtained, respectively. It was observed that lower precursor concentrations generally result in the formation of nanoparticles with smaller diameters, as precursors affect the crystallinity of the particles. Likewise, the higher precursor concentration may result in a higher nuclei density, which may lead to the formation of larger primary particles. These primary particles may lead to the agglomeration along the axis of the nanoparticle, which increases their size [36].

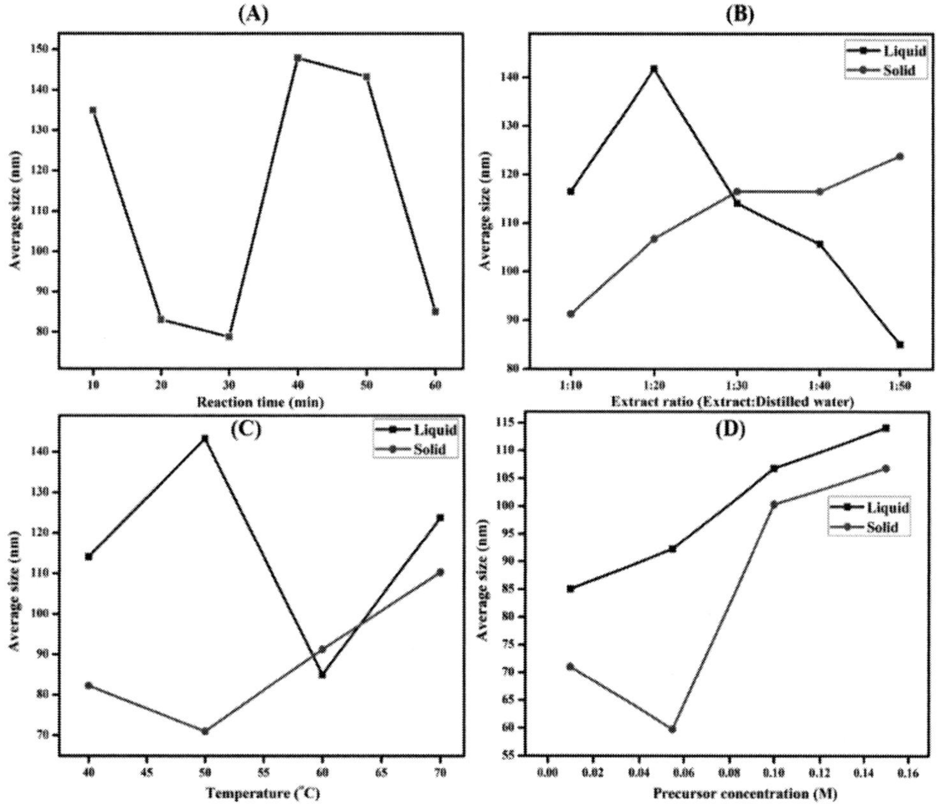

Figure 4. Optimization of MgO nanoparticle synthesis parameters (A) Reaction time, (B) Extract ratio, (C) Temperature and (D) Precursor concentration

3.5. Optimized parameters

The parameters such as reaction time, extract ratio, temperature and precursor concentration are optimized via OFAT approach. The optimized parameters are 30 min of reaction time, 60°C of reaction temperature and 0.01 M of precursor concentration for both Aloe (solid and liquid) extracts to yield smaller MgO nanoparticles. The optimized liquid extract to aqueous ratio is 1:50, whereas 1:10 ratio of solid extract: water was selected to obtain smaller MgO nanoparticles. Thus, it is evident that liquid and solid extract possesses significantly distinct phytochemicals, which alter their mechanism of nanoparticle formation. The MgO nanoparticles were synthesized using these optimized parameters for further characterizations.

3.6. Surface plasmon resonance analysis of MgO nanoparticles

Figure 5 (A) and (B) displays the UV and visible light absorbance spectra of the particles formed using liquid and solid Aloe latex powder extracts, respectively, magnesium nitrate as a precursor along with optimized parameters. The peak at ~280 nm in **Figure 5 (A) and (B)** represents the existence of oxide molecule in the particle. A miniscule peak at ~540 nm in **Figure 5 (A)** represents the presence of metal in the Aloe liquid extract yielded particles, which is absent in the particles formed by Aloe latex powder solid extract. Thus, it is evident that magnesium oxide is formed as colloidal particles in the liquid phytochemical extracts of Aloe latex powder. The results were similar to a previous study in which MgO nanoparticles are prepared via *Amaranthus tricolor* leaf extract and the optical absorbance of those nanosized particles are observed in the UV region (320 nm) [37]. The absence of peak in the visible region and the increase in a peak at 380 nm in **Figure 5 (B)** showed that hydroxide may be formed instead of oxide particles [38], while using Aloe solid latex powder extract. Hence, thermogravimetric analysis was performed to confirm the presence of metal hydroxide in the sample obtained from Aloe solid extracts.

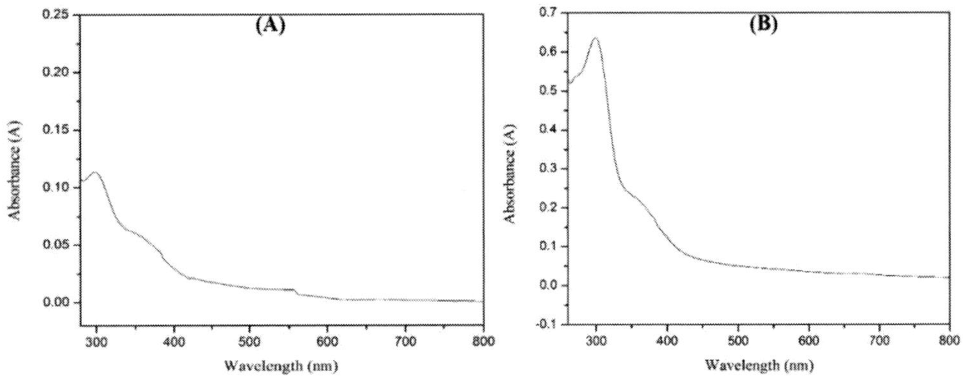

Figure 5. UV-Visible spectrum of (A) liquid extract and (B) solid extract of Aloe latex powder mediated MgO nanoparticles

3.7. Thermogravimetric analysis of solid Aloe extracts synthesized nanoparticles

Figure 6 shows the TG-DSC data of particles yielded by the Aloe latex powder solid extract. The weight loss percentage and its corresponding heat loss or gain will be beneficial in identifying the molecules present in the samples [39]. In the first degradation stage from 130 to 240°C, a weight loss of ~54% can be attributed to the evaporation of crystallizing water together with the impurities from phytochemicals present in the sample. In this process, most of the hydrogen molecule together with oxygen evaporates, which will lead to the transformation of magnesium hydroxide into MgO. The thermal degradation of transformation starts at 200°C, followed by the occurrence of melting transition process. The degradation curve is not that obvious as phytochemicals' impurities may have reduced the peak of crystallizing water evaporation. The second stage of weight loss which is approximately 4% from 230 to 410 °C was due to the breaking of bonds and transformation of MgO into another crystallite form. In literature, the second stage weight loss of 3.5% from 250 to 400°C is attributed to the crystallization of MgO nanoparticles [40]. The third stage weight loss of ~24% at 420 °C corresponds to the phase change of the MgO. There is a deflection in the DSC curve, corresponds to the weight loss as this may be due to the occurrence of phase change [41]. There is no further significant weight loss was observed at high temperature after 480 °C. Thus, it is evident that MgO nanoparticles can be formed via calcination and 350°C for 4 hours was selected as optimum calcination temperature for the MgO formation from TG-DSC curve.

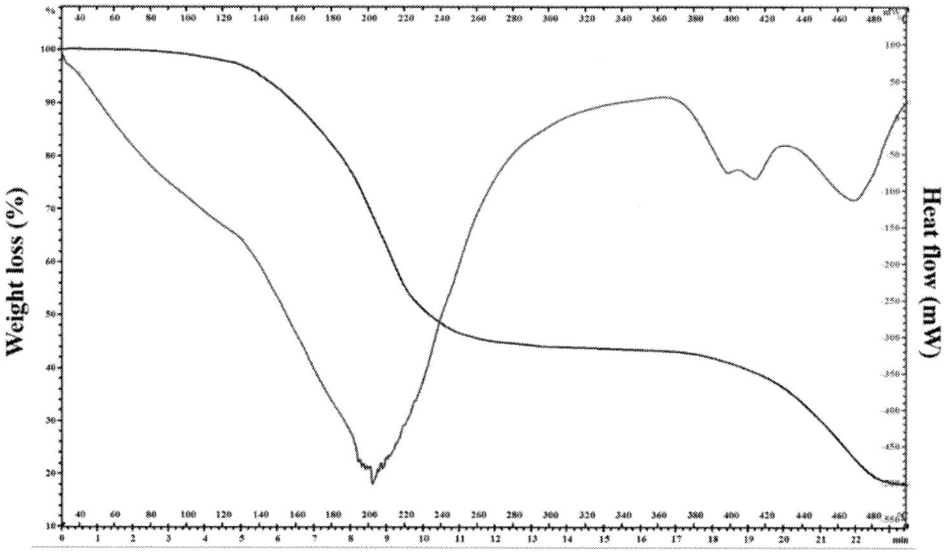

Figure 6. Thermogravimetric analysis of magnesium hydroxide nanoparticles synthesized via Aloe latex powder solid extract

3.8. Functional group analysis of optimized MgO nanoparticles

Figure 7 (A) and (B) shows the FTIR spectra of solid and liquid Aloe latex extract, respectively, whereas Figure 7 (C) and (D) displays the FITR spectra of MgO nanoparticles formed via the respective extracts. The FTIR spectral peaks present in the phytochemical extracts were depicted in the **Table 2**. The peaks at 3321.13, 2918.45, 2162.16, 1731.84, 1614.8, 1371.62, 1242.35, and 1316.45 cm^{-1} in solid extract indicate the existence of N-H stretching of amines, C-H stretches of alkanes, -C≡C- stretches of alkynes, C=O stretches of aliphatic esters, N-H bends of amines, N-O symmetric stretching of nitroalkane, the C-N stretch of aliphatic amines, and C-O stretches of alcohols, carboxylic acids, esters and ethers. Similarly, the peaks at 3277.79, 2929.43, 1715, 1372.18, 1242.02 and 1024.04 cm^{-1} in liquid extract indicate the existence of O-H stretches of the intramolecular hydrogen bond or phenols, C-H stretches of alkanes, C=O stretches ketones or saturated aliphatic, N-O symmetric stretches of nitroalkane, the C-N stretch of aliphatic amines and C-O stretching of alcohols, carboxylic acids, esters and ethers as reported by Kassama et al. (2015) [42]. The presence of these vibrational groups indicates the presence of phytochemicals such as flavanol, quercetin, Kaempeferol, myricetin and fisetin in both extracts [16]. However, the changes in the peak intensity and peak shifts indicates that the quantity and combination of these phytochemical are different in solid and liquid extracts.

In Figure 7 (C) and (D), most of the peaks present in Figure (A) and (B) are absent, which represents the modification of functional group for the formation of MgO. Peaks corresponding to alcohol group, alkyne, primary amine and carboxylic acid are present in the MgO nanoparticles samples which may be formed from the phytochemicals and act as capping agent for the stabilization of nanosized particles. The peaks at 1065.82 and 872.64 Cm^{-1} is due to the presence of Mg-O vibrational mode in the solid extract yielded particles [26], whereas the peak at 872.64 Cm^{-1} is miniscule and 1065.82 Cm^{-1} is too narrow in liquid extract yielded particles [43]. The vibration group related to metal oxide in the liquid extract yielded particles may not be detected by the FTIR spectroscopy as it is suspended in the liquid. **Table 3** is the comparison of functional groups present in the extract and MgO nanoparticles. It is evident from Table 3 that the transformation of functional groups in the phytochemicals, during the synthesis is the significant factor that contributes to the formation of nanoparticles.

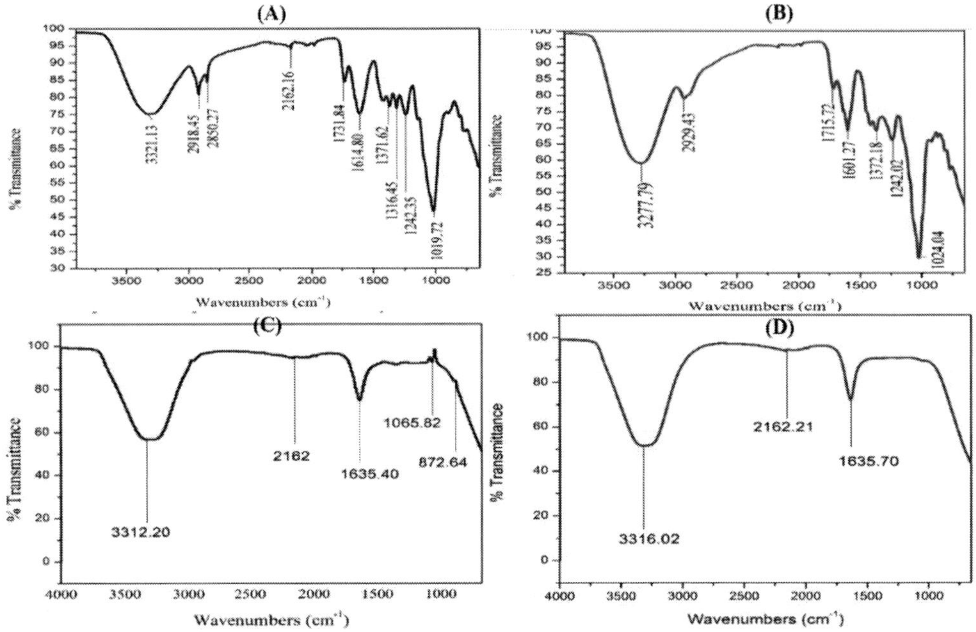

Figure 7. FTIR spectra of (A) Aloe latex solid extract, (B) Aloe latex liquid extract, (C) MgO nanoparticles from solid extract and (D) MgO nanoparticles from liquid extract

Table 2. FTIR peak values and functional groups of solid and liquid extracts of *Aloe* latex powder

Solid extract		Liquid extract	
Peak values	**Functional groups**	**Peak values**	**Functional groups**
3321.13	N-H (amines)	3277.79	O-H stretch (alcohols, phenols)
2918.45	C-H stretch (alkanes)	2929.43	C-H stretch (alkanes)
2850.27	C-H stretch (alkanes)	1715.00	C=O stretch (ketones, saturated aliphatic)
2162.16	-C≡C- stretch (alkynes)	1601.27	N-H (amines)
1731.84	C=O stretch (aliphatic esters)	1372.18	N-O asymmetric stretch (nitroalkane)
1614.80	N-H(amines)	1242.02	C-N stretch (aliphatic amines)
1371.62	N-O asymmetric stretch (nitroalkane)	1024.04	C-O stretch (alcohols, carboxylic acids, esters, ethers)
1316.45	C-O stretch (alcohols, carboxylic acids, esters, ethers)		-
1242.35	C-N stretch (aliphatic amines)		
1019.72	C-O stretch (aliphatic esters)		

Table 3. Comparison of functional groups in the extracts and phytochemical extract synthesized MgO nanoparticles

Frequency (cm^{-1})	Bond	Functional groups	ALSE	ALLE	ALSE synthesized MgO	ALLE synthesized MgO

3500-3200	O-H stretch, H bonded	Alcohols, phenols	-	+	-	-
3400-3250	N-H stretch	Amines	+	+	+	+
3000-2850	C-H stretch	Alkanes	+	+	-	-
2260-2100	-C≡C- stretch	Alkynes	+	-	+	+
1750-1735	C=O stretch	Esters, saturated aliphatic	+	-	-	-
1715	C=O stretch	Ketones, saturated aliphatic	-	+	-	-
1650-1580	N-H bend	1° amines	+	+	+	+
1360-1290	N-O symmetric stretch	Nitro compounds	+	+	-	-
1250-1020	C-N stretch	Aliphatic amines	+	+	+	-
1320-1000	C-O stretch	Alcohols, carboxylic acids, esters, and ethers	+	+	-	-
850-550	C-Cl stretch	Alkyl halides	-	-	+	-

+ = present, - = absent, ALSE = Aloe latex solid extract, ALLE = Aloe latex liquid extract

3.9. Morphology analysis of optimized MgO nanoparticles

Figure 8 shows the transmission electron micrographs of Aloe latex liquid extract (Figure 8 (A), (B) and (C)) synthesized MgO nanoparticles with optimized parameters and Aloe latex solid extract (Figure 8 (D), (E) and (F)) synthesized MgO nanoparticles with optimized parameters and calcinated at 350°C for 4 h. All the images reveal that the morphology of MgO nanoparticles formed via both Aloe extracts were spherical in shape. Figure 8 (A) show that the liquid extract synthesized MgO nanoparticles are agglomerated into a micro-sized spherical structure in the copper mesh grid used for TEM analysis. This emphasizes that the nanoparticles are needed to be dispersed in a more relevant solvent to de-agglomerate them, before using them for further applications. However, the size of the nanoparticle at 50 nm scale is noted to be in the range between 25-35 nm. The micrograph of solid extract synthesized MgO nanoparticles is different from the liquid extract synthesized counterparts, as the crude solid extract has led to the formation of translucent nanoparticles, which makes it tedious for the electrons to transmit through the particles. Thus, centered-dark field was used to analyze solid extract synthesized and calcinated MgO nanoparticles. The calcination provides has led to minimum agglomeration and the size of the nanoparticles are in the range of 39-60 nm. It is noteworthy that the liquid extract synthesized MgO nanoparticles are highly beneficial than solid extract counterparts, with respect to yield, monodispersity, rapid synthesis and less energy requirement aspects. The size of the liquid extract synthesized MgO nanoparticles are similar to our previous reports using different plant leaf extracts [37, 43, 44] and are smaller than the MgO nanoparticles fabricated via chemical synthesis approaches [26, 45, 46].

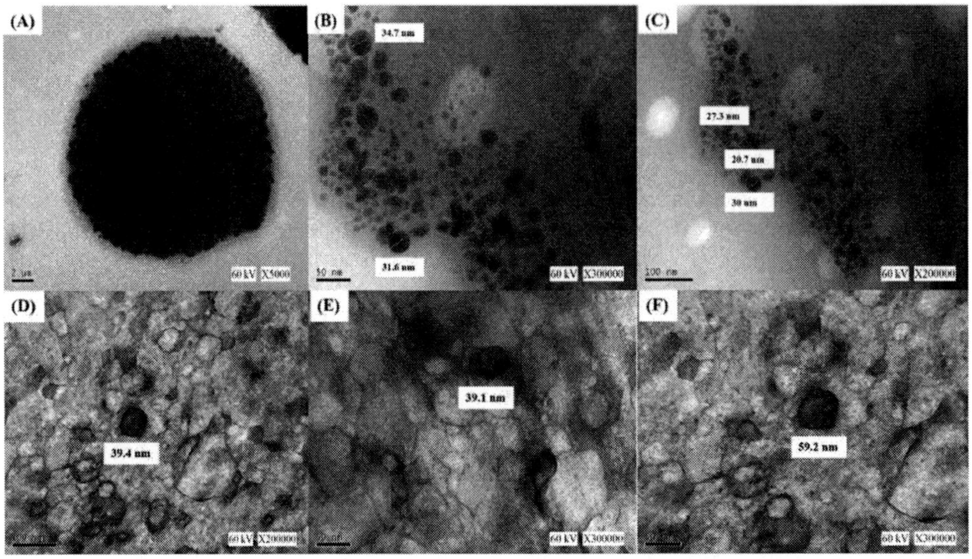

Figure 8. Transmission electron micrograph of liquid extract synthesized (A), (B) and (C) and solid extract synthesized MgO nanoparticles

4. Conclusion

The present work reports the synthesis of MgO nanoparticles via phytochemicals that are extracted in a novel method from *Aloe barbadensis* latex and magnesium nitrate as precursor. The latex is a pseudo-leaf, which is a unique structure with water content in Aloe and hence, two types are phytochemical extracts such as liquid and solid are obtained from them. Both were utilized for the fabrication of MgO nanoparticles and the parameters such as reaction time, extract ratio (dilution), temperature and precursor concentration. The optimized parameters are 30 min of reaction time, 60°C of reaction temperature and 0.01 M of precursor concentration for both Aloe (solid and liquid) extracts to yield smaller MgO nanoparticles. The optimized liquid extract to aqueous ratio is 1:50, whereas 1:10 ratio of solid extract: water was selected to obtain smaller MgO nanoparticles. The solid extract yielded magnesium hydroxide instead of MgO, which was evident from UV-visible spectroscopy analysis and thus, the calcination was performed at 350°C (4 h) for those samples, where the calcination temperature was selected from the thermogravimetric analysis results. Further, FTIR analysis confirmed the presence of MgO nanoparticles in both liquid and solid extracts synthesized samples along with alcoholic and carboxylic acid functional groups from phytochemicals that acts as reducing and stabilizing agent for the nanoparticle formation. Furthermore, TEM analysis confirmed that the spherical shaped MgO nanoparticles are formed from both extracts, where liquid extract yielded 25-35 nm sized particles and solid extract yielded 39-60 nm sized particles. Thus, the phytochemical extracts from *Aloe barbadensis* are proven to be beneficial in smaller sized MgO nanoparticles, which will be useful for various biomedical and pharmaceutical applications.

References

[1] Rukh, S., et al., *Antibacterial activity of magnesium oxide nanostructures prepared by hydrothermal method.* Asian Journal of Nanosciences and Materials, 2019. **2**(4): p. 425-430.

[2] Sharmila, G., et al., *Green fabrication, characterization of Pisonia alba leaf extract derived MgO nanoparticles and its biological applications.* Nano-Structures & Nano-Objects, 2019. **20**: p. 100380.

[3] Pudukudy, M., et al., *One-pot sol-gel synthesis of MgO nanoparticles supported nickel and iron catalysts for undiluted methane decomposition into COx free hydrogen and nanocarbon.*

Applied Catalysis B: Environmental, 2017. **218**: p. 298-316.

[4] Ding, Y., et al., *Nanoscale magnesium hydroxide and magnesium oxide powders: control over size, shape, and structure via hydrothermal synthesis.* Chemistry of materials, 2001. **13**(2): p. 435-440.

[5] Rao, K.G., et al., *Structural properties of MgO nanoparticles: synthesized by co-precipitation technique.* International Journal of Science and Research, 2014. **3**(12): p. 43-46.

[6] Aslani, A., et al., *Solvothermal synthesis, characterization and optical properties of ZnO, ZnO–MgO and ZnO–NiO, mixed oxide nanoparticles.* Applied Surface Science, 2011. **257**(11): p. 4885-4889.

[7] Mangalampalli, B., N. Dumala, and P. Grover, *Toxicity assessment of magnesium oxide nano and microparticles on cancer and non-cancer cell lines.* The Nucleus, 2019: p. 1-15.

[8] El-Sayyad, G.S., F.M. Mosallam, and A.I. El-Batal, *One-pot green synthesis of magnesium oxide nanoparticles using Penicillium chrysogenum melanin pigment and gamma rays with antimicrobial activity against multidrug-resistant microbes.* Advanced Powder Technology, 2018. **29**(11): p. 2616-2625.

[9] Raliya, R., et al., *Synthesis of MgO nanoparticles using Aspergillus tubingensis TFR-3.* Journal of Bionanoscience, 2014. **8**(1): p. 34-38.

[10] Sharma, G., R. Soni, and N.D. Jasuja, *Phytoassisted synthesis of magnesium oxide nanoparticles with Swertia chirayaita.* Journal of Taibah University for Science, 2017. **11**(3): p. 471-477.

[11] Suresh, J., et al., *Green synthesis and characterization of hexagonal shaped MgO nanoparticles using insulin plant (Costus pictus D. Don) leave extract and its antimicrobial as well as anticancer activity.* Advanced Powder Technology, 2018. **29**(7): p. 1685-1694.

[12] Sushma, N.J., et al., *Facile approach to synthesize magnesium oxide nanoparticles by using Clitoria ternatea—characterization and in vitro antioxidant studies.* Applied Nanoscience, 2016. **6**(3): p. 437-444.

[13] Reynolds, T., *Aloes: the genus Aloe.* 2004: CRC press.

[14] Grindlay, D. and T. Reynolds, *The Aloe vera phenomenon: a review of the properties and modern uses of the leaf parenchyma gel.* Journal of ethnopharmacology, 1986. **16**(2-3): p. 117-151.

[15] Sánchez-Machado, D.I., et al., *Aloe vera: Ancient knowledge with new frontiers.* Trends in Food Science & Technology, 2017. **61**: p. 94-102.

[16] Kumar, A., et al., *Prevalence and severity of periodontal diseases in type 2 diabetes mellitus of Bareilly region (India).* Int J Med Sci Public Health, 2013. **2**(1): p. 77-83.

[17] Ozsoy, N., E. Candoken, and N. Akev, *Implications for degenerative disorders: Antioxidative activity, total phenols, flavonoids, ascorbic acid, β-carotene and β-tocopherol in Aloe vera.* Oxidative medicine and cellular longevity, 2009. **2**(2): p. 99-106.

[18] Tippayawat, P., et al., *Green synthesis of silver nanoparticles in aloe vera plant extract prepared by a hydrothermal method and their synergistic antibacterial activity.* PeerJ, 2016. **4**: p. e2589.

[19] Radha, M.H. and N.P. Laxmipriya, *Evaluation of biological properties and clinical effectiveness of Aloe vera: A systematic review.* Journal of traditional and complementary medicine, 2015. **5**(1): p. 21-26.

[20] Pandey, A. and S. Singh, *Aloe Vera: A Systematic Review of its Industrial and Ethno-Medicinal Efficacy.* International Journal of Pharmaceutical Research & Allied Sciences, 2016. **5**(1).

[21] Sangeetha, G., S. Rajeshwari, and R. Venckatesh, *Green synthesis of zinc oxide nanoparticles by aloe barbadensis miller leaf extract: Structure and optical properties.* Materials Research Bulletin, 2011. **46**(12): p. 2560-2566.

[22] Kumar, P.P.N.V., et al., *Green synthesis of copper oxide nanoparticles using Aloe vera leaf extract and its antibacterial activity against fish bacterial pathogens.* BioNanoScience, 2015. **5**(3): p. 135-139.

[23] Rao, K.G., et al., *Green synthesis of TiO2 nanoparticles using Aloe vera extract.* Int. J. Adv. Res. Phys. Sci, 2015. **2**(1A): p. 28-34.

[24] Mehrabian, S., et al., *A STUDY OF THE ANTIMUTAGENIC EFFECTS OF DIFFERENT*

EXTRACTS OF ALOE VERA LEAF GEL AND LATEX USING AMES TEST. ARAK MEDICAL UNIVERSITY JOURNAL (AMUJ), 2012. **15**(2 (61)): p. 100-106.

[25] Hu, K. and D.J. McClements, *Fabrication of biopolymer nanoparticles by antisolvent precipitation and electrostatic deposition: Zein-alginate core/shell nanoparticles.* Food Hydrocolloids, 2015. **44**: p. 101-108.

[26] Jeevanandam, J., Y.S. Chan, and M.K. Danquah, *Calcination-Dependent Morphology Transformation of Sol-Gel- Synthesized MgO Nanoparticles.* Chemistry select, 2017. **2**(32): p. 10393 - 10404.

[27] Vatsha, B., et al., *Effects of precipitation temperature on nanoparticle surface area and antibacterial behaviour of Mg (OH) 2 and MgO nanoparticles.* Journal of Biomaterials and Nanobiotechnology, 2013. **4**(04): p. 365.

[28] Gopalakrishnan, D., D. Damien, and M.M. Shaijumon, *MoS2 quantum dot-interspersed exfoliated MoS2 nanosheets.* ACS nano, 2014. **8**(5): p. 5297-5303.

[29] Sahu, P.K., P.K. Sahu, and D.D. Agarwal, *Role of basicity and the catalytic activity of KOH loaded MgO and hydrotalcite as catalysts for the efficient synthesis of 1-[(2-benzothiazolylamino) arylmethyl]-2-naphthalenols.* RSC Advances, 2015. **5**(85): p. 69143-69151.

[30] Christensen, L., et al., *Biosynthesis of silver nanoparticles using murraya koenigii (curry leaf): an investigation on the effect of broth concentration in reduction mechanism and particle size.* Adv Mat Lett, 2011. **2**(6): p. 429-434.

[31] Ali, K., et al., *Aloe vera extract functionalized zinc oxide nanoparticles as nanoantibiotics against multi-drug resistant clinical bacterial isolates.* Journal of colloid and interface science, 2016. **472**: p. 145-156.

[32] Ismail, A.A., et al., *Multilayered ordered mesoporous platinum/titania composite films: does the photocatalytic activity benefit from the film thickness?* Journal of Materials Chemistry, 2011. **21**(21): p. 7802-7810.

[33] Ahmadi, M., M.R. Ghasemi, and H.H. Rafsanjani, *Study of different parameters in tio2 nanoparticles formation.* Journal of Materials Science and Engineering, 2011. **5**(1): p. 87.

[34] Al-Qubaisi, M.S., et al., *Cytotoxicity of nickel zinc ferrite nanoparticles on cancer cells of epithelial origin.* International journal of nanomedicine, 2013. **8**: p. 2497.

[35] Ozel, F., H. Kockar, and O. Karaagac, *Growth of iron oxide nanoparticles by hydrothermal process: effect of reaction parameters on the nanoparticle size.* Journal of Superconductivity and Novel Magnetism, 2015. **28**(3): p. 823-829.

[36] Muangban, J. and P. Jaroenapibal, *Effects of precursor concentration on crystalline morphologies and particle sizes of electrospun WO3 nanofibers.* Ceramics International, 2014. **40**(5): p. 6759-6764.

[37] Jeevanandam, J., Y. San Chan, and Y.H. Ku, *Aqueous Eucalyptus globulus leaf extract-mediated biosynthesis of MgO nanorods.* Applied Biological Chemistry, 2018. **61**(2): p. 197-208.

[38] Jeevanandam, J., Y.S. Chan, and M.K. Danquah, *Effect of pH variations on morphological transformation of biosynthesized MgO nanoparticles.* Particulate Science and Technology, 2019.

[39] Mallick, L., S. Kumar, and A. Chowdhury, *Thermal decomposition of ammonium perchlorate—A TGA–FTIR–MS study: Part I.* Thermochimica acta, 2015. **610**: p. 57-68.

[40] Jaison, J., S. Balakumar, and Y. Chan. *Sol–Gel synthesis and characterization of magnesium peroxide nanoparticles.* in *IOP Conference Series: Materials Science and Engineering.* 2015. IOP Publishing.

[41] Topnani, N., S. Kushwaha, and T. Athar, *Wet synthesis of copper oxide nanopowder.* International Journal of Green Nanotechnology: Materials Science & Engineering, 2010. **1**(2): p. M67-M73.

[42] Kassama, L.S., A.J. Kuponiyi, and T. Kukhtareva, *Comparative effects of aloe vera (aloe barbadensis) water vs ethanol extracts on the physicochemical properties and stability of silver nanoparticles synthesize.* American International Journal of Contemporary Research, 2015. **5**(2): p. 30-39.

[43] Jeevanandam, J., Y.S. Chan, and M.K. Danquah, *Biosynthesis and characterization of MgO*

nanoparticles from plant extracts via induced molecular nucleation. New Journal of Chemistry, 2017. **41**: p. 2800-2814.

[44] Hii, S.Y., J. Jeevanandam, and Y.S. Chan, *Plant mediated green synthesis and nanoencapsulation of MgO nanoparticle from Calotropis gigantea: Characterisation and kinetic release studies.* Inorganic and Nano-Metal Chemistry, 2019. **48**(12): p. 620-631.

[45] Jeevanandam, J., Y.S. Chan, and M.K. Danquah, *Effect of Gelling Agent and Calcination Temperature in Sol–Gel Synthesized MgO Nanoparticles.* Protection of Metals and Physical Chemistry of Surfaces, 2019. **55**(2): p. 288-301.

[46] Wong, C.W., et al., *Response Surface Methodology Optimization of Mono-dispersed MgO Nanoparticles Fabricated by Ultrasonic-Assisted Sol–Gel Method for Outstanding Antimicrobial and Antibiofilm Activities.* Journal of Cluster Science, 2019: p. 1-23.

The effect of the spacing and angle arrangement of the droplet on the spreading area in simultaneous multiple droplets deposition

K Y Pau and A N Sung*

Faculty of Engineering and Science, Curtin University Malaysia, CDT 250, 98009 Miri, Sarawak, Malaysia

E-mail: sung.aun@curtin.edu.my

Abstract. Simultaneous multiple droplet deposition involves dispensing of multiple droplets at the same instant on the substrate. A newly defined parameter, arrangement angle is considered as one of the factor that influencing the coalescence of the multiple droplets. However, the effect of such factor on the coalescence is largely neglected in the past. The paper aims to investigate the effect of droplet centre-to-centre distance and arrangement angle on the spreading area of four droplet deposited simultaneously using simulation method. The project starts with single droplet deposition and following by simultaneous multiple droplet deposition. For single droplet deposition, the spreading area of 80% wt aqueous glycerin droplet on the glass substrate were obtained using simulation and experimental methods. A correlation function was introduced to match the simulation result with the experiment result. For simulations of multiple droplet deposition, three droplet centre-to-centre distances as 2.22 mm, 3 mm and 4 mm and three droplet arrangement angles as 90°, 110° and 120° were studied. The larger the droplet spacing, the larger the spreading area for arrangement angle of 90°. The spreading area are largest for spacing of 4 mm, followed by 2.22 mm and 3 mm for arrangement angle of 110° and 120°. The arrangement angle is proven as a factor that affect the spreading area.

1. Introduction

Droplet deposition are classified into vapour and liquid phases. Liquid-phase methods such as inkjet printing, spray coating and drop-casting involve deposition of multiple droplets on a solid substrate. Multiple-droplet drop-casting (MDDC) is simultaneous deposition of highly ordered droplets with the equal size and spacing on the substrate surface. The technique is a simpler and more controllable compared to spray coating in producing continuous thin films.

Controlling the single or multiple-droplet impingement process is important for deposition process. The deposition mechanism of single droplet has been studied in the past researches. When a droplet impacts with a solid surface, the droplet can spread, rebound or splash [1]. A droplet can spread over the surface without splashing and rebounding. The maximum spreading of droplet is affected by several parameters such as droplet initial diameter, surface tension, viscosity, impact velocity and wettability of the substrate surface. These parameters are related to Reynolds and Weber numbers [2]. Lee, Derome, Guyer and Carmeliet [3] found that viscosity and surface tension play a significant effect on maximum spreading ratio of the impinging droplet. Surface roughness and type of surface have minor effect on maximum spreading ratio when the impact velocity is low. When impact velocity increases, spreading velocity increases as well, which results in a larger maximum spread [4].

When two or more droplets overlap, coalescence of droplets occurs which forms a larger droplet or thin film. The neighbouring droplets must be in contact and coalesce during deposition process to prevent discontinuities in dispensing pattern [5]. There are several factors that affect the coalescence of droplets, which include the parameters affecting deposition mechanism of single droplet and the arrangement of the droplets such as centre-to centre distance and the arrangement angle between the droplets. Lee, Kim, Chandra and Yoon [6] had conducted experiment to study the effect of viscosity on the coalescence of sessile droplets. Graham, Farhangi and Dolatabadi [7] studied the coalescence of falling droplet and sessile droplet on solid surface with different surface wettability. It was found that the maximum spreading length of the coalesced droplet decreased when the surface hydrophobicity increased.

Several researchers have studied the effect of droplet spacing on the coalescence of droplets. Castrejón-Pita, Betton, Kubiak, Wilson and Hutchings [8] studied the effect of droplet spacing on the coalescence of two droplets on a substance. A droplet was deposited adjacent to another droplet resting on the substrate. Dalili A, Chandra, Mostaghimi, Fan and Simmer [9] conducted experiment to investigate the effect of droplet spacing on the coalescence of glycerin-water droplets. The droplets were deposited sequentially on a flat steel plate at a frequency of 1 Hz to form a straight line or square pattern. Even though abovementioned researches studied the effect of droplet spacing on the coalescence of droplets, the droplets are deposited subsequently instead of at the same instant.

Simultaneous multiple droplet deposition involves dispensing of multiple droplets at the same instant to form a pattern. Eslamian and Soltani-Kordshuli [10] studied the surface roughness of poly(3,4-ethylenedioxythiophene)-polystyrene sulfonate (PEDOT:PSS) drop-cast films upon drying on a bare glass and a fluorine-doped tin oxide (FTO)-coated glass with different imposed vibration time. The experiment involved equally spaced and equally sized droplets deposited onto the surface at the same time. Thus, there are limited study on the multiple droplet deposition that droplets are deposited at the same instant.

Besides droplet spacing, a newly defined parameter, arrangement angle is considered as an influencing factor to the coalescence. Figure 1 illustrates the arrangement angle θ. The effect of arrangement angle on the coalescence is neglected in the past.

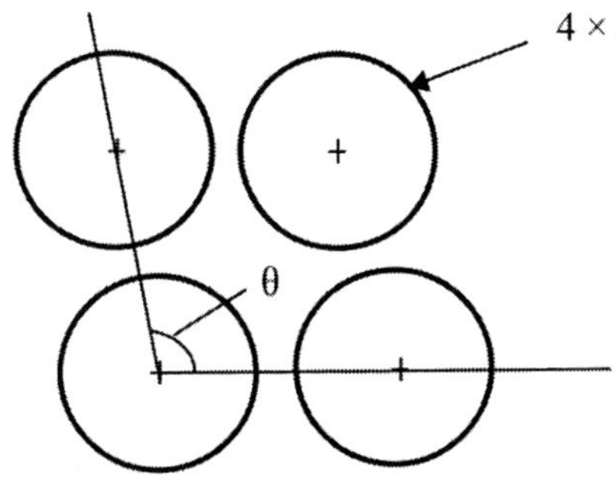

Figure 1. The arrangement angle.

The droplet deposition and coalescence phenomena were realized using numerical simulation method like marker-and-cell [11], Volume of Fluid [12], Lattice Boltzmann Method [13], Smoothed Particle Hydrodynamics [14] and Coupled Eulerian Lagrangian [15]. For Fluid Structure Interaction, Coupled Eulerian Lagrangian (CEL) method is a simple approach to model the interaction between fluid and solid. In Johansson and Ollar [15], a model of the sealing process was generated by applying CEL approach and the applicability of the method was discussed. For droplet deposition, Zhu, Kamnis and Gu [16] conducted a numerical study on impingement of molten and semi-molten ceramic using CEL method.

This research project focuses on how droplets centre-to-centre distance and arrangement angle affect the dispensing area through simulation study. Only the material properties of 80% wt aqueous glycerin droplet and glass substrate are input to the simulation to fix the fluid viscosity and the surface wettability. Four droplets are deposited simultaneously onto the glass substrate. The droplet centre-to-centre distance of 2.22 mm, 3 mm and 4 mm and arrangement angle of 90°, 110° and 120° were investigated. Nine sets of combination of the variables were investigated. For each combination, eight spreading area data were obtained for eight different spreading time range from 0-0.1 second. Thus, a total of 72 data were obtained in the study.

IOP Conf. Series: Materials Science and Engineering **943** (2020) 012031 doi:10.1088/1757-899X/943/1/012031

2. Methodology

The research methodology is divided into two stages, which are single droplet deposition and simultaneous multiple droplet deposition. The methodology of the research is summarized in figure 2. For single droplet deposition, experiment and simulation works were conducted. The experimental data were obtained to verify the simulation data. After verification, simulations of four aqueous glycerin droplets impacting and coalescing with each other on the glass substrate were conducted.

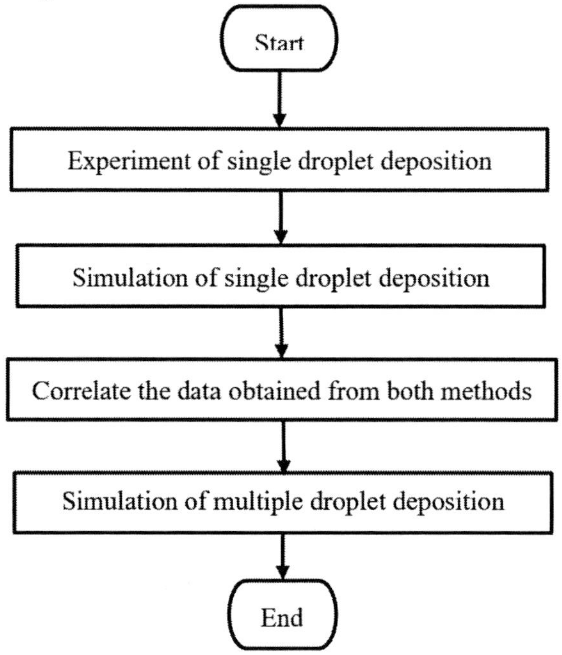

Figure 2. Flow chart of methodology.

2.1. Experiment on Single Droplet Deposition

The experiment was conducted by impacting an 80% wt aqueous glycerin droplet on a glass substrate. The density and viscosity of the aqueous glycerin droplet are 1205.45 kg/m^3 and viscosity of 0.047 kg/m s respectively [17]. The sound velocity is assumed to be similar to that of pure glycerin, which is 1920 m/s. The apparatus set-up of the experiment is shown in figure 3. A flat rectangular glass plate was placed on a horizontal platform of a retort stand. A microlitre syringe filled with aqueous glycerin solution was fixed on the clamp above the glass substrate. The needle tip of the syringe was positioned 20 mm away from the glass substrate. An aqueous glycerin droplet was formed and separated from the needle tip of the syringe once the droplet size was sufficiently large. The high speed camera Photron FASTCAM Mini UX50 was used to capture the droplet deposition. The high speed camera FASTCAM Mini UX50 has an operational pixel of 1280×1024 and exposure time frame per second (fps) can be increased up to 160000fps with reduced pixel. The optical lens is Nikkor Lens, Model: AF Zoom-Nikor 24-85mm f/2.4-4D IF. The experiment was repeated three times to obtain the average results.

The recordings of the deposition process were analyzed using Photron FASTCAM Viewer 4.0 software. The image of droplet before impacting the substrate and the image of droplet deposited on the substrate at the spreading time of 0.1 second are shown in figure 4(a) and figure 4(b) respectively. The scale of the image was calculated by comparing the distance between the needle tip of the microsyringe and the glass substrate surface in the image to the actual distance of 20 mm. The droplet initial diameter was measured based on the falling droplet as shown in figure 4(a). The droplet spreading diameter was measured from the image of droplet on the substrate. Deposited area is calculated using the spreading diameter by assuming the droplet spreads in circular pattern.

2.2. Simulation of Single Droplet Deposition

Simulia Abaqus v6.14 Student Edition software was used to apply finite element method to simulate the droplet impingement behavior with respect to time during impact with glass substrate. CEL method was used to create a model of single droplet impingement on solid surface. The steps of simulation of single droplet deposition is summarized in figure 5. There are several assumptions are made to simplify the simulation. First, the aqueous glycerin droplet is assumed to remain in spherical shape before it contacts with the glass substrate surface. It is assumed that glycerin droplets have zero initial velocity at the tip of the microsyringe and the effect of air resistance is negligible. The droplet is assumed to spread in circular pattern in deposition.

The 3D models of droplet, glass substrate and Eulerian instance were generated using Abaqus/CAE in Step 1. A hollow sphere with diameter of 2.2 mm was constructed as a droplet reference instance. The diameter of the sphere is equal to the average droplet initial diameter obtained from the experiment. A cuboid of 6 mm × 6 mm × 0.01 mm was created as the glass substrate. A cuboid of 6 mm × 6 mm × 3 mm was generated as an Eulerian instance that defines the fluid region. The Eulerian instance is required to be large enough to cover the glass substrate and the droplet reference instance.

In Step 2, the material properties were assigned to the parts created in Step 1. The substrate was assigned with glass material properties, such as density of 2500 kg/m^3, Elasticity modulus of 50 GPa and Poisson's ratio of 0.22. Assignment of material properties is not required for the droplet reference instance since the part only functions as a container to be filled with aqueous glycerin solution and the instance will be suppressed during the analysis. For 80% wt aqueous glycerin solution, the Eulerian instance was assigned with density of 1205.45

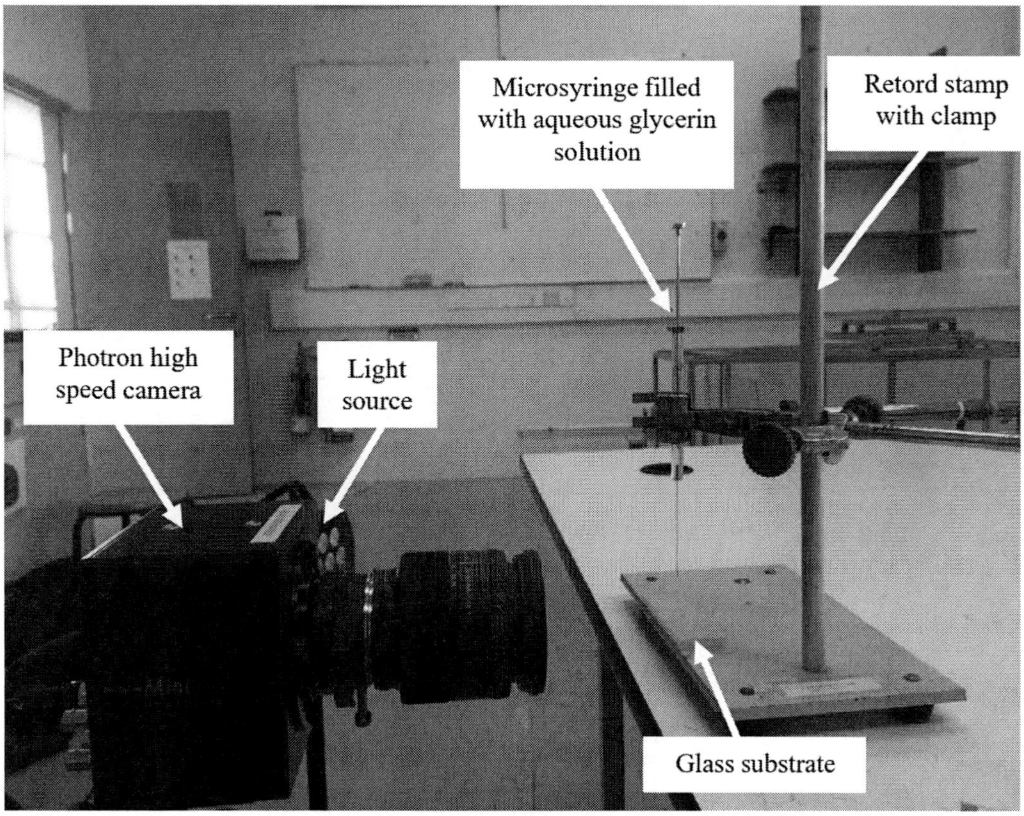

Figure 3. Apparatus set up of the experiment.

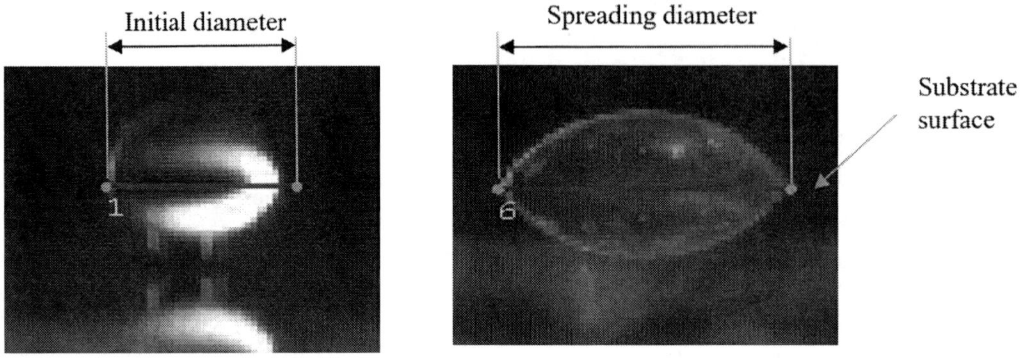

(a) (b)

Figure 4. (a) The falling droplet (b) The droplet on the substrate at the spreading time of 0.1 second.

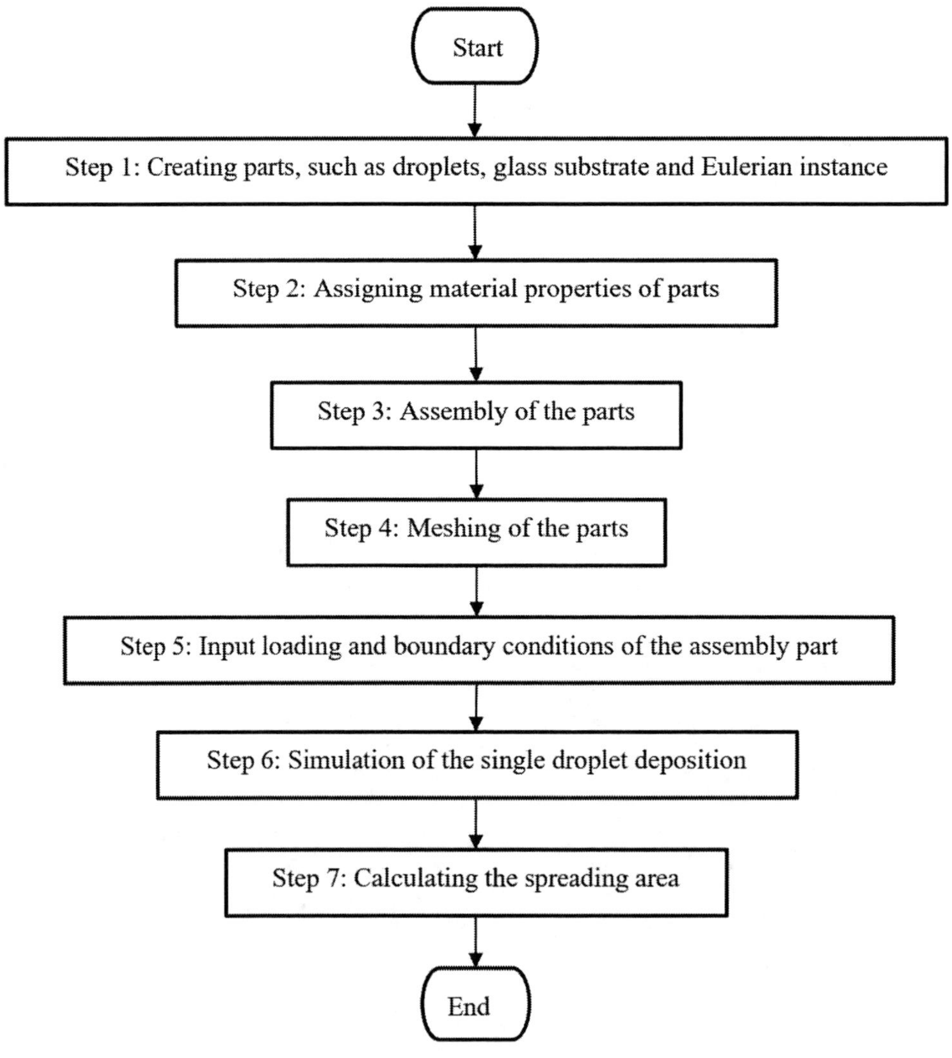

Figure 5. Flow chart of simulation of single droplet deposition.

kg/m^3, viscosity of 0.047 kg/m.s, speed of sound of 1920 m/s and defined with an Equation of State (EOS). In Abaqus, linear $U_s - U_p$ Hugonoit form is used to express the equations of state. The linear Hugonoit slope coefficient, s and material constant, Γ_o were both set to zero.

Step 3 involves the assembly of the three parts, such as droplet reference instance, glass substrate and Eulerian instance as shown in figure 6. The droplet reference instance and the glass substrate were positioned

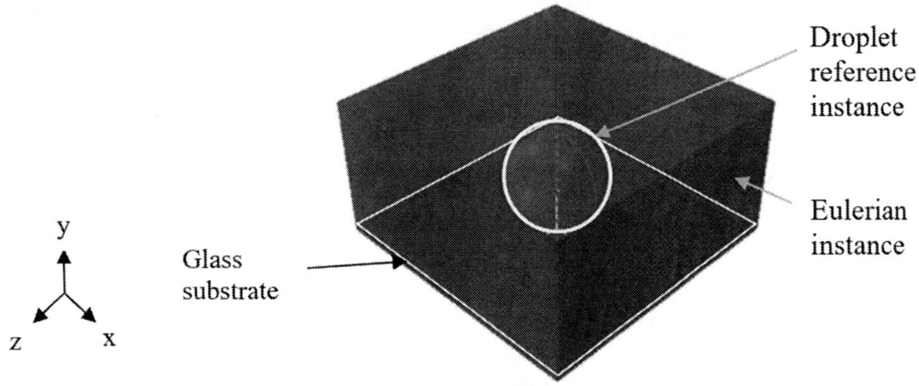

Figure 6. Assembly of droplet reference instance, glass substrate and Eulerian instance.

within the Eulerian instance to ensure that the fluid material can interact with the glass surface. The droplet reference instance and glass substrate are positioned 0.1 mm apart. The initial velocity of the droplet reference instance was calculated as the impact velocity V using the equation $V = \sqrt{2gh}$ where g is the gravitational acceleration and h is the falling distance. The impact velocity was calculated to be 0.6264 m/s. The reduce of distance between droplet reference instance and glass substrate from 20 mm as in the experiment with zero initial velocity to 0.1 mm with 0.6264 m/s initial velocity significantly decreased the computational time.

Step 4 involves meshing of the parts. It is suggested by the Simulia general guidelines that the mesh size of the Eulerian part should be at least smaller by a factor of three compared to the smallest feature of interest in Lagrangian mesh [15]. The Eulerian instance was meshed by an approximate global seed size of 0.1 mm using 8-node linear Eulerian brick with reduced integration and hourglass control. The glass substrate was meshed by an approximate global seed size of 0.6 mm and linear in three dimensions. The reduced integration was activated to reduce calculation time [18]. Element deletion was deactivated to prevent element from being removed by the impact of droplet. The droplet reference instance was not required to be meshed since it would be suppressed during the analysis. Finer mesh was not required for glass substrate part since the part was only for interaction with the aqueous glycerine droplet and the mesh was only for continuing the numerical processes such as assigning contact properties, load and boundary conditions. The number of elements in glass substrate and droplet reference instance are 10800 and 100 respectively.

In Step 5, the loads and boundary conditions of the assembly part were defined. The initial fluid location was defined inside the droplet reference instance. To implement fluid-structure interaction, two contact properties were created, which were tangential behaviour and normal behaviour. For tangential behaviour, the contact was given a penalty by applying friction coefficient of glass, which is 0.9. The assembly part was subjected to gravity load of 9.81 m/s^2 in negative y direction as shown in figure 6. The glass substrate was fixed at its original position throughout the simulation. The discrete field created in the Step 4 was assigned with the material properties of aqueous glycerin solution. A node set of Eulerian instance which consisted of 115351 nodes was created. Next, initial velocity of 0.6264 m/s in negative y direction was applied to the Eulerian instance.

In Step 6, a step time of 0.1 second was used to simulate the deposition process. The double precision was used to produce more accurate results. In Step 7, The spreading diameter of the droplet reference instance over time after in contact with the glass substrate were interpreted and analyzed using ImageJ software. The image scale was obtained by comparing the length of glass substrate in the image to the length of 6 mm. The spreading area was calculated from the obtained spreading diameter by assuming the droplet spreads in circular pattern.

2.3. Simulation of multiple Droplet Deposition

Nine sets of numerical simulations were carried out for multiple droplet deposition based on different combinations of droplet spacing and arrangement angle. The droplet spacing values chosen are 2.22 mm, 3 mm, and 4 mm. The arrangement angles selected are 90°, 110° and 120°. For each set of simulation, data were generated for 0.005 second, 0.01 second, 0.02 second, 0.04 second, 0.06 second, 0.08 second, 0.1 second.

The steps of simulation of multiple droplet deposition is similar to those of single droplet deposition. Instead of constructed a hollow sphere, in Step 1, four hollow spheres with the diameter of 2.2 mm were constructed. To make sure that there is enough space for the droplets to spread on the glass substrate, the substrates and Eulerian instances were created with the length and width varies with the droplet spacing. For droplet spacing of 2.22 mm, 3 mm and 4 mm, the length and width of the substrate and Eulerian instance are 12 mm x 12 mm, 15 mm x 15 mm and 20 mm x 20 mm respectively. The height of the substrate and Eulerian instance remaining unchanged.

In meshing of the parts in Step 4, the four aqueous glycerin droplets are merged together to form one part to ease the discrete field assignment process. The original instances are suppressed through this merging step. For meshing of Eulerian instance and glass substrate, the global seed size used also varies for three different sizes of Eulerian instance and glass substrate. The global seed size and number of elements for each sizing are tabulated in table 1. In Step 6, the process were simulated for 0.005 second, 0.01 second, 0.02 second, 0.04 second, 0.06 second, 0.08 second, 0.1 second.

3. Comparison of Experimental and Numerical Results of single droplet deposition

The experimental and numerical spreading areas for single droplet deposition were compared and verified. A correlation function was introduced to compensate the deviation found between both results. A graph of experimental and simulated spreading area against the spreading time was illustrated in figure 7. The time zero refers to the time right before the impact.

The spreading area data obtained from experiment and simulation method can be represented by equation (1) and equation (2) respectively:

$$\ln y_{\text{simulation}} = \ln 24.511 + 0.0783 \ln x \quad (1)$$

$$\ln y_{\text{experiment}} = \ln 9.0739 + 0.0435 \ln x \quad (2)$$

where $y_{\text{simulation}}$ is the spreading area obtained from simulation method, $y_{\text{experiment}}$ is the spreading area obtained from experiment, and x is the spreading time. The deviation between the data obtained from experiment and simulation method may be caused by the limitation of CEL method and on modelling droplet deposition process. Physical phenomena such as surface behaviors and material behaviors might not be correctly represented using CEL technique. These include adhesion force of droplet to glass surface, surface tension of the droplet impinged and cavitation.

Table 1. Global seed size and number of elements.

Droplet Spacing (mm)	Part	Global seed size (mm)	Number of elements
2.22	Eulerian instance	0.15	128000
	Glass substrate	1.20	100
3.00	Eulerian instance	0.15	200000
	Glass substrate	1.50	100
4.00	Eulerian instance	0.20	150000
	Glass substrate	2.00	100

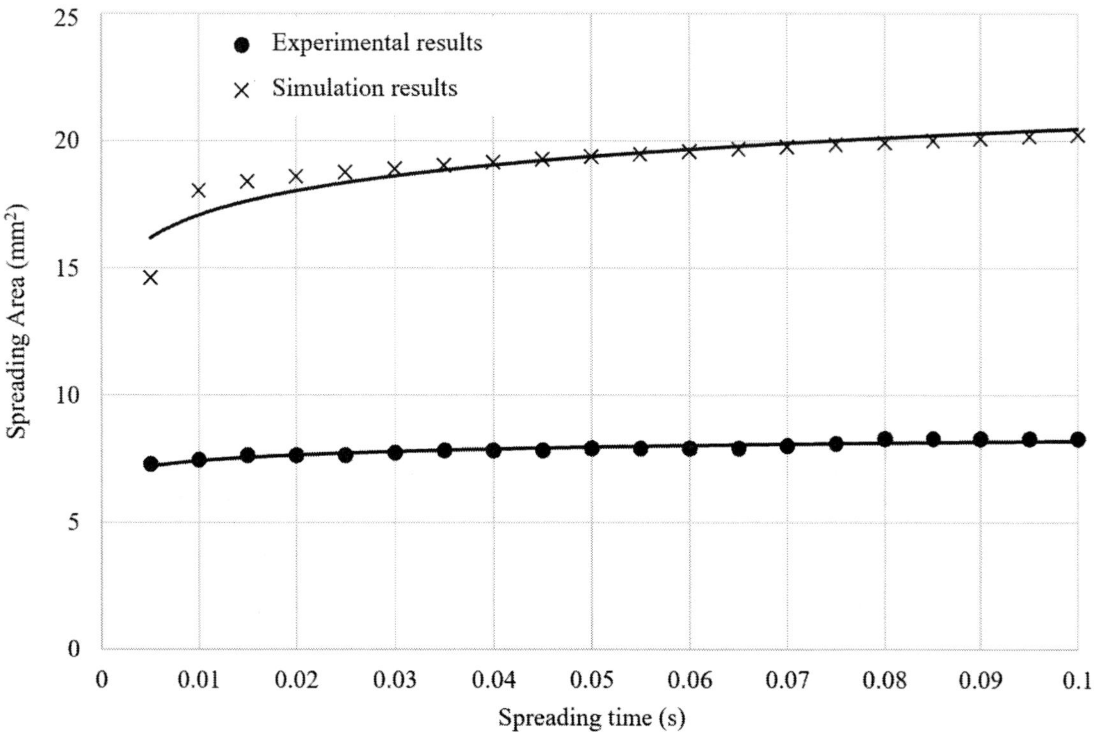

Figure 7. The spreading area for spreading time 0 to 0.1 second for single droplet deposition.

A correlation function was introduced to match the simulation results to the experimental results. The correlation function is given as

$$\ln y'_{\text{simulation}} = \ln y_{\text{simulation}} - 0.9937 - 0.0348 \ln x \quad (3)$$

where $y'_{\text{simulation}}$ is the spreading area that close to the spreading area obtained from experiment. To verify equation (3), five spreading time data were used to calculate the spreading area $y'_{\text{simulation}}$ based on equation (1) and equation (3). The calculated results were compared with those obtained from equation (2). Both results are shown in table 2. Based on table 2, the absolute percentage deviation of the spreading area obtained from correlation function and experiment are 0.002% only. Thus, the correlation function is comparable to the experiment and it was used to calculate the spreading area based on the simulated data obtained from multiple droplet deposition.

Table 2. The absolute percentage deviation of the results obtained from the correlation function and experiment.

Spreading time, x (second)	Spreading area, $y'_{\text{simulation}}$ (mm^2) based on correlation function	Spreading area, $y_{\text{experiment}}$ (mm^2) based on experiment	Absolute percentage deviation of spreading area (%)
0.02	7.65416	7.65401	0.002%
0.04	7.88846	7.88830	0.002%
0.06	8.02883	8.02867	0.002%
0.08	8.12993	8.12977	0.002%
0.10	8.20923	8.20907	0.002%

4. Simulated spreading area for multiple droplet deposition

The spreading area versus spreading time for different combinations of droplet spacing and arrangement angle was shown in figure 8. All the lines start from the origin where the spreading area is zero for spreading time of zero. The second spreading area data collected at 0.005 second for all the case. The second spreading area are more than 40 mm^2 for all the case. Therefore, the y-axis is starts from 40 mm^2.

Based on figure 8, the spreading area increased rapidly from 0 to 0.013 second. The slope of the increment decreased and slowly reached the equilibrium as the time went on. This was explained in Dorey [19] that the droplet spreading process could be divided into two stages, which were initial spreading and secondary wet spreading. Initial spreading of droplet depended on the kinetic energy of the impacting droplet which happened in the scale of microseconds. Secondary wet spreading depended on Ohnesorge number, which was influenced by surface tension, inertial and viscous forces. This type of spreading required longer time to reach equilibrium and drying effect might have occurred. The initial spreading occurred before 0.013 second in which rapid spreading of aqueous glycerin droplet occurred. After that, secondary wet spreading took place where the deposited droplet spread slowly on the glass substrate.

An array of droplets has more complicated interaction mechanisms compared to two-droplets interaction due to increasing in number of neighbouring droplets. During coalescence, more interface energy is released by multiple droplets interface, thus increasing the amount of kinetic energy converted. Higher amount of kinetic energy leads to larger deformation of the droplet interface during interaction and splashing of droplets may occur. Due to the existence of additional neighbouring droplets, the time required for contact of droplets to occur is also affected.

It was explained by Zhou, Loney, Fedorov, Degertekin and Rosen [20] that for an array of droplets, the interactions generated more forms of final equilibrium shape compared to two-droplet interaction. The interaction mechanisms were highly sensitive to impact conditions and droplet spacing. For small droplet spacing, the inertial force due to impact on solid surface led to coalescence of fast-moving droplet boundaries. High kinetic energy was gained from the coalescence of the droplet interface. Table 3 shows the droplet deposited pattern at 0.10 second for different combinations of droplet spacing and arrangement angle. It was observed that for image of droplet spacing of 2.22 mm and arrangement angle of 90° in table 3, the droplets had a tendency to spread out in the direction of neck growth due to high kinetic energy generated from coalescence. When droplet spacing was 3 mm, there were larger rooms for spreading which exhausted the inertial force gained from impact during the spreading of droplets. Coalescence of droplets occurred much later compared to droplet spacing of 2.22 mm. Less kinetic energy was gained and thus the droplets spread more steadily and uniform as shown in image of droplet spacing of 3 mm and arrangement angle of 90° in table 3. The droplet spacing of 4 mm generates largest spreading area compared to droplet spacing of 2.22 mm and 3 mm as shown in figure 8. From table 3, however, there are void between the droplets that are not covered up for arrangement angle of 90° and 110°. This is due to when the droplet spacing was further increased to 4 mm, coalescence of droplets could not be done fully due to lack of kinetic energy to energize the coalescence.

5. Effect of Droplet Spacing on Spreading Area

The effect of droplet spacing on the spreading area for three different arrangement angles at spreading time 0.1 second are illustrated in figure 9. The spreading area increased when the droplet spacing increased is only valid for 90° droplet arrangement angle. The finding is supported by the result of two droplets deposition

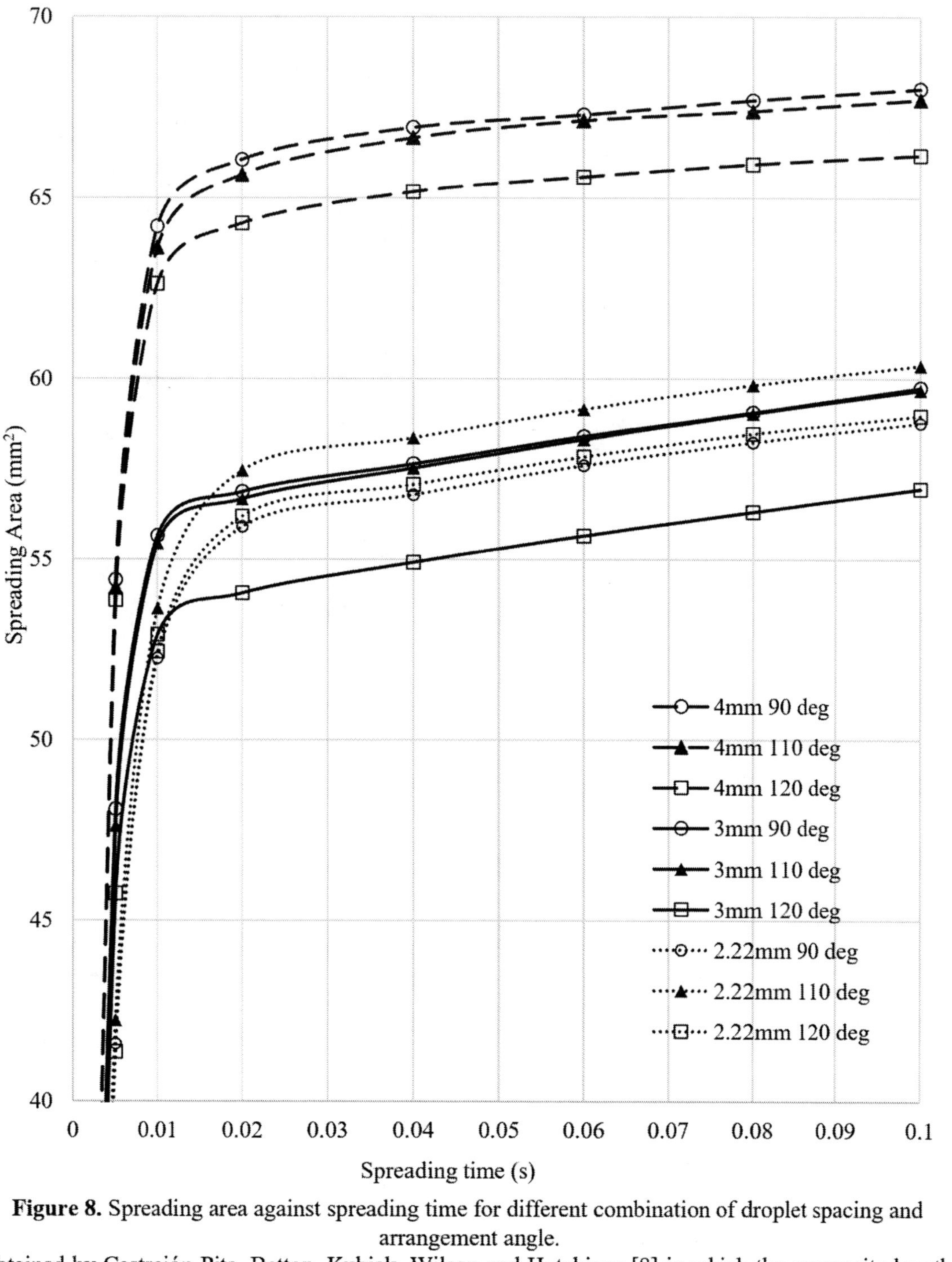

Figure 8. Spreading area against spreading time for different combination of droplet spacing and arrangement angle.

obtained by Castrejón-Pita, Betton, Kubiak, Wilson and Hutchings [8] in which the composite length increased as the droplet spacing increased. It was observed that the spreading area for droplet spacing of 4 mm is larger, followed by those for droplet spacing of 2.2 mm and 3 mm for arrangement angle of 110° and 120°.

Table 3. Deposited pattern of deposited droplet at 0.10 second for different combinations of droplet spacing and arrangement angle.

Arrangement angle, θ (°)	Droplet spacing, L (mm) 2.22	3.00	4.00
90			
110			
120			

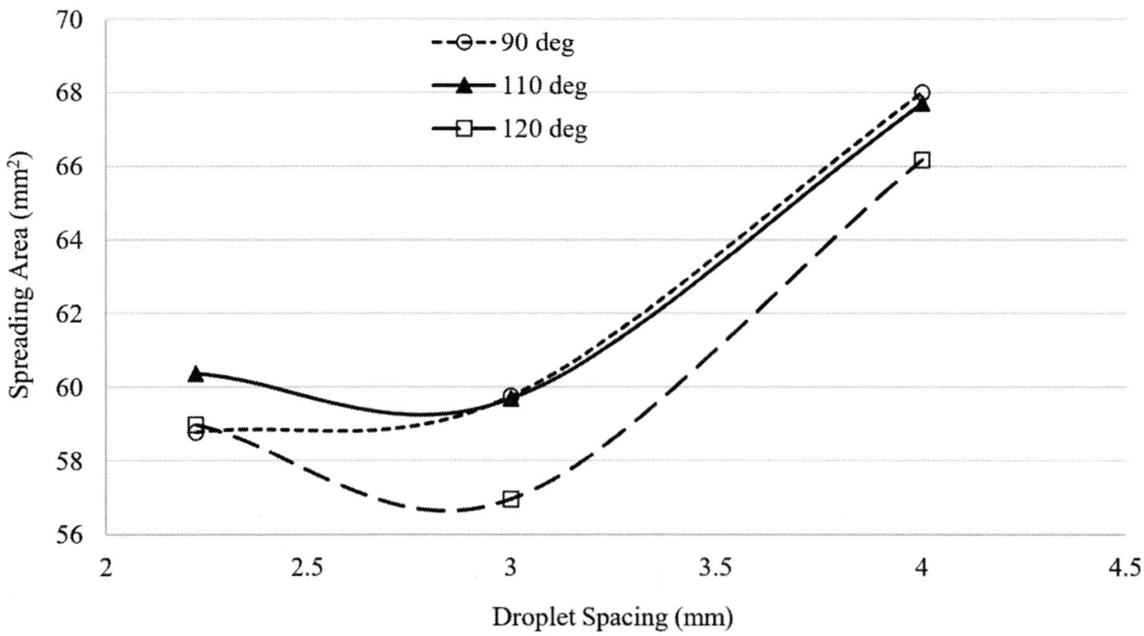

Figure 9. The spreading area vs droplet spacing for three different arrangement angles.

6. Effect of Arrangement Angle on Spreading Area

The effect of arrangement angle on spreading area is shown in figure 10. When the arrangement angle was increased from 90° to 110°, droplets with spacings of 3 mm and 4 mm displayed a slight reduction on spreading area while only droplets with spacing of 2.22 mm gave increment in spreading area as compared to that with 90° arrangement angle. For 120° arrangement angle, droplets of all three spacings exhibited reduction in spreading area.

For most of the research, the deposition of an array of droplets was done using arrangement angle of 90°. By changing the arrangement angle of an array of droplets, different deposition patterns can be observed. When the arrangement angle of droplets increases to 110° and 120°, the spacing between the diagonal droplets changed as shown in figure 11. Time for coalescence of droplets differed from each other and thus different kinetic energy was gained for each droplets. This led to more complicated interaction mechanisms compared to ordinary multiple droplet deposition.

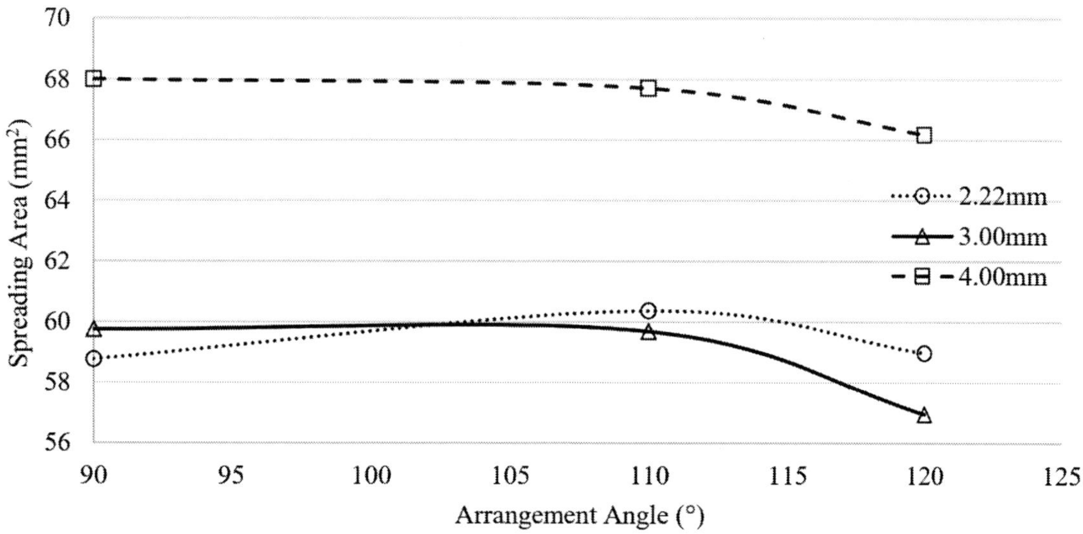

Figure 10. Comparison plot of spreading area vs arrangement angle for three different droplet spacings.

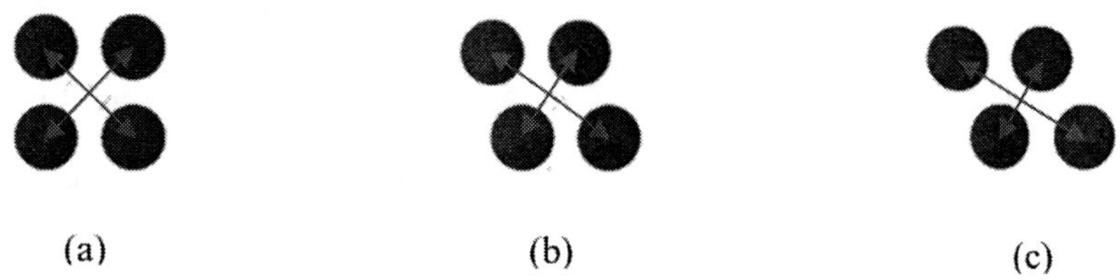

Figure 11. Diagonal distance of droplets for arrangement angle of (a) 90°, (b) 110° and (c) 120°.

7. Conclusion

The mechanisms of simultaneous multiple droplet deposition were studied using finite element analysis. The research was divided into two stages. For first stage, simulation on single droplet deposition was conducted and was verified using experiment. A correlation function was developed to match the simulation result with the experimental result. For second stage, nine sets of numerical simulations of simultaneous multiple droplet deposition were conducted by using an array consisting of four aqueous glycerin droplets with different combinations of droplet spacing and arrangement angles.

Based on the results obtained, it was found that the spreading area increased as the droplet spacing increased is only valid for arrangement angle of 90°. For arrangement angle of 110° and 120°, the spreading area for droplet spacing of 4 mm is the largest, followed by those for droplet spacing of 2.2 mm and 3 mm. The arrangement angle is proven as a factor that affect the spreading area.

References

[1] Bolleddula D A, Berchielli A, and Aliseda A 2010 Impact of a heterogeneous liquid droplet on a dry surface: Application to the pharmaceutical industry *Advances in Colloid and Interface Science* no. **159 (2)**144-159

[2] Li R, Ashgriz N and Chandra S 2010 Maximum Spread of Droplet on Solid Surface: Low Reynolds and Weber Numbers *Journal of Fluids Engineering* no. **132** (6):061302-061302-5.

[3] Lee J B, Derome D, G Guyer and J Carmeliet 2016 Modeling the Maximum Spreading of Liquid Droplets Impacting Wetting and Nonwetting Surfaces *Langmuir* no. **32** (5):1299-1308.

[4] Šikalo Š, Tropea C, Marengo M, and Ganić E N 2002 Spreading of Droplets on Horizontal Surfaces *New and Renewable Technologies for Sustainable Development* (Boston, MA: Springer US) p 683-692

[5] Farhangi M M 2012 *Numerical study of droplet coalescence and droplet shedding on surfaces with various wettabilities* Master's thesis, Concordia University

[6] Lee M W, Kim N Y, Chandra S and Yoon S S 2013 Coalescence of sessile droplets of varying viscosities for line printing *International Journal of Multiphase Flow* no. **56**:138-148.

[7] Graham P J, Farhangi M M, and Dolatabadi A 2012 Dynamics of droplet coalescence in response to increasing hydrophobicity *Physics of Fluids* no. **24** (11):112105.

[8] Castrejón-Pita J R, Betton E S, Kubiak K J, Wilson M C T and Hutchings I M 2011 The dynamics of the impact and coalescence of droplets on a solid surface *Biomicrofluidics* no. **5** (1):014112.

[9] Dalili A, Chandra S, Mostaghimi J, Fan H T C and Simmer J C 2014 Formation of liquid sheets by deposition of droplets on a surface *Journal of Colloid and Interface Science* no. **418**:292-299.

[10] Eslamian M and Soltani-Kordshuli F 2018 Development of multiple-droplet drop-casting method for the fabrication of coatings and thin solid films *Journal of Coatings Technology and Research* no. **15** (2):271-280.

[11] Harlow F H and Shannon J P 1967 The splash of a liquid drop *Journal of Applied Physics* no. **38** (10):3855-66.

[12] Bussmann M, Mostaghimi J, and Chandra S 1999 On a three-dimensional volume tracking model of droplet impact *Physics of Fluids* no. **11** (6):1406-17.

[13] Zhou W, Loney D, Fedorov A G, Degertekin F L, and Rosen D W 2014 Lattice Boltzmann simulations of multiple-droplet interaction dynamics *Physical Review E* no. **89** (3):033311.

[14] Yang X, and Kong S C 2018 3D Simulation of Drop Impact on Dry Surface Using SPH Method *International Journal of Computational Methods* no. **15** (03):1850011.

[15] Johansson H and Ollar J 2011 *Polymer Flow in Package Sealing Process - A Coupled Eulerian-Lagrangian Numerical Approach* Master's thesis, Lund University, Sweden.

[16] Zhu, Z, Kamnis S, and Gu S 2015 Numerical study of molten and semi-molten ceramic impingement by using coupled Eulerian and Lagrangian method *Acta Materialia* no. **90**:77-87.

[17] Glycerine Producers' Association 1963 *Physical properties of glycerine and its solutions* Glycerine Producers' Association.

[18] Naib S 2015 *Modeling the influence of water droplet impacts on steam turbine blade surfaces* Master's thesis, Institute of Materials Testing, Materials Science and Strength of Materials (IMWF) Universitaet Stuttgart.

[19] Dorey, R A 2011 *Ceramic thick films for MEMS and microdevices*: William Andrew.

[20] Zhou W, Loney D, Fedorov A G, Degertekin F L and Rosen D W 2015 Shape evolution of multiple interacting droplets in inkjet deposition *Rapid Prototyping Journal* no. **21** (4):373-385

IOP Conf. Series: Materials Science and Engineering **943** (2020) 012032 doi:10.1088/1757-899X/943/1/012032

The Effect of Fertilizer on *Epipremnum Aureum* in Phytoremediating Soil Contaminated with Crude Oil

K H D Tang*, and H T J Chai

Faculty of Engineering and Science, Curtin University Malaysia, CDT 250, 98009 Miri, Sarawak, Malaysia

E-mail: daniel.tang@curtin.edu.my

Abstract. *Epipremnum aureum* is a common ornamental foliage which is used mainly for indoor decoration. It demonstrates tolerance to low light environment and the ability to remove atmospheric chemicals such as gasoline and formaldehyde. A previous screening study showed the ability of *Epipremnum aureum* to remove crude oil from soil. Further to the screening study, this study examined the effect of fertilization on phytoremediation of crude oil-contaminated soil by *Epipremnum aureum Epipremnum aureum* was grown in pots with soil contaminated with 5%, 10% and 15% of crude oil by weight respectively, with and without the addition of fertilizer. In line with previous study, *Epipremnum aureum* showed the ability to remove crude oil contaminants. The plant demonstrated highest crude oil removal rate in pots with 15% contamination. Application of fertilizer enhanced the removal of crude oil in all experimental pots by Week 6, hence increasing the rate of crude oil decline over the experimental duration. It is deemed that application of fertilizer provides nutrients to promote plant growth and proliferation of roots which enhance rhizodegradation of crude oil in soil. This study extends on the existing knowledge that *Epipremnum aureum* can phytoremediate crude-oil contaminated soil by characterizing its tolerance to crude oil and its response to fertilization, which is crucial to its practical application in phytoremediation. Phytoremediation presents a promising yet cost-effective measure in treating contaminated land. Further study can examine the optimum fertilizer concentration for phytoremediation and the tolerance level of *Epipremnum aureum* to crude oil contamination.

1. Introduction

Intensifying anthropogenic activities ensuing population rise and the increase of industrial and agricultural activities have resulted in widespread pollution of the water, air and land. Land pollution is often manifested as soil contamination due to the presence of xenobiotic chemicals [1]. These chemicals frequently comprise petroleum hydrocarbons, polynuclear aromatic hydrocarbons, pesticides and heavy metals [2]. The detection of microplastics in terrestrial ecosystems such as soil also suggest contamination of soil by these fine plastic fragments which can be hard to remove [3]. Contamination of soil with petroleum hydrocarbon has been a persistent environmental issue the global communities are still combating. Activities such as onshore drilling, refining and storage of crude oil leave behind significant amount of petroleum hydrocarbons. Accidental spillage and leakage of petroleum products during transportation also causes entry of petroleum hydrocarbons into the terrestrial ecosystems in addition to improper disposal of petroleum products [2, 4]. It was estimated that the Europe alone has 2.5 million potentially contaminated sites of which approximately 342000 sites have been declared contaminated. The contamination was attributed mainly to heavy metals and petroleum products, indicating that petroleum products are one of the major soil contaminants [5].

Soil contamination by crude oil and petroleum products has a number of implications. Crude oil can alter the physical properties of soil by adsorbing on soil surfaces, thus, affecting porosity and permeability of soil [6]. Carbon-rich and nitrogen-deficient crude oil can also alter organic contents, pH, conductivity as well as carbon to nitrogen and carbon to phosphorus ratios of soil, which potentially affect the soil suitability for agriculture [7]. Nickel and vanadium commonly found in petroleum products can result in eco-toxicity following their release with petroleum products into soil [8]. Besides, crude oil contamination can result in modification of microbial population and enzyme system in soil [4]. Soil contaminated with crude oil has been shown to have lower germination rate, resistance to pests [9] and fertility due to decreased availability of nitrogen and phosphorus for the growth of crops [10]. Crude oil consists of a mixture of aliphatic, aromatic and organometallic constituents, many of which are harmful to human health, particularly the polycyclic aromatic hydrocarbons. These compounds may exhibit carcinogenic, mutagenic and teratogenic effects as they find their ways to the bodies of human and animals [11].

With widespread contamination of soil by petroleum products, removal of these pollutants has attracted enormous research interest. Four major removal methods of these pollutants have been identified, i.e. physical remediation, chemical remediation, microbiological remediation and phytoremediation [1]. Physical remediation involves soil removal and replacement as well as treatment of soil with heat and microwave which can be energy and cost intensive [1]. Chemical remediation uses elution, solvent extraction and oxidation which may leave behind an array of chemicals harmful to the environment [12]. Microbiological remediation uses microorganisms to degrade petroleum products in soil and the activities of microorganisms often depend on the soil conditions as well as the presence of nutrients [13]. Microbiological remediation, phytoremediation and a mixture of both are collectively known as bioremediation. Bioremediation provides a cost-effective alternative to removal of petroleum products from soil. As microbiological remediation is subject to numerous variables that affect the performance of remediating bacteria, phytoremediation has gradually gained popularity [7].

Phytoremediation works based on the interactions between plants, microorganisms and environment. It removes contaminants from soil via simulating microbial activity in the root zone as the roots secrete nutrient-rich exudates which also may contain contaminants degrading enzymes, or via direct absorption and transformation of contaminants [9]. For crude oil removal, studies have suggested degradation or transformation by microorganisms at the root zone as the main mechanism [14]. Many studies have been conducted for phytoremediation of crude oil contamination either globally or locally. In the North American, plants such as flatpea, alfalfa, reed canarygrass and deertongue, among others, have been shown to reduce polychlorinated biphenyl [15]. *Bassia scoparia* [14], *Salicornia persica* [16] and castor bean [17] have been reported as potential candidates for remediation of soil contaminated with crude oil. Locally, Tang and Angela showed the ability of *Pteris vittata*, *Epipremnum aureum*, *Mucuna bracteata* and *Imperata cylindrica* to remove crude oil from soil [12]. This was the first study that demonstrated the phytoremediating ability of *Epipremnum aureum* and *Mucuna bracteata* for crude oil.

Subsequently, a study was conducted by Tang and Law [18] to further investigate the ability of *Mucuna bracteata* in phytoremediating soil contaminated with crude oil up to 20% with and without application of fertilizer. The study demonstrated survival of *Mucuna bracteata* up to 15% crude oil contamination and better removal of crude oil in all pots where fertilizer was applied. This was in line with other studies pointing to the positive effect of fertilizer on phytoremediation [19, 20, 21]. However, no further studies of similar nature have been conducted for *Epipremnum aureum* which is a hardy, low-cost and widely available ornamental plant. This study therefore aims to examine the tolerance of *Epipremnum aureum* to different levels of crude oil contamination and the effect of fertilizer application on its ability to phytoremediate.

2. Methods

2.1. Soil preparation and plant cultivation

Loamy soil for cultivation of *Epipremnum aureum* was purchased and sieved through 50mm mesh to remove large organic debris and rock fragments. The soil was separated into parts of 2kg for both the control and experimental pots. Fertilizers were not applied to the plants in control pots while for experimental pots, the plants were fertilized.

Soil for control and experimental pots was mixed with crude oil at concentrations of 5%, 10% and 15% respectively by weight, as uniformly as was practicable. Crude oil used for the experiment was Miri medium sweet crude with 0.08% sulphur and an API of 32.3°. The fertilizer used was liquid NPK at a ratio of 10-8-7 fortified with 1% zinc, 1% magnesium and micronutrients of unspecified concentrations. Six control and experimental pots were prepared as follows:

Pot A: 2kg soil + 5% crude oil (without fertilizer application)

Pot A1: 2kg soil + 5% crude oil (with fertilizer application)

Pot B: 2kg soil + 10% crude oil (without fertilizer application)

Pot B1: 2kg soil + 10% crude oil (with fertilizer application)

Pot C: 2kg soil + 15% crude oil (without fertilizer application)

Pot C1: 2kg soil + 15% crude oil (with fertilizer application)

Healthy foliage of *Epipremnum aureum* weighing approximately 50g each, with comparable size was planted in polybags for 1 week prior to transplant to the pots above. Growth of plants in the pots were observed over the experimental duration of 6 weeks. The pots were placed at area exposed to sufficient sunlight but shaded from rain to minimize redistribution of crude oil due to rain splashes. The pots were watered twice daily by means of sprayer with similar amount of water to ensure sufficient moisture for *Epipremnum*'s survival. The volume of water used during each spraying was approximately 50ml 20g soil was collected weekly for analysis of the crude oil and moisture contents, and the pH.

2.2. Soil analysis

20g of soil from each pot was collected for analysis weekly after the transplant for 6 weeks. The soil was divided into two parts of 10g. The first 10g was tested for its moisture content using the ASTM D2974 method by evenly spreading it out on a petri dish and drying it in an oven at 125°C overnight until a constant weight was obtained [18]. The weight of soil before and after drying was recorded and the moisture content calculated with the following formula:

$$\text{Soil moisture (\%)} = 100 - \left(\frac{Weight\ of\ dry\ soil}{Weight\ of\ wet\ soil}\right) X\ 100\% \qquad (1)$$

The remaining 10g of soil was air-dried at room temperature for 8 hours and sieved with 2mm mesh. 5g of the sieved soil was mixed with water at a ratio of 1:1, stirred and left for 30 minutes before being tested for pH with a pH meter [22]. The purpose of moisture and pH analyses was primarily to ensure that these parameters were monitored for optimal plant growth. 2g of the sieved air-dried soil was transferred to a separating funnel and mixed with 10ml of hexane. The mixture was agitated to strip the crude oil from the soil and was left to stand until the soil had completely settled before filtration to obtain the filtrate containing crude oil extract. The filtrate was subsequently reconstituted with hexane to a final volume of 10ml before being transferred to a cuvette bound for UV-Vis Spectrophometer [12, 18]. The wavelength of the spectrophotometer was set at 360nm [22].

3. Results and discussion

Physiological condition of the plants was observed for 6 weeks. First two weeks after the transplant, the plants in all pots showed signs of yellowing and wilting as they were probably still adapting to the soil in the pots and the crude oil therein. After two weeks, the plants showed improved physiological conditions with greener leaves. Plants on potted soil contaminated with 5% crude oil showed better health compared to those on soil contaminated with 10% and 15% crude oil, wither fertilized or unfertilized. Generally, it can be concluded that *Epipremnum aureum* can survive in soil contaminated with 15% of crude oil. The limit of tolerance has not been identified in this study. Therefore, it is uncertain if *Epipremnum aureum* can survive higher levels of crude oil contamination.

3.1. Crude oil removal

Epipremnum aureum exhibits the ability to remove crude oil in all pots and the rate of crude oil removal, as indicated by the gradients of the trend lines, showed a rising trend as the level of crude oil contamination increased (see Figures 1 to 3). Application of fertilizer in all instances increased the rate of crude oil removal with more significant increase seen in Figure 2 and Figure 3. This was parallel to the findings of Tang and Law that fertilizer application increased the rates of crude oil reduction in soil planted with *Mucuna bracteata* [18]. Similar findings were reported by Kai et al. that soil amendment with perlite and higher basal fertilizer rates increased the removal rates of total petroleum hydrocarbon by *Zinnia hybrida* in soil contaminated with crude oil [19]. It is deemed that application of fertilizer provides nutrients to promote plant growth and proliferation of roots which enhance rhizodegradation of crude oil in soil [20].

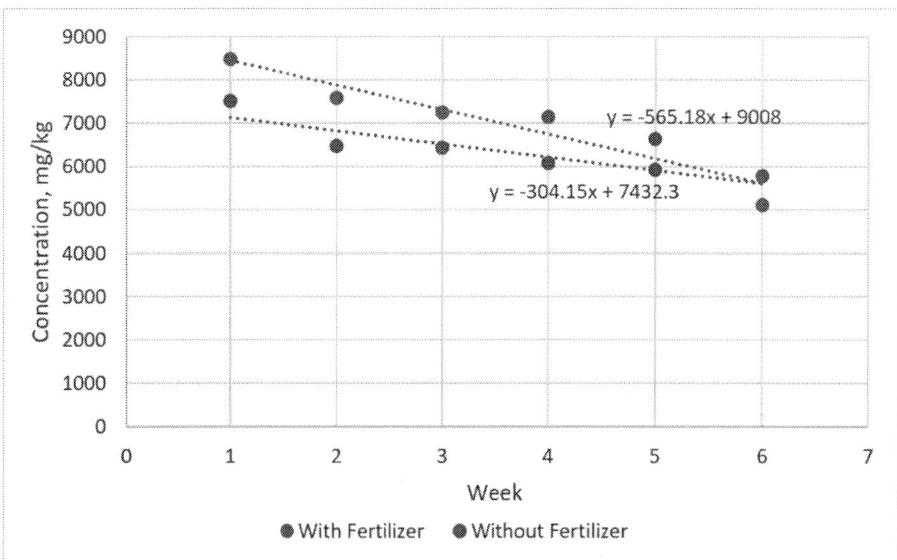

Figure 1: Weekly concentration of crude oil in soil contaminated with 5% crude oil

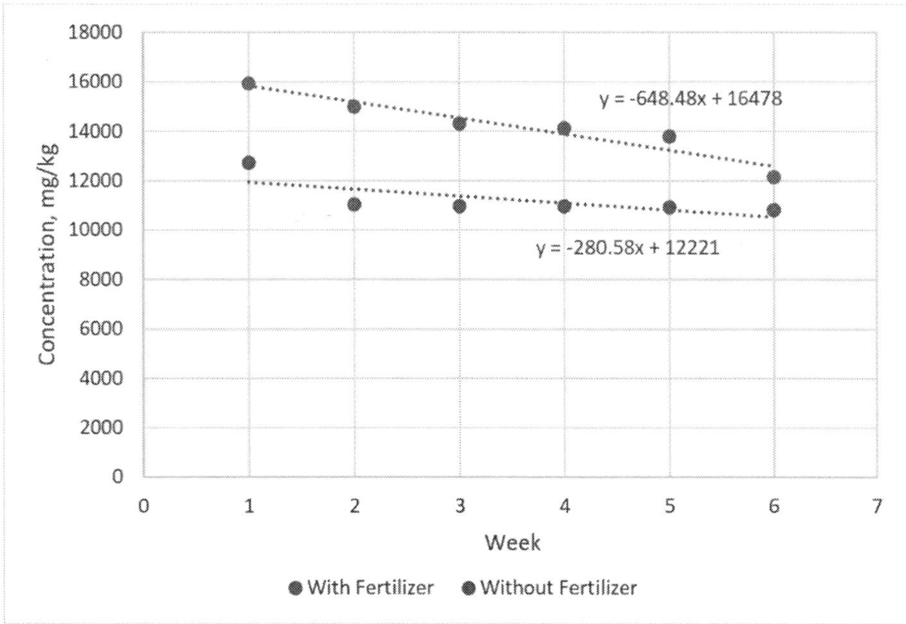

Figure 2: Weekly concentration of crude oil in soil contaminated with 10% crude oil

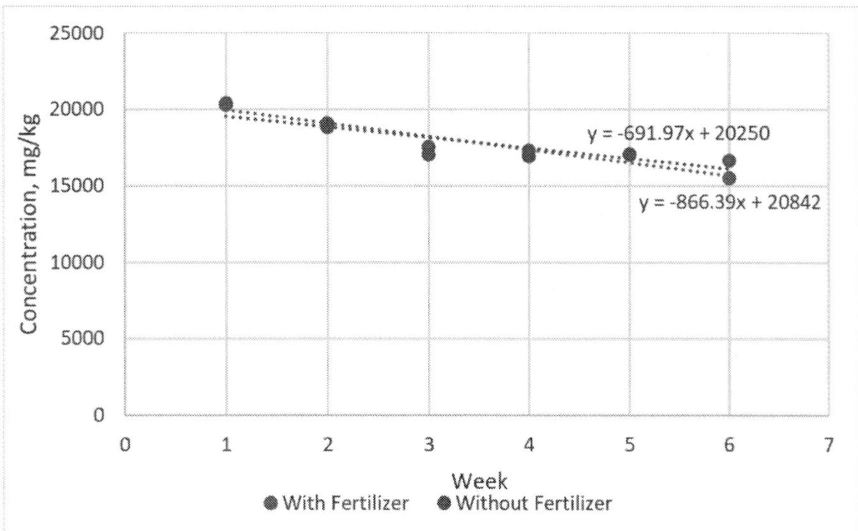

Figure 3: Weekly concentration of crude oil in soil contaminated with 15% crude oil

Percent removal of crude oil at Week 6 was calculated by expressing the difference between the crude oil concentration in Week 6 and that in Week 1, in relation to the crude oil concentration in Week 1 as percentage. The percent of crude oil removed was the highest in potted soil contaminated with 5% of crude oil where fertilizer was applied. Percent of crude oil removed was generally observed to be higher in pots with fertilizer applied than those without fertilizer application (Figure 4). In comparison to *Mucuna bracteata* [18], *Epipremnum aureum* showed significantly higher removal of crude oil in fertilized pots and unfertilized pots (a difference of approximately 29% and 14% respectively) with soil contaminated with 5% crude oil over 6 weeks. At 10% and 15% contamination, the percent crude oil removed by *Epipremnum aureum* was also higher than *Mucuna bracteata* by the range of 3% to 8% [18]. This agreed with the screening study of local plants for crude oil phytoremediation by Tang and Angela which revealed that *Epipremnum aureum* performed better than *Mucuna bracteata* in terms of the amount of crude oil removed over 6 weeks at 5% contamination [12].

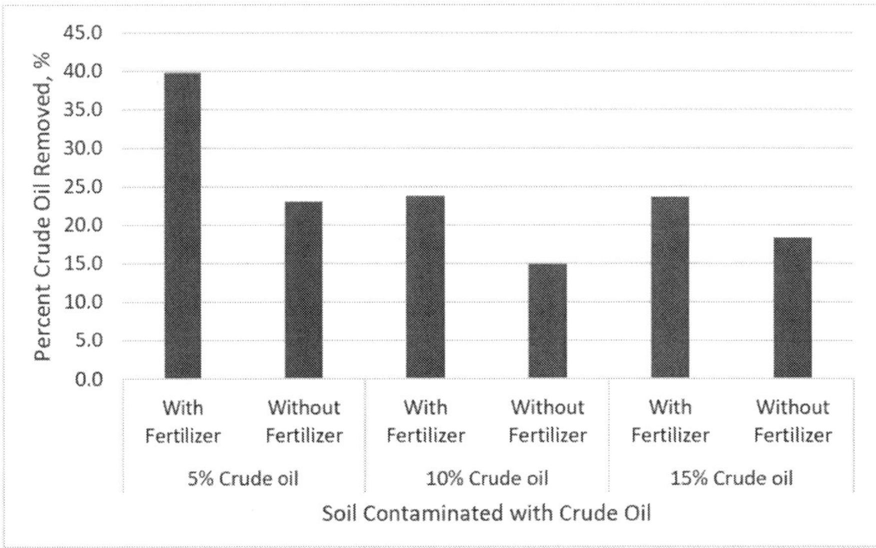

Figure 4: Percent crude oil removed from potted soil at different levels of contamination

3.2. Soil pH

The soil pH of all pots fluctuated within a narrow range of 5.8 to 6.35 (Table 1), indicating that there was no significant variation of the pH within the experimental duration and variability of phytoremediation due to the influence of pH was minimized. The pH of soil used in this study is weakly acidic. Soil pH influences nutrient availability and toxicity [23]. Commercially sourced soil commonly has pH in the range of 5.5 to 7.0 which is suitable for most crops [20].

Table 1. pH of soil contaminated with 5%, 10% and 15% crude oil respectively.

Week	5 % Crude Oil		10% Crude Oil		15% Crude Oil	
	Without Fertilizer	With Fertilizer	Without Fertilizer	With Fertilizer	Without Fertilizer	With Fertilizer
1	6.04	6.02	6.23	6.13	6.31	6.22
2	6.02	5.81	6.27	6.17	5.79	6.25
3	6.24	6.07	6.13	5.99	5.98	6.35
4	6.24	6.01	6.20	6.13	5.99	6.28
5	6.19	6.08	5.97	6.08	6.30	6.22
6	6.17	6.02	5.99	6.08	6.37	6.34

3.3. Soil Moisture

The soil moisture is shown Table 2. The purpose of monitoring the moisture is to ensure optimum moisture for healthy growth of *Epipremnum aureum.* Soil moisture was potentially affected by a number of factors such as weather conditions and humidity of air which could affect loss of moisture from soil via evaporation [22]. In view of variable weather conditions, humidity and evapotranspiration by plant, the watering frequency was adjusted accordingly to prevent wilting of plants. This led to variability of soil moisture shown in Table 2. The overall increasing trend of soil moisture was largely attributed to increasing watering frequency as the plants showed signs of wilting in the first two weeks of transplant, in addition to the hot weather in the subsequent weeks.

At the same watering frequency and volume, it was found that generally soil contaminated with higher amount of crude oil tended to retain slightly more moisture as crude oil had been reported to slow down evaporation of water from soil [18].

Table 2. Soil moisture contents.

Week	Moisture Content (%)					
	5% Crude Oil		10% Crude Oil		15% Crude Oil	
	Without Fertilizer	With Fertilizer	Without Fertilizer	With Fertilizer	Without Fertilizer	With Fertilizer
1	25.63	26.73	25.31	28.68	27.14	23.41
2	33.69	39.21	37.96	41.57	45.46	48.11
3	35.51	32.88	38.50	32.51	43.21	46.26
4	41.81	32.63	49.06	39.06	58.73	54.44
5	43.34	41.73	47.93	38.12	58.73	64.47
6	65.58	63.91	57.44	57.06	64.80	61.22

3.4. Limitation

The study has few inherent limitations. First, homogenous mixing of crude oil with soil might not have been attained though mixing was rigorously carried out. Second, watering though conducted in a controlled manner by spraying, might cause redistribution of crude oil in the soil. Third, soil was sampled at a fixed distance from the plant but the extent of root growth and penetration in the soil could be different, affecting the extent of phytoremediation. Fourth, the uptake hence the effect of fertilizer reaching the plant could not be tracked. Time was also a constraint in this study which did not permit longer experimental duration.

4. Conclusion

This study demonstrates that *Epipremnum aureum* can survive in soil contaminated with 5, 10 and 15% of crude oil and its ability to phytoremediate the contaminated soil. The rates of crude oil removal by *Epipremnum aureum* at all levels of contamination were increased by fertilizer application. This study contributes to the identification of *Epipremnum aureum* as a low-cost and hardy candidate for phytoremediation of soil contaminated with crude oil. This finding adds to the known ability of *Epipremnum aureum* to rhizofilter Co-60 and Cs-137 as well as its ability to clean indoor air [24, 25].

5. References

[1] Yuniati M D 2017 Bioremediation of petroleum-contaminated soil: A review. *IOP Conf. Ser.: Earth and Environmental Science* **118**(2018) 012063

[2] Wang S, Xu Y, Lin Z, Zhang J, Norbu N and Liu W 2017 The harm of petroleum-polluted soil and its remediation research. *AIP Conference Proceedings* **1864** 020222

[3] Zhang G S and Liu Y F 2018 The distribution of microplastics in soil aggregate fractions in southwestern China. *Sci. Total Environ.* **645**, 12-20

[4] Deng R Y 2014 *Microbiological monitoring and evaluation of compound pollution of petroleum and heavy metal in slated soils* (Shandong: Shan Dong University)

[5] EEA (European Environment Agency 2019 Data and Maps EEA, Progress in Management of Contaminated Sites. Available online: http://www.eea.europa.eu (accessed on 10 August 2019).

[6] Marinescu M, Toti M, Tanase V, Carabulea V, Plopeanu G and Calciu I 2010 An assessment of the effects of crude oil pollution on soil properties. *Ann. Food Sci. Technol.* **11** 94–99

[7] Li L Liu M and Zhao J L 2011 Current status and prospects of microbial remediation of soil pollution in oil fields in northern Shaanxi. *Chinese Journal of Soil Science* **42**(4) 1011-1013

[8] Dindar E F, Şağban F O T and Başkaya, H S 2015 Variations of soil enzyme activities in petroleum-hydrocarbon contaminated soil. *Int. Biodeter. Biodegr.* **105**, 268-275

[9] Shan B Q, Zhang Y T, Cao Q L, Kang Z Y, Li S Y 2014 Growth responses of six leguminous plants adaptable in Northern Shaanxi to petroleum contaminated soil. *Environmental Science* **35**(3) 1125-1130

[10] Liao J Q, Wang J, Huang Y. 2015 Bacterial community features are shaped by geographic location, physicochemical properties, and oil contamination of soil in main oil fields of China. *Microb Ecol.* **70**(2) 380-389

[11] Cocârţă D, Stoian M and Karademir A 2017 Crude oil contaminated sites: Evaluation by using risk assessment approach. *Sustainability* **9**(8), 1365

[12] Tang K H D and Angela J 2019 Phytoremediation of crude oil-contaminated soil with local plant species. *IOP Conf. Ser.: Mater. Sci. Eng.* **495** 012054

[13] Abbasian F, Lockington R, Mallavarapu M and Naidu R 2015 A comprehensive review of aliphatic hydrocarbon biodegradation by bacteria. *Appl. Biochem. Biotechnol* **176**(3), 670-699

[14] Hegazy A K, Mohamed N H, Moustafa Y M and Hamad A A 2015 Phytoremediation of soils polluted with crude petroleum oil using *Bassia scoparia* and its associated rhizosphere microorganisms. *Int. Biodeterior. Biodegr.* **98** 113-120

[15] Chekol T, Vough L R and Chaney R L 2004 Phytoremediation of polychlorinated biphenyl-contaminated soils: The rhizosphere effect. *Environmental International* **30**(6) 799-804

[16] Ebadi A, Sima N A K, Olamaee M, Hashemi M and Nasrabadi R G 2018 Remediation of saline soils contaminated with crude oil using the halophyte *Salicornia persica* in conjunction with hydrocarbon-degrading bacteria. *J. Environ. Manage.* **219** 260-268

[17] Olivares A R, Carrillo-González R, González-Chávez M C A and Hernández R M S 2013 Potential of castor bean (*Ricinus communis L.*) for phytoremediation of mine tailings and oil production. *J. Environ. Manage.* **114** 316-323

[18] Tang K H D and Law Y W E 2019 Phytoremediation of soil contaminated with crude oil using *Mucuna Bracteata*. *Research in Ecology* **1**(1), 29-30

[19] Kai T, Ikeura H, Ozawa S and Tamaki M 2018 Effects of basal fertilizer and perlite amendment on growth of *Zinnia* and its remediation capacity in oil-contaminated soils. *Int. J. Phytoremediat.* **20**(12), 1236-1242

[20] Neumann G and Römheld V 2012 Rhizosphere chemistry in relation to plant nutrition *Marschnar's Mineral Nutrition of Higher Plants* (3rd ed.) ed P Marschnar (New York: Elservier) 347–368

[21] Ronen E 2007 Micro-elements in agriculture. *Practical Hydroponics and Greenhouses* **July/August** 39–48

[22] Njoku K L, Akinola M O and Oboh B O 2008 Does crude oil affect the pH, moisture and organic content of soils? *Ecology, Environment and Conservation Paper* **14**(04) 731-736

[23] Tang K H D 2016 Ecological footprint analysis of electricity and water usage for Curtin University Sarawak. *The 33rd Annual Pan-Pacific Conference*, Miri, Sarawak, Malaysia

[24] Kil M, Kim K J, Cho J and Park C 2008 Formaldehyde gas removal effects and physiological responses of *Fatsia japonica* and *Epipremnum aureum* according to various light intensity. *Korean J. Hortic. Sci.* **26**(2) 189-196

[25] Kamel H A, Eskander S B and Aly M A S 2007 Physiological response of Epipremnum aureum for cobalt-60 and cesium-137 translocation and rhizofilgration. *Int. J. Phytoremediat.* **9**(5) 403-417

A Review on One-Way Concrete Wall Under Axial Loading

T Z H Ting[1], M E Rahman[1*] and H H Lau[2]

[1]Department of Civil & Construction Engineering, Faculty of Engineering and Science, Curtin University, 98009 Sarawak, Malaysia
[2]Faculty of Engineering, Computing and Science, Swinburne University of Technology, 93350 Sarawak, Malaysia

E-mail: merahman@curtin.edu.my

Abstract. As concrete wall panel is gaining popularity in precast construction industry, better insight into its behaviour is of paramount importance. Considerable research has been carried out on concrete wall under axial loading. Most of the published literatures have proposed design equations for wall design purpose. In this aspect, the paper is aimed to review the design equations of one-way concrete wall. The limitations of existing design equations are subsequently identified.

1. Introduction

Concrete wall panel is becoming increasingly popular as a load bearing structural member to transfer load from roof to foundation. Previously, concrete wall panel was mostly viewed as a partition member to isolate the internal building interspace as well as to provide protection against external environment [1]. Due to the extremely conservative design approach provided in the early version of concrete design standards, concrete wall panel was mostly designed with little consideration of structural capacity to sustain load.

With better understanding of the behavior of concrete wall panel as load carrying structural member due to substantial research, design guidelines have been gradually developed and improved. More reliable data has proven that the early version of concrete design standard underestimated the capability of concrete wall panel to be used as structural member [2-4]. Nevertheless, all the current standards allow concrete wall panel to be designed as structural member.

Reinforced concrete wall panels are now considered as important load transferring structural members in building structures. The function of concrete wall panel is determined by the chosen structural system as well as the corresponding type of loading. For wall subjected to wind and seismic loadings, it behaves as a shear wall to resist lateral loading. Concrete wall panel can be designed as load bearing wall if it resists vertical loading only, without being subjected to bending.Thus, the main objective of this paper is to review the current equations from published literature. The features, as well as the associated limitations, of the equations are discussed.

2. Research on Load Bearing Wall and Design Equation

In this section, design equations for concrete load bearing walls from the published literatures are reviewed. The experimental features from previous researchers, derived design equations as well as the associated limitations are also discussed. These semi-empirical equations are generally derived by correlating the experimental data with the equations based on scientific theory.

2.1. Leabu (1959)

Meanwhile, Leabu [5] conducted theoretical study based on allowable compressive stress concept to derive an equation for concrete wall panel under concentric loading. The equation is presented as (1).

$$P_u = 0.2f'_c\left[1-\left(\frac{H}{40t}\right)^3\right] \quad (1)$$

This equation only takes the effect of slenderness ratio into account without considering other factors. It is noticed that the factor of this equation is extremely low which will yield very conservative result. This equation is quite similar to the early version of ACI 318 (1963) empirical equation with the exception that a higher factor of 0.225 is being used in ACI 318 (1963) equation.

2.2. Oberlender (1975)
Oberlender [6] conducted a comprehensive study on 54 wall panels with slenderness ratio varied from 8 to 28 and aspect ratio from 1 to 3.5. The compressive cylinder strength of wall panels also varied from 28 to 42MPa. Minimum reinforcement ratios of 0.0033 and 0.0047 were provided for vertical and horizontal direction respectively. The wall panels were tested while hinged at both the top and bottom. They were subjected to uniformly distributed concentric loading and eccentric loading at $1/6^{th}$ of wall thickness. For the wall loaded under concentric loading, the authors noticed crushing failure for wall panel with slenderness ratio less than 20 while buckling failure was observed for slenderness ratio at 28. Combinations of crushing and slight buckling were observed for slenderness ratios of 20 and 24. It was noticed that buckling failure was observed for slenderness ratio of greater than 20 when the wall panel loaded eccentrically. The authors proposed an empirical equation based on their research findings as (2).

$$P_u = \varphi 0.6f'_u A_c\left[1-\left(\frac{H}{30t}\right)^2\right] \quad (2)$$

2.3. Pillai and Parthasarathy (1977)
Pillai and Parthasarathy [7] focused on testing of reinforced concrete wall panel with one layer of mesh embedded in its mid thickness. A total of 18 panels with slenderness ratios which ranged from 5 to 30, aspect ratios from 0.57 to 3.0 and wall thickness values of 40 to 80mm had been tested. The tests were carried out while hinged at both the top and bottom and under eccentric loading at t/6. The proposed studies were to verify the accuracy of theoretical calculation of extended column concept and empirical equation from ACI 318 (1971). The author noticed that the ultimate strength of wall panels estimated from column theory method with moment magnification as well as empirical equation were very conservative. The authors further recommended an empirical equation derived from their research results, shown as (3). The only difference of this proposed equation from that of ACI 318 (1971) was that the slenderness ratio term was modified to 50 instead of 40. It still has all the limitations of the later.

$$P_u = \varphi 0.57f'_c A_c\left[1-\left(\frac{H}{50t}\right)^2\right] \quad (3)$$

2.4. Kripanarayanan (1977)
Meanwhile, Kripanarayanan [8] conducted a theoretical analysis of reinforced concrete walls which were 200, 250 and 300mm in thickness while the reinforcement ratio was varied from 0.0015 to 0.01. The panels were load at the eccentricity of t/6. The authors proposed an empirical equation, (4), which consists of two functions, F_1 and F_2.

$$P_u = F_1 f'_c A_g F_2 \quad (4)$$

where F_1=0.55 and $F_2 = \left[1-\left(\frac{H}{32t}\right)^2\right]$.

The authors stated that substantial increase of wall capacity was observed when reinforcement ratio was increased from 0.0075 to 0.01. However, the wall axial capacity did not increase significantly with minimum reinforcement (0.0025) and therefore the contribution by reinforcement was neglected in the equation. Since the proposed F_2 did not yield realistic capacity estimation for wall with pin-ended supports, the authors recommended to include k factor in F_2 and it was modified to $F_2 = \left[1-\left(\frac{kH}{40t}\right)^2\right]$.

2.5. Zielinski et al. (1983)

Zielinski, Troitski [9] carried out an experimental study on five full scale ribbed walls under both concentric loading and eccentric loading of t/6. These wall panels were reinforced with two layers of mesh and had the geometry of 13.5 as slenderness ratio and 2.25 as aspect ratio. The authors compared the results of un-ribbed and ribbed wall and concluded that the inclusion of ribs can improve the stiffness and rigidity of wall panels. The authors proposed a design equation based on their research findings as (5).

$$P_u = \varphi 0.55 f'_c A_c \left[1 - \left(\frac{H}{40t}\right)^2\right] [1 + \rho_m (m - 1)] \tag{5}$$

The proposed equation is only applicable to the wall with slenderness ratio up to 40 and the applied load must be at eccentricity less than t/6. One of the features of the proposed equation is that the effect of reinforcement ratio is included.

2.6. Saheb and Desayi [2]

Saheb and Desayi [2] conducted a number of investigations into the effects of aspect ratio, slenderness ratio, vertical and horizontal reinforcement ratio on load bearing capacity of reinforced concrete wall under one-way action. A total of 24 samples of concrete grade 20 were studied. The slenderness ratio was varied from 9 to 27 while aspect ratio varied from 0.67 to 2. The vertical reinforcement ratio was varied from 0.0017 to 0.0085 while horizontal reinforcement varied from 0.002 to 0.005. Similar to other researchers, the wall panels were loaded at 1/6th of wall thickness with hinges at both the top and bottom. From their research outcome, the authors concluded that the ultimate strength of wall panel decreased with increasing of aspect ratio from 0.67 to 2.0. Also, the ultimate strength of wall panel decreased non-linearly with increasing of slenderness ration from 9 to 27. The authors also found that the ultimate strength of wall panel increased linearly with vertical reinforcement ratio but the increase of horizontal reinforcement ratio had negligible effect. The authors introduced aspect ratio effect and modified the design equations from ACI equation and Zielinski, Troitski [9]. The new equations are divided into two parts and are shown as (6) and (7) respectively.

$$P_u = 0.55\left[f'_c A_c + (f_y - f'_c) A_s\right]\left[1 - \left(\frac{kH}{32t}\right)^2\right]\left[1.2 - \frac{h}{10L}\right] \text{ for } \frac{H}{L} < 2 \tag{6}$$

$$P_u = 0.55\left[f'_c A_c + (f_y - f'_c) A_s\right]\left[1 - \left(\frac{kH}{32t}\right)^2\right] \text{ for } \frac{H}{L} \geq 2 \tag{7}$$

It can be seen that reinforcement has been considered in the proposed equation and the term representing slenderness ratio effect is similar to that in ACI equation. The authors demonstrated that the proposed equations could estimate the ultimate strength of reinforced concrete wall panel conservatively by using the test data available in the published literature. However, the proposed equations can only be used for wall with the slenderness ratio up to 32 and are limited to eccentricity of t/6.

2.7. Fragomeni and Mendis (1996)

Fragomeni and Mendis [10] performed a series of research to study normal and high strength concrete wall panels. 16 samples were tested as one-way wall while 4 samples were tested as two-way wall at the eccentricity of t/6. The slenderness ratio was varied from 12 to 25 while aspect ratio was varied from 2 to 5. The compressive strength of concrete was in the range of 32.9 to 67.4 MPa. Minimum reinforcement was provided in both vertical and horizontal directions. From the research outcome, the author observed that wall panel of H/t less than 20 failed by crushing while wall panel with H/t equal or greater than 20 failed by buckling. Comparing the normal strength and high strength wall panels of the same dimensions, high strength concrete wall panel exhibited more brittle failure under one-way action. The authors concluded that the failure mode of reinforced concrete wall panel under axial loading is governed by the concrete strength, slenderness ratio, and the amount of reinforcement provided. By comparing the experimental results with those calculated using AS equation, the authors concluded that the implementation of AS 3600 equation was unsafe for high strength concrete wall panel, especially when only minimum amount of reinforcement was provided. Thus, the equation underestimates the wall capacity.

The authors modified the design equation of AS 3600 by including high strength concrete parameters. The equations are shown in (8) and (9) respectively. They also incorporated an effective height factor from German Code DIN 1045 (1988) into the proposed equations in order to account for the effect of side support.

$$\varphi N_u = \varphi(t_w - 1.2e - 2e_a)0.6f_c' \text{ for} 20 \leq f_c' \leq 50 \tag{8}$$

$$\varphi N_u = \varphi(t_w - 1.2e - 2e_a)35(1 + (f_c' - 50)/80) \text{ for } 50 \leq f_c' \leq 80 \tag{9}$$

Although the investigation was only limited to compressive strength up to 70MPa, the author stated that the proposed equation was valid for wall panel with compressive strength up to 80MPa due to the conservative approach taken and the use of proper reduction factor.

2.8. Doh and Fragomeni (2005)

Doh and Fragomeni [4] conducted a series of comprehensive research on normal and high strength concrete walls through both experimental and numerical modelling studies. A total of 18 wall panels were tested experimentally with 6 of them tested as one-way wall while 12 of them tested as two-way wall. All the samples were tested at the eccentricity of t/6. The slenderness ratio was varied from 25 to 40 while aspect ratio varied from 1 to 1.6. Minimum reinforcement ratio of 0.0031 was provided in both vertical and horizontal directions. The authors proposed an empirical design equation by modifying the design equation of AS 3600 and the one proposed by Fragomeni and Mendis [10]. The proposed equation (10) shows that wall strength does not increase proportionally with concrete strength but is represented by $f_c'^{0.7}$.

$$\varphi N_u = \varphi 2f_c'^{0.7}(t_w - 1.2e - 2e_a) \tag{10}$$

The equation is applicable for wall with slenderness ratio greater than 30 and it accounts for the effect of side support. It also accounts for the effect of non-linear increase of wall strength with increase of concrete strength. However, the proposed equation still does not consider the effect of lightweight aggregates.

2.9. Ganesan et al. (2012)

Ganesan, Indira [11] investigated the behaviour of steel fiber reinforced self-compacting concrete (SFRSCC) and steel fiber reinforced concrete (SFRC) wall panels in one-way action. A total of 16 wall samples were tested. The slenderness ratios of 12 to 30 and aspect ratios of 0.75 to 1.5 were studied. A single layer of reinforcement was provided with reinforcement ratios of 0.0088 and 0.0074 for vertical horizontal directions respectively. Loads were applied at the eccentricity of t/6. All the wall panels were tested while pinned at both the top and bottom. From their research, the authors concluded that the inclusion of steel fibers were able to improve the cracking behaviour and ductility of concrete wall. The effects of slenderness ratio and aspect ratio were similar to normal type of concrete wall. (11) was proposed by the authors to calculate the ultimate axial strength of SFRSCC wall panels based on their experimental investigation. This equation includes the effect of slenderness ratio, aspect ratio and reinforcement ratio.

$$P_u = 0.56[f'_c A_g + (f_y - f'_c)A_s]\left[1 + \left(\frac{h}{29t}\right) - \left(\frac{H}{26t}\right)^2\right]\left[1 - \left(\frac{h}{11L}\right)\right] \tag{11}$$

2.10. Ganesan et al. (2013)

For more recent research, Ganesan, Indira [12] conducted a series of test on 20 samples of reinforced normal concrete and geopolymer concrete wall panels. The slenderness ratio was varied from 12 to 21 while aspect ratio was varied from 1 to 1.875. All the wall panels were tested with pins at both the top and bottom. Loads were applied at the eccentricity of t/6. The author noticed that for the same wall geometry, geopolymer concrete wall panel exhibited more softening behaviour when compared to normal concrete wall panel. Normal concrete wall panel showed steeper slope in load versus lateral deflection when compared to geopolymer concrete wall panel. The authors explained that content of finer particles in the matrix of geopolymer concrete resulted in more ductile behaviour. Attempts were made by the authors to compare the experimental results with those calculated using design equations from the published literature. The authors concluded that these equations are very conservative in predicting the load bearing capacity of geo polymer concrete wall. The authors proposed an equation based on the experimental results to predict the load bearing capacity of both geopolymer and normal concrete wall as (12). Similar assumptions of Saheb and Desayi [2] were adopted in the derivation of the equation. This equation considered the parameters such as slenderness ratio, aspect ratio and reinforcement ratio. However, the accuracy of this equation is debatable since the equation is derived from only 20 samples. Case study was not conducted to compare the proposed equation with other experimental results to prove the reliability of the equation.

$$P_u = 0.59\left[f'_c A_g + \left(f_y - f'_c\right)A_s\right]\left[1 + \left(\frac{h}{40t}\right) - \left(\frac{H}{30t}\right)^2\right]\left[1 - \left(\frac{h}{18L}\right)\right] \quad (12)$$

3. Conclusion

All these studies provide important insights into the development of reinforced concrete load bearing wall and its design equation. It is noticed that most of the research is concentrated on normal and high strength concrete walls. Hence, there is limited research on axial behaviour of lightweight concrete wall. It is also noted that majority of the researchers prefer to modify or derive the design equation based on ACI equation. These researchers concentrated on modifying the factors to account for geometric effects, which are slenderness ratio and aspect ratio. However, it is obvious that these simplified equations have not taken into consideration of all the material non-linearity effects. Further improvement of simplified equation can be made through the consideration of factors such as non-linear compressive strength increment and lightweight aggregates.

4. References

[1] Popescu, C., et al., *Concrete walls weakened by openings as compression members: A review.* Engineering Structures, 2015. **89**: p. 172-190.

[2] Saheb, S.M. and P. Desayi, *Ultimate strength of RC wall panels in one-way in-plane action.* Journal of Structural Engineering, 1989. **115**(10): p. 2617-2630.

[3] Fragomeni, S. and P. Mendis, *Applicability of current ACI318 wall design formula for high strength concrete walls.* Advances in Structural Engineering, 1999. **2**(2): p. 103-108.

[4] Doh, J.-H. and S. Fragomeni, *Evaluation of experimental work on concrete walls in one and two-way action.* Australian journal of structural engineering, 2005. **6**(1): p. 37-52.

[5] Leabu, V.F. *Problems and performance of precast concrete wall panels.* in *Journal Proceedings.* 1959.

[6] Oberlender, G.D., *STRENGTH INVESTIGATION OF PRECAST REINFORCED CONCRETE LOAD-BEARING WALL PANELS.* 1975.

[7] Pillai, S.U. and C. Parthasarathy, *Ultimate strength and design of concrete walls.* Building and Environment, 1977. **12**(1): p. 25-29.

[8] Kripanarayanan, K. *Interesting aspects of the empirical wall design equation.* in *Journal Proceedings.* 1977.

[9] Zielinski, Z., M. Troitski, and E. El-Chakieh, *Bearing capacity tests on precast concrete thin-wall ribbed panels.* PCI Journal, 1983. **28**(3): p. 88-103.

[10] Fragomeni, S. and P. Mendis, *Improved axial load formulae for normal and high strength reinforced concrete walls.* Australian Civil Engineering Transactions, 1996. **38**(2/3/4): p. 71.

[11] Ganesan, N., P. Indira, and S.R. Prasad, *Strength and behavior of SFRSCC and SFRC wall panels under one-way in-plane action*, in *High Performance Fiber Reinforced Cement Composites 6.* 2012, Springer. p. 279-286.

[12] Ganesan, N., P. Indira, and A. Santhakumar, *Prediction of ultimate strength of reinforced geopolymer concrete wall panels in one-way action.* Construction and building materials, 2013. **48**: p. 91-97.

Acknowledgments

The authors acknowledge funding from Curtin Malaysia Research Institute (CMRI).

IOP Conf. Series: Materials Science and Engineering **943** (2020) 012034 doi:10.1088/1757-899X/943/1/012034

Performance analysis of a modified Savonius hydrokinetic turbine blade for rural application

Armand Z. Anthony and Sukanta Roy

Department of Mechanical Engineering, Curtin University Malaysia, CDT 250, 98009, Miri, Sarawak.

E-mail: sukanta.roy@curtin.edu.my

Abstract. For rural areas situated in the deep topography of Sarawak, the distance from energy grids and plants are situated too far from the power appliances. Hence, sustainable alternative method of generating energy for the people should be put into consideration. For rural areas rich with riverine streams, the employment of hydrokinetic turbines (HKT) would be a good alternative to meet their daily requirements. The HKTs are fundamentally understood as a turbine that harnesses the kinetic motion of freestreams without the need of head. In the preceding years, the development of HKT turbines has undergone several breakthroughs in terms of efficiency and design. However, at low water depths cluttered with debris, it makes the implementation of certain HKT orientations limited. In such cases, the Savonius style SKTs would be one of the promising deployments in riverine systems. In this paper, a newly developed Savonius blade profile by Roy & Saha was simulated in a fluid medium of water through the CFD software StarCCM+. The two-dimensional unsteady Reynolds Averaged Navier Stokes equations was solved using the shear stress transport k-ω turbulence model at a Reynolds number of $0.71\text{-}1.88\times10^5$ and water speed ranging from 0.3-0.8 m/s. The maximum power coefficient generated by the modified blade was 0.43 at TSR=1.0 and Re=1.88×10^5.

1. Introduction

Energy plays a fundamental role and critical position in the social, economic, cultural and political sectors that moves the progress of civilization by sustaining all the needs of society. The demand has been too extensive in the past few decades, that the ever-increasing need of the community is not matched by the limited supply of energy resources, due to its over exploit to meet human demands. In the South Asian domain, most countries today still rely on the conventional energy sources that are depletable and non-renewable. It is straightforward to theorize that renewable energy is the better option as an energy source due to the outlying advantages. However, for the past century, civilization has an over reliance to fossil fuels as it is used as the main energy supply. The current projection of total electricity production is still dominated by conventional sources. In some of the developed countries such in US, UK and Europe, the trend of non-conventional energy generation is slowly increasing recently [1]. Hence, a regime for the incorporation of renewable energy is vital for the preservation of natural resources. The 2015 Paris Agreement had presented global ambitions to achieve stability between anthropogenic emissions by sources and removal of sinks of greenhouse gases by the second half of this respective century. The ambition of the Paris Agreement was to cultivate the increase in global average temperature to well below 2.0°C above the preindustrial levels while also enforcing the temperature limit of 1.5°C [2]. A solution to accommodate all these goals are to reduce the consumption of non-renewable sources and integrate a large-scale renewable energy supply in energy systems [3]. Hence, the motivation for this study is to facilitate the movement of implementing a contemporary green energy solution for the rural communities of Sarawak.

In the current state, the people located in the deep topography of Sarawak still rely on small diesel-powered generators and solar panels. However, the cost to maintain and operate the diesel generators are not cost efficient for the low-income remote communities. The fuel to power the generators are required to be transported over vast distances by river or by terrain. Further, the solar panels along with their electrical components were tested to be too fragile for the operation in the tropical rainforest as it susceptible to corrosion [4]. Hence, to alleviate the strain encountered by the villagers which are situated close to the riverbanks, an alternative form of energy generation method such as hydrokinetic turbines was considered. The feasibility of incorporating hydrokinetic turbines in Sarawak is due to the land's geographic characteristics. Sarawak is predominantly divided into nine districts which are Kuching, Samarahan, Sri Aman, Sarikei, Sibu, Bintulu, Kapit, Miri and Limbang. All of which are interconnected due to Sarawak being one of the largest river networks where the main rivers are the Sarawak river, Rajang river, Saribas river and the Lupar river [5].

Hydrokinetic turbines produce electricity by harnessing the flow of water in a river or stream, where turbines rotate the generator and capture the kinetic energy of the flow. Hydrokinetic turbines operate on the similar theory of wind power generation where the major significance of the respective configuration is the density of the fluid. For hydrokinetic turbines, the working fluid medium is water and it is approximately 800 times denser than air. Herein, this implies that the energy generated by hydrokinetic energy is far greater than the energy generated by wind turbines with similar geometry and flow speeds [6]. A hydrokinetic turbine operating at a rated speed of 2-3 m/s can produce four times the energy of a wind turbine with similar rated wind speed [7]. The power density comparison of hydrokinetic turbines and wind turbines conducted by Yuce & Muratoglu [7] found that the power density of a hydrokinetic turbine operating at flow speed 2 m/s is equal to the wind turbine operating at 16 m/s.

Hydrokinetic turbines can be classified as three separate subgroups; horizontal axis, cross flow and vertical axis. These subgroups are established on the arrangement of rotor axis with respect to the flow direction. Horizontal axis turbines or alternatively called an axial flow turbine, operates on the principle that the turbines rotate about the axes parallel to the incoming flow. These turbines predominantly utilize propeller type rotors. There exist two variations of horizontal axis hydrokinetic turbines (HAHKT), straight axis and inclined axis. HAHKT utilize the similar concept of aircraft wings, propeller blades and wind turbines [8]. The advancement of wind turbines and marine propellers have contributed greatly to the design and performance of most HAHKT in ocean and tidal energy extraction. There exist a multitude of literature on the respective configurations. However, the performance of HAHKT reduces substantially in riverine conditions. The presence of suspended particles or materials (eg; fish, debris, rocks, branches and twigs) could also be a problem for horizontal flow turbine as it could possibly damage the respective turbine [9].

The vertical axis hydrokinetic turbines (VAHKT) operate on the principle that the turbines axes are orientated orthogonal to the incoming water flow while being parallel to the water surface [10]. There exist three main configurations of the vertical axis hydrokinetic turbines in modern day research and commercial use. The three main configurations are classified as: (a) Darrieus Type Turbine, (b) Gorlov Turbine, and (c) Savonius Turbine. The VAHKT has been chosen to be the configuration of interest for this study due to the capability of having a generator configuration above water level, versatility towards changes of flow direction, horizontal configuration of shaft which allows turbines to sweep wide-shallow channels, low cost blade configuration, no need for yawing mechanism for reversing tidal flow and a marginally lower impact to aquatic life due to lesser noise, lower rotational speed and large hollow space. Furthermore, riverine currents offer lower energy content as opposed to ocean current and therefore, the turbines should have simple resolutions for the generators and rotor speed control. The generators and power transmission assembled at the surface of the water and this allows the configurations to ease in terms of design, operation and maintenance of the system [11].

2. Literature Review

Known as the simplest turbine configurations amongst the three other vertical turbine arch-types. The Savonius turbine can be described as a turbine with two semi-circular buckets cut from the top to bottom and placed sideways along the cutting plane. The cross section of the turbine forms an indistinguishable "S" letter [12]. The respective turbine operates on the principle of drag force difference between the concave side and convex side of the blade which exhibits excellent starting capabilities. However, Savonius turbines suffer from low aerodynamics efficiencies when compared to HAHKT and Darrieus turbines. The performance of a Savonius turbine in water is far efficient in than its original counterpart. Savonius turbines operates at lower tip speed ratios when placed in a denser fluid such as water, due to the greater drag force imposed by water. Since drag force is a product of density, it is only logical that the drag encountered in water is greater than that encountered in air. Furthermore, the pressure gradient that is subjected in water is adversely different than that of air. Where in water, the hydrostatic pressure gradients increase with the depth in the fluid [13]. Therefore, the analysis of hydrokinetic Savonius turbines are significantly different than that of wind energy generation due to the governance of gravity on the pressure difference [14].

There is a plethora of studies carried out devoted in improving the performance of Savonius rotors. The issue that engulfs the performance of conventional Savonius turbines are the large variations of static torque [10]. Numerous studies correlate the enhancement of performance of Savonius turbines with rotor blade profile, where research have varied the Savonius rotors from having twisted blade, helical blade and alternate blade geometry as opposed to the norm semi-circular shape. The prevalent criteria for a desired blade profile would be cost effectiveness and design simplicity yet offering a reasonable amount of coefficient of power. Therefore, a wide range of literature has been reviewed prior conducting this study [14-22]. It is worth noting, that the performance of certain blade profiles has yet to be tested in a working fluid medium of water and would serve as an additional motivation for this research. The summarization of literature along with the key aspects are shown in Table 1.0.

Table 1. Summary of Savonius turbine performance

Author(s)	**Methodology**	**Parameters Investigated**	**Maximum C_P**
Mabrouki, Driss and Abid **[16]**	Experimental (Hydrodynamic Test Bench)	Internal OR=0-0.3	Optimum OR=0.3 $C_{Pmax} = 0.327$
Golecha, Eldho and Prabhu **[17]**	Experimental (Open-Channel)	Double Deflector Plate Deflection Angle=15° – 60°	$C_{Pmax} = 0.35$
Patel, Bhat, et al. **[14]**	Experimental (Laboratory Channel and Irrigation Canal)	Varying AR and OR	Optimum OR=0.1-0.15 $C_{Pmax} = 0.23$
Roy and Saha **[15]**	Experimental (Low speed wind tunnel)	Modified Blade Profile	$C_{Pmax} = 0.31$
Current Study	CFD Simulation	Modified Blade Profile	$C_{Pmax} = 0.431$

According to an experimental analysis conducted by Roy and Saha [15], altering the blade profile of a 2-bladed Savonius wind turbine has shown significant increase in the turbine performance. The altered blade profile was described as a newly developed blade profile that was generated from numerous experimental studies on Bach and Benesh type turbines by altering the geometric arcs, overlap distances and dimension of blade profile. The experimental analysis conducted was done to compare the inherent performance of several altered blade profiles with the inclusion of the newly developed blade profile, to that of conventional Savonius wind turbine blade profile. The respective experimental testing was conducted in an open-type low-speed wind tunnel at Reynolds number, ranging from 6.0×10^4 to 1.5×10^5 with reasonable blockage correction factor of 4-9%. The highest obtainable coefficient of power generated by the newly developed blade profile was 0.31 at Reynolds number, $Re=1.2\times10^5$ corresponding to TSR of 0.82 [15]. In this study, the modified Savonius blade profile was integrated into a fluid medium of water and the performance was assessed in order to determine the feasibility of incorporating the respective blade for rural area deployment.

3. Numerical Methodology

3.1. Performance parameters

The performance of the Savonius HKTs are calculated in terms of averaged power and moment coefficient as shown:

$$TSR = \frac{\omega R}{V} \tag{1}$$

$$C_{M\,average} = \frac{T}{0.5\rho A V^2 R} \tag{2}$$

$$C_{P\,average} = \frac{T\,\omega}{0.5\rho A V^3} = \frac{T}{0.5\rho A V^2 R}\,\frac{R\omega}{V} = C_{M\,average} \times TSR \tag{3}$$

where ρ is the density of water [kg/m^3], V is the free stream water speed [m/s], A is the swept area of the turbine [m^2], TSR is the tip speed ratio, R is the radius of the turbine [m], ω is the rotational speed of the turbine [rad/s], T is the average moment generated on the turbine [Nm] respectively.

3.2. Mathematical model

The prevalent criterions for the utilization of a reliable CFD model are appropriate turbulence model, suitable definition of the spatial discretization refinement level and optimal time step dimension which ensure the capture of all relevant time scales [18]. Hence, a fully turbulent SST k-ω model with two additional equation for triggering the correct onset of the laminar to turbulent boundary layer transition is employed in this following thesis. Generally, the k-ω turbulence model can be separated as two entities: standard k-ω and SST k-ω. For the case of this research, the SST k-ω is known as the combination of a k-ω and k-ε turbulence model such that the k-ω model is incorporated in the inner regions of a boundary layer while the k-ε model is incorporated in the free shear flow outside a boundary layer. This allows the SST k-ω an optimal choice for the research as it allows for the utilization for low velocity conditions without any external energy input [19]. The SST k-ω can be expressed as:

$$\frac{D\rho k}{Dt} = \tau_{ij}\frac{\partial u_i}{\partial x_j} - \beta^*\rho\omega k + \frac{\partial}{\partial x_j}\left[(\mu + \sigma_k\mu_t)\frac{\partial k}{\partial x_j}\right] \tag{4}$$

$$\frac{D\rho\omega}{Dt} = \frac{\gamma}{v_t}\tau_{ij}\frac{\partial u_i}{\partial x_j} - \beta\rho\omega^2 + \frac{\partial}{\partial x_j}\left[(\mu + \sigma_\omega\mu_t)\frac{\partial \omega}{\partial x_j}\right] + 2\rho(1-F_1)\sigma_{\omega 2}\frac{1}{\omega}\frac{\partial k}{\partial x_j}\frac{\partial \omega}{\partial x_j} \tag{5}$$

$$\text{Where, } F_1 = \tanh\left(arg_1{}^4\right) \tag{6}$$

$$arg_1 = min\left(max\left(\frac{\sqrt{k}}{0.09\omega y};\frac{500v}{y^2\omega}\right);\frac{4\rho\sigma_{\omega 2}k}{CD_{k\omega}y^2}\right) \tag{7}$$

$$CD_{k\omega} = max\left(2\rho\sigma_{\omega 2}\frac{1}{\omega}\right)\frac{\partial k}{\partial x_j}\frac{\partial \omega}{\partial x_j}, 10^{-20} \tag{8}$$

$$\text{and } v_t = \frac{a_1 k}{max(a_1\omega;\Omega F_2)} \tag{9}$$

$$\text{Where, } F_2 = \tanh\left({arg_2}^2\right) \tag{10}$$

$$arg_2 = max\left(2\frac{\sqrt{k}}{0.09\omega y};\frac{500v}{y^2\omega}\right) \tag{11}$$

Two sets of constants which are generally used are as followed:

First Set (φ_1) – Wilcox	Second Set (φ_1) – Standard $k-\varepsilon$
$\sigma_{k1} = 0.5$, $\sigma_{\omega 1} = 0.5$, $\beta_2 = 0.0750$,	$\sigma_{k2} = 1.0$, $\sigma_{\omega 2} = 0.856$, $\beta_2 = 0.0828$,
$\beta^* = 0.09$, $k = 0.41$,	$\beta^* = 0.09$, $k = 0.41$,
$\gamma_1 = \frac{\beta_1}{\beta^*} - \frac{\sigma_{\omega 1}k^2}{\sqrt{\beta^*}}$	$\gamma_1 = \frac{\beta_2}{\beta^*} - \frac{\sigma_{\omega 1}k^2}{\sqrt{\beta^*}}$

3.3. Computational domain

For the computational modelling of this study, a rectangular computational domain is utilized to simulate the flow condition of an ideal riverine system. An enlarged computational domain is used in the following simulation to mitigate the wall boundary and blockage effect that would regularly affect the performance of turbines. The large computational domain also minimizes the impact of boundary condition uncertainties [20]. Figure 1 illustrates the computational domain of the modified Savonius turbine in two-dimensions confined by an upstream, far field downstream and symmetry type boundary. The respective boundary is circumscribed by a no-slip wall boundary condition on the blades along with a rotating mesh region. The inlet velocity and pressure outlet velocity are located 6D and 8D from the turbine center respectively. The diameter of the turbine (D=0.21) is used based on literature. The respective domain is discretized by unstructured polyhedral mesh with the inclusion of the boundary layer region. A 3-D CAD geometry is imported to StarCCM+ in order to generate a polyhedral mesh with a very small z-directional thickness (0.05 m). Generally, polyhedral meshes are reliable, quick to converge, generated easily in StarCCM+ and are more efficient than tetrahedral meshes. Once the mesh refinement is accepted, it is then projected into a two-dimensional plan for the generation of solution. The requirement for the two-dimensional is a very fine structed grid generated on the blade surface with an overall thickness of 1 mm along with a wall normal growth of 1.2 mm. A secondary level of mesh refinement is applied at the surrounding of the rotor blades to accurately replicate the flow fields inside the rotational radius of a turbine, which incurs vortex shedding. The surrounding of the rotor blades will be discretized with a minimal cell size of 3.15 mm. Furthermore, a third level of refinement is set on the outer circular zone and the level of the downstream refinement will be kept constant up to eight times the of the blade chord length to represent the wake flow. Lastly, a fourth level of mesh refinement will be applied to the far wake region with a minimal sized cell of 0.053D for minimized velocity dissipations. The velocity of the water at the inlet will be varied from 0.3 m/s to 0.8 m/s which corresponds to Re=0.707-1.88 $\times$ 10^5 in order to assess the hydrodynamic performance of the modified Savonius hydrokinetic turbine. The y+ value of the boundary layer was kept constant to ensure the same boundary layer resolution where values were less than or equal to 1. Meanwhile, the outlet is set at constant pressure P=0 Pa which adheres to the large computational domain size.

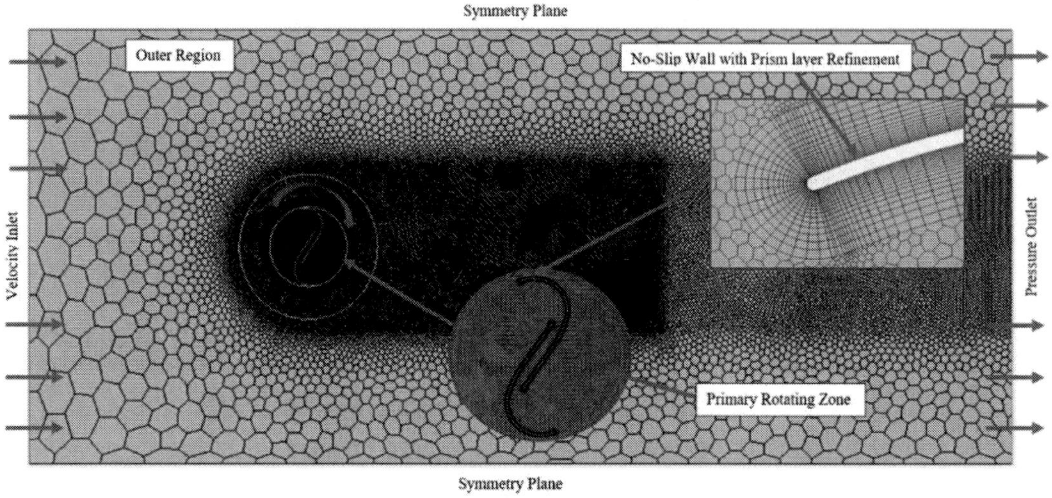

Figure 1. Overall computational domain

4. Validation

The quality of the mesh is mainly influenced by the distributions and size of the grid. The grid sensitivity analysis test conducted in Figure 2 shows that at least 86,000 polyhedral mesh is required in order to obtain less than 1% of errors compared to the finest grid that corresponds to a cell size consistency of 0.1 mm at the blade surface. The respective meshing method is based on the literature published by Roy and Ducoin [21]. The convergence of the computational simulations is noted to be achieved after a few initial fluctuations. The quasi steady time convergence of the simulations is achieved at approximately 1.5 seconds. The standards for the continuity, momentum and the two transport equations from SST $k - \omega$ turbulence model is defined by RMS values less than 10^3. For this study, the total cell count of polygonal mesh is 91, 091 cells which accounts for 0.5% of error.

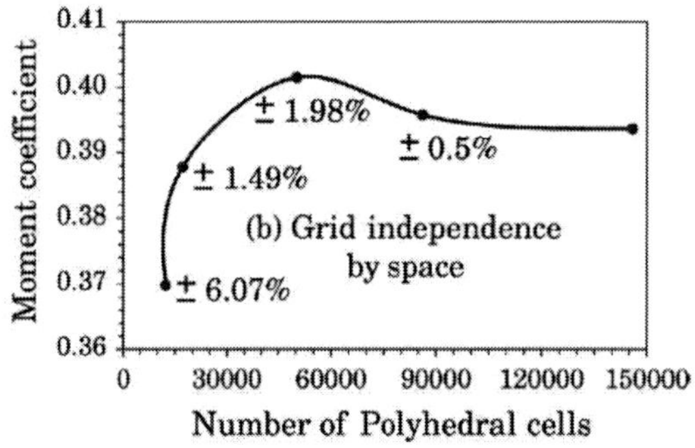

Figure 2. Grid independence test conducted by Roy & Ducoin [21]

5. Result and Discussion

5.1. Average moment and power coefficient at varying water speeds

In this numerical assessment, the performance of the modified Savonius turbine blade profile is analysed with respect to the averaged moment coefficients and power coefficients, at varying TSR. The goal is to assess the feasibility of the modified Savonius turbine blade profile when subjected to a fluid medium of density 997.561 kg/m^3. Based on figures 3 and 4, the maximum average power coefficient, $C_{p\ max}$ at the lowest water speed was found to be 0.38064 occurring at TSR=1.0 while the maximum average moment coefficient was found to be 0.53406 occurring at TSR=0.4. The trend of the power coefficient is comparable to that of literature published by the Kerikous and Thévenin

(2019), where it generates peak Savonius turbine power coefficient capabilities at TSR above 0.79. However, at elevated TSR ranges the performance of the Savonius turbine degrades after reaching a certain capstone TSR. The overall maximum power coefficient, $C_{p\ max}$ generated by the modified Savonius blade profile was 0.42603 at TSR 1.0 while the maximum overall moment coefficient was 0.6429 at TSR of 0.4 at water speed of 0.8 m/s. The minimum and maximum power coefficients occur at the lowest and highest water speeds respectively. Hence, there is a correlation that the water speeds play a role in determining the power output of the turbine. This inclination at water speed of 0.8 m/s can be attributed to the optimum decrease of moment coefficient. It is observed that the moment coefficient decrease is an s-like pattern, which stipulates the magnitude of torque generated at this respective speed is higher than average. Hence, the optimum operational speed and TSR can be drawn to be 0.8 m/s at TSR of 1.0. The highest average coefficient of power at TSR=1.0 would later be used to compare to conventional semi-circular Savonius blade profile. Based on figure 4, the progression of average power coefficient against TSR increases abruptly as opposed to the other water speeds.

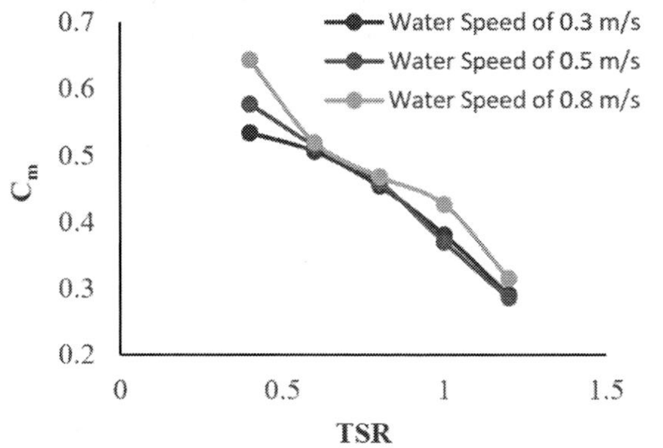

Figure 3. Average moment coefficient corresponding to TSR

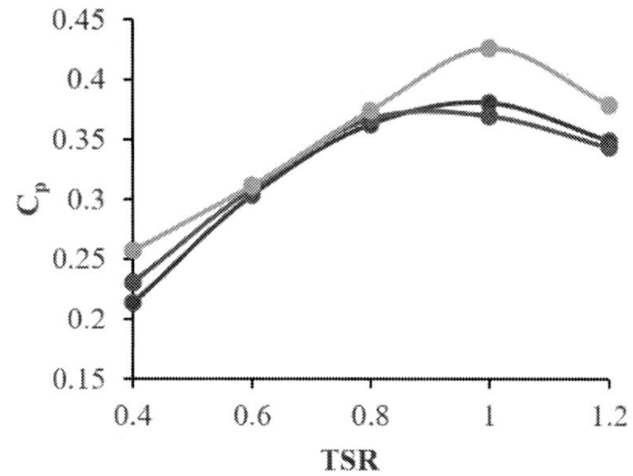

Figure 4. Average power coefficient corresponding to TSR

5.2. Significant static pressure and velocity magnitude contour at water speed 0.8 m/s (TSR=1.0)

The significant pressure and velocity contour that is considered occurs at water speed of 0.8 m/s. The relevance of the respective water speed is due to its peak power generation at TSR of 1.0. Coincidently, the correlation of elevated water speed to turbulence scale is also accounted for. For a higher water speed of 1.0 m/s, the peak power generation occurs at TSR of 1.2 and further degrades past the capstone TSR. In real-time conditions, it is not practical for the turbine to achieve the maximum power coefficient at a higher TSR. It implies that the turbine would require a substantial amount of rotational speed for it to achieve peak performance and would therefore defeat the purpose of implementing a VAHKT. Hence, the performance at water speed of 0.8 m/s is discussed.

The performance of the modified Savonius turbine at water speeds of 0.8 m/s is the most efficient amongst other water speeds. It can be observed from figure 5 that the maximum average power coefficient generated was 0.42603 at TSR=1.0. Based on the static pressure contour, at rotational angle of 180° it is observed such that the pressure subjected on the suction side of the advancing blade is relatively high. This attributes to the self-starting capabilities of the Savonius turbine. Further, at rotational angles of 120° and 150° the convex side of the advancing blade exhibits high negative pressure which aids in the rotation of the turbine.

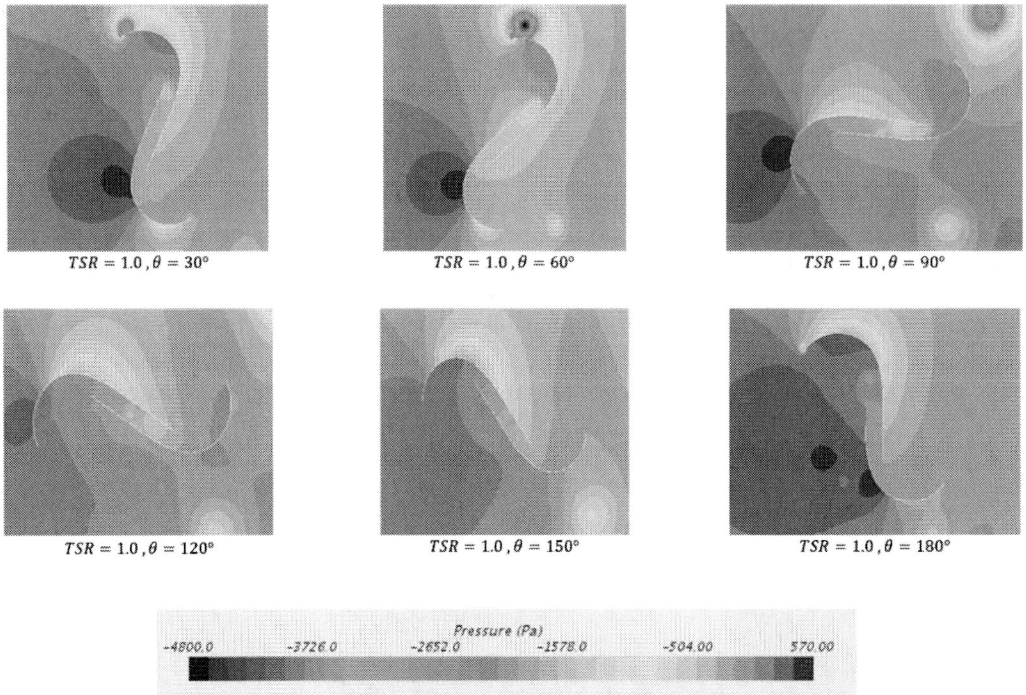

Figure 5. Static pressure contour at water speed 0.8 m/s (TSR=1.0)

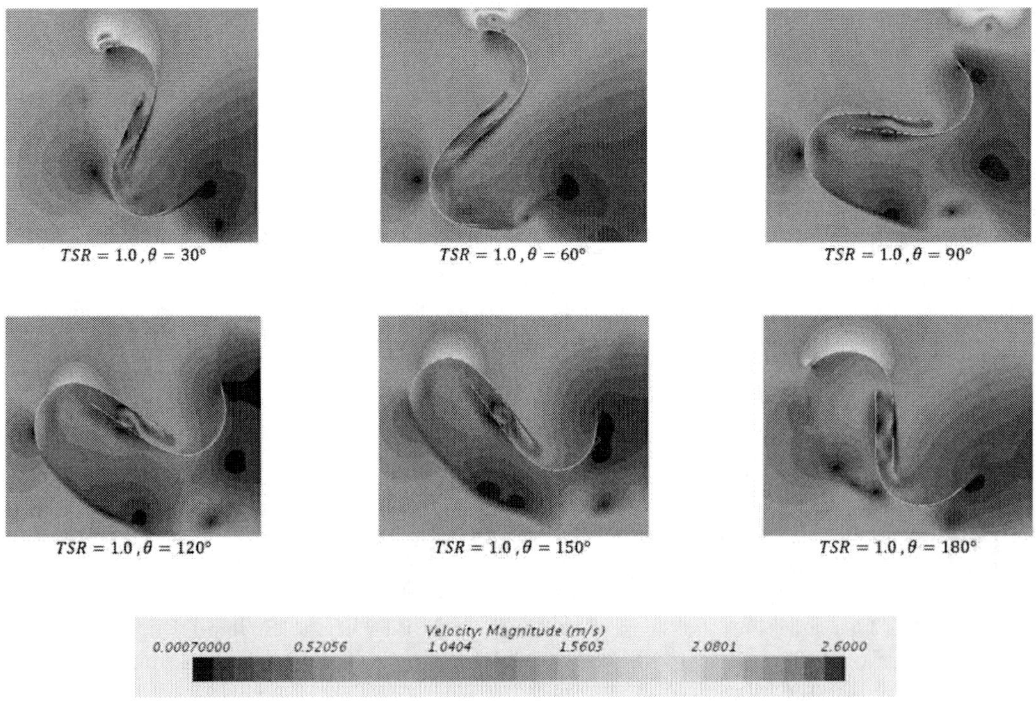

Figure 6. Highlighted velocity magnitude contour at water speed 0.8 m/s (TSR=1.0)

At TSR= 1.0 for rotational angle $\theta = 90°$, it can be observed from figure 6 that an excited wake region occurred along the convex side of the advancing blade which abruptly increased the velocity at the specific region. This is phenomenon is only present at TSR=1.0, as is regarded the contributory force to the increase turbine performance. According to Roy and Saha [22], the respective accelerating region near the trailing edge of the returning blade, increases the occurrence of air incidence on the concave side of the advancing blades. For this case, it is similar in nature with the inclusion that the incidence of air is substituted with water. Further, at angles of 120° and 150° similar effects of an excited wake regions occur along the leading edge towards the trailing edge of the returning blade which aids in accelerating the water to strike the concave side of the advancing blade. It is also noted that at angles 120° and 150° that a formation of vortices within the gap with occurs. The formation of vortices may or may not have an adverse impact on the overall performance of the turbine and may be considered in the future scope of this study.

6. Conclusion

The present work aims at improving the current performance of Savonius-style turbines with the implementation of a modified blade profile configuration at relatively low water speeds as it would be used for rural area deployment. Computational simulations have been performed to evaluate the performance of the modified Savonius blade profile which had never been tested in a medium other than air. The results of the computational simulations yielded that the modified Savonius blade profile exhibit excellent power generating capabilities as the water speeds increased. The modified Savonius blade profile also conforms to the norm of the majority of Savonius where its performance degrades after TSR=1.0. Hence, the feasibility of implementing the respective turbine blade profile in rural areas is viable. However, further testing should be conducted with respect to the future scope for the modified blade to be viable in real-time application. The modified Savonius HKT exhibit improved power generation capabilities at low water speeds up to 0.8 m/s. Based on the scope of this study, the influence of water speed at alternative TSR was tested and the results yielded that modified Savonius blade profile could generate a maximum average Cp value of 0.42603 at water speeds 0.8 m/s. The optimized performance of the modified Savonius HKT was found to occur at water speeds 0.8 m/s at TSR of 1.0. Also, the comparison of the modified Savonius blade profile to conventional semi-circular Savonius blade profile yielded an improved performance of 19.43% and modified Savonius blade exhibits excellent operational performance capabilities at low TSRs, all the modified Savonius turbines generated average Cm values above 0.5. Therefore, employment of the modified Savonius HKT is feasible for the employment in rural applications where water speed ranges from 0.3 m/s to 0.8 m/s. As a future research, the different designs will be studied for the hydro-kinetic applications.

References

[1] United States Energy Information Administration, "Annual Energy Outlook 2019 with Projections to 2050," United States Energy Information Administration, Washington, 2019.

[2] W. Steffen, J. Rockström, R. Richardson, T. M. Lenton, C. Folke, D. Liverman, C. P. Summerhayes, A. D. Barnosky, S. E. Cornell, M. Crucifix, J. F. Donges, I. Fetzer, S. J. Lade, M. Scheffer, R. Winkelmann and H. . J. Schellnhuber, "Trajectories of the Earth System in the Anthropocene," *Proceedings of the National Academy of Sciences,* vol. **115(33)**, pp. 8252-8259, 2018.

[3] K. Hansen, C. Breyer and H. Lund, "Status and perspectives on 100% renewable energy systems," *Energy,* vol. **175**, pp. 471-480, 2019.

[4] M. Anyi, B. Kirke and S. Ali, "Remote community electrification in Sarawak, Malaysia," *Renewable Energy,* vol. **35**, pp. 1609-1613, 2010.

[5] Z. Jamaluddin, E. Azrulhisham, M. Azri, N. M. Zain, S. Yusoff, M. Anyi and C. Peter, "Potential of Portable Hydrokinetic Energy Turbine Design for Remote Communities in Sarawak," *Journal of Adv Research in Dynamical & Control Systems,* vol. **10(9)**, p. 121, 2018.

[6] H. J. Vermaak, K. Kusakana and S. P. Koko, "Status of Micro-Hydrokinetic River technology in

rural applications: A review of literature," *Renewable and Sustainable Energy Reviews,* vol. **29**, pp. 625-633, 2014.

[7] M. I. Yuce and A. Muratoglu, "Hydrokinetic energy conversion systems: A technology status review," *Renewable and Sustainable Energy Reviews,* vol. **43**, pp. 72-82, 18 March 2015.

[8] N. D. Laws and B. P. Epps, "Hydrokinetic energy conversion: Technology, research, and outlook," *Renewable and Sustainable Energy Reviews,* vol. **57**, pp. 1245-1259, 2016.

[9] M. Khan, G. Bhuyan, M. Iqbal and J. Quaicoe, "Hydrokinetic energy conversion systems and assessment of horizontal and vertical axis turbines for river and tidal applications: A technology status review," *Applied Energy,* vol. **86(10)**, pp. 1823-1835, 2009.

[10] A. Kumar and R. Saini, "Performance parameters of Savonius type hydrokinetic turbine – A Review," *Renewable and Sustainable Energy Reviews,* vol. **64**, pp. 289-310, 2016.

[11] K. A. Ismail and T. P. Batalha, "A comparative study on river hydrokinetic turbines blade profiles," *International Journal of Engineering Research and Applications,* vol. **5(5)**, pp. 1-10, 2015.

[12] S. J. Savonius, "The S-Rotor and Its Application," *Mechanical Engineering,* vol. **53(5)**, pp. 333-338, May 1931.

[13] F. M. White, Fluid Mechanics, McGraw-Hill, 2003.

[14] V. Patel, G. Bhat, T. I. Eldho and S. V. Prabhu, "Influence of Overlap Ratio and Aspect Ratio on the Performance of Savonius Hydrokinetic Turbine," *International Journal of Energy Research,* vol. **41**, pp. 829-844, 2016.

[15] S. Roy and U. K. Saha, "Wind tunnel experiments of a newly developed two-bladed Savonius-style wind turbine," *Applied Energy,* vol. **137**, pp. 117-125, 2015.

[16] I. Mabrouki, Z. Driss and M. S. Abid, "Performance Analysis of a Water Savonius Rotor: Effect of the Internal Overlap," *Sustainable Energy,* vol. 2, no. 4, pp. 121-125, 2 July 2014.

[17] K. Golecha, T. Eldho and S. Prabhu, "Influence of the deflector plate on the performance of modified Savonius water turbine," *Applied Energy,* vol. **88 (9)**, pp. 3207-3217, 2011.

[18] S. Mauro, S. Brusca, R. Lanzafame and M. Messina, "CFD modeling of a ducted Savonius wind turbine for the evaluation of the blockage effects on rotor performance," *Renewable Energy,* vol. **141**, pp. 28-39, 2019.

[19] M. Basumatary, A. Biswas and R. Misra, "CFD analysis of an innovative combined lift and drag (CLD) based modified Savonius water turbine," *Energy Conversion and Management,* vol. **174**, pp. 72-87, 2018.

[20] N. Thakur, A. Biswas, Y. Kumar and M. Basumatary, "CFD analysis of performance improvement of the Savonius water turbine by using an impinging jet duct design," *Chinese Journal of Chemical Engineering,* vol. **27 (4)**, pp. 794-801, 2019.

[21] S. Roy and A. Ducoin, "Unsteady analysis on the instantaneous forces and moment arms acting on a novel Savonius-style wind turbine," *Energy Conversion and Management,* vol. **121**, pp. 281-296, 2016.

[22] S. Roy and U. K. Saha, "Computational study to assess the influence of overlap ratio on static torque characteristics of a vertical axis wind turbine," *Procedia Engineering,* vol. **51**, pp. 694-702, 2013.

IOP Conf. Series: Materials Science and Engineering **943** (2020) 012035 doi:10.1088/1757-899X/943/1/012035

Numerical Analysis of Exhaust Air Energy Extractor for Cooling Tower Applications

Enderaaj Singh, Sukanta Roy and Yam Ke San

Department of Mechanical Engineering, Faculty of Science and Engineering, Curtin University Malaysia

Email: sukanta.roy@curtin.edu.my

Abstract. Over the years, the need of renewable energy has been accelerated to minimise the reliance on using the fast depleting fossil fuel. Research on wind turbines to harness wind power is widely increasing especially with the progression of vertical axis wind turbines (VAWT) for the locations, where conventional horizontal axis wind turbines are not effective. The ability to provide renewable energy by generating efficient power is vital in combatting climate change due to increase in global energy demand thus, leading to this present study of extracting wind energy from a cooling tower. In this paper, a three dimensional numerical investigation has been conducted on the straight bladed Darrieus type VAWTs with S1046 airfoils. The implicit unsteady Reynolds-Averaged Navier-Stokes equations are solved using shear stress transport k-ω turbulent model. The VAWT model is placed at the outlet of the cooling tower to calculate power output with an aspect ratio of 0.88. The results show that a maximum power coefficient of 0.36 is obtained at a tip speed ratio of 3.0.

1. Introduction

Increase in population density with rapid commercial and social development have been causing the need of increasing energy generation to meet the ever-growing energy demand. Fossil fuels, a form of non-renewable energy is still largely the source of energy today and is expected to remain contributing 84% of the total energy demand in 2030 [1] as it provides cheaper and more reliable energy solution. Too much dependency on this carbon-based fossil fuel sources leads to high emission of carbon gases to the atmosphere, hence causing the global warming issues. The fossil fuel combustion which emits carbon dioxide (CO_2) excessively to the environment with 2012 recording 32.3 billion metric tons of CO_2 emission and is expected to continue rising reaching 43.2 billion metric tons by 2040 [2]. Moreover, if this growing energy demand continues to be met by fossil fuel at the current rate, chances for the fossil fuel resource to deplete completely in a shorter amount of time is high. Policies and global climate deal such as The Paris Agreement 2015, bringing all nations together for a common cause to keep the global rise in temperature capped at 2°C and further reduce it to 1.5°C in the future by constant effort in reducing greenhouse gases emissions with developed countries providing aid to share the burden with developing countries [3]. Thus, production of cleaner and sustainable energy sources has become increasingly crucial and need continuous focus to explore different alternative solutions. The available resources from International Renewable Energy Agency showed that there has been continuous increase in renewable energy capacity growth over the last 5 years [4]. However, it is still not enough to supply most of the increasing energy demand as the rate of growth is not fast enough to offset the effect of worldwide growing population and techno-economic and social needs.

In the above aspect, wind turbines are fund to be one of the most promising sources of renewable and sustainable energy generation, in large, medium, or small-scale applications [5]. Based on the axis of rotation, these wind turbines are classified into two broad categories, horizontal axis wind turbines (HAWT) and vertical axis wind turbines (VAWT). Many years of research and development focused on the HAWTs allowed possibility to widely utilize it for efficient commercial electricity generation in open field area with steady winds [6]. However, the rise in electricity needs and limited land space due to urbanization has directed more research to be done on the VAWTs to provide better solution to harness wind energy in areas where the HAWT is not suitable to be used. Advantages of VAWTs include simple design, low cut-in speed, lower noise level, adaptability to turbulent flows, flexibility of application in urban areas and its ease of maintenance. Being omnidirectional, i.e. capturing wind from any direction without yawing mechanism are added factor that give rise to research interest in this concept of wind turbines [7-8]. Based on the dominating aerodynamic forces, VAWTs can be further classified into lift type Darrieus wind turbines and drag type Savonius wind turbines [9]. Over the years, the evolution of the Darrieus wind turbines has led to two existing used designs: the curved eggbeater and the straight bladed H-rotor type. The better advantages of H-Darrieus turbines over its counterpart include a higher power coefficient and wider application capability that resulted it to be further researched and modified into designs such as fixed-pitch H-rotor, helical H-rotor, tilted H-rotor and also combined with Savonius type rotor [8]. The VAWTs are progressively being used in rural and urban areas to generate electricity to lessen the reliance on fossil fuel as an effort to battle climate issues. Ability to adapt to urban wind environment and being smaller in size allows the VAWTs to be installed at places such as on top of tall buildings, road side dividers, railway tracks and on top of lamp posts for local electricity generation [8]. However, locations with low speed wind conditions are unable to utilize the conventional way of wind power generation. An alternative approach of extracting power can be done by harnessing unnatural or artificial wind sources, such as exhaust air system of the cooling tower, at a consistent speed of about 5m/s to 16m/s [10] depending on the type and sizes of the cooling tower. Both atmospheric and mechanical induced cooling tower are used worldwide to remove heat waste from applications such as in thermal power plants, nuclear plants, oil refineries, chemical plants and heating, ventilation and air conditioning (HVAC) systems to cool buildings. The availability of these accelerated wind source through the cooling tower outlet for energy generation using wind turbines set off a new prospect to continue increasing the wind energy contribution in global renewable energy mix.

Studies expanding the scope of VAWTs is actively ongoing, and one interesting study on the prospect of generating clean energy by extracting wind energy coming out from the exhaust outlet of a cooling tower has been proposed by Chong et al. [11]. However, the reported VAWTs struggled with poor efficiency in converting wind energy to useful electrical energy forcing continuous effort to study and experiment the different parameters of turbines with diffuser plates and guide vanes through experimental and numerical analysis [11-15].

Till last two decades, the straight-bladed H-Darrieus wind turbines (HDWT) with its simple design, construction and manufacturing appealed many authors to do experiment on its geometric features in effort to optimize and improve the characteristics, producing the best performance out of this turbine. However, the application these turbines for exhaust air-energy extraction is very much limited to few studies. So, the present study is focused on a three-dimensional modelling of a wind energy extractor system by placing the HDWT at the outlet of a cooling tower and hereby analyze its performance for this application.

2. Methodology

2.1. Key performance parameters

The key design parameters which are used in this study are tip speed ratio, aspect ratio, power coefficient, moment coefficient, as described below with the standard equations.

IOP Conf. Series: Materials Science and Engineering **943** (2020) 012035 doi:10.1088/1757-899X/943/1/012035

Tip speed ratio (TSR), is the ratio of blade tip translational speed ($u = \omega R$) to the wind velocity (V_∞), where ω and R is the angular velocity and radius of the turbine respectively.

$$TSR = \frac{\omega R}{V_\infty} \quad (1)$$

Aspect ratio (AR) is the ratio of blade length (H) to the radius of the turbine (R), as expressed below:

$$AR = \frac{H}{R} \quad (2)$$

Moment coefficient, C_m and power coefficient, C_p are the parameters of interest to measure the performance of turbines, which are a function of wind velocity, V_∞, mechanical torque, T, output mechanical power, P and the turbine's swept area, A.

$$Swept\ area, A = H \times D \quad (3)$$

$$C_m = \frac{2T}{\rho ARV_\infty^2} \quad (4)$$

$$C_p = \frac{P_{turbine}}{P_{available}} = \frac{2P}{\rho AV_\infty^3} = C_m \times TSR \quad (5)$$

2.2. Cooling tower and turbine parameters

A proper dimensioning of the cooling tower is needed as a computational domain for the simulation to replicate the air flowing out of the tower's outlet as close as possible. Actual cooling tower parameters which was scaled down by factor of 200 used in a study by Liu et al. [16] was adapted in the present study. The parameters and dimensions of the cooling tower used as the shape of the computational domain are shown in figure 1 and table 1, respectively. The turbine selected for this study is a 3-bladed HDWT with S1046 airfoil with a turbine diameter of 456 mm and chord length of 45.6 mm. The aspect ratio is calculated as 0.88.

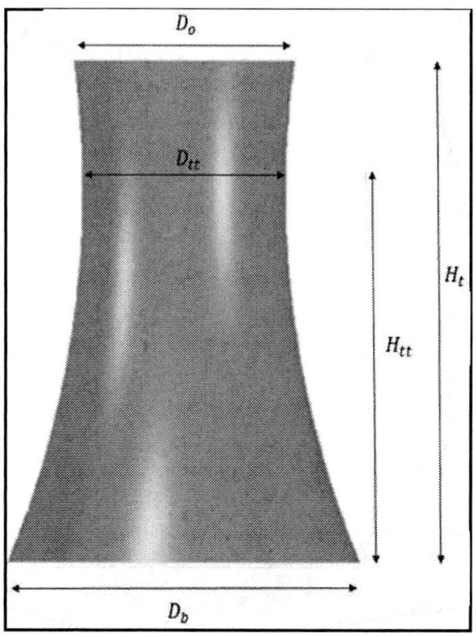

Figure 1. Cooling tower parameters.

Table 1. Cooling tower dimensions.

Parameter	Symbol	Dimension (mm)
Tower height	H_t	1175 mm
Base diameter of tower	D_b	910 mm
Outlet diameter of tower	D_o	570 mm
Throat diameter of tower	D_{tt}	525 mm
Height of throat from base	H_{tt}	881 mm

2.3. Computational domain & boundary conditions

The computational domain is shown in figure 2. The domain and HDWT is designed in a computer-aided design (CAD) software, Solidworks and imported into Star CCM+. The domain is divided into two regions, a rotating region consists of HDWT and a stationary region. The stationary region consists two control volume, a cooling tower control volume, and a cuboidal outer domain. The centre of HDWT is placed at $0.5D$ (D is the diameter of the turbine) from the cooling tower outlet at downstream. Cooling tower domain dimensions was set as mentioned in section 2.2 while the cuboidal domain dimensions were created large enough to avoid wall blockage and reverse flow effect as shown in figures 2 and 3.

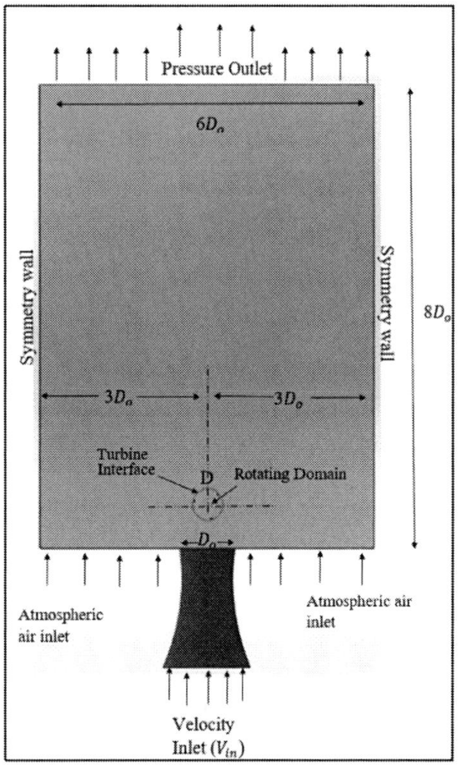

Figure 2. Computational domain

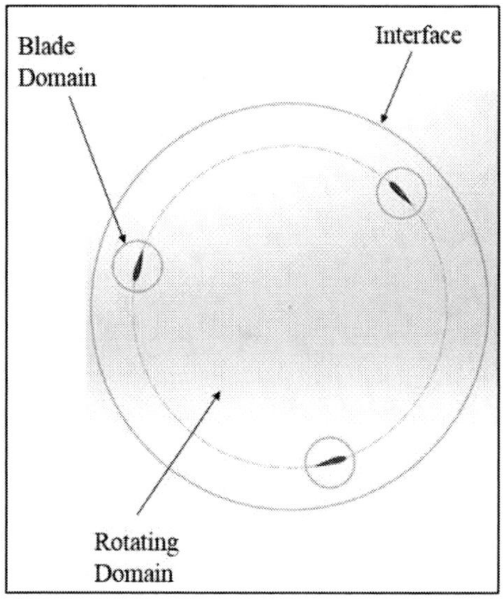

Figure 3. Turbine rotating computational domain

2.4. Meshing topology

In this study, polyhedral mesh is generated with boundary layers applied to the blade surface with growth rate of 1.1, and thickness of 3mm to achieve wall Y^+ value close to 1, as shown in figures 4 and 5.

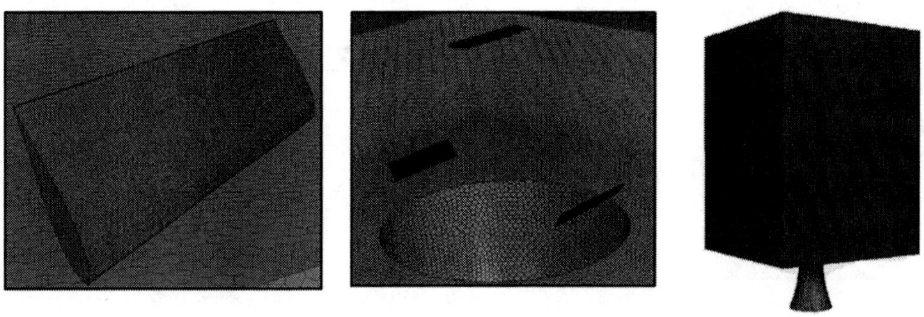

Figure 4. Three-dimensional computational domain.

2.5. Physics conditions

In this numerical study, three-dimensional implicit unsteady Reynolds-Averaged Navier-Stokes (URANS) equations were solved with shear stress transport (SST) k-ω turbulent model solver. The air density and dynamic viscosity was set constant at 1.18415 kg/m^3 and 1.85508 x 10^{-5} Pa.s respectively. Figure 5 illustrates the acceleration of the wind due to the converging cooling tower shape will result in higher wind speed at the outlet so, a inlet wind speed of 3.7 m/s was set such that the velocity obtained at the outlet of the cooling tower would be 9 m/s. Rotational speeds in the range of 47.37 rad/s – 421.05 rad/s were assigned to obtain TSRs in the range of 1.5 to 3.5. The time step was set to get azimuthal rotation increment of 5° which is acceptable for this study due to computational time constraints.

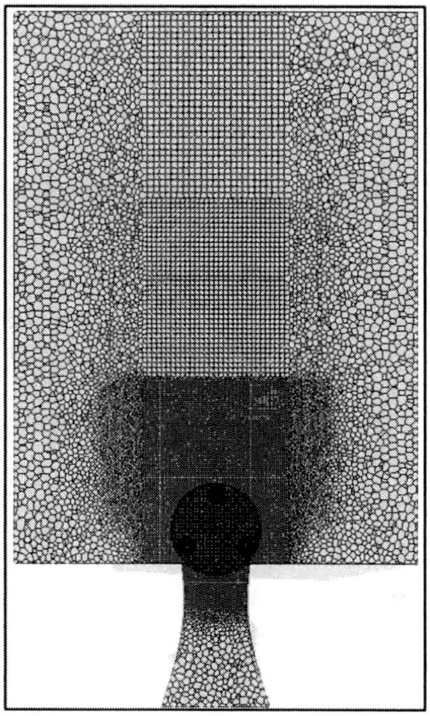

Figure 5. Sectional view of full domain

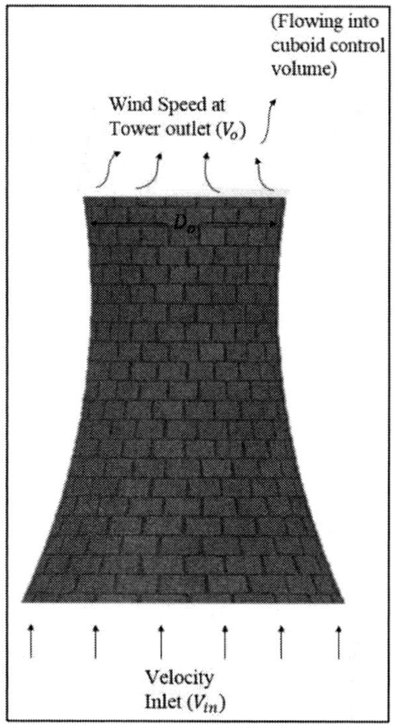

Figure 6. Cooling tower domain.

2.6. Governing equations

The SST k-ω model shows satisfactory efficiency and accuracy and has been used widely for turbulence modelling of VAWT. The mathematical formula associated with this turbulence model are as below:

Continuity equation:

$$\frac{\partial \rho}{\partial t} + \frac{\partial(\rho U_i)}{\partial x_i} = 0 \tag{6}$$

Momentum equation:

$$\frac{\partial(\rho U_i)}{\partial t} + \frac{\partial(\rho U_i U_j)}{\partial x_j} = -\frac{\partial p}{\partial x_i} + \frac{\partial\left(\tau_{ij} - \rho\overline{U_i U_j}\right)}{\partial x} \tag{7}$$

Where, the viscous tensor is expressed as:

$$\tau_{ij} = \mu\left(\frac{\partial U_j}{\partial x_i} + \frac{\partial U_i}{\partial x_j} - \frac{2}{3}\delta_{ij}\frac{\partial U_i}{\partial x_i}\right) \tag{8}$$

Transport equation for turbulent kinetic energy (k):

$$\frac{\partial(\rho k)}{\partial t}+\frac{\partial(\rho \bar{u}_i k)}{\partial x_i}=P_k+D_k+\frac{\partial}{\partial x_i}\left[(\mu+\sigma_k\mu_t)\frac{\partial k}{\partial x_i}\right] \tag{9}$$

Specific dissipation rate (ω):

$$\frac{\partial(\rho \omega)}{\partial t}+\frac{\partial(\rho \bar{u}_i \omega)}{\partial x_i}=P_\omega+D_\omega+\frac{\partial}{\partial x_i}\left[(\mu+\sigma_\omega\mu_t)\frac{\partial \omega}{\partial x_i}\right] \tag{10}$$

Where production terms are:

$$P_k=\mu_t\bar{S}^2-\frac{2}{3}\rho k\frac{\partial(\bar{u}_i)}{\partial x_i}-\frac{2}{3}\mu_t\left(\frac{\partial(\bar{u}_i)}{\partial x_i}\right)^2 \tag{11}$$

$$P_\omega=\rho\gamma\bar{S}^2-\frac{2}{3}\rho\gamma\omega\frac{\partial(\bar{u}_i)}{\partial x_i}-\frac{2}{3}\rho\gamma\left(\frac{\partial(\bar{u}_i)}{\partial x_i}\right)^2 \tag{12}$$

Where the term:

$$\bar{S}_{ij}=\frac{1}{2}\left(\frac{\partial\bar{u}_i}{\partial x_j}+\frac{\partial\bar{u}_j}{\partial x_i}\right) \tag{13}$$

Destruction terms:

$$D_k=-\rho\beta^* k\omega \tag{14}$$

$$D_\omega=-\rho\,\beta\omega^2 \tag{15}$$

Model coefficients are, σ_k, σ_ω, β^* and β, and μ is the dynamic viscosity of the fluid.

3. Model Validation

Accuracy of CFD simulations depends on a few factors and one of it is the quality of the grid. To accurately generate a 3D mesh for URANS simulation, two grid independence tests were conducted, one with only the cooling tower and cuboidal control volume with wake-refinement at the center stretching from the cooling tower outlet up to the pressure outlet were meshed, and second, with the turbine placed at the outlet of the cooling tower as described previously and refinement around the HDWT rotating domain and blade domain. Lastly, the final computational grid was validated using NACA 0021 VAWT against previous experimental and numerical results. The results from second grid convergence study (with turbine case) is shown in figure 7. Finally, 5.6 million cells were considered where the accuracy is assured with the difference in C_m value is less than 1% from the subsequent value.

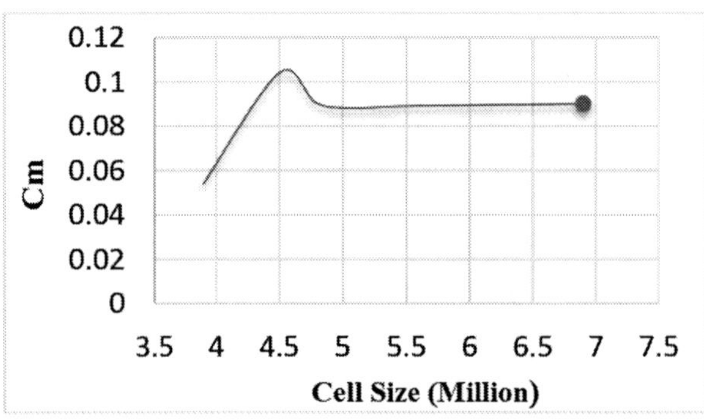

Figure 7. Grid independence study.

For validation, a three bladed NACA 0021 turbine with parameters shown in table 2 are used form the experimental data of Castelli [29] and simulation result from Hashem and Mohammed [23]. Figure 8 shows a good agreement of present numerical study to that of literature to allow the use of further study using this computational grid.

Table 2. Validation parameters.

Physics setup	Values
Airfoil profile	NACA 0021
Inlet wind speed (V_{in})	9.00 m/s
Turbine diameter (D)	0.515 m
Turbine span (H)	0.200 m
Solidity ratio (σ)	0.25
Tip speed ratio (TSR)	1.44 - 3.30

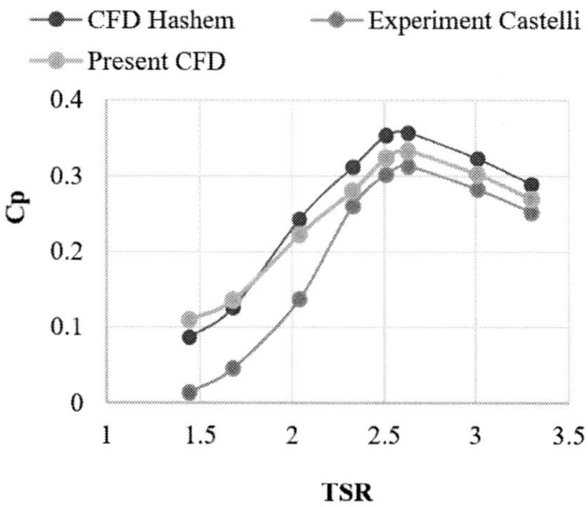

Figure 8. Computational model validation

4. Results and Discussion

Following the development of computational model and validation, a series of numerical simulations were conducted with three-bladed HDWT using S-1046 airfoils for the blades. In this paper, the simulation results using aspect ratio of 0.88 has been reported at TSR in the range of 1.5 to 4.0. Figures 9 and 10 shows the moment coefficient and power coefficient plots for the mentioned HDWT model, placed at half the diameter of turbine from the outlet of the cooling tower.

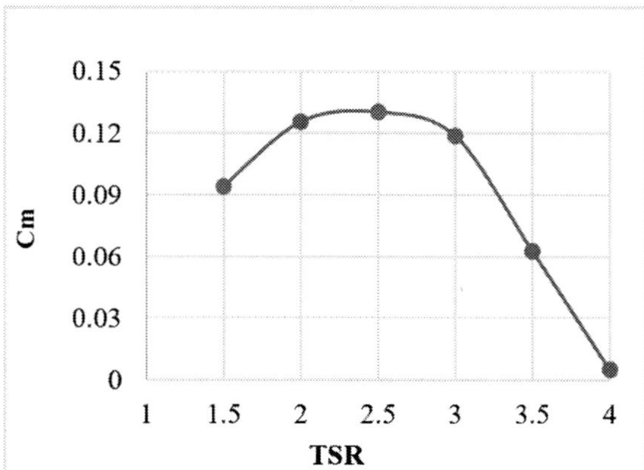

Figure 9. Moment coefficient around HDWT

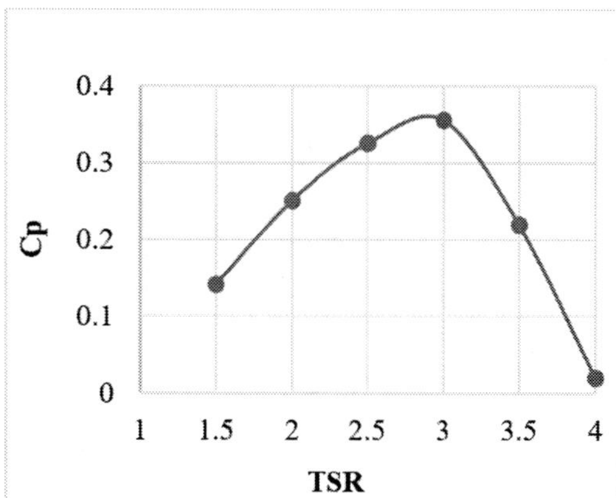

Figure 10. Power coefficient around HDWT

From the obtained results of moment coefficient, the power coefficient values are calculated by multiplying with the respective TSR values. It has been observed at TSR of 3.0, the maximum power coefficient is 0.36 for the tested model of HDWT. Further, for the complete turbine rotations, the instantaneous moments and power coefficients are analyzed with respect to 15-degree interval of azimuthal angle, as shown in Figures 11 and 12. Velocity contour of the air flow for the turbine at chosen critical azimuthal angles of θ=30°, 60°, 90° and 120° at low TSR of 2.0 and high TSR of 3.5 are shown in figure 13 and 14 respectively to analyse the flow characteristics and compare the difference in performance.

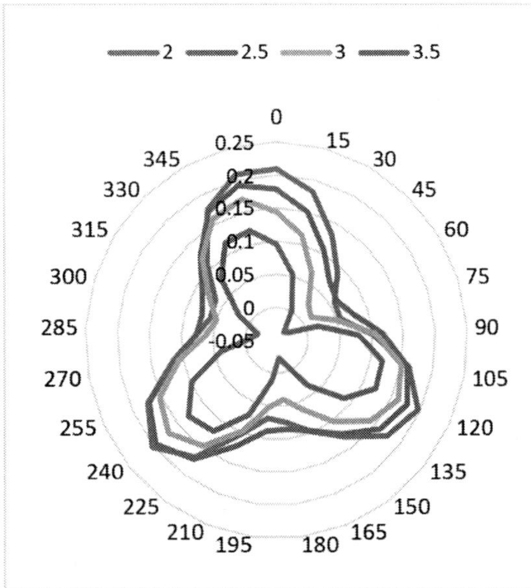

Figure 11. Instantaneous coefficient of moment versus rotor blade azimuth angles at TSR 2.0, 2.5, 3.0 and 3.5.

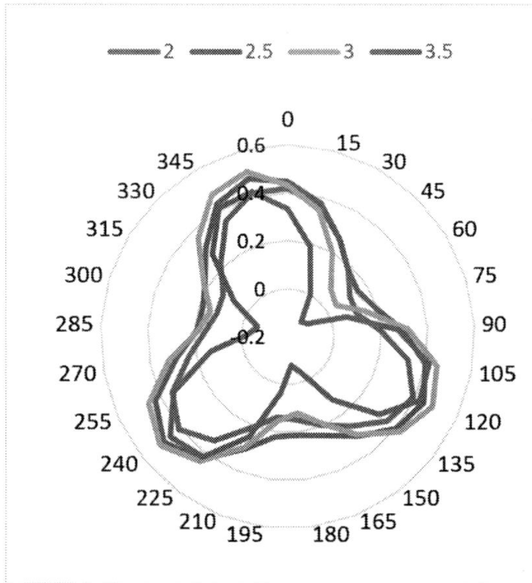

Figure 12. Instantaneous coefficient of power versus rotor blade azimuth angles at TSR 2.0, 2.5, 3.0 and 3.5.

The radar plot shows that the turbine has better moment coefficient at lower TSR of 2.0, 2.5 and 3.0 as it produce positive moment for a complete turbine rotation while at TSR of 3.5, it shows much lower moment coefficient with negative moment at some azimuth angles. Overall, TSR 3.0 gives highest power coefficient at most azimuth angles as seen in Figure 12.

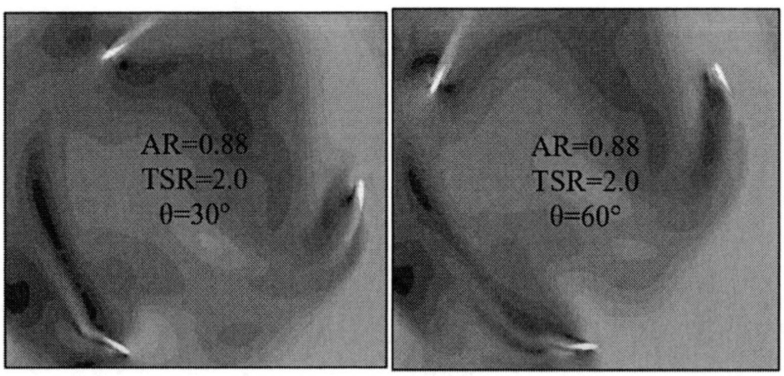

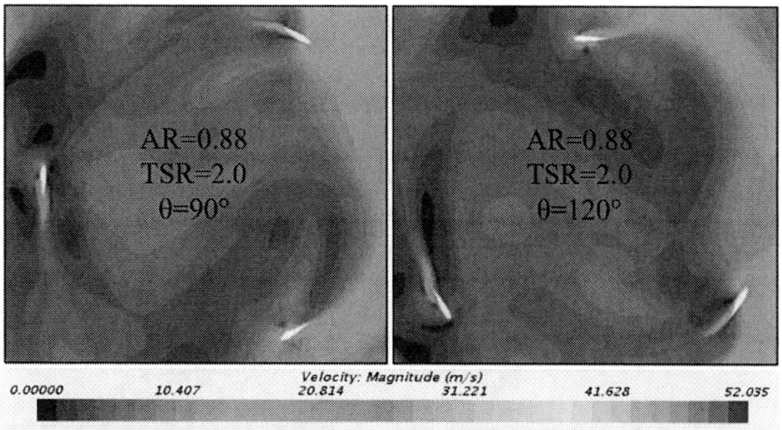

Figure 13. Velocity magnitude contour of AR=0.88 at TSR=2.00.

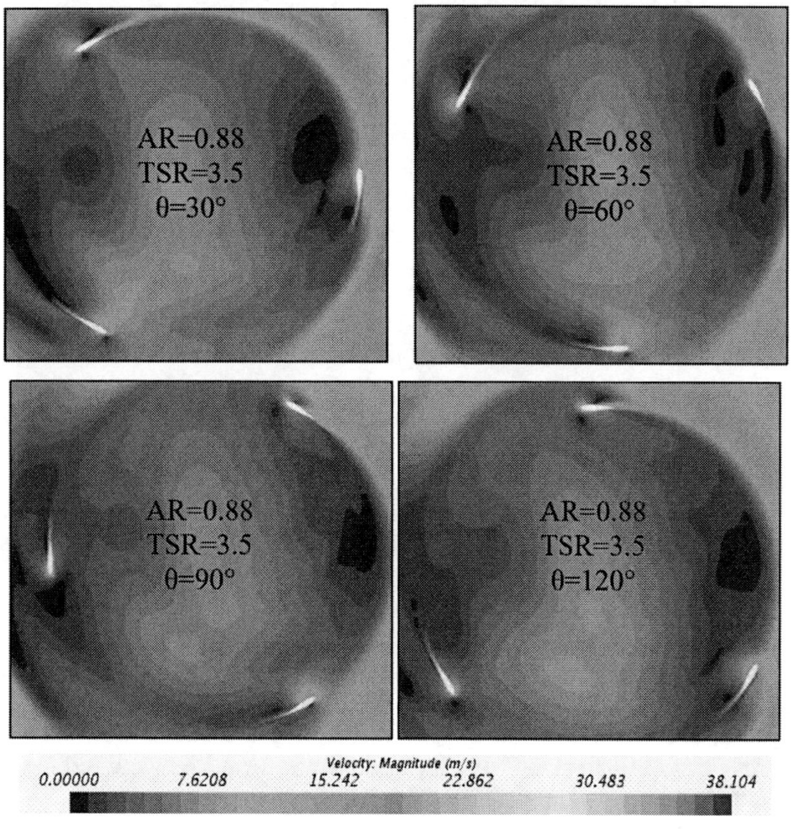

Figure 14. Velocity magnitude contour of AR=0.88 at TSR=3.50

4.1 Velocity contour at low TSR=2.00. There is presence of high wind velocity of 35 m/s to 50 m/s building up around the turbines. The turbine gave lowest C_m at azimuth angles between 60° to 90° as flow separation of high wind velocity occurs at blades 1 and 3 as can be seen in figure 13. At azimuth position of 60°, higher wind velocity located at the lower part of blades 2 and 3 for both the 0.88 aspect ratio turbine results in lesser lift generation. At 120°, the high velocity difference between the top and bottom side of the leading-edge turbine blades produces high pressure difference with lower pressure region above the blades allowing greater lift generations for the turbines at this azimuth angle. Overall, highest wind velocity regions are concentrated at the right side of the turbines as the air exiting the cooling tower is guided by the anticlockwise motion of the turbine towards the right side.

4.2 Velocity contour at high TSR=3.50. Figure 14 shows the minimal difference of pressure between top and bottom of the blades which obstructs lift generation for this high TSR turbine. The presence of wake regions from the blades at this TSR affects the performance slightly compared to lower TSR. This explains the negative moment generated at azimuth angles of 60° for the blades. The turbine still produces positive moment as it gives high difference in wind velocity of about 14-21 m/s between the top and bottom of blades 2 and 3 at positions 90° to 120° resulting in pressure difference which generates lift. The velocity contour as shown in figure 14 suggest the same comparison whereby TSR higher than 3.0 does not give any better performance compared to the lower TSR of 2.0, 2.5 and 3.0.

5. Conclusion

A computational analysis of wind turbine placed at cooling tower exhaust has been conducted using a straight-bladed H-Darrieus wind turbine (HDWT) with S1046 airfoil. With a turbine aspect ratio of 0.88, a maximum power coefficient of 0.36 is obtained at TSR 3.0. However, this is just a preliminary study focused on the computational mesh and model development for the three-bladed HDWT in a cooling tower application. The successful validation also signifies the outcome of the study. However, several studies being conducted on the aspect ratio and optimising the location of the turbine. The effect of guide vanes and diffusers to further improve the performance will also be investigated in future work.

Acknowledgement

The authors would like to thank Curtin University Malaysia for providing the facility, especially the Star CCM+ software license to carry out the numerical study. Sincere gratitude is also dedicated to Ministry of Higher Education (MOHE) Malaysia for the financial support by awarding the Fundamental Research Grant Scheme (FRGS Grant: FRGS/1/2018/TK10/CURTIN/03/1) to carry out this project.

References

[1] S. Shafiee and E. Topal, "When will fossil fuel reserves be diminished?," *Energy Policy*, vol. **37** (1), pp. 181–189, 2009.
[2] A. Kibria, S. B. Akhundjanov, and R. Oladi, "Fossil fuel share in the energy mix and economic growth," *International Review of Economics & Finance*, vol. **59 (C)**, PP. 253–264, 2019.
[3] A. Savaresi, "The Paris agreement: A new beginning?," *Journal of Energy & Natural Resources Law,* vol. **34(1)**, pp. 16-26, 2016.
[4] IRENA, "Renewable capacity highlights," 2019.
[5] S. Roy and U.K. Saha, "Numerical investigation to assess an optimal blade profile for the drag based vertical axis wind turbine." *ASME 2013 International Mechanical Engineering Congress & Exposition, Proceedings (IMECE)*, pp. V06AT07A084, 2013.
[6] N. Ma, H. Lei, Z. Han, D. Zhou, Y. Bao, K. Zhang, L. Zhou, and C. Chen, "Airfoil optimization to improve power performance of a high-solidity vertical axis wind turbine at a moderate tip speed ratio," Energy, vol. **150(1)**, pp. 236–252, 2018.
[7] S. Roy, P. Mukherjee and U.K. Saha, "Aerodynamic performance evaluation of a novel Savonius-style wind turbine under an oriented jet. ASME 2014 Gas Turbine India Conference, *ASME Proceedings:GT Cycle Innovations, Renewable Applications*, pp. V001T08A001, 2014.
[8] R. Kumar, K. Raahemifar, and A. S. Fung, "A critical review of vertical axis wind turbines for urban applications," *Renewable and Sustainable Energy Reviews*, vol. **89**, pp. 281–291, 2018.
[9] S. Roy and U.K. Saha, "Performance analysis of Savonius-style wind turbines under concentrated and oriented jets," ASME Turbo Expo 2014 Turbine Technical Conference and Exposition, *ASME Proceedings: Wind Energy*, pp. V03BT46A027, 2014.
[10] J.C. Hensley, "*Cooling Tower Fundamentals.* Compiled from the knowledge and experience of the entire SPX Cooling Technologies staff. SPX Cooling Technologies," Inc. Overl. Park.

Kansas USA, 2009.

[11] W.T. Chong, S.C. Poh, A. Fazlizan, S.Y. Yip, C.K. Chang, and W.P. Hew, "Early development of an energy recovery wind turbine generator for exhaust air system," *Applied Energy*, vol. **112**, pp. 568–575, 2013.

[12] W.T. Chong, S.Y. Yip, A. Fazlizan, S.C. Poh, W.P. Hew, E.P. Tan, and T.S. Lim, "Design of an exhaust air energy recovery wind turbine generator for energy conservation in commercial buildings," *Renewal Energy*, vol. **67**, pp. 252–256, 2014.

[13] W.T. Chong, W.P. Hew, S.Y. Yip, A. Fazlizan, S.C. Poh, E.P. Tan, and H.C. Ong, "The experimental study on the wind turbine's guide-vanes and diffuser of an exhaust air energy recovery system integrated with the cooling tower," *Energy Conversion and Management*, vol. **87**, pp. 145–155, 2014.

[14] S. Tabatabaeikia, N.N.B.N. Ghazali, W.T. Chong, B. Shahizare, N. Izadyar, A. Esmaeilzadeh, and A. Fazlizan, "Computational and experimental optimization of the exhaust air energy recovery wind turbine generator," *Energy Conversion and Management*, vol. **126**, pp. 862–874, 2016.

[15] A. Fazlizan, W. T. Chong, S. Y. Yip, W. P. Hew, and S. C. Poh, "Design and experimental analysis of an exhaust air energy recovery wind turbine generator," *Energies*, vol. **8**, no. 7, pp. 6566–6584, 2015.

[16] Z. Liu, C. Zhang, and T. Ishihara, "Numerical study of the wind loads on a cooling tower by a stationary tornado-like vortex through LES," Journal of Fluids and Structures, vol. **81**, pp. 656–672, 2018.

[17] I. Hashem and M. H. Mohamed, "Aerodynamic performance enhancements of H-rotor Darrieus wind turbine," *Energy*, vol. **142**, pp. 531–545, 2018.

[18] M. Raciti Castelli, A. Englaro, and E. Benini, "The Darrieus wind turbine: Proposal for a new performance prediction model based on CFD," *Energy*, vol. **36(8)**, pp. 4919–4934, 2011.

Aerodynamic Analysis of Active Trailing-Edge Flaps and Passive Surface Roughness on NACA 4412 Airfoil

Thanesh A/L Manickam and Sukanta Roy

Email: sukanta.roy@curtin.edu.my

Abstract. The focus of this study is to understand the effect of the flow control mechanism on NACA 4412 airfoil. Two configuration of flow control method is used in this investigation. The first configuration is, airfoil with only the trailing-edge flap. This is the active flow control method. The trailing-edge flap can be deflected to various angle accordingly. The second configuration is a combination of, active trailing-edge flap with a passive surface roughness at the leading-edge of the airfoil. The surface roughness is expected to reduce the flow separation at the leading-edge of the airfoil. This is expected to enhance the aerodynamic performance of the airfoil. Both configuration 1 and configuration 2 was simulated and the results compared with the base airfoil without any flow control mechanism. Based on the results, airfoil with configuration 1 performed better compared to the base airfoil. Airfoil with configuration 2, performed well but not better than the airfoil with configuration 1. Upon further analysis, key patterns emphasise airfoil with configuration 2 could perform better if the roughness element parameters further optimised. This airfoil to be used for wind turbine applications.

1. Introduction

Surge of 4 percent in electricity consumption globally, from the year 2017 to 2018 was recorded [1]. This is most obvious in Southeast Asian countries, as most are considered as developing nations and require more energy due to industrialization [2]. As most power plant uses fossil fuels, increase in greenhouse gases raises the environmental concerns. Countries such as Malaysia, Indonesia, Vietnam and Thailand have extensive coastlines which makes it suitable for wind energy utilization by erecting wind turbines. This can help in decarbonizing the generation of electricity. Optimized wind turbines can improve reliability of utilizing wind energy. To improve wind turbines performances, optimization of airfoil is required as it will help to generate a greater lift coefficient which is crucial to wind turbine blade characteristics [3].This paper aims to investigate the effect of active flow control method (configuration 1) and a combination of active with passive flow control method (configuration 2) on NACA 4412.

1.1 Flow Control Methods.

Trailing-edge flaps, a common component used as an active-flow control mechanism. Adjusting flaps on turbine blades to a sinusoidal cycle per revolution eliminates cyclic loading acting on wind turbines [4]. Flaps can also be used to reduce vortex down-wash which causes drag, by employing a morphed trailing-edge flap with an extended radial length [5]. As to improve lift to drag coefficient ratio, high chambered flap profile with a high deflection angle can be used [6]. Lift coefficient can also be improved by increasing the exposed surface area by employing extrados flexible trailing-edge flaps [7].

Passive flow control method is usually designed on the surface of an airfoil to reduce the flow separation that occurs at high angle of attack (AOA). Air slots, a common passive flow control feature. An airfoil with an air slot wind inlet at the leading-edge and outlet at the trailing-edge respectively improves the lift coefficient [8]. Without optimization of air slot feature, undesired drop in lift coefficient is produced [9]. Surface roughness element near the leading-edge of an airfoil is another passive flow control feature. This feature reduces the flow separation of an airfoil despite showing a reduction of lift force coefficient due to increase in skin friction [10]. By optimizing the roughness element's height and location near the leading-edge of the airfoil, an increase in the lift to drag coefficient ratio as well as reduction of flow separation on the airfoil was attained [11]. Another passive feature is attachment of oscillating micro-cylinders at the leading-edge. By optimizing oscillating frequency, improvement of lift coefficient is recorded [12].

Based on observation, a study of both this flow control method combined on an airfoil is insufficient. Therefore, this paper adapts the surface roughness element near the leading-edge of the airfoil as the passive flow control element and a simple trailing-edge flap as the active flow control mechanism. The surface roughness element is expected to reduce the flow separation that would occur in an airfoil at high angle of attacks. The simple trailing-edge flap is attached to increase the surface area. The flow around the airfoil and lift to drag ratio is analyzed.

2. Methodology

This section explains the selection of the airfoils, physical parameters used in this investigation, computational methodology, governing equation applied, airfoil modelling, simulation domain and boundary condition as well as the mesh generation.

2.1. Airfoil Selection

The airfoil selected as the base for the study is NACA 4412 airfoil. The airfoil has a chord length of 175mm and a width of 100mm. Plot of NACA 4412 is shown in figure 1. For the trailing-edge flap, NACA 64-008A airfoil was selected. The airfoil had a chord length of 6.25% of the base airfoil length. A plot of NACA 64-008 A is depicted in figure 2.

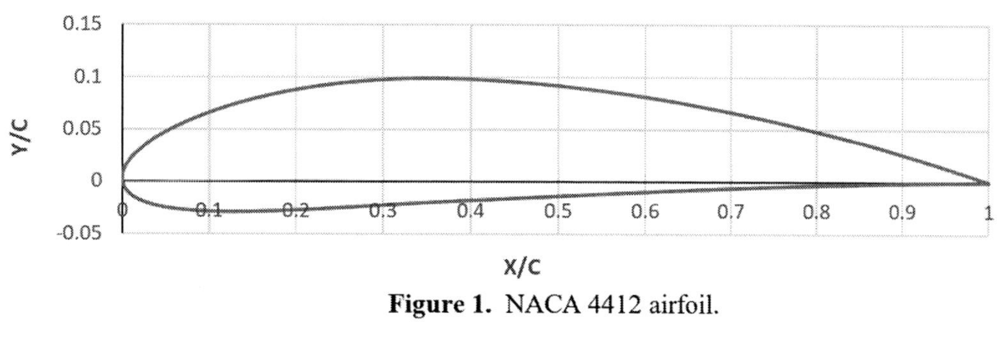

Figure 1. NACA 4412 airfoil.

2.2. Physical Parameters

The surface roughness element used at the leading-edge of the NACA 4412 airfoil is based on a previously optimised value [11]. Therefore, the ratio of height of the surface roughness element to airfoil chord length (k/C) is 0.003, making the roughness height 0.51mm. The location of surface roughness element on the airfoil is at 15% from the leading-edge up to 25% of the chord length. Other conditions are as summarised in the table 1 below.

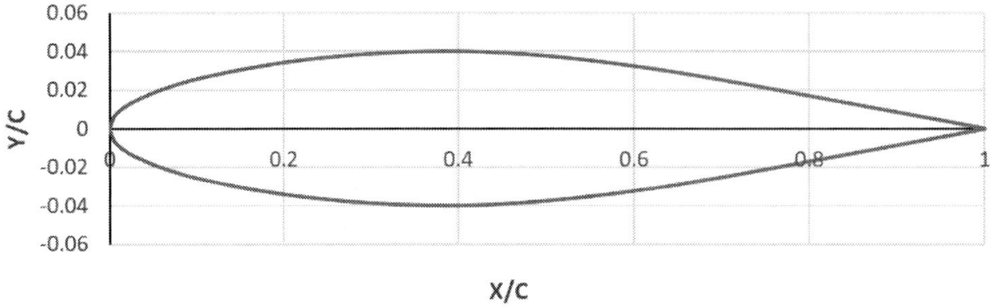

Figure 2. NACA 64-008A airfoil.

Table 1. Physical parameters used for investigation.

Features	**Values**
Angle of attacks (AOA)	0°, 5°, 10°, 15°
Inlet wind speed	6 m/s, 12m/s, 18 m/s
Deflection of trailing-edge flaps	0°, 5°, 10°,15°,

2.3. Computational Methodology

This investigation is carried out using computational fluid dynamic (CFD) approach. In Star-CCM+, a three-dimensional analysis using the Unsteady Reynold's-Averaged Naiver Strokes (U-RANS) model was created. A three-dimensional setup is necessary to provide an accurate result. This is due to the introduction of the roughness element on the surface of the airfoil. To understand the flow of near wall condition, $SST\ k-\omega$ turbulence model was selected for this investigation. Star-CCM+ solver was utilized in solving the governing equation of mass and momentum conservation.

2.4. Governing Equation.

Following equations applied to solve airflow problems around the airfoil. The main equation consists of continuity of flow equation and momentum equation.

Continuity equation:

$$\frac{\partial \rho}{\partial t}+\frac{\partial(\rho U_i)}{\partial x_i}=0$$

Momentum equation:

$$\frac{\partial(\rho U_i)}{\partial t}+\frac{\partial(\rho U_i U_j)}{\partial x_j}=-\frac{\partial p}{\partial x_i}+\frac{\partial(\tau_{ij}-\rho\overline{U_i U_j})}{\partial x}$$

Where, the viscous tensor is expressed as:

$$\tau_{ij}=\mu\left(\frac{\partial U_j}{\partial x_i}+\frac{\partial U_i}{\partial x_j}-\frac{2}{3}\delta_{ij}\frac{\partial U_i}{\partial x_i}\right)$$

Transport equation for turbulent kinetic energy (k):

$$\frac{\partial(\rho k)}{\partial t}+\frac{\partial(\rho \overline{u_i} k)}{\partial x_i}=P_k+D_k+\frac{\partial}{\partial x_i}\left[(\mu+\sigma_k\mu_t)\frac{\partial k}{\partial x_i}\right]$$

Specific dissipation rate (ω):

$$\frac{\partial(\rho\omega)}{\partial t}+\frac{\partial(\rho\bar{u}_i\omega)}{\partial x_i}=P_\omega+D_\omega+\frac{\partial}{\partial x_i}\left[(\mu+\sigma_\omega\mu_t)\frac{\partial\omega}{\partial x_i}\right]$$

Where the production terms:

$$P_k=\mu_t\bar{S}^2-\frac{2}{3}\rho k\frac{\partial(\bar{u}_i)}{\partial x_i}-\frac{2}{3}\mu_t\left(\frac{\partial(\bar{u}_i)}{\partial x_i}\right)^2$$

$$P_\omega=\rho\gamma\bar{S}^2-\frac{2}{3}\rho\gamma\omega\frac{\partial(\bar{u}_i)}{\partial x_i}-\frac{2}{3}\rho\gamma\left(\frac{\partial(\bar{u}_i)}{\partial x_i}\right)^2$$

Where the term

$$\bar{S}_{ij}=\frac{1}{2}\left(\frac{\partial\bar{u}_i}{\partial x_j}+\frac{\partial\bar{u}_j}{\partial x_i}\right)$$

Destruction terms:

$$D_k=-\rho\beta^*k\omega$$
$$D_\omega=-\rho\,\beta\omega^2$$

Model coefficients are,σ_k, σ_ω, β^* and β. Where, μ is the dynamic viscosity of the fluid

2.5. Airfoil Modelling.

The airfoil was modelled using SolidWorks software. A three-dimensional model of the base airfoil, airfoil with trailing-edge flap (configuration 1) and airfoil with combined trailing-edge flap with surface roughness (configuration 2) is modelled as depicted in figures 3, 4 and 5. The angle of deflection of the trailing-edge flap in this investigation is calculated relative to the chord line of the NACA 4412 airfoil as shown in figure 7.

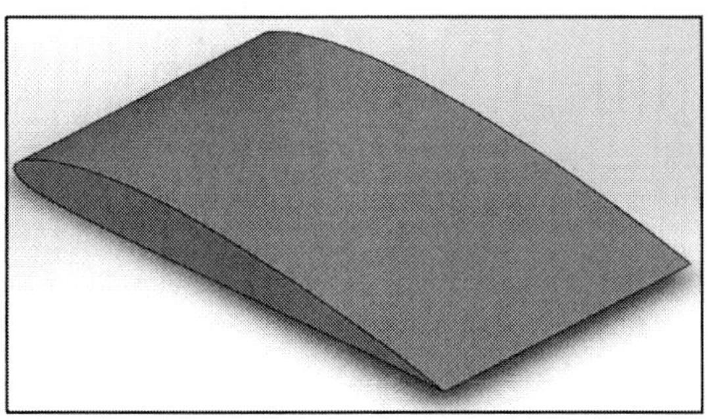

Figure 3. Base NACA 4412 airfoil.

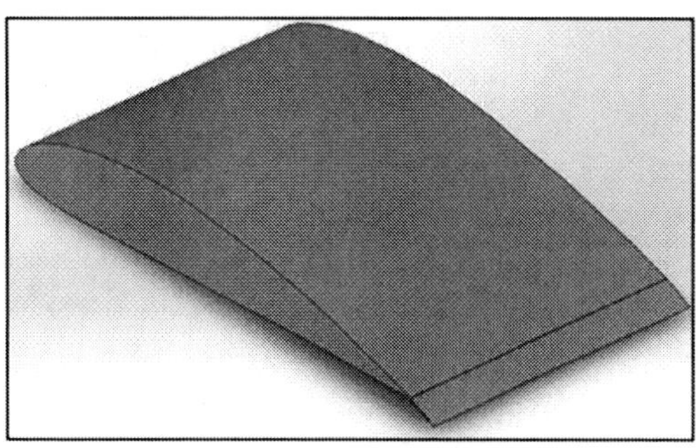

Figure 4. Configuration 1: trailing-edge flap.

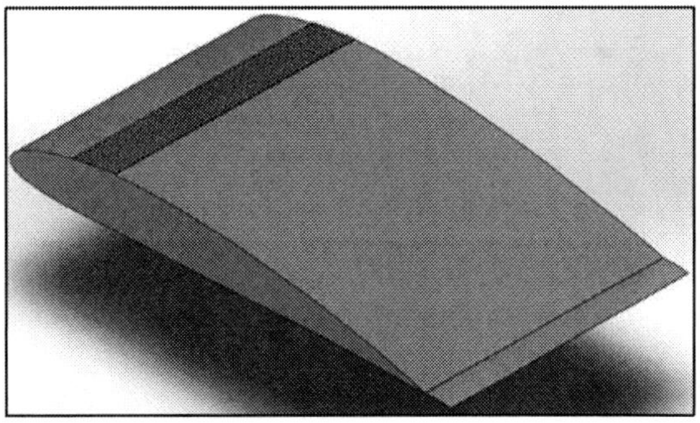

Figure 5. Configuration 2: airfoil with combined trailing-edge flap with surface roughness

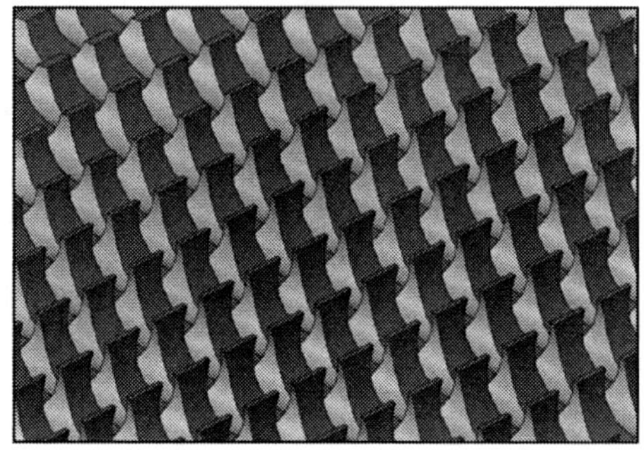

Figure 6. Zoomed view of Surface roughness element on configuration 2.

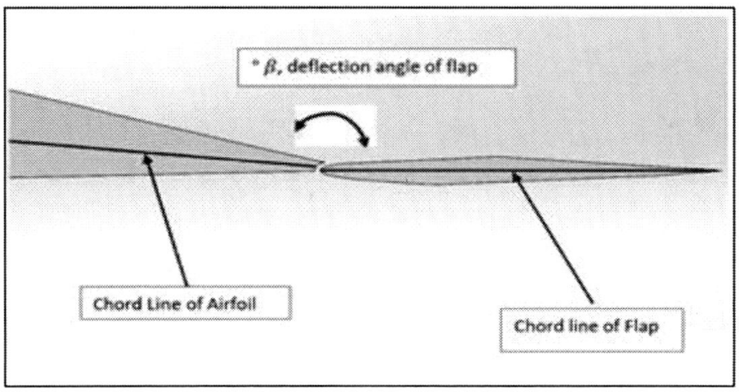

Figure 7. Angle of deflection.

2.6. Simulation Domain and Boundary Condition.

The domain was created in Star-CCM+. The wind inlet is 900mm away from the leading-edge of the airfoil. The wind outlet is 1800mm away from the trailing-edge of the airfoil. The height of the domain 2440mm and the width of the domain is 500mm. The wind inlet condition set to be velocity inlet and the wind outlet condition to pressure outlet. Both sides of the domain wall set to be interference with periodic translational to define a confined space. The domain was setup to be sufficiently large to avoid disturbance of flow over the airfoil from the induced inlet speed and remove confinement effects.

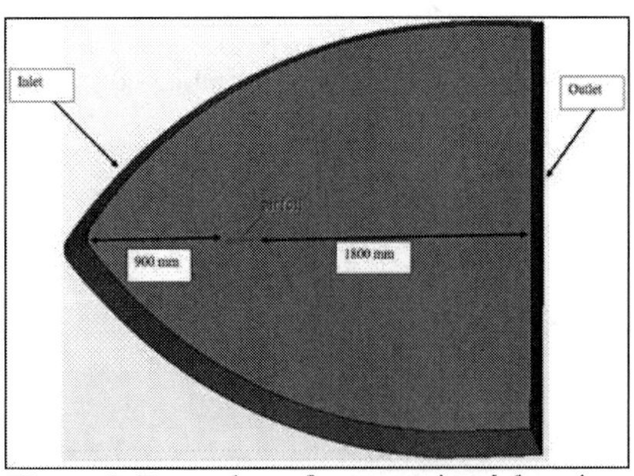

Figure 8. Front view of computational domain

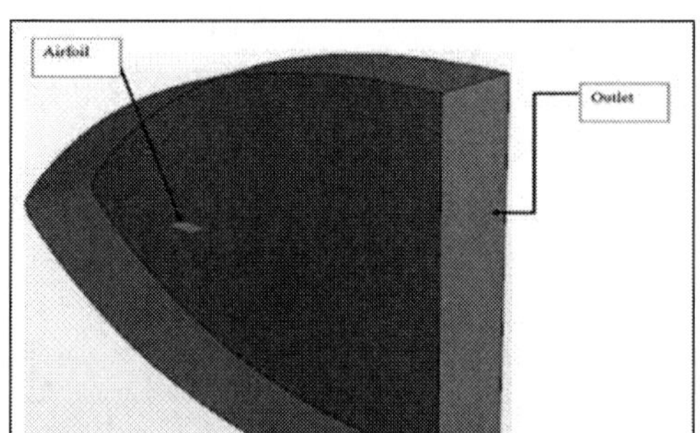

Figure 9. 3D view of the computational domain.

2.7. Mesh Generation.

A polyhedral mesh is used in this investigation with a wall Y^+ value close to 1. The mesh had a base value of 10mm and volume of refinement of 5mm. The mesh around the airfoil, refined to value of 3.1mm minimum size and 4.0mm target size. The mesh near the trailing-edge flap, refined with to minimum value of 0.05mm and a target value of 4.0mm. Finally, to accurately capture the flow changes near the surface roughness element, mesh around the surface roughness was set to be the finest with a minimum value of 0.05mm and a target size of 1.0mm. The number of prism layer used in this investigation is 17 with a stretching factor of 1.2mm for the first layer and 0.5mm for the subsequent layers.

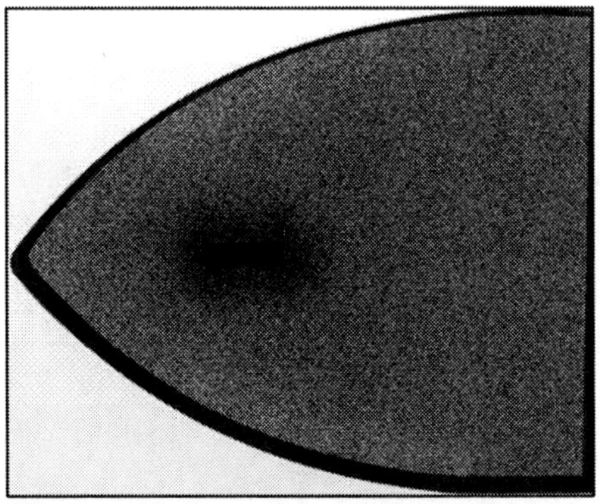

Figure 10. Overall mesh in the domain

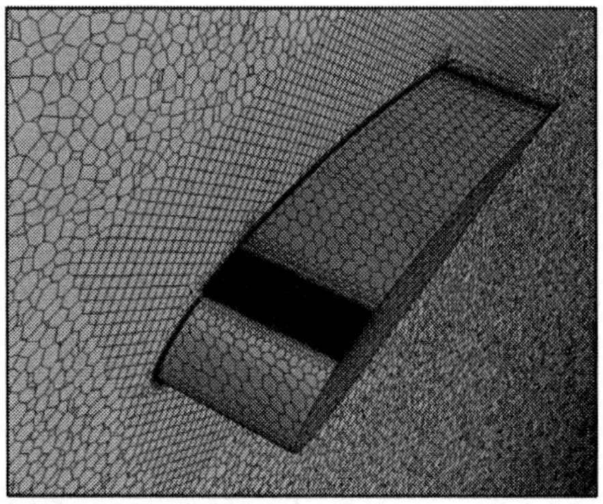

Figure 11. Mesh generation over airfoil

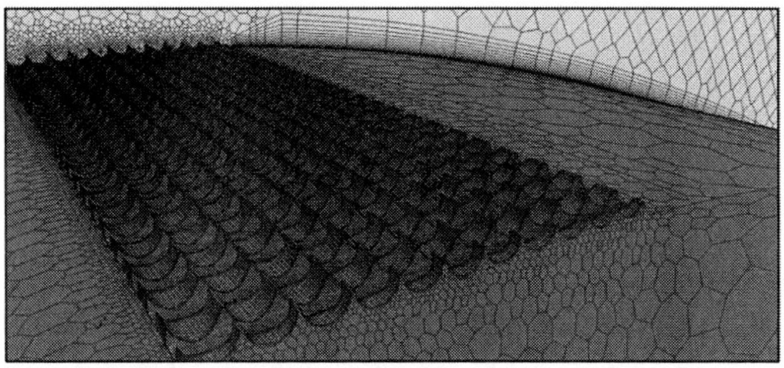

Figure 12. Surface roughness mesh generation.

2.8. Validation.

The validation study is initially conducted to ensure the computational setup is accurate and able to mimic the results of a published paper [7]. This would provide confidence to the results obtained later in this investigation. The obtained result is very closed to a published result. Slight variation in the results is acceptable. Thus, this validates the simulation setup used for the investigation to be adequate.

3. Results

3.1. Configuration 1.

From AOA of 0° up to 4°, it is observed that airfoil with flap with a deflection angle of 15° has the highest lift to drag coefficient ratio. Beyond AOA of 5°, a drop in lift to drag coefficient ratio is recorded and significantly reduces at angle of attack of 15°. At AOA of 0°, lift to drag coefficient ratio of 36.57

percent higher than the base airfoil is generated a significant increase in airfoil performance. But the value dropped below to 10.22 percent lower than of the base airfoil at a higher AOA of 15°.

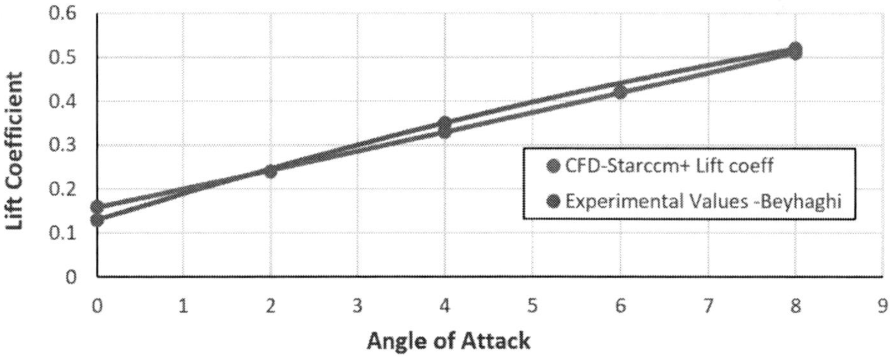

Figure 13. Validation of CFD Results with results obtained from Beyhaghi and Amano [7].

For airfoil with flap of 10° deflection angle, a lower lift to drag coefficient ratio is generated compared to the airfoil with flap of 15° deflection angle in a lower AOA of 0° up to 4°. Beyond an AOA of 4°, this airfoil generated a higher lift to drag coefficient ratio compared to the airfoil with flap deflection angle of 15°. At AOA of 0°, this airfoil generated lift to drag coefficient ratio higher compared to the base airfoil by 23.25%.

For airfoil with flap 5° deflection angle, lift to drag coefficient ratio higher than the base airfoil for a longer AOA from 0° up to 7° was generated. At the AOA of 6°, this airfoil managed to obtain the highest lift to drag coefficient ratio.

For airfoil with trailing-edge flap of 0° deflection angle, from an AOA of 0° up to 8° it had generated the lowest lift to drag coefficient ratio compared to the base airfoil and airfoil with other flap deflection angles. As the AOA is increased, lift to drag coefficient ratio generated became close to the value generated by the base airfoil. It can be summarised that, for airfoil with configuration 1 flow control mechanism the higher flap deflection angle performed well in a lower AOA. Airfoil with lower deflection angle performed better at higher AOA. The values of lift to drag coefficient ratio of airfoil with different flap angles in different angle of attack is plotted as shown in figure 14.

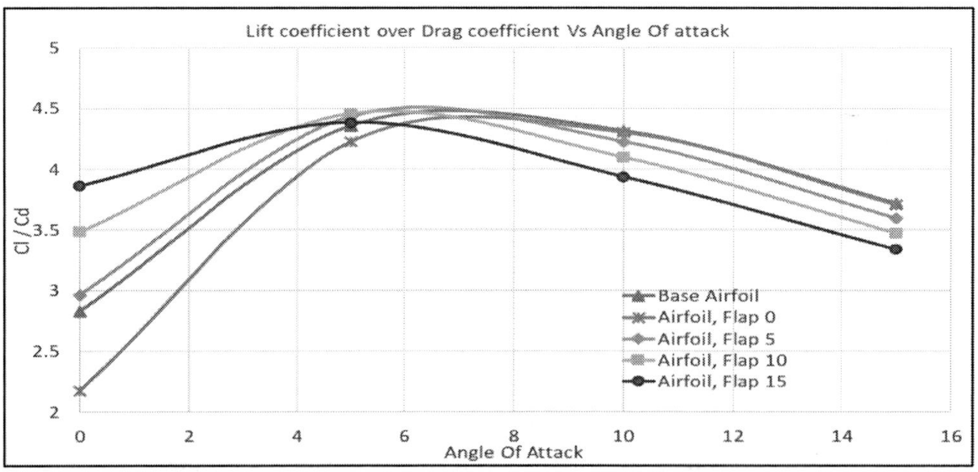

Figure 14. Lift to drag coefficient ratio for configuration 1.

3.2. Configuration 2.

For airfoil with a trailing-edge flap of 15° deflection angle, highest lift to drag coefficient ratio was obtained at lower AOA of 0° up to 4°. Drop in the lift to drag coefficient ratio was noticed beyond the AOA of 4°. The lift to drag coefficient ratio of 29.25 percent higher than the base airfoil recorded for the AOA of 0°.

For airfoil flap angle of 10°, it displayed a similar pattern of lift to drag coefficient ratio when plotted against the AOA. At lower AOA of 0° up to 4° it generated a lift to drag coefficient ratio higher than the base airfoil. Lift to drag coefficient ratio of 18.48 percent higher than the base airfoil was obtained for the AOA of 0°. At AOA of 4°, the lift to drag coefficient value begins to reduce. The lift to drag coefficient ratio continues to drop as the AOA is continuously increased.

For airfoil with flap angle of 5°, lift to drag coefficient ratio with small difference than that of the base airfoil was generated. Lift to drag coefficient ratio of 0.53 percent lower than the base airfoil was obtained at 0° AOA. This value is insignificant difference compared to the base airfoil value and can be said to have similar or a close aerodynamic performance of the base airfoil.

For airfoil with flap of 0°, the lowest lift to drag coefficient ratio was obtained. Lift to drag coefficient ratio of 24.62 percent lower than the base airfoil was obtained at 0° AOA and this is the lowest value generated by an airfoil studied in this investigation for configuration 2. Increment in the AOA showed that the lift to drag coefficient value gets closer to the value generated by the base airfoil.

To summarise the data obtained for configuration 2, it shows a similar pattern as the airfoil in configuration 1. This is because, as the deflection angle of the flap is increased it performs better in a lower AOA. Lower deflection angle of the flap has a better performance at high AOA. Although, the airfoil in configuration 1 and 2 has the same pattern when comparing the data of lift to drag coefficient at various AOA, configuration 2 had generated a lower lift to drag coefficient ratio compared to the airfoil in configuration 1. The lift to drag coefficient ratio against AOA is shown in figure 15.

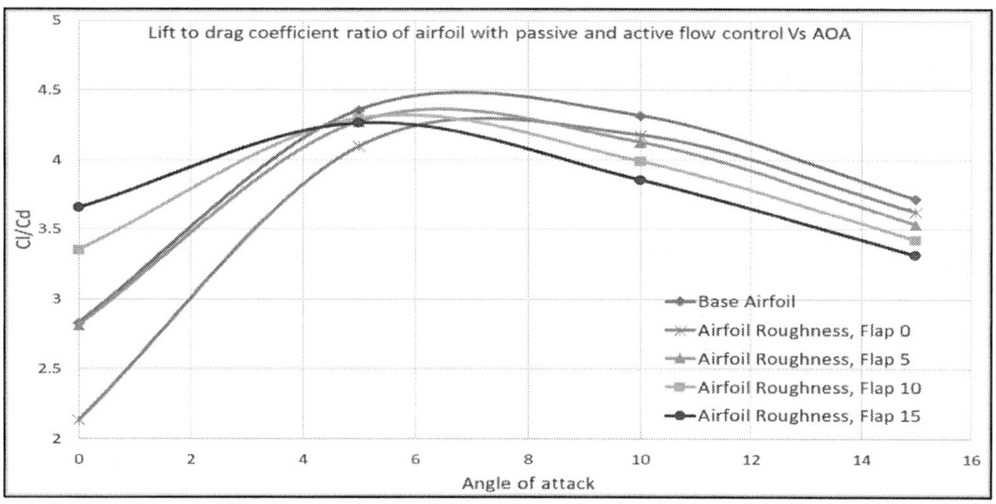

Figure 15. Lift to drag coefficient ratio for configuration 2.

3.3. Comparison Between Configuration 1 and Configuration 2.

Table 2 shows the value of lift to drag coefficient ratio obtained by airfoil with configuration 1 and 2 relative to the base airfoil in terms of percentage. The (+) symbolize the lift to drag coefficient ratio is higher than the base airfoil, where the (-) symbolize the lift to drag coefficient ratio is lower than the base airfoil.

In table 2, observing the difference between airfoil with configuration 1 and configuration 2, the airfoil with only active flow control method used for this investigation performed better in terms of aerodynamic efficiency.

Table 2. Comparison of Cl/Cd for airfoil with configuration 1 and 2.

AOA	Airfoil with configuration 1. (Cl/ Cd) %	Airfoil with configuration 2. (Cl/Cd) %
Flap 0°		
0	-23.1 %	-24.62 %
5	-3.0 %	-5.99 %
10	-0.32 %	-3.20 %
15	-0.3 %	-2.45 %
Flap 5°		
0	+ 4.84 %	-0.53 %
5	+ 1.81%	-1.88 %
10	-2.06 %	-4.40 %
15	-3.28 %	-4.89 %
Flap 10°		
0	+ 23.25 %	+ 18.48 %
5	+ 2.31 %	-1.38 %
10	-5.05 %	-7.55 %
15	-6.58 %	-7.88 %
Flap 15°		
0	+ 36.57 %	+ 29.25 %
5	+ 0.6 %	-2.11 %
10	-8.78 %	-10.63 %
15	-10.22 %	-10.83 %

4. Conclusion.

This paper investigated the effect of using both active and passive flow control method on NACA 4412 airfoil. It was observed that, a reduction of lift to drag coefficient ratio was observed for configuration 2. It was also observed that for airfoil with configuration 1, the lift to drag coefficient value improved relative to the base airfoil. With further optimisation of roughness element on NACA 4412, the lift to drag coefficient ratio can further improve along with its aerodynamic performance.

Based on the results obtained, it can be observed that the flow separation was high for the airfoil with configuration 2. This is due to the parameters related to the roughness element in the leading-edge of the airfoil. But, the drop in the lift to drag coefficient ratio in the airfoil with configuration 2 is not too far from the values obtained by the airfoil with configuration 1. To achieve a better lift to drag coefficient ratio for the airfoil with configuration 2, optimisation of the roughness element for the airfoil is required. With a proper optimisation of the roughness element, drag coefficient generated by the airfoil is expected to reduce.

Acknowledgement

The authors would like to thank Curtin University Malaysia for providing the facility, especially the Star CCM+ software license to carry out the numerical study. Sincere gratitude is also dedicated to Ministry of Higher Education (MOHE) Malaysia for the financial support by awarding the Fundamental Research Grant Scheme (FRGS Grant: FRGS/1/2018/TK10/CURTIN/03/1) to carry out this project.

IOP Conf. Series: Materials Science and Engineering **943** (2020) 012036 doi:10.1088/1757-899X/943/1/012036

References

[1] "World Power consumption | Electricity consumption | Enerdata." [Online]. Available: https://yearbook.enerdata.net/electricity/electricity-domestic-consumption-data.html. [Accessed: 31-May-2019].

[2] H. Reza Karimi, *Structural Control And Fault Detection Of Wind Turbine Systems*. 2018.

[3] N. C. Oltmann, D. Sobotta, and A. Hoffmann, "Load reduction of wind turbines using trailing edge flaps," *Energy Procedia*, vol. 136, pp. 176–181, 2017.

[4] E. Jost, T. Lutz, and E. Krämer, "A Parametric CFD Study of Morphing Trailing Edge Flaps Applied on a 10 MW Offshore Wind Turbine," *Energy Procedia*, vol. 94, no. January, pp. 53–60, 2016.

[5] H. Kamliya Jawahar, Q. Ai, and M. Azarpeyvand, "Experimental and numerical investigation of aerodynamic performance for airfoils with morphed trailing edges," *Renew. Energy*, vol. 127, pp. 355–367, 2018.

[6] C. Hafien and T. Ben Mbarek, "Reduced order model for the lift coefficient of an airfoil equipped with extrados and/or trailing edge flexible flaps," *Comput. Fluids*, vol. 180, pp. 82–95, 2019.

[7] S. Beyhaghi and R. S. Amano, "A parametric study on leading-edge slots used on wind turbine airfoils at various angles of attack," *J. Wind Eng. Ind. Aerodyn.*, vol. 175, no. December 2017, pp. 43–52, 2018.

[8] R. Belamadi, A. Djemili, A. Ilinca, and R. Mdouki, "Aerodynamic performance analysis of slotted airfoils for application to wind turbine blades," *J. Wind Eng. Ind. Aerodyn.*, vol. 151, pp. 79–99, 2016.

[9] F. R. Marzabadi and M. R. Soltani, "Sharif University of Technology Effect of leading-edge roughness on boundary layer transition of an oscillating airfoil," *Sci. Iran.*, vol. 20, no. 3, pp. 508–515, 2013.

[10] M. Serdar GENÇ, K. KOCA, and H. H. AÇIKEL, "Investigation of pre-stall flow control on wind turbine blade airfoil using roughness element," *Energy*, vol. 176, pp. 320–334, 2019.

[11] X. Shi, S. Xu, L. Ding, and D. Huang, "Passive flow control of a stalled airfoil using an oscillating micro-cylinder," *Comput. Fluids*, vol. 178, pp. 152–165, 2019.

Numerical analysis of erosion pattern on pipe elbow bend with swirling flows

K S Yam*, S Roy, V C C Lee, and M C Law

Department of Mechanical Engineering, Faculty of Sciences and Engineering, Curtin University Malaysia, 98009, Miri, Sarawak, Malaysia

Email: yamks@curtin.edu.my

Abstract. Abrasive solid erosion is a common issue faced in many industrial applications and can incur significant loss to production efficiency. In a piping system, the bends are generally the most vulnerable to the abrasive erosion due to the abrupt change of flow. Reducing the erosion at the bend is key to industries for safety purpose and ensure equipment longevity. This research focusses on the effectiveness of utilizing the swirling flow in reducing the erosion rate at the elbow bends. Numerical approaches are adopted to systematically evaluate the impact of the degree of swirling in the flow on the erosion reduction at the elbow. The results demonstrate the promising prospect of the swirling flows as a mechanism to control the erosion at the pipe elbow.

1. Introduction

Abrasive erosion are common issues encountered in numerous manufacturing and processing industries that employs pneumatic system to transport granular material. Abrasive erosion has detrimental effect to the industries as it causes more frequent production downtime for maintaining equipment. Abrasive erosion left unchecked can result in severe consequence, such as leakages of processing materials from the system that damages the equipment and contaminates the surrounding environment. Minimising erosion is important to industries due to the impact it has on the maintenance cost, and the risk management.

Typically, the pneumatic system's elbows are identified to be the component that is most vulnerable to abrasive erosion. As a result, numerous investigations were conducted to understand the mechanism of erosions at the elbow, and the factors that could impact the erosion rate. Various strategies were proposed to control the erosion rate. The previous studies focussed on the impact of the operating conditions on the erosion rate at the elbows, such as how the erosion is controlled by the pipe curvature ratio and diameter by El-Behery et al. [5], the effect of particle size and pipe material [6], the influence of particle impact angle and the material properties [10,13], impact velocity [11].

Researchers also have proposed various designs to alter the shape elbows and manipulate the flows at the bend. An overview of the designs was comprehensively summarized by Dhodapkar, Solt, and Klinzing [2]. The common fittings such as Tee-bends were proposed to counter erosion by creating a particle-particle impact zone. However, the design was less effective in reducing erosion [1]. The specialized bends are the latest innovative designs that exerts better control on the erosion. Gamma bend, pellbow bend and smart elbow bend are examples of the recently proposed models. Each offers unique physical mechanism on erosion reduction. The smart elbow bend utilizes a spherical chamber to form rotating particle shield that deflect incoming particles from impinging directly on the wall. There are evidences from the recent studies [2,11] on the effectiveness of the designs in reducing erosion compared to the standard elbow. Gamma bends and pellbow designs utilizes their customized shape to increase the area of the particle-particle impact zones.

This paper investigates the use of swirling inlet as a mechanism to reduce the erosion. Most previous investigations assume uniform flow inlets. Swirling flows as mechanism to reduce erosion was investigated recently [4], but a systematic study into the effect of swirling degree on the erosion was not performed before. The objective of this study is to numerically investigate the effect on the erosion pattern on a common elbow bend when the inlet flows are angled at a range 10 to 30° from the normal inlet.

2. Methodology

A common elbow pipe bend with particulate air flows from the inlet as shown in Fig. 1 is considered. The pipe has a diameter D of 0.0254m and the elbow has a radius to pipe diameter ratio of 1.5. The length of the inlet pipe was assigned to be twelves times the pipe diameter to allow the flow to be fully developed before it enters the bend. The flow is turbulent with a velocity of 34.1m/s and Reynolds number of over 60,000. The particulate mass flowrate is 0.0217 kg/s and the corresponding mass loading is 0.013. Four cases are considered here, namely a common uniform inlet that is parallel the pipe, and three swirling inlets that are angled at 10, 20 or 30° from the normal direction, as shown in Figure 2.

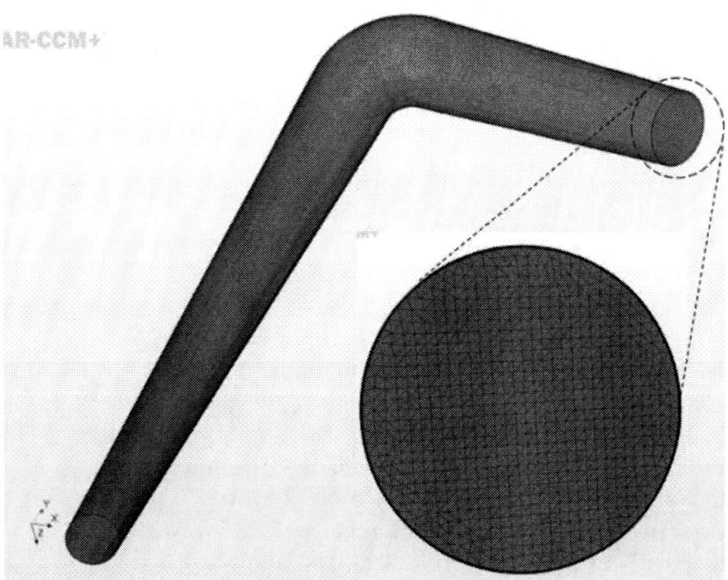

Figure 1 Geometry and meshing employed for the elbow pipe bend.

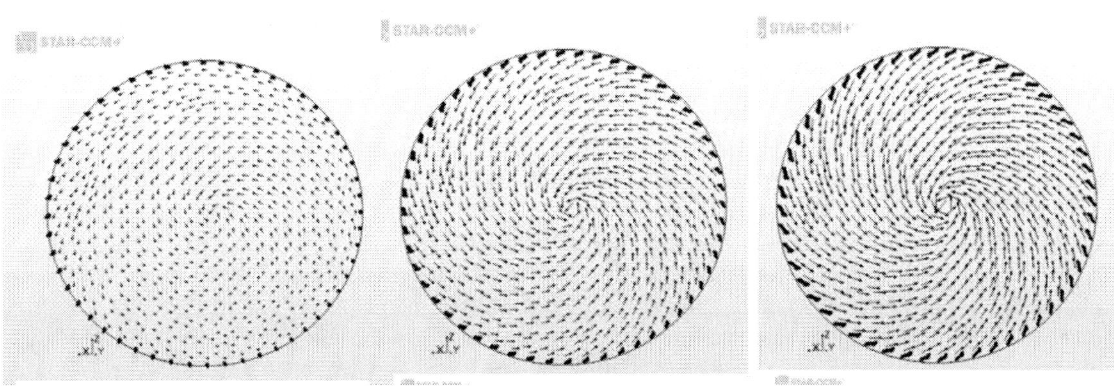

(a) Top View

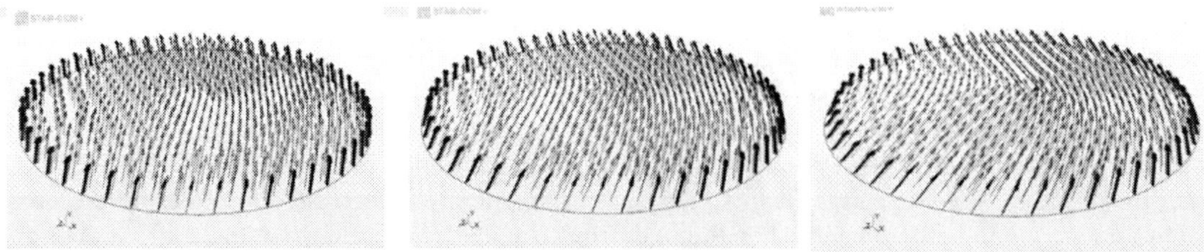

(b) Isometric View

Figure 2 Vector plot from the top view and side view on the inlet for case of 10° swirl (*left*), 20 ° swirl (center), and 30 ° swirl (right).

2.1 Computational Approaches

The analysis were performed using computer approaches. Commercial Computational Fluid Dynamics (CFD) code StarCCM v13 were employed to simulate the problem. In this computational method, the governing equations of the fluid flows were treated using the finite volume method whereby the equations are numerically integrated on a control volume basis into a set of discrete algebraic equations that are then solved iteratively using computer algorithm. The governing equations of the model used in this study is summarized in the Table 1. In this simulation, the second order discretization schemes are employed to numerically discretize the Reynolds-Averaged Navier Stokes (RANS) and the Turbulence model equations. The SIMPLE algorithm is chosen to iteratively solve the governing equations in the domain until the solver convergence criteria is achieved. The transport of the solid particles are resolved separately from the fluid flow transport. After the calculation on the airflows is completed, the transport of the solid particles is then solved using the particle Lagrange equation. Giving the low mass loading of the flow (less than 2%), one-way coupling is assumed between the particle and the air flows. Particles impacting on the wall are assumed to always rebound from the wall at the angles following Grant & Tabakoff model. The strength of the impact is estimated using the Oka et al. erosion model.

Table 1. Mathematical Models For Particulate Airflow

Physics	**Mathematical Model**
Airflow Transport	Reynolds-Averaged Navier Stokes (RANS)
Flow Turbulence	Standard K-Epsilon
Near Wall Treatment	Two-layer All y+
Particle Transport	Lagrange with 1-way coupling
Particle Drag Correlation	Schiller-Naumann
Particle Shear Lift	Sommerfeld [13]
Erosion model	Oka et al. [10]
Impact model	Grant-Tabakoff [7]

3. Model Validation

Simulations of different mesh density were simulated in order to determine the mesh independent solution. The effect of mesh density was analyzed based on the erosion profile along the pipe bends. Figure 3 shows that the erosion profile predicted using a lower mesh count of 1.2million closely matches with the simulation with almost triple mesh cells. This demonstrates that the simulation configuration using 1.2 million is sufficiently fine for the current problem. Also included in Figure 3 is the experimental result from previous study [9] for verifying the simulation model's accuracy. The overall comparison of the simulation model prediction with the experimental result is fairly good. The overall characteristics are well captured, i.e. an increasing erosion from the initial location, followed by an abrupt increase of erosion near to the peak at 50-degree position, and then a drastic drop of erosion after the peak.

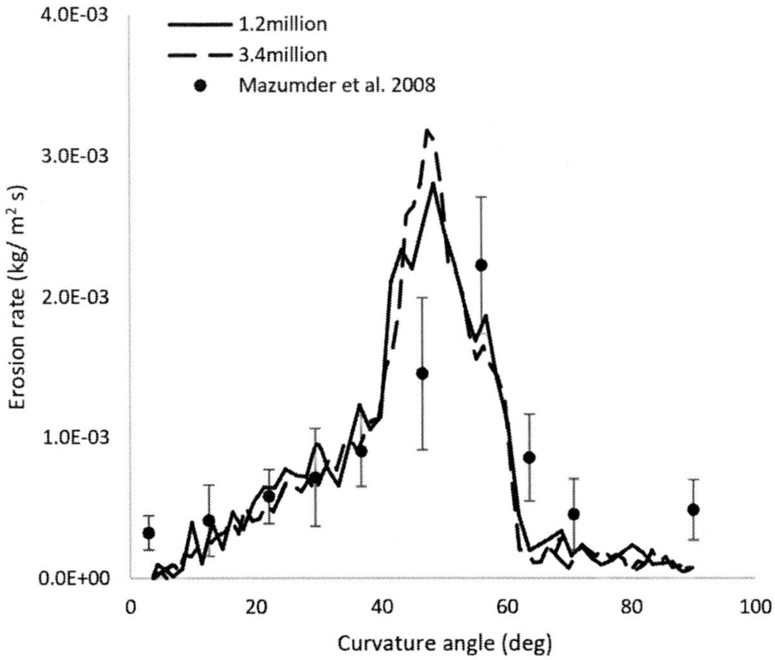

Figure 3. Erosion rate prediction along the centerline of the pipe bend for different mesh and the comparison against experimental result.

4. Result & Discussion

In a common configuration where there is no swirl, the erosion pattern is characterized by a small area of peak erosion at the middle of the pipe, with much lower erosion around the peak. This pattern is the result of the abrupt change of direction of the pipe that forces the solid particle to concentrate into a smaller stream and impact on the small area on the pipe as depicted by the particle trajectories plot on Figure 4(a). On the other hand, the continuous flows exhibit different profile from that of particle trajectories, i.e. the flows along the bend is more uniformly distributed. This observation is quite typical in gas-solid flows with high Reynolds number due to the weak coupling between the continuous flow and the solid particles. Solid particles cannot follow the continuous flow when there is abrupt change of direction.

When swirl of 10 degree is imposed on the flows, the flows along the bend are slightly changed and the erosion pattern on the bend is significantly changed. It can be seen that the peak is slightly diffused and is shifted slightly downstream and away from the centerline. The erosion around the peak is significantly reduced and the prominent V-shaped that is observed in the common scenario is also diffused out. The key observation is the reduction of the maximum erosion by 35% at the peak. The observation shows the effectiveness of the swirling flows in accommodating the solid particles in facing the abrupt change in the flow direction. The swirl allows "more time" for the particles to adapt to the flows changes. This is evidenced in the shift of the particle trajectories from a concentrated stream to a diffused stream, as depicted in the particle trajectories plot in Figure 4(b). Increasing the swirl to 20 degree further amplifies the effect of the swirling but only marginal improvement on the erosion reduction. The erosion peak is shifted further downstream, and a further reduction of 8% is achieved in the maximum erosion.

A significant shift in the flow dynamic occurs when the swirl is increased to thirty degree. At this condition, the swirl is sufficiently strong to transform the characteristics of the particle trajectories from a converging streamline into rotating flows. As a result, the erosion peak is eliminated and, instead, the erosions are diffused out over the entire pipe, with characteristics of a swirling flow. This effectively

IOP Conf. Series: Materials Science and Engineering **943** (2020) 012037 doi:10.1088/1757-899X/943/1/012037

reduces the maximum erosion by a hefty margin of 82%. The maximum erosion reduction is summarized in Figure 5.

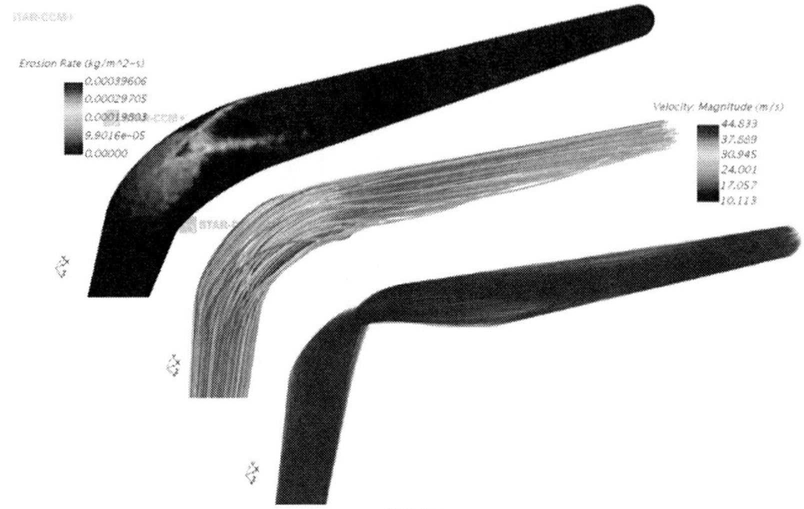

(a) 0°

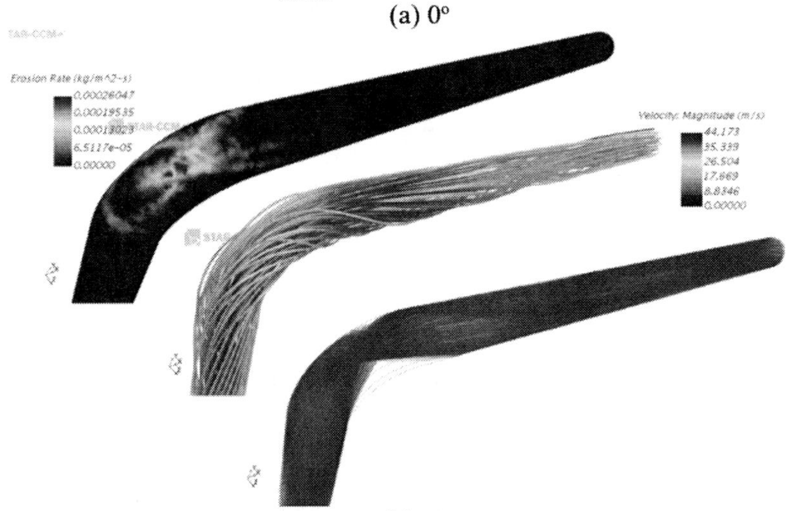

(b) 10°

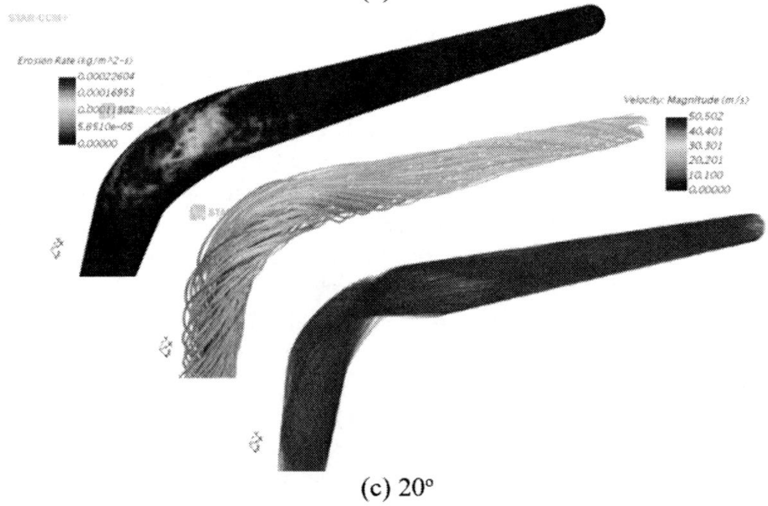

(c) 20°

IOP Conf. Series: Materials Science and Engineering **943** (2020) 012037 doi:10.1088/1757-899X/943/1/012037

(d) 30°

Figure 4. Erosion contour on the pipe (top), flow streamline (middle), and particle trajectories (bottom) for 10°, 20° and 30° swirling flows.

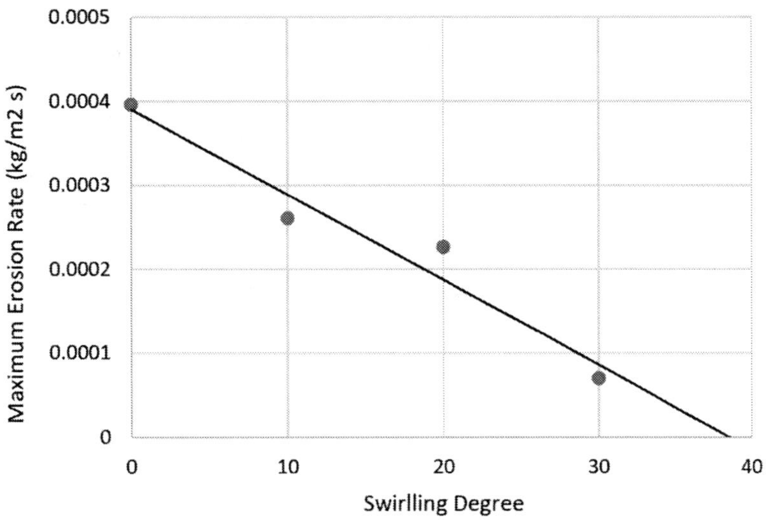

Figure 5. Maximum erosion rate as a function of swirling degree.

5. Conclusion

This study presents a numerical investigation on the erosion reduction on the elbow-typed bend using the mean of swirling inlet. The study shows that a significant 35% reduction in the maximum erosion is achieved when a mere 10 degree swirl is imposed on the flows. When the swirl is increased to 30 degree, the particle trajectories are completely transformed into a rotating flows that sees a significant 82% reduction is achieved in the maximum erosion. The study highlights the effectiveness and potential of the swirling flows as a mechanism to reduce erosion. Further study will investigate more diversified boundary conditions to ascertain the applications of the swirling in greater range of scenarios.

References

[1] Chen, X., McLaury, B.S. and Shirazi, S.A., 2006. Numerical and experimental investigation of the relative erosion severity between plugged tees and elbows in dilute gas/solid two-phase flow. Wear, 261(7-8), pp.715-729.

IOP Conf. Series: Materials Science and Engineering **943** (2020) 012037 doi:10.1088/1757-899X/943/1/012037

[2] Dhodapkar, Shrikant, Paul Solt, and George Klinzing. 2009. "Understanding bends in pneumatic conveying systems." *Chemical Engineering* 116 (4):53.

[3] Duarte, Carlos Antonio Ribeiro, Francisco José de Souza, and Vinicius Fagundes dos Santos. 2016. "Mitigating elbow erosion with a vortex chamber." *Powder Technology* 288:6-25.

[4] Duarte, C.A.R. and de Souza, F.J., 2017. Innovative pipe wall design to mitigate elbow erosion: A CFD analysis. Wear, 380, pp.176-190.

[5] El-Behery, Samy M, Mofreh H Hamed, KA Ibrahim, and MA El-Kadi. 2010. "CFD evaluation of solid particles erosion in curved ducts." *Journal of Fluids Engineering* 132 (7):071303.

[6] Gandhi, BK, and SV Borse. 2002. "Effects of particle size and size distribution on estimating erosion wear of cast iron in sand-water slurries." *Indian Journal of Engineering and Materials Sciences* 9 (6):480-486.

[7] Grant, G, and W Tabakoff. 1975. "Erosion prediction in turbomachinery resulting from environmental solid particles." *Journal of Aircraft* 12 (5):471-478.

[8] Lataste, J., Huilier, D., Burnage, H. and Bednář, J., 2000. On the shear lift force acting on heavy particles in a turbulent boundary layer. *Atmospheric environment, 34*(23), pp.3963-3971.

[9] Mazumder, Q.H., Shirazi, S.A. and McLaury, B., 2008. Experimental investigation of the location of maximum erosive wear damage in elbows. Journal of Pressure Vessel Technology, 130(1), p.011303.

[10] Oka, Y Isomoto, K Okamura, and T Yoshida. 2005. "Practical estimation of erosion damage caused by solid particle impact: Part 1: Effects of impact parameters on a predictive equation." *Wear* 259 (1):95-101.

[11] Peng, Wenshan, and Xuewen Cao. 2016. "Numerical simulation of solid particle erosion in pipe bends for liquid–solid flow." *Powder Technology* 294:266-279.

[12] San, Y.K., Thien, R. and Lee Chieng Chen, V., 2019. Numerical study on erosion of a pipe bend with a vortex chamber. *Particulate Science and Technology, 37*(2), pp.200-206.

[13] Sommerfeld, M. 2000. "Theoretical and Experimental Modelling of Particulate Flows," Technical Report Lecture Series 2000-06, von Karman Institute for Fluid Dynamics pp 20-23.

[14] Veritas, Det Norske. 2007. Recommended Practice RP O501 Erosive Wear in Piping Systems. Technical report, DNV RP O501–Revision 4.2., pp. i-iii, 1-39.

IOP Conf. Series: Materials Science and Engineering **943** (2020) 012038 doi:10.1088/1757-899X/943/1/012038

Recent Developments in City Logistics Research: A Literature Review

S Istiqomah[1*], Yuniaristanto[2] and W Sutopo[2]
[1] Master Program of Industrial Engineering, Faculty of Engineering, Universitas Sebelas Maret Surakarta, Indonesia
[2] Department of Industrial Engineering, Faculty of Engineering, Universitas Sebelas Maret Surakarta, Indonesia

Email: silviistiqomah23@gmail.com

Abstract. City logistics is an important problem of the modern urban logistics system that needs to be resolved. In this review, we can integrating economic, environmental, and social dimensions problem to analyze the relevance of urban logistic issues and to identify gaps. This problem require coorperation between several type of stakeholders. The stakeholders have different objectives and decision spaces. We identify some information and research, so this literature review can contribute to the evolution of knowledge management. The literature review can develop opportunities and challenges for understanding about development of city logistics.

1. Introduction

The city is a place for community activities concerning the economy, social activities, and the biggest politics. In the times, shipping goods became one of the important activities and had a significant impact on the condition of the community. A lot of community comments about the impact received from the city logistics, there are good impacts that can be beneficial, there are also negative impacts that make people's lives worse. With this problem, the development of urban logistics needs to be analyzed to sustain the life of the city and business activities that encourage the growth and development of the city.

City logistics is one of the most widely debated problems in most cities in the world. Current issues are related to phenomena such as urbanization or the revival of people's expectations of the level of well-being [1]. With existing developments, city logistics is influenced by several sectors, infrastructure, social, economic, and environment. With the linkages between sectors, this research is expected to be able to provide advice or development for integrated city logistics problems. Many things need to be regulated regarding city logistics in order to achieve mobility, sustainability, and life in the city [2]. The design and planning of city logitics produced will have an impact on people's welfare and good social conditions [3]. One of the most important things and has a significant influence on city logitics is stakeholders. There are some of the most important stakeholders, namely shippers, shippers, local governments and residents (consumers) who maintain different goals and perspectives towards their own goals [4]. Differences often occur between the interaction of private and public stakeholders [5]. Given the differences between several stakeholders, stakeholder behavior needs to be considered in evaluating city logistics activities [6]. City logistics is regulated with the aim to plan, manage, and control the movement of goods efficiently and effectively in networks between regions in a city to reach an optimal point by considering the results of integration of several different stakeholder goals [7]. The purpose of this paper is to present various aspects of urban logistics and current trends to provide a comprehensive view of this broad field of research. Then research trends in urban logistics are used to identify more aspects that are not being addressed.

2. Framework for The Review

The review framework of the results of this study is based in part on the analysis of the urban transport domain by Odgen in Anand (2012).The author explores several things about city logistics by studying stakeholders, descriptors, and the objectives associated with them. In addition, this review literature will explore the solution approaches and methods used to solve existing problems. In this chapter, we explain some problems related to city logistics. With the presence of various stakeholders related to city logistics, there are several perspectives used in developing this review.

Relationships that occur between stakeholders have different impacts for each stakeholder involved. In addition, there are several other impacts generated by the city logistics in the environmental and social sectors. The movement of goods is influenced by various parameters (eg Infrastructure conditions, industrial structures) and on the other hand, it affects different parameters (eg congestion, accessibility, pollution, etc.). Causes and effects of the movement of goods are also interdependent According to various market segments, the stakeholders that influence the domain of urban transportation are mainly 1) Producers & Shippers, 2) Freight Transport and Logistics Operators, 3) Receivers, 4) Residents, 5) Public Authorities, 6) Other Stakeholders. Private stakeholders seek to maximise profits, and there is thus an implicit need to achieve a reduction in transportation costs as well as an increase in sales, while keeping in mind that their customers expect a service that is both economical and reliable. In the discussion of the stakeholders involved, the original intent and purpose of engagement must be known, the method used, the participants, a summary of stakeholder concerns, expectations and perceptions, a summary of the discussion, and a list of strong outputs (decisions, actions, proposals, and recommendations).

Some researchers discuss and analyze various perceptions, needs, goals, and perspectives from stakeholders to address different problems in city logistics. The main purpose of the decisions they make is to increase the potential and actions that will be realized. According [38] relies on inter-city public transportation with rail-based by analyzing stakeholder goals and interests. According [37] tried to consider the key factors for city logistics by considering various stakeholder perspectives. According [35] point to the importance of considering diverse preferences among stakeholders for local policy makers with compilation preferences. According [36] focused exclusively on stakeholders' perspectives to design and evaluate last mile solutions. They first presented different criteria for three types of stakeholders: industrial, institutional and consumer; secondly developed stakeholders' interactions with respect to their interests and finally the proposed conceptual framework to demonstrate the key considerations of stakeholders.

3. Research Methodology

There are methods used to produce this paper, some methods obtained from the literature review have the same procedures adopted in various fields of research [8][9][10]. The present study adopts the eight step methodology proposed by [41] to conduct the SLR: (i) research problem formulation, (ii) literature search, (iii) data collection, (iv) quality assessment, (v) data analysis and synthesis, (vi) interpretation, (vii) presentation of results, (viii) and updating of the review. From the first stage, we formulated the problem to be solved. In this report, the problem to be solved is about city logistics. The second stage is looking for literature to add references in problem solving and problem viewpoints, in this stage some literature has been obtained from 2010 to 2020. in the third stage, there is no primary data taken from research, data obtained from several news , conditions and current research. For the fourth stage, it does not produce anything significant because the research conducted only compares the various studies that have been done to obtain solutions to the current problems. The next stage regarding analysis, this analysis is drawn from the various conclusions of each study that has been conducted. the final stage is the interpretation of results and conclusions from various studies that have been conducted to determine trends and gaps regarding city logistics.

The first thing to do is to determine the topic from the literature review that will be discussed. then look for papers and discussions that are relevant to the topic to be discussed. Then analyze the data that has been obtained to find out the objectives that can be done for the development and literature study of the paper that will be produced. The first step is to look for keywords that are relevant to city logistics with the scientific platform Scopus and Web of Science released since 2010. Then for supporting material obtained from several articles related to the citation relationship contained in the selected paper. From the selected paper, it can be the primary data used to discuss the development of city logistics. After several papers have been obtained, they will be chosen based on the topics to be discussed. The document filtered according to the review title and abstract of each existing paper. Last step, the document filtered again based on full paper that has a relationship with the topic to be discussed. Literature review data obtained from 2010 to 2020, filtered based on Review articles and Research articles, and filtered based on each sub-theme to be discussed, then selected articles that are full text review.

4. City Logistics Modeling

Some things that will be discussed in this literature study are about stakeholders, the relationship between stakeholders activities (descriptors), the purpose of each research conducted, the solution approach, and the method used in each problem solving.

4.1 Stakeholder

In this chapter discusses several sub-chapters that have a direct relationship with city logistics. The first thing that will be discussed is the stakeholders involved. in the development of urban logistics, there is a relationship between stakeholders involved. Knowledge is needed about the relationships between stakeholders and the roles of each stakeholder involved. The main stakeholders are shippers, transportation of goods, local authorities and residents (consumers) who have their respective goals and differ in the logistics scope of the city and these stakeholders have diverse perspectives on their own goals [4]. There are several difficulties discussed in the latest urban logistics development, the first difficulty is the high number of stakeholders involved from different circles and having their respective goals [11]. Often conflicting interactions usually occur between private and public stakeholders [5]. The next problem is the lack of real data on city logistics. there were some local authority stakeholders who were previously less interested in providing data because public stakeholders caused real data shortages in the context of city logistics [12].

City logitics involves several stakeholders in interactions, especially from the commodity, transportation and infrastructure sectors. The involvement of these stakeholders makes the attributes and behavior of each stakeholder need to be considered in the decision making and chosen business strategy. Stakeholders in the city transport domain can be divided into two categories: 1) public sector stakeholders which include traffic authorities, infrastructure authorities, cities, train terminals / port authorities etc. All these stakeholders can be referred to as "administrators". 2) Private sector stakeholders which include producers, suppliers, shippers, shippers, trucking companies, truck drivers, shop assistants, recipients, etc. So far the existing stakeholder classifications are only divided into senders, transporters and recipients.

4.2 Descriptors

Descriptors are the classification of activities and interactions among stakeholders [13]. In fact, this descriptor is a typical indicator that is observed, measured and analyzed to find out the results that can be used for the development of future urban goods movement. Analyzing the effects of several different descriptors provides insight into how the system works and how to modify the system to achieve the desired goals.

Table 1 Key urban freight and logistics stakeholders (Source : Bossche et al, 2017)

Producers & Shippers	Freight Transport and Logistics Operators	Receivers
1. Producing the goods. 2. Often located outside cities. 3. Some outsource transport and logistics operations. 4. Others operate own fleet	1. Visible face of urban freight logistics operations. 2. Provide transport and distribution services. 3. Highly heterogeneous group (small family companies up to major international transport companies). 4. Hired by producers, shippers or receivers. 5. Behaviour reflects the need to meet customers' requirements.	1. Highly heterogeneous group: small retailers, international retail chains, shopping centres, households, etc. 2. Each segment has specific demands (in terms of delivery time, transport conditions, pricing, etc.). 3. Expect high quality of service (reliability, flexibility, short transit times) at reduced prices.

Residents	Public Authorities	Other Stakeholders
1. May also be Receivers ecommerce). 2. Expect proper quality of life (no pollution, security, clean built environment, quiet green areas, charming leisure and shopping zones, etc.). 3. Expect access to a wide range of high-quality goods at affordable prices. 4. Expect customised and variety of goods.	1. Commonly municipalities and local agencies 2. Balance between promoting sustainable urban development and fostering economic growth. 3. Limited intervention in urban freight logistics activities.	1. Investors, Infrastructure providers, landowners, software providers, manufacturers, nongovernmental organisations 2. Not directly involved in the urban freight logistics operations.

Table 2 Urban freight descriptors (Source: adapted from Ogden in Anand, 2012)

Market	Descriptor	Market	Descriptor
Commodity	Freight Generation	Traffic Service	Traffic Design
	Comodity Flow		Traffic Flow
	Industry Structure		Pollution Level
Transport Service	Vehicle Loading	Infrastructure	Land Use
	Vehicle Design		Location
	Trip Generation		Building & Site Design
	Cost		Modal Transfer

4.3 Objective

The framework carried out at the initial stage is to determine the research objectives and aspects related to the themes to be discussed.The complexity of aspects related to urban logistical problems has been facilitated by the development of technology. We try to broaden our views on some of the problems at hand and some alternative solutions used in city logistics. City logistics has a less stable trend in its completion. In the settlement used, it needs renewal in regulating policies and explaining their impact on the community. So that this development takes time to be accepted among the people. Infrastructure was approved recently because it concerns the claimed logistics efficiency of the city [14].

Most of the goals to be achieved still consider one stakeholder, so that the results of city logistics research have positive and negative impacts that can be felt directly by the community. Therefore the objectives of each stakeholder must be accommodated with the latest models or research. This is done so that city logistics can work efficiently and effectively. By synchronizing several objectives, the researcher can draw a conclusion that shows that the results achieved can make the city better condition for each aspect and stakeholders involved.

4.4 Solution Approach

There are several approaches used in dealing with urban logistics problems. According [33] divide into 3 approaches, the approach used is the policy, technology and planning approach. To explore the results of this review, a social approach is also carried out in order to understand the impact of city logistics in society, both in terms of economy, environment, and also the development that will be accepted by the community. Understanding of each approach taken needs to be known, this is needed to facilitate the classification of objectives and models of existing problem solving.

a. Policy: This model analyzes the appropriateness of the regulations used in city logistics and analyzes the impact resulting from the policies that are already running or planned for the environment and related stakeholders.
b. Planning: This model analyzes infrastructure planning or things that will be implemented to support the city's logistics success.
c. Technology and information: This model analyzes the technology that is developing, the technology that will be used, and the technology that is suitable for use in solving problems related to city logistics. this development needs to be analyzed to determine the impact produced by the technology that will be used in the development of city logitistics.
d. Social: This model analyzes the social impacts resulting from various city logistics situations. several aspects that need to be considered so as not to cause harm or negative impacts on the community and also stakeholders associated with city logistics.

4.5 Method

There are several settlement methods used. Quantitative modeling methods, case studies, surveys, and experimental simulation. This method is a method that is widely used in solving existing city logistics problems. Some of the approaches taken aim to overcome various problems regarding city logistics, some of them only aim to minimize the negative impact resulting from a decision that is applied to city logistics. Multi-criteria analysis is used mostly as a qualitative method in city logistics, this is due to the large number of stakeholders involved with different interests. Various different goals to be achieved by each stakeholder make the method of completion and approaches so varied, this is done so that research can approach the real conditions. With the different types of problems, the solutions needed to suit each problem are needed. With a variety of ways of resolution, the resulting optimization will also vary. Therefore, the method used must also be in accordance with the problem to be solved. Neghabadi et al (2018) describe several methods that can be used to solve problems, this is explained in Figure 1.

Table 3 Description of Objective Problem (Source: Ogden and Woudsma in Anand et al (2015).)

IOP Conf. Series: Materials Science and Engineering **943** (2020) 012038 doi:10.1088/1757-899X/943/1/012038

Purpose	Description
Economy	Develop and improve the transport system towards improving the local, regional and national economy
	Focus on the trade sector that is directly related to the economic region
	Focus on ports and intermodal facilities
Efficiency	Minimization or cost of travel for transportation operations related to endroute travel, endpoint activity and energy
	Focus on congestion, transportation of goods, and the cost of transporting goods
	Lack of road networks including road design and geometry, maintenance, road boards, management of regional traffic and capacity associated with the main road
	Costs at the end point (endpoint costs) associated with loading and unloading, parking, activities at the terminal, operating hours and the way in and out to the location Energy costs related to vehicle speed, as well as character and delivery type
Safety on the road	Minimization of property damage, accidents and deaths related to accidents Focus on policies related to traffic management, road design, vehicle design, driver training and land use
Environment	Focus on agreed mitigation, air and vibration Threats received by large vehicles and activities taken in residential areas
Infrastructure and Management	Adjustment of education costs through regulations, price controls, taxation and investment
	Road construction and maintenance with the help of the freight forwarding sector
Urban Structure	Focus on the interaction between transport facilities and urban structures including the interaction between cargo and urban structure, the size of the city and the effect on transport costs, and the load as urban land users

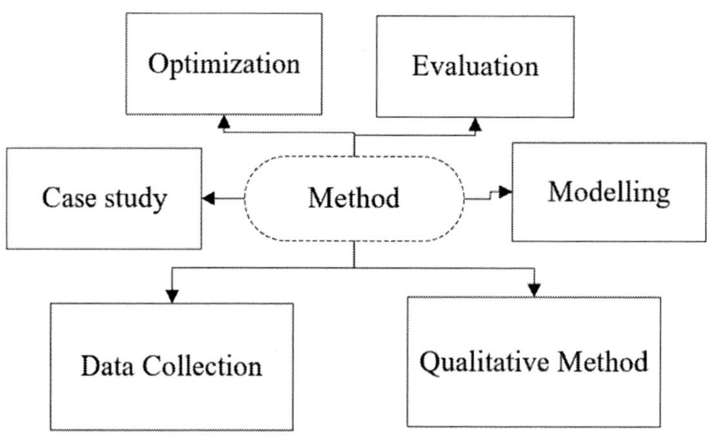

Figure 1. Method Category

5. Recent Development

All aspects of contribution have a relationship. Existing aspects have been active in their own fields, if these aspects have poor performance or poor decisions. if one of the stakeholders does not consider the impact that will occur on the stakeholder or the negative impact produced, then this needs to be justified so that the logistics conditions of the city are comfortable for all communities. The various negative social, environmental and economic impacts of city logistics can be managed by city leaders. Developments occurring in the industrial world "Industry 4.0 (fourth industrial revolution)" can change and synchronize business content and related environments, this can be one of the pioneers of a dynamic market structure. With technology, the core idea of Industry 4.0 has been to use information technology that has emerged to implement the Internet of Things (IoT) and services, so that business processes and engineering processes can be well integrated, making it easy to process and monitor the results of developments that have been made. in addition, in depth so that production operates in a way that is flexible, efficient, and environmentally friendly with low quality and cost [15].

The Era of Digitalization is the most important thing in a business today, it is used in realizing a reliable and sustainable future transportation system and supply of goods. Digitalisation basically has the sense of capturing analog signals and turning them into digital forms for the purpose of producing digital representations that can be stored or processed electronically. The digital logistics ecosystem is based on four main driving factors: technology, processes, organization and knowledge [16]. The goal of a sustainable digital logistics ecosystem is to rethink digital-based business models and redesign the way business processes along the supply chain lead to sustainable development and to balance sustainability in terms of economic, social and environmental dimensions and reflect the interconnection between them [17].

5.1 Stakeholder

There are several stakeholders who have a goal in developing city logistics. The objectives of each stakeholder are different, some are still related and some are very different. But almost all want the costs in city logistics to be minimal and have a slight negative impact on society. All stakeholders of the city are interested, They all have a common goal of making the city attractive (economic, social, traffic, ecology, culture and others) [18]. Administrators, residents, shippers, freight carriers and urban expressway operators are stakeholders who are considered to arrange the transportation of goods and logistics inside urban and metropolitan areas [19]

5.2 Descriptors

Night delivery, logistics providers can make lower costs and shorter delivery times, and from the public, lower emissions and traffic density during the day can be reduced [20]. using an e-grocery delivery system will help companies and consumers alike buy goods more easily, and the growth of e-grocery is more focused on strengthening customer loyalty and acquisition rather than economic benefits [21]. The scenario analysis produced by the simulation can be used to investigate the logistics and transportation of goods in relation to land use, delivery of demand and supply goods, logistics profiles and performance and impact of the road network [22]. City logistics measures that can be taken to regulate freight transport and logistics within urban and metropolitan areas, implementation measures requires a full assessment of social costs and who has to abide by them [19]. With limited land and increasing logistics delivery activities, Urban Underground Logistics is one of the solutions that needs to be studied and considered to get the optimal solution for several problems [23].

5.3 Method

To solve problems involving many stakeholders, the hybrid MCDM model is one of the methods used by combining the Delphi, AHP and SWARA methods [18]. Kayikci (2018) uses qualitative methods and connected semi-structured interviews to find out the benefits of the logistics digitalization process and test the impact of the sustainability of digitalization in logistics. According [42] used an agent-based model to investigate parking and shipping management systems together with UDCs. According [40] used a nested logit model based on a survey in Rome to study stakeholder reactions to existing urban transport policies. According [35] discuss a special agent approach to increasing the awareness and ability of decision makers to make better decisions. Liu et al (2018) use simulation as one of the methods to solve urban logistical problems, the method used is a systematic architecture of dynamic optimization based on real-time information for sustainable reverse logistics. with the different goals of the stakeholders, the best solution is needed to produce a good and structured decision, so that it can be obtained in the development of consensus-based strategies [24]. According [39] SCOR simulation is the most suitable method used in measuring the performance of several stakeholders involved in the newspaper industry supply chain case. Analysis of environmental problems and Macro situations (PEST-SWOT) is a strategic analysis method that combines SWOT and PEST to identify the advantages, losses, opportunities and threats that will be generated, first used to analyze the internal and external factors involved [23]. According [34] using agent-based modeling methods in order to know the interaction between the stakeholders involved and netlogo as one of the tools to solve the problem of distributing newspapers to consumers who are scattered. gaps that occur in the quality and price of beef commodities are resolved by the supply chain model using Mix Integer Linear Programming (MILP), this is done so that the government can make good decisions to improve the existing conditions [25].

5.4 Objective

Political regulations that can integrate various community needs in city logistics [18]. Saskia et al (2015) explained that with e-grocery, purchasing goods from sellers to consumers is more efficient and easier, because technology can cut processes that usually require a lot of time. Kayikci (2018) explains the advantages of digitizing city logistics in three sustainable dimensions (economy, environment and society) more easily and attractively. The simulation method used can solve several problems at once, can reduce the total logistics cost, energy consumption, and the total logistics distance, optimize the configuration of logistics resources, and achieve sustainable reverse logistics services [26].

5.5 Solution Approach

There are several approaches taken in this city logistics problem. technological, political, planning and social approaches. Political initiatives are made to be used by the public in order to regulate their logistics business activities and make them sustainable [27-28]. e-grocery is a process of technological innovation using the internet, explaining supplier-retailer relationships more easily [21]. The results showed that the use of digital technology and applications in logistics within FMCG companies and transportation service providers have a large sustainability impact, especially the sustainability impact of digitalization. [29]. Internet of Things technology is used to build a supportive environment to get real time data and accurate data that can support the development of city logistics [27]. The agent-based modeling approach makes it possible to represent the city logistics domain in its original form by modeling the city logistics entity as an autonomous agent [30]. The simulation is able to point out the relations existing among city logistics measures, decision-maker choice dimensions by using a multi-stage demand model and a discrete choice approach for each decision level [22]. Smart City logistics on cloud computing model (Cloud Smart City logistics) is one of the methods used to achieve goals in an effective and efficient way, in addition to the many negative impacts that can be minimized by the development that is implemented in logistics activities. [31]. GIS is one of the technology tools that can support decision making that can consider several stakeholders involved to reach good decisions [32].

6. Conclusions and Suggestions

There are several developments that make the distribution of city logistics easier to do. One of them is by involving technology, this will digitize the flow of goods delivery so that we can track the location of goods easily. In addition, the number of parts in the shipping groove that can be trimmed, this will make the goods sent will immediately arrive at the desired destination. To digitize this process, it may reduce the social and environmental impacts that will be caused. but further study is needed to discuss the optimization of the existence of technology for city logistics.

This literature study discusses the impact of digitalization in the field of logistics. in addition, the impact of adopting digitalization for the logistics process will be reviewed further. this research is a new approach to understanding the sustainability implications of logistical perspectives. many problems occur in the development of urban logistics. one consideration is the goal of stakeholders. many papers discuss stakeholder objectives, but many have not considered the multiple stakeholders involved. the goals of the multiple stakeholders involved can minimize problems that occur because of differences in the objectives to be achieved. Therefore, agent-based is one method that is done to achieve all the desired goals. The planned development is an underground logistics system to reduce pollution. but this needs more research to be ready to be implemented [23].

Smart logistics can be utilized in various forms. from services to consumers RFID (Radio Frequency Identification), GPS, cloud computing, and other information technology into the logistics process, this can have a good impact so that it becomes efficient and cost-effective. the concept used by smart logistics does not have a significant difference with logistics 4.0. Important components for smart logistics will be connected to data, sophisticated analytics, autonomous decisions, and IoT. Therefore, smart logistics will make the process between e-commerce markets - sellers - logistics companies - and end customers more effective and efficient. So there is a synergy of utilities that are integrated end-to-end and ultimately have an impact not only in terms of profit for business, but also for human life itself. The company utilizes shipping needs with a technology-based logistics network. For example with real time tracking updates, alternative pick up points and package tracking options that can be utilized by consumers.

One of the developments carried out is the process of improvement that exists in the city logitics. this process makes the process easier to do. Further research is expected to be able to accommodate all the objectives of the stakeholders involved, then to minimize the negative impacts generated. One thing that needs to be eliminated is about the increased congestion, air credit, economic and social society. With digitalization, it is hoped that all parties can feel it.

References

[1] Bozzo, R., Conca, A., Marangon, F. (2014). Decision Support System for City Logistics: Literature Review

[2] Taniguchi, E., Thompson, R. G., Yamada, T. (2014). Recent trends and innovations in modelling city logistics. Procedia-Social and Behavioral Sciences, 125, 4-14.

[3] Rao, C., Goh, M., Zhao, Y., Zheng, J. (2015). Location selection of city logistics centers under sustainability. Transportation Research Part D: Transport and Environment, 36, 29-44.

[4] Taniguchi, E., Thompson, R.G., Yamada, T. (2012). Emerging techniques for enhancing the practical application of city logistics models. Procedia-Social and Behavioral Sciences, 39, 3-18.

[5] Gonzalez-Feliu, J., Taniguchi, E., d'Arcier, B. F. (2013). Financing urban logistics projects. From public utility to public-private partnerships. Chapter in "Sustainable Urban Logistics: Concepts, Methods and Information Systems", Part of the series EcoProduction, pp 245-265

[6] Tamagawa, D., Taniguchi, E., Yamada, T. (2010). Evaluating city logistics measures using a multi-agent model. Procedia-Social and Behavioral Sciences, 2(3), 6002-6012.

[7] Amaral, R. R., Aghezzaf, E. H. (2015). City Logistics and Traffic Management: Modelling the Inner and Outer Urban Transport Flows in a Two-tiered System. Transportation Research Procedia, 6, 297-312.

[8] Behrends, S. 2016. "Recent Developments in Urban Logistics Research - A Review of the Proceedings of the International Conference on City Logistics 2009-2013." Transportation Research Procedia 12: 278-287.

[9] Gahm, C., F. Denz, M. Dirr, and A. Tuma. 2016. "Energy-efficient Scheduling in Manufacturing Companies: A Review and Research Framework." European Journal of Operational Research 248 (3): 744-757.
[10] Wee, B. V., and D. Banister. 2016. "How to Write a Literature Review Paper." Transport Reviews 36 (2): 278-288
[11] Maggi, E., Vallino, E. (2015) Simulating urban mobility and the role of public policies: the challenges of Agent Based Models.
[12] Russo, F., Comi, A. (2011). A model system for the ex-ante assessment of city logistics measures. Research in Transportation Economics, 31(1), 81-87.
[13] Anand, N., Quaka, H., Duina R.V., Tavasszy, L. (2012). City logistics modeling efforts: Trends and gaps - A review. The Seventh International Conference on City Logistics. Procedia - Social and Behavioral Sciences 39 (2012) 101 – 115
[14] Jha, M. K., S. Shariat, J. Abdullah, and B. Devkota. 2012. "Maximizing Resource Effectiveness of Highway Infrastructure Maintenance Inspection and Scheduling for Efficient City Logistics Operations." Procedia - Social and Behavioral Sciences 39: 831-844.
[15] S. Wang, J. Wan, D. Li, C. Zhang, Implementing smart factory of industrie 4.0: an outlook, International Journal of Distributed Sensor Networks (2016).
[16] M. Stuermer, G. Abu-Tayeh, T. Myrach, Digital sustainability: basic conditions for sustainable digital artifacts and their ecosystems, Sustain Sci 12 (2017) 247-262.
[17] K. van Marwyk, S. Treppte, 2016 Logistics Study on Digital Business Models, Roland Berger, White Paper, 2016.
[18] Tadic, S., Zecevic, S., Krstic, M. (2018). Assessment of the political city logistics initiatives sustainability. Transportation Research Procedia 30 (2018) 285-294
[19] Russoa, F., Comi, A. (2010). A classification of city logistics measures and connected impacts. Procedia Social and Behavioral Sciences 2 (2010) 6355-6365
[20] Tadić, S.; Zečević, S. 2015. Integrated planning aimed at sustainability city logistics solutions (in Serbian: Integrisano planiranje u cilju održivosti rešenja city logistike), Tehnika, in press.Saskia, S., Marei, N., Blanquart, C. 2015. Innovations in e-grocery and logistics solutions for cities. Transportation Research Procedia 12 (2016) 825 - 835. The 9th International Conference on City Logistics, Tenerife, Canary Islands (Spain), 17-19 June 2015
[21] Comi, A., Rosati, L. (2013). CLASS: a City Logistics Analysis and Simulation Support System. Procedia - Social and Behavioral Sciences 87 (2013) 321 - 337 1877-0428
[22] Chen, Z., Dong, J., Ren, Rui. 2017. Urban underground logistics system in China: Opportunities or challenges?. ScienceDirect Underground Space 2 (2017) 195-208
[23] Rubini et al, 2018
[24] Lupita, A., Rangkui, S.H., Sutopo, W., Hisjam, M. (2017). A supply chain model to improve the beef quality distribution using investment analysis: A case study. AIP Conference Proceedings 1902, 020003 (2017)
[25] Liu, S., Zhangb, G., Wanga, L. (2018). IoT-enabled Dynamic Optimisation for Sustainable Reverse Logistics. Procedia CIRP 69 (2018) 662 - 667
[26] Tadić, S.; Zečević, S.; Krstić, M. 2014b. City logistics initiatives aimed at improving sustainability within existing context of urban area (in Serbian: Inicijative city logistike u cilju poboljšanja održivosti unutar postojećeg konteksta urbane sredine), Tehnika, 61(3): 487-495.
[27] Awasthi, A., Chauhan, S.C. 2012. A hybrid approach integrating Affinity Diagram, AHP and fuzzy TOPSIS for sustainable city logistics planning. Applied Mathematical Modelling 36, 573-584
[28] Kayikci, Y. (2018). Sustainability impact of digitization in logistics. 15th Global Conference on Sustainable Manufacturing. Procedia Manufacturing 21 (2018) 782-789
[29] Anand, N., Duinb, J.H.R, V., Tavasszy, L. (2016). Framework for modelling multi-stakeholder city logistics domain using the agent based modelling approach. Transportation Research Procedia 16 (2016) 4 – 15

[30] Nowicka, K. 2014. Smart City logistics on cloud computing model. Procedia - Social and Behavioral Sciences 151 (2014) 266 – 281
[31] Guerlain, C., Cortina, S., Samuel, R. 2015. Towards a collaborative Geographical Information System to support collective decision making for urban logistics initiative. Transportation Research Procedia 12 (2016) 634 – 643
[32] Anand, N., Duina R.V., Quaka, H., Tavasszy, L. (2015). Relevance of City Logistics Modelling Efforts: A Review, Transport Reviews, 35:6, 701-719, DOI: 10.1080/01441647.2015.1052112
[33] Febriandini, I. F., Yuniaristanto, Sutopo, W., and Hisjam, M. (2019). Analysis daily newspaper distribution in Solo by Agent Based Simulation. IOP Conf. Series: Materials Science and Engineering 528 (2019) 012033
[34] Gatta, V., and E. Marcucci. 2016. "Stakeholder-specific Data Acquisition and Urban Freight Policy Evaluation: Evidence, Implications and New Suggestions." Transport Reviews 36 (5): 585-609.
[35] Harrington, T. S., J. Singh Srai, M. Kumar, and J. Wohlrab. 2016. "Identifying Design Criteria for Urban System 'Last-Mile' Solutions - A Multi-stakeholder Perspective." Production Planning & Control 27 (6): 456-476.
[36] Kiba-Janiak, M. 2016. "Key Success Factors for City Logistics from the Perspective of Various Groups of Stakeholders." Transportation Research Procedia 12: 557-569.
[37] Kordnejad, B. 2016. "Stakeholder Analysis in Intermodal Urban Freight Transport." Transportation Research Procedia 12: 750-764.
[38] Putri, A.S., Sutopo, W., Hisjam, M. (2017). Framework of Supply Chain Simulation Using SCOR Model in Newspaper Industry. 2017 IEEE International Conference on Industrial Engineering and Engineering Management (IEEM)
[39] Stathopoulos, A., Valeri, E. and Marcucci, E. (2012) Stakeholder reactions to urban freight policy innovation, Journal of Transport Geography, 22, 34-45.
[40] Thome, Antonio Marcio Tavares, Scavarda, Luiz Felipe, Scavarda, Annibal Jose, 2016. Conducting systematic literature review in operations management. Prod. Plann. Contr. 27 (5), 408e420.
[41] Wangapisit, O., Taniguchi, E., Teo, J.S.E. and Qureshi, A.G. (2014) Multi-agent systems modelling for evaluating joint delivery systems, ProcediaSocial and Behavioral Sciences, 125, 472-483.

IOP Conf. Series: Materials Science and Engineering **943** (2020) 012039 doi:10.1088/1757-899X/943/1/012039

Multi-Compartment Vehicle Routing Problem to Find the Alternative Distribution Route of Petroleum Product Delivery

I F Febriandini[1], Yuniaristanto[2], W Sutopo[3]

[1] Laboratory Assistant Logistic System and Business, Industrial Engineering Department, Universitas Sebelas Maret, Indonesia
[2,3] Research Group of Industrial Engineering and Techno-Economic, Industrial Engineering Department, Universitas Sebelas Maret, Indonesia

Email: izatulfitria23@gmail.com

Abstract. Distribution is an activity that the firm has to do to deliver their product to the customers. The product has to deliver to the customers on the right condition, at the right time, and with the minimal cost of delivery. This paper focuses to solve the planning distribution problem on petrol product delivery or Petrol Station Replenishment Problem. One of the ways to solve the planning distribution is VRP (Vehicle Routing Problem). VRP aims to find the minimal total cost delivery with some constraints depend on the factory. In this research, VRP is used by the distribution center to find the optimal route of distribution of petrol products. Using Multi-Compartment Vehicle Routing Problem (MC-VRP) that considers the demand for the petrol product, tank-truck capacity, distribution cost, and the number of the compartment. In this work, a mathematical formulation for MC-VRP is developed in the Mixed Integer Linear Programming (MILP) model. Using Ilog Cplex with branch and bound algorithm to solve this distribution problem with the minimum total distribution cost. There are 10 customers with different distances, different time deliveries and different travel costs in this research. The mathematic formulation produces the best alternative that is Rp 1.898.584 to deliver products from depot to the customers.

1. Introduction

Goods and service industries cannot be separated from distribution activities. Distribution activities are one of the strategy of the company to share their product to their customers. Distribution channels are a series of organizational participants who perform all the functions needed to deliver products or services from the company to the end customers [1].

One of the problems in the distribution process is selecting the distribution route or we mostly call it VRP (Vehicle Routing Problem). VRP itself has many variations in its development. In this paper, using one of the variants on the VRP, called MC-VRP (Multi-Compartment Vehicle Routing Problem). MC-VRP is the variant of the VRP that when selecting the route of the vehicle they enable to jointly transport product with different variances, thus it could reduce the number of visits to a store [2]. Martins also said that MC-VRP can be used to deliver many variations of the product, such as fuel distribution, grocery distribution (e.g. product with different temperature zones), and waste distribution (e.g. glass waste).

Another version of VRP is PSRP (Petrol Station Replenishment Problem). PSRP is a variation of VRP that studies the distribution of fuel products (Oil and Gas Company). The aim is to optimize the distribution of petroleum products to the petrol station (SPBU) using a limit number of a vehicle [3].

IOP Conf. Series: Materials Science and Engineering **943** (2020) 012039 doi:10.1088/1757-899X/943/1/012039

PSRP is a tool to find an alternative solution to distribute products from the company to its customers. This method already discussed in previous studies with different cases and different algorithms. There are so many kinds of the algorithm to do VRP, there are Tabu Search, Branch And Bound, Neighbourhood Search, Genetic Algorithms, etc.

Table 1. Previous Research.

No	Author	VRP Variation	Algorithm
1	Sevaux and Sorensen, 2008 [6]	VRP-Cluster	Floyd-Warshall
2	Surjandari, et.al, 2011 [3]	VRP-Capacitated, Time Window, Multi-Compartment, Loading-Unloading	Tabu Search
3	Benantar and Ouafi, 2012 [4]	VRP- Time Window, Multi-Compartment	Tabu Search
4	Battarra, et.al, 2014 [7]	VRP-Cluster, Capacitated	Branch and Cut & Branch and Cut and Price
5	Henke, et.al, 2015 [8]	VRP-Capacitated, Multi-Compartment	Neighborhood Search
6	Coelho and Laporte, 2015 [9]	VRP-Capacitated, Time Window, Multi-Compartment	Branch and Cut Algorithms
7	Belov and Slastnikov, 2017 [10]	VRP-Capacitated	Ant Colony
8	Pop, et.al, 2018 [11]	VRP-Cluster, Capacitated	Genetic Algorithms
9	Marc, et.al, 2018 [12]	VRP-Cluster, Capacitated	Genetic Algorithms
10	Henke, et.al, 2018 [13]	VRP-Cluster, Capacitated, Multi-Compartment	Branch and Cut Algorithms

This problem is based on the real study case of the oil and gas depot owned by PT. Pertamina, one of the Oil and Gas Company in Indonesia. This research held in PT. Pertamina Terminal Bahan Bakar (Depot of the petrol product) located in Boyolali, Central Java.

Based on the observation, TBBM Boyolali delivers the product to their customers, we call it SPBU (petrol station) every day. But, before the company delivers its product, they need to make distribution planning. Distribution planning is about the route of distribution and the delivery schedule. They have to make the distribution planning every day, based on the demand of the customers. So this research aims to minimize the total cost of distribution to deliver the petrol product.

2. Methodology

TBBM Boyolali delivers its product into 3 zones. Three of them be distinguished based on the distance from customers to the depot. Zone 1 for consumers with a distance of fewer than 30 km, zone 2 for customers between 30 to 60 km, and the last is zone 3 is for customers with a distance more than 60 km from the depot location. Figure 1 describes the distribution zones of TBBM Boyolali.

The company determines the shipping route in line 1, which is delivered in zone 2 at 01.00 - 09.00, followed by shipping on line 2, which is in zone 1 at 09.00-15.00, and the last one in line 3, which is delivered in zone 3 at 15.00 - end.

Table 2. Distribution Zones.

No	ZONE I		ZONE II		ZONE III	
	Area	**Total**	**Area**	**Total**	**Area**	**Total**
1	Boyolali	15	Salatiga	12	Sragen	1
2	Salatiga	3	Semarang	3	Magetan	16
3	Surakarta	24	Sragen	24	Wonogiri	14
4	Karanganyar	12	Klaten	8	Purwodadi / Blora	13
5	Sukoharjo	15	Karanganyar	8	Salatiga	1
6	Klaten	16	Sukoharjo	3	Ngawi	16
7	Sragen	1			Pacitan	5
					Semarang	14
	TOTAL	**86**		**58**		**80**

The business process of petrol product distribution in TBBM Boyolali are:

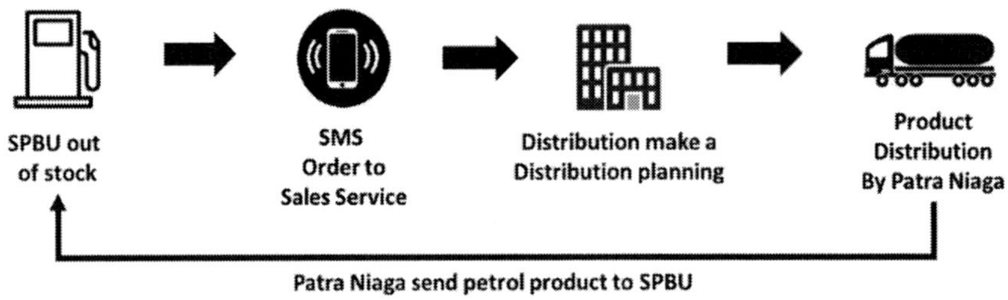

Figure 1. Distribution Zones.

SPBU (gas stations) that run out of stock will order via SMS (Short Message Service) to Boyolali TBBM Sales Services one day before petrol products are sent to the gas station. The Sales Service will send the results of the product order recapitulation to the Distribution Division. Distribution Division then makes a planning distribution that will be given to Patra Niaga. PT Patra Niaga was launched in 2004 to focus on the downstream oil and gas (O&G) business [Patra Niaga]. Distribution Division will make a different distribution planning every day, this depends on the requests that go to the Sales service the day before.

The product sent by TBBM Boyolali are RON 88, RON 95, RON 95 and bio solar. There are 241 customers (SPBU/ Petrol Station) around the 3 zones. 103 tank trucks have a capacity of 16 kiloliters, 24 kiloliters, and 32 kiloliters in TBBM Boyolali explained in table 3. Whereas to meet consumer demand every day, TBBM Boyolali uses around 95 tank truck, the rest of the tank cars are not operating (maintenance). Each compartment has 8.000 liters on its capacity.

Table 3. Several vehicles and their capacity.

Capacity (kiloliter)	Number of Compartment	Number of Vehicles	Total Compartment	Total Capacities
16	2	30	32	480
24	3	52	72	1248
32	4	21	128	672
TOTAL		103	232	2.400

All of these vehicles are leased from the Patra Niaga Company, which is a subsidiary of PT. Pertamina. Patra Niaga only focuses on the distribution of petrol products. Each truck has a different number of compartments, depending on the capacity of the vehicle. Each compartment has a capacity of 8 kiloliters. So, consumers who want to order petrol products must order with multiple capacities of 8 kiloliters.

2.1. Mathematic Formulation

Assumptions:

- The delivery cost per kilometer is known and constant.
- Delivery cost Rp 2.920,- per kilometer [3].
- Several trips can be assigned to the same vehicle.
- Only study 10 samples of location in zone 3, from 241 real customer location.
- Only use 5 tank trucks with capacity: 32 kiloliters.
- SPBU (gas station) only order the amount of fuel that is a multiple of 8 kiloliters (according to the capacity of each compartment).

Model. In this paper, a mathematical formulation for MC-VRP is developed in the Mixed Integer Linear Programming (MILP) model. The model below is written based on the research conducted by Benantar 2012 [4] and 2016 [5] which was conducted in the petroleum industry.

The MC-VRP problem is formulated by a multi-product, single depot, with time windows.

Notations of parameters used for the tank trucks assignment and trip route are as follow:

N	:	Number of customers $N = 10$
V	:	$N \cup \{0\}$ Where 0 represent the depot
K	:	Number of vehicle k that can be used $K = 5$
P	:	Number of product $P = 3$ {Premium, Pertamax, Pertalite}
Q	:	Number of compartment $Q = 3$
d_{ip}	:	The demand of customer i for the product p
C_k	:	The capacity of vehicle k
c_{qk}	:	The capacity of compartment q for the vehicle k
d_{ij}	:	Distance between customer i and customer j
t_{ijk}	:	Travel time of vehicle k from customer i to customer j
a_i	:	Earliest start time for servicing the customer i
b_i	:	Latest start time for servicing the customer i
C_{ij}	:	The cost spent by the vehicle k from customer i to customer j

Then, the decision variable are:

$$x_{ijk} = \begin{cases} 1 : \; if\ i\ precede\ j\ in\ the\ route\ of\ vehicle\ k \\ 0 : \; otherwise \end{cases}$$

$$y_{ikp} = \begin{cases} 1 : \; if\ and\ only\ if\ customer\ i\ receives\ product\ p\ from\ vehicle\ k \\ 0 : \; otherwise \end{cases}$$

S_{ik} = Specifies the arrival time at i when serviced by vehicle k. If vehicle k does not service i, s_{ik} has no meaning and consequently, its value is considered irrelevant.

So, the mathematic formulation is:
Objective Function:

$$min \sum_{k\in\{K_1 \cup K_2\}} \sum_{i\in V} \sum_{j\in V} c_{ij}\, x_{ijk} \quad (1)$$

The objective function in formula (1) is to minimize the total cost of distribution of all routes.

$$\sum_{i\in V, i\neq h} x_{ihk} - \sum_{i\in V, j\neq h} x_{hjk} = 0, \qquad h \in N, \qquad k \in \{K_1 \cup K_2\} \quad (2)$$

$$\sum_{j\in V, i\neq j} \sum_{k\in K} x_{ijk} \leq 1, \qquad i \in N \quad (3)$$

$$\sum_{i\in V, i\neq j} \sum_{k\in K} x_{ijk} \leq 1, \qquad j \in N \quad (4)$$

$$\sum_{j\in N} x_{j0k} = 1, \qquad k \in \{K_1 \cup K_2\} \quad (5)$$

$$\sum_{i\in N} x_{i0k} = 1, \qquad k \in \{K_1 \cup K_2\} \quad (6)$$

$$\sum_{j\in N} d_{jp}\, y_{jkp} = c_k^q, \qquad k \in \{K_1 \cup K_2\}, \qquad q \in \{Q_1 \cup Q_2\}, \qquad p \in P \quad (7)$$

$$y_{jkp} \leq \sum_{i\in V} x_{ijk}, \qquad j \in N, \qquad k \in \{K_1 \cup K_2\}, \qquad p \in P \quad (8)$$

$$s_{ik} + t_i + t_{ij} - M_{ij}(1 - x_{ijk}) \leq s_{jk}, \qquad i \in N, j \in V, i \neq j, \qquad k \in \{K_1 \cup K_2\} \quad (9)$$

$$a_i \leq s_{ik} \leq b_i, \qquad i \in V, \qquad k \in K \quad (10)$$

$$x_{ijk} \in \{0,1\}, \qquad i,j \in V, \qquad k \in \{K_1 \cup K_2\} \quad (11)$$

$$y_{ipk} \in \{0,1\}, \qquad i \in V, \qquad k \in \{K_1 \cup K_2\}, \qquad p \in P \quad (12)$$

$$s_{ik} \geq 0, \qquad i \in N, \qquad k \in \{K_1 \cup K_2\} \quad (13)$$

Constraint (2) states that the vehicle arrived at each customer has to leave it. Constraint (3) and (4) states that the vehicle will not return to the previous customer. Constraint (5) shows that the start of the route for each vehicle must be depot and (6) and the end of the route must be depot. Constraint (7) ensure that the capacity of the compartment is not exceeded. Constraint (8) shows that if customer i is not visited by vehicle k, so the value of y_{ikp} is zero. Constraint (9) represents the relationship between the starting time of the service to one customer and the departure time from its predecessor. M is an arbitrarily large value. And constraint (10) states that the service takes place at each customer concerning the time window. Decision variables stated in constraint (11), (12) and (13).

3. Result and Discussion

There are 10 customers and using 5 vehicles in this research. Each customer located in Semarang, Central Java, Indonesia. Each customer has their code to describe the petrol station. There are:

Table 4. List of customers.

No.	Code of Petrol Station	Address
1	44.505.01	Jl. Raya Klepu Karangjati Ungaran
2	44.505.02	Jl. Diponegoro Ungaran
3	44.505.03	Kel. Karangjati Kec. Bregas
4	44.505.04	Ds. Jatijajar Dk. Tegal Panas Bergas
5	44.505.05	Ds. Randugunting Kec. Klepu
6	44.505.06	Jl. A Yani Kel. Ungaran
7	44.505.07	Jl. Karangjati Pringapus Ungaran
8	44.505.08	Jl. Diponegoro 204
9	44.505.09	Jl. Halmahera Kel. Gendangan Kec. Ungaran
10	44.506.01	Ds. Harjosari Bawen

The input data that been used for this research are distance, time deliveries and cost of delivery from the depot to each customer. Distance is obtained from the real distance by google maps. The distance matrix from each customer shows in table 5.

Table 5. The distance for each customer and depot.

Matrix of Distance (KM)											
	DEPOT	1	2	3	4	5	6	7	8	9	10
DEPOT	-	78	67	61	61	58	65	62	65	67	57
1	78	-	5	1	4	6	7	1	6	6	6
2	67	5	-	7	8	11	2	7	1	2	10
3	61	1	7	-	3	6	7	2	7	5	5
4	61	4	8	3	-	2	11	5	8	10	2
5	58	6	11	6	2	-	13	6	11	11	1
6	65	7	2	7	11	13	-	8	1	19	11
7	62	1	7	2	5	6	8	-	7	7	5
8	65	6	1	7	8	11	1	7	-	3	10
9	67	6	2	5	10	11	19	7	3	-	10
10	57	6	10	5	2	1	11	5	10	10	-

From the distance, we can calculate the cost per km by multiple the distance with traveling cost per kilometer.
Table 6 shows the travel cost from depot to 10 customers.

Table 6. The travel cost for each customer and depot.

Matrix of Cost (Rp)											
	DEPOT	1	2	3	4	5	6	7	8	9	10
DEPOT	0	228636	195932	178704	178996	168192	190092	179872	190968	194180	166732
1	228636	0	14600	2920	11096	17812	21608	4088	18688	17228	16644
2	195932	14600	0	20440	24528	31244	7008	21024	4088	6424	30076

3	178704	2920	20440	0	9636	16352	20148	6132	20148	15768	15184
4	178996	11096	24528	9636	0	6716	32412	13432	23944	28032	5548
5	168192	17812	31244	16352	6716	0	36792	17812	30952	32412	2336
6	190092	21608	7008	20148	32412	36792	0	23360	3796	56648	32412
7	179872	4088	21024	6132	13432	17812	23360	0	20440	19272	13432
8	190968	18688	4088	20148	23944	30952	3796	20440	0	9344	29492
9	194180	17228	6424	15768	28032	32412	56648	19272	9344	0	28032
10	166732	16644	30076	15184	5548	2336	32412	13432	29492	28032	0

The vehicle needs 40 minutes to load all of the products to the compartments, then the vehicle needs 40 minutes to unload the product to each petrol station. To minimize the total cost of delivery, this problem solved by Ilog Cplex 12.7.1. with branch and cut algorithm. Using 5 available vehicles with 4 compartments or 32.000 liter capacity. Each compartment has 8.000 liter capacity. Each compartment carries only one kind of product.

To solve the problem using Ilog Cplex, we need to insert the mathematic formulation such as decision variables, constraints, and objective function into the software. If 0 represent the depot, so the result is:

Table 7. Alternatives solution in Cplex.

Vehicle	Route	Code Delivery	Time Windows
1	0 – 5 – 10 – 0	Depot – 44.505.05 – 44.506.01 – Depot	15.00 – 16.27 – 17.29 – 19.56
2	0 – 3 – 4 – 0	Depot – 44.505.03 – 44.505.04 – Depot	15.00 – 16.30 – 17.39 – 20.02
3	0 – 1 – 7 – 0	Depot – 44.505.01 – 44.505.07 – Depot	15.00 – 16.18 – 17.27 – 19.53
4	0 – 9 – 2 – 0	Depot – 44.505.09 – 44.505.02 – Depot	15.00 – 16.38 – 18.19 – 20.30
5	0 – 8 – 6 – 0	Depot – 44.505.08 – 44.505.06 – Depot	15.00 – 17.00 – 18.18 – 20.50

Based on table 7, the best cost is Rp 1.898.584. The result of the mathematic formulation produces 5 routes from 5 available vehicles. For example, vehicle 1 must distribute the petrol product to customer 44.505.05 and 44.506.01. The vehicle will start shipping at 15.00, then at 16.27 the vehicle will arrive in point 44.505.05. Then the vehicle will unload the product and then continue to deliver the product to the last customer 44.506.01, and arrive at 17.29. After unloading, the vehicle must come back to the depot and arrive at 19.56.

4. Conclusion

Pertamina TBBM Boyolali should distribute their product every day based on the customer need. From table 7, we knew that the best cost for this research is Rp 1.898.584. Using 5 available vehicles with the result of each vehicle are vehicle 1: Depot – 44.505.05 – 44.506.01 – Depot, vehicle 2: Depot – 44.505.03 – 44.505.04 – Depot, vehicle 3: Depot – 44.505.01 – 44.505.07 – Depot, vehicle 4: Depot – 44.505.09 – 44.505.02 – Depot, and vehicle 5: Depot – 44.505.08 – 44.505.06 – Depot. So the company could minimize their distribution cost for all routes by choosing the best route. This model could be applied for all routes, so the company could reduce the distribution cost.

Acknowledgments

This research is supported by the Institute for Research and Community Service, Universitas Sebelas Maret with Hibah Penelitian Unggulan Terapan UNS (PUT-UNS) Research Program (Contract No. 516/UN27.21/PP/2019).

5. References

[1] Tjiptono, F. Pemasaran Jasa – Prinsip, Penerapan, dan Penelitian, Andi Offset, Yogyakarta. (2014).

[2] Martins, S., et.al. Product-Oriented Time Window Assignment for a Multi-Compartment Vehicle Routing Problem. European Journal of Operation Research. (2019)

[3] Surjandari, I., et.al. Petrol Delivery Assignment with Multi-Product, Multi-Depot, Split Deliveries and Time Windows. International Journal of Modeling and Optimization, Vol. 1, No, 5, p. 375-379. (2011).

[4] Benantar, A. and Ouafi, R. Optimization of Vehicle Routes : An Application to Logistic and Transport of the Fuel Distribution. 9th International Conference on Modeling, Optimization and Simulation, Bordeaux France. (2012).

[5] Benantar, A., Ouafi, R., & Boukachour, J. A petrol station replenishment problem: new variant and formulation. Logist. Res. (pp. 1-18). Springerlink.com. (2016).

[6] Sevaux, M., & Sorensen, K. Hamiltonian Paths in Large Clustered Routing Problem. EU / MEeting 2008, (pp. 1-7). Troyes, France. (2008).

[7] Battara, M., Edorgan, G., & Vigo, D. Exact Algorithms for the Clustered Vehicle Routing Problem. Operation Research, 58-71. (2014).

[8] Henke, T., Speranza, M. G., & Wascher, G. The multi-compartment vehicle routing problem with flexible compartment sizes. European Journal of Operational Research, 730-743. (2015).

[9] Coelho, L. C., & Laporte, G. Classification, Models and Exact A`lgorithms for Multi-Compartment Delivery Problems. European Journal of Operational Research, 854-864. (2015).

[10] Belov, A., & Slastnikov, S. Modeling of Multi Depot Vehicle Routing Problem for Petroleum Products. Lobachevskii Journal of Mathematics, 884-887. (2017).

[11] Pop, P., & Chira, C. A Hybrid Approach based on Genetic Algorithms for Solving the Clustered Vehicle Routing Problem. IEEE Congress on Evolutionary Computation (CEC), 1421-1426. (2014).

[12] Marc, A. H., Fuksz, L., Pop, P. C., & Danciulescu, D. A decomposition-based method for solving the clustered vehicle routing problem. UK: Oxford University Press. (2017).

[13] Henke, T., Speranza, M. G., & Wascher, G. Abranch-and-cut algorithm for the multi-compartment vehicle routing problem with flexible compartment sizes. Annals of Operations Research (pp. 1-18). Germany: Springer. (2018).

IOP Conf. Series: Materials Science and Engineering **943** (2020) 012040 doi:10.1088/1757-899X/943/1/012040

Determination of Routes for Daily Newspaper Product Distribution with Saving Matrix Methods

N A F P Adam*, I P Sari, A Tasya, W Sutopo, and Yuniaristanto
Department of Industrial Engineering, Faculty of Engineering, Universitas Sebelas Maret Surakarta, Indonesia

Emails: intan.permata.sari02@gmail.com

Abstract. The newspaper industry has several distinctive characteristics that make it stand out from other industries due to tight production and delivery times. The results of a case study in one of the newspaper companies in the city of Surakarta, found several problems, one of which is the slow distribution process due to the determination of the wrong route. The problem of distribution is the problem of the Vehicle Routing Problem (VRP). Vehicle Routing Problem is a problem related to determining the optimal distribution route by involving several obstacles to serve a number of agents at the request of each agent. One of the obstacles of the Vehicle Routing Poblem is the capacity of the vehicle. This study applies the Vehicle Routing Problem using the Nearest Neighbor method to optimize the distribution route. The Nearest Neighbor method is used to design routes based on the next closest distance. Distribution efficiency is evaluated based on the total distance, time and load borne by the vehicle fleet. The results showed that the distribution route was divided into two routes with a total distance traveled as far as 27.3 km and a total time of 1 hour 30 minutes in accordance with the time windows policy that is applied daily Solopos. Based on the use of these methods, it can save company expenses of Rp. 134,160.00 per day or Rp. 1,609,920.00 per month in product distribution for the Surakarta City area.

1. Introduction

The globalization and free trade era have erased the regional border in the trading system, resulting in a sharply increasing competition in the business sector. The easy access of information provided by various media causes the competition in business sector to improve. One example of the media is newspaper [Basriati, 2015]. The newspaper industry has several distinctive characteristics that make it stand out from other industries. The tight delivery time makes it impossible to save the inventories in a very short time for production and distribution. On one hand, there are pressures from the newsroom to foster the production process as late as possible to input last minute news, and on the other hand there are also pressures from the production and distribution departments to begin the production as early as possible [Furnes, 2014]. The supply chain of newspaper industries mostly consists of content editing, manufacture, and distribution. The manufacture comprises of pre-press, printing, and packing. While the distribution consists of delivery to distribution centers, drop-off points or retailers and customers [Furnes, 2014].

Generally, the production and distribution problems in the paper supply chain are the production planning integration and distribution to minimize the total costs [Aqidawati, Sutopo & Hisjam, 2018].

Companies have their own distribution deadlines. In this case, transportation plays an important role in the supply chain. Even though there are suppliers who conduct delivery in the same market area, coordination between two or more suppliers is seldom to happen [Intan, Sutopo & Hisjam, 2017]. There is a research that can build a model to optimize the profits of supply chain by including uncertainty in supply chain negotiations, transportation problems, and plannig of product allocation from suppliers to buyers by taking into account the time value of money (TVM) [Sutopo, Erliza & Heryansyah, 2016]. Most daily newspapers are published in the morning, evening newspapers were once quite common but now they are hard to find. It is common for newspapers to guarantee the delivery of daily newspapers before 06.00, and fulfilling the delivery deadlines is very important. As a result of time pressure, most of the production and distribution are carried out at night and often done in a short time [P3L, 2006].

The case study results in a newspaper company in Surakarta city illustrate some problems, one of which is the slow distribution process due to the incorrect route determination [Saputra, Sutopo & Hisjam, 2020]. Location and routing are very complex problems. The most important reason is that there are many alternative location models combined with a large number of routing models. Another reason is that different horizontal planning for location, routing decisions, and the fact that location problems are in contrast to routing problems require request aggregation [Klose & Drexl, 2005]. The distribution problem is caused by a Vehicle Routing Problem (VRP) problem. It is the problem of finding a route with minimum cost from a depot to customers with a number of different requests [Amri, Rahman & Yuniarti, 2014].

In order to solve the VRP problem, this research utilizes saving matrix method to minimize the cost as well as the transportation time. In addition, nearest neighbor method is used to determine the delivery routes. Other researchers have developed the VRP problem solving, for example the integer programming (Kulkarni and Bhave, 1985), mixed integer programming (Longo and Aragao, 2004), tabu search (Fermin and Roberto, 2004), genetic algorithm (Baker and Ayechew, 2003), simulated annealing (Tavakkoli, 2005), ant colony (Bell and McMullen, 2004), etc. Utilizing the exact optimization approach such as integer programming will take a long computation time, especially for large problems (if there are a large number of points served). Therefore, the authors choose the heuristic approach through the algorithm which is used to provide the optimum solution by learning, so it is faster than the exact optimization approach. Furthermore, one oftenly-used heuristic approach is the saving matrix. It is a method that minimize the route taken by considering the existing obstacles. In the initial case at PDAM Bandarmasih, it is proven that the saving matrix method can minimize time and transportation costs [Darmawan, 2011]. The saving matrix method is able to provide results regarding the fleet deployment based on their capacity load in a number of delivery areas based on the biggest savings [Ikfan & Masudin, 2013].

This article aims to solve the problems of route determination in the newspaper company in Surakarta city so the time and cost minimization as well as the best delivery route can be obtained.

2. Research Method

One classification of route-determining and scheduling problems is the Vehicle Routing Problem (VRP). The VRP problem is an integer programming problem, where a set of routes for a vehicle fleet sourced coming from one or several depots should be determined to serve the geographically-dispersed customers (www.neo.lcc.uma.es; Marinakis, 2012). The purpose of VRP is to deliver the product to a group of customers whose demand whose demands are already known by spending a minimum cost, also starting and ending at one or more depots. The output of this problem is a low cost and feasible route for each vehicle (the route is a sequence of locations that must be visited with an indication of requiring service). One interesting problem in the discussion of Vehicle Routing Problem is the Capcitated Vehicle Routing Problem (CVRP). CVRP is a case of determining the route of a K vehicle that aims to minimize the total distance covered by all routes, which will meet the capacity of vehicle Q and serve each customer. The following constraints will exist:

a. Each route will start and end at one depot.
b. Each customer will only be visited by one vehicle.

IOP Conf. Series: Materials Science and Engineering **943** (2020) 012040 doi:10.1088/1757-899X/943/1/012040

c. The total demand for each route cannot exceed the capacity of vehicle Q.
The CVRP formulations (Fukusawa, 4), are:
a. Purpose Function

$$\min \sum_{e \in E} C_e x_e$$

b. Constraint Function

$$\sum_{e \in \delta(\{i\})} x_e = 2 \qquad \forall i \in V+$$

$$\sum_{e \in \delta(\{0\})} x_e = 2K$$

$$\sum_{e \in \delta(S)} x_e \geq 2.k(S), \qquad \forall S \in V+$$

$$xij \in \{0,1\}, \forall \{i,j\} \in E$$

Note:
V : set node,
E : set egde / arc,
V' / V+ : {1…n} set node of customer,
N : customer,
ce : non negative values in each arc {i,j} ∈ δ E, can be the cost, distance or time, which are assumed to be symmetrical, where cij = cji,
xe : the number of vehicles passing through the arc {i,j} ∈ E, is equal to 0 if not passed by the vehicle and is equal to 1 if passed by the vehicles
∀ : all,
K : number of vehicles,
δ(S) : cut set defined in S,
S : {S ⊆ V', S ≥ 2},
k(S) : the minimum number of vehicles with Q capacity needed to fulfill customers' demand in S, the value is {d(S) / Q},
d(S) : indication of the total node requests in the subset S ⊆ V', where q(S) = ΣiЄS qi ,
Q : vehicle capacity,
qi : non negative demand on V ', where each V' vertex has non negative demand qi on condition of 0 < qi ≤ Q.

Saving Matrix method is used to determine the product's distribution route in the marketing area by determining the distribution route to be travelled and the number of vehicles based on the capacity of the vehicles in order to obtain the shortest route and minimum transportation costs. There are steps or several algorithms that should be taken in the matrix saving method. Below are the steps to complete the Saving Matrix method:

Step 1: Determining the distance matrix

The distance matrix states the distance between each pair of customers that should be visited. The distance can be determined utilizing google earth application, google maps, or manual calculation with the speedometer on the vehicle used. Vo Vi Vj Vn C0i

TABLE 1. General Form of a Distance Matrix

	Vo	...	Vi	...	Vj	...	Vn
Vo	0						
...		0					
Vi	Coi		0				
...				0			
Vj	Coj				0		
...						0	
Vn	Con		Cin		Cjn		0

Step 2: Determining the Saving Matrix

The savings matrix shows the savings that occur when two customers are combined. If S (x, y) states the distance saved, for example the trip from the center or starting point in a route by distance from depot to customer 1 and from customer 1 back to depot plus the distance from depot to customer 2 and then back to depot) subtracted by (distance from depot to customer 1 to customer 2 plus distance from customer 2 to depot), then the equation to find the amount of savings is:

TABLE 2. General Form of Savings Matrix

	Vo	...	Vi	...	Vj	...	Vn
Vo	-						
...		-					
Vi	Sii		-				
...				-			
Vj	Sij		Sij		-		
...						-	
Vn	Sin		Cin		Sjn		-

Step 3: Combining customers into vehicle travel routes

In this step, the customers are divided into a vehicle travel route by considering the customers and the vehicle capacity used. A route is considered feasible if the total number of requests from all customers does not exceed the capacity of the vehicle and the number of requests from one customer can be accommodated by one vehicle. The procedure used for grouping the customers is based on the largest savings matrix value. So, the first thing to do is sorting the largest saving matrix value until the capacity of the vehicle used can accommodate all demands. If the maximum capacity is reached, the procedure will be repeated until all customers are allocated in a route.

Step 4: Determining the order of customers

The order of customers is determined using farthest insert, nearest insert, Nearest Neighbor, Greedy, and Sweep procedures. The procedure which has the least distance result will be chosen.

- Farthest Insert Method

This method starts from the farthest customers and continued by putting other customers to the route that has the biggest mileage increase using the following formula :

IOP Conf. Series: Materials Science and Engineering **943** (2020) 012040 doi:10.1088/1757-899X/943/1/012040

- Nearest Insert Method

Beginning from the depot, this method adds the closest customers to complete the route. Afer there is no more closest customer to other customers, the last point visited by the vehicle will be inserted until all customers are visited.

- Nearest Neighbor Method

Starting from the depot, this method adds the closest customers to complete the route. At each step, the route is built by adding the customers closest to the last point visited by the vehicle until all consumers are visited.

- Greedy Method

This method is one type of algorithm which uses a problem solving approach by finding a temporary maximum value in each step or the local maxium. Greedy algorithms usually provide solutions that are close to the optimum value in a fairly quick time.

- Sweep Method

The principle used in the sweep method is to divide refugee posts into several clusters and then optimize the routes for each cluster.

First, all the vehicles are empty. Then, beginning with the first vehicle, this method inserts the unvisited nearest customers (Nearest Neighbour) into the route one by one. During the customer insertion process, the vehicle may not exceed its maximum capacity. Then the same process is carried out for the next vehicles, until all vehicles are full or all customers are visited.

The algorithm advantage of Nearest Neighbor is that it has shorter iteration which gives optimum result to solve the combinatorial optimization problems. Therefore some trips that use the Nearest Neighbor method can be used as an initial route that can make improvements for other methods.

The algorithm lack of Nearest Neighbor is when the point is more than 20, then the calculation will need quite a long time so it tries to find a way to get good, but not the best results. However, some cities that are not too far away can be travel and then visited at the end which results in further distance and more expensive costs.

The steps of the Nearest Neighbor method are as follows:

Step 1

Starting from the depot, then look for unvisited customers who have the shortest distance from the depot as the first location.

Step 2

Other customers who have the closest distance to the previously selected customer and the delivery number does not exceed the vehicle's capacity.

a. If there is a customer selected as the next customer and there is remaining vehicle capacity, return to step (2).
b. If the vehicle has no remaining capacity, return to step (1).
c. If no location is selected because the number of delivery exceeds the capacity of the vehicle, return to step (1). It starts again from the depot and visits unvisited customers who have the closest distance.

Step 3

If all customers have been visited exactly once then the algorithm ends

3. Result and Discussion

Previous researches are one of the authors' references in conducting research so that the author can enrich the theories used. From the previous researches, the authors do not find research with the same title as the authors' research title. However, the authors put forward some research as references to enrich the study material in the authors' research. Below are the previous researches in the form of journals related to the research conducted by the authors.

Table 3. Previous Research

Name of Researcher	Research Title	Research Results
Rahmawati, Nazaruddin, & Sari, 2014.	*Usulan Model dalam Menentukan Rute Distribusi untuk Meminimalkan Biaya Transportasi dengan Metode Saving Matrix di PT. XYZ*	The sub formation in the proposed route using the saving matrix method produces fewer sub routes than the distribution route applied by the company, where there are 7 proposed sub routes and there are 14 sub routes used by the company. The savings matrix can save a distribution cost of IDR 309,725,-

Difference : besides using the number of requests variable, transportation cost variable, and distance variable, Rahmawati, Nazaruddin and Sari's research used the capacity variable for each vehicle. Meanwhile, the variables used by the authors of this research are number of requests, coordinates of distribution points, distances between warehouses & agents, vehicle rental costs variables.

Name of Researcher	Research Title	Research Results
Sodikin, 2014.	*Penentuan Rute Distribusi Produk yang Optimal dengan Memperhatikan Faktor Kecepatan Kendaraan guna Meningkatkan Efisiensi Penggunaan BBM*	The initial distance is 1092.4 km. After saving matrix is used by considering factors of vehicle's speed, the distance travelled for product distribution reduces to 454.2 km. 638.2 km of distance can be saved in route 1. The effectiveness obtained through the use of saving matrix method by considering the vehicle speed is 58.42%. The fuel cost saved is IDR 400,717,- or distribution cost efficiency of 57.45%.

Difference : the research carried out by Sodikin (2014) focused on determining the optimal product distribution route by considering the vehicle's speed factor based on the distance from the warehouse to the retailer and the distance between retailers in order to improve fuel efficiency.

Name of Researcher	Research Title	Research Results
Indrawati, 2014.	*Penentuan Rute Distribusi Produk yang Meminimumkan Jarak Tempuh dan Jumlah Kendaraan dengan Pendekatan Metode Heuristik Vehicle Routing Problem*	The product distribution problem modeled in VRPTW can provide a sequence that has taken into account the obstacles such as time and vehicle capacity. By knowing the total distance, the Mantab Rasa SME can consider implementing the route obtained from the data processing. Reducing the number of vehicles used from 3 type B vehicles to only 2 vehicles can save 1 type B vehicle.

Difference : the research conducted by Indrawati (2014) utilized nearest to depot method to determine the distribution route. Meanwhile, the authors of this research use farthest insert, nearest insert, nearest neighbor, greedy, and sweep methods to determine the route.

In this research, observation of the daily distribution of Solopos in May 2018 was conducted. The initial data used is the distance between agents and the number of requests for each agent shown in Table 3 with a transport car that has the capacity of 2000 copies of newspaper . The company has 10 permanent agents with various daily demands for each agent ranging from 115 to 625 in May 2018. The everyday distribution of Solopos newspaper begins at 3:00 a.m. and ends at 6:00 a.m. with a service duration of 1 minute for 50 copies of newspaper. The coordinates of each agent can be determined using Google Maps as shown in Table 5.

TABLE 4. List of Agents, Addresses & Number of Demands

No.	Code	Name of Agent	Address	Demand (copies)
1	A1	TB Matahari	Keprabon	625
2	A2	Surya 1 Agency Solo	Keprabon	190
3	A3	ABC Agency	Keprabon	294
4	A4	Wahyu Agency	Keprabon	350
5	A5	Mandira/Kencana Agency	Laweyan	280
6	A6	ABA Agency	Timuran	122
7	A7	Sheva Agency	Nonongan	587
8	A8	Momokl Agency	Cemari	115
9	A9	Asih Agency	Cemari	115
10	A10	Margono 2 Agency	Laweyan	315
		TOTAL		**2993**

TABLE 5. Information of Distribution Point Coordinate

No.	Code	Location Coordinate		Demand (copies)
		X Coordinate	Y Coordinate	
1	A1	-7.569067	110.824795	625
2	A2	-7.545586	110.779160	190
3	A3	-7.545586	110.779160	294
4	A4	-7.550793	110.817880	350
5	A5	-7.553314	110.820476	280
6	A6	-7.553469	110.820597	122
7	A7	-7.568512	110.823682	587
8	A8	-7.567927	110.817130	115
9	A9	-7.572134	110.823997	115
10	A10	-7.573876	110.818691	315

The company spends operational costs that comprise of vehicle operating costs and fuel costs. Since the company uses third party services, the company only needs to pay a vehicle rental fee of Rp 650,000.00 per month for each vehicle and also the fuel costs.

The saving matrix method is used to minimize the route. The method can form a route and sequence of stop points in one route. The initial step taken is identifying the distance matrix between warehouse to each agent and the distance between stores by using the coordinates presented in Table 5.

TABLE 6. Distance Between Warehouse and Agent

	Warehouse	A1	A2	A3	A4	A5	A6	A7	A8	A9	A10
A1	7.6	0.0									
A2	0.2	7.7	0.0								
A3	0.3	7.7	0.1	0.0							
A4	5.3	3.6	5.1	5.5	0.0						
A5	5.1	3.3	5.7	5.6	0.2	0.0					
A6	5.1	3.3	5.7	5.6	0.2	0.1	0.0				
A7	7.7	0.2	6.9	6.8	2.8	2.8	2.4	0.0			
A8	6.8	1.7	6.5	6.4	2.8	2.8	2.4	0.8	0.0		
A9	7.7	0.4	8.8	8.7	4.7	4.7	4.3	0.5	2.8	0.0	
A10	7.5	1.9	7.7	7.6	3.6	3.6	3.2	1.1	0.9	0.9	0.0

The next step is identifying the savings matrix assuming that each agent will be visited by one truck exclusively. In other words, there will be 10 different routes with one destination each. The following example shows the calculation of the distance saving from warehouse to store 1 to store 2.

$$S(x,y) = J(G,x) + J(G,y) - J(x,y)$$
$$S(x,y) = 7.6 + 0.2 - 7.7$$
$$= 0.1km$$

So, the distance saving from store 1 to store 2 is 0.1 km. By using the same formula, the saving results are as follows :

TABLE 7. Results of Distance Savings Calculation

	Warehouse	A1	A2	A3	A4	A5	A6	A7	A8	A9	A10
A1	7.6	0.0									
A2	0.2	0.1	0.0								
A3	0.3	0.2	0.4	0.0							
A4	5.3	9.3	0.4	0.6	0.0						
A5	5.1	9.4	-0.4	-0.2	10.2	0.0					
A6	5.1	9.4	-0.4	-0.2	10.2	10.1	0.0				
A7	7.7	**15.1**	1.0	1.2	10.2	10.0	10.4	0.0			
A8	6.8	12.7	0.5	0.7	9.3	9.1	9.5	**13.7**	0.0		
A9	7.7	**14.9**	-0.9	-0.7	8.3	8,1	8.5	**14.9**	11.7	0.0	
A10	7.5	**13.2**	0.0	0.2	9.2	9.0	9.4	**14.1**	**13.4**	**14.3**	0.0

Based on the distance saving calculation, the next step is allocating the agent to the vehicle or route with the initial assumption of 10 different routes. However, the agents allocation can be combined with the limit of existing vehicles' capacity. The combination starts from the largest savings in order to maximize savings.

Based on calculations in allocating stores into the routes or vehicles, the results of calculation of store allocation into the route or delivery vehicle are known to be carried out with 2 shipping routes:

Route 1 : Agent 8, Agent 9, Agent 10, Agent 1, and Agent 7 (load 1757 copies)

Route 2 : Agent 2, Agent 3, Agent 4, Agent 5, and Agent 6 (load 1236 copies)

After the route allocation is made, the next step is determining the order of delivery by using Farthest Insert, Nearest Insert, Nearest Neighbour, Greedy, and Sweep methods. It is done because the agents in the route group are still random so they need to be sorted. It is assumed that each agent travel the route of Warehouse-Agent-Warehouse, then the delivery is sorted based on the method and the route as seen in Table 8. Then, route 1 and 2 are formed with the help of google maps. Finally, the Solopos distribution route is made, as seen in table 7.

TABLE 8. Recapitulation of Distance and Delivery Order Based on Method

Route	Order of Delivery	Method	Total Distance (km)
Route 1	Warehouse – A7 – A8 – A9 – A10 – A1 – Warehouse	Farthest Insert	17.06
	Warehouse – A8 – A7 – A9 – A1 – A10 – Warehouse	Nearest Insert	17.08
	Warehouse – A8 – A7 – A9 – A1 – A10 – Warehouse	Nearest Neighbor	17.08
	Warehouse – A8 – A7 – A1 – A9 – A10 – Warehouse	Greedy	16.06
	Warehouse – A8 – A1 – A7 – A9 – A10 – Warehouse	Sweep	17.06
Route 2	Warehouse – A4 – A5 – A6 – A3 – A2 – Warehouse	Farthest Insert	11.04
	Warehouse – A2 – A3 – A4 – A5 – A6 – Warehouse	Nearest Insert	10.07
	Warehouse – A2 – A3 – A4 – A5 – A6 – Warehouse	Nearest Neighbor	10.07
	Warehouse – A2 – A3 – A4 – A5 – A6 – Warehouse	Greedy	10.07
	Warehouse – A3 – A2 – A4 – A5 – A6 – Warehouse	Sweep	10.09

From the calculation by using Farthest Insert, Nearest Insert, Nearest Neighbor, Greedy, and Sweep result in minimum total distance. Based on table 8, route 1 have a total distance of 16.6 km from the Greedy calculation method. Route 2 have a minimum distance of 10.7 km based on Nearest Insert. Nearest Neighbour, and Greedy methods. So, the order of distances selected by the method is presented in Table 9 below. The route can be applied provided that there are no changes to agents or current traffic signs. In addition, the results obtained only cover the area of Surakarta city.

TABLE 9. Recapitulation of Distribution Route Distance

Route	Order of Delivery	Total	Distance	Time
Route 1	Warehouse – A8 – A7 – A1 – A9 – A10 – Warehouse	1757	16.06	60 minutes
Route 2	Warehouse – A2 – A3 – A4 – A5 – A6 – Warehouse	1236	10.07	30 minutes

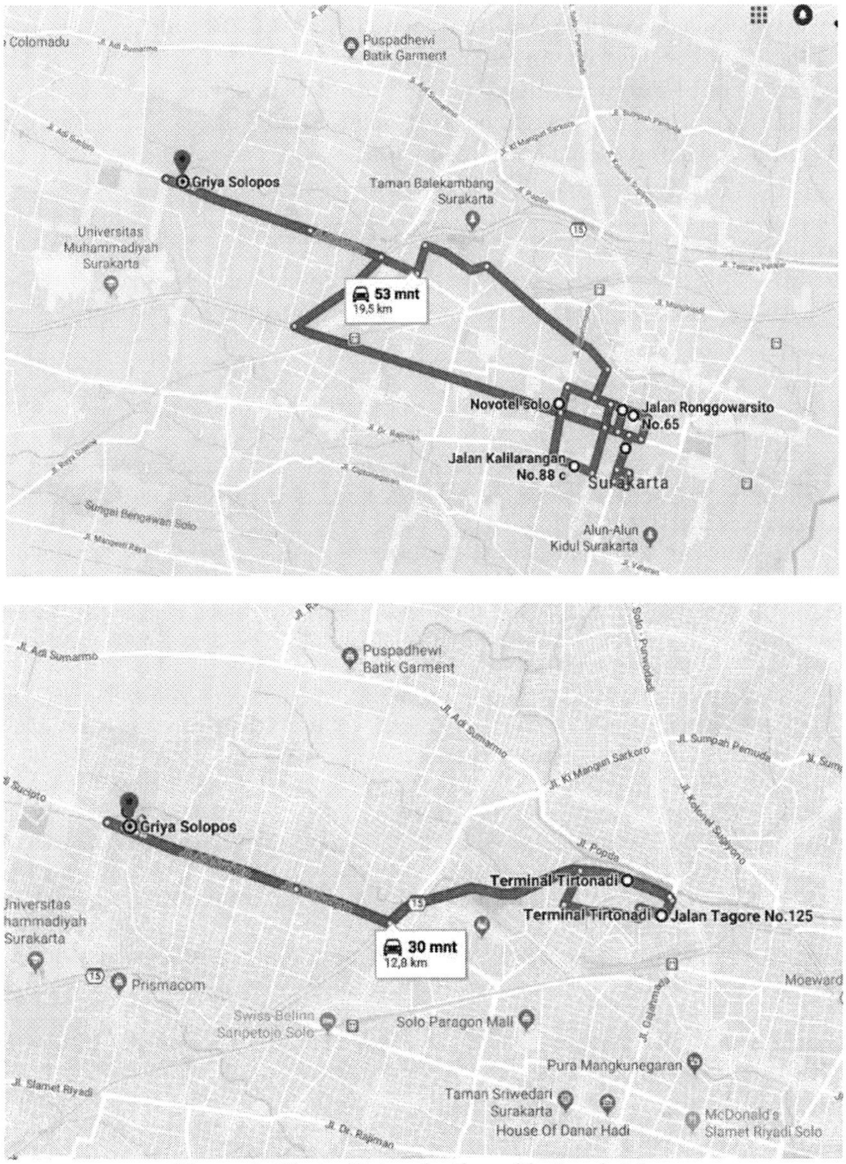

Figure 1. Cluster Determination with Google Maps

From the data collected, the next step is calculating the cost by comparing the expenses incurred by the company. Based on the results obtained, the company can save 4.3 km every day. Since the distribution process in this area takes place every day, the company can save a distance of 129 km monthly. Assuming that 1 litre of fuel can cover a distance of 10 km, and 1 litre of Pertamax fuel costs IDR 10,400, then the company can save up to IDR 134,600 each month and IDR 1,609,920 each year. The savings are considered quite large considering that the company does not need to invest.

4. Conclusions

Based on the Vehicle Routing Problem solving with Time Windows using saving matrix method, there are two routes for Solopos distribution optimization. The total travel time of both routes is 1 hour 30 minutes according to the time windows policy implemented by the company. By using saving matrix method, the company can save up to IDR 134,600 each month and IDR 1,609,920 each year for the product distribution in Surakarta.

This article can be developed by combining qualitative solving methods. In addition, this research still uses distance and operational costs in processing the data. It will be better if future research considers the traffic level and delivery time:

1. Solopos can use the routes provided in this research to reduce the delivery expenses in distributing their product.
2. The final results of processing those methods are not the most optimal results. So, this research can be developed by using other methods of distribution.

References

[1] Abadi, C., Susanty, S., Adianto, H. 2014. Penentuan Rute Kendaraan Distribusi Produk Roti Menggunakan Metode Nearest Neighbor dan Metode Sequential Insertion. Jurnal Online Institut Teknologi Nasional, 1(3), 152-163

[2] Amri, M., Rahman, A., & Yuniarti, R. (2014). Penyelesaian Vehicle Routing Problem Dengan Menggunakan Metode Nearest Neighbor. JRMSI Student Journal UB.

[3] Arinal, F., Imran, A., & Fitria, L. 2013. Penentuan Rute Kendaraan Pengangkutan Sampah dengan Menggunakan Metode Nearest Neighbour (Studi Kasus PD Kebersihan Kota Bandung). Jurnal Online Institut Teknologi Nasional, 1(1), 22-32.

[4] Aqidawati, E. F., Sutopo, W., & Hisjam, M. (2018). The Integration of Production-Distribution on Newspapers Supply Chain for Cost Minimization using Analytic Models: Case Study. IOP Conf. Series: Materials Science and Engineering, 319.

[5] Chandra, A. 2018. Optimasi Jalur Distribusi dengan Metode Vehicle Routing Problem (VRP). Jurnal Manajemen Transportasi & Logistik, 5(2)

[6] Darmawan, A. 2011. Penentuan Rute Terpendek Pada Pengiriman Air Bersih dengan Menggunakan Metode Optimal. PKN. FT-UMM, Malang.

[7] Furnes, K. K. (2014). Supply chain optimization in the newspaper industry. Industrial Economics and Technology Management.

[8] Hutasoit, C. S., Susanty, S., & Imran, A. 2014. Penentuan Rute Distribusi Es Balok Menggunakan Algoritma Nearest Neighbour dan Local Search (Studi Kasus di PT. X). Jurnal Online Institut Teknologi Nasional, 2(2), 268-276

[9] Intan, C., Sutopo, W., & Hisjam, M. (2017). Sales Forecasting Newspaper with ARIMA: A case Study. The 3rd International Conference on Industrial, Mechanical, Electrical, and Chemical Engineering, 1931.

[10] Ikfan, N., & Masudin, I. (2013). Penentuan Rute Transportasi Terpendek Untuk Meminimalkan Biaya Menggunakan Metode Saving Matriks. JITI, 12(2), 165-178.

[11] Jakaria, R. B., & Rosid, M. A. 2017. Implementasi Algoritma Greedy Pada Metode Transportasi Dengan Menggunakan Vam Dalam Pendistribusian Produk. Spektrum Industri, 15(1), 1 – 119

[12] Hayati, E. N., & Yohanes, A. 2014. pencarian rute terpendek menggunakan algoritma greedy, Seminar Nasional, 2337-4349

[13] Klose, A., & Drexl, A. (2005). Facility location models for distribution system design. European Journal of Operational Research, 162(1), 4-29.

[14] Marinakis. (2012). Multiple Phase Neighborhood Search-GRASP for the Capacitated Vehicle Routing Problem. Expert Systems with Applications, 39(8), 6807–6815.

[15] P3L. (2002) .Newspaper Supply Chains. http://www.p3logistics.com/papers.html

[16] Pujawan, I. N. 200. Supply Chain Management, 2nd ed., Surabaya, Indonesia: Guna Widya.

[17] Sarjono, H. 2014. Determination of Best Route to Minimize Transportation Costs Using Nearest Neighbor Procedure. Applied Mathematical Sciences, 8(62), 3063 – 3074

[18] Saraswati, R., Sutopo, W., & Hisjam, M. (2017). Penyelesaian Capacitated Vechile Routing Problem Dengan Menggunakan Algoritma Sweep Untuk Penentuan Rute Distribusi Koran : Studi Kasus. Jurnal Manajemen Pemasaran, 11(2)

IOP Conf. Series: Materials Science and Engineering **943** (2020) 012040 doi:10.1088/1757-899X/943/1/012040

[19] Saputra, I. W., Sutopo, W., & Hisjam, M. (2017). Optimization of Distribution Channel Vehicle Routing Problem with Time Windows using Differential Evolution Algorithm A case Study in Newspaper Industry. Asia-Pacific Management and Business Application, 10, 1-10.
[20] Sianipar, M., Fu'ani, D., Sutopo, W., & Hisjam, M. 2017. Penentuan Rute Kendaraan Menggunakan Metode Clark And Wright Saving Heuristic (Studi Kasus : Pt. Sinar Sosro). Performa, 16(2), 143-151
[21] Sutopo, W., Erliza, A., & Heryansyah, A. (2016). Optimizing Supply Chain Collaboration Based on Agreemenet Buyer-Supplier Relationship with Network Design Problem. Makara J. Technology, 20, 114-120.

Supply Chain Performance Analysis With Data Envelopment Analysis in TBBM PT. Pertamina Boyolali

Z Hanif[1*], P Kristy[2], M W D Utami[3], Yuniaristanto[4] and W Sutopo[5]
[1,2,3] Undergraduate Program of Industrial Engineering, Faculty of Engineering, Universitas Sebelas Maret Surakarta, Indonesia
[2] Department of Industrial Engineering, Faculty of Engineering, Universitas Sebelas Maret Surakarta, Indonesia

Email: hendrahanif@gmail.com

Abstract. PT. PERTAMINA TBBM Boyolali is a supplier of fuel oil in the form of premium, pertamax and diesel. The company should distribute fuel products using tank cars with operating areas covering Boyolali, Surakarta, Karanganyar, Sukoharjo, Sragen, Klaten, Wonogiri, Ngawi, Pacitan, Magetan, Ungaran, Semarang, Salatiga, Purwodadi, and Blora. An efficient supply chain is needed to provide the right products and services, with the specifications needed, in the right place and to the right customers. This article aims to find out and analyze the efficiency that occurs at operational TBBM Pertamina Boyolali and to find out the variables that most influence supply chain performance to be analyzed and improved so that improvements in supply chain performance can be more focus and optimal. The Decision-Making Units (DMUs) stated in this article are the period of supply chain activities from January 2018 to November 2018 at TBBM Pertamina Boyolali, the number of DMU is 11 DMUs. Supply Chain Operation Reference (SCOR) method with responsiveness, flexibility, and reliability attributes is used to measure Supply chain performance. To analyze supply chains efficiency, this article applies input-oriented DEA CCR (Charnes, Cooper, and Rhodes) model which considers linear relationships in its input and output. Data Envelopment Analysis (DEA) is a strong tool for measuring the efficiency of the DMUs and does not need to normalize indicator values that have different dimensions. The software used for processing data in this article is Banxia Frontier Analyst 4.2. This article provides suggestions for improving cash-to-cash cycle time, lead time, and flexibility for more efficient fuel distribution.

1. Introduction

At present, the supply chain has become a very important problem in companies when effective supply chain management leads to high-performance supply chains. The supply chain has involved many companies working together in such a way that the competition created is not between companies but between supply chains. The complexity which tends to develop and the increasing involvement of various stakeholders in the supply chain has shown the need for performance measurement and re-engineering of the business processes that have been traversed [1]. In the distribution of petroleum, supply chain costs up to 40% of the total cost of refining and distribution, so that effective management and optimization of the chain is very important in improving the performance of logistics and SCM.

TBBM Boyolali is one part of the Marketing Operation Region IV - PT Pertamina, which in its operation has used the Terminal Automation System (TAS). TBBM PT. Pertamina Boyolali is located on Jalan Solo - Semarang km 18. The Boyolali BBM Terminal receives fuel supplies from the Lomanis Transit Terminal through multi-product pipes using the CY2 pipeline. The length of the CY2 pipe is 246 km with an average flow rate of 300 kiloliter/hour and the total volume of the pipe contents is 18,098 kiloliter. TBBM PT. Pertamina Boyolali receives fuel products (Premium, Pertamax and Solar) from the Lomanis Transit Terminal through pipelines, stores in storage tanks, and distributes using tank cars with operating areas including Boyolali, Surakarta, Karanganyar, Sukoharjo, Sragen, Klaten, Wonogiri, Ngawi , Pacitan, Magetan, Ungaran, Semarang, Salatiga, Purwodadi, Blora.

Data Envelopment Analysis (DEA) has proven to be a strong tool in measuring the efficiency of Decision Making Units (DMUs). Also, this method is not necessary to normalize indicator values with different dimensions. The term DMU is referred to every production unit with certain inputs and outputs. In the literature on the evaluation of operational performance, certain DEA models are widely applied by many researchers. Tajbakhsh & Hassini (2015) [1] and Fathi & Saen (2018) [2] evaluated sustainability in the distributive chain of transportation companies. Huang, (2018) [3] assessing the performance of tourism supply chains with N-DEA. Boudaghi & Farzipoor Saen (2018) [4] carried out the development of the DEA model to predict the removal of suppliers in the supply chain. Goodarzi & Saen (2018) [5] analyzed the phenomenon of the bullwhip effect in the supply chain using DEA. Saputri, Sutopo, Hisjam & Ma'aram (2019) [6] analyzed sustainable agri-food supply chain performance measurement model for GMO and Non-GMO.

In this study, we use the input-oriented DEA-CCR model. This model is a DEA model that was developed by Charnes, Cooper, & Rhodes in 1978 [7]. This model is also called the Constant Return to Scale (CRS) model, which considers linear relationships in the input and output. Every increase in the input will result in a proportional and constant increase in output. This can mean that efficiency will not change on a scale regardless of the unit operating. Input-oriented measurement is the identification of inefficiencies through the possibility to reduce inputs without changing the output value [8]. This article aims to find out and analyze the efficiency that occurs at operational TBBM Pertamina Boyolali and to find out the variables that most influence supply chain performance to be analyzed and improved so that improvements in supply chain performance can be more focus and optimal.

Based on the literature review by Najmi, Gholamian, & Makui (2013) [9], the development of performance analysis methods are as follows.

Table 1. The Development Of Performance Analysis Methods

Number	Method	Reference	Number	Method	Reference
1	Simulation	Persson & Olhager, 2002 [10]	13	DEA	Wong, 2009) [22]
2	Delphi	Bottani & Bigliardi, 2010 [11]	14		Xu, Li, & Wu, 2009 [23]
3	AHP	Chan, 2003 [12]	15		Peng Wong & Yew Wong, 2008 [24]
4		Yang, 2009 [13]	16		Peng Wong & Yew Wong, 2007 [25]
5		Wadhwa, Varma, & Deshmukh, 2008 [14]	17		F. Yang, Wu, Liang, & Bi, 2011 [26]
6		Bhagwat & Sharma, 2009 [15]	18		Hamid, Hakim, Lin, & Choy Chong, 2009 [27]
7		Drzymalski, Odrey, & Wilson, 2010 [16]	19		Parkan & Wang, 2007 [28]
8		Bhagwat & Sharma, 2007 [17]			
9		Askariazad & Wanous, 2009 [18]	20	Hybrid	Berrah & Clivillé, 2007 [29]
10		Bhagwat, Chan, & Kumar Sharma, 2008 [19]	21		Jia & Ierapetritou, 2003 [30]

IOP Conf. Series: Materials Science and Engineering **943** (2020) 012041 doi:10.1088/1757-899X/943/1/012041

Number	Method	Reference	Number	Method	Reference
11	Heuristic	Ho, 2007 [20]	22		Kodali & Soni, 2010 [31]
12		Angerhofer & Angelides, 2006 [21]	23		Chen & Larbani, 2005 [32]

Based on the State of the Art, the DEA method is used by 8 studies from 24 studies. Research with the DEA method is quite popular for performance analysis.

2. Quantitative Approach

Supply Chain Operation Reference (SCOR) method with responsiveness, flexibility, and reliability attributes is used to measure Supply chain performance. To analyze supply chains efficiency, this article applies input-oriented DEA CCR (Charnes, Cooper, and Rhodes) model which considers linear relationships in its input and output. Data Envelopment Analysis (DEA) is a strong tool for measuring the efficiency of the DMUs and does not need to normalize indicator values that have different dimensions. The software used for processing data in this article is Banxia Frontier Analyst 4.2. The Distribution Channel of PT. TBBM Boyolali is shown in Figure 1.

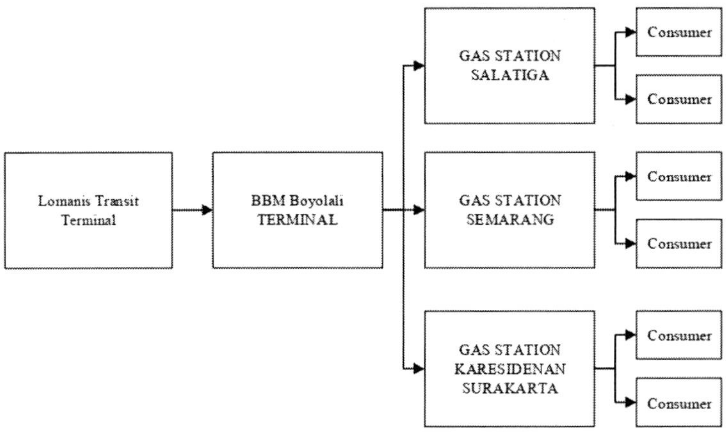

Figure 1. Distribution Channel of PT. TBBM Boyolali

The Decision-Making Units (DMUs) stated in this article are the period of supply chain activities from January 2018 to November 2018 at TBBM Pertamina Boyolali, the number of DMU is 11 DMUs. The classifications of DMUs are shown in Table 2.

Table 2. The classifications of DMUs in 2018

Month	DMU
January	DMU 1
February	DMU 2
March	DMU 3
April	DMU 4
May	DMU 5
June	DMU 6
July	DMU 7
August	DMU 8
September	DMU 9
October	DMU 10
November	DMU 11

Variables used based on the method of Supply Chain Operation Reference (SCOR) are responsiveness, flexibility, reliability, and financial measures. Responsiveness is seen from the cash-to-cash cycle time and lead time (order fulfillment cycle). Cash-to-cash cycle time is used to measure the time starting from the incoming order until the product reaches the consumer. Cash-to-cash cycle time is calculated based on days sales outstanding (DSO) added days of inventory (DOI) then minus days payable outstanding (DPO). Increasing DPO, reducing DSO and DOI can minimize the value of the cash-to-cash cycle time. DSO size shows existing collectible receivables. The longer the collectible receivables are recorded, then the company's financial flows will be disrupted. DOI shows inventory turnover speed. The DPO shows the number of hours the debt is paid to the supplier. Lead time is the period required from the time of order until the ordered item has been received.

Flexibility is the tenacity of the supply chain and the ability to adapt to changes in market demand to maintain supply chain competitive advantage. Flexibility is defined as the ability to respond to consumer requests and overtake competitors. Flexibility is one of the determining factors in the market competition [33]. The input and output variables used in this article are shown in Table 3.

Table 3. Input and Output Variables

No	Variabel	Sub-criteria Variabel	Unit	Type of Criteria	Exlanation
1	Responsiveness	Cash-to-cash cycle time	Hours	Input	The time needed from the fuel is bought by TBBM until it is purchased by the customer
		Lead time	Hours	Input	The time needed by suppliers to fulfill orders
2	Flexibility	Flexibility	Hours	Input	The company's ability to respond to changes in company orders
3	Reliability	Compliance with standards	%	Output	Percentage of suitability of the product entering the gas station
		Order fulfillment	%	Output	Persentase pemenuhan pesanan konsumen
		Delivery performance	%	Output	Percentage of timeliness in order fulfillment

Calculations performed to obtain the output variable are as follows.

a. Conformance to standard (%)

$$\frac{\text{Number of products received - Number of products rejected}}{\sum receipt}\times 100\% \quad \text{.....................Equation (1)}$$

b. Order fulfillment (%)

$$\frac{\sum \text{order completion}}{\sum \text{order receipt}}\times 100\% \quad \text{.......... Equation (2)}$$

c. Delivery performance (%)

$$100\% - \left(\frac{\text{Receipt date - date of appointment}}{\text{Lead Time}}\times 100\%\right) \quad \text{..........Equation (3)}$$

Calculation of Efficiency Value was done by Banxia Frontier Analyst 4.2 software. The following equation with the input-oriented DEA-CCR model.

a. Objective function

$$MAXh_n \sum_{i=1}^{m} u_r y_{rj} \quad \text{..........Equation (4)}$$

b. Constrain

$$\sum_{i=1}^{m} v_i x_{ij} = 1 \qquad \text{.....Equation (5)}$$

$$\sum_{r=1}^{t} u_r y_{rj} - \sum_{i=1}^{m} v_i x_{ij} \leq 0, j = 1,\ldots..,n \qquad \text{.....Equation (6)}$$

$$u_r, v_i > 0, r = 1,\ldots \ldots.,t, i = 1,\ldots\ldots,m \qquad \text{.....Equation (7)}$$

h_n = DMU relative efficiency
t = number of outputs
m = number of inputs
u_r = weight of output r
v_i = weight of input i
y_{rj} = value of the fourth output from DMU to j
x_{ij} = the first input value from the DMU to j

3. Result and Discussion

Data input and output chain variables are shown in table 3. Data analysis was carried out with the help of Banxia Frontier Analyst 4.2 software. In the table 4, 5, and 6, there are 3 conditions with 3 different colors. green for efficient conditions with a value of 100%, yellow for Ammber conditions with a value of 95% -99.99% (with improvement potential) and red for conditions under 95% [34]. Table 4 shows data of input variable and output variable of 1st Supply Chain Alternative

Table 4. Data Input Variable And Output Variable of 1st Supply Chain Alternative

Month	Input			Output		
	Cash-to-cash Cycle Time	Lead Time	Flexibility	Compliance with standards	Order fullfilment	Delivery performance
	Hour	Hour	Hour	%	%	%
1	83	24	11	96	100	93
2	96	22	9	100	100	90
3	92	20	9	98	100	88
4	87	19	12	96	100	90
5	54	32	18	100	97	95
6	81	22	11	99	100	98
7	90	24	12	100	95	95
8	93	20	10	100	100	100
9	93	19	11	99	100	85
10	92	20	14	100	100	98
11	84	20	11	98	99	88

Table 5 shows data of input variable and output variable of 2nd Supply Chain Alternative

Table 5. Data Variable Input Dan Variable Output of 2nd Supply Chain Alternative

Month	Input			Output		
	Cash-to-cash Cycle Time	Lead Time	Flexibilit y	Compliance with standards	Order fullfilment	Delivery performance
	Hour	Hour	Hour	%	%	%
1	90	17	11	96%	98%	97%
2	90	18	9	99%	95%	91%
3	92	15	8	98%	100%	95%
4	90	15	8	100%	97%	100%
5	85	18	8	96%	95%	100%
6	85	19	9	97%	98%	100%
7	90	20	7	199%	100%	96%

8	90	15	12	98%	100%	97%
9	80	18	11	100%	99%	100%
10	80	18	10	97%	97%	92%
11	88	18	10	99%	97%	96%

Table 6. Eficiency Score for each DMU

DMU	1st Alternatif			2nd Alternatif		
	Score	Eficiency	Condition	Score	Eficiency	Condition
1	98.60%	Yes	Green	96.2%	No	Amber
2	100.00%	Yes	Green	96.6%	No	Amber
3	100.00%	Yes	Green	100.00%	Yes	Green
4	100.00%	Yes	Green	100.00%	Yes	Green
5	100.00%	Yes	Green	100.00%	Yes	Green
6	100.00%	Yes	Green	100.00%	Yes	Green
7	88.10%	No	Red	100.00%	Yes	Green
8	100.00%	Yes	Green	100.00%	Yes	Green
9	100.00%	Yes	Green	100.00%	Yes	Green
10	99.40%	No	Amber	100.00%	Yes	Green
11	100.00%	Yes	Green	94.20%	No	Red

In the table 4, 5, and 6, there are 3 conditions with 3 different colors. green for efficient conditions with a value of 100%, yellow for Ammber conditions with a value of 95% -99.99% (with improvement potential) and red for conditions under 95% [34]. The data on the efficiency value of each DMU in the 1st Alternative has the lowest efficiency value in July 2018 of 88.1%. From the overall data obtained the efficient value of green at 9 DMU is in January, February, March, April, May, June, August, September, November. Ammber conditions in October, as well as Red's condition in July.

The lowest efficiency value was obtained in November at 94.20% for the 2nd Alternative. In the table, there were 8 green DMU conditions, namely in March, April, May, July, June, August, September, and October. Amber's condition in January, and February and Red's condition in November

The average value in alternative 1 obtained at 98.74% and in alternative 2 at 98.82%. hence based on this condition it is still possible to make improvements based on a potential improvement to increase the efficiency value to 100%.

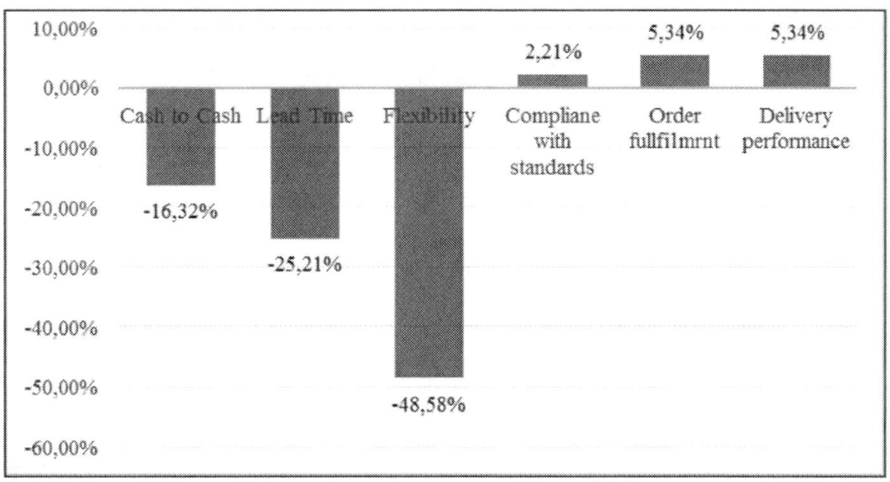

Figure 2. Total Potential improvements of 1st Alternative

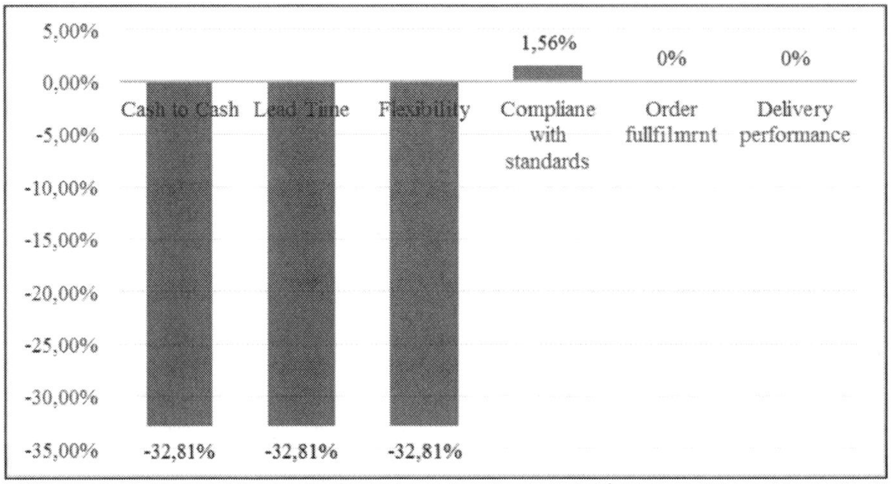

Figure 3. Total Potential improvements of 2nd Alternative

Figure 4 is a total diagram of potential improvement resulting from the processing of the average potential improvement of each DMU. This data is used to base the improvement and managerial implications of the company.

Overall, the supply chain performance in alternatives 1 and 2 is quite good with the results of calculations on amber conditions. To get the optimal value in alternatives 1 and 2, it is necessary to improve input performance by decreasing the actual value of the existing input variable. The repairs are carried out in several ways.

1. Cash-to-cash cycle time
   Improvement In 1st alternative, improvements in the cash-to-cash cycle time can be done by educing the value of the input variable by 16.32% so that the actual value can reduce from 86 hours to 72 hours. In 2nd alternative, there is a reduction of 32.81% at the actual value so that the actual time of 87 hours is reduced to 58 hours. The action that can be taken is by updating the agreement or payment method between Pertamina and SPBU to facilitate payment at the end of the deadline or can use the credit payment system. This will greatly influence the increase in the value of efficiency if an improvement in the cash-to-cash cycle time factor is made.
2. Lead time improvement
   Improvement of lead time is done by reducing the input value by 25.21% in 1st alternative and by 32.81% in 2nd alternative. So the actual value in 1st alternative can be reduced from 22 hours to 16 hours and in 2nd alternative from 17 hours to 12 hours. Steps that can be taken to reduce lead time can be done by setting a better delivery schedule with the order forecasting process before entering the order and reviewing the shipping hours where the previous delivery is served from 12 pm to 9 pm.
3. Flexibility Improvement
   Improvement of time flexibility is done by reducing the input value by 45.18% in 1st alternative and by 32.81% on 2nd alternative. So that the actual value in 1st alternative can be reduced from 11 hours to 6 hours and in 2nd alternative from 9 hours to 7 hours. Steps taken to support this change can be done by preparing a truck that specifically serves orders that require fast supply needs and a fast ordering process with a payment delay system.

An efficiency value of 100% must be achieved for each output variable in the supply chain. This value can be achieved by improving the actual value. Improvements are made by reducing input values with the results of the input variable in the 1st alternative are cash-to-cash cycle time of 72 hours, lead time of 16 hours and flexibility of 6 hours and for input variables in the 2nd alternative are cash-to-cash 59 hours cycle time, 12 hours of lead time, and 7 hours of flexibility. The input variables that most influence the supply chain efficiency value on the 1st alternative and 2nd alternative are lead time and flexibility. So that lead time and flexibility are the most influential factors to be improved in this case.

4. Conclusion

PT. PERTAMINA TBBM Boyolali should distribute fuel products using tank cars with operating areas covering Boyolali, Surakarta, Karanganyar, Sukoharjo, Sragen, Klaten, Wonogiri, Ngawi, Pacitan, Magetan, Ungaran, Semarang, Salatiga, Purwodadi, and Blora. The Decision-Making Units (DMUs) stated in this article are the period of supply chain activities from January 2018 to November 2018 at TBBM Pertamina Boyolali, the number of DMU is 11 DMUs. This article provides suggestions for improving cash-to-cash cycle time, lead time, and flexibility for more efficient fuel distribution. Improvements using the DEA-CCR model in the case of Pertamina's Fuel Distribution can be used to increase efficiency by changing input values for 3 variables, namely cash-to-cash cycle time, lead time, and flexibility. This article has certain limitation that should be overcome in order to provide in deep analysis on the performance measurement analysis. For further research will complete the research with Balance Score Card for more balance set of performance measures.

References

[1] Tajbakhsh, A., & Hassini, E. (2015). A data envelopment analysis approach to evaluate sustainability in supply chain networks. Journal of Cleaner Production, 105, 74–85.

[2] Fathi, A., & Saen, R. F. (2018). A novel bidirectional network data envelopment analysis model for evaluating sustainability of distributive supply chains of transport companies. Journal of Cleaner Production, 184, 696–708.

[3] Huang, C. (2018). Assessing the performance of tourism supply chains by using the hybrid network data envelopment analysis model. Tourism Management, 65, 303–316.

[4] Boudaghi, E., & Farzipoor Saen, R. (2018). Developing a novel model of data envelopment analysis–discriminant analysis for predicting group membership of suppliers in sustainable supply chain. Computers & Operations Research, 89, 348–359.

[5] Goodarzi, M., & Saen, R. F. (2018). How to measure bullwhip effect by network data envelopment analysis? Computers & Industrial Engineering.

[6] Saputri, V., Sutopo, W., Hisjam, M., & Ma'aram, A. (2019). Sustainable Agri-Food Supply Chain Performance Measurement Model for GMO and Non-GMO Using Data Envelopment Analysis Method. Applied Sciences, 9(6), 1199.

[7] Charnes, A., Cooper, W. W., & Rhodes, E. (1978). Measuring the efficiency of decision making units. European Journal of Operational Research, 2(6), 429–444.

[8] Yeni, S., & Siswanto, N. (2005). Penerapan Data Envelopment Analysis dalam Pemilihan Supplier dan Perbaikan Performansi Supplier. In Dalam Prosiding Seminar Nasional Manajemen Teknologi II (Vol. 200, hal. 1–9).

[9] Najmi, A., Gholamian, M. R., & Makui, A. (2013). Supply chain performance models: A literature review on approaches, techniques, and criteria. Journal of Operations and Supply Chain Management; Vol 6, No 2 (2013): July - December.

[10] Persson, F., & Olhager, J. (2002). Performance simulation of supply chain designs. International Journal of Production Economics, 77(3), 231–245.

[11] Bottani, E., & Bigliardi, B. (2010). Performance measurement in the food supply chain: a balanced scorecard approach. Facilities, 28(5/6), 249–260.

[12] Chan, F. T. S. (2003). Performance Measurement in a Supply Chain, 534–548.

[13] Yang, J. (2009). Integrative performance evaluation for supply chain system based on logarithm triangular fuzzy number-AHP method. Kybernetes, 38(10), 1760–1770.

[14] Wadhwa, S., Varma, S., & Deshmukh, S. G. (2008). Evaluating petroleum supply chain performance: Application of analytical hierarchy process to balanced scorecard. Asia Pacific Journal of Marketing and Logistics, 20(3), 343–356.

[15] Bhagwat, R., & Sharma, M. K. (2009). An application of the integrated AHP-PGP model for

performance measurement of supply chain management. Production Planning & Control, 20(8), 678–690.

[16] Drzymalski, J., Odrey, N. G., & Wilson, G. R. (2010). Aggregating performance measures of a multi-echelon supply chain using the analytical network and analytical hierarchy process. International Journal of Services, Economics and Management, 2(3–4), 286–306.

[17] Bhagwat, R., & Sharma, M. K. (2007). Performance measurement of supply chain management using the analytical hierarchy process. Production Planning & Control, 18(8), 666–680.

[18] Askariazad, M., & Wanous, M. (2009). A proposed value model for prioritising supply chain performance measures. International Journal of Business Performance and Supply Chain Modelling, 1(2–3), 115–128.

[19] Bhagwat, R., Chan, F., & Kumar Sharma, M. (2008). Performance measurement model for supply chain management in SMEs. Int. J. of Globalisation and Small Business (Vol. 2).

[20] Ho, C.-J. (2007). Measuring system performance of an ERP-based supply chain. International Journal of Production Research, 45(6), 1255–1277.

[21] Angerhofer, B. J., & Angelides, M. C. (2006). A model and a performance measurement system for collaborative supply chains. Decision Support Systems, 42(1), 283–301.

[22] Wong, W. P. (2009). Performance evaluation of supply chain in stochastic environment: using a simulation based DEA framework. International Journal of Business Performance and Supply Chain Modelling, 1(2–3), 203–228.

[23] Xu, J., Li, B., & Wu, D. (2009). Rough data envelopment analysis and its application to supply chain performance evaluation. International Journal of Production Economics, 122(2), 628–638.

[24] Peng Wong, W., & Yew Wong, K. (2008). A review on benchmarking of supply chain performance measures. Benchmarking: An International Journal, 15(1), 25–51.

[25] Peng Wong, W., & Yew Wong, K. (2007). Supply chain performance measurement system using DEA modeling. Industrial Management & Data Systems, 107(3), 361–381.

[26] Yang, F., Wu, D., Liang, L., & Bi, G. (2011). Supply chain DEA : production possibility set and performance evaluation model, 195–211.

[27] Hamid, A., Lukman Hakim, I., Lin, B., & Choy Chong, S. (2009). Measuring supply chain performance among public hospital laboratories. International Journal of Productivity and Performance Management, 59(1), 75–97.

[28] Parkan, C., & Wang, J. (2007). Gauging the performance of a supply chain. International Journal of Productivity and Quality Management, 2(2).

[29] Berrah, L., & Clivillé, V. (2007). Towards an aggregation performance measurement system model in a supply chain context. Computers in Industry (Vol. 58).

[30] Jia, Z., & Ierapetritou, M. (2003). Mixed-Integer Linear Programming Model for Gasoline Blending and Distribution Scheduling. Industrial & Engineering Chemistry Research, 42(4), 825–835.

[31] Kodali, R., & Soni, G. (2010). Internal benchmarking for assessment of supply chain performance. Benchmarking: An International Journal, 17(1), 44–76.

[32] Chen, Y.-W., & Larbani, M. (2005). Simulating the performance of supply chain with various alliances. International Journal of Advanced Manufacturing Technology (Vol. 25).

[33] Groover, M. P. (2002). Automation, Production Systems, and Computer-integrated Manufacturing 2nd ed.20022 Automation, Production Systems, and Computer-integrated Manufacturing 2nd ed. . Pearson Education – Prentice-Hall, 2001. 856 pp., ISBN: 0-13-088978-4. Assembly Automation (Vol. 22).

[34] Hussain, A. & Jones, M. (2010). An Introduction to Frontier Analyst 4. United Kingdom: Banxia Frontier Ltd.

IOP Conf. Series: Materials Science and Engineering **943** (2020) 012042 doi:10.1088/1757-899X/943/1/012042

Redesign Production Layout Using Dedicated Storage Method: Case Study of PT.Solo Grafika Utama

C Dianto[1*], F Widiandoko[2], D Rahmanasari[3], Yuniaristanto[4], W Sutopo[5]
[1,2,3] Undergraduate Program of Industrial Engineering, Faculty of Engineering, Sebelas Maret University Surakarta, Indonesia
[4,5] Department of Industrial Engineering, Faculty of Engineering, Sebelas Maret University Surakarta, Indonesia

Email: cornelius.dianto@gmail.com

Abstract. PT. Solo Grafika Utama is one of the companies in the field of newspaper production. The newspaper industry has strict production characteristics. Therefore, it requires a good factory layout in order to create effectiveness and production efficiency. The warehouse layout in PT Solo Grafika Utama still deficient in the application of material handling in warehouses so that it often raises problems. The problem is that the product arrangement in the warehouse is not optimal because there are still products stored in an wrong place. This article aims to provide an effective and efficient proposal to improve warehouse layout in PT Solo Grafika Utama using the dedicated storage method. Dedicated Storage is chosen to improve warehouse arrangement and minimize the distance of product displacement. After the proposed improvement, a better warehouse layout is obtained. In the proposed layout can also minimize the product displacement distance of 23% from the initial distance. Thus it can be said that the proposed layout is better than the initial layout and can be proven to be able to facilitate the handling of materials in the warehouse so that it can make newspaper production more effective and efficient.

1. Introduction

To be able to compete with businesses today, companies must have the ability to continue to grow and always have change. In a company, many factors affect the sustainability of the production process including production planning, production layout planning, receiving, shipping, warehousing, etc (Wingjosoebroto, 2009). In general, a well-planned factory layout will also determine efficiency and will maintain the survival or success of an industry. One important layout plan is the warehousing system. Warehouse is an important part of production in managing the flow of goods, information, and costs. Warehouse is a place to store materials, raw materials, equipment, and finished products. The purpose of the warehousing system itself is to manage and store goods that are ready to be distributed so that the goods can be received on time by consumers. So that a good warehousing system will produce efficiency and effectiveness of the production process (Permana, 2014).

PT. Solo Grafika Utama is a printing company in Surakarta that has a main product, Solopos newspaper. As a local newspaper in the Solo area, SOLOPOS has the potential to promote in the Surakarta, Boyolali, Sokoharjo, Wonogiri, Sragen, and Klaten regions which are the main distribution areas. Solopos general daily production process at PT. Solo Grafika Utama based on orders from its parent company, PT. Solopos script. But along with the development of the company, now PT. Solo Grafika Utama has produced other products, such as tabloids, Student Worksheets (LKS), magazines or orders to print other general daily

The problem faced by PT Solo Grafika Utama is that the company does not have clear arrangements regarding product layout, storage position arrangement, and product arrangement in the warehouse so that the warehouse arrangement pattern becomes irregular. Therefore, it is necessary to redesign the product storage layout using the dedicated storage method. This method, arranging products by placing one product in one storage location only. This placement is based on a comparison of the activities of each product and space requirements. (Abdullah, 2009) say that the advantage of this method is to propose improvements to the layout of the product warehouse that is more flexible and more effective, minimizes transportation distance, saves material transfer, and makes it easier to arrange goods in the warehouse. While the disadvantage of this method is the low space utilization because the location of the product cannot be changed or used by other products even though the location is empty. The problem at PT Solo Grafika Utama's warehouse was solved by using a dedicated storage method to redesign the warehouse layout, determine the needs of the warehouse area, and determine the distance of initial material handling and proposals. With this research the warehouse is expected to be more tidy and orderly so as to facilitate material handling activities in product storage and retrieval, as well as facilitate operators in the product search process.

2. Methodology

This section contains the flowchart of research conducted in PT. Solo Grafika Utama :

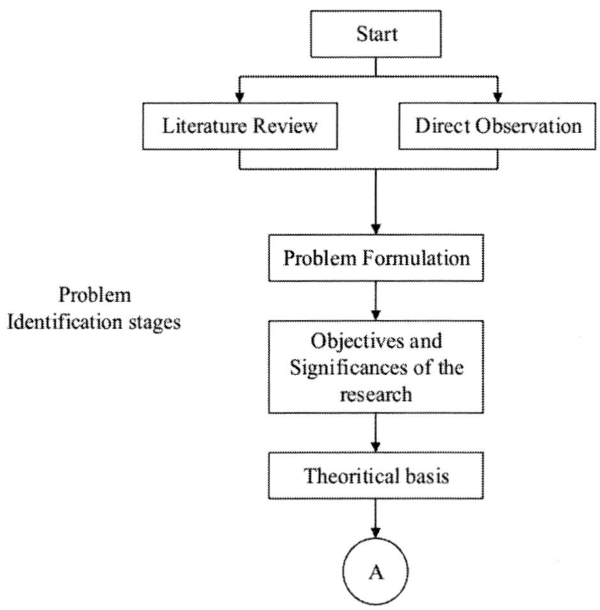

Figure 1 Flowchart of the Research

IOP Conf. Series: Materials Science and Engineering **943** (2020) 012042 doi:10.1088/1757-899X/943/1/012042

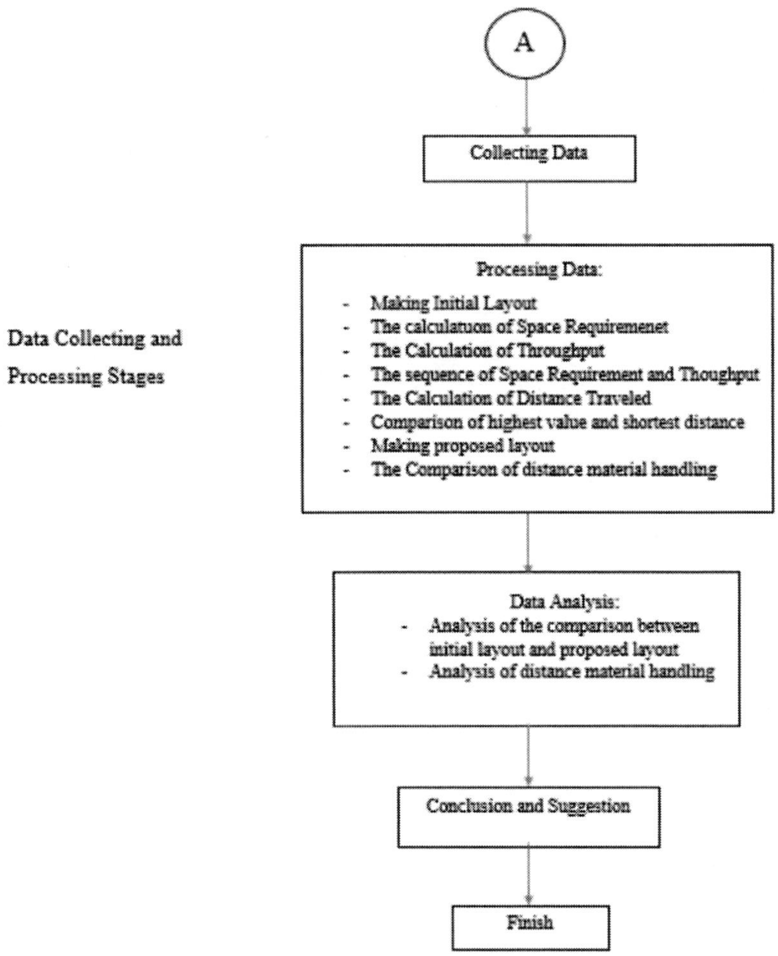

Figure 1 Flowchart of the Research (continued)

The research method is carried out in several stages so that the discussion can be structured and the research objectives are achieved. Before doing research, a literature study is conducted on the object to be studied. Furthermore, direct observations were made in the production section of PT. Solo Grafika Utama, Surakata. These observations are generally carried out in the entire factory area, and specifically in the area in the printing production section. This observation aims to determine the course of the newspaper printing process from raw materials into finished products. At the time of direct observation, interviews were conducted with manpower and foremen who were directly involved in the production process. The next step is to collect the data needed. Data obtained by making observations directly on the work system on the production floor and interviews with business owners, foremen, and operators, as well as by taking from company data sources.

Data that has been collected at the observation stage is then processed according to the data processing stage. Making the initial layout of the company as a reference in this study, then proceed with the calculation of space requirements, throughput, and distance traveled. The proposed improvement method used is the dedicated storage method. This method is done by sorting the product by comparison of the largest throughput and space requirements with the smallest distance.

The analysis step by comparing the initial layout with the proposed layout. The analysis also compares the distance of the old material handling with the proposal. The final stage of this study is drawing conclusions from the processing and analysis of data that has been done to answer the research objectives. Then there are suggestions given to improve the layout of the production floor at PT. Solo Grafika Utama and the next research.

3. Results and Discussion

Problems faced by the company are the irregularity in the layout of the warehousing system and the lack of effectiveness in handling material handling. The layout of the production floor can be a factor in the lack of material handling. Existing space also becomes a limitation in managing goods in the production floor. Space shortages occur when demand exceeds warehouse storage capacity. Additional space requirements are met by considering existing storage space. The design of repairs is done by analyzing the layout of the initial warehouse then designing improvements to the layout of the warehouse. Finally, a discussion of the results of the comparison of the material handling distance of the initial layout and proposals.

3.1 Initial Warehouse Layout

At present the problem faced by the company is that the layout of the product warehouse still does not have definite rules, meaning that the product will be placed in any location that is still empty or with the least number of stacks regardless of the dimensions of the product in one slot. Circumstances like this will allow the need for space for storage to be even greater. It will also lead to a longer search process, thereby inhibiting material handling activities and product delivery times to consumers.

The company's service system that serves purchases either from regular customers or customers who at any time make an order makes the frequency of ordering products difficult to predict precisely. So the company also does not have the right order lead time. This condition causes the lead time to fluctuate because in reality there are products with a storage period of more than 1 year. Meanwhile, there are also other products that process the entry and exit of it very quickly even less than 1 day. The initial layout of the production floor at PT. Solo Grafika Utama as follows.

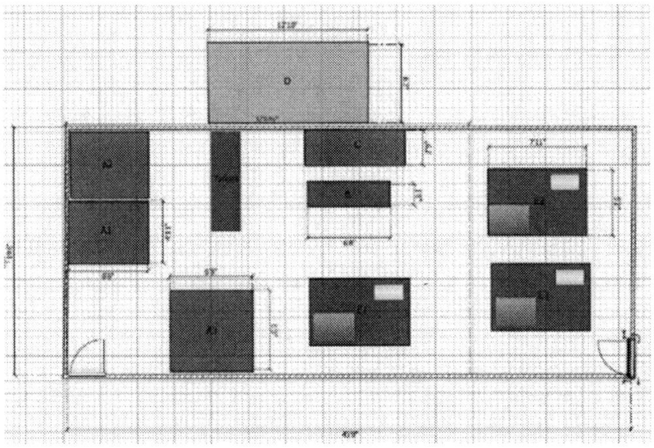

Figure 2. Initial Warehouse Layout

The calculation of the total distance of material handling the initial conditions are as in table 1.

Table 1. The Calculation of Distance Traveled

No	No. Blok	Distance	Product Name	T/S	Reach
1	A1	700			
	A2	1800	Tinta BW News Proc Yellow	43,71	15.500
	A3	600			
2	A1	700			
	A2	1800	CD 78,5 cm/45 gsm	21,01	15.500
	A3	600			
3	A1	700			
	A2	1800	Tinta Black SP-H	20,83	15.500
	A3	600			
4	B	3100	Plate Fuji CTP FDT 889 x 586 x 0,30	15,65	15.500
5		3100	Plate Fuji CTP FDT 889 x 608 x 0,30	15,21	15.500
			Sum of Reach		**77.500**

3.2 Proposed Warehouse Layout

The product arrangement in the proposed warehouse layout uses a dedicated storage method which is a product storage layout method based on the activities of entering and leaving the product in the warehouse with the shortest distance to the point I / O. The use of this method aims to facilitate the work of operators in the activities of receiving and shipping products because the goods or products already have a fixed slot (fixed location) in the warehouse. In addition, the distance to be traveled by each product can be ascertained and the time needed in the process of storing and taking the product will be smaller. So it will be known whether the existing warehouse area is sufficient, less, or exceeds the needs. Another reason for choosing this method is that there are many types of products with different product dimensions. So with the application of this method it is hoped that a better arrangement of products takes into account the path of movement of people and goods. This will have an impact on the loading and unloading process which becomes shorter and the process of finding the desired product will also be faster and easier, and will eventually be able to minimize the total distance of material handling. The steps for the dedicated storage method are:

3.3 Space Requirement (S)

Space requirement is the number of places or areas occupied by one type of product. Space requirment is needed to find out the storage capacity in the available slots or storage areas, so that it will be known how many slots are needed to store one type of product and then to calculate whether the number of slots available in the warehouse can be sufficient or not. The need for space on the production floor is one factor that is an obstacle to effective manual handling on a production floor. The condition of a company will change because of the influence of technology, changes in products, changes in the level of demand, and organizational structure in the future. These uncertain conditions cause a facility designer to face difficulties in determining true space requirements (Tompkins, 2003). Table 2 shows the space requirements for each type of product

Table 2. The Calculation of Space Requirement

No	Product Name	Amount per Bundle	Average Product Acceptance	Block Capacity	Space Requirement Teoritis	Space Requirement
1	Plate Fuji CTP FDT 889 x 586 x 0,30	200	4.938	39.500	1,65	2
2	Plate Fuji CTP FDT 889 x 608 x 0,30	200	4.813	38.500	1,60	2
3	Tinta Black SP-H	75	3.250	26.000	2,17	3
4	CD 78,5 cm/45 gsm	35	2.750	30.000	2,68	3
5	Tinta BW News Proc Yellow	150	6.850	54.800	2,28	3

3.4 Throughput (T)

Throughput is a dynamic activity measurement or storage, which shows the flow in storage. Measurement of product reception and delivery activities in the warehouse is used in the calculation of throughput which will later be used as a reference in the placement of products in the available slots in the warehouse. The priority of placing the product to the entrance and exit is based on the highest throughput value, followed by the second highest throughput value and so on. This is intended to facilitate the process of loading and unloading products by placing products whose activities are dense close to the point I / O so as to minimize the distance of material handling. Table 5 shows the throughput of each product.

Table 3. The Calculation of Throughput

No	Product Name	Average Product Acceptance	Average Product Delivery	Goods Transported	Throughput
1	Plate Fuji CTP FDT 889 x 586 x 0,30	4.938	4.450	300	31,29
2	Plate Fuji CTP FDT 889 x 608 x 0,30	4.813	4.312,5	300	30,42
3	Tinta Black SP-H	3.250	3.000	100	62,5
4	CD 78,5 cm/45 gsm	2.750	3.500	115	63,04
5	Tinta BW News Proc Yellow	6.850	6.262,5	100	131,13

3.5 Product Ranking Based on Comparison of Throughput and Storage

The process of product placement using the dedicated storage method is done by ranking the products based on a comparison of throughput to space requirment (storage). Ranking is done by sorting the largest T / S value up to the smallest value, and in the end it will be able to know the priority ordering of the products to be arranged. Product ranking based on comparison of throughput and storage is shown in table 4.

Table 4. The Calculation of T/S

No	Product Name	Space Requirement	Throughput	T/S
1	Plate Fuji CTP FDT 889 x 586 x 0,30	2	31,29	15,65
2	Plate Fuji CTP FDT 889 x 608 x 0,30	2	30,42	15,21
3	Tinta Black SP-H	3	62,5	20,83
4	CD 78,5 cm/45 gsm	3	63,04	21,01
5	Tinta BW News Proc Yellow	3	131,13	43,71

3.6 Calculation of Distance Traveled Between Each Storage Slot with I / O Point

Travel distance calculation is done starting from I / O point to each storage area (slot). Calculations are performed using the rectilinear distance method, where distances are measured along a path using a line perpendicular to each other towards the midpoint of each storage area. After obtaining the distance of each slot, the next step is to sort from the shortest distance to the farthest from the I / O point. Ordering aims so that the product with the largest T / S can be placed in the slot nearest the I / O point, then proceed with the product with the second largest T / S placed in the next slot and so on until all products are accommodated in the available slots.

3.7 Sorting T / S Values Descending

The proposed warehouse layout uses the dedicated storage method, where the laying process is done based on the ranking of products against T / S starting from the largest to the smallest placed at the shortest distance to the farthest from the I / O point. This rule is done to improve the warehousing system so that operators will find it easier to carry out the search process, because storage of products in slots is only for products in one type. The product placement with the highest T / S value in the slot with the smallest distance is shown in table 5 and the proposed warehouse layout is shown in Figure 3.2.

Table 5. Sorting T/S Values Descending

No	Product Name	Amount per Bundle	Average Product Acceptance	Block Capacity	Space Requirement Teoritis	Space Requirement
1	Plate Fuji CTP FDT 889 x 586 x 0,30	200	4.938	39.500	1,65	2
2	Plate Fuji CTP FDT 889 x 608 x 0,30	200	4.813	38.500	1,60	2
3	Tinta Black SP-H	75	3.250	26.000	2,17	3
4	CD 78,5 cm/45 gsm	35	2.750	30.000	2,68	3
5	Tinta BW News Proc Yellow	150	6.850	54.800	2,28	3

3.8 Calculation of Proposed Improvements using Dedicated Storage

In dedicated storage, a special storage location is assigned for each item to be stored. One of the advantages of dedicated storage is the efficiency of data handling because the data collection is fixed from the items that will go into storage. The proposed layout design based on the largest T / S value has the smallest mileage, so ordering the T / S for each product is sorted from largest to smallest and ranks the blocks based on the shortest mileage. The proposed layout based on dedicated storage is as follows:

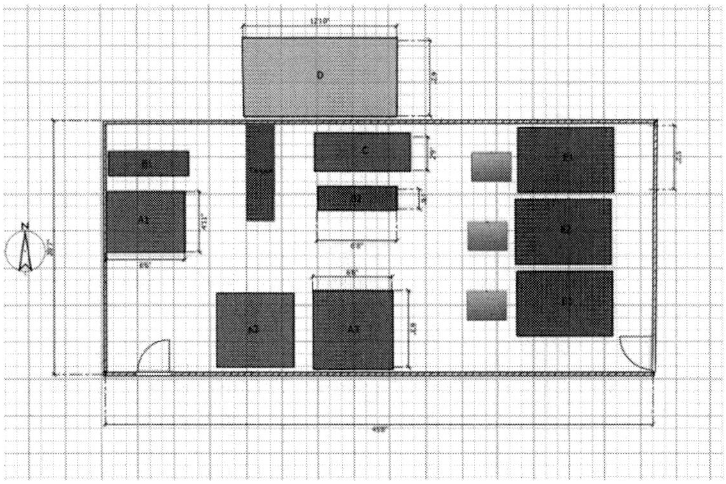

Figure 3S. Proposal Warehouse Layout

There was a change in the inventory storage area (A) because in the comparison of T / S values, the inventory area has the largest value so that it is near the I / O point. Space requirements in the raw material warehouse (B) also increase based on the calculation of space requirements and are placed not far from the inventory area. The distance traveled calculation is based on the proposed production floor layout as follows.

Table 6. The Calculation of Distance Traveled Based on the Proposed Layout

No	No. Blok	Distance	Product Name	T/S	Reach
1	A1	700	Tinta BW News Proc Yellow		
	A2	600		43,71	12.500
	A3	1200			
2	A1	700	CD 78,5 cm/45 gsm		
	A2	1200		21,01	12.500
	A3	600			
3	A1	700	Tinta Black SP-H		
	A2	1200		20,83	12.500
	A3	600			
4	B1	800	Plate Fuji CTP FDT 889 x 586 x 0,30	15,65	11.000
	B2	1400			

5	B1	800	Plate Fuji CTP FDT 889 x 608 x 0,30	15,21	11.000
	B2	1400			
			Sum of Reach		**59.500**

Comparison of Material Handling Distance

After obtaining the results of the material handling distance of the initial layout and layout of the proposed improvements, then a comparison is done to find out the difference between the initial layout and layout of the proposed improvements

Table 7. Comparison of Initial and Proposed Layout Distances

Layout	Total Distance	Difference to Existing	Percentage Decrease in Distance
Initial Layout	77.500		
Proposed Layout	59.500	18.000	23%

Changes in the distance value on the new layout will affect the cost factors associated with material handling, such as the use of material handling tools and material handling life. A good placement will affect material handling activities, meaning that the reduction in distance will be proportional to the decrease in material handling activities and will be followed by a reduction in material handling operating costs. With the minimization of distance means companies can reduce material handling operating costs decreased. Another thing that is affected by this change is the movement of operators and the operation of the warehousing system. With the existence of a good planner, the operator will be easier to check and search for the desired product, so that the warehousing system is increasingly well controlled. These factors will affect the handling process, where later the handling process will be faster and more precise so that service to consumers will also increase.

4. Conclusion

Improvement of factory layout with the dedicated storage method was chosen to improve the arrangement of the warehouse and minimize the distance of product movement so that efficiency and effectiveness occur in the production process. From the results of the discussion it was found that there was a decrease in the total distance of material handling with the application of the dedicated storage method, where in the initial conditions the warehouse total material handling distance was 77,500 m, whereas in the proposed layout it was 59,500 m. This means that there is a decrease in the total material handling distance of 18,000 m from the initial conditions. From the comparative data analysis between the initial layout and the proposed layout it can be seen that there is a decrease in distance of 23%, this shows that the proposed layout has a smaller material handling cost. Thus it can be said that the layout of the proposal is better than the initial layout and can be proven able to facilitate the handling of materials in the warehouse so that it can make newspaper production more effective and efficient.

IOP Conf. Series: Materials Science and Engineering **943** (2020) 012042 doi:10.1088/1757-899X/943/1/012042

References

[1] Abdullah, Fikrie. 2009. Usulan Perbaikan Tata Letak Gudang Produk Jadi Dengan Menggunakan Metode Dedicated Storage di PT. Cahaya Kawi Ultra Polyintraco. Universitas Sumatera Utara: Medan.
[2] Apple, J. M. 1990. *Tata Letak dan Pemindahan Bahan*. Edisi Ketiga. Penerjemah : Nurhayati Mardiono. Bandung : ITB.
[3] Armin Klausnitzer, R. L. (2018). Optimal facility layout and material handling network design. *Computers and Operations Research*.
[4] Frazelle, E.H. 2002. Word Class Warehousing and Material Handling. McGraw Hil New York
[5] Gutta, P. R., Chinthala, V. S., Manchoju, R. V., MVN, V. C., & Purohit, R. (2018). A Review On Facility Layout Design Of An Automated Guided Vehicle In Flexible Manufacturing System. *Materials Today: Proceedings 5*, 3981-3986.
[6] Hadiguna, R. A. dan Setiawan, H. 2008. *Tata Letak Pabrik*. Yogyakarta : CV. Andi Offset.
[7] Hadiguna, R. A. dan Setiawan, H. 2008. *Tata Letak Pabrik*. Yogyakarta : CV. Andi Offset.
[8] Harmon, L. Roy. 1993. Warehouse Management System.3th Edition. Prentice Hall,New Jersey
[9] Horta, M., Coelho, F., & Relvas, S. (2016). Layout design modelling for a real world just-in-time warehouse. *Computers & Industrial Engineering 101*, 1-9.
[10] Maniya, K., & Bhatt, M. (2011). An alternative multiple attribute decision making methodology for solving optimal facility layout design selection problems. *Computers & Industrial Engineering 61*, 542-549.
[11] Hadiguna, R. A. dan Setiawan, H. 2008. *Tata Letak Pabrik*. Yogyakarta : CV. Andi Offset.
[12] Nurul Hayati Abdul Halim, N. Y. (2015). Effective Material Handling System for JIT Automotive Production Line. *Procedia Manufacturing 2*, 251-257.
[13] Permana, Hadi. 2014. Relayout Tata Letak Gudang Produk Jadi Baja Tulangan dengan Metode Dedicated Storage. Unversitas Sultan Ageng Tirtayasa.
[14] Prasetyo, Anindito Adi. 2011. Perancangan Tata Letak dan Sistem Informasi Gudang Obat PT. XY. Tugas Ahir Sarjana Teknik Industri Universitas Diponegoro. Semarang.
[15] Wingnjosoebroto, Sritomo. 2009. Tata Letak Pabrik dan Pemindahan Bahan. Penerbit Guna Widya:Surabaya.

IOP Conf. Series: Materials Science and Engineering **943** (2020) 012043 doi:10.1088/1757-899X/943/1/012043

Six Sigma Application to Minimize Castor 5 Inch Scrap Material in EOP Warehouse: A Case Study

A Fajrianto[1], M H I N Asnan[1], N A Khofiyah[2], W Sutopo[3], and Yuniaristanto[3]
[1]Department of Industrial Engineering, Faculty of Engineering, Universitas Sebelas Maret, Surakarta
[2]Master Program of Industrial Engineering, Faculty of Engineering, Universitas Sebelas Maret, Surakarta
[3]Research Group Industrial Engineering and Techno-Economic, Department of Industrial Engineering, Universitas Sebelas Maret, Surakarta
Jl. Ir. Sutami, 36 A, Surakarta, Indonesia

Email: fajriantoaldy@gmail.com

Abstract. Case study in this research at PT. Mega Andalan Kalasan, which one of the manufacturing companies engaged in manufacturing hospital equipment that often has problems regarding the quality of materials stored in warehouses. The problem is often found in the 5 Inch castor material that becomes scrap and incurs costs so that waste occurs. Based on data from April 2018 to September 2018, the percentage of scrap material is increasing every month from 1.44% to 1.77%. Due to these problems, the purpose of this research is to minimize the 5 Inch castor material which is a scrap to avoid waste costs in the Export Oriented Product Warehouse Unit of the company. The methodology in this study uses the Six Sigma method, namely by implementing the Define, Measure, Analyze, Improve, and Control (DMAIC) stages. The results obtained are the most dominant cause of scrap castor 5 Inch material is storage of material on a rack that is not organized with an RPN value of 240. Thus, it is necessary to make improvements to minimize the scrap castor 5 Inch material by conducting a routine briefing by the PIC Area Warehouse Unit Export Oriented Product before operational activities begin.

1. Introduction

Case study in this research at PT. Mega Andalan Kalasan (PT. MAK) Yogyakarta, where is a hospital manufacturing factory. One problem that is often encountered is the problem of material stored in the Warehouse Unit Export Oriented Product, which is often found in materials that are cast 5 Inch which experience scrap. Scrap material is material that is damaged in the production process [1]. Based on data from April 2018 to September 2018, the percentage of scrap material is increasing every month from 1.44% to 1.77%. The scrap material can come from returning the remaining material from the production department to the Export Oriented Product Warehouse Unit (remaining production), material handling errors, and discontinue products, and old designs.

Based on the explanation of the above problems, a study was conducted to minimize scrap 5-inch castor material in the Export Oriented Product Warehouse Unit of the company uses the Six Sigma method. Six Sigma is a structured methodology to improve processes that are focused on efforts to reduce process variances while reducing defects (products/services that are out of specification).

This research was only conducted in the Export Oriented Product Warehouse Unit section and the data used are scrap data of Castor 5 Inch material in April 2018 to September 2018 as a barrier and assimilation used in this study is a blocking factor on grading on Severity (S), Occurrence (O) , and Detection (D) does not exist. This study aims to minimize scrap 5 Inch castor material in the Export Oriented Product Warehouse Unit uses the Six Sigma method.

Six Sigma is also based on real data and facts which are used to perform a detailed analysis. It is based on continuous improvement of all aspects of functioning development in the organization as well as proactive management and cooperation without boundaries at every level in the enterprise. It should be underlined that it is not only an approach for solving the problems with manufacturing but also business processes [2].

Six Sigma is a well-structured, data-driven methodology for eliminating defects, waste or quality control problems of all kinds in manufacturing, service delivery, management, and other business activities [3]. According to [4], Six Sigma uses two major structured methods for improvement known as DMAIC (define-measure-analyse improve-control) and DMADV (define-measure-analyse design-verify). The highlight of the DMAIC method is the five-phased methodological framework that guides in conducting the improvement project [5].

The DMAIC methodology is used for improving production processes, successfully contributing to the reduction of the number of non-compliant products and reducing production costs [6]. Process improvements include reducing the occurrence of defects so that defects can be minimized as much as possible [7].

2. Methodology

This research method uses the stages in Six Sigma, namely DMAIC. According to [8], the DMAIC method is used to identify and eliminate waste that exists. Before identifying risk, researchers first recapitulate scrap data in the Warehouse section based on the SAP system then identified risks using the SIPOC diagram, the process of receiving residual production material. These data are then measured by risk using the Pareto diagram. According to [9], the Pareto diagram is a frequency distribution from the largest to the smallest data. So that the Pareto diagram is used to find out the most dominant causes of scrap to be resolved.

After the most dominant cause of risk, then a risk analysis was carried out by calculating the value of the Risk Priority Number (RPN) for the scrap material using the FMEA method based on the results of the brainstorming with the PIC Area Warehouse and PPIC. Through the FMEA method, the dominant causes of product defects will be known following the value of the RPN [10]. According to [11] the formula used to calculate RPN is:

$$RPN = Severity\ x\ Occurance\ x\ Detection \quad (1)$$

Then the risk evaluation of the highest RPN value is carried out by determining the proposed improvement using the Five Why Analysis method so that scrap material can be minimized. Each source of the problem will be determined what causes up to five stages of the question why so that the root causes can be found and solutions are sought [12]. The last step is to control the proposed improvements that have been given to minimize scrap material. Following is the flow of the research methodology used.

1. Determine (D)
   It is the first stage to discuss the problem, the process of determining internal and external objectives and requirements. Determining customer needs, determining goals and problems, forming teams, and determining sources are parts of the define phase. The tools used in define are SIPOC diagrams (Suppliers, Inputs, Processes, Outputs, and Customers), affinity diagrams, and relationship diagrams.
2. Measuring (M)

The purpose of improving the basics of repairs is basic. Measure is a step in collecting data, what is agreed to is to set performance standards. Important tools in this phase usually include trend charts, Pareto charts, process flow diagrams, and measurement of process capability (sigma level, or it can also be called a sigma process).

3. Analysis (A)
   It is a device that is carried out when doing root problems with data in the analysis. The results of the analysis can be used to create solutions in developing and improving the supporting processes. The tool used is FMEA (Failure Mode Effect and Analyze) causal diagram, Pareto diagram, flow chart, and control diagram. According to [13] Priority number RPN is the product rating severity, occurrence, and detection that shows the ranking of corrective action needs.
4. Improve (I)
   The step enhances the full understanding of the main causes that analyze in phase, with the intention of either controlling or overcoming these problems to achieve maximum results. Six Sigma tools commonly used in this phase include regression analysis, hypothesis testing, experimental design (DOE), and variance analysis (ANOVA).
5. Control (C)
   The stage of control on adopting DMAIC is about maintaining changes made in the improvement phase. The aim is to maintain acquisition, restore repairs to ensure success, plan repairs, and win approval documents, business processes, and required training records.

3. Result and Discussion

3.1 Define

Based on the results of the scrap data recapitulation in the Warehouse section of the company, it can be seen that the amount of scrap material in the section has increased from 1.44% to 1.77%. More details are explained in table 1.

Table 1. Scrap Recapitulation of 5 Inch Castor Material

Month	Warehouse Stock (Unit)	Total Scrap (Unit)	Presentation
April	7423	107	1.44%
May	7246	108	1.49%
June	7321	105	1.43%
July	7159	112	1.56%
August	7367	117	1.59%
September	7228	128	1.77%
Total	**43744**	**677**	
Average	**7291**	**113**	**1.55%**

In facilitating the depiction of the stages of the material acceptance process in the Warehouse Unit Export Oriented Product section until the material is stored in a Warehouse that can lead to the occurrence of scrap material, a SIPOC (Supplier-Input-Process-Output-Customer) diagram is used. Parts that can cause scrap material are found in the Warehouse operator section. More details are explained in figure 2.

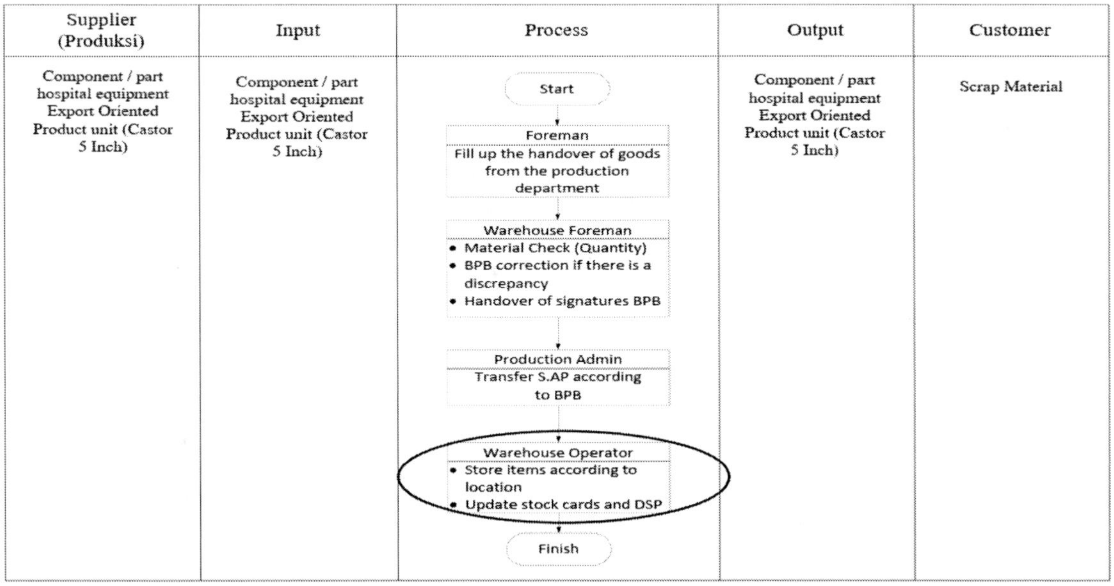

Figure 1. SIPOC Diagram of Acceptance of Production Remaining Materials in the Warehouse Section

3.2 Measure

Based on the recapitulation of data on the causes of the scrap material, a Pareto Diagram was made to determine the causes of scrap material that must be prioritized to be completed. Pareto diagram is a graph to map problems in decreasing frequency sequence [14]. Events are arranged according to their size, from the largest on the left to the smallest on the [15]. Figure 2 is a Pareto diagram of the data on the causes of the scrap material.

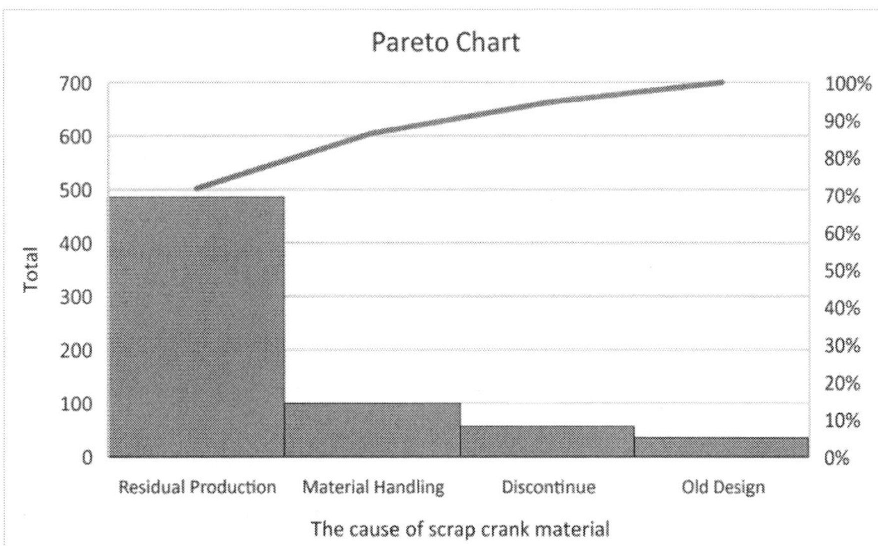

Figure 2. Pareto Diagram Causes of 5 Inch Castor Material in Warehouse Unit Export Oriented Product

Based on these results, the cause of the 5 inch castor material that becomes scrap which needs to be prioritized to be resolved in an effort to minimize the 5 inch castor material that becomes scrap in the Export Oriented Product Warehouse Unit section is the remaining production. This is because the amount of castor 5 inch material which becomes scrap in section Export Oriented Unit Product Warehouse caused by residual production of 485 units (71.6%).

3.3 Analyze

Based on these results, the cause of the 5-inch castor material that became a priority to be completed to minimize scrap material in the Warehouse Unit Export Oriented Product section is the remaining production. Then a risk analysis is carried out through 2 stages, namely risk assessment using the Failure Mode and Effect Analysis (FMEA) method and the probability determination of the causes of the 5-inch castor material that becomes scrap.

The following is a risk analysis table and calculation of the Risk Priority Number (RPN) that causes scrap 5-inch castor material for the period April 2018 to September 2018.

Table 2. Risk Analysis and Calculation of Risk Priority Number (RPN) Causes of 5 Inch Castor Material Scrap in Warehouse Unit Export Oriented Product Section

Process Description	Potential Failure Mode	Potential consequences of failure	S	Causes of potential failure	O	Detection Method	D	RPN
Storage of materials in the Warehouse Export Oriented Product	Production Waste	FIFO doesn't work well	6	Material storage on a shelf disorganized	8	Improve material arrangement	5	240
				Lazy operators place material following the provisions (oldest batch)	7	Conducting routine briefings and checkers must ensure the placement of returned items	5	210
				The operator is lazy to update the stock card	5	Doing routine briefings and checkers always checks every item that enters and exits from the shelf	4	120
				Conditions in the field do not match the Product Status Data (DSP)	4	Foreman is more regularly doing DSP updates	4	96
		Material is broken on the shelf	3	Material falls from the shelf	2	Routinely do pallet replacements that are not suitable for use	4	24
				The material was wet because the gutters leaked	2	Move material to dry place	2	12
				The material on the inside was hit by the material to be inserted into the rack	4	Operators are more careful and in no hurry to put material into a rack	4	48

According to [16], severity is the quantification of how serious the conditions are caused if a failure occurs which is consequently mentioned in the failure effect, this severity is made on 10 scales (1-10) which show the consequences that are not serious (1) to very serious (10). Occurrence is the level of probability of failure, the occurrence is made in 10 scales (1-10) which shows that very rare consequences (1) until it is very possible (10). While detection is the level of possibility of the failure of the control that has been given, this detection is made in 10 scales (1-10) which indicates the possibility of escaping from a small control (1) to the possibility of escaping large control (10). Based on the results of the risk analysis, the value of the largest RPN is the potential failure of the lazy operator to place the material following the provisions (the oldest batch). According to [17], RPN confirms the priority level of a failure. The RPN value is used to rank potential process failures [18].

3.4 Improve

After the root cause of the problem causing the scrap material, the next step is to find a solution to the problem. The steps in making improvements for minimizing scrap material use the Five Why Analysis method. The following is the Five Why Analysis of the 5-inch castor material that becomes scrap.

Table 3. Five Why's Analysis of Scrap of 5 Inch Castor Material in Warehouse Unit Export Oriented Product Section

Question	Information
Why is the storage of material on the shelf unorganized?	Because the operator is lazy to place the material following the provisions (the oldest batch) and the checker does not routinely check the suitability of the stock card with Product Status Data (DSP) and the placement of returns on the remaining production material.
Why do lazy operators place materials following the provisions (oldest batch) and checkers do not routinely check the suitability of stock cards with Product Status Data (DSP) and the placement of returns on the remaining production material?	Because this has become a habit for a long time, both from the operator and the checker in the Warehouse Unit Export Oriented Product section.
Why is the placement of material not following the provisions (oldest batch) and checking the suitability of the stock card with non-routine Product Status Data (DSP) has become a habit long ago?	Because of the lack of briefings every day before operational activities began long ago.
Why are there minimal briefings every day before operational activities begin?	Because the PIC Area Warehouse has not routinely conducted briefings before operational activities have begun and so far the briefing has only been carried out only if the audit will be held in the Warehouse section.
Why hasn't the PIC Area Warehouse routinely conducted a briefing before the operational activities began?	Because the PIC Area feels reluctant to direct workers to the Warehouse section because most workers in the Warehouse section are older / senior to themselves and there is a lack of supervision from the section head Warehouse regarding the habituation of the briefing.

Based on the description of the Five Why's analysis, the solution that can be recommended to minimize the cause of the 5-inch castor material that becomes scrap in the Warehouse Unit Export Oriented Product section is as follows:

a. The operator began to get used to placing the material following the provisions (the oldest batch).
b. The checker is more routine in checking the suitability of the stock card with Product Status Data (DSP) as well as the placement of returns for the remaining production material every time they make a shift.
c. Briefings are not only limited to the holding of an audit in the Warehouse section.
d. The PIC area Warehouse Unit Export Oriented Product is more routine in conducting briefings to workers in the Warehouse Unit Export Oriented Product section before operational activities begin.

e. The Head Warehouse Section is more routine in supervising the briefing by the PIC Area Warehouse Unit Export Oriented Product before operational activities begin.

3.5 Control

After making a corrective action in the form of a proposed improvement to the results of the Five Why 'analysis, the cause of scrap 5-inch castor material in the Warehouse Unit Export Oriented Product section, the next step is to control the results of improvements using the Six Sigma method. Without control of the results of these improvements, the repair process will not achieve the expected results. In addition, the control starts from the PIC Area Warehouse Unit Export Oriented Product and the Section Head Warehouse is very much needed in the effort to repair it.

4. Conclusion

Based on the results of each stage of Six Sigma, at the stage of defining the parts that can lead to the occurrence of scrap material are found in the Warehouse Unit Export Oriented Product operator. At the measuring stage, the cause of scrap material that must be prioritized for completion is the remaining production. In the analyse stage, the largest RPN value is the potential cause of material storage failure on an unorganized shelf. At the improve stage, the proposal given is addressed to the operator, checker, PIC Area Warehouse Unit Export Oriented Product, and Section Head Warehouse to make the proposal that has been given. While at the control stage, control efforts must be made to the results of improvements using the Six Sigma method that has been given.

The limitation of this study is that the researcher assumes that the blocking factor for giving values to Severity (S), Occurrence (O), and Detection (D) does not exist. So that the next study is expected to consider the disrupting factors (block) on the value of Severity (S), Occurrence (O), and Detection (D) so that the results of the assessment of the three factors obtained are more valid.

References

[1] Tebae, I., A., Karamoy, H., dan Runtu, T. (2016). Perlakuan Akuntansi Terhadap Sisa Bahan Pada PT. Multi Nabati Sulawesi. Jurnal EMBA, Vol. 4, hal. 1431-1441.

[2] Smętkowska. M and Mrugalska. B. 2018. Using Six Sigma DMAIC to improve the quality of the production process: a case study. SIM 2017 / 14th International Symposium in Management. Social and Behavioral Sciences 238 (2018) 590 – 596.

[3] Al-Qatawneh, L., Abdallah, A., and Zalloum, S. Z. (2019). Six Sigma Application in Healthcare Logistics: A Framework and A Case Study. Hindawi Journal of Healthcare Engineering Volume 2019, Article ID 9691568, 12 pages https://doi.org/10.1155/2019/9691568.

[4] Prasad. K. G. D, Subbaiah. K. V, and Padmavathi. G. 2012. Application of Six Sigma Methodology in an Engineering Educational Institution. Int. J. Emerg. Sci., 2(2), 222-237, June 2012 ISSN: 2222-4254

[5] Pyzdek. T and Keller. P. 2014. The Six Sigma Handbook, McGrawHill Education, New York, NY, USA, 4th edition, 2014.

[6] Zhang, M., Wang, W., Goh, T. N., and He, Z. (2015). Comprehensive Six Sigma application: a case study. Production Planning & Control The Management of Operations 26:3, 219-234, DOI: 10.1080/09537287.2014.891058.

[7] Heryadi, A. R. And Sutopo, W. (2018). Review of the Utilization of the DMAIC Analysis Methodology in Industry Garment. Seminar dan Konferensi Nasional IDEC ISSN: 2579-6429 2018 Surakarta, 7-8 Mei 2018.

[8] Gultom, S., Sinaga, T., S., dan Sinulingga, S. (2013). Studi Pengendalian Mutu Dengan Menggunakan Metode Lean Six Sigma Pada PT. XYZ. E-Jurnal Teknik Industri FT USU, Vol. 3, hal. 23-30.

[9] Utami, I., Solikhah, R., dan Istadi, I. (2012). Sintesa Katalis Super Asam SO42-/ ZnO Untuk Produksi Biodiesel Dari Minyak Kelapa Sawit. Jurnal Teknologi Kimia dan Industri, Vol. 1, hal. 69-78.

[10] Hariastuti, N., L., P. (2015). Analisis Pengendalian Mutu Produk Guna Meminimalisasi Produk Cacat. Seminar Nasional IENACO, hal. 268-275.

[11] Adrianto, W., dan Kholil, M. (2015). Analisis Penerapan Lean Production Process Untuk Mengurangi Lead Time Process Perawatan Engine (Studi Kasus PT. GMF Aeroasia). Jurnal Optimasi Sistem Industri, Vol. 4, hal. 299-309.

[12] Basuki, M., Susanto, R., B., dan Herianto, H., P. (2015). Analisis Risiko Kegiatan Bongkar Muat Sebagai Komponen Dwelling Time Di Pelabuhan. Seminar Nasional Sains dan Teknologi Terapan III, hal. 511-517.

[13] Indrawati, S., and Ridwansyah, M. (2015). Manufacturing Continuous Improvement Using Lean Six Sigma: An Iron Ores Industry Case Application. Industrial Engineering and Service Science 2015, IESS 2015.

[14] Soetjitro, P. (2010). Instrumen Total Quality Management (TQM) Sebagai Pilihan Alat Pengendalian. VALUE ADDED, Vol. 6, hal. 49-60.

[15] Rusdianto, A., S., Novijanto, N., dan Alihsany, R. (2011). Penerapan Statistical Quality Control (SQC) Pada Pengolahan Kopi Robusta Cara Semi Basah. J Agrotek, Vol. 5, hal. 1-10

[16] Putra, E., S. (2012). Sistem Pendukung Keputusan Dalam Mengurangi Stok Poy Downgrade Menggunakan Metode Six Sigma Dengan Tools FMEA pada PT. MGT. STIKOM

[17] Hanif, R., Y., Rukmi, H., S., dan Susanty, S. (2015). Perbaikan Kualitas Produk Keraton Luxury Di PT. X Dengan Menggunakan Metode Failure Mode and Effect Analysis (FMEA) dan Fault Tree Analysis (FTA). Jurnal Teknik Industri Itenas, Vol. 3, hal. 137-147.

[18] Puspitasari, N., B., dan Martanto, A. (2014). Penggunaan FMEA Dalam Mengidentifikasi Resiko Kegagalan Proses Produksi Sarung ATM (Alat Tenun Mesin) (Studi Kasus PT. Asaputex Jaya Tegal). Jurnal TI Undip, Vol. 9, hal. 93-98.

Redesign Determination of Plat Raw Material Supplier with Analytic Hierarchy Process (AHP) Approach: A Case Study

K A Hidayati[1*], A A Rosanti[1], N A Khofiyah[2], and W Sutopo[3]
[1]Department of Industrial Engineering, Faculty of Engineering, Universitas Sebelas Maret, Surakarta
[2]Master Program of Industrial Engineering, Faculty of Engineering, Universitas Sebelas Maret, Surakarta
[3]Research Group Industrial Engineering and Techno-Economic, Department of Industrial Engineering, Universitas Sebelas Maret, Surakarta
Jl. Ir. Sutami, 36 A, Surakarta, Indonesia

Email: hidayatikartika2@gmail.com

Abstract. PT. XYZ is a company in the newspaper industry. The selection of suppliers so far is only based on the price offered and the promised delivery speed. The price offered by the supplier is also influenced by the exchange rate of the rupiah against the dollar resulting in price fluctuations in the purchase price of raw materials. Even though the price of raw materials has increased, newspaper prices have remained the same, and to avoid this problem, the company has supplied raw materials from many suppliers. The company has made good cooperation with several suppliers, but does not yet have the right supplier selection method and can be applied with a good planning process. Therefore, companies need to select suppliers with the right methods so that raw material orders are optimally fulfilled and get the best suppliers who can work together in the long run. In this study, the determination of the best suppliers is carried out on the raw material of the plate as well as important criteria in the selection of these suppliers. Data collection is done using interviews, questionnaires, and literature studies. Based on the results of data collection, among others, the purchase price, delivery time, maximum number of orders, the level of disability, and the distance of the supplier with the company. Data processing uses one of the MCDM (Multi Decision Making Criteria) methods, namely AHP (Analytic Hierarchy Process). With the results of supplier one being designated as the best supplier. This can be known with the highest final value in the final AHP calculation, with a value of 3.95643, and the criteria most considered in determining suppliers is the bid price.

1. Introduction

Logistics system is part of the process of the supply chain (supply chain) that serves to plan, execute, control effectively, efficient procurement, management, storage of goods, services and information ranging from the starting point (point of origin) to the point of consumption (point of consumption) with the aim of meeting the needs of the consumer. One of the things that are important to any company that is doing inventory management. One of the activities within the inventory management is conducting the procurement of raw materials that are included in the activity of logistics systems. Raw materials imported from some suppliers so that the supplier be one external party that is important to the existence and sustainability of a manufacturing company. Inventory management has an impact on all fronts, primarily financial, operational, and marketing [1]. Concerning inventory, there is a different goal in a

IOP Conf. Series: Materials Science and Engineering **943** (2020) 012044 doi:10.1088/1757-899X/943/1/012044

variety of fields. On the part of Finance wants the level of inventories are low, while in operational marketing and want a high level of supplies so that consumers ' needs can be met. When the supplies could not be sufficient demand exists, the company will suffer losses. To keep the supply of raw material remains optimal, schedule the purchase of raw materials and a good relationship with the supplier to note. Valid supplier selection can help companies achieve the desired production outcomes. The company generally have different requirements in the selection of suppliers. So, companies need to do a periodic evaluation to ensure that raw materials and procurement by the standards set by the company.

PT. XYZ is a company engaged in the printing industry that produces newspapers, magazines and is CATEGORIZED in Surakarta. The complete results of the company's products are distributed to Central Java so that product quality and timeliness are becoming essential criteria that should always be maintained. This company has multiple suppliers as significant suppliers of raw materials for its production of the product to be. Supplier, which is owned by the company among other suppliers, suppliers of paper, and ink suppliers plat.

The selection of suppliers of raw materials is one of the activities in the supply chain management in pt. XYZ. The event is a strategic activity, due to the role of the supplier in determining the success of the company. If the raw materials provided by the supplier of poor quality, then this will affect the quality of the resulting product. When the supplier could not provide raw materials according to the needs of the company, then the production schedule will also be disrupted. So the selection of the supplier is essential to note. The selection of suppliers for this only based on the price offered and the speed of delivery is promised. The price offered by the suppliers and also influenced by the exchange rate of the rupiah against the dollar, which price fluctuations on the purchase price of raw materials. Although the price of raw materials prices increased, the newspaper remains the same, and to avoid the issue, PT. XYZ was supplying raw material plate from many suppliers. If companies still use many suppliers at once could cause problems among other things too hard in the setting of a schedule entered the warehouse, control of the supplier, as well as setting the payment orders of raw materials [1]. When the supplier is chosen, any common issues such as quality, quantity, and delivery time is not as promised when booking, so that interferes with the operations of the production process. Although PT. XYZ has conducted good cooperation with some suppliers, but does not yet have a proper supplier selection method and can be applied with a good planning process. Therefore, companies need to make a selection of suppliers with the right method for raw materials orders fulfilled optimally and get the best supplier who can work together in the long term. This research was conducted on the determination of the best supplier is done on a plate of raw materials as well as the essential criteria in the selection of the supplier.

2. Literature Review

AHP is a decision-making method that is developed to prioritize several alternatives when several criteria must be considered and allow decision-makers to construct complex problems into a hierarchy. However, the AHP method has weaknesses in the inability to accommodate the disguise or uncertainty (vagueness), and the inaccuracy of respondents in making choices [2]. In this study, the AHP method is used for decision making in developing a redesign for determining suppliers of plate raw materials.

AHP is one of the most popular MCDM tools for formulating and analyzing decisions, especially in operation management [3]. The AHP method is applied in a great variety of areas, including planning, selection of the best alternative, resource allocation, and conflict resolution, thus being the most applied method. This method is aimed at establishing priorities or weights to be attributed to different criteria and alternatives, which features a decision and, as a consequence, enables the choice of the most suitable alternative [4]. This model should be an instrument to easily and quickly state the purchasing situation, which is often hidden by ordinary production management and the customers' delivery time [5].

According to [6], to compare the Business process management and the analytic hierarchy process as the tools of process performance assessment. Conclusions are Business process management is helpful in process performance assessment because it offers a conceptual foundation. The analytic hierarchy process is a complement that makes the intuitions based on business process management rigorous.

IOP Conf. Series: Materials Science and Engineering **943** (2020) 012044 doi:10.1088/1757-899X/943/1/012044

According to [7], the study aims to identify and evaluate the barriers faced by Electrical and electronics (E&E) manufacturing companies in Malaysia in implementing material efficiency strategies. Conclusions are the identified barriers that are categorized into internal and external barriers. From the empirical analysis, a total of nine barriers that influence the implementation of material efficiency were shortlisted.

Therefore, from the two case studies in this paper, we found confidence in using the AHP method in answering goals and finding solutions in this study.

3. Methodology

The methods used in this research include the study of literature, data collection, and then analyze the data regarding the determination of the supplier.

a) Literature Study
   Literature studies were done by a literature review regarding the problems which occurred in the procurement cycle with a method of Analytical Hierarchy Process (AHP).
b) Data collection
   Data in the study obtained from interviews, questionnaires, and observation. Interview conducted by Assistant General Manager. Then a detailed questionnaire distributed to respondents during the three parts of the production, logistics, and procurement.
c) Data analysis the data analysis used IE using the method of the Analytical Hierarchy Process (AHP).

In the method of AHP, three principles must be observed, i.e., [8]:

1. Principles of drafting hierarchy
2. The principle of determining priority
3. The principle of logical consistency

The steps in the performance analysis with the AHP method: first do a recap of the calculation of the questionnaire from the three respondents. Then make a multiple comparison chart of each criterion and the supplier. After that, the table using aggregating WGCM (Weighted Geometric Mean Complex). Next, do the normalization, eigenvector performs calculations, test the consistency of the data by searching for the value of the CR, and the selection of suppliers and criteria based on the eigenvalue.

4. Result and Discussion

The criteria used to determine the best supplier is obtained based on the study of literature and previous research. The criteria consist of the purchase price, delivery time, the addition error or reduction of the amount of the booking, the level of disability, and the distance with supplier companies. A detailed questionnaire distributed to respondents during the three parts of the production, logistics, and procurement.

Table 1. The Data Information Of The Respondents

No	Name	Parts	Weight
1	Respondents 1	Production	33.33%
2	Respondent 2	Procurement	33.33%
3	Respondent 3	Logistic	33.33%

4.1 Pairwise Comparison Between Supplier and Delivery Criteria

For each of the criteria and alternatives, the pairwise comparison that compares each element with other elements. At each level of the hierarchy in pairs so that the obtained values the importance of the element.

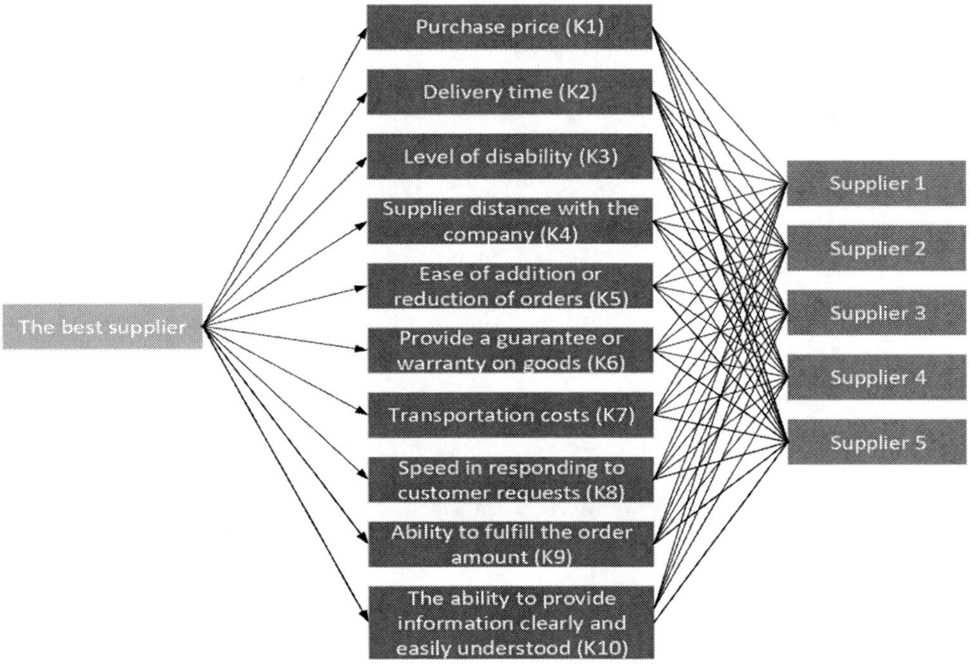

Figure 1. The Structure of The Hierarchy Of The Selection Criteria For Supplier

The criteria used to determine the best supplier are obtain based on literature studies and previous research. Criteria consist of price purchase, delivery time, defect rate, supplier distance from the company, and the ease of adding or reducing the number of orders, giving a guarantee or warranty on goods, transportation costs, speed in terms of responding to customer requests, the ability to fulfill the number of orders, the ability to provide information clearly and easily understood.

From the results of a questionnaire and interviews conducted on the part of the production, logistics, and procurement is done recapitulation as in table 3 for the criteria and table 5 for the supplier.

Table 2. Supplier Selection Criteria

Criteria	Information
K1	Price of bid
K2	Delivery time
K3	Rate of Disability
K4	Supplier distance with companies
K5	Ease of adding or decrease in the number of orders
K6	Guarantee or warranty against goods
K7	Transportation cost
K8	Speed in response to customer requests
K9	Fulfillment abilities to the number of orders
K10	The ability to provide information clearly and easily understood

Table 3. WGMC (*Weighted Geometric Mean Complex*) Criteria

Criteria	K1	K2	K3	K4	K5	K6	K7	K8	K9	K10
K1	1,000	2,333	1,667	1,667	1,667	1,000	2,333	1,000	1,667	3,000
K2	0,429	1,000	1,000	1,000	1,667	1,667	1,000	1,667	1,667	1,000
K3	0,004	1,000	1,000	3,000	1,667	2,333	1,667	2,333	1,667	1,667
K4	0,004	1,000	0,333	1,000	1,667	1,000	2,333	2,333	1,667	1,000
K5	0,004	0,004	0,004	0,004	1,000	2,333	1,000	1,000	1,000	1,667

K6	1,000	0,004	0,429	1,000	0,429	1,000	1,667	3,000	1,000	1,000
K7	0,429	0,007	0,004	0,429	1,000	0,004	1,000	1,000	1,667	2,333
K8	1,000	0,004	0,429	0,429	1,000	0,333	1,000	1,000	1,667	1,000
K9	0,004	0,004	0,004	0,004	1,000	1,000	0,004	0,004	1,000	2,333
K10	0,333	1,000	0,004	1,000	0,004	1,000	0,429	1,000	0,429	1,000
Total	**6,590**	**8,740**	**7,257**	**10,724**	**11,695**	**12,267**	**13,029**	**14,933**	**13,429**	**16,000**

Table 4. Supplier Code

Code	Supplier
S01	Supplier Plat 1
S02	Supplier Plat 2
S03	Supplier Plat 3
S04	Supplier Plat 4
S05	Supplier Plat 5

Table 5. WGMC between Suppliers on any criteria

Supplier Code	S01	S02	S03	S04	S05
S01	1	3	1.6667	2.3333	1
S02	0.333333	1	3	1.6667	2.3333
S03	0.6	0.33	1	3	1.667
S04	0.428571	0.6	0.33	1	2.3333
S05	1	0.428571	0.2727	0.428571	1

4.2 Normalization Of Multiple Matrices

The normalization of the result obtained from the Division of each value in the column to the-i and j lines with total value columns to-i. The column number of the summation of results obtained from each line j, while the vector of weights derived from the average value of normalizing each line to a-j, as shown in tables 6 and 7.

4.3 Calculation Of The Eigen Vector

Eigen Vector calculation is performed by multiplying the vector with the initial matrix of weights. Eigen Vector calculation the following criteria.

$$\begin{vmatrix} 1 & \dots & \dots\dots & \dots. & 3 \\ 0.4285 & \dots & \dots & \dots & \dots \\ 0.6 & \dots & \dots & \dots & \dots \\ \dots\dots & \dots\dots & \dots. & \dots & \dots \\ 3 & 1 & \dots & \dots & 1 \end{vmatrix} \times \begin{pmatrix} 0.15855 \\ \dots\dots \\ \dots. \\ \dots. \\ 0.0668 \end{pmatrix} = \begin{pmatrix} 1,655 \\ \dots. \\ \dots. \\ \dots. \\ 1,136 \end{pmatrix}$$

Based on the calculation of the above matrix, Eigen Vector values obtained at each such criteria in table 8.

Table 6. Eigen Value Criteria

Criteria	Eigen Value
K1	1.655
K2	1.135
K3	1.589
K4	1.150
K5	0.964
K6	1.024
K7	0.785
K8	0.799
K9	0.821

K10	1.136

Eigenvalue calculations based on the results, take the top 5 criteria based on the eigenvalue. The percentage of the respective percentage of the criteria described in the Pareto diagram below.

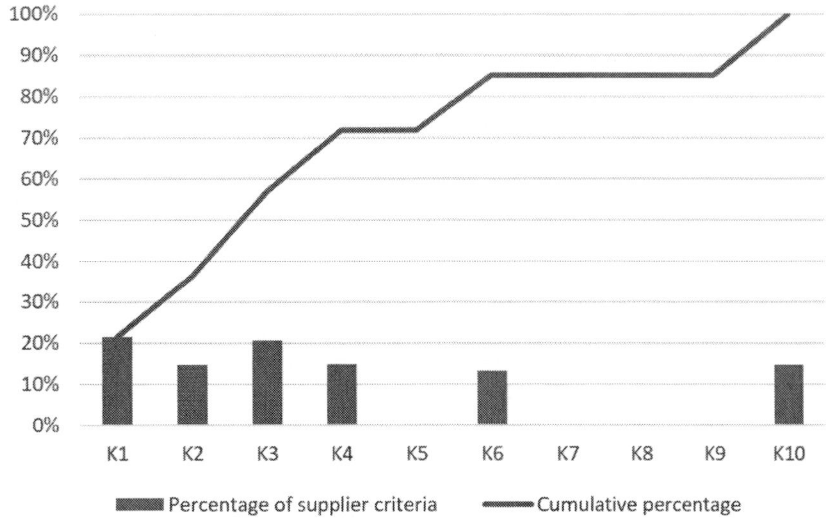

Figure 2. Pareto Supplier Selection Criteria

The following is a calculation of the Eigen Value for suppliers.

$$\begin{vmatrix} 1 & 3 & 1.67 & 2.33 & 1 \\ 0.333 & 1 & 3 & 1.67 & 2.33 \\ 0.6 & 0.333 & 1 & 3 & 3.67 \\ 0.42857 & 0.6 & 0.33 & 1 & 2.33 \\ 1 & 0.42 & 0.27 & 0.428 & 1 \end{vmatrix} \times \begin{pmatrix} 0.29925 \\ 0.23749 \\ 0.22217 \\ 0.12739 \\ 0.1137 \end{pmatrix} = \begin{pmatrix} 3.95643 \\ 1.48135 \\ 1.27995 \\ 0.7507 \\ 0.6336 \end{pmatrix}$$

Table 7. Eigen Value Supplier

Supplier	Eigen Value
S01	3.95643
S02	1.48135
S03	1.27995
S04	0.73749
S05	0.62992

Table 6 explains the normalization of comparisons twos between criteria. Normalization is obtained from the results of the division each value in the i-the column and j-row with the total value in the i column. The following is a normalization table for the recapitulation of questionnaire data.

Table 8. Multiple Matrix normalization between Criteria

Criteria	K1	K2	K3	K4	K5	K6	K7	K8	K9	K10	Total	Weight Vector
K1	0,152	0,267	0,230	0,155	0,143	0,082	0,179	0,067	0,124	1,302	1,585	0,159
K2	0,065	0,114	0,138	0,093	0,143	0,136	0,077	0,112	0,124	0,434	1,064	0,106
K3	0,091	0,114	0,138	0,280	0,143	0,190	0,128	0,156	0,124	0,104	1,468	0,147
K4	0,091	0,114	0,046	0,093	0,143	0,082	0,179	0,156	0,124	0,434	1,091	0,109

K5	0,091	0,069	0,083	0,056	0,086	0,190	0,077	0,067	0,074	0,104	0,896	0,090
K6	0,152	0,069	0,059	0,093	0,037	0,082	0,128	0,201	0,074	0,434	0,957	0,096
K7	0,065	0,001	0,083	0,040	0,086	0,049	0,077	0,067	0,124	0,146	0,737	0,074
K8	0,152	0,069	0,059	0,040	0,086	0,027	0,077	0,067	0,124	0,434	0,762	0,076
K9	0,091	0,069	0,083	0,056	0,086	0,082	0,046	0,402	0,074	0,146	0,772	0,077
K10	0,051	0,114	0,083	0,093	0,051	0,082	0,033	0,067	0,032	0,434	0,668	0,067

Tabel 9. Multiple Matrix normalization between Suppliers

Supplier	S01	S02	S03	S04	S05	Jumlah	Weight Vector
S01	0,297	0,560	1,845	0,277	0,097	1,496	2,078
S02	0,099	0,187	0,478	0,198	0,226	1,187	1,649
S03	0,178	0,062	0,159	0,356	0,355	1,111	1,543
S04	0,127	0,112	0,053	0,119	0,226	0,637	0,885
S05	0,297	0,080	0,043	0,051	0,097	0,568	0,790

4.4 Test the Consistency of The Data

The next step, namely, to calculate the Consistency Index (CI) based on the current Eigen Value and Weighting Vector. The steps are as follows:

- Scoring λ

$$\lambda = \frac{\sum EV}{VB} \quad (1)$$

- Determine the Consistency Index (CI)

$$CI = \frac{(\lambda - n)}{(n-1)} \quad (2)$$

- Determine CR

$$CR = \frac{CI}{RI} \quad (3)$$

The results of the calculation will be the value obtained by CI CR. Data is said to be consistent if the value CR≤0,1.

Tabel 10. The results of the calculation of the λ on the criteria

EV	VB	EV/VB	λ
1.65546	0.158547	10.44141	11.27111
1.135223	0.106384	10.67099	
1.588768	0.146818	10.82135	
1.149782	0.109062	10.54243	
0.963762	0.089639	10.75157	
1.023914	0.095664	10.70324	
0.784969	0.073655	10.6547	
0.798912	0.076241	10.47873	
0.820786	0.077187	10.63366	
1.136317	0.066802	17.01029	
Total		112.7111	

Based on table 10, CR values obtained on Supplier of 0.0974, which means that the value of CR ≤ 0.1 then the consistent assessment.

4.5 Determination of Supplier and Criteria

Determination of supplier based on the eigenvalue. The selected supplier is a supplier to the value of the highest eigenvalue. Likewise, with the criteria considered in determining supplier eigenvalue derived from the value of the highest. The selected supplier is a supplier of 1 with the highest value i.e., and the eigenvalue amounted to 3.956. Table 11 below explains the selected suppliers

Table 11. Supplier Selected

Supplier	Eigen Value
S01	3.95643
S02	1.48135
S03	1.27995
S04	0.73749
S05	0.62992

Criteria to consider in determining the best supplier is the first criteria, namely price bidding, with eigenvalues of 1.655. Next up is the level of disability, the distance with supplier companies, the ability to provide information in a clear and easy to understand, and delivery time can be seen in table 12.

Table 12. The criteria taken into consideration

Criteria	Eigen Value
K 1	1.165
K 2	1.135
K 3	1.589
K 4	1.150
K 5	0.964
K 6	1.024
K 7	0.785
K 8	0.799
K 9	0.821
K 10	1.136

4.6 Analysis and Interpretation of Result

- Analysis of the Criteria Hierarchy Structure

The hierarchical structure is a hierarchical model to show criteria that influence supplier determination. Based on a hierarchical structure, the criteria used in determining suppliers are price purchase, delivery time, defect rate, supplier distance from the company, the ease of adding or reducing the number of orders, giving a guarantee or warranty on goods, transportation costs, speed in terms of responding customer demand, the ability to fulfill the number of orders, the ability to provide information clearly and easily understood. Criteria were obtained from literature studies and previous research.

- Analysis of the Questionnaire Recapitulation

The questionnaire was given to 3 respondents each from the production, procurement, and logistics. Based on the calculation of the eigenvalue, priority selection criteria are obtained suppliers based on the highest to the lowest eigenvalue. Criteria consider the bid price, the level of disability, and the supplier's distance to the company.

- Supplier Determination Analysis

The suppliers used in the research of plate suppliers are because of the plate components have high price fluctuations. Based on the results of the eigenvalue calculation, The selected supplier is supplier 1, with the largest eigenvalue, which is equal to 3.9564. Then the next supplier is supplier 2 with an eigenvalue of 1.4813, and the last is supplier 3 with an eigenvalue of 1.27995. Then the last supplier 4 and 5.

5. Conclusion

The supplier selection method has been carried out using an analytical hierarchy process (AHP) approach. Based on processing and data analysis that has been done in the previous section, then the conclusions that can be drawn are set the supplier one as the best. It can know with the highest end of

IOP Conf. Series: Materials Science and Engineering **943** (2020) 012044 doi:10.1088/1757-899X/943/1/012044

the value in the calculation of the end of the AHP, the value of 3.95643, and the most criteria considered in determining supplier is the bid price. Suggestions for further research need to be considered or take into account the costs in selecting suppliers such as optimization on
storage costs with several suppliers that have been ranked.

References

[1] Taufik, R., Sumantri, Y., & Tantrika, C. F. (n.d.). Implementation Of Ready Mix Raw Material Supplier Selection Based On Ahp And Topsis Methods Integration. 1067-1076.

[2] Wardayanti, A., Zakaria, R., Sutopo, W., and Louhenapessy, B. L. (2018). Supplier Selection Model of the Lithium-ion Battery using Fuzzy AHP and Analysis of BOCR. International Journal of Sustainable Transportation Technology Vol. 1 No. 1, April 2018, 1-8.

[3] Gupta, S., Dangayach, G. S. Singh, A. M., Raod, P. N. (2015). Analytic Hierarchy Process (AHP) Model for Evaluating Sustainable Manufacturing Practices in Indian Electrical Panel Industries. XVIII Annual International Conference of the Society of Operations Management (SOM-14). DOI: 10.1016/j.sbspro.2015.03.216.

[4] Tramarico, C. L., Mizuno, D., Salomon, V. A. P, Marins F. A. S. (2015). Analytic Hierarchy Process and Supply Chain Management: a bibliometric study. Peer-review under responsibility of the Organizing Committee of ITQM 2015 DOI: 10.1016/j.procs.2015.07.005.

[5] Bianchini, A., Benci, A., Pellegrini, M., and Rossi, J (2018). Supply chain redesign for lead-time reduction through Kraljic purchasing portfolio and AHP integration. Benchmarking: An International Journal, Vol. 26 No. 4, 2019, pp. 1194-1209. DOI 10.1108/BIJ-07-2018-0222

[6] Gorbanev, I., Cortes, I., and Londoño, S. A. (2017). Redesign of the Attention Process of Patients with Rheumatologic Diseases: Assessing the Performance with the Analytic Hierarchy Process. Hindawi Advances in Operations Research Volume 2017, Article ID 6367037, 6 pages https://doi.org/10.1155/2017/6367037

[7] Ho, F. H., Abdul-Rashid, S. H., and Ghazilla, R. A. R. (2016). Analytic Hierarchy Process-Based Analysis to Determine the Barriers to Implementing a Material Efficiency Strategy: Electrical and Electronics' Companies in the Malaysian Context. Sustainability 2016, 8, 1035; DOI:10.3390/su8101035.

[8] Ngatawi, & Setyaningsih, I. (2011). Analisis Pemilihan Supplier Menggunakan Metode Analytic Hierarchy Process (AHP). Jurnal Ilmiah Teknik Industri, 7-13.

IOP Conf. Series: Materials Science and Engineering **943** (2020) 012045 doi:10.1088/1757-899X/943/1/012045

Aggregate Planning Method as Production Quantity Planning and Control to Minimizing Cost

I Nugraha[1*], M Hisjam[2] and W Sutopo[2]

[1] Master Program of Industrial Engineering, Faculty of Engineering, Universitas Sebelas Maret Surakarta, Indonesia

[2] Department of Industrial Engineering, Faculty of Engineering, Universitas Sebelas Maret Surakarta, Indonesia

Email: isna.nugraha13@gmail.com

Abstract. Production planning is very important in a production because it is related to the amount of fluctuating consumer demand in each period. If there is a product shortage will cause consumer demand cannot be fulfilled so the company loses profits and commitment from customers who have trusted. However, if the excess production in a certain period there will be a buildup of products that are generated and will result in high storage costs of finished products and waste of product storage. Products that often occur in fluctuating demand is wheat flour products. Based on observations at PT. Bulog, the authors provide an analysis to overcome the problem of fluctuating demand at PT. Bulog with aggregate planning methods. Based on forecasting results that have been obtained in the previous period wheat flour demand data at PT. Bulog, we will determine the quantity and the future production time for wheat flour products. Production planning is a very important activity of all production activities of a company. Based on the calculation of aggregate planning that has been done using the company's demand and production data PT. Bulog in the previous period found that the best method that can be used in the company's production process is the chase strategy method, this method is used by increasing or reducing the number of workers based on the number of consumer demand. By using a chase strategy, the costs used by the company are less when compared to other methods. Costs to be incurred based on the selection of the best method for the product "Tepung Terigu Kita" is IDR 341,857,000.00.

1. Introduction

Industry in the world has grown and developed rapidly. Along with its development, competition between companies will be increasingly fierce, especially in the food industry in Indonesia. With increasing competition, of course the company will further improve the quality of its management in order to survive in the competition. One of them is to improve the continuity of production in order to meet consumer demand in a timely manner, also with minimal production costs [1].

Perum Bulog is a state-owned public corporation in the field of food logistics. Bulog's sales of "KITA" products are related to Bulog's current position which must be able to support itself through commercial business. One of the food products from Bulog's business activities is wheat flour with the trademark "Terigu Keluarga Indonesia" "Terigu Kita" is produced and distributed by Perum BULOG along with companies that work with Perum Bulog.

Along with the increasingly familiar products that have been produced by PT. Bulog and with continued product development, the demand for these products has also increased. The main problem in increasing productivity in anticipation of increasing market demand is the limited production capacity. To anticipate the possibility of inadequate production capacity, planning for capacity needs must be done as well as possible with minimal costs. So there is not enough capacity and the company gets the optimal profit in every production.

Production planning is very important in a production because it is related to the amount of fluctuating consumer demand in each period. If there is a product shortage will cause consumer demand cannot be fulfilled so the company loses profits and commitment from customers who have trusted. However, if the excess production in a certain period there will be a buildup of products that are generated and will result in high storage costs of finished products and waste of product storage.

Products that often occur in fluctuating demand is wheat flour products. Based on observations at PT. Bulog, the authors provide an analysis to overcome the problem of fluctuating demand at PT. Bulog with aggregate planning methods. Based on forecasting results that have been obtained in the previous period wheat flour demand data at PT. Bulog, we will determine the quantity and the future production time for wheat flour products. Production planning is a very important activity of all production activities of a company [2]. Production planning activities begin with forecasting to find out what and how much needs to be produced in the future [3]. Production forecasting aims to estimate demand for the company's goods or services. But almost all companies cannot always adjust their production levels to real changes in demand. Therefore, companies develop rational plans that show how they will respond to the market..

Aggregate planning is the process of planning the quantity and setting the output time for a certain period of time [4] (usually between three months to one year) through adjusting the levels of production, employee, inventory and other controllable variables [5]. Aggregate planning allows companies to arrange the optimal utilization of company resources, in order to achieve effective and efficient capacity that is made based on forecast demand in the future [6]. Effective which means harmony between planning and the results obtained, while efficient means being able to produce a certain output with the available resources to a minimum. Therefore, it is necessary to make a production plan that can accommodate these goals properly.

2. Methodology

Data is collected by survey, observation, interview and documentation techniques to view and use reports in the company. The second method is the study of literature which aims to solve problems that have been formulated based on previous research and other information. Analyzing problems that occur in the company using several methods. The first method is demand forecasting using linear regression method with the smallest tracking signal calculation. The second method consists of three alternative strategies (level, chase and flexible) for aggregate planning. Sales data for the past three years (May 2016 - April 2019) is to determine demand forecasting for the next 5 months period (May 2019 - September 2019). Data is processed using Microsoft Excel software.

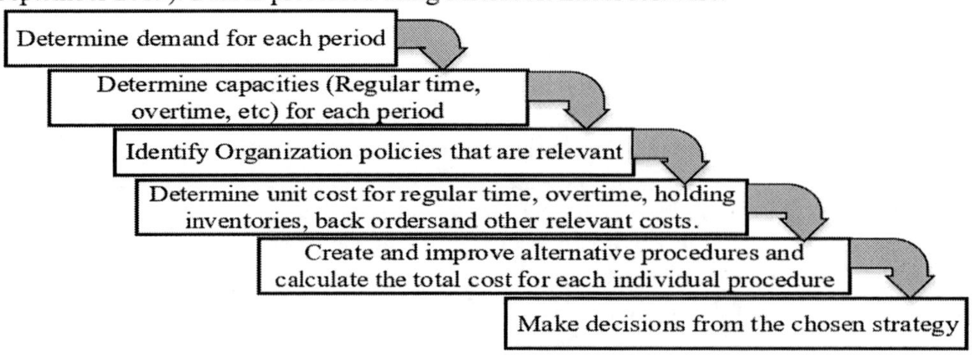

Figure 1. Research Method

Some of the data needed to calculate aggregate planning are:

1. Product demand data for 3 years
2. Data on available resources throughout the period of the aggregate production planning
3. Data related company policies, such as; labor regulations
4. Data on various related costs, such as: costs of recruitment and firing cost, overtime and unemployment costs, inventory storage costs, subcontracting costs, part-time labor costs, inventory costs or re-ordering.

The objectives to be achieved are:

1. Able to plan production, inventory, and resources that are stable against fluctuations in demand.
2. Able to determine a proper production planning strategy and can minimize the total production costs.
3. Understanding the aggregate planning process and capacity in an industry.

The steps taken are:

1. Perform calculation of chase strategy, level strategy and flexible strategy.
2. Analyzing the calculation of chase strategy, level strategy and flexible strategy.
3. Take decisions from the best strategy.

3. Literature Review

3.1 Definition Of Aggregate Planning

Aggregate planning also known as aggregate scheduling is an approach usually carried out by operations managers to determine the quantity and time of production in the medium term (usually between 3 and 18 months ahead) [5]. Aggregate planning can be used in determining the best way to meet the predicted demand by adjusting the value of production, labor level, inventory level, overtime work, subcontracting level, and other variables that can be controlled [7]. Scheduling decisions regarding the formulation of monthly and quarterly plans that prioritize the problem of matching productivity with fluctuating demand [8]. Therefore, aggregate planning is included in the medium-term plan [9].

3.2 Purpose Of The MRP

Basically the purpose of aggregate planning is to try to obtain an optimal solution to costs or benefits in the planning period [10]. However, there are other strategic issues that may be more important than low costs. The strategic problems in question include reducing labor level problems, reducing inventory levels, or meeting higher levels of service [11]. For manufacturing companies, the aggregate schedule aims to link the company's strategic goals with the production plan, but for service companies, aggregate scheduling aims to connect targets to the schedule of workers.

3.3 Characteristics Of Aggregate Planning

Aggregate planning means combining appropriate resources into the whole period [12]. With the prediction of demand, facility capacity, inventory level, labor size, and interconnected inputs, planners must choose the level of output for a facility for the next 3 to 18 months. In aggregate planning, the production plan does not describe per product but concerns how many products will be produced regardless of the type of product.

4. Results & Discussion

4.1 . Data of Observation Results

The following information is obtained from the data collection results in the company shown in the table below:

Table 1. Demand Data for Wheat Flour Products

Period	Demand	Period	Demand	Period	Demand
1	8649	13	7249	25	5813
2	9727	14	8348	26	9023
3	9537	15	6689	27	8881
4	9181	16	8107	28	6499

Period	Demand	Period	Demand	Period	Demand
5	7209	17	6238	29	9431
6	5915	18	9722	30	6809
7	7330	19	8602	31	9267
8	5259	20	9208	32	5824
9	9504	21	6063	33	5004
10	9902	22	8243	34	6728
11	6230	23	5869	35	5006
12	6522	24	9129	36	5191

Source: Data of PT. Bulog 2019

Table 2. Costs

Raw Material	6500	/unit
Holding costs	600	/unit/mont
Stockout costs	12500	/unit/mont
Hiring and training cost	5000	/worker
Layoff costs (firing)	11500	/worker
Overtime cost	7000	/hour /worker
Worker costs	5000	/hour

Source: Data of PT. Bulog 2019

Table 3. Other Cost

Initial inventory	225	unit
Productive time / worker / day	7.5	hour
Paid work hours / workers	8	
The number of workers	40	worker
Production time	0.75	hour /unit

Source: Data of PT. Bulog 2019

4.2 Results Of Strategy

Table 4. Forecasting Results

Demand (unit)	May	June	July	August	September	Average
	6605	6554	6503	6453	6401	6504

Source: Data Processed

Table 4. shows that demand information for each period from May to September 2019 is 6504 units. Data obtained from forecasting using the Centered Moving Average method performed in the previous calculation.

Table 5. Calculation of Data Production Need in 2019

	May	June	July	August	September
Number of days	25	24	24	25	26
Hour / worker / month	187.5	180	180	187.5	195
Unit/ worker	250	240	240	250	260
IDR / worker / month	1,000,000	960,000	960,000	1,000,000	1,040,000
Maximum Production Capacity	10000	9600	9600	10000	10400

Source: Data Processed

4.3 Calculation with Level Strategy

In this strategy, companies have limited production capacity and a fixed number of workers. The amount of production is fixed and the inventory that arises can be used to meet the excess demand for the product for a certain period. The calculation formula for the strategy level is shown below [18]:

a. Demand was obtained from the forecasting demand in the past.
b. Get the initial inventory from the information. In this strategy, the initial inventory is constant.

c. Net requirement = demand - initial inventory
d. Production = net requirement
e. Worker needs = production / (unit / worker)
f. Workers = workers' needs
g. Lastest inventory = net requirement – Production
h. Surplus = inventory
i. Shortage = net requirement – production
j. Raw material costs = raw material costs per unit x of production
k. Worker costs = number of workers x costs / workers / month
l. Holding cost = Surplus x storage costs

Table 6. Calculation of Level Strategy

Month	May	June	July	August	September
Number of days	25	24	24	25	26
Hour / worker / month	187.5	180	180	187.5	195
Unit / worker	250	240	240	250	260
IDR / Worker / month	1000000	960000	960000	1000000	1040000
Demand	6,605	6,554	6,503	6,453	6,401
Initial Inventory	225	124	0	1	52
Net requirement	6,380	6,430	6,503	6,452	6,349
Worker	40	40	40	40	40
Production	6,504	6,504	6,504	6,504	6,504
Lastest Inventory	124	0	1	52	155
Surplus	124	0	1	52	155
Shortage	0	0	0	0	0

Source: Data Processed

Table 7. Calculation of Production Costs

Costs	May	June	July	August	September	Costs
Workers (IDR)	IDR40,000,000	IDR38,400,000	IDR38,400,000	IDR40,000,000	IDR41,600,000	IDR198,400,000
Raw Material (IDR)	IDR42,276,000	IDR42,276,000	IDR42,276,000	IDR42,276,000	IDR42,276,000	IDR211,380,000
Holding cost (IDR)	IDR74,400	IDR0	IDR600	IDR31,200	IDR93,000	IDR199,200
Stockout cost (IDR)	IDR0	IDR12,500	IDR0	IDR0	IDR0	IDR12,500
Total						**IDR409,991,700.00**

Source: Data Processed

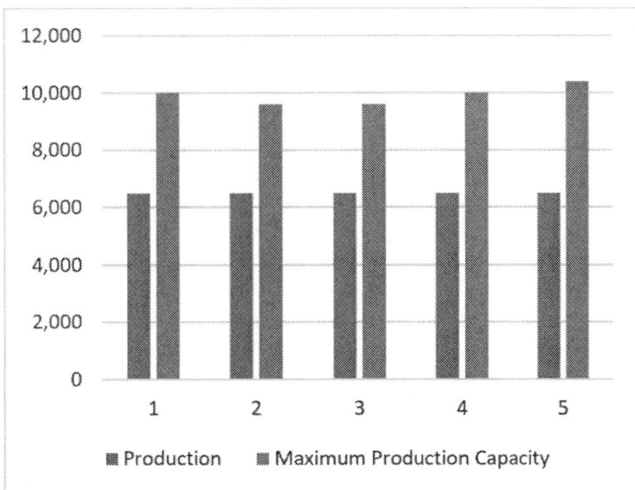

Figure 2. Level Strategy Chart

Table 6 explains the results of calculations with level strategy. From the calculation, it is found that the initial inventory obtains from the previous information which is 225 units, specifically for the beginning of the period in May, while for the following month it is obtained from the difference between production and net requirements. For net requirement obtained from the amount of demand reduced by initial inventory. Obtained that the net requirement for the May period is 6380 units, June 6430 units, July 6503 units, August 6452 units and September 6349 units. Whereas for production is obtained from multiplication between workers and units per worker, so the results for January to September are equal at 6504 units. For the initial inventory with a surplus amount is the same for each period. At this level, there is no shortage because the amount of net requirement is less than the amount of production, so the company does not lack production. And then, table 7 explains the results of the calculation of production costs level strategy. Figure 2 shows the level strategy graph. The graph includes the number of production and maximum production capacity. Maximum production capacity is always the same as the number of productions in each period. For the amount of production, which is always the same as the maximum production capacity in each period.

4.4 Calculation with Chase Strategy

In this strategy the capacity and amount of production have the variety of demand according to each period. The Chase Strategy calculation formula is shown below [18]:

a. Demand was obtained from the forecasting demand in the past
b. The initial inventory of the month is obtained from information. The next one comes from the last inventory in the previous period.
c. Net requirement = demand - initial inventory
d. Production = net requirement
e. Worker needs = production / (unit / worker)
f. Employed = worker needs – available workers
g. Fired = available workers - conditional workers
h. Workers = workers' needs
i. Last inventory = net requirement - production
j. Raw material costs = raw material costs per unit x production
k. Worker costs = number of workers x costs / workers / month
l. Costs are employed = costs of being fired / workers x number of reinstating workers
m. Costs fired = costs of being fired / workers x number of workers fired

TABLE 8. Chase Strategy Calculation

Month	May	June	July	August	September
Number of days	25	24	24	25	26
Hour / worker / month	187.5	180	180	187.5	195
Unit / worker	250	240	240	250	260
IDR / Worker / month	1000000	960000	960000	1000000	1040000
Demand	6,605	6,554	6,503	6,453	6,401
Initial Inventory	225	0	0	0	0
Net requirement	6,380	6,554	6,503	6,453	6,401
Production	6,380	6,554	6,503	6,453	6,401
Worker needs	26	28	28	26	25
Hired	0	2	0	0	0
Fired	14	0	0	2	1
Workers	26	28	28	26	25
Lastest inventory	0	0	0	0	0

Source: Data Processed

TABLE 9. Calculation of Production Costs

Cost	May	June	July	August	September	Costs
Raw Material (IDR)	IDR 41,470,000.00	IDR 42,601,000.00	IDR 42,269,500.00	IDR 41,944,500.00	IDR 41,606,500.00	IDR 209,891,500.00
Worker (IDR)	IDR 26,000,000.00	IDR 26,880,000.00	IDR 26,880,000.00	IDR 26,000,000.00	IDR 26,000,000.00	IDR 131,760,000.00
Hiring cost (IDR)	IDR 0.00	IDR 10,000.00	IDR 0.00	IDR 0.00	IDR 0.00	IDR 10,000.00
Firing cost (IDR)	IDR 0.00	IDR 0.00	IDR 0.00	IDR 23,000.00	IDR 11,500.00	IDR 195,500.00
Total						IDR **341,857,000.00**

Source: Data Processed

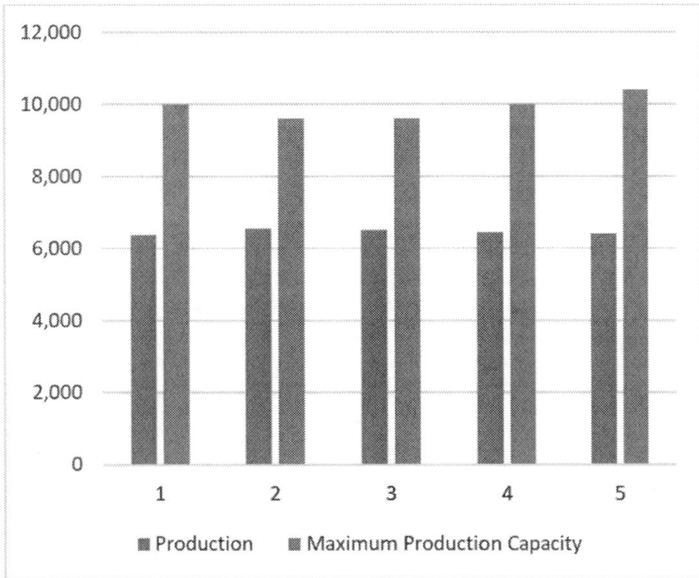

Figure 3. Chase Strategy Chart

Table 8 explains the results of calculations with chase strategy. From the calculation, it is found that the initial inventory is obtained from previous information, namely 225 units since the beginning of the period in May, while for other periods are zero because there is no inventory. The net requirement is obtained from the total demand reduced by the initial inventory so that it is obtained that the net need for May 6380 units, June 6554 units, July 6503 units, August 6453 units and September 6401 units. The amount of production is also equal to the net requirement. Workers need for the period from May to September are between 25 and 28 workers. Due to the initial number of workers of 40 people, while only requiring 26 people, there were 14 workers who were fired in the May period, so for the next period the number of workers has been adjusted to the number of workers' needs. Because the amount of net requirement is equal to the amount of production, there is no inventory in all periods. And then, Table 9 explains the results of the calculation of production costs chase strategy. Figure 3 shows the chase strategy graph. The graph includes the number of production and maximum production capacity. Maximum production capacity is always higher than the amount of production in each period.

4.5 Calculation with Flexible Strategy

This strategy is a combination of level strategy and chase strategy. For example, a company has sufficient production capacity, then the utility of production facilities is used as a reference in determining aggregate planning. In this case the number of workers is permanent, but the working hours of each worker have variability adjusted to the level of product demand. The flexible strategy calculation formula is shown below [18]:

a. Demand was obtained from the forecasting demand in the past
b. The initial inventory of the month is obtained from information. The next one comes from the last inventory in the previous period.
c. Net requirement = demand - initial inventory
d. In flexible strategy, Net requirement only available in the first period.
e. Production = worker x unit / worker
f. Worker Needs = from the initial number of workers (constant number for each period)
g. Lack of production = net requirement - production
h. Surplus = production - net requirement
i. Overtime Production = shortage

j. Overtime / work hours = (overtime production x production time) / number of days / number of workers
k. Raw material costs = raw material costs per unit x production
l. Worker costs = number of workers x costs / workers / month
m. Holding cost = Surplus x storage costs
n. Along with the cost of time = the number of overtime hours x the cost of overtime number x workers.

TABLE 10. Workforce and Overtime Strategy

Month	May	June	July	August	September
Number of days	25	24	24	25	26
Hour / worker / month	187.5	180	180	187.5	195
Unit / worker	250	240	240	250	260
IDR / Worker / month	1000000	960000	960000	1000000	1040000
Demand	6,605	6,554	6,503	6,453	6,401
Initial Inventory	225	0	0	0	0
Net requirement	6,380	6,554	6,503	6,453	6,401
Worker	40	40	40	40	40
Production	10,000	9,600	9,600	10,000	10,400
Shortage of production	0	0	0	0	0
Surplus	3,395	3046	3097	3547	3,999
Overtime Production	0	0	0	0	0
Overtime / worker hours	0	0	0	0	0

Source: Data Processed

TABLE 11. Calculation of Production Costs

Cost	May	June	July	August	September	Costs
Workers (IDR)	IDR 40,000,000.00	IDR 38,400,000.00	IDR 38,400,000.00	IDR 40,000,000.00	IDR 41,600,000.00	IDR 198,400,000.00
Raw material (IDR)	IDR 65,000,000.00	IDR 62,400,000.00	IDR 62,400,000.00	IDR 65,000,000.00	IDR 67,600,000.00	IDR 322,400,000.00
Holding cost (IDR)	IDR 2,037,000.00	IDR 1,827,600.00	IDR 1,858,200.00	IDR 2,128,200.00	IDR 2,399,400.00	IDR 10,250,400.00
Biaya Lembur (IDR)	IDR 0.00	IDR 0.00	IDR 0.00	IDR 0.00	IDR 0.00	IDR 0.00
Total						IDR **531,050,400.00**

Source: Data Processed

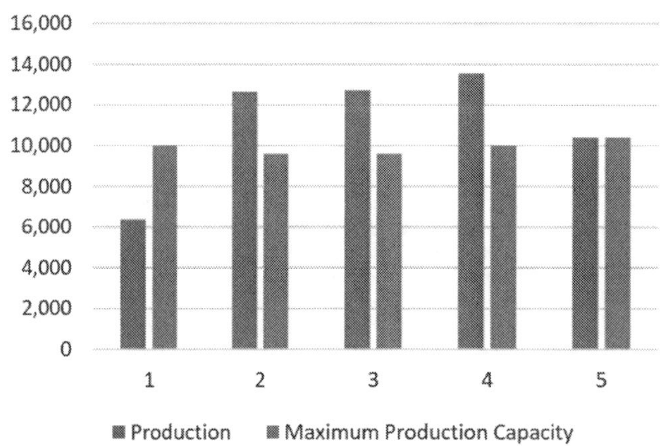

Figure 4. Flexible Strategy Chart

Table 10 explains the results of calculations with flexible strategy. From the calculation, it is found that the initial inventory is obtained from previous information, namely 225 units since the beginning of the period in May, while for other periods are zero because there is no inventory. The net requirement is obtained from the total demand reduced by the initial inventory so that it is obtained that the net requirement for May 6380. The amount of production for the flexible strategy is not the same as the net requirement. Workers' needs for the period from May to September are the same, namely 40 workers. The amount of production obtained from multiplying the number of workers by units per worker. The result is 10000 units in May; 9600 units in June and July; 10000 units in August; and 10400 units in September. For production shortages do not occur in each period because the number of net requirement is less than the amount of production. Because of this the company had a surplus of 3,395 units in May; 3046 units in June; 3097 units in July; 3547 units in August, and 3999 units in September. There is no shortage production so there is no overtime production from May to September, so there are no overtime hours per day. And then, Table 11 explains the results of the calculation of production costs with flexible strategy. Figure 4 shows the flexible strategy graph. The graph includes the number of production and maximum production capacity.

4.6 Comparison of Three Strategies

After calculating costs based on three strategies, we have a recapitulation and comparison of three strategies. The recapitulation cost will be displayed in the following table 12:

TABLE 12. Recapitulation Cost

Strategy	Regular production	Overtime production	Sub-contracted	Shortage	Inventory	Cost
Level Strategy	√	-	√	-	√	**IDR 409,991,700.00**
Chase Strategy	√	-	√	-	-	**IDR 341,857,000.00**
Flexible Strategy	√	-	√	-	√	**IDR 531,050,400.00**

Source: Data Processed

Table 12 shows the cost comparison of the three strategy costs. First, the level strategy, the number of workers is fixed at 40 people. By maintaining the amount of workers, a relatively constant level of production will be obtained. If there is a surplus of production, then the surplus is used to cover the production shortage in another month. The level strategy is handling the inventory to meet the demand in the next month. Level strategy plans in which the aggregate production level remains the same for a period of time (constant output). Scheduling levels maintain output levels, production levels, or constant employment levels in the planning horizon. The total cost for level strategy is IDR 409,991,700.00.

Second, chase strategy is a variation in the level of workers, the amount of workers used follows the level of demand for production. Where, sometimes in each month there is a reduction or increase in the number of workers. Chase strategy manufacture products according to the demand at that time so it does not use any inventory handling. Chase strategy aligns production to meet the demand with minimum inventory level. So that the total cost for a variety of workers is IDR 341,857,000.00.

Finally, flexible strategy. This strategy has a fixed workers, but the production level uses the lowest demand forecast every month. When demand increases, the company makes overtime production. The addition of working hours is limited to 4 to 5 hours per day. Thus, it has the largest total production costs among other strategies. In this strategy, there is no shortage of supply so that demand can be met in the planning period and there is a handling charge of inventory. The total production cost of the flexible strategy is IDR 531,050,400.00.

From this table, it can be seen that the lowest cost is the second strategy, the chase strategy. With the most economical cost because in this strategy the company produces only in the first period. The capacity and amount of production determined has variability that matches the variability of the number of product demands in each period. This strategy is used to minimize and stabilize inventory levels [19]. This method is also used by increasing or reducing the number of workers based on the number of consumer demand [20] [21]. To produce an appropriate and efficient strategy, the company must consider the condition of the company such as the company's production capacity. To implement chase strategy, companies must carefully estimate the workdays of the following year, and determine the worker to be employed, both trained and skilled, with untrained or unskilled workers to optimize the costs of hired or fired workers.

4.7 Approach to Strategy Implementation

If product demand variability during the planning period, there are two alternative approaches, namely:

a. Managing Demand [22, 23]

This approach is done to regulate the level of demand, that is if the demand is smaller than capacity or otherwise. This approach can be done by regulating several factors, namely: price strategy, promotion, reservation and backlog.

b. Adjust capacity [22, 23]

Balancing the required resources is done by changing several factors that affect production capacity, namely: Appointment and dismissal of workers, overtime (overtime work), part time labor, inventory, subcontracting, and setting the engine capacity level.

This approach can be chosen based on two strategies, namely:

a. Pure Strategy [24, 25]

Pure strategy is a strategy by making changes to one of several factors that affect production capacity and performance, such as: appointment and dismissal of workers, subcontracting, and reservation.

b. Mixed Strategy [24, 25]

This strategy is a combination of two or more factors used in a pure strategy by considering company policies and costs incurred.

5. Conclusions

Based on the calculation of aggregate planning that has been done using the company's demand and production data PT. Bulog in the previous period found that the best method that can be used in the company's production process is the chase strategy method, this method is used by increasing or reducing the number of workers based on the number of consumer demand. By using a chase strategy, the costs used by the company are less when compared to other methods. Costs to be incurred based on the selection of the best method for the product "Tepung Terigu Kita" is IDR 341,857,000.00.

References

[1] Gasversz and Vincent 2004 *Production Planning and Inventory Control Berdasarkan Sistem Terintegrasi MRP II dan JIT Menuju Manufakturing 21* Jakarta: PT Gramedia Pustaka Utama

[2] Dejonckheere J, Disney S M, Lambrecht M and Towill D R 2003 *Production Planning and Control* **14** pp 497–516

[3] Djordjevica I, Petrovicb D and Stojica G 2019 *Computers in Industry* **110** pp 48-63

[4] Noegraheni E and Nuradli H 2016 *Binus Business Review* **7** pp 39-45

[5] Heizer J H and Render B 2015 *Manajemen Operasi: Manajemen Keberlangsungan dan Rantai Pasokan (11th ed.)* Jakarta: Salemba Empat

[6] Jamalnia A, Yang J B, Feili A, Xu D L and G. Jamali 2019 *The International Journal of Advanced Manufacturing Technology* **102** pp 159–181

[7] Rasmi S A B, Kazan C and Türkay M 2019 *Computers & Industrial Engineering* **132** pp. 348-360

[8] Goli A, Tirkolaee E B, Malmir B, Bian G B and Sangaiah A K 2019 *Computing* **101** pp 499–529

[9] Cheraghalikhani A, Khoshalhan F and Mokhtarib H 2019 *International Journal of Industrial Engineering Computations* **10** pp 309-330

[10] Jamalnia A, Yang J B, Xu D L, Feili A and Jamali G 2019 *Journal of Manufacturing Systems* **50** pp 146–162

[11] Demirel E, Özelkan E C and Lim C 2018 *International Journal of Production Economics* **202** pp. 45-58

[12] Thompson S D and Davis W J 1990 *IEEE Transactions on Systems, Man, and Cybernetics* **20** pp 1000-1012

[13] Missbauer H 2002 *International Journal of Production Research* **40** pp 699-718

[14] Kogan K and Khmelnitsky E 1995 *Computers and Industrial Engineering* **28** pp 851-859

[15] Gen M, Tsujimura Y and Ida K 1992 *Computers and Industrial Engineering* **23** pp 117-120
[16] Zapfel G and Missbauer H 1993 *European Journal of Operational Research* **67** pp 297-320
[17] Axsater S and Jonsson H 1984 *European Journal of Operational Research* **17** pp 338-350
[18] Sukendar I and Kristomi R 2008 *Prosiding Seminar Nasional Teknoin* pp c107- c112
[19] Baroto and Teguh 2002 *Perencanaan dan Pengendalian Produksi* Jakarta: Ghalia Indonesia
[20] Chinguwa S, Madanhire I and Musoma T 2013 *International Journal of Science and Research (IJSR), India Online* **2** pp 370-383
[21] Rahmadhani A, Rahman A and Tantrika C F M 2014 *Jurnal Rekayasa dan Manajemen Sistem Industri* **2** pp. 1192-1202
[22] Cahyono D D 2007 *Perencanaan Produksi Disagregasi Dengan Pendekatan Reguler Knapsack Method Pada Produk Mini Boom ZX 25 YYZX22B dan Mini Boom ZX 30 YYZX30B* Retrieved from http://library. gunadarma.ac.id
[23] Octavianti I A, Setyanto N W and Tantrika C F M 2013 *Jurnal Rekayasa dan Manajemen Sistem Industri* **1** pp. 264-274
[24] Radwan A and Aarabi M 2011 *Proceedings of the 2011 International Conference on Industrial Engineering and Operations Management,* Kuala Lumpur, Malaysia
[25] Simamora B H and Natalia D 2014 *International Business Management* **8** pp. 353-356

IOP Conf. Series: Materials Science and Engineering **943** (2020) 012046 doi:10.1088/1757-899X/943/1/012046

The Dynamic Simulation Model of Local Soybean Competitiveness Policy to Support the Soybean Price Stabilization

I Nugraha[1], W Sutopo[2], M Hisjam[2] and N Oktyajati[2]

[1] Master Program of Industrial Engineering, Faculty of Engineering, Universitas Sebelas Maret Surakarta, Indonesia
[2] Research Group Industrial Engineering and Techno-Economic, Department of Industrial Engineering, Universitas Sebelas Maret, Surakarta

Email: isna.nugraha13@gmail.com

Abstract. Fluctuations in soybean commodity prices become a crucial issue in Indonesia every year. In this study, it is proposed to conduct a research on the development of a dynamic simulation model for fulfilling the logistics of soybeans, which aims to look at the competitiveness of local soybeans and to support soybean price stability (case study: in Central Java Province). Researchers focus on examining the relationship between variables in the system for fulfilling soybean logistics in Central Java, namely food needs (demand), total soybean availability (supply) and soybean prices using a dynamic system framework. This study also aims to identify factors in increasing local soybean competitiveness towards meeting the needs of consumers with dynamic system simulations to increase local soybean competitiveness in the national market and stabilize soybean prices that can benefit both parties, namely producers and consumers. Based on simulation results, to look at the competitiveness of local soybeans and to support soybean price stability for the next 20 years using software powersim studio 10, the government needs to take action as follows: (1) Dummy monopoly bulog intervention in soybean import price and soybean producer prices; (2) Increasing soybean import tariffs 20%; (3) Increasing soybean local prices with predicting soybean price conditions.

1. Introduction

Fluctuations in soybean commodity prices become a crucial issue in Indonesia every year [1]. Fluctuations in food prices always make it difficult for food producers and consumers and for the economy as a whole, because the direction and development, especially the end result is difficult to predict beforehand, so that economic growth is disrupted. Therefore, almost all countries in the world always try to overcome this fluctuation through food price stabilization. Based on simple logic, food price stabilization is an effort to create a climate that encourages food distribution to benefit producers and help consumers, so that distribution management is not only based on market mechanisms that are very dynamic and dependent on various factors and policies, because the government is obliged to manage food availability as long as time.

One of the industrial commodities made from agricultural raw materials that have the ability to compete is soybean as a raw material for the processed soybean industry. Soybean has become a culture in Indonesian society in the household economy of farmers, food consumption, needs and national food trade [2]. Utilization of soybean seeds in addition to direct consumption, is also an industrial raw material, such as fermented products such as tempeh, soy sauce, tauco, oncom, soy fresh and nonfermented products including tofu, milk, soy cheese, yuba and artificial meat.

Soybean needs increase every year, along with increasing population growth, public awareness of nutrition, which is characterized by increased consumption of soybeans per capita and the growth of the processed soybean product industry. Based on data released by BPS, the average per capita consumption of soybeans in Central Java is 10 kg / year or 0.1 quintal / year and continues to increase along with the increasing population in Central Java. The current problem is the demand for soybeans continues to increase, but cannot be matched by domestic production. To fulfill this, imports are increasing every year. Since 1975 Indonesia's position has shifted from exporting countries to importing soybeans [3]. This is because soybean demand is so fast, while production develops slowly due to low local soybean productivity [4].

In the last decade, there has been a decline in the amount of production due to soybean price problems that affect farmers' decisions in producing [5]. The government has made various efforts to increase the production and fulfillment of soybean needs since 1986. But the target of soybean production has not been achieved due to various problems [4]. Low local soybean production causes insufficient local soybeans to meet the demand of the soybean processing industry. This causes the dependence of soybean processing industries on imported soybeans [6]. Central Java contributed 13.95% of Indonesian soybean production (an average production of 123.54 thousand tons). Empirically the increase in soybean prices is caused by the element of scarcity on the supply side, or local soybean production is slower than demand [7] [8]. According to the Indonesian Ministry of Trade, more than 90% of Indonesia's soybean needs are still met from imports. Large depending on soybean consumption, especially in Java, will have an impact on dependence on imports if there is no significant increase in domestic production [9].

In this study, it is proposed to conduct a research on the development of a dynamic simulation model for fulfilling the logistics of soybeans, which aims to look at the competitiveness of local soybeans and to support soybean price stability (case study: in Central Java Province). Researchers focus on examining the relationship between variables in the system for fulfilling soybean logistics in Central Java, namely food needs (demand), total soybean availability (supply) and soybean prices using a dynamic system framework. This study also aims to identify factors in increasing local soybean competitiveness towards meeting the needs of consumers with dynamic system simulations to increase local soybean competitiveness in the national market and stabilize soybean prices that can benefit both parties, namely producers and consumers.

Dynamic system frameworks can be used to analyze models and scenarios to improve system performance because of the ability to represent physical flows and information, based on feedback control information that is continuously transformed into decisions and actions [10]. The main objective of this research is to see how to analyze and find alternative solutions quickly and as expected to see the stability of soybean prices over the next 20 years using software powersim studio 10. The results of the model scenario can be used as supporting material for government and stakeholder decisions in developing strategies for implementing increased local soybean competitiveness in the national market and stabilizing soybean prices. Thus, stakeholders can utilize the results of this simulation to make strategic decisions.

2. Basic Theory

2.1 Price Stabilization Policy

Soybean demand from year to year is increasing, but Indonesia has various problems such as the small production of soybeans, on average only reached around 40 percent so that it meets the shortage of the soybean supply through imports [11]. In 2013, the Government issued a policy of Soybean Price Stabilization Program (SHK Program) to stabilize soybean prices at the farmer level and at the craftsman level simultaneously. This is done by the government considering that since August 2012, domestic soybean prices have increased and soybean prices are difficult to control. The Ministry of Agriculture's Food Security Agency is directly involved in the formulation of SHK Program policies, from the process of drafting regulations, to the mechanism of policy implementation, including determining the purchase price of soybeans at the farm level. The soybean price stabilization program (SHK) is a regulation on soybean purchases from farmers, soybean imports and soybean sellers to tofu and tempeh sellers. The SHK policy was issued on May 28, 2013 through the Minister of Trade Regulation No.23 / 2013 about Soybean Price Stabilization program which is the implementation of Perpres No.32 / 2013 about assignment to Perum Bulog for safeguarding soybean price and distribution. Through the soybean price stabilization program, the government is trying to regulate soybean trade through purchasing soybeans.

2.2 System Dynamics

A dynamic system is a framework that focuses on systems thinking by using a feedback loop and taking some additional structural steps and testing it through a computer simulation model [12]. The process of comparing outputs with inputs in a system, also known as a feedback system, is a general concept [13]. A statistical correlation between a pair of variables is taken from existing data in a state that these variables have a relationship with other variables in the system and all of them change simultaneously [14]. Stocks and flows are the main components of a dynamic system. A stock represents the storage of some type of information or entity (such as money or population) in the system. Flows define the rate of change in stocks and adds more than the type of information or entity to stocks, removes some to be replaced elsewhere in the system, or removes from the system. System dynamics can be used as an analytical tool to evaluate the impact of short-term and long-term policy. The final goal of the simulation model creation is validation of models and scenarios decisions. The purpose of the validation is that the model created can certainly approach the original and credible system. The credibility of the model can be expressed from the results of the verification and validation of the model. Credible models can be simulated using computer-assisted predictions to see results quickly.

3. Previous Research

By using a dynamic system framework for fulfilling food logistics to support soybean price stability in Central Java to illustrate price stability and to formulate several future policy scenarios. Research related to local soybean competitiveness and soybean price stabilization in Indonesia ever conducted by previous researchers using various methods including a system dynamics approach. The strategy to create soybean competitiveness [15] through a policy of combining soybean price increases and import tariffs of 20% will stimulate producers to increase their harvest and production areas. The research also studies and establishes factors in increasing local soybean competitiveness through policy simulations. Other studies [16] include several additional conditions such as soybean imports, sales, import volumes, import taxes, soybean import demand, dollar exchange rates, etc. The research also studies about the scenario of stabilization soybean prices to increasing farmer productivity through the implementation of policies related to import soybeans.

4. Research Steps

This research consists of four stages: data collection, analysis of existing conditions, design of computer model and perform the process simulation with scenarios. Figure 1 shows the chart of the research methodology to solve the problem. The variables involved can be identified by conducting in depth interviews and study the literature conducted in previous research.

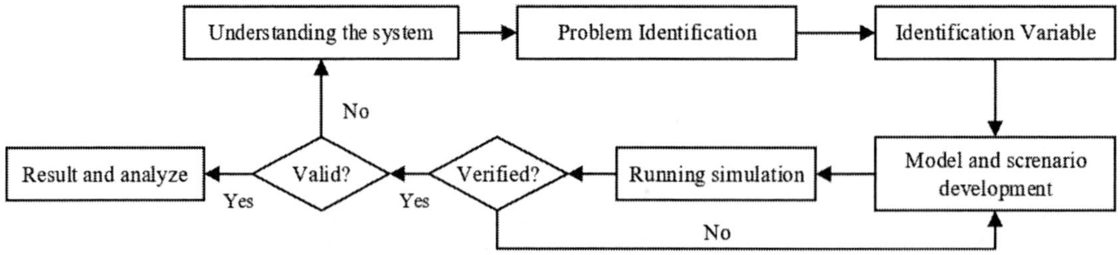

Figure 1. Research Method

4.1 Research Information and Data

This study uses secondary data from BPS and the Department of Industry and Trade Central Java Province, also related to research that has been conducted in the Central Java region.

4.2 Analysis of Existing Condition

Soybean trading system, which is dominated by importers, often has an impact on the instability of soybean prices at the community level, both producers, in this case, tofu and tempe producers, consumers or the wider people. Soybean dependence on imported products also affects domestic prices due to fluctuations in soybean prices on the international market. This condition causes soybeans to influence changes in inflation [17]. In addition to the availability of land, the availability of farmers who want to grow soybeans is also important. Farmers do not want to grow soybeans if they cannot benefit from agriculture [18]. Farmers will lose if people prefer to import soybeans because the price is cheaper. Lower selling prices will encourage farmers to use their land for cultivation of other food [19]. The establishment of government purchase prices does not help if the soybean import price is still cheaper than the soybean local prices. To overcome this problem, it is necessary to keep the soybean import price under control.

Price instability will also make farmers switch crops other than soybeans. Many factors affect price instability, one of which is price and quality competition with import soybeans [20]. Soybean local prices are influenced by soybean producer level prices, import prices and volumes, productivity and soybean local prices in the previous year. Soybean producer prices are influenced by production, import volumes, soybean consumption, dummy monopoly Bulog and soybean producer prices in the previous year. Import volumes are influenced by soybean international prices, production, consumption, population and import prices. Soybeans import price is influenced by international prices, dollar exchange rate, import tariffs, dummy monopoly Bulog and import prices of the previous year. In a systems approach, these conditions can be explained in the Causal Loop Diagram below (Fig. 2). Causal loops are made to describe the interaction or cause-effect relationships of the main variables that will be created in the model. Causal relationships in the model can be positive and negative. Figure 2 shows the CLD of this study.

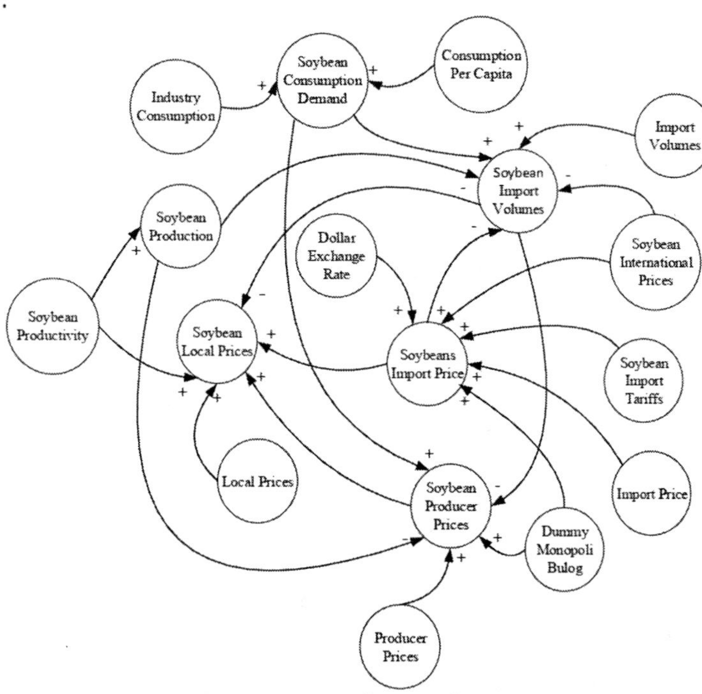

Figure 2. Causal Loop Diagram

4.3 System Dynamics Model

Data modeling is carried out to verify and validate the relationships between variables (model formulation) to suit the model with the actual system. Based on a causal loop diagram, a dynamic model of local soybean competitiveness can be developed to support soybean price stability to the stock flow diagram. It can be seen in the Figure 3 base model of local soybean competitiveness to support soybean price stability using software powersim studio 10 explained as follows (Fig. 3):

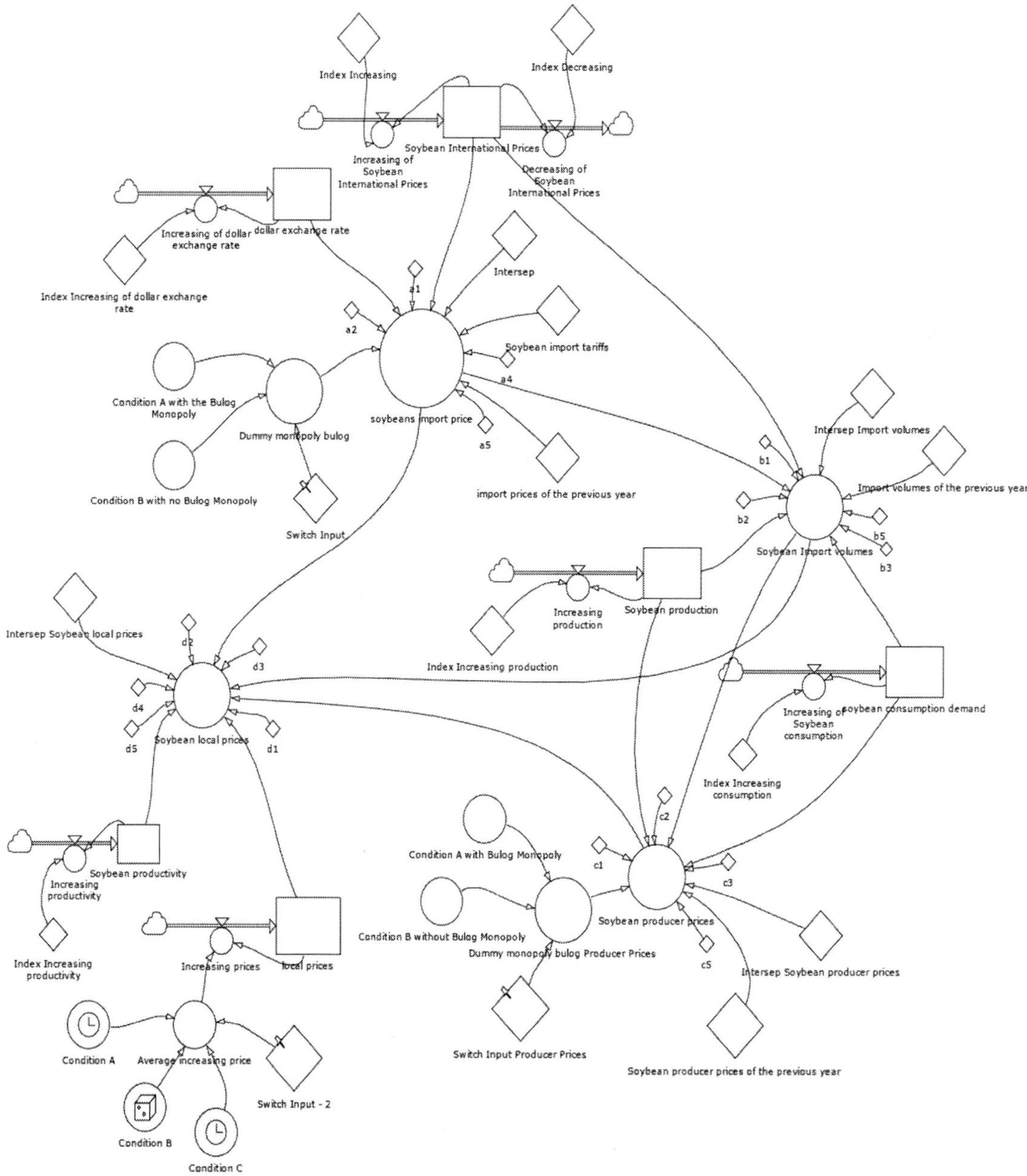

Figure 3. Causal Loop Diagram

4.4 Model Simulations

The base model is a basic model that will later be developed for analysis. The initial step is to identify related variables that affect the main system. Model simulations are carried out to obtain the results and behavior of the system during the simulation period. The simulation is run by entering the values of the input parameters and changing the structure of the model if necessary. The simulation period used for the model is from 2018 to 2038. To see the competitiveness of local soybeans and to support soybean price stability in the next 20 years, the following scenarios are carried out: (1) The scenario without dummy monopoly Bulog intervention; (2) Scenarios of with dummy monopoly bulog intervention on soybeans import price and soybean producer prices; (3) Scenarios for increasing soybean import tariffs 20%; (4) Scenario of increasing soybean local prices.

4.4.1 Result of Model Simulation

The simulation process is done using software powersim studio 10. The data used came from secondary data from BPS and the Department of Industry and Trade Central Java Province, also related to research that has been conducted in the Central Java region.

4.4.1.1 Scenario without dummy monopoly bulog intervention.

The variable without dummy monopoly Bulog has an effect on lower import prices and more import volumes. If the soybean import price is cheap, it will disadvantage farmers because local soybeans are valued cheaply, so they cannot cover farming costs. This also shows that without the monopoly on soybean trade by Bulog, the effect on soybean local price has decreased from the previous year. Increasing soybean import volume will reduce the soybean local price. If Bulog does not monopolize soybeans, the soybean import price is getting lower. As a result, many soybean processing industries use soybean import. This resulted in sluggish local soybean production.

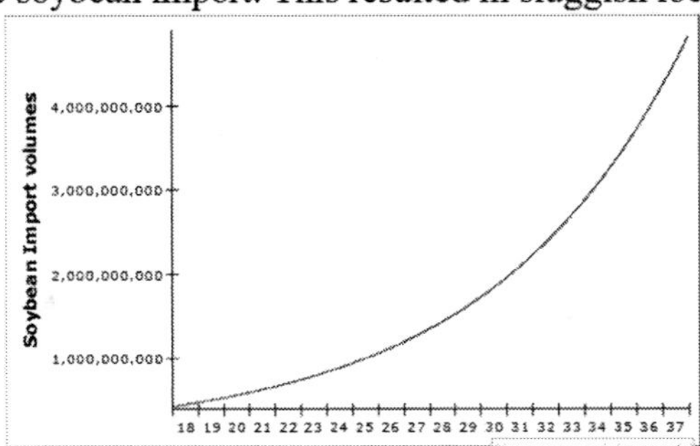

Figure 4. Soybean Import Volume

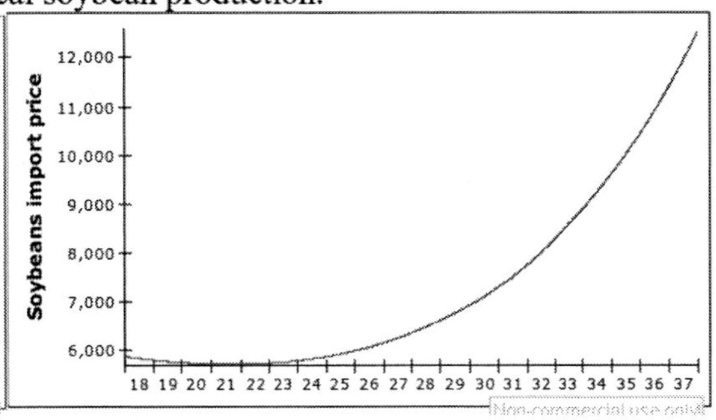

Figure 5. Soybeans Import Price

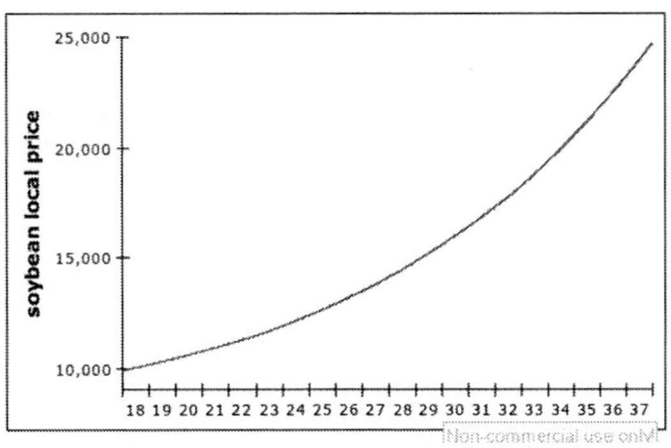

Figure 6. Soybean Local Price

4.4.1.2 Scenarios of with dummy monopoly bulog intervention on soybeans import price and soybean producer prices.

The variable dummy monopoly Bulog has an effect on increasing soybean local price, decreasing import volumes and increasing import prices. This shows that the monopoly Bulog causes the soybean local price to increase. Bulog has a role in stabilizing domestic soybean prices, through the procurement, storage and distribution of soybeans. The aim is to ensure the availability of soybean, especially for the

IOP Conf. Series: Materials Science and Engineering **943** (2020) 012046 doi:10.1088/1757-899X/943/1/012046

tofu and tempeh industry, especially for Kopti members. If Bulog manages soybeans, a quota policy can be implemented, so that import volumes can be controlled.

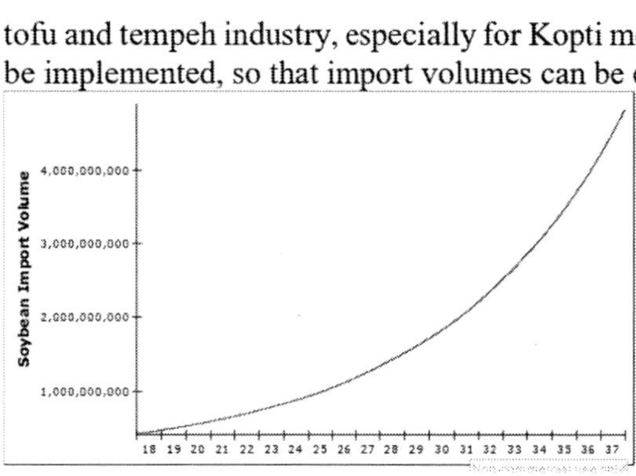

Figure 7. Soybean Import Volume

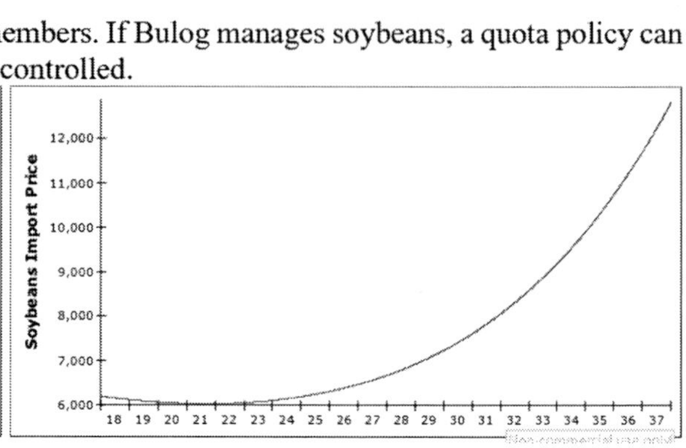

Figure 8. Soybeans Import Price

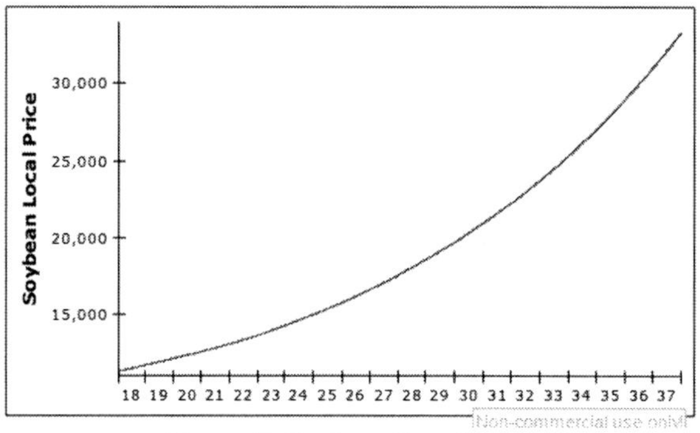

Figure 9. Soybean Local Price

4.4.1.3 Scenarios for increasing soybean import tariffs 20% and dummy monopoly Bulog intervention.

The government is currently implementing an import tariff policy 0.5% for soybean commodity to prevent the amount of import soybeans and effort to increase domestic soybean production. By raising import tariffs 20%, so there is an increase in the soybean import prices. This shows that the increase in the soybean import prices will stimulate an increase in the soybean local price, so farmers are interested in planting soybeans again, the implication of this is an increase in soybean production. This is consistent with observations at the industry level that soybean local price provisions are based on market mechanisms by following soybean import prices with soybean local price slightly above import prices.

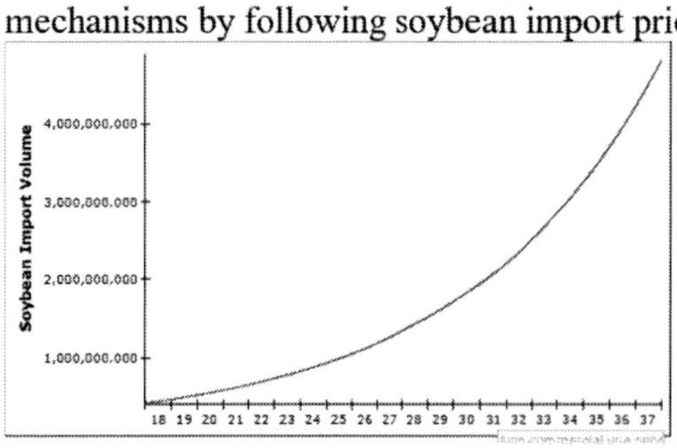

Figure 10. Soybean Import Volume

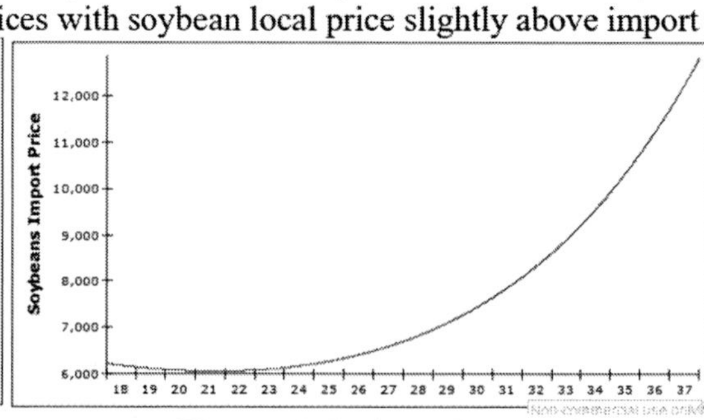

Figure 11. Soybeans Import Price

4.4.1.4 Scenario of increasing soybean local prices.

By raising the soybean local prices properly, it is estimated that farmers will benefit properly and consumers buy at affordable prices. If the government tries to increase local soybean production and farmer income, it is necessary to have a policy to determine the soybean base price that benefits farmers.

Predict soybean prices with 2 conditions: (1) Condition B: Soybean prices have increased randomly to normal distribution with a mean of 8.1% per year and a standard deviation of 1% per year; (2) Condition C: The average price increase in the form of a cycle with a price increase of 8.1% per year and a change of 5% per year in 5 years.

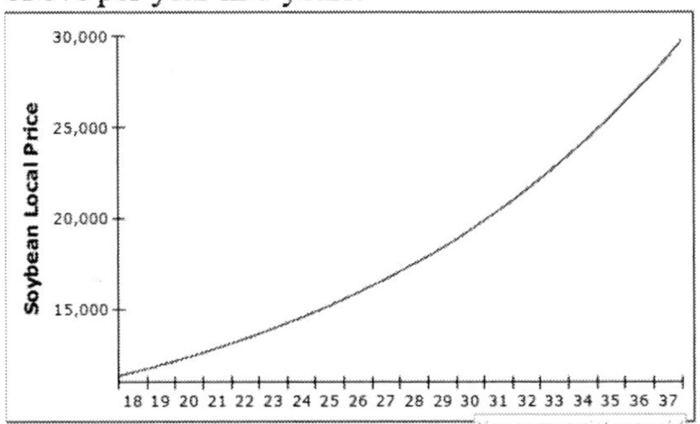

Figure 12. Soybeans Import Price Condition B

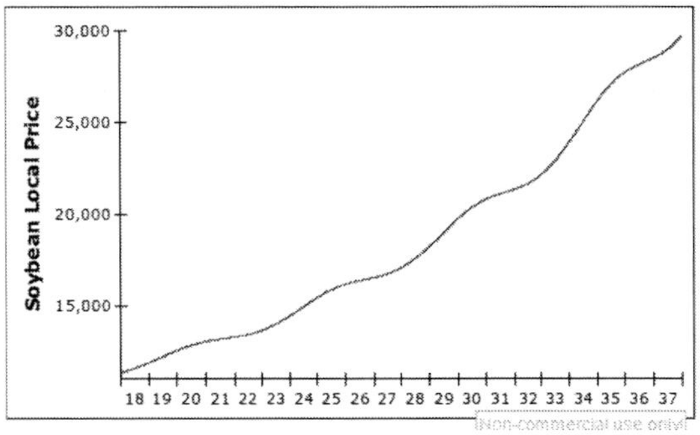

Figure 13. Soybeans Import Price Condition C

5. Conclusions

Based on simulation results, to look at the competitiveness of local soybeans and to support soybean price stability for the next 20 years, the government needs to take action as follows:

1. Dummy monopoly bulog intervention on soybeans import price and soybean producer prices. The variable dummy monopoly Bulog has an effect on increasing soybean local price, decreasing import volumes and increasing import prices.
2. Increasing soybean import tariffs 20%. This shows that the increase in the soybean import prices will stimulate an increase in the soybean local price, so farmers are interested in planting soybeans again, the implication of this is an increase in soybean production.
3. Increasing soybean local prices with predicting soybean price conditions. By raising the soybean local prices properly, it is estimated that farmers will benefit properly and consumers buy at affordable prices.

References

[1] Zainuri A, Wardhono A, Sutomo and Ridjal J A 2015 *Economics and Informatics* **7** pp 99-106
[2] Sudaryanto T dan Swastika D K S 2007 *Forum Agro Ekonomi (FAE)* **12** pp 1-27
[3] Amang B 1996 *Ekonomi kedelai di Indonesia* IPB Press Bogor
[4] Suryana A 2005 *Prospek dan arah pengembangan kedelai* Badan Penelitian dan Pengembangan Pertanian Departemen Pertanian Jakarta.
[5] Swastika D K S, Nuryanti S and Sawit M H 2007 *Teknik Produksi dan Pengembangan* pp 28-44
[6] Zakiah 2012 *Mimbar* **28** pp 77-84
[7] Elizabeth R 2007 *Pusat Analisis Sosial Ekonomi dan Kebijakan Pertanian* pp 165-173
[8] Zakaria, Amar K, Wahyuning K and Kustiari R 2010 *Jurnal Agro Ekonomi* **28** pp 21-37
[9] Hadipurnomo T 2000 *Tesis Institut Pertanian Bogor*
[10] Suryani E, Chou S Y and Chen C H 2009 *Expert Sistems with Applications* **37** pp 2324-2339
[11] Wulandari S S 2017 *JOM FISIP* **4** pp 1-15
[12] Forrester J W 1994 *System Dynamics Review Summer* **10** pp 1-14
[13] Zulkarnaini 2005 *Jurnal Teknik SIMETRIKA* **4** pp 356-359
[14] Mukti E T, Sjafruddin A and Kusumawati A 2014 *FSTPT Internasional Symposium* **2** pp 1045-1053
[15] Handayani D, Bantacut T, Munandar J M and Budijanto S 2007 *Jurnal Teknologi Industri Pertanian* **19** pp 7-15
[16] Hasan N, Suryani E and Hendrawan R 2015 *Procedia Computer Science* **72** pp 605-612
[17] Sutopo W, Bahagia S N, Cakravastia A and Arisamadhi T M A 2011 *Proceeding of the 2011 IEEE IEEM* pp 298-302

IOP Conf. Series: Materials Science and Engineering **943** (2020) 012046 doi:10.1088/1757-899X/943/1/012046

[18] Oktyajati N, Hisjam M and Sutopo W 2018 *The 3rd International Conference on Industrial, Mechanical, Electrical, and Chemical Engineering AIP Conference Proceedings* **1931**
[19] Utomo H, Udisubekti C M and Nurhadi S 2005 *Prosiding Seminar Nasional Manajemen Teknologi II* pp A91-A98
[20] Firdaus M 2015 *Journal of Basic and Applied Scientific Research* **5** pp 105-110

IOP Conf. Series: Materials Science and Engineering **943** (2020) 012047 doi:10.1088/1757-899X/943/1/012047

The Selection of Lithium Battery raw Materials by Environmental, Economic, and Social Sustainable

AI Sholichah[1], M Hisjam[2], W Sutopo[3]

[1]Master Program of Industrial Engineering, Faculty of Engineering, Universitas Sebelas Maret, Surakarta, Indonesia

[2,3]Departement of Industrial Engineering, Faculty of Engineering, Universitas Sebelas Maret, Surakarta, Indonesia

Email: aisyahnanik95@gmail.com

Abstract. The development of technology is done to advance knowledge and simplify life, and to develop the issues that are around. The development of electric vehicles must be encouraged by the development of batteries used to run these vehicles. The batteries used are usually made from lithium, because it has a strong electrical shelf life, not easily damaged, fast charging power, and so forth. This study evaluates and selects the best raw materials for lithium batteries from three selected raw materials, namely Lithium Cobalt Oxide (LCO), Lithium Iron Phospate (LFP), and Lithium Nickel Manganese Cobalt Oxide (NMC), because these materials are often used to make good lithium batteries and have some things in common. The approach used to choose the right standard material by knowing the life cycle of each material. This life cycle consists of three elements namely environmental, social, and economic. An environmental approach to find out whether the material used will have an impact on the surrounding environment such as air, water pollution and so forth. An economic approach to find out which materials are more economical so as not to burden the company and consumers who use them. The social approach is carried out to find out information about the material has been known by the consumer related to applicable standards.

1. Introduction

Problems in this world are increasingly complex and complicated to be solved with just one point of view. The population of the earth is currently faced with environmental problems that are crisis, such as global warming, disruption of the ozone layer and tropical rain forests, water and air pollution, and social problems such as poverty in developing countries, inter-religious conflicts, and so on [1]. The problem of air pollution is often found throughout the world. Developing countries have higher levels of air pollution. The impact of air pollution can disrupt human health, plants, aesthetics and comfort, or damage property. One of the causes of air pollution is smoke from burning vehicles. Because most vehicles at the moment the fuel does not burn completely on the engine which causes harmful smoke. There are 3 hazards caused by vehicle fumes namely: vehicle exhaust gases are casinogenic, damage to the circulatory system, and trigger damage to the respiratory system [2].

Seeing from this problem, an electric vehicle was made to replace today's commercial vehicles. The development of electric vehicle technology is very fast and has spread widely throughout the world. The more development of electric vehicle technology, the more developed batteries for electric vehicles. Batteries are energy sources that can convert the stored chemical energy into electrical energy that can be used like an electronic device, so there is no need to connect the power cord to the

Published under licence by IOP Publishing Ltd

terminal to be able to activate electronic devices. Each battery consists of a positive terminal (cathode) and negative terminal (anode). The output of the electric current from the battery is direct current or also known as DC current (Direct Current). At this time lithium batteries have been found as a good battery for electric vehicles [3].

Lithium-Ion batteries began to be developed in 1912. Li-ion batteries are batteries that can be removed like laptop batteries, smartphones and tablets. Li-ion batteries have different compounds, such as Lithium Cobalt Oxide (LCO), Lithium Manganese Oxide (LMO), Lithium Nickel Manganese Cobalt Oxide (NMC), Lithium Iron Phospate (LFP), Lithium Nickel Cobalt Aluminum Oxide (NCA)), and Lithium Titanate (LTO) [4]. Lithium batteries continue to be developed to suit electric vehicles and not only in terms of their development that must be considered, but the life cycle of lithium batteries also needs to be considered [5].

Life cycle analysis is an analysis of environmental impacts during the product life cycle from the extraction of raw materials to the end of their useful life. The environmental life cycle assessment (LCA / ELCA) has developed into an established tool and is widely used to assess the environmental impacts caused by products. Because it consists of all relevant 'product life' phases, such as production, use and disposal, which provide broad assessments and comprehensive results. By using three approaches namely economics (ELCA), environment (LCA), social (SLCA), and Life Cycle Costing (LCC) can arrange equipment to assess the sustainability of a product's life cycle [6]. This idea is in accordance with the 'three pillars' model of sustainability which considers environmental, economic, and social problems as three dimensions for modern sustainable concepts. Sustainability is the endurance of a system and process [7]. The principle of a sustainable development is the first environmentally sustainable where the extraction of raw materials from natural environmental resources for human needs for these resources has limitations. So that sustainable environment is implemented to protect the environment from over-exploitation and utilization of resources. Lithium batteries are used to reduce air pollution, one of the causes is conventional vehicle smoke, and to find out which lithium material is better and healthier for the environment [8].

Second, sustainable social which guarantees social justice in the distribution of wealth and services. Knowing the problems that occur in social related to the selection of raw materials to be used in batteries. Awareness of the social problems that often occur in the economic sector in certain countries, helps in the selection of supplies with more efficient social standards. In addition, the estimation of social problems helps producers anticipate social problems that might occur during future supply chain distribution, and prepare actions to solve them [9].

Third is a sustainable economy which maintains stable economic growth by restructuring productive systems to save energy resources. Where this aspect is used to assess the economic performance of the product. There are three perspectives taken, namely from the point of view of customers, producers and larger entities such as public authority or society as a whole. This aspect also influences the previous aspects, which are environmental and social. To avoid this, external effects need to be considered in determining the most cost-effective option for the community [10].

Based on the types of lithium raw materials used to make selected 3 raw materials which are often used to make batteries, namely Lithium Cobalt Oxide (LCO), Lithium Nickel Manganese Cobalt Oxide (NMC), and Lithium Iron Phospate (LFP). This research was conducted to determine the differences in the life cycle of the battery raw material viewed in terms of environmental, social, and economic.

2. Methodology

This type of research is a qualitative descriptive study, the research used to analyze data by describing or describing collected data that is applicable to the public or generalization (Sugiyono, 2008). Data obtained from previous studies that discuss lithium batteries for electric vehicles. This research discusses the raw materials of Lithium Cobalt Oxide (LCO), Lithium Nickel Manganese Cobalt Oxide (NMC), and Lithium Iron Phospate (LFP) to use lithium batteries, as well as the raw materials used as raw materials for lithium batteries. Then do an analysis of three life cycle cycles (ELCA, LCC, SLCA) to improve the performance of the raw materials to be used. Where offers several advantages such as battery cycle perspective, cycle perspective, and manage useful software and databases.

Figure 1. The sustainble Process

Then compare the results of the analysis of the three materials that will be used in making lithium batteries. Is it in accordance with what is desired and how the level of security and economics of these raw materials.

3. Results and Discussion

This research describes lithium-ion batteries made from Lithium Cobalt Oxide (LCO), Lithium Nickel Manganese Cobalt Oxide (NMC), and Lithium Iron Phospate (LFP). The following table explains the slight differences in raw material for batteries:

Table 1. Data of raw materials for batteries

Spesification	Lithium Cobalt Oxide (LCO)	Lithium Iron Phosphate (LFP)	Lithium Nickel Manganese (NMC)
Voltage, nominal	3.60 V	3.20V	3.60V, 3.70V
Specific energy (capacity)	150-250 Wh/kg	90-120 Wh/kg	150-220 Wh/kg
Charge (C-rate)	0.8 C, 1C maximum, 4.20V peak, 3 hr charge	1C typical, 3.65V peak, 3 hr charge time	1C, 4.20V peak, 3 hr charge time
Discharge (D-rate)	1C, 2.5V cut off	25-30C continuous, 2V cut off	2C continuous, 2.5V cut off
Cycle life	500-1000 related to depth of discharge, load, and temperature	1000-2000 related to depth of discharge and temperature	1000-2000 related to depth of discharge and temperature
Thermal runaway	150 °C (302°F) full charge promotes thermal runaway	270°C (518°F) very safe battery even fully charged	210°C (410°F), high charge promotes thermal runaway
Applications	Mobile phones, laptops, cameras	Portable and stationary needing high load currents and endurance	E-bikes, medical device, Evs, Industrial
Comments	Very high specific energy, limited specific power, cobalt is expensive	The very flat voltage discharge curve, but low capacity. One of safest Li-ions. Elevated self-discharge.	Provides high capacity and high power. Serves as Hybrid cell. This chemistry is often used to enhance Li-manganese.

The high specific energy makes Li-cobalt the popular choice for mobile phones, laptops and digital cameras. The battery consists of a cobalt oxide cathode and a graphite carbon anode. The cathode has a layered structure and during discharge, the lithium ions move from the anode to the cathode. The flow reverses on charge. The drawback of Li-cobalt is a relatively short life span, low thermal stability and limited load capabilities (specific power). Li-cobalt should not be charged and discharged at a current higher than its C-rating. This means that an 18650 cell with 2,400mAh can only be charged and discharged at 2,400mA. Forcing a fast charge or applying a load higher than 2,400mA causes overheating and undue stress. For an optimal fast charge, the manufacturer recommends a C-rate of 0.8C or about 2,000mA. The performance of Li-cobalt in terms of specific energy or capacity that relates to runtime, specific power or the ability to deliver high current, safety, performance at hot and cold temperatures, life span reflecting cycle life and cost.

Lithium Nickel Manganese Cobalt Oxide (NMC)
One of the most successful Li-ion systems are a cathode combination of nickel-manganese-cobalt (NMC). Similar to Li-manganese, these systems can be tailored to serve energy cells or power cells. NMC in an 18650 cell for moderate load condition has a capacity of about 2,800mAh and can deliver 4A to 5A; NMC in the same cell optimized for specific power has a capacity of only about 2,000mAh but delivers a continuous discharge current of 20A. The secret of NMC lies in combining nickel and manganese. An analogy of this is table salt in which the main ingredients, sodium and chloride, are toxic on their own, but mixing them serves as seasoning salt and food preserver. Nickel is known for its high specific energy, but poor stability; manganese has the benefit of forming a spinel structure to achieve low internal resistance but offers a low specific energy. Combining the metals enhances each other strengths. NMC is the battery of choice for power tools, e-bikes and other electric powertrains. The cathode combination is typically one-third nickel, one-third manganese and one-third cobalt, also known as 1-1-1.

Lithium Iron Phosphate (LFP)
Li-phosphate is more tolerant to full charge conditions and is less stressed than other lithium-ion systems if kept at high voltage for a prolonged time. As a trade-off, its lower nominal voltage of 3.2V/cell reduces the specific energy below that of cobalt-blended lithium-ion. With most batteries, cold temperature reduces performance and elevated storage temperature shortens the service life, and Li-phosphate is no exception. Li-phosphate is often used to replace the lead acid starter battery. Four cells in series produce 12.80V, a similar voltage to six 2V lead acid cells in series. Vehicles charge lead acid to 14.40V (2.40V/cell) and maintain a topping charge. Time will tell how durable Li-Phosphate will be as a lead acid replacement with a regular vehicle charging system. Cold temperature also reduces the performance of Li-ion and this could affect the cranking ability in extreme cases.

4. Conclusion
Based on this research, it is known that lithium NMC and LFP batteries are both the most stable lithium batteries, but some projects may require a higher energy density so they adapt lithium NMC and some projects require a longer life cycle so they adjust LFP batteries. For LCO has been just to small production like robot, games, and others. Price for NMC and LFP is mostly same because they have combination material and for each ingredient the price is mostly same. For the environmental approach phosphate, manganese and iodide has different characteristics. Such as phosphate is more likely to be cold, manganese tends to be hot because it is made of metal, and iodide tends to be room temperature. And for social people or consument mostly know LFP and NMC, because NMC almost same with Nickel Cobalt Aluminium (NCA). Because NCM just needs 100 cell battery and for LFP needs 200 cell battery.

5. References
[1] Segar,B.N. 2015 *Education for Sustainable Development (ESD) sebuah Upaya Mewujudkan Kelestarian Lingkungan* Social Science Education Journal
[2] Klopffer, W. 2002 *Life-Cycle Based Method for Sustainable Product Development* Int. J. LCA 8(3) 157-159
[3] Hello sehat Online 20/06/2019 *3 Bahaya Utama yang Ditimbulkan Asap Knalpot Kendaraan* https://hellosehat.com/hidup-sehat/fakta-unik/3-bahaya-asap-knalpot-bagi-kesehatan/
[4] Valdicvidia, S. Soenneman, G. Et. Al 2011 *Toward a Life Cycle Sustainbility Assesment* UNEP/SETAC Life Cycle Initiative
[5] Online 04/11/2019 Battery University **https://batteryuniversity.com/learn/article/types_of_lithium_ion**

[6] Benjamin, R. 2016 *Assesment of Sustainability Issue for The Selection of Materials and Technologies during Product Design: a Case Stusy of Lithium-Ion Batteries for Electric Vehicles* Int. J. Interact Des Manuf 10:217-227
[7] Benoit, C., Mazjin, B., et al, 2009 *Guidelines for Social Life Cycle Assesment of Products* UNEP/SETAC Life Cycle Initiatve
[8] U.S., 2013 *Geological Survey* Mineral Commodity Summeries
[9] Reuter, B., Riedl, J., Hamacher, T., Lienkamp, M., 2014 *Future Resource Availability for The Production of Lithium-Ion Vehicle Batteries* In: Proeeding of The Coference on Future Automotive Technology Munich German
[10] Morley, N., Eatherley, D., 2008 *Material Security: Ensuring Resource Availability for The UK Economy* Knowledge Transfer Ntworks Chester

A Literature Review: Commercialization Study of Electric Motorcycle Conversion in Indonesia

A Habibie[1] and W Sutopo[2]
[1] Master Program of Industrial Engineering, Faculty of Engineering, Universitas Sebelas Maret Surakarta, Indonesia
[2] Department of Industrial Engineering, Faculty of Engineering, Universitas Sebelas Maret Surakarta, Indonesia

Email: achmad.habibie.mail@gmail.com

Abstract. The number of motorcycles in Indonesia increased very significant in Indonesia at the last 10 years, reaching 300%. Automatically, its will make carbon pollution in Indonesia increasing. Electric motorcycle is one that can be used to remove carbon pollution. Indonesian government very concern about this by issuing Presidential Regulation Number 55 of 2019 for push and accelerate the electrification all vehicle in Indonesia including motorcycle. The goal of this study is to see the development of research on converting Internal Combustion Engine (ICE) motorcycles into electric motorcycle, both in the world and in Indonesia. The method used is quantitative descriptive statistics through Scopus web. The results of this research are expected to be a reference for research in Indonesia that want to be immediately prepare for electric motorcycle conversion. In this research also give taxonomy and classification based on Goldsmith Model as a technology commercialization model and SWOT analysis for implementation electric motorcycle conversion program in Indonesia.

1. Introduction

The number of motorcycles in Indonesia increased from 2007 to 2017 significantly. In 2007 the number of motorcycles in Indonesia is 41 million units. In 2017 the number of motorcycles in Indonesia become more than 113 million units. Based on this trend, it is not impossible that the number will continuously increase in years forward.

With the increasing number of motorcycles, automatically carbon emissions will also increase. Whereas at present the governments in all countries are reducing gas emissions from transportation in their respective countries [1]. As a national action, the government of Indonesia plans to suppress carbon gas emissions. It is stated in Presidential Regulation Number 61 of 2011 [2]. The action is designed from transportation sector and targets a reduction of $cO2$ by 0.038 up to 0.056 gigatons during 2010-2020. In 2019, through Presidential Regulation Number 55 Indonesia's government push for accelerate the electrification all vehicle in Indonesia [3]. Transportation sector is the one of the most pollution source in the world [4]. One of the contributors of carbon gas emissions from the transportation sector is Internal Combustion Engine (ICE) motorcycles. The usage of electrical technology is one that can be used on ICE motorcycles to reduce carbon emissions. This study will discuss the conversion ICE motorcycle into an electric motorcycle in Indonesia.

It is called electric motorcycle because electricity used as a replacement of fuel in ICE motorcycle [5]. Electric vehicles have low noise level and high energy saving [6]. The implementation of electric vehicles will reduce gas emissions from transportation and also will support the development of electricity establishment [5]. Electric motorcycle conversion is a new technology for Indonesia. Goldsmith commercialization model is one of the most effective models to identified technology commercialization. The Goldsmith commercialization model is a model that integrates technical aspects, markets and business process elements for commercialization into a sequential concurrent activity matrix [7].

2. Method

The methodology in this research is study of literature through the page www.scopus.com. Scopus is the largest abstract database, citations and literature for tracking and analyzing research [8].
The flow of research as follows:

- Search at the Scopus for the keywords *"electric motorcycle"* from first paper at 1989 until August of 2019.
- Scopus data analysis.
- Paper taxonomy and classification based on goldsmith model.
- SWOT analysis for implementation electric motorcycle conversion in Indonesia.

3. Results and Discussion

3.1. *Data of Scopus*

From searching for data through the Scopus web, there are 286 studies about electric motorcycles. The following is the result of the searching:

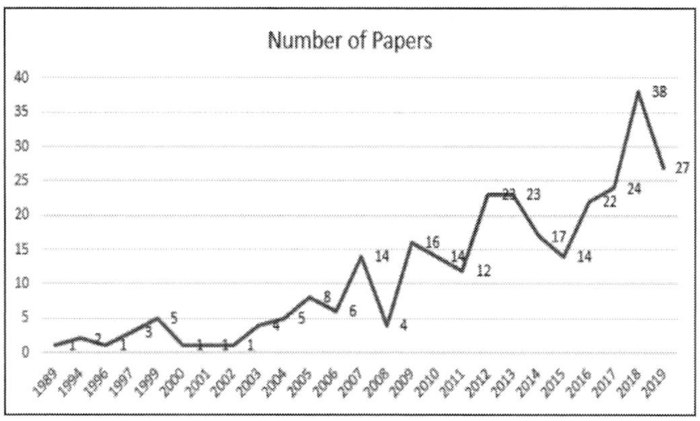

Figure 1. Number of Papers Based on Years

Figure 1 shows that research on electric motorbikes began in 1989. The research was conducted in Japan with the title "The concept of a future electric vehicle and the development of an electric motorcycle". This paper studied about comparation between electric vehicle performance and ICE vehicle performance [9]. After 2004, research on electric motorcycles began to be consistently studied and after 2010 research about electric motorcycle increased. It is showed by the number of studies presented in more than ten papers each year. In early 2019, there are already 27 papers presented.

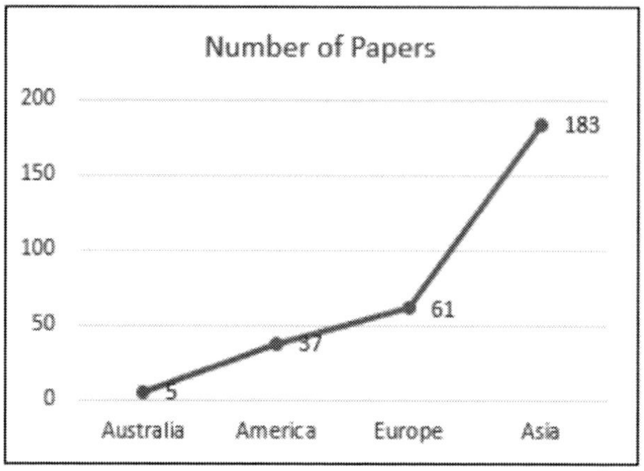

Figure 2. Number of Papers Based on Continents

From the figure 2, it shows that Asia has the biggest courage in studying about electric motorcycle. Asia has presented 183 papers about it. Europe becomes runner up by presenting 61 papers and America has presented 37 papers. Australia become the lowest continent that studied about it. Australia has presented only 5 papers.

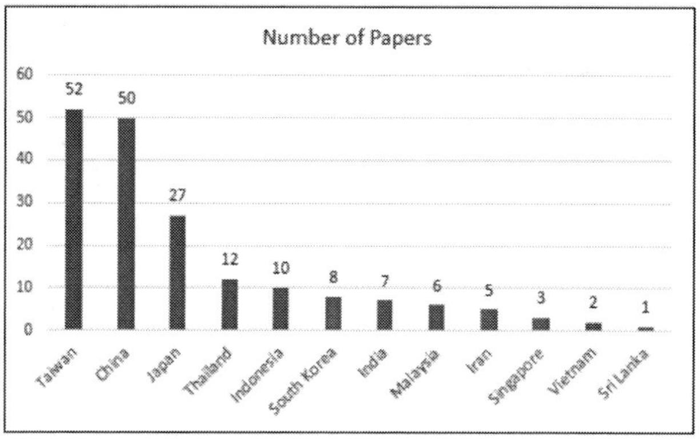

Figure 3. Number of Papers in Asia

Asia is the greatest continent in doing study about electrical motorcycle. About 183 studies have been released in Scopus. Based on Figure 3, the country that release the greatest amount is Taiwan with 52 papers. Recently, Taiwan becomes model country that is already implementing electric vehicle as on their mass transportation. China becomes runner up with 50 papers released and the third is Japan with 27 papers. In South East Asia, Thailand becomes the greatest country that release 12 papers. Indonesia has released 10 papers about electric motorcycle. Malaysia has released 6 papers, Singapore has released 3 papers, and the last Vietnam has released 2 papers.

3.2. Research Classification

There are 286 studies about electric motorcycle. It will be classified to 6 topics. The following result of classification shown in the table below:

Table 1. Classification of Paper

Topic of Paper	Number of Papers
Product Design and Simulation Study	144
Feasibility and Market Study	56
Environmental Issue Study	38
Implementation Impact Study	34
Supply Chain Study	9
Ergonomic Issue Study	5

Based on the Table 1, the biggest topic is Product Design and Simulation Study. The number of papers in this topic is 144 papers or more than 50% from all topic. The second topic is Feasibility and Market Study, the papers released is 56 papers or 20%. Environmental Issue Study about electric motorcycle has released 38 papers or 13%, Implementation Impact Study has released 34 papers or 12%, and Supply Chain Study has released 9 papers or 3%. The smallest topic is Ergonomic Issue Study, the papers in this topic only 5 papers or 2%.
There are some previous papers in each topic. The following papers shown in the table below:

Table 2. Previous Paper

Title and Author	Topic of Paper	Content of Paper
A study of electric motorcycle (Matsuda, 2014)	Product Design and Simulation Study	In this research, developed the prototypes of electric motorcycle about technical design and parts suppliers. The developed electric motorcycle has the equivalent with ICE motorcycle 250cc and can travel distances of up to 100km. The result of this paper is prototype and simulation with environmental loads such as water problem, cold and heat situation, shock problem, and impact of an accident [10].
A Study of Function Control in the Electric Motorcycle (Kawai et al., 2015)	Product Design and Simulation Study	This paper shows that control in electric motorcycle simpler than ICE motorcycle. This study develops the prototype electric motorcycle having manual transmission. The problem points of electric motorcycle which put electric propulsion motor and manual transmission together at first tested [11].
Design and Analysis of Chassis for an Electric Motorcycle (Jeyapandiarajan et al., 2018)	Product Design and Simulation Study	This paper designs an electric motorcycle chassis with structurally stable. This chassis simulation with frame forces and geometries suspension. This paper also makes other material comparation to find the best suitable component for electric motorcycle chassis (strength, cost, and weight) [12].

To produce or convert: A case for large scale electric motorcycle conversion in Singapore (Teo et al., 2014)	Feasibility and Market Study	This paper discusses about the electric motorcycle economic feasibility and viability. In this research, simulated by producing per year of 2500 motorcycles and produced by 20 people. The result is electric motorcycles more economical than ICE motorcycles. In a short period, this cost savings can offset the motorcycle price [13].
Techno-economic analysis of lithium-ion battery for motorcycle development (Kurniyati, et al., 2016)	Feasibility and Market Study	This paper explores about development lithium-ion battery technology for motorcycles in Indonesia. This technology development is conducted by reviewing the technical aspects, market aspects, and aspects of the business. The financial analysis that used in this paper are the analysis of NPV, PP, and IRR [14].
Electric vehicles, air pollution, and the motorcycle city: A stated preference survey of consumers' willingness to adopt electric motorcycles in Solo, Indonesia (Guerra, 2019)	Feasibility and Market Study	This paper discusses the replacement for ICE motorcycle. Five scenarios given to respondents what they selected between electric motorcycle, ICE motorcycle, and no motorcycle based on speed, price, charge time, and range. The results is a convincing market in Solo for electric motorcycle. However, the motorcycle must be low price and powerful [15].
Exploring the determinants of consumers' WTB and WTP for electric motorcycles using CVM method in Macau (Zhu et al., 2019)	Feasibility and Market Study	This study discussed the willingness to pay (WTP) factors and willingness to buy (WTB) factors consumers for electric motorcycle. The results indicate the respondents don't quite understand about electric motorcycle. When interesting electric motorcycle, the respondents more concern about actual cost of electric motorcycle (repair fee, tax incentives, sale price, charging fee, etc.). The load capacity and driving speed have just a little attention [16].
Environmental impact of converted electrical motorcycle (Xuan et al., 2014)	Environmental Issue Study	This study discusses the environmental impact ICE sports motorcycle conversion to electric sports motorcycle. This conversion will reduce gas emissions. However, impacts and energy sources during post-conversion usage must be considered. The result of conversion will reduce gas emissions and life cycle energy used [17].

ICE Motorcycle and Electric Motorcycle: Environmental and Economic Analysis (Tuayharn et al., 2015)	Environmental Issue Study	This research studies the comparation using electric motorcycle and ICE motorcycle in Thailand. The fuel consumption of ICE motorcycle's is 1.94 Lge/100km with CO2 released equivalent to 43.59 g/km. On the other side, the electric motorcycle consumed 0.27 Lge/100km with CO2 released is at 20.67 g/km if use coal generated electricity, and 10.7 g/km if use natural gas generated electricity. Total cost of electric motorcycle save up to 20% over 10 years [18].
Environmental impact of electric motorcycles: Evidence from traffic noise assessment by a building-based data mining technique (Sheng et al., 2016)	Environmental Issue Study	ICE motorcycles make a high noise because its traffic at the city. The result of this study indicated that electric motorcycle can reduce the human risk to high traffic noise in the city until 22.7% than before [19].
Study of the Energy Efficiency and Greenhouse Emissions from Motorcycles Powered by Electric and Internal Combustion Engines (Daemme et al., 2017)	Environmental Issue Study	This paper discusses about comparation ICE motorcycle and electric motorcycle in Brazil. The result indicated the difference of gas emissions significantly between both of them. The efficiencies of ICE motorcycles are 13%-17% and for electric motorcycle showed 47% [20].
Energy and emission impacts of liquid fueled engines compared to electric motors for small size motorcycles based on the Brazilian scenario (de Assis et al., 2019)	Environmental Issue Study	This paper considers environmental impact for evaluation of greenhouse gases in Brazil. The result shows that electric motorcycles have more energy saving than ICE motorcycle. Gas emissions from electric motorcycle are not dangerous for nature, but the impact from battery must be considered further [21].
Failure safety method according to battery's state-of charge condition in electric motorcycle (Lee and Kim, 2014)	Implementation Impact Study	This paper explores about electric motorcycle battery safety based on the status of battery. This study simulated by Automotive Simulation Models (ASM) program and used Model Based Design (MBD). The result is if Li-ion battery discharges until 10% will have a significant problem [22].
Mobile phone use among motorcyclists and electric bike riders: A case study of Hanoi, Vietnam (Truong et al., 2016)	Implementation Impact Study	This paper explores the factors related to mobile phone used by motorcycle rider and electric bike riders. This paper also studies their prevalence. The method of this research is observation survey at 12 cross sectional sites. The result of prevalence when riding

		and using a mobile phone was 8.4% (95% CI: 8.06-8.74%) [23].
Evaluation and selection of suitable battery for electrics motorcycle in Taiwan - application of fuzzy multiple attribute decision making (Tzeng et al.,1997)	Supply Chain Study	This paper describes that general qualities of electric motorcycles are lower than the qualities of ICE motorcycle. The motors and batteries technology are the main reason in this problem. The method of this study is TOPSIS method, fuzzy theory, analytic hierarchy process, and Borda method. The results show that the best of seven alternative is Lithium battery, and the second one is the Nickel-Zinc battery [24].
Electric vehicle charging strategy for isolated systems with high penetration of renewable generation (Clairand, 2018)	Supply Chain Study	This paper explores an electric vehicle charging strategy given the Renewable Energy Sources (RES) high penetration. The result proposes take CO2 emissions tax. If mostly used by fossil fuel, it will be high taxed. And if mostly used RES power source, it will be low taxed [25].
Sustainable supply chain planning for swap battery system: Case study electric motorcycle applications in Indonesia (Prianjani et al., 2019)	Supply Chain Study	This time, electric motorcycle in Indonesia have a low progress. The chance for accelerate is planning sustainable supply chains by regarding the triple bottom line, such as environmental, social, and economic. This literature study used as the method to measure the swap battery system supply chain and create a sustainable supply chain scheme for electric motorcycle [26].
Impact of electric motorcycles on distribution network and design of a charger for electric motorcycles (Wijesekera and Binduhewa, 2019)	Supply Chain Study	This paper shows the comparation of charging pattern between electric motorcycle and electric car because of their charging time difference. And automatically the collective power profile will be different. The ability of slow and fast charging on electric motorcycle are showed in this paper [27].
The study of ergonomic evaluation and critical design factors for electric scooter (Hsu et al., 2016)	Ergonomic Issue Study	This paper shows the electric scooter design has 23 factors and has its own priorities. Furthermore, while riding the electric scooter, its postures have been evaluated. The result indicates the significant factors are comfort and convenience, the design of vehicle control, dimensions of vehicle space and performance, and the design of dashboard [28].

Work-related road safety: The impact of the low noise levels produced by electric vehicles according to experienced drivers (Pardo et al., 2019)	Ergonomic Issue Study	This paper shows that the low noise levels of electric vehicle could take a new safety risk even though it seen as positiveness previously. This study explores the experience of electric vehicle drivers and hybrid vehicle drivers. The total of respondents in Malaga, Spain is 95 drivers. The result indicates "medium" risk situations at low speeds with pedestrians because of the low noise [29].

3.3. Taxonomy and classification based on Goldsmith Model

Goldsmith model can describe well about technology commercialization [7]. Goldsmith commercialization model consists of three phases consisting of the concept phase, the development phase and the growth phase. In each phase of the commercialization model analyses the technical, market and business aspects so that the commercialization process can be carried out effectively. The combination of these three phases and aspects forms a matrix model that described by Figure 4. There are 6 papers described in Step 1-Technology Analysis, 12 papers explored Step 2-Market Needs Assessment, 38 papers studied in Step 3-Venture Assessment. Step 4-Technical Feasibility at the most paper by 86 papers. There are still no paper at Step 17-Market Diversification and Step 18-Business Maturity.

Stage	Technical	Market	Business
Concept Phase			
Stage 1 Investigation	Step 1 Technology Analysis	Step 2 Market Needs Assessment	Step 3 Venture Assessment
	6 papers	*12 papers*	*38 papers*
Development Phase			
Stage 2 Feasibility	Step 4 Technical Feasibility	Step 5 Market Study	Step 6 Economic Feaseibility
	86 papers	*18 papers*	*15 papers*
Stage 3 Development	Step 7 Engineering Prototype	Step 8 Strategic Market Plan	Step 9 Strategic Business Plan
	51 papers	*10 papers*	*3 papers*
Stage 4 Introduction	Step 10 Business Start Up	Step 11 Pre Production Prototype	Step 12 Market Validation
	5 papers	*5 papers*	*3 papers*
Growth Phase			
Stage 5 Growth	Step 13 Production	Step 14 Sales and Distribution	Step 15 Business Growth
	5 papers	*2 papers*	*3 papers*
Stage 6 Maturity	Step 16 Production Support	Step 17 Market Diversification	Step 18 Business Maturity
	14 papers	*0 papers*	*0 papers*

Figure 4. Taxonomy Based on Goldsmith Model

3.4. SWOT Analysis for Implementation Electric Motorcycle Conversion in Indonesia

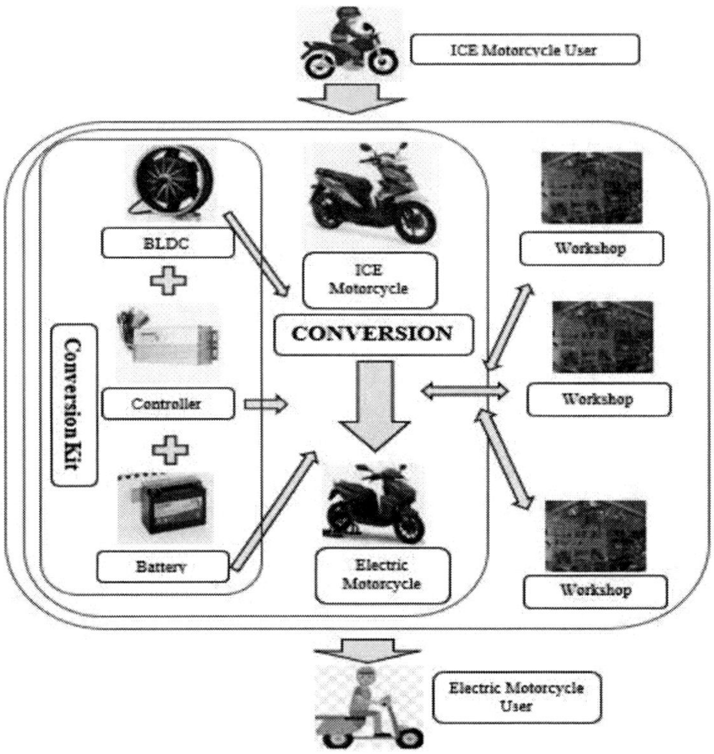

Figure 5. Electric Motorcycle Conversion Diagram

Electric motorcycle conversion is creating electric motorcycle by change the power source of ICE motorcycle. ICE motorcycle can be converted to electric motorcycle by adding Conversion Kit. The conversion kit consists of Brushless Direct Current Motor (BLDC), controller, and battery. The function of BLDC is generating the power for motorcycle. BLDC converts supplied electrical energy into mechanical energy. BLDC usually installed at rear wheel of motorcycle. The controller function as set the power from BLDC to motorcycle power. The last part of conversion kit is battery. Battery is the power source of electric motorcycle, if it installed greater battery capacity, automatically it will be produced greater power for electric motorcycle. The main part of conversion kit is battery because battery is the most expensive than the other. The diagram of electric motorcycle conversion showed by Figure 5. The differentiation between electric motorcycle conversion, ICE motorcycle, and electric motorcycle showed by Table 3.

Table 3. Motorcycle Difference

Description	ICE Motorcycle	Electric Motorcycle Conversion	Electric Motorcycle
Power source	Gasoline	Gasoline converted to electric	Electric
Classification	Engine capacity	Battery capacity	Battery capacity
Motorcycle drive	Engine	Engine converted to BLDC	BLDC
Expense	Less efficient	More efficient	More efficient
Environment	Less green friendly	Greener friendly	Greener friendly
Maintenance	More maintenance	Less maintenance	Less maintenance

Many consumers in Indonesia not have attention yet about green product such as electric motorcycle. Even though, electric motorcycle is more environmentally friendly than ICE motorcycle. Electric motorcycle is 53% more green friendly than ICE motorcycle [18]. Moreover, implementation of electric motorcycle can reduce traffic noise. Electric motorcycle can reduce high traffic noise until 22.7% at the city [19].
In Indonesia, ICE motorcycle consumers have very attention about economic value. The following result of economic comparation shown in the figure below:

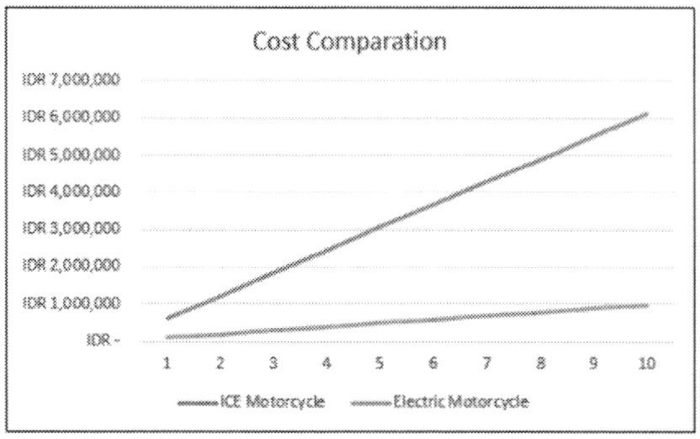

Figure 6. Cost Comparation ICE Motorcycle VS Electric Motorcycle

Based on Figure 6, could be seen that electric motorcycle more efficient than ICE motorcycle. The power source of motorcycle comparation between gasoline and electrical shown on Figure 5. In the end of the 10th month, electric motorcycle can save until 84% compared with ICE motorcycle.

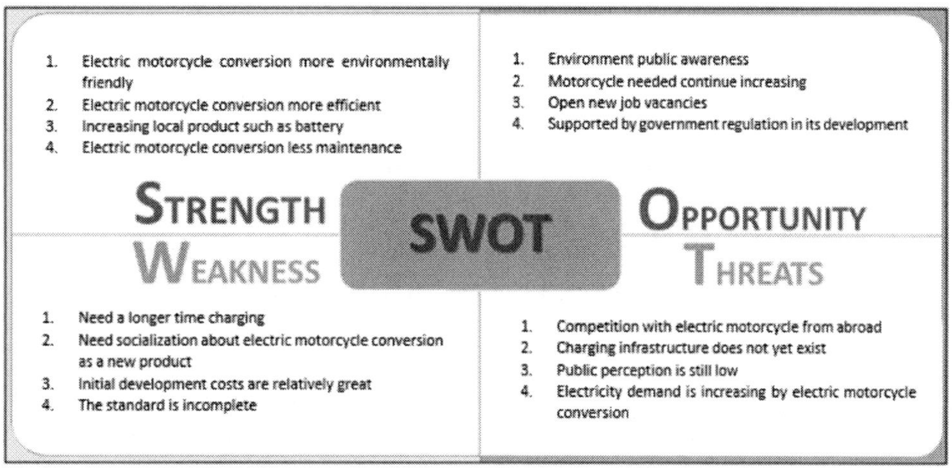

Figure 7. SWOT Analysis of Electric Motorcycle Conversion

Figure 7 above shows the SWOT analysis of implementation of electric motorcycle conversion in Indonesia. The explanation of each showed below:

3.4.1. Strength

- This conversion program will have an impact in environmental, economic, and social. As explained above electric motorcycle can reduce gas emission pollution until 53%. This electric motorcycle also can reduce 22.7% of high traffic noise. This conversion program is expected reduce energy dependence from non-renewable resources.

- From economic perspective, electric motorcycle is more efficient than ICE motorcycle. Consumers can save their budget for gasoline at ICE motorcycle until 84% by using electric motorcycle at the end of the 10th month. This conversion program can grow the new industry such as BLDC industry, controller for electric motorcycle industry, and battery industry as the main part of electric motorcycle. Based on Figure 4, private workshop as people's economy can grow by converting ICE motorcycle to electric motorcycle.
- There is some local product that can be used in electric motorcycle conversion such as battery. Battery is the main part of electric motorcycle conversion because battery is the most expensive than the other parts. By developing electric motorcycle conversion, automatically it will be increasing battery production as the main part from local product.
- Electric motorcycle conversion has a different maintenance with ICE motorcycle. In ICE motorcycle, it must take periodic maintenance and service such as oil change, engine cleaning, etc. But in electric motorcycle conversion, it doesn't need to do that because there is no engine. So electric motorcycle conversion is less maintenance than ICE motorcycle.

3.4.2. Weakness

- Fast charging technology is not ready in Indonesia. It will be a weakness of electric motorcycle conversion. Now, the charging time of battery is 5-6 hours. But if fast charging used, it just only takes 2-3 hours. Fast charging technology must be faster developed in Indonesia for supporting electric motorcycle conversion program.
- Socialization is one of the most important in this conversion program. Good socialization to the community will make this program succeed quickly. The socialization for example gas emissions awareness, government regulation, safety awareness of electric motorcycle, gas emissions awareness, etc. Generally, Indonesia's people confused trying new technology and by this socialization expected to receive it more quickly.
- As a new product and new technology, electric motorcycle conversion will take great initial development cost. But it will be paid off with healthier environment. Indonesia's government has also prepared a great budget for researching and developing it.
- Indonesia doesn't yet have a standard about electric motorcycle. International regulation standard of electric motorcycle is UN R136 "about uniform provisions concerning the approval of vehicles of category L with regard to specific requirements for the electric power train". This regulation standard consists of Electrical Safety and Rechargeable Electrical Energy Storage System (REESS) Safety. The others international standard related to electric motorcycle shown in the table below:

Table 4. International Standard Related to Electric Motorcycle

Topic	International Standard
Electrification safety (vehicle)	ISO 13063
Charging systems	IEC 60335-2-29(Rev.), IEC 61851-3, ISO 18246
DC charging connectors	IEC 62196-4
Cell testing & safety	IEC 62660-1,-2
Cell safety	IEC 62660-3
Battery testing & safety	ISO 18243
Electricity consumption	ISO 13064-1
Vehicle performance	ISO 13064-2

3.4.3. Opportunity

- Environment public awareness is increasing recently. Carbon emissions has disrupted public health. Around the world is also actively promoting about this environmental issue. Electric motorcycle conversion will be faster accepted because this awareness.

- Public needs for motorcycle transportation are increasing. It showed by motorcycle sales are increasing every year. From this, it will be advantageous for electric motorcycle conversion as an option motorcycle that more environmentally friendly.
- With the growth of new industries and private workshop, automatically will require labor and will open new job vacancies. This is one of the social impacts from this conversion program. It will This will reduce the number of unemployed people in Indonesia.
- Indonesia's government at least has released 7 regulation related to electric vehicle, such as Regulation Number 30 of 2007 about Energy [30], Presidential Regulation Number 61 of 2011 about National Action Plan for Reducing Greenhouse Gas Emissions [2], Government Regulation Number 79 of 2014 about National Energy Policy [31], Government Regulation Number 14 of 2015 about National Industrial Development Master Plan 2015-2035 [32], Regulation Number 16 of 2016 about Ratification of the Paris Agreement to The United Nations Framework Convention on Climate Change [33], Presidential Regulation Number 22 of 2017 about National Energy General Plan [34], and the last is Presidential Regulation Number 55 of 2019 about Acceleration of Electric Vehicle Program Based on Battery for Road Transportation [3]. Based on those regulation, Indonesia's government is very focus about conversion of electric vehicle in Indonesia. Government's expectation at 2035, at least 30% of road transportation use electric energy.

3.4.4. Threats

- Electric motorcycle had already been developed in the other country. Electric motorcycle product from other country automatically will become a competitor electric motorcycle conversion in Indonesia. The advantages of electric motorcycle conversion are public don't need purchase a new motorcycle but they can convert their old ICE motorcycle to electric motorcycle.
- The main problem before implementation electric motorcycle conversion is infrastructure. The infrastructure of electric motorcycle such as fast charging technology, public electric charging station, battery swap technology, etc. Indonesia's government must be serious and fast in infrastructure development so the people can trust more and faster too in making decision for electric motorcycle conversion.
- Public perception about electric motorcycle conversion is still low. Generally, public will think more before decide use a new product and new technology. But it can be prevented with socialization and advertisement about electric motorcycle conversion.
- If electric motorcycle conversion increasing automatically electricity demand will increase too. Government must be prepared and forecast the electricity demand so problem in electricity will not happen in the future.

4. Conclusion

From this study, it was found that electric motorcycle conversion includes new technology and is still highly developed. This literature review expected provide benefits in adding references to all parties involved in this conversion program. The future research about electric motorcycle conversion still very open such as feasibility and market study for each city in Indonesia, infrastructure study for supporting electric motorcycle conversion, implementation impact study, environmental study by using electric motorcycle in Indonesia, etc. For the government must be serious for electric motorcycle conversion so as not to lag behind other countries. Socialization to the public about electric motorcycle conversion should be intensive so that the public understands about this product. Indonesian government also should accelerate infrastructure development such as fast charging technology, public electric charging station, battery swap technology. Tax deduction for companies involved can also accelerate its growth. Indonesian National Standards (SNI) for electric motorcycle conversion must be formalized immediately.

References

[1] Bonilla O and Donald M N 2010 *Engineering Management Journal* 22:3 pp 34-44
[2] Presidential Regulation Number 61 of 2011 the Republic of Indonesia concerning National Action Plan for Reducing Greenhouse Gas Emissions *(Peraturan Presiden No. 61 Tahun 2011 tentang Tentang Rencana Aksi Nasional Penurunan Emisi Gas Rumah Kaca (RAN GRK)* in Bahasa Indonesia
[3] Presidential Regulation Number 55 of 2019 the Republic of Indonesia concerning Acceleration of Electric Vehicle Program Based on Battery for Road Transportation *(Peraturan Presiden No. 55 Tahun 2019 tentang Percepatan Program Kendaraan Berbasis Listrik (KBL) Berbasis Baterai untuk Transportasi Jalan)* in Bahasa Indonesia
[4] European Environment Agency 2010 *The Contribution of Transport to Air Quality –TERM 2012: Transport Indicators Tracking Progress*
[5] Ahmad A, Khan Z, Alam M, and Khateeb S 2018 *Smart Science* Vol 6 No 1 pp 36-53
[6] Song Y, Li J, Ji G, and Xue Z 2016 *China International Conference on Electric Distribution*
[7] Goldsmith H R 2003 *Arkansas Small Business and Technology Development Center*
[8] Elsevier 2017 *Scopus Content Coverage Guide*
[9] Shimizu H, Iikura Y, Naitoh M, and Ono M 1989 *SAE Technical Papers*
[10] Matsuda Y 2014 *SAE Technical Papers*
[11] Kawai D, Matsuda Y, and Murase T 2015 *SAE Technical Papers*
[12] Jeyapandiarajan P, Kalaiarassan G, Joel J, Shirbhate R, Felix Telare F, and Bhagat 2018 *Materials Today: Proceedings*
[13] Teo B S, Howe A S S, Inna S C R, Henz M, and Dieter Weigl J 2014 *World Electric Vehicle Symposium and Exhibition*
[14] Kurniyati I, Sutopo W, Hisjam M, and Astuti R W 2016 *Lecture Notes in Engineering and Computer Science* Vol 2 pp 751-755
[15] Guerra E 2019 *Transportation Research Part D: Transport and Environment* Vol 68 pp 52-64
[16] Zhu L, Song Q, Sheng N, and Zhou X 2019 *Energy Policy* pp 64-72
[17] Xuan P Y, Henz M, and Dieter W J 2014 *2013 World Electric Vehicle Symposium and Exhibition*
[18] Tuayharn K, Kaewtatip P, Ruangjirakit K, and Limthongkul P 2015 *SAE Technical Papers*
[19] Sheng N, Zhou X, and Zhou Y 2016 *Science of the Total Environment* pp 73-82
[20] Daemme L C, Penteado R, Schneider P S, Da Rocha B P, Piccoli B D S, Errera M R, and Corrêa S M 2017 *SAE Technical Papers*
[21] De Assis Brasil Weber N, da Rocha B P, Smith Schneider P, Daemme L C, and de Arruda Penteado Neto R 2019 *Energy* pp 70-79
[22] Lee T H and Kim B W 2014 *International Journal of Control and Automation* Vol 7 pp 41-48
[23] Truong L T, Nguyen H T T, and De Gruyter C 2016 *Accident Analysis and Prevention* Vol 91 pp 201-215
[24] Tzeng Gwo-Hshiung, Chen June-Jye, Teng Jin-Di, and Pan Jyh-Sheng 1997 *Journal of the Chinese Institute of Industrial Engineers* Vol 14 pp 319-331
[25] Clairand J M, Rodríguez-García J, and Álvarez-Bel C 2018 *Energies* Vol 11
[26] Prianjani D, Sutopo W, Hisjam M, and Pujiyanto E 2019 *IOP Conference Series: Materials Science and Engineering* Vol 495
[27] Wijesekera A and Binduhewa P 2019 *IEEE Region 10 Humanitarian Technology Conference, R10-HTC*
[28] Hsu C H, Liu B S, Jhuang S L, and Li Y C 2016 *International Conference on Applied System Innovation, IEEE ICASI*
[29] Pardo-Ferreira M D C, Rubio-Romero J C, Galindo-Reyes F C, and Lopez-Arquillos A 2019 *Safety Science*
[30] Regulation Number 30 of 2007 the Republic of Indonesia concerning Energy *(Undang-Undang No. 30 Tahun 2007 tentang Energi)* in Bahasa Indonesia

[31] Government Regulation Number 79 of 2014 the Republic of Indonesia concerning National Energy Policy *(Peraturan Pemerintah No. 79 Tahun 2014 tentang Kebijakan Energi Nasional)* in Bahasa Indonesia
[32] Government Regulation Number 14 of 2015 the Republic of Indonesia concerning National Industrial Development Master Plan 2015-2035 *(Peraturan Pemerintah No. 14 Tahun 2015 tentang Rencana Induk Pembangunan Industri Nasional Tahun 2015-2035)* in Bahasa Indonesia
[33] Regulation Number 16 of 2016 the Republic of Indonesia concerning Ratification of the Paris Agreement to The United Nations Framework Convention on Climate Change *(Undang-Undang No. 16 Tahun 2016 tentang Pengesahan Persetujuan Paris atas Konvensi Kerangka Kerja Perserikatan Bangsa-Bangsa Mengenai Perubahan Iklim)* in Bahasa Indonesia
[34] Presidential Regulation Number 22 of 2017 the Republic of Indonesia concerning National Energy General Plan *(Peraturan Presiden No. 22 Tahun 2017 tentang Rencana Umum Energi Nasional)* in Bahasa Indonesia

Performance Measurement in Supply Chain Using SCOR Model in The Lithium Battery Factory

Yuniaristanto*, **N Ikasari, W Sutopo, R Zakaria**

Research Group of Industrial Engineering and Techno-Economic, Universitas Sebelas Maret, Surakarta

Email: yuniaristanto@ft.uns.ac.id

Abstract: The research aim is to measure the performance of supply chain in the UNS Lithium Battery Factory using the SCOR model in order to determine an improvement plan for the company. The battery supply chain performance is measured using the SCOR model with following steps: (i) identifying supply chain performance metrics, (ii) validating metric level 1 as KPI, (iii) performance measurements using metrics level 1 and level 2 and (iv) normalization using Snorm de Boer formula. Based on the monitoring system of the performance indicators, the performance of reliability and cost is categorized in good while the performance of responsiveness, agility, and asset management efficiency is categorized average.

1. Introduction

The UNS Lithium Battery Factory is the first lithium battery manufacturing with independent production processing in Indonesia and has been able to produce commercially with a capacity of 1,000 cells per day. The UNS Lithium Battery Factory produces 2 types of lithium batteries, namely Lithium Ferro Phosphate (LFP) and Nickel Cobalt Aluminum (NCA) which have a voltage of 3.2 volts and 3.6 volts, with a capacity of 1400 mAH and 2700 mAH respectively. This lithium battery is marketed through the business to the business scheme, where the battery buyer company will process it further by assembling the battery cell into a battery module. This module can be applied to electric motors, electric cars, and others. Lithium Battery Factory manufactures lithium batteries with production systems both Make To Order [1].

The UNS Lithium Battery Factory has just been running commercially for 2 years so it can be categorized as a startup company. The company will strive to achieve the best performance in order to be able to compete and survive in the battery industry. Performance measurements need to be conducted on the supply chain system of the UNS lithium battery factory. In this case, the battery supply chain involves several entities consisting of 2 suppliers of raw materials, a shipping company from supplier to factory, UNS Lithium Battery Factory, freight delivery company from the factory to customer and customer itself.

Supply chain management (SCM) is defined as an approach that is used to integrate all supply chain entities efficiently so that production and distribution of goods are in the right amount, the right location, and the right time to minimize overall system costs and fulfill all needs at every level [2]. There are several methods to measure the performance of the supply chain, including the Balanced Scorecard, Activity Based Costing, Multi-Criteria Analysis, Economic Value Added, Life-Cycle Analysis, and the Supply Chain Operations Reference (SCOR) model. This study uses SCOR to measure the supply chain performance of the UNS Lithium Battery Factory.

SCOR was introduced by the Supply Chain Council (SCC), an independent, non-profit global company interested in implementing and developing sophisticated supply chain technology in management systems and practices. SCOR is a reference model with standard terminology and processes that serves as a benchmark for operational measurements by creating a portfolio of priority improvements related to the company's financial statement to improve performance and revenue [3]. The SCOR model excels because of its balanced approach as a reference model with the ability to integrate business process reengineering, benchmarking, and best practices into supply chain frameworks in various attributes [4].

Since the company or researchers have never measured the company performance, especially the supply chain area while the company has been operating for 2 years. So that it is important to measure supply chain performance for the company. Measuring supply chain performance at the UNS Lithium Battery Factory by using SCOR 11.0 to support developing companies, design goals, evaluate performance and determine future steps. Based on the previous description, the purpose of this research is to measure the performance of the supply chain in the UNS Lithium Battery Factory using the SCOR model in order to determine an improvement plan so that the company can survive and compete with other competitors.

2. Literature Review

SCOR is a reference model used to measure supply chain performance. The SCOR model was developed by the SCC which is a global non-profit consortium by having methodologies, diagnostics and benchmarking tools to help companies develop rapid improvements in the supply chain activities. SCC developed the SCOR to analysis and compare the supply chain process and performance. The SCOR model was developed by providing a self-assessment and comparison methods of supply chain processes and performance as a standard for SCM in a cross-industry scope. This model presents a business process framework, performance indicators, best practices and techniques to support collaboration between several parties in the supply chain in order to increase the supply chain efficiency [5].

The SCOR model is a supply chain term, which can be applied in various contexts to design, define, reconfigure various types of business activities. The application of the SCOR model within certain limits is quite flexible and can be adjusted to increase productivity to meet the needs of consumers [5]. The scope of SCOR consists of six core processes, namely plan, source, make, deliver, return, and enable. The six core processes function as follows [6]:

a. Plan describes activities related to prepare a plan of the supply chain operation. The plan includes collecting needs and information on existing resources, balancing demand and resources to determine capabilities.
b. Source is the procurement process of products to fulfill demand. Sources include shipment scheduling, receiving, inspecting and payments authorization of products, supplier selection, and supplier evaluation.
c. Make describes activities related to material conversion or the content preparation of services. This activity can operate based on make-to-stock or make-to-order. Make includes production scheduling, production, testing quality, managing work-in-process, and maintaining production facilities.
d. Deliver describes activities related to the preparation and fulfillment of customer orders. Deliver includes processing customer orders, selecting shipping companies, handling finished products in the warehouse and sending invoices to customers.
e. Return is the process of receiving a product return for various arguments. Returns include identifying product conditions, requesting a return defect authorization, scheduling returns and making returns.
f. Enable describes activities related to supply chain management. Enable includes business rules, performance management, data management, resource management, facility management, contract management, supply chain network design, managing compliance with regulations and risk assessment.

SCOR covers all customer interactions, all material transactions, and all market transactions but does not represent every business activity. The SCOR model is not intended for sales and marketing, product development, research and development, and post-shipment customers [7]. At present, the SCOR model 11.0 is the last revision made and released in 2012. The revision of the SCOR model was conducted by SCC members because changes must be done to facilitate the practical use of the model. The change process of the SCOR Model was managed by the voluntary efforts of SCC members. The SCC relies on the contribution of members to upgrade knowledge about the supply chain by identifying, researching and validating these changes, then creating consensus on proposed changes [6].

The performance of SCOR focuses on understanding two elements, namely attributes and metrics. Attributes are the categorization of metrics used to describe specific strategies. The attribute itself cannot be measured and used to direct strategic [8] SCOR recognizes five attributes, namely reliability, responsiveness, agility, cost, and asset management efficiency. The attributes of reliability, responsiveness, and agility are consumer-oriented attributes, while the attributes of cost and asset management efficiency are internal-oriented attributes [9].

The SCOR model has been widely applied to measure the performance of company supply chains. Goparvar & Seifbarghy measured supply chain performance from Iranol Oil Company (IOC) by mapping supply chain processes level 1, level 2 and level 3 [10]. Then conducted a level 3 process analysis based on SCOR best practices and carried out improvement projects and the prioritization using TOPSIS. Seifbarghy, Akbari, & Sajadieh measured supply chain performance at steel companies by identifying supply chain processes at levels 1, 2 and 3 and analyzing level 3 based on best practices and conducting improvement projects and the prioritization using TOPSIS [11]. Erkan & Ugur measured supply chain performance in the Turkish manufacturing industry with the SCOR Model and Enterprise Resources Planning (ERP) approaches [12]. Lee, Tzong-Ru, Shiu, & Sivakumar measured supply chain performance in the manufacturing industry in Taiwan with the SCOR Model and Information, Communication, & Technology (ICT) [13]. Immawan, Marimin, Arkeman, & Maulana designed a sustainable supply chain system on the MTS-MTO system with a case study of the batik industry in Solo Indonesia [14].

3. Research Method

At this stage, the battery supply chain performance is measured using the SCOR model. The steps taken are as follows.

3.1. Identify supply chain performance metrics

The performance of SCOR focuses on understanding two elements, namely performance attributes and metrics. Attributes cannot be measured because they are used to set strategic direction. Each attribute can be measured through several metrics. Metrics are the standard for measuring the performance of a supply chain or process. SCOR has three predetermined metric levels, namely metrics level 1 to level 3. This study uses performance measurements based on metrics level 1 and level 2. Metric level-1 is diagnostic for the entire supply chain known as Key Performance Indicators (KPI) and metric level-2 measures as diagnostics for metric level-1. Diagnostic relationships help to identify the root causes or causes of performance gaps for metric level-1 [6].

3.2. Validate KPI

The questionnaire was prepared based on the results of the identification of metrics that have been conducted on the lithium battery supply chain. It aims to find out what performance metrics that represent the performance of the UNS Lithium Battery supply chain. The KPI validation questionnaire is a questionnaire with closed questions with two responses, namely valid if the metric is relevant or in accordance with the conditions and invalid if the metric is not relevant or not in accordance with company conditions. Supply chain performance metrics that have been validated by respondents are referred to as KPI so that performance can be measured.

3.3. SCOR performance measurements using metrics level 1 and level 2
Each KPI consisting of metrics level 1 and level 2 is calculated based on company data and observations of researchers. Calculations are conducted using the formulation written in the SCOR guidelines.

3.4. Normalization (Snorm de Boer)
Normalization level 1 metric will use %. Calculation of normalization using the Snorm de Boer formula:

$$\text{Larger is better, Snorm} = \frac{S_i - S_{min}}{S_{max} - S_{min}} \times 100\%$$

$$\text{Lower is better, Snorm} = \frac{S_{min} - S_i}{S_{max} - S_{min}} \times 100\%$$

where: S_{min}, S_{max}, S_i is the minimum value, the maximum value, the actual value of the performance attributes respectively.

4. Result and Discussion

The metrics used are metrics level 1 and level 2 contained in each performance attribute. Then the level 1 and level 2 performance metrics are identified on the five performance attributes. Performance metrics level 1 and level 2 that have been identified are then validated using a questionnaire to respondents (Site Manager). The KPI validation questionnaire is a questionnaire with closed questions with available answers that are valid and invalid. Valid also means whether the metric is important to measure or not for the company. The validity test results there are 26 valid KPIs and 15 non-valid KPIs.

Table 1. Supply Chain Performance Measurement Results

Attributes	Metric L-1 (Code)	Metric L-2 (Code)	Score	Unit	Total
Reliability (RL)	*Perfect Order Fulfillment* (RL.1.1)	RL.2.1	100	%	80.75%
		RL.2.2	85	%	
		RL.2.3	100	%	
		RL.2.4	95	%	
Responsiveness (RS)	*Order Fulfillment Cycle Time* (RS.1.1)	RS.2.1	0	hour	4,896 hours
		RS.2.2	4,704	hours	
		RS.2.3	192	hours	
Agility (AG)	*Upside Supply Chain Flexibility* (AG.1.1)	AG.2.1	0	hour	1,056 hours
		AG.2.2	1,008	hours	
		AG.2.3	48	hours	
	Upside Supply Chain Adaptability (AG.1.2)	AG.2.6	0	%	0%
		AG.2.7	0	%	
		AG.2.8	0	%	
Cost (CO)	*Total Cost to Serve* (CO.1.001)	CO.2.002	0.33	%	74.08%
		CO.2.003	66.67	%	
		CO.2.004	6.7	%	
		CO.2.006	0.42	%	
		CO.2.008	74.08	%	
Asset Management Efficiency (AM)	*Cash-to-Cash Cycle Time* (AM.1.1)	AM.2.1	46	days	47 days
		AM.2.2	1	days	
		AM.2.3	0	day	
	Return on Fixed Assets (AM.1.2)	AM.2.5	7.2	%	7.2%
	Return on Working Capital (AM.1.3)	AM.2.6	0	%	3.63%
		AM.2.7	0	%	
		AM.2.8	7.14	%	

Then all of the validated KPIs are measured. The performance value of metric level 1 is an accumulation of the performance value of metric level 2. Measurements are made based on data from the company and interviews and the results can be obtained as in table 1.

Because the metric level 1 is still in different units such as percent, day, and hour, it is necessary to equalize the units in order to calculate the performance of the supply chain as a whole. Normalization metric level 1 (KPI) will use units of percent. Metric level 1 among order fulfillment cycle time, upside supply chain flexibility, and cash to cash cycle time are not in percent. In this measurement, each metric value is converted into a range of 0% to 100%. Zero is the worst value and one hundred is the best value. Normalization is done by using the Snorm de Boer formula. The results of normalization can be seen in the following table.

Table 2. Normalization Results for Performance Values

Attributes	Score
Reliability	80.75 %
Responsiveness	50%
Agility	50%
Cost	74.08%
Asset Management Efficiency	64.63%

Based on the monitoring system of the performance indicators [15], the performance of reliability and cost is included in the Good category while the attribute performance of responsiveness, agility, and asset management efficiency is included in the average category.

The performance of lithium battery manufacturers in the average category is still understandable because the company has been operating for 2 years. However, this situation is not sufficient to be able to compete with companies that have existed. So that it needs improvement action so that company performance can be improved. Company performance can be improved by referring to existing best practices. Best practices are current, structured, and repeatable practices that had a proven on supply chain performance. Based on the results of the supply chain performance measurement, the company should focus its improvement on the attributes that have the lowest value, namely responsiveness and agility.

The responsiveness attribute can be improved based on best practice as follows:

a. The variability of the order fulfillment cycle time is reduced so that it can be predicted more accurately.
b. Evaluating the performance of cargo delivery. Improvement in the quality of on-time delivery and in good conditions can increase company satisfaction.

The agility attribute can be improved based on best practice as follows:

a. Use Lean principles to increase the agility of the production run or production cycle in order to modify production planning, finished goods, and work-in-process inventory. This can be combined with a reduction of production batch size and pull-based production.
b. Applying Just in Time strategy to reduce in-process inventory and associated carrying costs.
c. Strategic sourcing is needed to improve and evaluate the purchasing activities of the company.
d. Use the supply chain visibility system that allows the supplier to see requirements and inventory levels at the customer site.

5. Conclusion

Based on the results of performance measurements, the supply chain performance of the UNS Lithium Battery Factory has not been maximized because the total value obtained is 65.13% or included in the average category. Attributes with low values are responsiveness, agility, and asset management efficiency, the values of which are 50 (average), 50 (average) and 64.63 (average) and require further improvements to increase supply chain performance of the UNS Lithium Battery Factory. In further research, it is necessary to do supply chain mapping and benchmarking with similar companies so that recommendations for improvement are more appropriate and can be a reference for companies to determine targets based on SCOR metrics. Other research that can be done is to integrate SCOR with multi-criteria decision analysis to measure overall supply chain performance based on the priority weights of the five performance attributes.

Acknowledgments
This work is supported by Non-Tax State Revenue, Universitas Sebelas Maret with Penelitian Unggulan Terapan (PUT) Grant (Contract No. 516/UN27.21/PP/2019).

References

[1] Unit Produksi Baterai Lithium Ion UNS 2019 *Profil Perusahaan Unit Produksi Baterai Lithium Ion UNS.* (in Indonesia)
[2] Simchi-Levi D, Kaminski P and Simchi-Levi E 2000 *Designing and managing the supply chain: concept, strategies, and case studies* (Irwin McGraw-Hill)
[3] Meyr H, Rohde J and Stadtler H 2002 Basics for modeling *Supply Chain Management and Advanced Planning* ed H Stadtler and C Kilger (Berlin: Springer) pp 45–70
[4] Aramyan L H, Ondersteijin C J M, van Kooten O and Oude A G J M 2006 Performance Indicators in Agri-Food Production Chains. in: Quantifying The Agrifood Supply Chain *Agribusiness: An International Journal* **22**(2) p 47 – 64
[5] Supply Chain Council [SCC] 2014 *Supply Chain Operations Reference Model Revision 11.0 Handbook*
[6] Supply Chain Council [SCC] 2012 *Supply Chain Operations Reference Model Revision 11.0.*
[7] Supply Chain Council [SCC] 2001 *Supply Chain Operations Reference Model Overview of Version 5.0*
[8] Paul J 2014 *Panduan Penerapan Tranformasi Rantai Suplai dengan Model SCOR® 15 Tahun Aplikasi Praktis Lintas Industri* (Jakarta: PPM Manajemen in Indonesia)
[9] Sutopo W, Maryanie D I and Yuniaristanto 2015 Evaluation of Valuable Chain in Palm Oil Industry Based on SCOR Model: A Case Study *Int. J. Logistic Systems and Management* **21**(2) p 229 – 244
[10] Goparvar M and Seifbarghy M 2009 Application of SCOR Model in An Oil- Producing Company *Journal of Industrial Engineering* **2**(4) p 59 – 69
[11] Seifbarghy M, Akbari M R and Sajadieh M S 2010 Analyzing The Supply Chain Using SCOR Model in a Steel Producing Company *Proceedings of the 40th International Conference on Computers & Industrial Engineering* p 1-6
[12] Erkan T E and Ugur B 2011 Supply Chain Performance Measurement: A Case Study about Applicability of SCOR Model in A Manufacturing Industry firm *International Journal of Business and Management Studies* **3**(1) p 381-390
[13] Lee T, Shiu Y, and Sivakumar P 2012 The Applications of SCOR in Manufacturing: Two Cases in Taiwan *Procedia Engineering* **38** p 2548-2563
[14] Immawan T, Marimin, Arkeman Y, and Maulana A 2015 Suistainable Supply Chain Management for Make to Stock-Make to Order Production Typology Case Study: Batik Industry in Solo Indonesia *European Journal of Business and Management* **7**(11) p 94-106
[15] Trienekens J H and Hvolby H H 2000 Performance Measurement and Improvement in Supply Chains *Proceedings of the 3rd CINet Conference* p 399-408

IOP Conf. Series: Materials Science and Engineering **943** (2020) 012050 doi:10.1088/1757-899X/943/1/012050

Supply Chain Performance Analysis Using Hybrid Overall Equipment Effectiveness and Discrete Event Simulation (Case Study: Futura Energy Nusantara)

Z A Fadhlurrahman[1*],Yuniaristanto[2], M Hisjam[2], R Zakaria[2]

[1], Undergraduate Program of Industrial Engineering, Faculty of Engineering, Universitas Sebelas Maret, Surakarta
[2], Research Group of Industrial Engineering and Techno Economic, Department of Industrial Engineering, Universitas Sebelas Maret, Surakarta

Email: zufarasyraff@gmail.com

Abstract. Abstract. Futura Energy Nusantara is the first manufacturer of lithium-ion battery cell production in Indonesia. Because highly demand, the company must fill the order and increase production Productivity is one of the most fundamental and important determinants of system production. Productivity measurement help identify the problems and find solutions to improve system performance. OEE (Overall Equipment Effectiveness) is one method used for effective time manufacturing systems related to the existence of equipment in the production system. The Element of OEE is Availability Rate (A), Performance Efficiency (P), and Quality Rate (Q). The study aims to test through the simulation production system in Lithium-Ion Batteries (LIB), by considering factor OEE, with discrete event simulation system. As a result, this study is a model to applicable, for the company to improve Productivity

1.Introduction

The production process is the transformation of the process or a set of activities that transform a collection of some inputs (human resources, materials, energy, information, etc.) Into output products (finished products or services) that have added value. Productivity is one of the most basic and important determinants of system production. Productivity measurement helps the company to identify the problems and find solutions to improve system performance. [3]. In the manufacturing industry, production productivity is the main thing that needs to be improved. Increased productivity can be realized by the existence of a production process that runs effectively and efficiently. The production process carried out certainly requires production resources such as humans, machinery and materials.[7] nowadays to be competitive, manufacturers must provide excellent reliability and the best quality of their equipment at competitive prices. To have a very reliable machine to ensure the smooth manufacturing process, many organizations have implemented TPM as a tool that makes it possible to maximize the effectiveness of the equipment. [12]. Futura Energy Nusantara is the first manufacturer of lithium-ion batteries in Indonesia. The target of production must 80.000 battery cells. With a total of 20,000 cells for lithium iron phosphate (LFP) batteries and 60,000 cells for nickel cobalt aluminum (NCA) cell batteries [5]. The agreed period of work is seven months with the fulfillment of a weekly period of 4000 cells per week. In February 2018 PT Pertamina (persero) to collaborate in the formulation and production optimization study consultant services prototype of a lithium-ion battery, in this project, has succeeded in formulating and improving the production capacity of up to 15,000 cells per month with a floor area of 700m2 involving 20 employees. The OEE is one of the approaches that can calculation of overall equipment/machine effectiveness and can simplify the complexity of production problems into simple information. OEE can systematically analyze the production process and identify potential problems in influencing the work of the machine. There are three basic metrics of the OEE such as performance, availability, and rate of quality [2] OEE can measure performance, identify increased opportunities and focus on improvement efforts in areas related to equipment efficiency and effectiveness [9]. OEE calculates how well the performance of the production system is relative to the designed capacity. During the operation period. A survey reported that OEE is often used as a means to increase company output because it focuses on quality, productivity, and the use of equipment simultaneously [1]. OEE aims to reduce or eliminate speed loss in a lean manufacturing environment, consider two types of uncertainties, namely in the speed of production and cessation of measurement of duration, which is used in calculating OEE components. [13]

In this study, we use the input-oriented Overall Equipment Effectiveness OEE-DES hybrid.

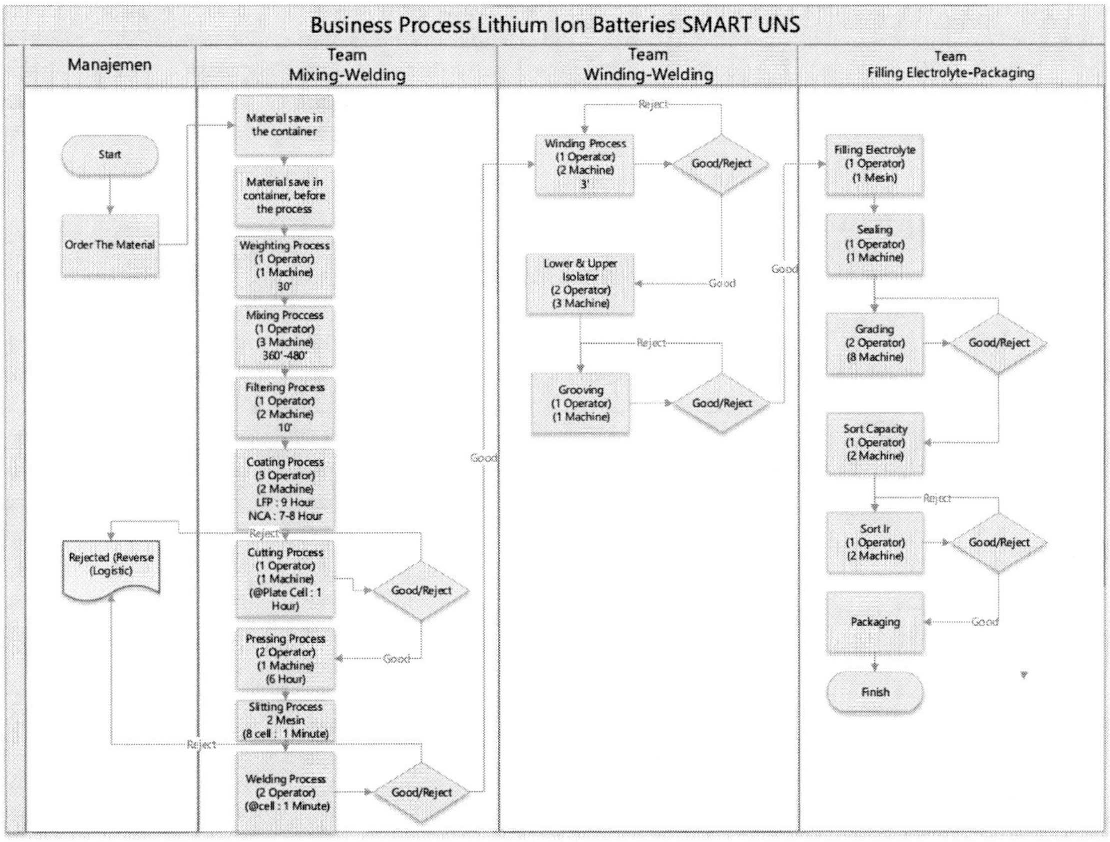

Figure 1. Business Process Lithium Ion Batteries SMART UNS

In making the lithium battery production process, it is divided into 3 people in charge of production, the first is mixing-welding which includes the weighting process, which is the weighing process of the main ingredients. This process is carried out with 1 operator 1 machine. Second is the mixing process, this process is the process of mixing several main ingredients. This process is carried out with 1 operator and 1 machine. Third is the screening process, namely the eliminated filtering material, which is carried out with 1 operator and 1 machine, the fourth is the coating process, namely plate making, consisting of 2 namely the cathode and anode, this process is carried out by 3 operators and 2 machines, the fifth is cutting process, namely the process of cutting the plate, into the size of a battery. Sixth is the process of pressing, which is the process of applying pressure to the layer that has been cut. Seventh is the process of slitting, which is then followed by the rolling process, which is the welding process carried out by 2 operators in turn. The second person in charge is winding-welding, which is given a winding process, namely wrapping the anode and cathode into the battery shell with 1 operator and 2 machines. Then start with the installation of the top and bottom insulators, namely the installation on the top and bottom of the insulator cover by 2 operators and 3 machines. After installation is complete, a grooving process is carried out, which is the process of making grooves done by 1 operator and 1 machine. The third person in charge is the electrolyte-packaging process, after the material and the battery compiler have finished, then the electrolyte filling is carried out, which is carried out by 1 operator and 1 machine if the electrolyte content process is finished the sealing process is carried out by 1 operator and 1 machine. After that the grading process is carried out, namely the process of testing the battery price by 2 operators and 8 machines. After the cost test is successful, then a capacity sorting will be completed with sorting per ir done by 1 operator and 2 machines. When all processes have been completed, the last meal is packaging, that is, packaging the batteries, and the batteries are ready to be sent.

2. Quantitative Approach

2.1 Overall Equipment Effectiveness

In table 1 is the data collected and processed using the method OEE data processing is performed to determine the level of effectiveness of use machines just for 6 process from 18 process of lithium-ion batteries, there are pressing, slitting, welding, winding, screw & spot welder, and filling electrolyte & punching

Table 1. Element of OEE SMART UNS in January – June 2019

Process	Available Time (Minutes)	Planned Down Time (Minutes	Loading Time (Minutes)	Unplanned Down Time (Minutes)	Operating Time (Minutes)	Processed Amount (Cell)	Ideal Cycle Time (Minutes/Cell)	Defect Amount (Cell)
Pressing	42596	3205	39391	1687	37704	60288	0,384	1463
Slitting	37510	3562	33948	1617	32331	58826	0,527	2732
Welding	40164	3227	36937	2077	34860	56094	0,529	969
Winding	38744	4119	34625	2159	32466	55125	0,421	9204
Screw & Lid Welding	38134	3236	34898	2003	32895	45921	0,498	470

Filling electrolyte & Punching	36788	3896	32892	2094	30798	45451	0,613	540

Overall equipment effectiveness (OEE) is the one measure of TPM (Total Productive Maintenance)to evaluate the effectiveness of equipment or machine. The function of OEE is to measure reveals the hidden costs that cause a considerable amount of production loss. The OEE consists of three parameters are availability, performance, and quality. It can be calculated as follows [13]

$$OEE = Availability\ x\ Performance\ x\ Quality \tag{1}$$

2.1.1 Availability.

Availability is the percentage level availability of machine or time scheduled production that compared with the amount of time producing. Which is not included in availability is Scheduled maintenance, planned downtime, or equipment trials. This can help a plant manager to identify whether machine downtime issues are part of a known calendar, or if there is a more serious problem. [10]

The Example of calculate availability of Pressing Machine as follows :

$$Availability = \frac{Operating\ Time}{Loading\ Time} x100\% \tag{2}$$

$$= \frac{Loading\ Time - Down\ Time}{Loading\ Time}$$

$$= \frac{(Total\ Available\ Time - Planned\ Time\ own\ Time) - Down\ Time}{Load\ Time} x\ 100\%$$

$$= \frac{(42596 - 3205) - 1687}{3931} = 95{,}72\%$$

2.1.2. Performance

Performance is the percentage level of the total parts that produced on the machine to the production rate of the machine. This can help the manager to indicate problems with the machine itself if has recurring efficiency issues. [10]

The Example of calculate Availability of Pressing Machine as follows :

$$Performance = \frac{Processed\ Amount\ x\ Ideal\ Cycle\ Time}{Operating\ Time} x100\% \tag{3}$$

$$Performance = \frac{60288\ x\ 0{,}384}{37704} x100\% = 61{,}40\%$$

2.1.3 Quality.

Quality indicates the level of ability of the machine or equipment inside produces a product that complies with predetermined standard specifications. It can help the manager to compare consistency between individual machines, different manufacturers as well as machine, specifications and even individual operators [8]

The Example of calculate Quality of Pressing Machine as follows

$$Quality\ Rate = \frac{Processed\ Amount\ x\ Defect\ Amount}{Processed\ Amount} x100\% \tag{4}$$

$$Quality\ Rate = \frac{60288 - 1463}{60288} x100\% = 97{,}57\%$$

Total OEE of Pressing Machine as follows :

$$OEE = 95{,}72\%\ x\ 61{,}40\%\ x\ 97{,}57\% = 57{,}34\%$$

3. Results and Discussion

Based on the calculation results in table 1.data analysis was carried out with OEE Methode of 6 process lithium-ion batteries.

Table 2. The Result of OEE SMART UNS in January – June 2019

Process	Availability (%)	Performance Efficiency (%)	Quality Rate (%)	OEE (%)
Pressing	95,71	61,40	97,57	57,34
Slitting	95,23	95,88	95,35	87,07
Welding	94,75	85,12	98,27	80,33
Winding	94,27	71,48	83,30	55,83
Screw & Lid Welding	93,79	69,52	98,97	64,85
Filling Electrolyte & Punching	93,79	90,46	98,81	83,69
OEE	**93,79**	**78,97**	**95,38**	**71,52**

After we calculate the 6 process with OEE element, we get the result.

3.1 Availability
Pressing Process has high availability than the other process with 95.71% because, operators and materials are always there, and has small downtime. In the filling electrolyte & Punching is the lowest availability with 93.79% because the cell must be waited for the process before and has most downtime

3.2 Performance Efficiency
Slitting process has the highest Performance with 95.88% because the process has the smallest losses idling and minor stoppages. The pressing process has the lowest with 61.40%, because the process has the difference between actual production capacity with an ideal run time very far

3.3 Quality
Screw & Welding process has the Highest Quality with 98.97% because, in this process has a small defect product, then the other process. In the Winding process, has the lowest Quality with 83.30%, because in this process has big defect product, cause a combination of several materials to become a cell battery

The result of total availability is 93.70%, total performance efficiency is 78.97%, and the quality rate is 95.38%. And the total OEE is 71.52%

We get the lower OEE is Winding Process with 55.83% because the machine mostly has long time set up, and get stuck in the middle of the process, besides that, this station has lacked the operator to operate this machine.
To identification furthermore, we make model flowchart simulation. The simulation model get 6 processes of Lithium-Ion Batteries with Discrete Event Simulation to know the model simulation with any logic software

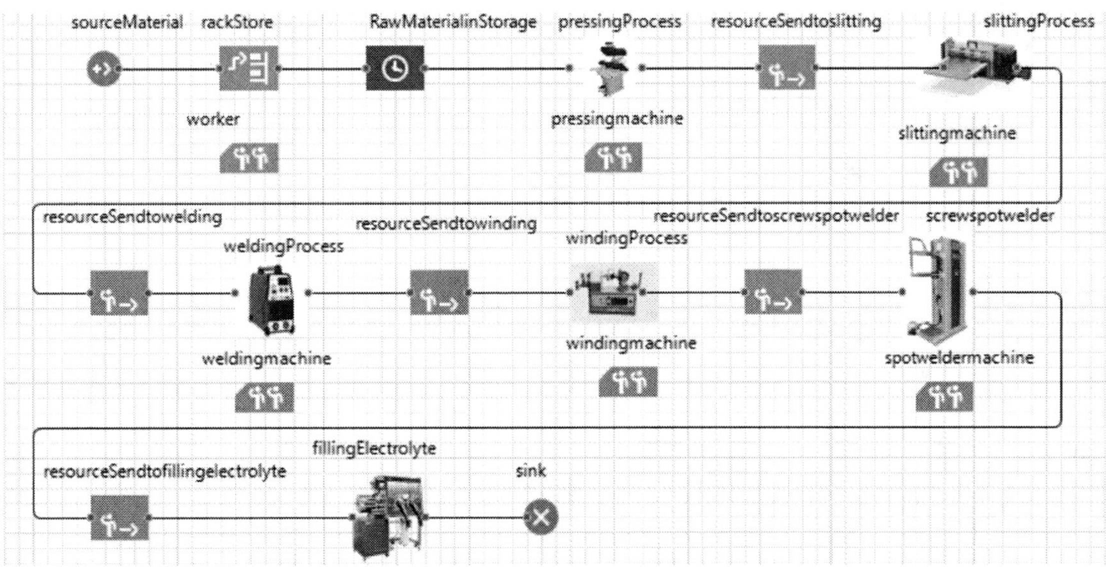

Figure 2. Any logic Flowchart

We use Any Logic to make the design model. We make the model with a discrete event method. Utilizes blocks that describe a specific state in the process. There are numerous different blocks in the software library, as a source, delay, resource send to, resource pool, rack store, sink, etc. The blocks are interconnected by connectors. Each block, depending on the type, contains defined properties, such as name; type; delay time; capacity; maximum capacity; agent location; agent type, etc.

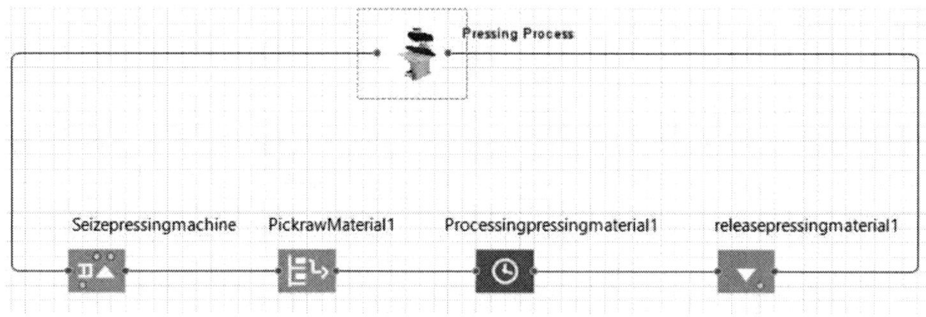

Figure 3. Pressing Process Flowchart

4. Conclusion

Futura Energy Nusantara is the first manufacturer of lithium-ion battery cell production in Indonesia. Because highly demand, the company must fill the order and increase production Productivity is one of the most fundamental and important determinants of system production. To evaluate the production process, we use OEE to find the effectiveness of equipment/machines. Overall equipment effectiveness (OEE) is to evaluate the effectiveness of equipment or machine.

After we calculate the OEE, we get the lower OEE is Winding Process with 55.83% because the machine mostly has a long time set up, and gets stuck in the process, besides that, this station has lacked the operator to operate this machine. And we get that the total OEE is 71,52%, need to improve the score to get 85% to become a world-class manufacturer

This study also provided a discrete system simulation flowchart model using any logic software.

This article has certain limitations, that should be overcome to provide in the deep analysis of the performance measurement analysis. For further research will complete the research with ORE method for accurate performance measures, and run the simulation

Acknowledgments

This research is supported by Institute for Research and Community Service, Universitas Sebelas Maret with Hibah Penelitian Unggulan Terapan UNS (PUT-UNS) Research Program (Contract No. 516/UN27.21/PP/2019).

IOP Conf. Series: Materials Science and Engineering **943** (2020) 012050 doi:10.1088/1757-899X/943/1/012050

References

[1] Bamber, C, Castka, P, Sharp, J. and Motara, Y 2003 *Cross-functional team working for overall equipment effectiveness*, Journal of Quality in Maintenance Engineering, Vol. **9** No. 3, pp. 223-239.

[2] Benjamin, M.S.J., Marathamuthu, M.S. and Murugaiah, U. (2015), *The use of 5-WHYs technique to eliminate OEE's speed loss in a manufacturing firm*, Journal of Quality in Maintenance Engineering, Vol. **21** No. 4, pp. 419-435.

[3] Bon, A.T. and Lim, P.P. (2011), *Implementation of total productive maintenance (TPM) in automotive industry*, IEEE Symposium on Business, Engineering and Industrial Applications, pp. 163-166

[4] Borris, Steve. *Total Productive Maintenance*. USA : The McGraw-Hill Companies (2006)

[5] C.A Mursid, W.Sutopo, Yuniaristanto. (2019), *Master Productions Scheduling with goal programming on lithium battery production:case study,* IOP Conference Series : Materials Science and Engineering 2019 **495** 012046

[6] Dawood, L.M and Abdullah, Z.H (2018), *Study Impact of Overall Equipment and Resource Effectiveness onto Cement Industry*. Journal of University of Babylon, Engineering Sciences, Vol.**26**, No.3: 2018.pp 187-198

[7] Eswaramurthi, K.G., & Mohanram, P.V., (2013), *Improvement of Manufacturing and Evaluation of Overall Resource Effectiveness*, American Journal of Applied Sciences. **10**

[8] Fabio De Felice, Stanislao Monfreda, Antonella Petrillo, Dr.Ing. Maria, 2013 ," Operations Management" This Work is Licensed Under A Creative CommonsShare Alike 4.0 International License, Original Source: Intech

[9] Kumar, J., Soni, V.K. and Agnihotri, G. (2014), *Impact of TPM implementation on Indian manufacturing industry*, International Journal of Productivity and Performance Management, Vol. **63** No. 1, pp. 44-56

[10] Kalpande S.D 2014 *OEE-an Effective Tool for TPM Implementation- A Case Study*, Proceedings of 8th International Quality Conference, Center for Quality, Faculty of Engineering, University of Kragujevac, pp.521-526.

[11] Nachiappan, R.M. and Anantharaman, N. (2006), *Evaluation of overall line effectiveness (OLE) in a continuous product line manufacturing system*, Journal of Manufacturing Technology Management, Vol. **17** No. 7, pp. 987-1008.

[12] Nguyen Truong Son, Salwa Hanim Abdul Rashid , 2011,*Using the Value Stream Mapping and Overall Equipment Effectiveness to Improve Productivity: A Literature Review", Proceedings of the Fourth Regional Conference on Manufacturing*, Malaysia,pp.1-9.

[13] Sonmez, V., Caner, M., Ozlem, T. and Testik, M. (2018), *Overall equipment effectiveness when production speeds and stoppage durations are uncertain*, The International Journal of Advanced Manufacturing Technology, Vol. **95** Nos 1-4, pp. 121-130

[14] Yu, C. (2016), *Airline productivity and efficiency: concept, measurement, and applications*, in Bitzan, J.D., Peoples, J.H. and Wilson, W.W. (Eds), Airline Efficien S

IOP Conf. Series: Materials Science and Engineering **943** (2020) 012051 doi:10.1088/1757-899X/943/1/012051

Strategy Business of Battery Swap for Electric Vehicle Using Business Model Canvas

AI Sholichah[1], W Sutopo[2]

[1]Master Program of Industrial Engineering, Faculty of Engineering, Universitas Sebelas Maret, Surakarta, Indonesia
[2] Department of Industrial Engineering, Faculty of Engineering, Universitas Sebelas Maret, Surakarta, Indonesia

Email: aisyah_nanik95@gmail.com

Abstract. Electric vehicles have developed rapidly in this global era. Many countries have replaced conventional vehicles into electric vehicles. Electric vehicles need batteries as electric reservoirs. The way to charge batteries for electric vehicles is only through charging and it takes a long time, so the idea of battery exchange grows to simplify battery changes in electric vehicles. This study aims to design a business for battery exchange with a canvas model business approach consisting of 9 important factors such as customer segments to map the market, channels to facilitate information provided to consumers, value propositions or set prices according to consumers by providing some advantages from battery replacement products, customer relationships or maintaining good relationships with consumers, revenue streams provide or provide other services in addition to battery changes, key resources to establish cooperation with raw material suppliers or banks to facilitate and provide discounts to consumers, key activities describe the activities to be carried out in the company, key partners give discounts to consumers who want to become members or regular customers, and the cost structure makes a breakdown of funds or costs from raw materials to commercial costs such as promotions, water, electricity, and so forth. The business model that has been formed will be the basis for designing a busines plan consisting of a management profile and financial analysis. Researchers conducted qualitative research, namely data collection is done by searching from the previous literatre. This research and analysis is to design a battery exchange business that can be done in the future.

1. Introduction

Business is an organization that provides goods or services that aim to earn profits. Business developments in Indonesia have developed that provide high competitiveness for each company. So companies are required to pay more attention to the environment [1]. Some research on business or strategy development uses the Strength, Weakness, Opportunities, Threats (SWOT) and Business Model Canvas (BMC) methods.

The business model canvas is a logical business model of how an organization creates, delivers and capture value (Osterwalder, 2010). The business model canvas can also be called a strategy in management in the form of a visual chart consisting of 9 elements that can comprehensively explain a business process. These 9 elements consist of customer segments (customers who have the potential to produce products or come from a problem that can be solved by the business), value propositions (value added that will make a business look attractive and different from other businesses such as innovation or solutions that are offer and become the company's main advantage), channel (a business media used to deliver solutions that can be offered to consumers in the form of websites, applications, online advertisement, or sales person), customer relationships (knowing how to keep business in touch with the customers subscribers in the form of newsletters, after sales services, etc.), revenue streams (costs and revenue which need to describe the financials of a company such as products / services that can provide revenue), key activities (creating value preposition of the company), key resources (corporate strategy in the form of raw materials, infrastructure what is needed), and key partnerships (a business needs a partner that can be a supplier, vendor, agency, and the like), cost structure (what costs are needed to run the whole business activity) [2]. Managing costs efficiently will make the business being lived more economical and can minimize the risk of loss and can determine the right value proposition for customers.

he transportation sector is a major contributor to environmental pollution and dependence on oil, because the majority of vehicles are based on internal combustion engines. The development of technology in vehicles in recent years and to reduce greenhouse gas emissions, electric vehicles are one of the chosen paths. The future of mobility is interesting because there are many new players in the automotive industry who want to embrace electric mobility in a variety of products and services to overcome the energy crisis and climate change [3]. The commercialization of electric vehicles mash has not yet developed widely because there are still consumers who think that the cost is expensive or high, limited products, and the uncertainty of electric charging stations. Although the battery charging station is quite a lot, but the time to charge the battery is quite long and makes consumers uncomfortable. Then a battery exchange station was made, which only exchanges empty batteries for batteries that are fully charged making it easier for consumers without having to wait long enough [4]. The ownership of a leased battery is different from the ownership of the vehicle. Battery degradation due to charging or usage, maintenance, and labor costs will be carried out by those who have the batteries. While these battery customers simply pay a service fee. This is one way to reduce the cost of very expensive costs to customers [5]. Based on this development, it is possible for a business opportunity to come up with a battery station with a business model that is competitive and technological innovation in the ecosystem as a whole and is compatible with other business strategies, and is likely to succeed and last for a long time. The business model has six functions such as value propositions, customer segments, value chains, cost structures, formulation of competitive strategies, and the position of corporate strategies in the value network [6].

he electric battery used for electric motorbike vehicles is usually made from lithium. These batteries have been packaged into one like a battery. Where it is easier and more practical in assembling the motor. When this battery runs out of power it can be recharged but with a long time allocation until the battery is fully charged or by the method of charging directly with electricity or in the charger or using the battery swap method [4]. Swap has the meaning of reducing it by exchanging batteries without having to wait longer and easier [7]. By exchanging batteries, the electricity network system does not need to be reinvested. Advances in battery technology will also make the system better. The advantages of exchanging batteries such as getting a full charge faster than vehicles that charge for hours, do not need electricity to increase high, charging, vehicle ownership and the batteries can be separated so as to reduce consumer risk.

his research was conducted to expand the design of business models in the business ecosystem in battery swaps for motor vehicles by illustrating or making a mature business plan by using the Business Model Canvas to reduce obstacles or the impact of losses that would occur before conducting the business and using a feasibility analysis to determine whether the business is worth doing.

2. Methodology

This type of research is a qualitative descriptive study, research that is used to analyze data by describing or describing collected data that is applicable to the public or generalization (Sugiyono, 2008). The type of data used in this study is qualitative data. Datacualitative is data that are not stated in numbers or numbers. The data collection method is used to search for data on electric battery companies in the world such as Tesla, Wanxiang, and BYD Shenzen. Furthermore, the Business Model Canvas (BMC) data will be divided based on 9 elements, namely customer segments, customer relationships, customer channels, revenue structures, value propositions, key activities, key resources, cost structures, and key partners in the business model canvas to create a battery swap business strategy.

By using the BMC that has been made, can then perform financial analysis projections. Financial analysis is a matter related to investment estimates, operating and maintenance costs, working capital requirements, sources of financing, income estimates, and calculation of investment criteria. There are 2 elements needed to make a financial analysis, projection, namely revenue stream (revenue) and cost structure (cost). By using these 2 elements, you can financially analyze the feasibility of the battery swap business.

3. Result and Discussion

The following is the design of the results of the business model of the canvas for the exchange of electric vehicle batteries:

Customer Segment

Segmentation is the activity of dividing markets into different groups of buyers. Market segmentation can be done based on geographical, demographic, psychographic, and consumer behavior (Osterwalder, 2010) Looking at examples from several companies such as testla, WanXiang, OEMs, and others, so that the market segment for battery swaps in Indonesia is selected in terms of its best class for electric vehicles, battery swaps are advances in the replacement of batteries in electric vehicles, and there is still a lack of battery swap business. Nearly 72% of Indonesians use motorbikes for transportation because they are considered faster, fuel efficient, cheaper and more sporty.

Seeing from the many motorcycle users in Indonesia especially in big cities, the possibility of enthusiasts for electric motorcycles is also large, so the need for batteries also increases. Meanwhile, if you have to charge an electric motor battery requires around 2-3 hours and there is time inefficiency, it is necessary to exchange batteries to make it easier to replace an empty battery. The price for electric motorcycle vehicles can also compete with other conventional motorcycle vehicles, because the price of batteries is not included in it.

Value Proporsitions

The value that will be given to consumers will be divided into a number of things like the first newness where products for exchanging electric vehicle batteries in Indonesia are still underdeveloped so that the facilities provided to customers must be satisfactory, information and equipment are very clear, and the availability of battery exchange stations in Indonesia overall. Second, the performance provided is fast service, easy service, clear information, and equipment that is not lacking, and service that is both simple and easy. Third, the price on the battery exchange is relatively cheaper because the customer only needs to buy the electricity costs without having to buy the batteries back. Fourth, risk reduction is by giving guarantees to consumers related to products used so as to give a good impression to consumers.

Channel

Online and place of service is a means to provide services / products to consumers who need services when faced with obstacles during use. As for the swap battery replacement, there is a battery swap area that is easily accessible to customers. And notice of how to use is given online or in the battery swap area.

Customer Relationships

To maintain a good relationship with consumers, a company usually provides a service such as a company's brand, company reputation, discounts, guarantees and customer service. By providing these servants hoping to strengthen the relationship between the company and consumers personally.

Revenue Streams

Focusing on the automotive segment revenue, the energy generation and storage segment and the battery swap application. This battery swap application covers how you use, notices, components, and so on about the battery swap information. To provide consumer interest, prices can be set according to the price of electricity or the manufacturing process or see existing companies.

Key Activities
To achieve good, friendly, fast and innovative services the company makes updates to its design, carries out maintenance for charging stations, builds charging stations in strategic locations, develops and maintains software or technology on battery swaps, and maintains sales and marketing for batteries. swap.

Key Resource
The most important key resources are people, systems and equipment. People here mean labor that runs the presentation of services. A business or business is largely influenced by the interaction between labor, systems and instruments. If the system can work and the workforce is also satisfying, the consumer perception of the company is very good. So the company provides services to consumers such as battery systems, effective coolers, technological knowledge, inverters, automatic and technicians.

Key Partnerships
Raw material is an important element in the process of making and providing services. From a financial point of view, raw material also influences profitability. If you get good and suitable raw materials, the electrical content will last long and the battery can be used several times. If you get cheap raw material prices, you can get high profits. So it is suggested that the company has a partnership with fund holders, major suppliers, transportation companies, and the government to facilitate the operation of the company and get high profits.

Cost Structure
The cost structure can be determined from key resources, key activities, and key partners in the business model canvas. The costs required are divided into 2 types, namely the first variable cost where costs incurred are costs such as electricity, water, battery installation facilities, overtime employees. In the variable there is a cost savings in a company's expenses such as water that is used only. The second fixed cost is the cost of the building because it is the largest cost incurred.

Based on the discussion of the model design above, we obtain the elaboration of 9 elements of the battery canvas replacement business model as follows:

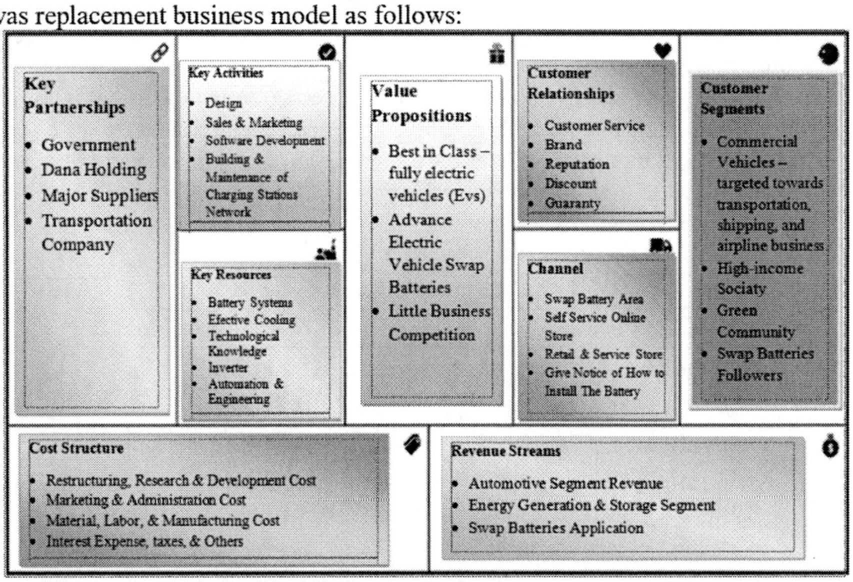

Figure 1. Business Model Canvas for business strategy of battery swap

The marketing plan must be known by the public by providing promotions and clear information about exchanging batteries for electric motor vehicles.

Financial Analysis
Financial analysis is important, used to calculate capital expenditures, and the estimated costs required to open a new business, these are the elements contained in financial analysis:

Initial investor projections
Starting a battery exchange business for electric vehicles that is motorcycles requires investment in land and buildings, renovation of buildings, investment in equipment and supplies at the battery exchange station. The following is the total initial investment needed for the business of exchanging electric motor vehicle batteries of Rp. 119,000,000.

Table 1. The initial investment

No.	Initial Investment	Price (Rp)
1.	Land and Building	50,000,000
2.	Renovation	4,000,000
3.	Equipment and Supplies	32,000,000
4.	Storefront	3,000,000
5.	Decor	30,000,000
	Total	**119,000,000**

Sales projections
To get a profit or profit, an income from a business that is obtained from sales made by the exchange of batteries and services provided is Rp. 160,000,000 with a replacement price of Rp. 5,000 and a damaged battery is purchased for Rp. 15,000,000. If there are 50 customers who come to charge a battery, the company will get a payback period in the 16th month.

Projected costs
Cost projections are needed in financial analysis because with the projected costs it can be seen that new income projections for battery exchange can reach Rp. 180,000,000.

Projection of income statemen
To find out how much profit or loss is obtained, it must be sought how much total income earned each month and then reduced by the amount of expenditure incurred each month. Because the business for exchanging batteries is still new in Indonesia, the benefits are still uncertain at the beginning. However, this business has a high profit, because the development of electric vehicles, especially motorcycles, has been growing rapidly in Indonesia.

Table 2. The payback periode

Payback Periode	Net Income (Rp)
Mounth 1	1,500,000
Mounth 2	3,000,000
Mounth 3	4,500,000
Mounth 4	6,000,000
Mounth 5	7,500,000
Mounth 6	9,000,000
Mounth 7	7,500,000

Mounth 8	6,000,000
Mounth 9	7,500,000
Mounth 10	6,000,000
Mounth 11	6,000,000
Mounth 12	6,000,000
Mounth 13	7,500,000
Mounth 14	9,000,000
Mounth 15	9,000,000
Mounth 16	9,000,000

4. Conclusion

With the business model canvas, it can be described that there are 9 elements that must be done to create a battery swap business strategy, namely customer relationships, customer segments, channels, key activities, key partnerships, cost structures, key resources, revenue streams, and value propositions. Battery swaps must have marketing objectives and the right place to install battery swaps so that the right partner is needed. To carry out all activities oriented to the value created by potential consumers, the battery swap must prioritize raw materials and key resources in human resources, systems, and renewable technology equipment. Thus reducing the risks that will occur in the future.

To improve the driving performance of the company, it must cooperate. Establish cooperation not only with companies that provide raw materials, but also with other companies that can reduce production costs to provide high profits. The battery swap business in Indonesia also still has many opportunities and has low business competition.

From the canvas business model that has been designed to analyze the finances in the battery replacement business for electric motor vehicles, that the company gets a payback period in the 16th month. Based on this research, it can be seen that the business model canvas can be a tool or a simple medium to produce alternative corporate strategies in knowing what is needed to do a business.

References

[1] M. E. Porter 1992 *Keunggulan Bersaing: Menciptakan dan Mempertahankan Kinerja Unggul* Erlangga
[2] Wisnu Dewobroto November 2012 *Penggunaan Business Model Canvas Sebagai Dasar untuk Menciptakan Alternatif Strategi Bisnis dan Kelayakan Usaha* ResearchGate
[3] Chew Vee Kuan 2016 *Business Dynamic Evaluation of Battery Swapping in Electric Vehicle Systems* Keio Associated Repository of Academic Resources Japan
[4] Sarkar R. M. Pandzic Hrvoje Ortega Vazquez A. M. 2013 *electric Vehicle Battery Swapping Station: Business Case and Optimization Model* University of Washington Seattle, Washington
[5] Sarkar R. M. 2010 *Electric Vehicles as Grid Resources* University of Washington, Washington
[6] H. Chesbrough and R. S. Rosenbloom 2002 *The Role of The Business Model in Capturing Value from Innovation: Evidence from Xerox Corporation's Technology Spin-off Companies* Industrial and Corporate Change, vol. 11 no. 3 pp. 529-555
[7] A. Osterwalder Y. Pigneur and C. L. Tucci 2005 *Clarifying Business Models: Origins, Present, and Future of The Concept* Communication of The Association for Information Systems vol. 15

Framework to assess technology readiness of lithium-ion battery to fulfill the safety standard: IEC 62660-3:2016 (case study: smart lithium battery of UNS product)

K Fachriansyah[1*], F Fakhma[2] and E Pujiyanto[3]

[1] Department of Industrial Engineering, Sebelas Maret University, Surakarta, Indonesia
[2,3] Quality System Laboratory, Department of Industrial Engineering, Sebelas Maret University, Surakarta, Indonesia

Email: kfachri16@gmail.com

Abstract. Electric vehicles are one of the ways to solve the problem of air. Electric vehicles have a very vital component which is the battery as a source of power for other components. The type of battery that has bes performance for electric vehicles is lithium-ion battery. Reliability and safety present significant challenges in lithium-ion batteries. Thus, lithium product standards are important to guarantee the quality of lithium battery products. The International Electrotechnical Commission (IEC) published a lithium battery standard product regulation in document IEC 62660-3:2016, that provide a basic level of safety test methodology and criteria with general versatility, serves a function in common primary testing of lithium-ion cells to be used in variety of battery systems. This paper aims to identify the technological readiness of Universitas Sebelas Maret (UNS) lithium-ion enterprise to fulfill the lithium battery product standards by modifying the Quality Functional Deployment (QFD). Several production processes in technoware and humanware are not compatible with lithium battery standard, but all production processes in infoware and orgaware are compatible with lithium battery standard.

1. Introduction

The source of air pollution in the world is motorized vehicles [1]. One of the ways to overcome the problem of air pollution is to present an electric vehicle. The battery is a vital component of electric vehicles, because the battery is a source of power from the components of the electric vehicles [2]. The various types of batteries used in electric vehicles, and lithium batteries have the highest performance compared to other types of batteries [3]. Battery safety is a rather complex and sophisticated problem, and the safety of lithium-ion battery in vehicle is apriority of automotive industry [4]. The availability of international standards regarding secondary lithium-ion cell safety requirements have been developed by world organizations, the one of them is International Electrotechnical Commission (IEC) published IEC 62660-3:2016 standard. The standard is concerning in secondary lithium-ion cells for electric vehicles on safety requirements. Nevertheless, achieving a good level of safety and reliability is now a goal for global large-scale distributed energy management systems [5].

In addition to meet the standard criterias for the battery production process, the readiness of technological readiness in the production process must also be considered. Many people avoid technology because they are uncomfortable, not ready to use technology [6][7]. Technology readiness refers to the human tendency to use new technology to achieve goals both at work and in everyday life

[8]. In accordance with the systematic approach to technology, the components of technological system include technoware, humanware, infoware, and orgaware must be kept out of date.

This study purpose to identify each stage of the production in Universitas Sebelas Maret (UNS) lithium battery plant against the standard lithium battery product. furthermore, to provide recommendations for improvement component technologies to fulfill the quality requirements of IEC 62660-3:2016. This study was conducted in the production process at Universitas Sebelas Maret (UNS) battery plant. The UNS lithium plant sells lithium battery using make-to-order production system. The name of lithium battery product commercialized by UNS is SMART UNS-Lithium Battery. The specifications of SMART UNS-Lithium Battery are presented in **Table 1**.

Table 1. SMART UNS-Lithium Battery product specification

Cathode material	LFP	NCA	NMC
Material	$LiFePO_4$	$LiNiCoAlO_2$	$LiNiMnCoO_2$
Electrical conductivity (V)	3.4	3.6	3.6
Current capacity (Ah/Kg)	160	180	180
Energy density/weight (Wh/Kg)	544	648	648
Energy density/volume (Wh/L)	1,953	3,110	3,110

2. Methodology

Notation

j	: Index for production process	k	: Index for the lithium standard
HL	: Lithium's humanware sophistication for production process number k	HS	: Standard humanware sophistication for production process number k
TL	: Lithium's technoware sophistication for production process number k	TS	: Standard technoware sophistication for production process number k
IL	: Lithium's infoware sophistication for production process number k	IS	: Standard infoware sophistication for production process number k
OL	: Lithium's orgaware sophistication for production process number k	OS	: Standard orgaware sophistication for production process number k
AHL	: Lithium's aggregate humanware sophistication to fulfill lithium standard number j	AHS	: Standard aggregate humanware sophistication to fulfill lithium standard number j
ATL	: Lithium's aggregate technoware sophistication to fulfill lithium standard number j	ATS	: Standard aggregate technoware sophistication to fulfill lithium standard number j
AIL	: Lithium's aggregate infoware sophistication to fulfill lithium standard number j	AIS	: Standard aggregate infoware sophistication to fulfill lithium standard number j
AOL	: Lithium's aggregate orgaware sophistication to fulfill lithium standard number j	AOS	: Standard aggregate orgaware sophistication to fulfill lithium standard number j
W_j	: Weight importance of production process number k	C_{jk}	: Correlation value between production process number k with lithium standard number j
NW	: The normalized matrix on planning	W_k	: Weight importance of standard number j

According to Noor, Quality Function Deployment (QFD) is a methodology that translates the needs and desires of consumers into a product/service design that has certain technical requirements and quality characeristics [9]. This method is identify the Voice of Customer (VOC) [10]. QFD consists of six main parts, namely the Voice of Customer (VOC), relationship matrix, technical response, technical correlation matrix, technical matrix, and planning matrix [11]. Then QFD is modified to fit the research conducted, Voice of Customer (VOC) is represented by lithium battery product standard, relationship matrix, technical response is represented by production process, technical matrix is represented by technological readiness, and planning matrix is represented by technological component standard benchmark. The technical correlation matrix is not include in this study because the production process is assumed to be independent of each other. The original QFD is presented in **Figure 1** and modificated QFD is presented in **Figure 2**.

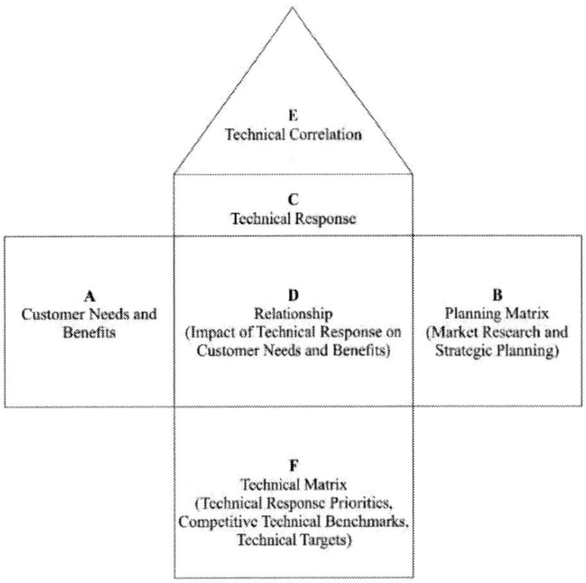

Figure 1. Original QFD

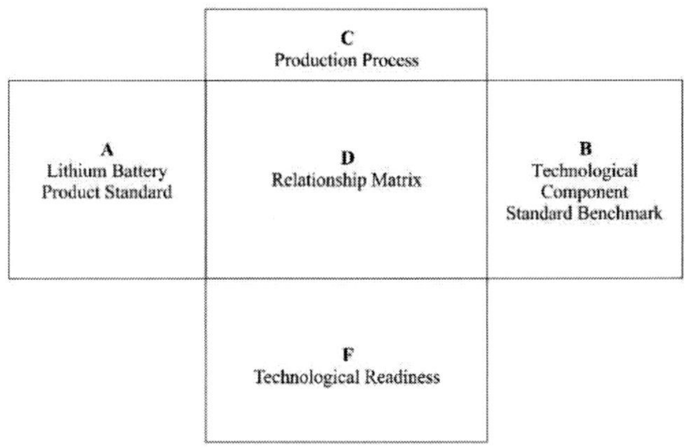

Figure 2. Modified QFD

The steps of methodology is presented in Figure 3. First step is identify lithium product standard. After that, lithium battery production processes are identified. Third, the existing sophistication of technology component (technoware, humanware, infoware, and orgaware) in every process is measured to get TL, HL, IL, and OL. Fourth, experts are asked to give TS, HS, IS, and OS value. TS, HS, IS, and OS value is minimum sophistication level of IEC's quality requirement that needs to be fulfilled. Fifth, the relationship between every production process with the quality requirement in IEC 62660-3:2016 is evaluated. Sixth, the importance between production production processes is evaluated by using paiwise comparison technique. Seventh, ATL, AHL, AIL, AOL, ATS, AHS, AIS, and AOS are calculated based on relationship matrix, importance weighting TL, HL, IL, OL, TS, HS, IS, and OS. Thus, the gap between ATL with ATS, AHL with AHS, AIL with AIS, and AOL with AOS.

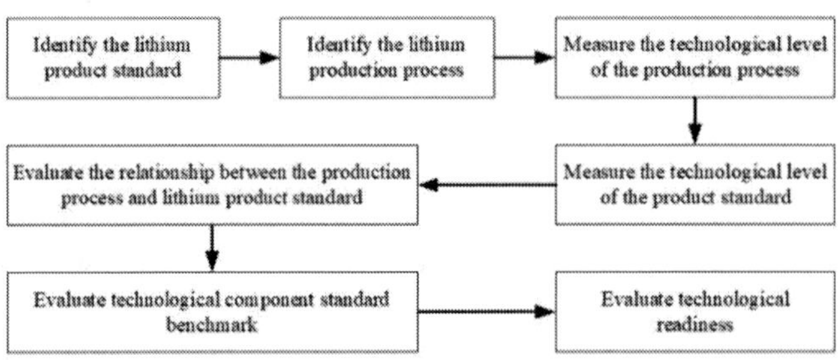

Figure 3. Methodology steps

3. Results

The process of producing lithium batteries is based on survey results can be identified are (1) mixing, (2) coating, (3) pressing, (4) slitting, (5) nickel tab welding, (6) winding, (7) welding into sleeves, (8) grooving, (9) cap welding, (10) electrolyte filling, (11) punching, (12) grading, (13) sorting, and (14) delivery.

To get TL, HL, IL, and OL score, a degree sophistication degree from UNESCAP (1989). The model has range score from 1 to 9, which the basic technological component has score 1 and the most advanced technological components has a score 9 with field survey and interview was taken to get TL, HL, IL, and OL score. Furthermore, two experts from electric vehicle technical committee and lithium battery researcher was asked to give TS, HS, IS, and OS to every production process. ATL, AHL, AIL, AOL, ATS, AHS, AIS, and AOS score is recapitulated in **Tabel 2**.

To find correlation between lithium battery standard *k* and production process *j,* a correlation matrix is used. A correlation value of 0, 1, 3 and 9 shows that the correlation between lithium battery standard *k* and production process *j* have no, weak, medium, and high correlation respectively. Two experts was asked to give the correlation value (C_{kj}) in relationship matrix. The correlation value is recapitulated in **Tabel 3**.

Every production process has a different importance weighting to create overall lithium battery product. the importance weighting between production process *j* is estimated by pairwise comparison technique, with two experts make the comparison judgment. In **Table 4** is recapitulation of the pairwise comparison (W_j) technique.

Table 2. ATL, AHL, AIL, AOL, ATS, AHS, AIS, and AOS score

Production Process / Tech. Components	*1*	*2*	*3*	*4*	*5*	*6*	*7*	*8*	*9*	*10*	*11*	*12*	*13*	*14*
TL	4.5	5.5	4.0	4.0	2.0	5.5	2.5	3.0	2.5	4.5	3.0	5.5	6.0	1.5
HL	4.0	5.5	4.0	4.0	4.0	4.5	4.0	4.0	4.5	5.0	4.0	4.0	5.0	4.5
IL	4.5	4.5	4.5	4.0	4.5	4.5	4.0	4.5	4.0	5.0	4.0	5.0	4.5	4.0
OL	5.0	5.0	4.0	3.5	4.5	5.5	3.5	4.0	3.0	5.0	3.5	4.0	3.5	3.5
TS	4.0	4.5	4.0	3.5	3.0	4.5	3.0	3.0	3.0	3.5	3.5	5.0	4.0	1.5
HS	3.0	3.5	3.0	3.0	4.5	3.5	3.5	3.0	3.5	4.5	3.0	5.0	3.5	2.5
IS	4.0	4.5	4.0	4.0	4.0	4.5	4.0	4.0	4.0	4.5	4.0	4.5	4.0	3.0
OS	3.0	3.5	3.0	3.0	3.0	3.5	3.0	3.0	3.0	3.5	3.0	3.0	3.0	2.5

Table 3. Correlation value in realtionship matrix

No. *j* / No. *k*	*1*	*2*	*3*	*4*	*5*	*6*	*7*	*8*	*9*	*10*	*11*	*12*	*13*	*14*
1	9.0	9.0	3.0	1.0	9.0	9.0	5.2	1.0	3.0	9.0	1.0	0	0	3.0
2	9.0	9.0	3.0	1.0	9.0	9.0	5.2	3.0	3.0	9.0	1.0	0	0	3.0
3	9.0	9.0	3.0	1.0	9.0	9.0	5.2	3.0	3.0	9.0	1.0	0	0	3.0
4	9.0	9.0	1.7	1.0	9.0	9.0	9.0	1.0	3.0	9.0	1.0	0	0	1.0
5	9.0	9.0	1.7	1.0	9.0	9.0	9.0	1.0	3.0	9.0	1.0	0	0	1.0
6	9.0	9.0	1.7	1.0	9.0	9.0	9.0	1.0	3.0	9.0	0	3.0	0	1.0
7	9.0	9.0	1.7	1.0	9.0	9.0	9.0	1.0	3.0	9.0	0	3.0	0	1.0
8	9.0	9.0	1.7	1.0	9.0	9.0	9.0	1.0	3.0	9.0	0	3.0	0	1.0
9	9.0	9.0	1.7	1.0	9.0	9.0	9.0	1.0	3.0	9.0	0	30	0	1.0

Table 4. Importance weighting of the production process

No. j	1	2	3	4	5	6	7	8	9	10	11	12	13	14
Importan	15.8	16.4	10.4	4.4	6.0	11.2	5.3	4.7	4.6	8.8	3.3	3.9	2.8	1.7
ce (%)	4	8	1	7	9	1	0	3	8	9	2	7	1	1

Lithium's technoware, humanware, infoware, and orgaware gap is defined as when raw technoware lithium is lower than raw technoware standard, raw humanware lithium is lower than raw humanware standard, raw infoware lithium is lower than raw infoware standard, and raw orgaware lithium is lower than raw orgaware standard respectively. Before determining the gap, we need to calculate ATL, AHL, AIL, AOL, ATS, AHS, AIS, and AOS by Equations (1)-(8). Technoware, humanware, infoware, and orgaware gap is identified when the value between raw lithium and raw standard is negatif. If the value between raw lithium and raw standard is positive, the production process meets lithium standard. Raw technoware lithium, raw humanware lithium, raw infoware lithium, raw orgaware lithium, raw technoware standard, raw humanware standard, raw infoware standard, and raw orgaware standard can be calculated by Equations (9)-(16). But before calculated the gap, we need to estimate the importance weighting between lithium standard by using pairwise comparison based on two experts. The result of importance weighting between lithium standard is showed in **Table 5** and the result of calculation by Equation (9)-(16) is showed in **Table 6, Table 7, Table 8, and Table 9**.

$$ATL_j = \sum_k W_k \, x \, TL_{k,j} \quad (1)$$
$$AHL_j = \sum_k W_k \, x \, HL_{k,j} \quad (2)$$
$$AIL_j = \sum_k W_k \, x \, IL_{k,j} \quad (3)$$
$$AOL_j = \sum_k W_k \, x \, OL_{k,j} \quad (4)$$
$$ATS_j = \sum_k W_k \, x \, TS_{k,j} \quad (5)$$
$$AHS_j = \sum_k W_k \, x \, HS_{k,j} \quad (6)$$
$$AIS_j = \sum_k W_k \, x \, IS_{k,j} \quad (7)$$
$$AOS_j = \sum_k W_k \, x \, OS_{k,j} \quad (8)$$
$$Raw\ Technoware\ Standard = \sum_k C_{k,j} \, x \, W_k \, x \, TS_{k,j} \, x \, NW_j \quad (9)$$
$$Raw\ Humanware\ Standard = \sum_k C_{k,j} \, x \, W_k \, x \, HS_{k,j} \, x \, NW_j \quad (10)$$
$$Raw\ Infoware\ Standard = \sum_k C_{k,j} \, x \, W_k \, x \, IS_{k,j} \, x \, NW_j \quad (11)$$
$$Raw\ Orgaware\ Standard = \sum_k C_{k,j} \, x \, W_k x \, OS_{k,j} \, x \, NW_j \quad (12)$$
$$Raw\ Technoware\ Lithium = \sum_k C_{k,j} \, x \, W_k \, x \, TL_{k,j} \, x \, NW_j \quad (13)$$
$$Raw\ Humanware\ Lithium = \sum_k C_{k,j} \, x \, W_k x \, HL_{k,j} \, x \, NW_j \quad (14)$$
$$Raw\ Infoware\ Lithium = \sum_k C_{k,j} \, x \, W_k \, x \, IL_{k,j} \, x \, NW_j \quad (15)$$
$$Raw\ Orgaware\ Lithium = \sum_k C_{k,j} \, x \, W_k x \, OL_{k,j} \, x \, NW_j \quad (16)$$

Table 5. Importance weighting of the lithium standard

No. k	1	2	3	4	5	6	7	8	9
Importance (%)	5.77	5.53	5.95	17.50	9.86	17.88	11.27	5.20	21.04

Table 6. Technoware gap

Production Process	*1*	*2*	*3*	*4*	*5*	*6*	*7*	*8*	*9*	*10*	*11*	*12*	*13*	*14*
Raw Technoware Standard	4.00	4.50	0.87	0.39	3.00	4.50	2.78	0.41	1.00	3.50	0.17	0.92	0	0.22
Raw Technoware Lithium	4.50	5.50	0.87	0.44	2.00	5.50	2.32	0.41	0.00	4.50	0.15	1.02	0	0.22
Gap	0.50	1.00	0	0.06	-1.00	1	-0.46	0	-0.17	1.00	-0.02	0.09	0	0

Table 7. Humanware gap

Production Process	*1*	*2*	*3*	*4*	*5*	*6*	*7*	*8*	*9*	*10*	*11*	*12*	*13*	*14*
Raw Humanware Standard	3.00	3.50	0.65	0.33	4.50	3.50	3.24	0.41	1.17	4.50	0.15	0.92	0	0.37
Raw Humanware Lithium	4.00	5.50	0.87	0.44	4.00	4.50	3.71	0.55	1.50	5.00	0.20	0.74	0	0.67
Gap	1.00	2.00	0.22	0.11	-0.50	1.00	0.46	0.14	0.33	0.50	0.05	-0.18	0	0.30

Table 8. Infoware gap

Production Process	*1*	*2*	*3*	*4*	*5*	*6*	*7*	*8*	*9*	*10*	*11*	*12*	*13*	*14*
Raw Infoware Standard	4.00	4.50	0.87	0.44	4.00	4.50	3.71	0.55	1.33	4.50	0.20	0.83	0	0.45
Raw Infoware Lithium	4.5	4.5	0.98	0.44	4.50	4.50	3.71	0.61	1.33	5.00	0.20	0.92	0	0.60
Gap	0.50	0	0.11	0	0.50	0	0	0.07	0	0.50	0	0.09	0	0.15

Table 9. Orgaware gap

Production Process	*1*	*2*	*3*	*4*	*5*	*6*	*7*	*8*	*9*	*10*	*11*	*12*	*13*	*14*
Raw Orgaware Standard	3.00	3.50	0.65	0.33	3.00	3.50	2.78	0.41	1.00	3.50	0.15	0.55	0	0.37
Raw Orgaware Lithium	5.00	5.00	0.87	039	4.50	5.50	3.24	0.55	1.00	5.00	0.17	0.74	0	0.52
Gap	2.00	1.50	0.22	0.06	1.50	2.00	0.46	0.14	0	1.50	0.02	0.18	0	0.15

From the result of gap calculation, in the technoware gap there are four production processes are not sophisticated enough to fulfill the lithium battery standard, there are nickel tab welding process, welding into sleeves process, cap welding process, and punching process. In the humanware gap there are two production processes are not sophisticated enough to fulfill the lihtium battery standard, there are nickel tab welding process and grading process. In the infoware and orgaware gap that compatible to fulfill all lithium battery standard.

4. Analysis

The result show that all technological components in production process are partially compatible with lithium battery standard. In another hand, several production processes in technoware and humanware are not compatible with battery lithium standard, but the production processes in infoware and orgaware are compatible with battery lithium standard. There are several reasons why lithium battery technoware and humanware do not sophisticate enough to fulfill the standard. The first reason is standards that are still considered less important for their business. The second is because the business is relatively new, so it still needs to be adjusted to existing standards. The last is lack of knowledge and financial to conduct training and improve the sophistication of machines and tools. Those three reasons are main problems for lithium battery plant to improve their compatibility with standard.

Production process No. 1, 2, 5, 6, 7, and 10 has high correlation value with lithium battery standard. Thus, improvement of production process No. 1, 2, 5, 6, 7, and 10 have high impact to fulfill lihtium battery standard requirement.

References

[1] Siami L, Sofyan A and Frazila Russ B 2014 Perhitungan penurunan beban emisi pencemaran udara dari pembangunan jalur tol jorr w2 DKI Jakarta. *J. Teknik Lingkungan* **20** 2.

[2] Afif Muhammad T, Pratiwi and Ilham Ayu P 2015 Analisa perbandingan baterai lithium-ion, lithium polymer, lead acid dan nickel-metal hydride pada penggunanaan mobil listrik-

review. *J. Rekasaya Mesin* **6** 2.

[3] Wang Yang N 2019 Power battery performance detection system for electric vehicles. *Procedia Computer Science* **154** 759-763.

[4] Hollmotz L 2013 Safety of lithium ion batteries in vehicles – state of the art, risk, dan trends. *23rd International Technical Conference on the Enhanced Safety of Vehicle (ESV): Research Collaboration to Benefits Safety of All Road Users.*

[5] Liao Z, Zhang S, Li K, Zhang G, and Habetler T G 2019 A survey of method for monitoring and detecting thermal runaway of lithium-ion batteries. *Journal of Power Sources* **436**

[6] Lin J S C and Hsieh P L 2007 The influence of technology readiness on satisfaction and behavioral intentions toward self-service technologies. *Computers in Human Behavior* **23** 1597–1615.

[7] Meuter M L, Ostrom A L, Bitner M J and Roundtree R 2003 The influence of technology anxiety on consumer use and experiences with self-service technologies. *Journal of Business Research* **56** 899–906.

[8] Smith R and Sharif N 2007 Understanding and acquiring technology assets for global competition. *Technovation* **27** 643-649.

[9] Noor Asep Muhamad 2000 Pengembangan komputer software quality function deployment *Procceding Pertemuan Ilmiah BKSTI.*

[10] Suhendar E and Suroto 2014 Penerapan metode quality function deployment (QFD) dalam upaya peningkatan kualitas pelayanan akademik pada UB. *Faktor Exacta* **7** 372-386.

[11] Cohen L 1995 *Quality Function Deployment : "How To Make QFD For You".* Menlo Park: Addison-Wesley Publishing Company.

[12] International Electrotechnical Commission, "IEC 62660-3:2016 Secondary Lithium-Ion Cells for Propulsion of Electric Road Vehicles Part 3: Safety Requirement", 2016.

IOP Conf. Series: Materials Science and Engineering **943** (2020) 012053 doi:10.1088/1757-899X/943/1/012053

Sound Energy Approach in The Use of LRAD as A Bird Deterrent Device in Sam Ratulangi Manado International Airport, North Sulawesi, Indonesia

B Basuki* and M R Palupi

Center for Research and Development of Human Resources, National Standardization Agency, Indonesia.

Email: budhyb@bsn.go.id

Abstract. To fulfil the airport safety requirements as stated in International Civil Aviation Organization document, ICAO 9137-3 AN part 3 – Airport Services Manual, Wildlife control and reduction, especially the bird stike control programme, the Sam Ratulangi International Airport is decided to apply the auditory method to control the birds in aerodrome area. Study on this need have been carried out using the sound energy approach, more specifically on inverse square law theory in free field. The theory said that by doubling the distance between the speaker and the sound level meter will decrease in sound pressure level of around 6 dB. From the field measurement results, the measured sound pressure level were 85 dB and 79 dB at the distance of 50 m and 100 m from the speaker. Both levels and distance were used in trials of repelling birds at 1600 Hz to 3150 Hz in frequencies. The use of long range acoustic device, LRAD speakers, that genereate 156 dB at a distance of 1 m in front of the speaker can effectively reach longer distances.

1. Introduction

Bird strike is the impact of touching two high-speed materials which have very large differences in characteristics, where birds are soft in contact with rigid fuselage, resulting in nonlinear material behavior, high strain rate, and very large deformation [1]. The incidence of bird disturbance is reported to increase every year mainly due to environmental conservation and protection programs that are increasingly stringent so that it supports the growth of populations of species of birds. Bird disturbance can occur at varying heights and 41% of bird disturbance events in America occur during take-off or landing [2].

International Civil Aviation Organization (ICAO) document 9137-AN section 3, Wildlife Control and Reduction requires the prevention of bird disturbance at airports to be met for flight safety [3]. Flight safety is regulated in the Law of the Republic of Indonesia No. 1 of 2009 concerning Aviation, and reporting procedures for bird attacks in airports and surrounding areas are regulated in General Director of Civil Aviation Regulation No. KP.468 in 2011 [4,5].

The techniques for controlling bird disturbances at airports can be categorized as auditory, visual, chemical, exclusion, habitat modification, and annihilation. Auditory is usually done using the acoustic method by recording the sound of predatory animals and emitting predator animal sounds through loudspeakers at airports [6]. Another method that developed by acoustic researchers in Indonesia is to use the bird communication frequencies which are estimated also be the frequencies of bird sensitive hearing. These frequencies are then emitted through loudspeakers to a group of birds [7].

As the one of International Airports in Indonesia, the Sam Ratulangi International Airport must fulfil the airport safety requirements as stated in regulations above. The Sam Ratulangi International Airport is decided to apply the auditory method to control the birds in aerodrome area by using the hearing sensitive frequencies method. The airport puts some long range acoustic device, LRAD speakers to repel the birds at its aerodrome area.

This paper reports about the study of the effective range of speaker area by using the sound energy approach, more specifically on inverse square law theory in free field and the effectiveness of repelling birds by using LRAD speakers in Sam Ratulangi International Airport.

2. The Inverse Square Law Theory[8]

Radiation of sound energy in all directions with power (power) with a constant sound / fixed, sound spread with intensity (power per unit area) that is spread evenly to all the direction forming the surface of the ball (spherically) is increasingly expanding with a radius of d, 2d, 3d, 4d and so on. With the intensity of the sound continues to decrease the additional distance of the ball radius where the surface area of the ball is 4πr2. Addition of the distance doubled from d to 2d will decrease the intensity to a quarter (1/4 of it), addition of the distance tripled to 3d intensity decreases to 1/9, addition of the distance fourfold ie 4d intensity will decrease by 1/16 . The sound intensity of an open free area with the addition of the distance from the point of the sound source will decrease in proportion to the inverse square of the distance, confirm the inverse square law.

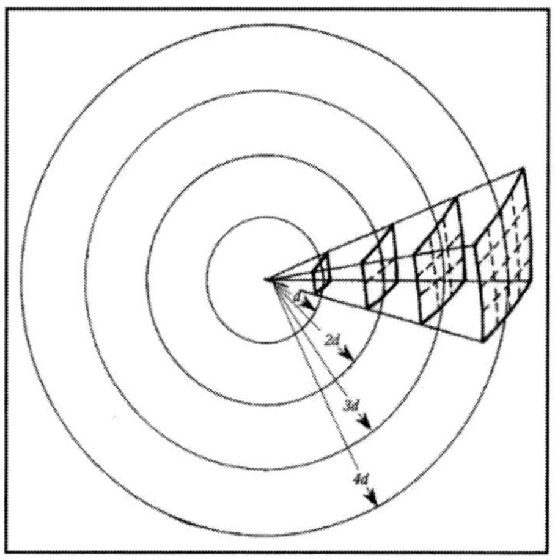

Figure 1. A series of illustrative characteristics of sound in an open free area spreading in all directions and forming a ball (spherically). Sound intensity is inversely proportional to the square of the distance from the point of the sound source

Sound intensity, ie power per unit area is a parameter that is not easy to measure, which is much easier to measure is sound pressure. The intensity is proportional to the pressure squared: I ~ p2 ~ 1 / r2, r is the distance from the sound source, that p ~ 1 / r, then the inverse square law for sound intensity becomes the inverse distance law for sound pressure, this means that the amount of sound pressure will vary with distance. The amount of sound pressure at a certain distance from a sound source point in open space if measured at different distances then the sound pressure will be inversely proportional to the change in distance from the measuring point. As shown by Figure 2 the sound pressure level (SPL) in the decibels is mapped and varies with distance, there is a decrease in sound pressure level of 6 dB for each doubling the distance of the measuring point to the sound source.

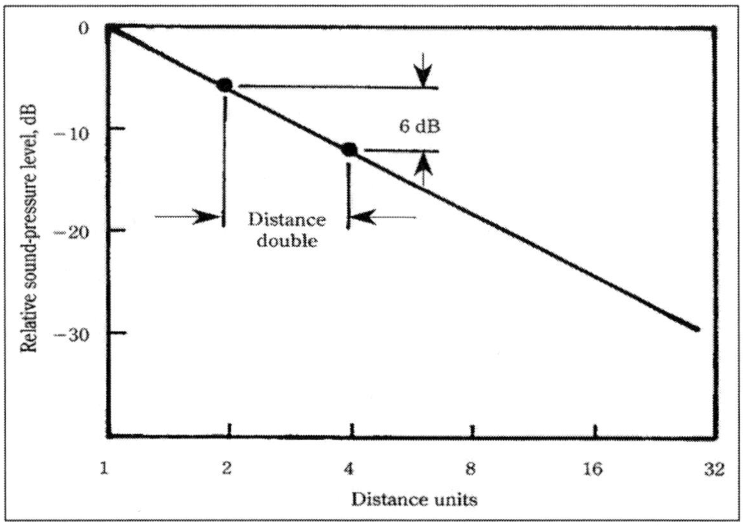

Figure 2. The inverse square law for sound intensity becomes the inverse distance law for sound pressure. The sound pressure level decreases by 6 dB for every double addition of distance

This calculation can be shown by equation (1) below, for example the sound pressure level at the position closest to the sound source is L1 with distance d1 from the sound source. Then in the next position the sound pressure level is L2 with distance d2 from the sound source then:

$$\boldsymbol{L2 = L1 - 20log\frac{d2}{d1}} \text{ dB} \quad (1)$$

thus the difference in sound pressure level of the two measuring points with the distance from the sound source d1 and d2 respectively:

$$\boldsymbol{L2 - L1 = 20log\frac{d2}{d1}} \text{ dB} \quad (2)$$

3. Bird Hearing Frequencies [9]

Birds communicate with each other using sounds produced by the birds themselves. Birds have hearing sensitivity 10 x more sensitive than humans. By using the analogy that birds are living things, there are sensitive areas of hearing birds that can be subject to sound at certain sound pressure levels that will touch the threshold of pain so that they feel uncomfortable.

Bird ear is divided into three parts, namely the outer ear, the middle part, and the inner part. The outer ear consists of a tube (ear canal) which will deliver sound waves to the eardrum (eardrum). Behind the eardrum there is the middle ear which consists of a single bone called collumella which is responsible for transforming the vibrational signal into electrical pulses and is transmitted to the inner ear called the cochlea. The inner ear is filled with fluid and nerve cells will deliver electrical pulses to the brain for processing. The results of processing electrical pulses by the brain is in the form of sound perceptions such as loud weak, rhythm, high-low, distance, sound position and so forth. With this ability, living things like humans and birds can focus their attention and respond or act on the sounds that are heard in their ears.

4. Measurement Results

The background noise in Sam Ratulangi International Airport aerodrome was measured to find out the minimum level should be emitted from the speakers by using the inverse square law theory. The minimum level at the furthermost of speaker perimeter range must be at least 10 dB higher than the highest LAeq of background noise. The highest LAeq of background noise is 75.8 dB measured. Thus, the minimum level should be 85.8 dB or rounded to 86 dB at the furthermost of speaker perimeter range.

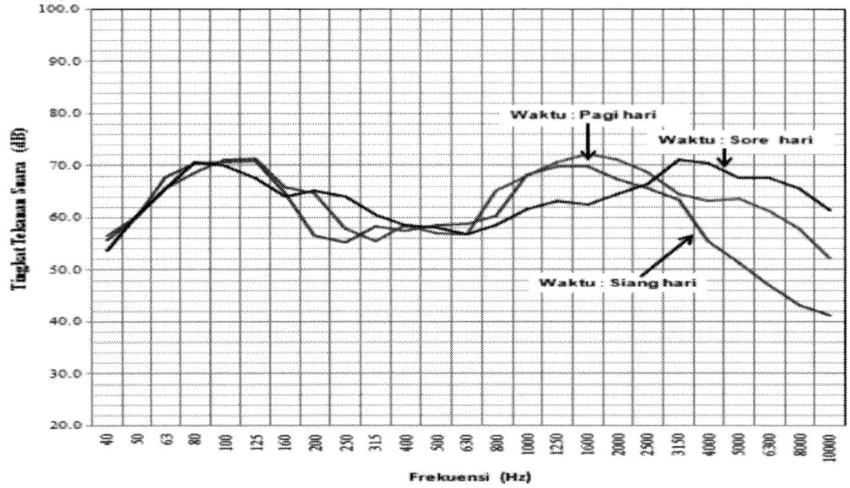

Figure 3. Background Noise Measurement Results at Sam Ratulangi International Airport measured on May 29, 2016. Red line is the morning measurement, blue line is the noon measurement, and black line is afternoon measurement.

The aircraft noise and birds communication frequencies were measured and both were compared to see the frequency distribution. The aircraft noise about 40 Hz to 250 Hz. Meanwhile the bird communication frequencies about 1600 Hz to 4000 Hz.

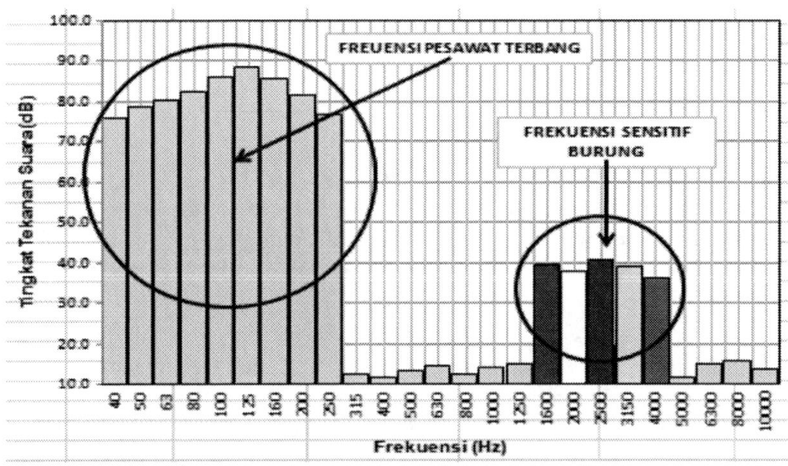

Figure 4. Comparison between aircraft noise (silver bar) and bird communication frequencies (colours) measured at Sam Ratulangi International Airport.

After obtaining the above data, a field trial was carried out to repel birds. The first experiment was carried out by emitting 85 dB of sound at a distance of 50 m, frequencies of 1600 Hz, 2000 Hz, 2500 Hz, 3150 Hz and 4000 Hz. The output frequency is set to automatic swipe per frequency for 2 minutes to see bird behaviours. The second experiment was carried out in the same way to "shoot" a flock of birds within 100 m without changing sound amplification. In this second experiment, the sound level at 100 m is 79 dB measured.

Figure 5. Buffalo egrets (bubulcus ibis) fly away from the speaker when given sound exposure in a bird's sensitive hearing frequency area with a loudness level of 85 dB.

In the first experiment, when the sound signal was at 1600 Hz, the birds had not yet responded. The birds were still carrying out their activities, looking for food in between the grass. When the sound signal moved to frequencies of 2000 Hz and 2500 Hz, most birds simultaneously flew away from the origin of the sound. When the signal moved to 3150 Hz, the remaining birds flew all away from the speaker perimeter area and leave a clean area of the bird.

In the second experiment, when the sound signal at frequencies of 1600 Hz to 2000 Hz the birds at the target location did not respond at all. Then when the frequency shifted at a frequency of 2500 Hz, 3150 Hz and 4000 Hz, there were only some birds that flew but not far away. The flew birds landing back at another location within 10 meters of their original position before. This indicates that the exposure of acoustic waves with a level of 79 dB has no disruptive effect on the bird's ears even though the frequency is in accordance with the communication frequency of the bird.

5. Discussion

The distance of Sam Ratulangi International Airport runway is 2650 m and 45 m width. If the perimeter range of speaker is 100 m, then the airport will need about 13 fixed speakers to be put near the runway to clear the runway area from birds. From the experiments and equation (2), the ideal sound pressure level at the furthest perimeter area is about 85 dB. It means that at 100 m should reach 85 dB, and it means at 1 m in front of speaker should be 119 dB measured.

The amount of 13 fixed speakers could be reduce by expanding the perimeter area to 500 m. It only need 2 fixed speakers to be put near the runway, and the rest unreach area can be cleaned by the mobile speaker system. For the fixed speaker, it should meet the need of emitting sound at 1 m in front of speaker about 139 dB measured in order to reach 85 dB at 500 m.

The officer who in charge to maintain the bird deterrent speakers should be equipped with ear muff to protect his hearing. Based on the Indonesia Ministry of Health regulation No 70/2016, the threshold of noise could accept by human ear is 14.06 seconds for 118 dB sound pressure level and 0.11 second for 139 dB [10].

Based on the measurement results, experiments results, and the calculation of sound pressure level that should be emitted from speaker, Sam Ratulangi International Airport decided to use long range acoustic device (LRAD) as the speaker for the bird deterrent system. It installed on a pickup car as the mobile version of bird deterrent system. The specification of LRAD is a speaker capable of producing loud and clear sound, this device is able to produce up to 156 dB sound in a distance of 1 meter with a flat frequency response graph that indicate that this device relatively could emit the stable sound [11,12].

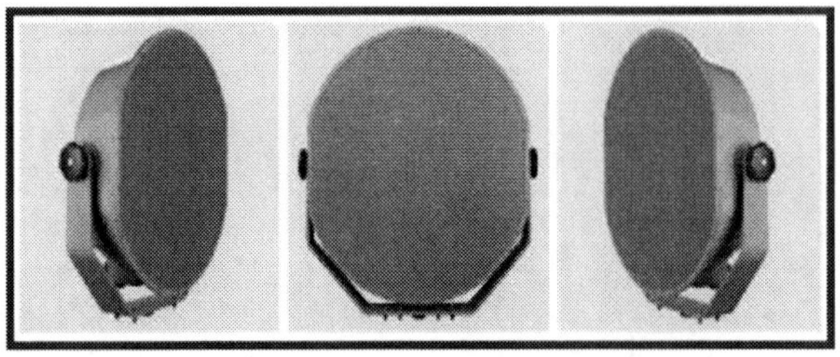

Figure 6. Speaker Model Type Long Range Acoustic Device (LRAD) is a speaker capable of producing loud and clear sound, this device is able to produce up to 156 dB sound in a distance of 1 meter

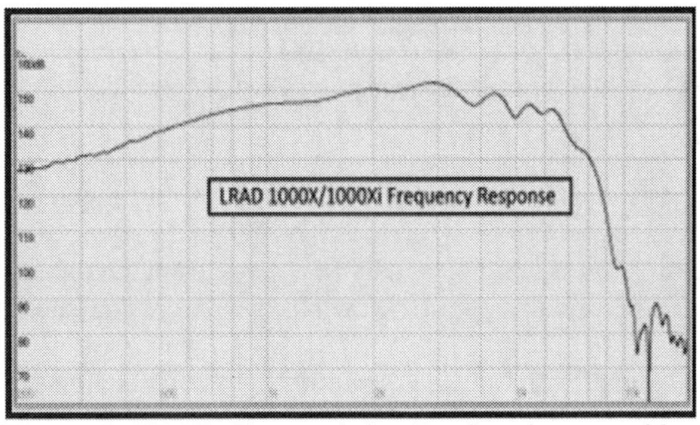

Figure 7. LRAD Characteristics Speaker: have a wide frequency range capable of producing maximum intensity in the frequency range of 3 kHz to 10 kHz.

Figure 8. LRAD Speaker that already operated at Sam Ratulangi International Airport as as the mobile version of bird deterrent system

6. Conclusion

The sound energy approach especially by using inverse square law theory in free field can assist to determine the speaker specification needed as a bird deterrent device in Sam Ratulangi International Airport. In addition, the data needed to determine the system specifications are background noise, aircraft noise frequencies, bird species and communication frequencies in the aerodrome area.

From the field measurement results, the measured sound pressure level were 85 dB and 79 dB at the distance of 50 m and 100 m from the speaker. Both levels and distance were used in trials of repelling birds at 1600 Hz to 3150 Hz in frequencies. The 85 dB more effective to make the bird flew away from the speaker.

If the perimeter area of the speaker is about 500 m, the speaker should meet 139 dB at 1m in front of speaker in order to reach 85 dB at 500 m. This 85 dB meet the 10 dB above of the highest LAeq background noise, that about 75.8 dB.

Acknowledgments

Authors wishing to acknowledge the management of Sam Ratulangi International Airport in Manado, North Sulawesi, PT Angkasa Pura I (Persero), the Safety Management System team. PT Angkasa Pura I Sam Ratulangi International Airport, Manado, North Sulawesi, the Acoustic & Vibration Metrology Subdivision of the LIPI Metrology Research Center (currently known as the National Standard Laboratory for Acoustic & Vibration Measuring Units of the National Standardization Agency), assistance from colleagues: Achmad Suwandi, Dewi Malia, Rukmana, and Ahnan Ma'ruf (Alm).

Contribution Statement

IOP Conf. Series: Materials Science and Engineering **943** (2020) 012053 doi:10.1088/1757-899X/943/1/012053

The authors of this paper stated that both authors were the main contributors to the writing of this paper.

References

[1] Hedayati R., Sadighi M. 2015. *Bird Strike 1st Edition. An Experimental, Theoretical and Numerical Investigation.* Woodhead Publishing.

[2] Martono A. 2007. *Gangguan Burung di Bandara dan Sekitarnya.* Bahan Paparan di Hadapan Civitas Akademika Fakultas Kedokteran Hewan Universitas Airlangga.

[3] International Civil Aviation Organization. 2012. *Doc 9137. Airport Services Manual. Part 3 Wildlife Control and Reduction.* ICAO.

[4] Indonesian Government. 2009. *Law of the Republic of Indonesia No. 1 of 2009 concerning Aviation.* Republic of Indonesia.

[5] General Director of Indonesia Aviation. 2011. *Regulation No. KP.468/2011 about Bird Strike Report Procedure in Airport.* Republic of Indonesia.

[6] Bishop, J. 2003. *Review of international research literature regarding the effectiveness of auditory bird scaring techniques and potential alternatives.* https://pdfs.semanticscholar.org/52d1/7a806a7156c45b3f50bf7ea8eb7b918ac4ab.pdf

[7] Akil, Husein A. 2003. *Metode Akustik untuk Menghalau Jenis Burung yang Terdapat di Sekitar Landasan Pacu Bandara Soekarno Hatta*. Prosiding Seminar Nasional Akustik dan Teknik Tata Suara.

[8] Everest Alton F. 2001. *Master Handbook of Acoustics*, Fourth Edition, McGraw-Hill, pages 41-87.

[9] Linda Pesek DVM. 2003. *The Avian Ear – Part I Anatomy*, http://www.birdsnways.com/wisdom/ww71eiv.htm

[10] Indonesian Ministry of Health regulation. 2016. No 70/ 2016: Standard and Requirements of Industrial work environment health. Republic of Indonesia.

[11] American Technology Corp. 2005. *Long Range Acoustic Devices*, Power Point Presentation, September 2005, San Diego, CA: at http: /www.atcsd.com/

[12] LRAD, *LRAD Product Guide*, https: //www.lradx.com/product_categories/long-range-acoustic-devices/

IOP Conf. Series: Materials Science and Engineering **943** (2020) 012054 doi:10.1088/1757-899X/943/1/012054

A Closed-loop Supply Chain Inventory Model Considering Limited Number of Remanufacturing Generation and Environmental Investigation

W A Jauhari[1*], R D Septian[1], P W Laksono[1], A R Dwicahyani[2]
[1] Department of Industrial Engineering, Universitas Sebelas Maret, Surakarta, Indonesia
[2] Department of Industrial Engineering, Adhi Tama Institute of Technology, Surabaya, Indonesia

E-mail: wakhidjauhari@gmail.com

Abstract. This study presents a mathematical inventory model in the Closed-loop Supply Chain (CLSC) system. We investigate a CLSC system that consists of a retailer, a manufacturer and a 3rd party collection dealer. The model aims to minimize the annual joint total inventory cost for the three parties and determine several decision variables, including number of shipments, manufacturer shipment lot size and number of remanufacturing generations. The number of remanufacturing generations is one of our concerns since we consider a type of product that could not be remanufactured in an unlimited number of time. Thus, how many times the product should be remanufactured becomes important. In this study, we also investigate the environmental impacts in term of carbon emissions, energy usage, as well as annual disposed items. The model can guide companies in the related industry to plan a system of CLSC inventory management with environmental considerations.

1. Introduction

A supply chain provides a network for business players and suppliers working together to deliver goods, services and information to consumers to be more effectively and efficiently [1]. Within a good planning and design of supply chain management, the company could achieve what we call by an "optimal supply chain system". There are two streams in supply chain management, namely open-loop supply chain and closed-loop supply chain. Open-loop supply chain is noticed by no product returns from consumer to initial producer [2]. Whereas closed loop supply chain is characterized by the existence of product returns managed by the manufacturer or supplier by performing the recovery process on the returned product prior to resale. The recovery process can be categorized into repair, refurbishing, remanufacturing, cannibalization and recycling [3]. The categorization of the recovery process depends on the type of product and the treatment needed to recover the product.

In a supply chain system, inventory need to be controlled effectively and efficiently. Inventories play a role in eliminating the risk of delays and material deficiencies. Excessive inventory will be resulting in waste due to high storage and maintenance costs. While the lack of inventory leads to shortages that may disrupt the activities of production and distribution. There are many prior studies that focus on the development of an optimal inventory modelling in reverse logistics, including [4], [5], [6], [7], [8], [9]. [10].

On the other hand, environmental issues that continue to be the world's attention have forced manufacturing companies to carry out sustainable production and distribution activities. As discussed formerly, a closed-loop supply chain is a sustainable operation that enables manufacturer to recover returned products from the market for future resale. The recovery of used items can reduce waste of end-of-life products in the market [11]. In addition, it can also reduce the use of raw materials from nature. Therefore, the application of closed loop supply chain in the world of manufacturing industry is closely related to sustainable supply chain and environmental considerations.

El Saadany *et al.* [7] developed a single-echelon inventory model considering a number of limited remanufacturing generations. The number of remanufacturing generations indicates the number of times a product can be remanufactured. This relaxes the general assumption regarding products that can be continuously remanufactured at an unlimited number of times. The number of remanufacturing generations depends on the amount of the R&D investments made on the product. The higher number of remanufacturing generations, the lesser the waste would be. Bazan *et al.* [8] improved the model of [7] to consider carbon emissions, energy, and transportation costs. Dwicahyani *et al.* [10] then developed the model of [8] by including more parties among the supply chain, which are a distributor that acts as both retailer and collector of used items.

Based on the above discussion, many studies have accommodated the recovery process of used items. However, the company who collects used items from the market is always the manufacturer or the retailer itself. Moreover, there has been no research to include inspection of used items that returned from the market before they went through the recovery process. In this study, we developed a three-echelon closed-loop supply chain inventory model consisting of a manufacturer, a retailer and a 3rd party collection dealer as a collector of used items from the market. We consider a limited number of remanufacturing generations [7] and also includes freight costs as well as carbon emissions costs generated from the regular production process, remanufacturing process and transportation activities.

2. Model development

Here, we examine a CLSC system that consists of a retailer, a manufacturer and a used item collector. We seek to get the optimal solution regarding several inventory decisions, such as the shipment lot size, number of shipments per year, and also the number of remanufacturing generations. The material flow among those parties are depicted in Figure 1.

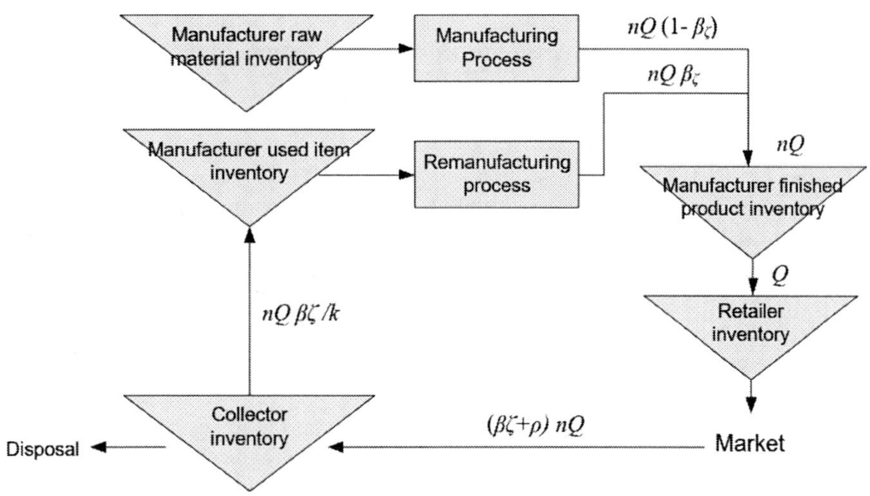

Figure 1. The CLSC system investigated in the model.

In this model, we consider the concept of remanufacturing generations firstly proposed by [7]. By this concept, items that classified as recoverable will go through several number of remanufacturing generations before they are completely disposed as waste. This study intends to determine the number of remanufacturing generations along with the determination of shipment lot size and number of shipment between investigated parties so the total profit can be maximized.

We consider two components of cost related to environmental implications, which are cost of carbon emission and energy usage. We adopt the functions of both costs from [8]. The following component of costs are considered in the development of model's objective function:

a. Holding costs at retailer, manufacturer (finished product inventory, recoverable item inventory, and raw material inventory), and collector

b. Ordering costs for retailer and manufacturer
c. Set-up costs of manufacturing and remanufacturing processes
d. Manufacturer's raw materials purchasing cost
e. Manufacturer's used items purchasing cost
f. Used item collection cost
g. Cost for remanufacturing
h. Cost for manufacturing
i. Annual R&D investment cost
j. Collector's waste disposal cost
k. Collector's inspection cost
l. Finished goods (from manufacturer to retailer) and return items (from collector to manufacturer) transportation cost
m. GHG emissions cost from transportation activities
n. GHG emissions cost from manufacturing and remanufacturing processes
o. Energy usage cost for manufacturing and remanufacturing processes

The objective of this model is to minimize all of the above costs and determine the following decision variables:

k : number of shipments from collector to manufacturer, w.r.t $k \geq 1$ and an integer
n : number of shipments from manufacturer to retailer, w.r.t $n \geq 1$ and an integer
Q : shipment lot size from manufacturer to retailer (unit)
ζ : number of remanufacturing generations, w.r.t $\zeta \geq 0$ and an integer

The parameters used in this model following with their notations and constraints are listed as follows:

β : nominal proportion of used items returned for remanufacturing, when the number of remanufacturing generations is unlimited (ζ=∞), with $0 \leq \beta < 1$;
β_ζ : actual proportion of used items returned for remanufacturing, when the number of remanufacturing generations is limited ($\zeta \geq 0$), with $\beta_\zeta = 1 - \left[\frac{(1-\beta)}{(1-\beta^{\zeta+1})}\right]$;
H_1 : holding cost per unit per unit of time at manufacturer finished product inventory ($/unit/year);
H_2 : holding cost per unit per unit of time at retailer inventory ($/unit/year);
H_3 : holding cost per unit per unit of time at collector inventory ($/unit/year);
H_4 : holding cost at manufacturer recoverable item inventory ($/unit/year);
H_5 : holding cost at manufacturer raw material inventory ($/unit/year);
A_1 : set-up cost of remanufacturing ($/set-up);
A_2 : set-up cost of manufacturing ($/set-up);
S_1 : ordering cost for manufacturer recoverable inventory ($/order);
S_2 : ordering cost for retailer inventory ($/order);
P_m : retailer purchasing cost ($/unit);
P_b : collector used item collection cost ($/unit);
C_{re} : remanufacturing cost per unit ($/unit);
C_{mn} : manufacturing cost per unit ($/unit);
a_i : emissions function parameter for process i (ton year^2/unit^3), with $i=\{p, r\}$ where $i=p$ for manufacturing activity and $i=r$ for remanufacturing activity;
b_i : emissions function parameter for for process i (ton year/unit^2), with $i=\{p, r\}$ where $i=p$ for manufacturing activity and $i=r$ for remanufacturing activity;
c_i : emissions function parameter for process i (ton/unit), with $i=\{p, r\}$ where $i=p$ for manufacturing and $i=r$ for remanufacturing;
$C'0$: the required energy at the machine to manufacture one unit (KWh/unit);
$C'1$: the required energy per year when manufacturing is idle (KWh/year);
$C''0$: the required energy at the machine to remanufacture one unit (KWh/unit);
$C''1$: the required energy per year when remanufacturing is idle (KWh/year);
d : annual demand rate (units/year);
v : annual remanufacturing rate (units/year), with $v > d$;
γ : annual manufacturing rate (units/year) with $\gamma > d$;
C_{GHG} : carbon emission cost ($/ton CO_2);
c_{en} : cost of energy ($/KWh);
c_w : disposal cost per unit ($/unit);
c_{is} : inspection cost ($/unit);
ρ : collector defective item proportion, with $0 \leq \rho \leq 1$;

F_t : fixed cost per truck per trip ($/truck);
t_c : truck capacity (units/truck);
g_t : number of gallons required per truck per distance travelled (gallons/truck);
e_t : amount of GHG emissions produced from one gallon of diesel-truck fuel (ton/gallon);
θ : investment increment factor that governs the ratio of investment for each remanufactured generation, with $0 \leq \theta < 1$;
c_{inv} : annual investment in the R&D process to, theoretically, be able to remanufacture it for an indefinite number of time ($/year)

Here are several assumptions that used to develop the model:

1. Manufacturing and remanufacturing processes are perfect
2. Remanufacturing rate is less than the demand rate ($v < d$)
3. Collector will inspect all of its items at the beginning of each cycle
4. Demand (d) is deterministic
5. Equal shipment lot size for each party
6. Lead time for transportation and production processes are not considered (equals to zero)
7. Remanufactured items have the same quality as the new items (as-good-as-new-one)
8. Used items return from the market have uniform quality.

The formulation of cost functions are referred to the model of [7], [8], [9], and [10]. The objective function is to minimize the total inventory related costs among the supply chain.

A. Retailer Inventory Cost Function

Inventory cost related to the retailer (TIC_{re}) consists of holding cost (HC_{re}), ordering cost (SC_{Re}), transportation cost (TC_{re}) and carbon emission cost from transportation activity ($CGHG_{re}$). The on-hand stock level at retailer inventory is depicted in Figure 2.

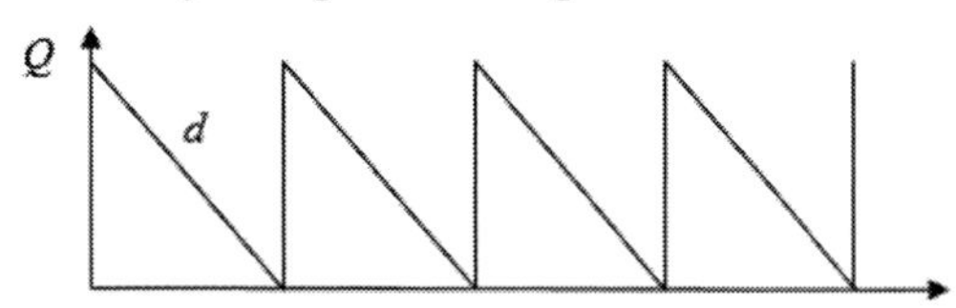

Figure 2. On-hand stock level at retailer inventory.

The function of TIC_{re} is given by:

$$TIC_{re} = H_2\frac{Q}{2} + S_2\frac{d}{Q} + \frac{dF_t}{tc} + g_t e_t c_{GHG}\left(\frac{d}{tc}\right) \quad (1)$$

B. Manufacturer Inventory Cost Function

Figure 3 shows the on-hand stock level at finished products, recoverable items and raw materials inventory. As mentioned previously, the holding costs belongs to manufacturer (HC_m) include the cost of storing the finished products, recoverable items and raw materials. The annual inventory cost for manufacturer (TIC_m) is consists of holding costs (HC_m), ordering cost (OC_m), set-up cost (SC_m), production costs (MC_m), transportation cost (TC_m), R&D investment cost (C_{inv}), carbon emission cost ($CGHG_m$), and energy usage cost (EC_m).

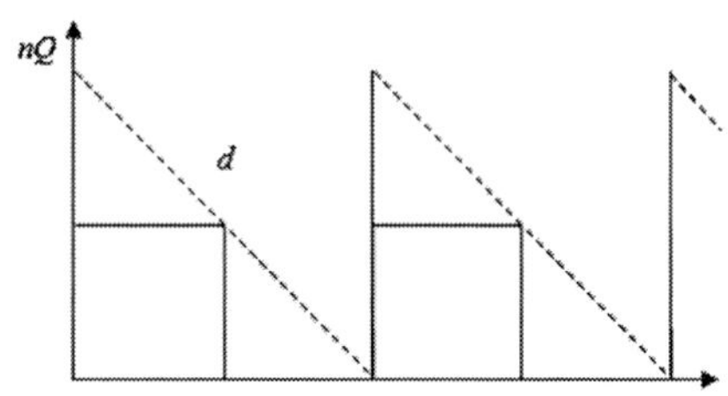

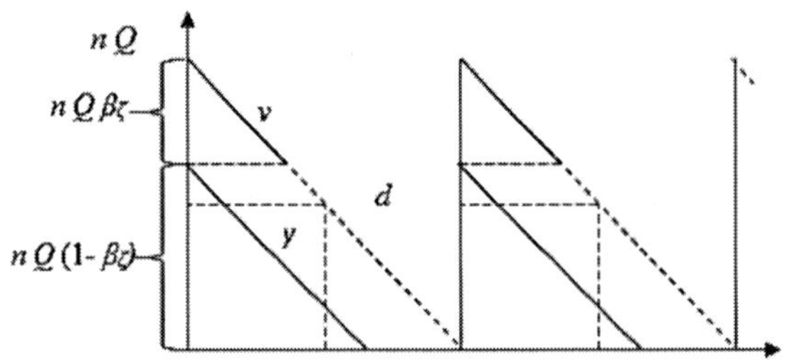

(b)

Figure 3. On-hand stock level at manufacturer finished product inventory (a) and raw material-recoverable item inventory (b).

TIC_m is formulated as:

$$TIC_m = HC_m + OC_m + SC_m + MC_m + TC_m + C_{inv} + CGHG_m + EC_m$$

$$= \left(H_1 \frac{n}{2d} + H_4 \frac{nQ}{2}\left(1 - \frac{1-\beta}{1-\beta^{\zeta+1}}\right)^2 + H_5 \frac{nQ}{2}\left(\frac{1-\beta}{1-\beta^{\zeta+1}}\right)^2\right) + \left(S_1 k \frac{d}{nQ}\right) + \left(A_1 \frac{d}{nQ}\left(1 - \frac{1-\beta}{1-\beta^{\zeta+1}}\right) + A_2 \frac{d}{nQ}\left(\frac{1-\beta}{1-\beta^{\zeta+1}}\right)\right) + \left(d\left(C_{re}\left(1 - \frac{1-\beta}{1-\beta^{\zeta+1}}\right) + C_{mn}\left(\frac{1-\beta}{1-\beta^{\zeta+1}}\right)\right)\right) + \left(\frac{dF_t}{t_c}\left(1 - \frac{1-\beta}{1-\beta^{\zeta+1}}\right)\right) + c_{inv}(1 - e^{-\zeta\theta}) + c_{GHG}\, d\left((c_r - b_r v + a_r v^2)\left(1 - \frac{1-\beta}{1-\beta^{\zeta+1}}\right) + (c_p - b_p \gamma + a_p \gamma^2)\left(\frac{1-\beta}{1-\beta^{\zeta+1}}\right) + \left(\frac{g_t e_t}{tc}\left(1 - \frac{1-\beta}{1-\beta^{\zeta+1}}\right)\right)\right) + c_{en}\, d\left(\left(C_0'' v + \frac{C_1'' v}{v}\right)\left(1 - \frac{1-\beta}{1-\beta^{\zeta+1}}\right) + \left(C_0'\gamma + \frac{C_1'\gamma}{\gamma}\right)\left(\frac{1-\beta}{1-\beta^{\zeta+1}}\right)\right) \quad (2)$$

C. Collector Inventory Cost Function

Inventory cost related to the collector (TIC_c) consists of holding cost (HC_c), collection cost (PC_c), inspection cost (C_{ins}) and waste disposal cost (WC_c). Figure 4 shows the on-hand stock level at collector used item inventory.

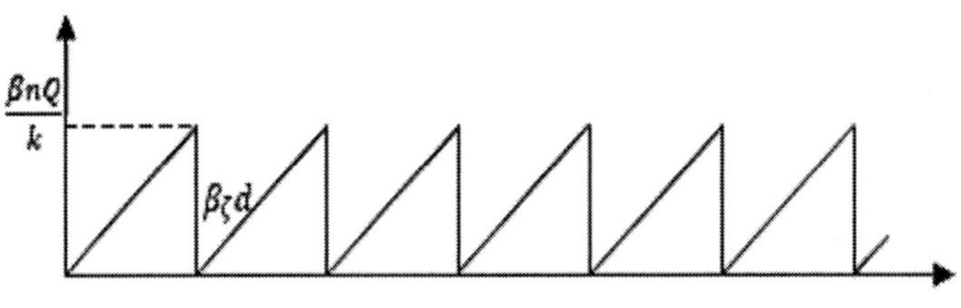

Figure 4. On-hand stock level at collector used item inventory.

The function of TIC_c is given by:

$$TIC_c = H_3 \frac{nQ}{2k}\left(1 - \frac{1-\beta}{1-\beta^{\zeta+1}}\right) + d\,(P_b + c_{isc} + c_w)\left(\left(1 - \frac{1-\beta}{1-\beta^{\zeta+1}}\right) + \rho\right) \quad (3)$$

Finally, we get the function of annual joint total cost for the three parties as given by

$$AJTC(n, Q, k, \zeta) = TIC_r + TIC_m + TIC_c \quad (4)$$

3. Solution Procedure

The optimization problem of the proposed model is to determine the values of n, Q, k and ζ that minimize the annual inventory related costs of all parties, $AJTC(n, Q, k, \zeta)$. We get the optimal solutions of Q by taking the first partial derivative of Equation (4) with respect to Q, and then setting the result to zero. The optimal value of Q denoted by Q^* is derived as follows.

$$Q^*(n, k, \zeta) = \sqrt{\frac{2dk\left(\beta^{1+\zeta}-1\right)\left(A_2(\beta-1) - A_1\beta + A_1\beta^{1+\zeta} - nS_3 + nS_3\beta^{1+\zeta} + kS_1\left(\beta^{1+\zeta}-1\right)\right)}{n\left(H_2 k\left(\beta^{1+\zeta}-1\right)^2 + n\left(\beta\left(\beta^{\zeta}-1\right)\left(H_3\left(\beta^{1+\zeta}-1\right) + H_4 k\beta\left(\beta^{\zeta}-1\right)\right) + H_5 k(\beta-1)^2\right)\right)}} \quad (5)$$

IOP Conf. Series: Materials Science and Engineering **943** (2020) 012054 doi:10.1088/1757-899X/943/1/012054

With the same procedure, we obtain n^* and k^* as given by Equation (6) and (7) respectively.

$$n^*(Q,k,\zeta)=\sqrt{\frac{2d^2k\left(\beta^{1+\zeta}-1\right)\left(A_2(\beta-1)+A_1\beta\left(\beta^{\zeta}-1\right)+kS_1\left(\beta^{1+\zeta}-1\right)\right)}{Q\left(H_1k\left(\beta^{1+\zeta}-1\right)^2+dQ\left(\beta\left(\beta^{\zeta}-1\right)\left(H_3\left(\beta^{1+\zeta}-1\right)+H_4k\beta\left(\beta^{\zeta}-1\right)\right)+H_5k(\beta-1)^2\right)\right)}} \tag{6}$$

$$k^*(n,Q,\zeta)=\sqrt{\frac{H_3n^2Q^2\beta}{2\,d\,S_1}\left(\frac{\beta-1}{\beta^{\zeta+1}-1}\right)} \tag{7}$$

The optimal values of Q^*, n^*, k^* and ζ^* that minimize $AJTC(Q, n, k, \zeta)$ with $1\leq\zeta\leq10$ are derived by the following solution procedure.

Step 1 Set $\zeta=1$

Step 2 Substitute $k^*(n, Q)$ in Equation (7) into Equation (6), by solving the equation we then obtain the function of $n^*(Q)$

Step 3 Substitute $n^*(Q)$ into Equation (5) and solve the equation. We then obtain the function of Q^*

Step 4 Compute $n^*(Q)$ with $Q=Q^*$

Step 5 Compute $k^*(n, Q)$ with $Q=Q^*$ and $n=n^*$

Step 6 Repeat steps 1 to 5 with $\zeta=\zeta+1$

Step 7 Find the value of ζ that gives the minimum value of $AJTC(Q^*, n^*, k^*, \zeta^*)$, and set as ζ^*.

By solving the optimization problem analytically, we guarantee a global optimal solution for the problem.

4. Numerical Example, Discussion, and Analysis

To illustrate the model, we provide a numerical example using a set of data collected from the previous studies ([4], [8], and [10]). The parameters used to illustrate the model are:

β = 0.67	C_{mn} = \$20/unit	C_{GHG} = \$2/ton CO_2
d = 100 units/year	C_{re} = \$10/unit	C_0' = 57.96 Kwh/unit
γ = 150 units/year	P_s = \$10/unit	C_1' = 1,855,744 Kwh/year
v = 100 units/year	P_c = \$8/unit	C_0'' = 18.9 Kwh/unit
H_1 = \$50/unit/year	P_b = \$5/unit	C_1'' = 605,110 Kwh/year
H_2 = \$60/unit/year	c_w = \$5/unit	c_{en} = \$0.000928/Kwh
H_3 = \$25/unit/year	ρ = 0.1	c_{isc} = \$2/unit
H_4 = \$40/unit/year	a_r = 8.33×10^{-8} ton.year^2/unit^3	c_{in} = \$500/year
H_5 = \$40/unit/year	b_r = 0.0002 ton.year/unit^2	θ = 0.2
A_1 = \$100/batch	c_r = 1.4 ton/unit	F_t = \$100/truck
A_2 = \$200/batch	a_p = 3×10^{-7} ton.year^2/unit^3	t_c = 20 unit/truck
S_1 = \$30/order	b_p = 0.0012 ton.year/unit^2	g_t = 375 gallons/truck
S_2 = \$50/order	c_p = 1.4 ton/unit	e_t = 0.1008414 ton/gallon

We solve the problem using the procedure given in Section 3. The solutions for the given problem are: 3 shipments/cycle with the size of 13 units/shipment from manufacturer to retailer; 2 shipments/cycle from collector to manufacturer; 3 production cycles/year; 2 remanufacturing generations. By setting Q, n, k, and ζ at the suggested values, the supply chain system minimizes its annual cost at \$7,580.16.

More generation of remanufacturing will eventually end up in a higher amount of CO_2 emission from the remanufacturing process and yet lower amount of CO_2 emission from the manufacturing process (see Figure 5a). The more generations of remanufacturing will increase the amount of product returned to the system and the amount of product produced through remanufacturing process. Emissions generated from the manufacturing process remains decreased since the amount of products produced by the manufacturing activities is smaller. Lowest CO2 emission is obtained at $\zeta=1$, and highest CO2 emission is obtained at $\zeta = 10$. The value of $\zeta=1$ indicates that the product is only manufactured 1 time in its lifetime. Therefore, few emissions are generated from the remanufacturing process. In addition, there is only one transportation activity to deliver the used items from the market, resulting in less amount of CO_2 emission generated from related activity.

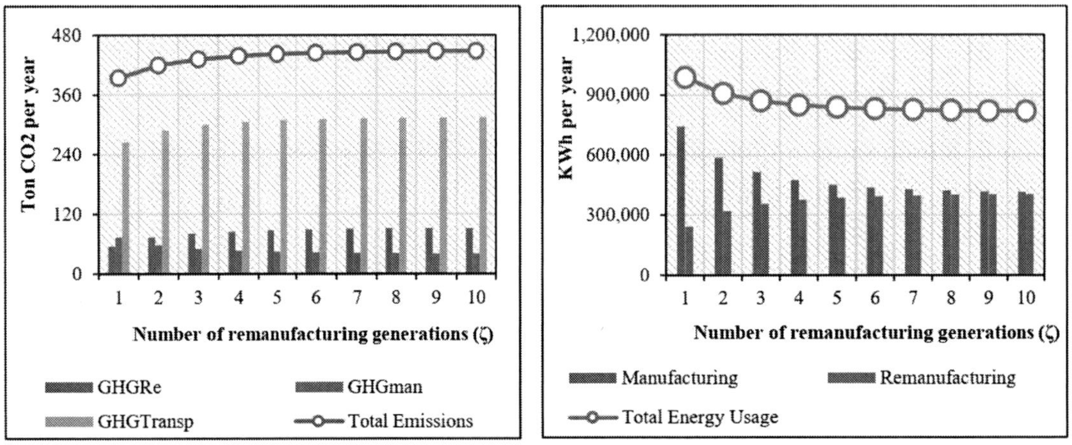

Figure 5. The amount of ton CO_2 emissions per year (a) and amount of energy (KWh) per year at different values of ζ.

Manufacturing energy usage gradually decreases as the number of remanufacturing generation increases (see Figure 5b). Whereas, the increased number of remanufacturing generations actually increases the amount of energy used to perform remanufacturing process. The lowest usage of energy came when the number of remanufacturing generations is at its highest. When ζ=10, the system reduce a significant amount of energy usage from manufacturing process because the energy required to produce one unit of product in the manufacturing process is higher than remanufacturing.

We understand that less number of remanufacturing generations will give less CO_2 emissions, but highest energy usage. Therefore, companies should reconsider what criteria to prioritize in order to investigate environmental effects. If the company is focusing more in reducing the CO_2 emissions, then the decision will be not to remanufacture more. Yet, if the company is more concerned to reduce the amount of energy usage, the decision should be to remanufacture more. However, this should also be considered in conjunction with the costs and other parameters associated and experienced by the company. Hence in the end, the decision can actually provide a best solution that gives lowest cost to the system.

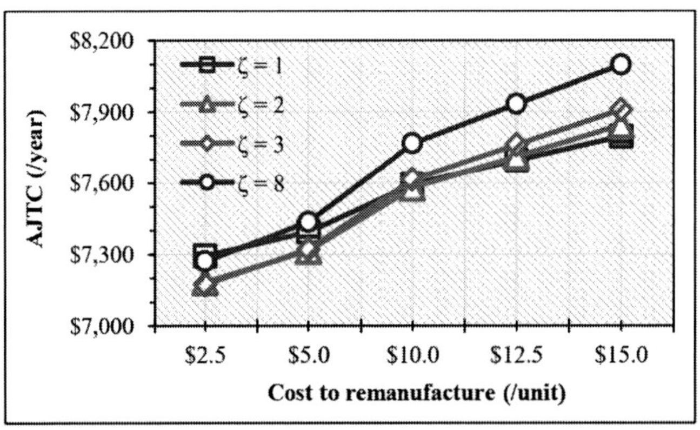

Figure 6. The effects of c_{re} on AJTC for $\zeta = \{1, 2, 3, 8\}$.

Figure 6 shows the effects of different c_{re} on *AJTC* for $\zeta=\{1, 2, 3, 8\}$. The interesting finding is that, the optimal value of ζ will be increased from 2 to 3, when the cost to remanufacture is decreased to $2.5/unit. In addition, when the cost to remanufacture is increased to $12.5/unit, the optimal value of of ζ will be decreased from 2 to 1. It suggests that remanufacturing variable cost actually affects the optimal number of remanufacturing generations. Higher c_{re} will be resulting in lower ζ, while lower c_{re} will be resulting in higher ζ. Here, we can take a note that when the cost to remanufacture one unit of item is expensive, the system should consider designing its product to be less remanufacturable. It will be not profitable if there are more number of items that will go through the remanufacturing process.

5. Conclusion

This study develops a CLSC inventory model consisting a manufacturer, a retailer, and a used item collection dealer. Previous works on this problem have not yet included the third party collection dealer as one separate party among the reversed supply chain. Here, we investigate limited number of remanufacturing generations and consider carbon emission and energy effects as component of inventory costs. The results explain that in order to obtain minimum integrated supply chain inventory cost, the system should design a product in such a way so that it could be remanufactured at a desired number before it undergoes disposal. However, designing a product to be highly remanufacturable does not necessarily become a good strategy, since it actually gives higher cost to the system.

Future works for the related problems should include more recovery activities, such as disassembly and remanufacturing of individual components as well as assemblies. Another challenge is to consider that not all collected items have the same quality. Since it comes from the different consumers, the remaining quality of each item may not be uniformly distributed. As stated by [12] uncertainties in product acquisition may affect both manufacturing and remanufacturing strategies. It is explained that the return rate of used item will actually be possible to be controlled if the company performs a specific acquisition effort. As of it will affect the quality, quantity, cost, and lead time of the return item management activity.

References

[1] Akdogan S M and Coskun A 2012 Drivers of reverse logistics activities: an empirical investigation *Int. Stra. Manag. Conf. Procedia-Soc. and Behav. Scie.* **58** 1640.

[2] Pujawan I N 2008 Supply Chain Management (2nd ed), Surabaya: Guna Widya

[3] Thierry M, Salomon M, Nunen J V and Wassenhove L V 1995 Strategic issues in product recovery management *Californa Manag. Rev.* **37** 114.

[4] Mitra S 2009 Analysis of two echelon inventory system with returns *The Int. J. Manag. Sci.* **37** 106.

[5] Chung S, Wee H M and Yang P 2008 Optimal policy for a closed-loop supply chain inventory system with remanufacturing *J. Math. and Comp. Modelling* **48** 867.

[6] Yuan K F and Gao Y 2010 Inventory decision-making models for a closed-loop supply chain system *Int. J. Prod. Res.* 6155.

[7] El Saadany A M A, Jaber M Y and Bonney M 2013 How many times to remanufacture? *Int. J. Prod. Econ.* **143** 598.

[8] Bazan E, Jaber M Y and El Saadany A M A 2015 Carbon emissions and energy effects on manufacturing-remanufacturing inventory model *Comp.and Indust. Eng.* **88** 307.

[9] Dwicahyani A R, Jauhari W A and Jonrinaldi 2017 A regular production-remanufacturing inventory model for a two-echelon system with price-dependent return rate and environmental effects investigation *J. Physics: Conf. Series* **855**.

[10] Dwicahyani A R, Jauhari W A, Rosyidi C N and Laksono P W 2017 Inventory decisions in a two-echelon system with remanufacturing, carbon emission, and energy effects *Cogent Eng.* **4** 1.

[11] Fleischmann M, Bloemhof-Ruwaard J M, Dekker R, Van der Laan E, Van Nunen J A A E E and Van Wassenhove L N 1997 Quantitative models for reverse logistics: A review *European J. Oper. Res.* **103** 1.

[12] Lechner G and Reimann M 2014 Impact of product acquisition on manufacturing and remanufacturing strategies *Prod. and Manufac. Res.* **2** 831.

Lot-Sizing Decisions in Manufacturer-Retailer Inventory System under Carbon Emissions Reduction

W A Jauhari
Industrial Engineering Department, Universitas Sebelas Maret, Surakarta, Indonesia

Email: wachid_aj@yahoo.com

Abstract. This paper develops a mathematical model for supply chain inventory system of a manufacturer and a retailer under stochastic demand and carbon reduction policy. The objective of the model is to obtain the decisions regarding inventory management in supply chain system by minimizing the joint total cost. The model contributes to the current stochastic joint lot-sizing literature by allowing the investigation of the impact of carbon cap policy to inventory decisions. The carbon mainly emitted from several activities in the supply chain system, including the production, transportation and storage activities. The regulator, for example the government, uses a carbon cap policy to restrict the amount of emissions generated from the system. As we use strict carbon cap, the carbon emissions generated from the system must not exceed the carbon emission level set by regulator. A list of procedure is proposed to obtain the optimal solutions, which are ordering lot, safety factor, number of shipments and production rate. A numerical instance is provided to show the applicability of the model and to explore the impact of different levels of carbon cap on inventory decisions.

1. Introduction

Environmental responsibility has become an important factor that must be considered by manager for doing the modern business. The environmental regulations issued by government and the increased of customer's awareness regarding environmental problems are encouraging the supply chain parties to lessen the generations of carbon emissions from their operations. Operational managements, such as inventory, shipment and ordering decisions, have proven to be best way to lessen the carbon emissions influence [1]. Thus, scholars have considered carbon emissions in developing the inventory management for supply chain system. The developed models not only concern to minimize total inventory cost but also try to minimize the negative impact of carbon emissions generated from the supply chain operations.

An inventory model dealing with carbon emissions was firstly developed by Wahab et al. [2]. They developed an integrated inventory model for domestic and international supply chain systems and investigated the impact of carbon emissions on inventory decisions. Bonney and Jaber [3] developed an inventory model and proved that the carbon emissions reduction can be gained through increasing the delivery lot size and reducing the frequency of deliveries. Konur [4] investigated heterogeneous truck types and carbon cap mechanism in an EOQ model. Battini et al. [5] studied the impact of carbon emissions on inventory and transportation decisions. Schaefer and Konur [6] developed inventory model addressing the continuous review policy, transportation and carbon emissions with two different objectives. Benjaafar et al. [7] and Chen et al. [8] studied joint economic lot sizing problem (JEL under carbon emission reduction. They considered some mechanisms, which are carbon emission tax, inflexible cap and cap-and-trade, to restrict the carbon emitted from the systems. Jaber et al. [10] addressed emission tax and penalty in a manufacturer-retailer system. Jauhari et al. [11] included carbon emission costs in their JELP model. They proposed single-retailer single-manufacturer inventory model in which the deliveries are made under unequal-shipment policy and the demand is deterministic.

Some scholars have also investigated the carbon emissions reduction in stochastic inventory models. Wangsa [12] proposed a supply chain inventory model containing a manufacturer and a retailer under probabilistic demand and carbon reduction. An incentive and penalty is proposed to lessen the carbon levels emitted from industrial and transportation operations. Jauhari (2018) considered imperfect items and carbon tax in developing a mathematical model for two-level inventory system. The Manufacturing process is flawed and the lead time is formulated under stochastic function. Jauhari et al. (In press) studied a JELP with imperfect items, fuzzy demand, inspection errors and service level constraint. They adopt the incentive and penalty mechanism proposed by Wangsa (2017) to cut down the amount of carbon emitted from manufacturing and transportation operations.

Our review to the JELP literature shows that the research concerning with the impact of strict carbon policy, adjustable production rate and imperfect production on inventory decisions is rarely discussed in the paper. The published papers mostly focused on discussing the carbon tax and fixed production rate under perfect production. Therefore, this paper propose a mathematical model considered a situation in which the emissions generated by supply chain is restricted by a carbon cap regulated by government. In addition, defective items are accounted in holding cost and emission cost.

2. Problem description

In this research, a mathematical inventory model for a manufacturer and a retailer system is proposed. The retailer faces a stochastic demand, which follows normal distribution, from end customers and uses a continuous review policy to control his inventory level. The retailer makes an order of nQ units from the manufacturer and incurs ordering cost A for each order. The unsatisfied demand is considered to be partially backordered. To satisfy the demand of retailers, the manufacturer produces a batch of size nQ units with a rate of P ($P>D$) per setup and incurs a production setup cost K for each manufacturing cycle. The delivery of Q units is done from the manufacturer to the retailer over n times. The lead time is assumed to be dependent on lot size, setup time and transportation time.

The manufacturer's production process is flawed, hence, generating a portion of defective items during production time. The lot received by the retailer will be inspected to categorize the quality of items. The defective items obtained during inspection process will be sent back to the manufacturer at the end of delivery cycle. The manufacturer can determine the production capacity flexibly. It is assumed that the production rate can be adjusted between P_{min} and P_{max}. The emissions are investigated in the model and computed from production, delivery and storage activities. The regulation to reduce the carbon emission, namely strict carbon cap, is applied in the system.

To formulate the mathematical model, we employ the following notations

D	end customer's demand (units/year)
σ	standard deviation of end customer's demand (units/year)
A	replenishment cost per order (\$/order)
F	delivery cost per shipment (\$/shipment)
h_b	retailer's carrying cost (\$/unit/year)
h_v	manufacturer's carrying cost (\$/unit/year)
π	backorder cost per unit backordered (\$/unit)
π_0	marginal profit per unit for retailer (\$/unit)
θ	percentage of backorder items
Y	the defect rate
w	warranty cost per unit item (\$/unit)
s	inspection cost per unit item (\$/unit)
x	inspection rate (units/year)
S_R	fixed carbon emission for retailer (kg CO_2)

S_M	fixed carbon emission for manufacturer (kg CO_2)
z_R	carbon emission generated from items kept in the retailer (kg CO_2/unit)
z_M	carbon emission generated from items kept in the manufacturer (kg CO_2/unit)
W_r	carbon emission emitted from keeping items (kg CO_2/unit)
C	carbon cap (kg CO_2)
a_1	fixed production cost which is independent of production rate ($)
a_2	variable production cost which is depended on production rate ($/unit)
P	production rate in units per unit time (units/year)
K	setup cost ($/setup)
n	number of shipments
Q	delivery lot from manufacturer to retailer (units)
k_1	safety factor of first shipment
k_2	safety factor of shipment 2,……,n
T_w	setup and transportation time (year)
T_s	transportation time (year)

3. Model formulation

As mentioned above, we propose a JELP for manufacturer-retailer system under stochastic environment, flexible production rate and carbon reduction. The objective of the proposed mathematical model is to search the optimal solutions which are, delivery lot size, safety factor and number of shipments so the total cost is minimized. The total cost of supply chain consists of two costs, which are retailer cost and manufacturer cost. The development of each cost is provided in the sub section below

3.1 Retailer cost

Consider the customer's demand occurred at the retailer follows a normal distribution with mean $DL(Q)$ and standard deviation $\sigma\sqrt{L(Q)}$. The lead time of the first delivery is constructed by addressing setup and transportation time and the next lead time is only transportation time considered in the model (Hsiao, 2008). Therefore, the safety stock for first shipment and next shipment can be formulated by the following equations

$$S_1 = k_1\sigma\sqrt{L(Q)} = k_1\sigma\sqrt{Q/_P + T_w} \quad (1)$$

$$S_2 = k_2\sigma\sqrt{T_s} \quad (2)$$

Equations (3) and (4) express the expected shortage of first shipment and the expected shortage of next shipment, respectively

$$ES_1 = \sigma\sqrt{Q/_P + T_w}\,\psi(k_1) \quad (3)$$

$$ES_2 = \sigma\sqrt{T_s}\,\psi(k_2) \quad (4)$$

where,

$$\psi(k_1) = f_S(k_1) - k_1\left[1 - F_S(k_{1,})\right] \quad (5)$$

$$\psi(k_2) = f_S(k_2) - k_2\left[1 - F_S(k_{2,})\right] \quad (6)$$

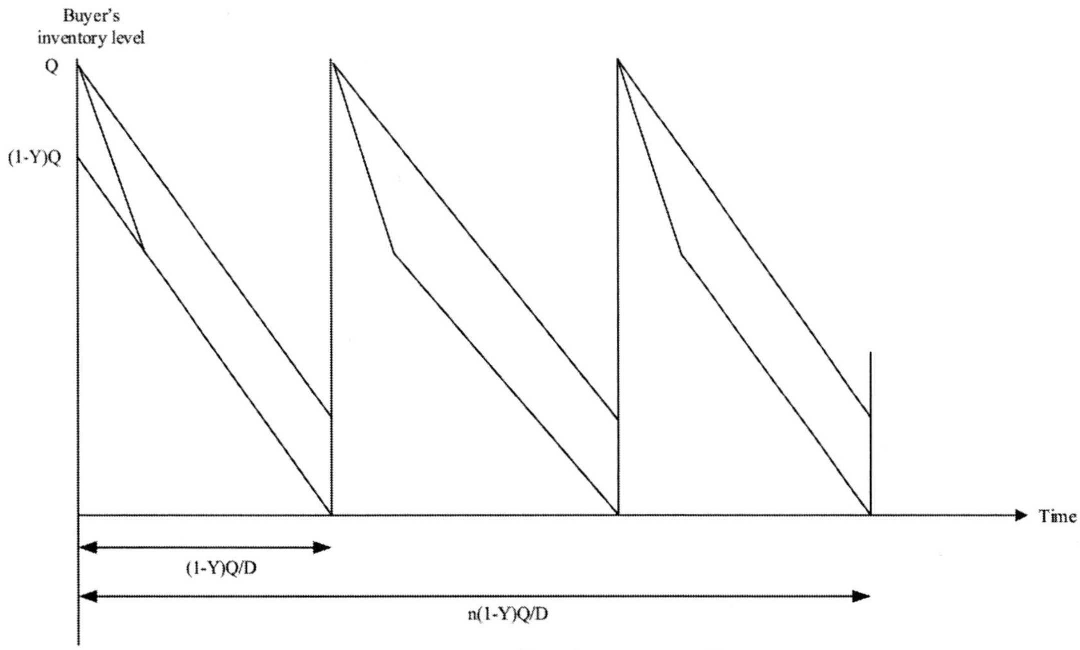

Figure 1. Retailer's inventory profile

Figure 1 demonstrate the behavior of retailer's inventory pattern. By considering this figure, the expected retailer's carrying cost per unit time is formulated as follows

$$h_R\left(E\left(\frac{Y}{1-Y}\right)\frac{QD}{x}+\frac{[1-E(Y)]Q}{2}+k_1\sigma\sqrt{Q/P+T_w}+(1-\theta)\sigma\sqrt{Q/P+T_w}\psi(k_1)\right) \tag{7}$$

The expected retailer's shortage cost per unit time is presented by the following expression

$$\left(E\left(\frac{1}{1-Y}\right)\frac{D\sigma}{nQ}(\pi+\pi_0(1-\theta))\right)\left[\sqrt{\frac{Q}{P}+T_w}\psi(k_1)+(n-1)\sqrt{T_s}\psi(k_2)\right] \tag{8}$$

The expected retailer's total cost per unit time, which is consisted of replenishment cost, delivery cost, inspection cost, carrying cost and backorder cost, is calculated by

$$\begin{aligned} ETCR &= E\left(\frac{1}{1-Y}\right)\frac{D}{nQ}(A+nF)+E\left(\frac{1}{1-Y}\right)sD \\ &+h_R\left(E\left(\frac{Y}{1-Y}\right)\frac{QD}{x}+\frac{[1-E(Y)]Q}{2}+k_1\sigma\sqrt{Q/P+T_w}+(1-\theta)\sigma\sqrt{Q/P+T_w}\psi(k_1)\right) \\ &+\left(E\left(\frac{1}{1-Y}\right)\frac{D\sigma}{nQ}(\pi+\pi_0(1-\theta))\right)\left[\sqrt{\frac{Q}{P}+T_w}\psi(k_1)+(n-1)\sqrt{T_s}\psi(k_2)\right] \end{aligned} \tag{9}$$

3.2 Manufacturer cost

The manufacturer's inventory level is computed the accumulated deliveries minus the accumulated manufacturer's production. The expected manufacturer's carrying cost per unit time is given by

$$h_R \frac{Q}{2}\left(n\left[1-E\left(\frac{1}{1-Y}\right)\frac{D}{P}\right]-1+E\left(\frac{1}{1-Y}\right)\frac{2D}{P}\right) \tag{10}$$

The production cost is also included in the model and formulated by following Khouja and Mehrez's (1994) formula, which is

$$E\left(\frac{1}{1-Y}\right)\left(\frac{a_1}{P}+a_2 P\right)D \tag{11}$$

The expected total cost of manufacturer, which is composed of production setup cost, production cost, warranty cost, carbon emission cost and carrying cost is expressed by the following equation

$$ETCM = E\left(\frac{1}{1-Y}\right)\frac{DK}{nQ}+E\left(\frac{Y}{1-Y}\right)wD+h_M \frac{Q}{2}\left(n\left[1-E\left(\frac{1}{1-Y}\right)\frac{D}{P}\right]-1+E\left(\frac{1}{1-Y}\right)\frac{2D}{P}\right) \\ +E\left(\frac{1}{1-Y}\right)\left(\frac{a_1}{P}+a_2 P\right)D \tag{12}$$

3.3 Joint total cost

As we described above that the proposed model considers carbon emission reduction policy, namely strict carbon cap. This means that the carbon emitted from supply chain activities (transportation, production, storage) is restricted by carbon emission level (carbon cap) determined by the regulator. For transportation activity, the emissions are calculated from delivering the items from manufacturer to retailer and delivering defective items from retailer to manufacturer. Here we adopt two types of emission, namely fixed emission and variable emissions. The emissions generated from transportation activity are given by

$$EM_{Trans} = E\left(\frac{1}{1-Y}\right)\frac{DS_R}{Q}+E\left(\frac{1}{1-Y}\right)z_R D+E\left(\frac{1}{1-Y}\right)\frac{DS_M}{Q}+E\left(\frac{Y}{1-Y}\right)z_M D \tag{13}$$

The emissions generated from production activity are given by

$$EM_{Prod} = E\left(\frac{1}{1-Y}\right)(y_1 P^2 - y_2 P + y_3)D \tag{14}$$

The emissions generated from storage activity are formulated by the expression below

$$EM_{Stor} = W_r\left(E\left(\frac{Y}{1-Y}\right)\frac{QD}{x}+\frac{[1-E(Y)]Q}{2}+k_1\sigma\sqrt{Q/P+T_w}+(1-\theta)\sigma\sqrt{Q/P+T_w}\psi(k_1)\right) \\ +W_r\frac{Q}{2}\left(n\left[1-E\left(\frac{1}{1-Y}\right)\frac{D}{P}\right]-1+E\left(\frac{1}{1-Y}\right)\frac{2D}{P}\right) \tag{15}$$

Therefore, by using equation $M = E\left(\frac{1}{1-Y}\right)$, the proposed problem can be written as

$$EJTC(n,Q,k_1,P) = \frac{MD}{nQ}(A+nF)+MsD \\ +h_R\left(\frac{MQD}{x}-\frac{QD}{x}+\frac{[1-E(Y)]Q}{2}+k_1\sigma\sqrt{Q/P+T_w}+(1-\theta)\sigma\sqrt{Q/P+T_w}\psi(k_1)\right) \\ +\left(\frac{MD\sigma}{nQ}(\pi+\pi_0(1-\theta))\right)\left[\sqrt{\frac{Q}{P}+T_w}\psi(k_1)+(n-1)\sqrt{T_s}\psi(k_2)\right]+\frac{MDK}{nQ} \\ +wDM-wD++h_M\frac{Q}{2}\left(n\left[1-\frac{MD}{P}\right]-1+\frac{2MD}{P}\right)+M\left(\frac{a_1}{P}+a_2 P\right)D \tag{16}$$

Subject to

$$EM_{Trans}+EM_{Prod}+EM_{Stor} < C$$

$$P_{min} \le P \le P_{max}$$

IOP Conf. Series: Materials Science and Engineering **943** (2020) 012055 doi:10.1088/1757-899X/943/1/012055

4. Solution procedure

To obtain the proposed mathematical model's solution, we search the first partial derivatives of $EJTC(n,Q,k_1,P)$ with the respect to k_1, Q and P, respectively and set the results to zero, we then obtain the equations below

$$\frac{[1-F_s(k_1)]+(n-1)\left[1-F_s\left(k_1\sqrt{\frac{Q/P+T_w}{T_s}}\right)\right]}{F_s(k_1)+\theta[1-F_s(k_1)]}=\frac{h_b nQ}{MD(\pi+\pi_0(1-\theta))} \tag{17}$$

$$Q=\sqrt{\frac{2MD\left\{\left[\frac{A+K}{n}+F\right]+\frac{(\pi+\pi_0(1-\theta))\sigma\left(\sqrt{Q/P+T_w}\,\psi(k_1)+(n-1)\sqrt{T_s}\,\psi(k_2)\right)}{n}\right\}}{h_M\left\{n\left(1-\frac{MD}{P}\right)-1+\frac{2MD}{P}\right\}+h_R[1-E(Y)]+\frac{2h_RMD}{x}-\frac{2h_RD}{x}+\frac{h_Rk_1\sigma}{P\sqrt{Q/P+T_w}}+\frac{h_R(1-\theta)\sigma\psi(k_1)}{P\sqrt{Q/P+T_w}}+\frac{MD(\pi+\pi_0(1-\theta))\sigma\psi(k_1)}{nPQ\sqrt{Q/P+T_w}}}} \tag{18}$$

$$P=\sqrt{\frac{a_1MD+MQh_MD-\frac{MQh_MnD}{2}+\left(\frac{h_Rk_1\sigma Q}{2\sqrt{Q/P+T_w}}+\frac{h_R(1-\theta)\sigma\psi(k_1)Q}{2\sqrt{Q/P+T_w}}+\frac{\frac{MD\sigma}{n}(\pi+\pi_0(1-\theta))\psi(k_1)}{2\sqrt{Q/P+T_w}}\right)}{a_2MD}}$$

$$=\sqrt{\lambda} \tag{19}$$

If λ is negative, expression (19) is not feasible. Thus, we use $P=P_{min}$ if $\lambda<0$. The equation P should be reformulated as

$$P=Max\{P_{min},Min(P_{max},\sqrt{\lambda})\} \tag{20}$$

An iterative procedure, depicted in figure 2, is suggested to obtain the model's solusions.

5. Numerical example

To show the applicability of the proposed model, we provide the following data for parameters: D=1,000 units/year, σ=25 units/year, F=\$50/delivery, A=\$100/order, K=\$400/setup, h_M=\$1/unit/year, h_R=\$3/unit/year, π=\$50/unit, π_0=\$150/unit, θ=0.5, P_{min}=1,500 units/year, P_{max}=5,000 units/year, x=7,000 units/year, a_1=\$2,500, a_2=\$0.0004/unit, S_R=1,300 $kgCO_2$, S_M=1,300 $kgCO_2$, z_R=10.4 $kgCO_2$/unit, z_M=10.4 $kgCO_2$/unit, y_1= 0.0003 $kgCO_2$.year^2/unit^3, y_2=1.2 $kgCO_2$.year/unit^3, y_3=140 $kgCO_2$/unit, C=170,000 $kgCO_2$, w=\$2/unit, s=\$0.5/unit, T_w=0.05, T_s=0.04. The pdf for the defect rate is taken to be

$$f(y)=\begin{cases}25, & 0\le y\le 0.04\\ 0, & otherwise\end{cases}$$

Thus, we can formulate the expected value of Y, which is

$E(Y)=\int_0^{0.04}25y\,dy=0.02$ and $E\left(\frac{1}{1-Y}\right)=\int_0^{0.04}\frac{25}{1-y}dy=1.02055.$

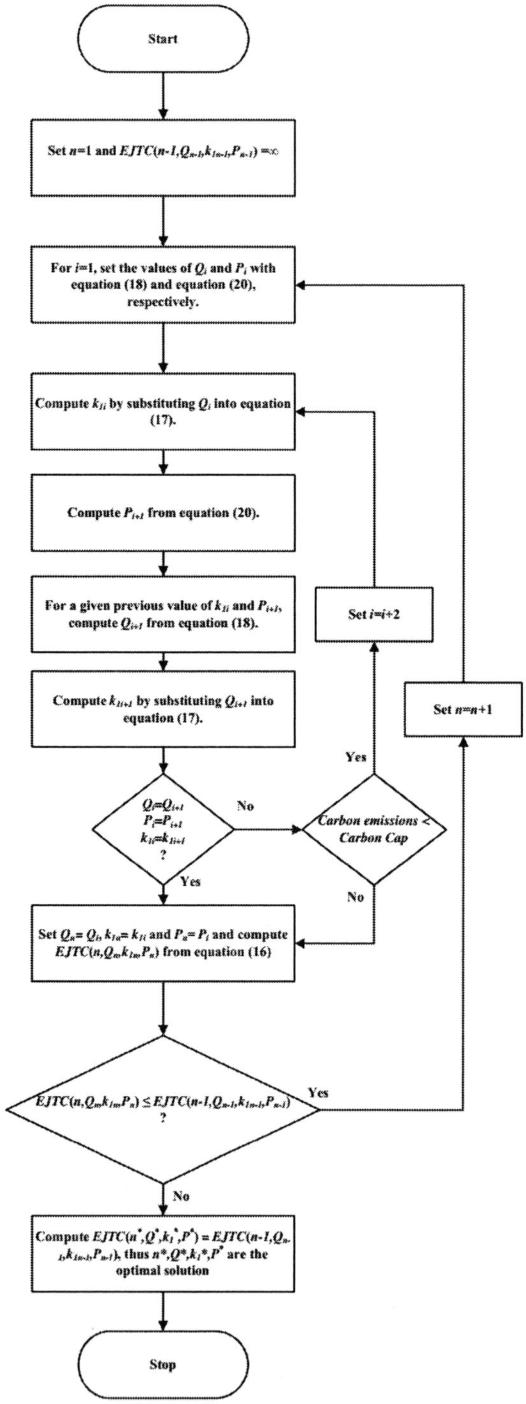

Figure 2. Proposed procedure

By applying the above iterative procedure, the solutions of the numerical example are as follows: the number of deliveries is 7, production rate is 2297 units/year, safety factor 1.89, delivery quantity 190 units and the total cost is \$3912.6. The emissions generated by vendor and buyer are 151,460 $kgCO_2$ and 18,395 $kgCO_2$. Figure 3 shows the influence of carbon cap changes on cost and carbon emissions. It is observed 3 that when the cap is increased the carbon emissions are increased as well. However, the parties cost is relatively decreased due to the increase in carbon cap. The influence of the changes of carbon cap on n and Q is presented in figure 4. It shows that when cap increases, n sharply increases while Q sharply decreases. It can be concluded that the impact of cap changes on decision variables is more significant than its impact on total cost.

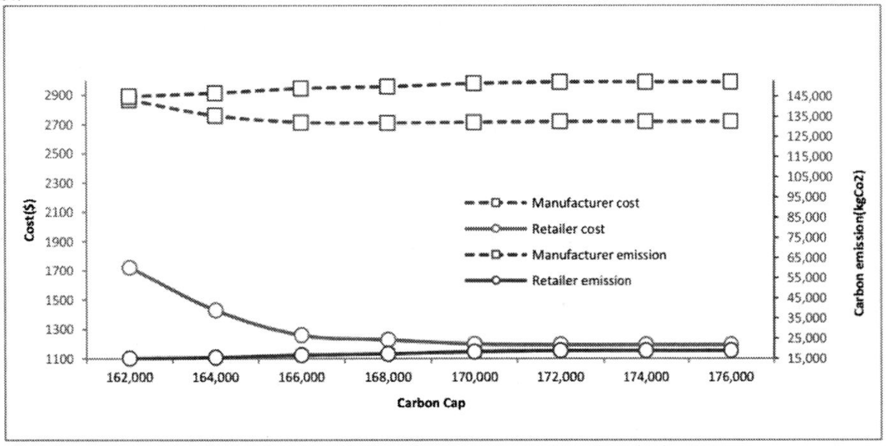

Figure 3. Impact of carbon cap on cost and emissions

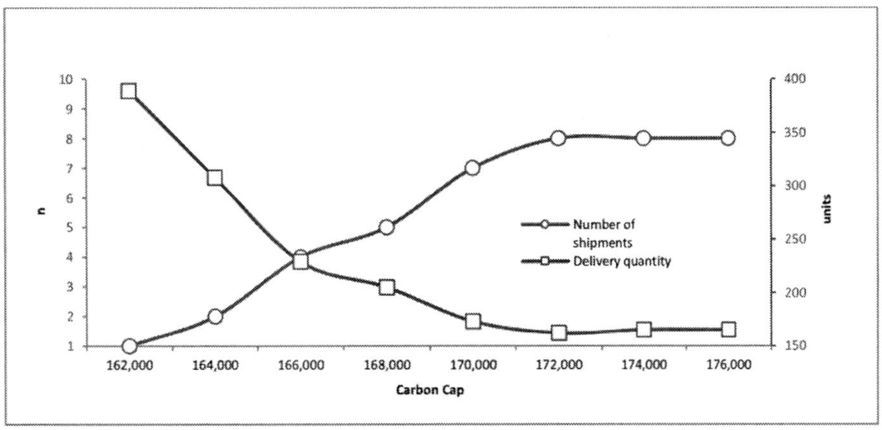

Figure 4. Impact of carbon cap on n and Q

6. Conclusion

In this paper, a JELP containing a manufacturer and a retailer is introduced to comply with carbon emission reduction regulated by a government. The manufacturing process is unreliable, thus producing some defective items during production run. Adjustable production rate is allowed in the proposed model which finally gives an opportunity to the vendor's manager to set its value flexibly to optimize the production capacity. Generally, the manufacturing, warehousing and transportation activities will produce a certain level of carbon emissions. A strict carbon cap is selected by the regulator as a mechanism to reduce the emission generated from the vendor system and buyer system. A mathematical inventory model is built to minimize total cost charged by supply chain and to ensure that the emission levels are lower than carbon cap. A numerical instance is presented to show the application of the model. Future research may focus on discussing the impact of another carbon policies regulated by government on inventory decisions. An extension to multi supply chain structure, such as multi-vendor multi-buyer, is also interesting to be investigated.

References

[1] Bouchery Y, Ghaffari A, Jemai Z, Dallery Y 2012 Including sustainability criteria into inventory models *European J. Oper. Res* **222** 229
[2] Wahab M I M, Mamun S M H and Ongkunaruk P 2011 EOQ models for a coordinated two-level international supply chain considering imperfect items and environmental impact *Int. J. Prod. Econ.* **134** 151
[3] Bonney M, and Jaber M Y 2011 Environmentally responsible inventory models: nonclassical models for a non-classical era *Int. J. Prod. Econ.* **133** 43
[4] Konur D, 2014 Carbon constrained integrated inventory control and truckload transportation with heterogeneous freight trucks *Int. J. Prod. Econ.* **153** 268
[5] Battini D, Persona A and Sgarbossa F 2014 A sustainable EOQ model: Theoretical formulation and applications *Int. J. Prod. Econ.* **149** 145
[6] Schaefer B, and Konur D 2015 Economic and environmental considerations in a continuous review inventory control system with integrated transportation decisions *Trans. Res. Part E* **80** 142
[7] Benjaafar S, Li Y, and Daskin M 2012 Carbon footprint and the management of supply chains: Insights from simple models *IEEE Trans. on Auto. Sci. and Eng.* **10** 99
[8] Chen X, Benjaafar S, Elomri A 2013 The carbon-constrained EOQ *Oper. Res. Let.***41** 172.
[9] Jaber M Y, Glock C H, El Saadany A M A 2013 Supply chain coordination with emissions reduction incentives Int. J. Prod. Res. **51** 69
[10] Jauhari W A, Pamuji A S and Rosyidi, C N 2014 Cooperative inventory model for vendor-buyer system with unequal-sized shipment, defective items and carbon emission cost *Int. J. Log. Sys. and Manag.***19** 163
[11] Wangsa I D 2017 Greenhouse gas penalty and incentive policies for a joint economic lot size model with industrial and transport emissions Int. J. Indus. Eng. Comput. **8** 453
[12] Hsiao Y C 2008 A note on integrated single vendor single buyer model with stochastic demand and variable lead time *Int. J. Prod. Econ.* **114** 294

A review on sustainable metrics for Sustainability Measurement in Supply Chain

V H LSaputri[1*], M Hisjam[2], and W Sutopo[2]

[1] Master Program of Industrial Engineering, Faculty of Engineering, Universitas Sebelas Maret Surakarta, Indonesia.
[2] Department of Industrial Engineering, Faculty of Engineering, Universitas Sebelas Maret Surakarta, Indonesia.

Email: hisjam@staff.uns.ac.id

Abstract: This research analyzes the performance measurement in sustainable supply chain using sustainable triple bottom line aspects, namely environmental dimension, social dimension, and economic dimension. The analysis process is accomplished for 40 articles, which are from Scopus website related to sustainable supply chain. The focus in analysis process is the sustainable dimension that has been used in the previous research. Furthermore, the other focuses are the analysis processes about the journal name, organizations that fulfill the performance measurement, the utilized approach, and the type of research. Then, in the final research, there will be explanation about the metric that is utilized to be a tool for performance measurement of sustainable supply chain in the organization.

1. Introduction

The sustainable supply chain management (SSCM) is a topic that is being discussed currently. The SSCM has been a cornerstone for every organization that is trying to achieve the continual goal. Now, it continually becomes a requirement from all of stakeholders, such as government, non-governmental organization, consumer, community activist, and global competition. The SSCM comes along with the increasing of concerns about the impacts of environment and social in supply chain, so that, all of related stakeholders forces the companies to minimize the negative impacts of their own supply chain system [1], [2]. The SSCM focuses on creating the three pillars of sustainable development that emphasize on social, economic, and environment study, namely Triple Bottom Line (TBL) [3].

A lot of research that related to SSCM has been accomplished in every aspect and perspective. Molamohamadi, Ismail, Leman, & Zulkifli [4] performed a research that integrated the sustainable concept in supply chain to create a SSCM framework. Bendul, Rosca, & Pivovarova [5] accomplished a research that connected the SSCM with the study of Pyramid Base to develop the integrated sustainable model perspective in sourcing, making and delivering area. Tajbakhsh and Hassini [6] evaluated the operation of supply chain that maximized the return of economic, minimized the impact of environment, and fulfilled the social expectation by using data envelopment analysis model. Barbosa-Póvoa, da Silva, & Carvalho [7] reviewed the SSCM research based on Operations Research (OR) perspective which the optimization model was applied in strategic level decisions by assessing the economic, the environment, and social aspect. Qorri, Mujkić, & Kraslawski [8] suggested the new conceptual framework and provided a short guidance to assess the SSCM by using multi-criteria decisions making method. Moreover, there are some research that conducted for sustainable wooden furniture industry using goal programming [9], [10], [11], [12], simulation [13], [14], and fuzzy Chance Constrained Model (CCM) [15].

A lot of SSCM research shows that SSCM has become a concern for many parties in the era of global competition today. For facing the global competition, the company needs to improve the competitive advantage. One of ways that can be used is by measuring and managing the sustainable performance of the company's supply chain effectively and efficiently [16]. Measuring and improving the company's performance certainly need a development of the supply chain performance measurement system. Performance metric and measurement tool/ method are integral part of system. Finally, it is important to analyze the performance measurement approach that can help the manager to focus on the main decision that is related to the company's SSCM [17].

Given the importance of measuring the SSCM's performance, this research is aimed to find out and summarize the approach and metric that is used for performance measurement of the company's supply chain that is related to "the sustainable" issue. Hereafter, the methodology that is applied in this research will be explained in the Chapter 2. Then, the understanding of existing SSCM literature and the detailed analysis of metric that is used in SSCM will be revealed in the Chapter 3. Lastly, the conclusion of the entire research, include the scope and the boundary of SSCM research study in the future will be explained in the Chapter 4.

2 Methodology

The method that was used in this study was by studying of the literature on the related research which discussed about SSCM. The literature study was conducted on five methodological steps from Tranfield *et al.* [18] as illustrated on Figure 1. The related articles was browsed on Scopus Website (www.scopus.com) that was one of the biggest database to examine and view trends of the related research over time.

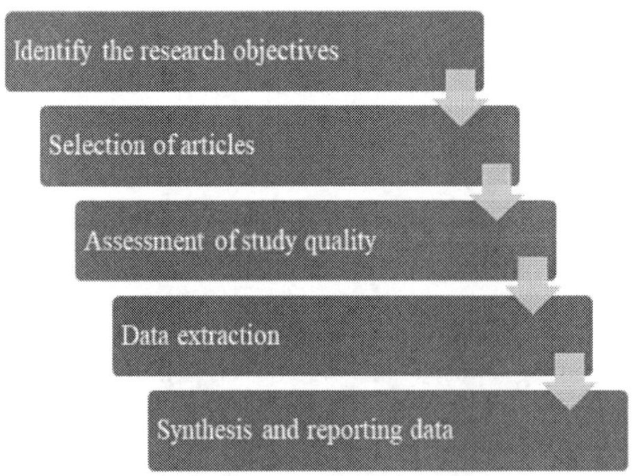

Fig. 1 Research Methodology

The collecting data process was started by browsing documents/ articles on Scopus database through the search field of title, abstract, and keyword using the search string as follows:

("sustainable" OR "sustainability" OR "green") AND ("environmental" OR "social) AND "supply chain" AND "performance" AND ("management" OR "measurement" OR "assessment" OR "evaluation") AND ("framework" OR "system" OR "method" OR "tool" OR "concept" OR "standard") AND ("metric" OR "metrics")

The article search process was only done for the articles that were published on Scopus from time range January 1, 2000 to June 20, 2018. As for the document type which was used was the journal articles that published in English, the other journal articles were not reviewed. The collecting data process was continued by reading the abstract of every article. The illustration of the article search process was presented on Figure 2. Based on the search, it could be recognized that there were 40 related article documents from Scopus. Then from those articles, it could be done the dimension analysis of Triple Bottom Line (TBL) that related to the sustainable and the performance measurement in supply chain.

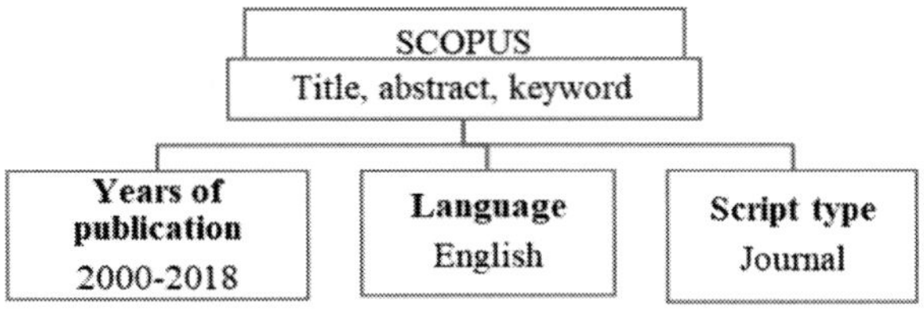

Fig. 2 The article serach process

3 Results and Discussions

3.1 Descriptive Analysis

The research study on SSCM performance measurement tends to increase every year. It can be proved based on the search result on Scopus that is illustrated on Figure 3 which shows that the research study on SSCM has increased in the last 5 years.

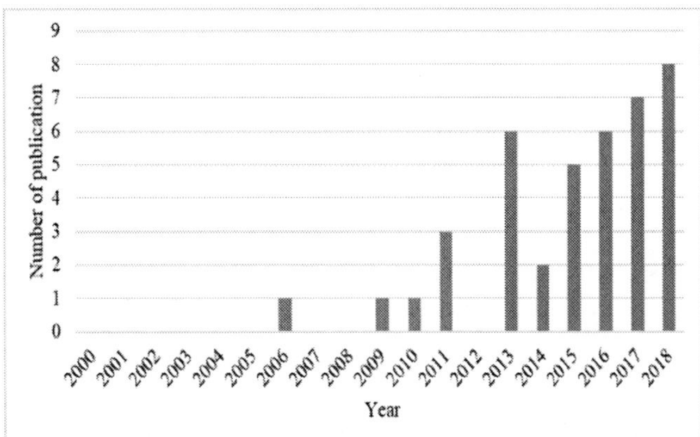

Fig. 3 Distribution of the number of research study by published year (n=40)

Those research studies have been scattered in several journals, as illustrated in Figure 4 which shows the top 10 of journal names. The most widely published journal about SSCM is Journal of Cleaner Production. Furthermore, there are several other journal names such as Energy Conversion and Management, Environmental Science and Technology, International Journal of Global Energy Issues, International Journal of Logistics Management, etc. They have at least one published article about SSCM.

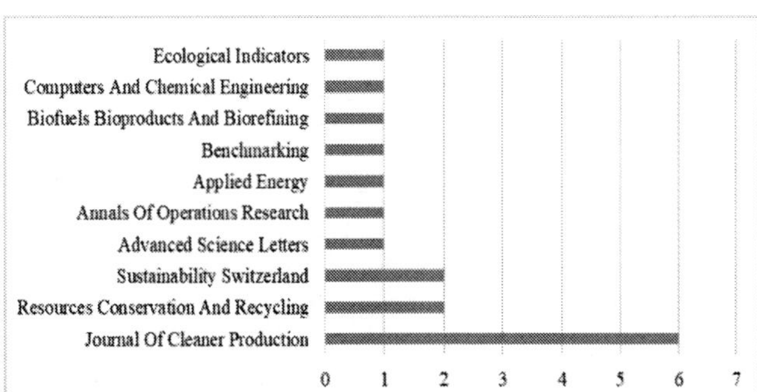

Fig. 4 Distribution of the number of research study by journal name

The articles that fit the criteria, which are articles that discuss about Sustainable Supply Chain Management (SSCM) are analyzed based on the industries studied, sustainable dimension, approach method for problem solving, and type of data processing that is used. The result of data analysis is showed in Figure 5, 6, 7 and 8.

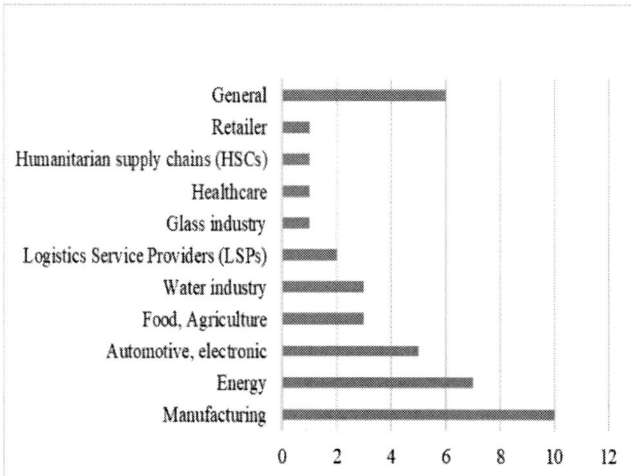

Fig. 5 Distribution of the number of research study by industries studied

The five industries that are used as the research object in SSCM is manufacturing [19], [20], [21], energy [22], [23], [24], automotive/electronic [25], [26], [27], food/agriculture [28], [29], [30], and water industry [31], [32], [33].

The dimension of sustainability measured that is including the dimension of the environment [34], [35], [36], social [37], economy [38], as well as its third integration about TBL [39], [40], [41]. From the result, can be seen that the dimensions of the many studied are the environment. It is in accordance with the discussion of SSCM and GSCM which further highlight the environmental dimension in every aspect of sustainability.

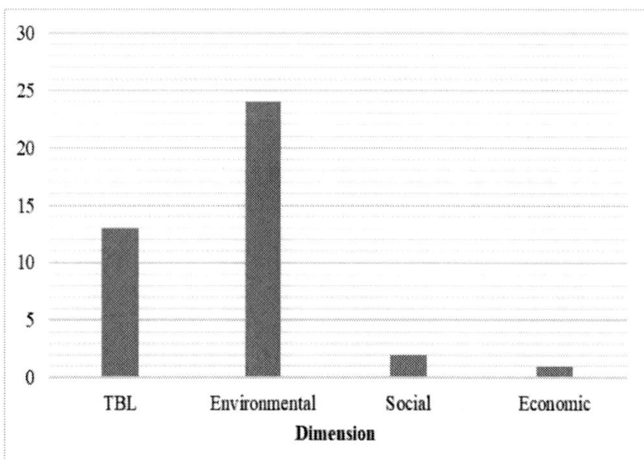

Fig. 6 Distribution of the number of research study by used sustainable dimension

In addition, five problem solving approaches that are widely used in research study include performance measurement frameworks [42], [43], LCA [44], [45], AHP [46], [47], MCDM [48], NSGA-II [49], [50], TOPSIS [51], and behavior-based [29], [38]. Then, for the type of research is mostly done by way of survey. In addition, there are other ways that can be done such as the formulation of mathematical models, literature reviews, case studies, and conceptual frameworks of a study of SSCM.

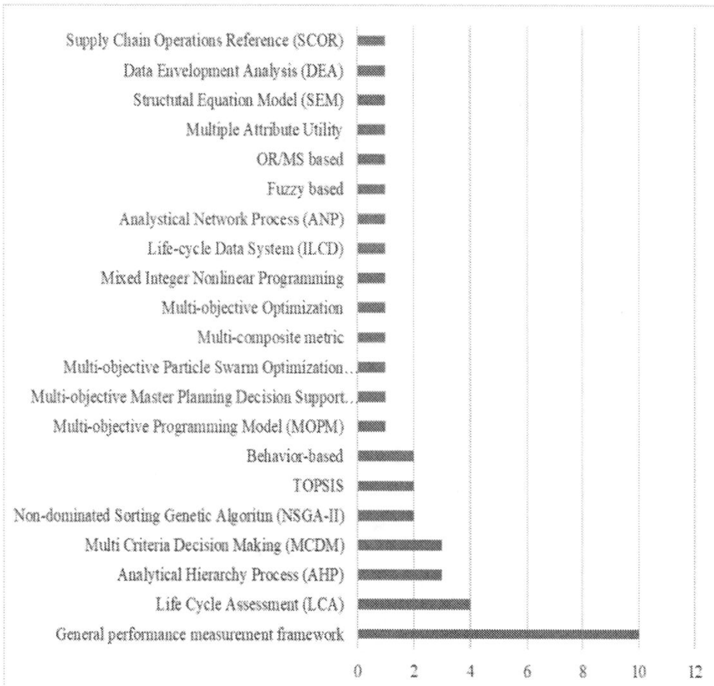

Fig. 7 Distribution of research study by approach method for proble solving

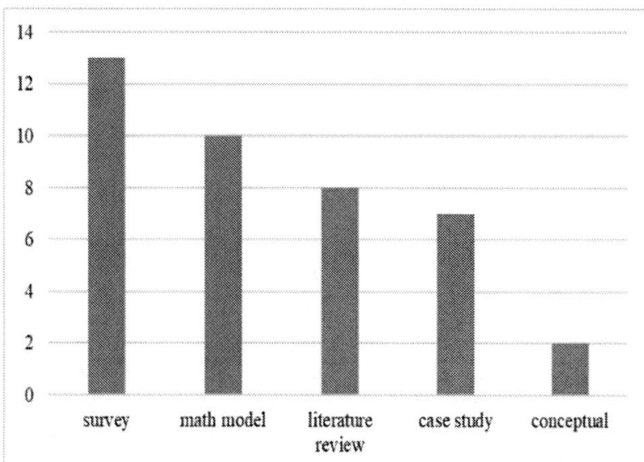

Fig. 8 Distribution of research study by research types

3.2 Sustainable metrics

The sustainability considers three dimensions or aspects, namely economic, social, and environmental or often termed a triple bottom line. For measuring the performance of sustainability in the supply chain, it can use sustainability metric. Where the metric can be categorized based on each aspect in the dimension of sustainability. Based on the literature which has been analyzed can be made a metric that can be used to measure the performance of a sustainable supply chain of an organization. The proposed metric is presented in Table 1.

IOP Conf. Series: Materials Science and Engineering **943** (2020) 012056 doi:10.1088/1757-899X/943/1/012056

Table 1. The Sustainability Metric

Dimension	The Sustainability Metric
Environmental	Nature resources, solid waste, air emission and water, transportation modes and transportation distance, global warming, biodiversity, acid deposition, landscape, ultimate waste disposal, product recycling ability, restoration of forest and eco-systems, minimization of deforestation, conservation of primary forest, use of GMOs, use of chemicals, pest control, and fertilizer
Social	Employment, salary, labour intensiveness or productivity, accident, risk, noise, welfare, education, resources, personal rotation rate, health and safety, human rights, legality verification, raw material/food security
Economical	Value added, output production, turnover, resources used as input, profitability, competitiveness, market shares, product durability, research and development efforts, business viability, economic stability, benefits sharing

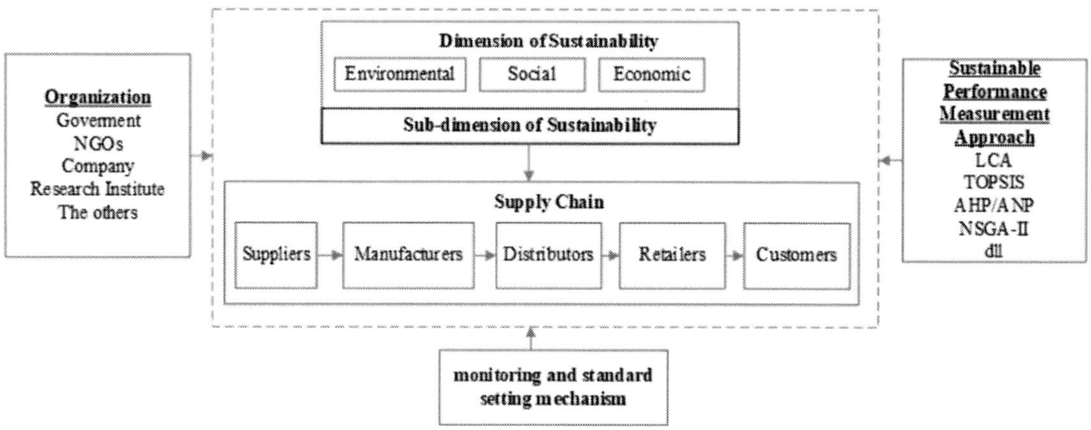

Fig. 9 The Framework for Sustainability Measurement of Supply Chain

Involved organizations in measuring sustainability performance of the supply chain may include governments, non-governmental organizations, corporations, media, research institutes, as well as the other parties that involved in decision-making related to the sustainability of an industry. The relevant parties can measure the sustainability performance based on metrics that are appropriate to the industry to be assessed. The approaches that can be used to perform the measurement namely, Life Cycle Assessment (LCA), Analytical Hierarchy Process (AHP), Multi Criteria Decision Making (MCDM), Non-dominated Sorting Genetic Algorithm (NSGA), TOPSIS, Behavior-based, Supply Chain Operations Reference (SCOR), Data Envelopment Analysis (DEA), Structural Equation Model (SEM), Fuzzy-based, and Multi-objective Optimization.

After measuring the performance of sustainability, the involved parties can perform a monitoring and standard setting mechanism, so that, the sustainability carried out by an industry can be monitored. The mechanism may include the application of the complaint procedure, trademark protection, label, assessment result publication, regular 3rd party assessment, 3rd party certification and accreditation processes. From all those processes can then be described a performance measurement framework of the supply chain as in Figure 9.

4. Conclusion

The review process of the previous literature has been done by analyzing based on the published year, the journal name, the used sustainable dimension, the industries that performs the performance measurement process, and the approach used in conducting the research. Then based on the results of the analysis, can be made a framework that can be used to measure the performance of supply chain sustainability in an industry. Where the framework describes the involved organization, the dimensions and sub-dimensions of sustainability that can be used, the approach that can be used as a method of performance measurement, as well as monitoring mechanism and standard measurement of the process of measuring the performance of the sustainable supply chain.

Acknowledgements

The research is supported by Institute for Research and Community Service, Sebelas Maret University with Hibah Penelitian Mandatory UNS (PM-UNS) Research Program (Contract No. 543/UN27.21/PP/2018).

References

[1] Delai, I. and Takahashi, S.: "Sustainability measurement system: A reference model proposal," Soc. Responsib. J., Vol. 7, pp. 438-471 (2011)
[2] Hassini, E., Surti, C., and Searcy, C.: "A literature review and a case study of sustainable supply chains with a focus on metrics," Int. J. Prod. Econ., Vol. 140, pp. 69-82 (2012)
[3] Joyce, A. and Paquin, R. L.: "The triple layered business model canvas: A tool to design more sustainable business models," Journal of Cleaner Production, Vol. 135, pp. 1474-1486 (2016)
[4] Molamohamadi, Z., Ismail, N., Leman, Z., and Norzima, Z.: "Developing aframework for sustainable supply chain management," Applied Mechanics and Materials, Vol. 564, pp. 661-666 (2014)
[5] Bendul, J. C., Rosca, E., and Pivovarova, D.: "Sustainable supply chain models for base of the pyramid," Journal of Cleaner Production, Vol. 160, pp. S170-S120 (2017)
[6] Tajbakhsh, A. and Hassini, E.: "A data envelopment analysis approach to evaluate sustainability in supply chain networks," Journal of Cleaner Production, Vol. 105, pp. 74-85 (2015)
[7] Barbosa-Póvoa, A. P., da Silva, C. and Carvalho, A.: "Opportunities and challenges in sustainable supply chain: An operations research perpective," European Journal of Operational Research, Vol. 268, No. 2, pp. 399-431 (2018)
[8] Qorri, A., Mujkić, Z. and Kraslawski, A.: A conceptual framework for measuring sustainability performance of supply chains," Journal of Cleaner Production, Vol. 189, pp. 570-584 (2018)
[9] Habibie, A., Hisjam, M., Sutopo, W., and Widodo, K. H.: "A relationship model between supplier and manufacturer for securing availability of teak log in export oriented furniture industry with sustainability considerations," Lecture Notes in Engineering and Computer Science Proceedings, pp. 1292-1297 (2012)
[10] Devi, R.A., Hisjam, M., Sutopo, W., and Widodo K.H.: "A relationship model between manufacturer and buyer for order fulfillment in export oriented furniture industry with sustainability considerations," Lecture Notes in Engineering and Computer Science Proceedings, pp. 1407-1411 (2012)
[11] Hisjam, M., Habibie, A., Sutopo, W., and Widodo K.H.: "A supplier-manufacturer model for procurement plan in export oriented furniture industry with sustainability considerations," Lecture Notes in Electrical Engineering Proceedings, pp. 247-260 (2013)
[12] Hisjam, M., Sutopo, W., Devi, R.A., and Widodo K.H.: "A manufacturer-buyer relationship model in export oriented furniture industry with sustainability considerations," Proceedings - Joint International Conference on Electric Vehicular Technology and Industrial, Mechanical, Electrical and Chemical Engineering, ICEVT 2015 and IMECE 2015, pp. 298-303 (2015)
[13] Putri, D.N.E., Hisjam, M., Sutopo, W., and Widodo, K.H.: "Simulation of supplier-manufacturer relationship model for securing availability of teak log in furniture industry with sustainability consideration," IEEE International Conference on Industrial Engineering and Engineering Management Proceedings, pp. 367-371 (2014)
[14] Herjuna, S.A.S., Hisjam, M., Putri, D.N.E., Sutopo, W., and Widodo K.H.: "A decision support application for teak log supplier to simulate scenarios of utilizing forest resources by considering the sustainability of teak log supply," Lecture Notes in Engineering and Computer Science Proceedings, pp. 938-941 (2015)
[15] Kurniawan, B., Hisjam, M., and Sutopo, W.: "Integration of production and supply chain strategic planning for renewable resources under sustainability considerations: Teakwood case study," IEEE International Conference on Industrial Engineering and Engineering Management Proceedings, pp. 433-437 (2011)

[16] Shepherd, C. and Günter, H.: "Measuring supply chian performance: current research and future directions," Int. J. Product. Perform. Manag., Vol. 55, pp. 242-258 (2006)
[17] Bai, C. and Sarkis, J.: "Determining and applying sustainable supplier key performance indicators," Supply Chain Manag. An Int. J., Vol. 19, pp. 275-291 (2014)
[18] Tranfield, D., Denyer, D. and Smart, P.: "Towards a methodology for developing evidence-informed management knowledge by means of systematic review," Br. J. Manag., Vol. 14, pp. 207-222 (2003)
[19] Alsaffar, A. J., Raoufi, K., Kim, K. Y., Kremer, G. E. O., and Haapala, K. R.: "Simultaneous consideration of unit manufacturing processes and supply chain activities for reduction of product environmental and social impacts," Journal of Manufacturing Science and Engineering, Transactions of the ASME, Vol. 138, No.10 (2016)
[20] Entezaminia, A., Heydari, M., and Rahmani, D.: "A multi-objective model for multi-product multi-site aggregate production planning in a green supply chain: Considering collection and recycling centers," Journal of Manufacturing Systems, Vol. 40, pp. 63-75 (2016)
[21] Egilmez, G., Kucukvar, M., and Park, Y. S.: "Supply chain-linked sustainability assessment of the US manufacturing: An ecosystem perspective," Sustainable Production and Consumption, Vol. 5, pp. 65-81 (2016)
[22] Garofalo,P., Campi, P., Vonella, A. V., and Mastrorilli, M.: "Application of multi-metric analysis for the evaluation of energy performance and energy use efficiency of sweet sorghum in the bioethanol supply-chain: A fuzzy-based expert system approach," Applied Energy, Vol. 220, pp.313-324 (2018)
[23] Sluka, C. and Peck, P.C.: "Stakeholder dynamics in the EU forest energy sector: Key issues to manage and ways forward," Biofuels, Bioproducts, and Biorefining, Vol. 9, No. 1, pp. 51-71 (2015)
[24] Halog, A.: "Models for evaluating energy, environmental and sustainability performance of biofuels value chain," International Journal of Global Energy Issues, Vol. 32, pp.83-101 (2009)
[25] Schoggl, J. P., Fritz, M.M.C., and Baumgartner, R. J.: "Toward supply chain-wide sustainability assessment: A conceptual framework and an aggregation method to assess supply chain performance," Journal of Cleaner Production, Vol. 131, pp. 822-835 (2016)
[26] Sahu, N. K., Datta, S., and Mahapatra, S. S.: "Green supplier appraisement in fuzzy environment," Benchmarking, Vol. 21, No. 3, pp. 412-429 (2014)
[27] Olugu, E. U., Wong, K. Y., and Shaharoun, A.M.: "Development of key performance measures for the automobile green supply chain," Resources, Conservation and Recycling, Vol. 55, No. 6, pp. 567-579 (2011)
[28] Mohammed, A., Setchi, R., Filip, M., Harris, I., and Li, X.: "An integrated methodology for a sustainable two-stage supplier selection and order allocation problem," Journal of Cleaner Production, Vol. 192, pp. 99-114 (2018)
[29] Lengnick, L., Miller, M., and Marten, G.G.: "Metropolitan foodsheds: a resilient response to the climate change challenge?," Journal of Environmental Studies and Sciences, Vol. 5, No. 4, pp. 573-592 (2015)
[30] Cao, L., Diana, J. S., Keoleian, G.A., and Lai, Q.: "Life cycle assessment of Chinese shrimp farming systems targeted for export and domestic sales," Environmental Science and Technology, Vol. 45, No. 15, pp. 6531-6538 (2011)
[31] Balfaqih, H., Nopiah, Z. M., Al-Nory, M. T., and Saibani, N.: "Sustainable framework for water desalination supply chain: Delphi-AHP technique," Advanced Science Letters, Vol. 23, No. 5, pp. 4721-4724 (2017)
[32] Hoekstra, A.Y.: "A critique on the water-scarcity weighted water footprint in LCA," Ecological Indicators, Vol. 66, pp. 564-573 (2016)
[33] Huysveld, S., Schaubroeck, T., De Meester, S., Sorgeloos, P., Van Langenhove, H., Van Linden, V., and Dewulf, J.: "Resource use analysis of Pangasius aquaculture in the Mekong Delta in Vietnam using Exergetic Life Cycle Assessment," Journal of Cleaner Production, Vol. 51, pp. 225-233 (2013)
[34] Tuni, A., Rentizelas, A., and Duffy, A.: "Environmental performance measurement for green supply chains: A systematic analysis and review of quantitative methods," International Journal of Physical Distribution and Logistics Management, doi: 10.1108/IJPDLM-02-2017-0062 (2018)
[35] Aymard, V. and Botta-Genoulaz, V.: "Normalisation in life-cycle assessment: consequences of new European factors on decision-making," Supply Chain Forum, Vol. 18, No. 2, pp. 76-83 (2017)
[36] Raoufi, K., Haapala, K.R., Jackson, K.L., Kim, K.-Y., Kremer, G.E.O., and Psenka, C.E.: "Enabling non-expert Sustainable Manufacturing Process and Supply Chain Analysis During the Early Product Design Phase," Procedia Manufacturing, Vol. 10, pp. 1097-1108 (2017)

[37] Alsaffar, A. J., Raoufi, K., Kim, K. Y., Kremer, G. E. O., and Haapala, K. R.: "Simultaneous consideration of unit manufacturing processes and supply chain activities for reduction of product environmental and social impacts," Journal of Manufacturing Science and Engineering, Transactions of the ASME, Vol. 138, No. 10, doi:10.1115/1.4034481 (2016)

[38] Kirchoff, J.F., Omar, A., and Fugate, B.S.: A "Behavioral Theory of Sustainable Supply Chain Management Decision Making in non-exemplar Firms," Journal of Supply Chain Management, Vol. 52, No. 1, pp. 41-65 (2016)

[39] Laguna-Salvadó, L., Lauras, M., Okongwu, U., and Comes, T.: "A multicriteria Master Planning DSS for a sustainable humanitarian supply chain," Annals of Operations Research, pp. 1-41 (2018)

[40] Gong, M., Simpson, A., Koh, L., and Tan, K.H.: "Inside out: The interrelationships of sustainable performance metrics and its effect on business decision making: Theory and practice," Resources, Conservation and Recycling, Vol. 128, pp. 155-166 (2018)

[41] Abidi, N., Bandyopadhayay, A., and Gupta, V.: "Sustainable supply chain management: A three dimensional framework and performance metric for Indian IT product companies," International Journal of Information Systems and Supply Chain Management, Vol. 10, No. 1, pp. 29-52 (2017)

[42] Ahi, P. and Searcy, C.: "An analysis of metrics used to measure performance in green and sustainable supply chains," Journal of Cleaner Production, Vol. 86, pp. 360-377 (2015)

[43] Halog, A. and Manik, Y.: "Advancing integrated systems modelling framework for life cycle sustainability assessment," Sustainability, Vol. 3, No. 2, pp. 469-499 (2011)

[44] Huysveld, S., Schaubroeck, T., De Meester, S., Sorgeloos, P., Van Langenhove, H., Van Linden, V., and Dewulf, J.: "Resource use analysis of Pangasius aquaculture in the Mekong Delta in Vietnam using Exergetic Life Cycle Assessment," Journal of Cleaner Production, Vol. 51, pp. 225-233 (2013)

[45] Neupane, B., Halog, A., and Lilieholm, R.J.: "Environmental sustainability of wood-derived ethanol: A life cycle evaluation of resource intensity and emissions in Maine, USA," Journal of Cleaner Production, Vol. 44, pp. 77-84 (2013)

[46] Mohammed, A., Setchi, R., Filip, M., Harris, I., and Li, X.: "An integrated methodology for a sustainable two-stage supplier selection and order allocation problem," Journal of Cleaner Production, Vol. 192, pp. 99-114 (2018)

[47] Entezaminia, A., Heydari, M., and Rahmani, D.: "A multi-objective model for multi-product multi-site aggregate production planning in a green supply chain: Considering collection and recycling centers," Journal of Manufacturing Systems, Vol. 40, pp. 63-75 (2016)

[48] Mansoornejad, B., Pistikopoulos, E.N., and Stuart, P.: "Metrics for evaluating the forest biorefinery supply chain performance," Computers and Chemical Engineering, Vol. 54, pp. 125-139 (2013)

[49] Rabbani, M., Navazi, F., Farrokhi-Asl, H., and Balali, M.H.: "A sustainable transportation-location-routing problem with soft time windows for distribution systems," Uncertain Supply Chain Management, Vol. 6, No. 3, pp. 229-254 (2018)

[50] Moslemi, S., Zavvar Sabegh, M.H., Mirzazadeh, A., Ozturkoglu, Y., and Maass, E.: "A multi-objective model for multi-production and multi-echelon closed-loop pharmaceutical supply chain considering quality concepts: NSGAII approach," International Journal of Systems Assurance Engineering and Management, Vol. 8, pp. 1717-1733 (2017)

[51] Sahu, N. K., Datta, S., and Mahapatra, S. S.: "Green supplier appraisement in fuzzy environment," Benchmarking, Vol. 21, No. 3, pp. 412-429 (2014)

Review of BIM literature and government initiatives to promote BIM in Malaysia

S S Sinoh[1], Z Ibrahim[1], F Othman[1] and N L N Muhammad[1]

[1] Department of Civil Engineering, Faculty of Engineering University of Malaya, Lembah Pantai, 50603, Kuala Lumpur, Malaysia

E-mail: saznizam@um.edu.my

Abstract. The purpose of this paper was to review the current state of building information modelling (BIM) literature in Malaysia. A detailed description of the history and definition of BIM was also presented. A literature search was carried out on the Web of Science Core Collection (WoS CC) on BIM research in Malaysia. Results showed that Malaysian output of BIM research peaked in 2016 and has shown a steady decline since. Most of the research from local affiliations were published in conference proceedings as opposed to academic journals. Also, a search among non-academic sources was conducted to reveal government efforts to promote BIM in Malaysia. Generally, the strategies used by the government to promote BIM include spreading awareness through roadshows, increasing competency through workshops and encouraging higher education institutes to include BIM in their curricula. Current focus should now be shifted to increasing collaboration among construction industry players using BIM as a medium for cooperation. Implications include the need for more peer-reviewed studies and a systematic nationwide study of BIM implementation in Malaysia to facilitate its growth on the national level.

1. Introduction

The construction industry in Malaysia contributed to 6.7% of the country's economy in 2017 and is critical for the development of social and economic infrastructure [1]. This industry is currently undergoing a transformation phase into widespread implementation of building information modelling (BIM). It is a move which is seen as a revolution in this industry and fully supported by the government of Malaysia. BIM is one of the latest technologies revolutionizing the architectural, engineering and construction industry worldwide. It goes beyond a technological advancement and represents a shift in the traditional building delivery process towards an integrated workflow. The aim of the study was to analyze the literature to identify BIM studies relating to Malaysia. This study also examined the efforts made by the Malaysian government to promote BIM.

1.1. Origins of BIM

Construction projects are heavily paper-based, and any documents related to the project must be printed out on physical media. In fact, most of the construction industry relies on physical specifications and drawings. The path to digitization in the construction industry began when the first software packages for computer-based design of buildings were developed in the 1970s. The concept of BIM was created in the early 1970s by Charles Eastman. He proposed the Building Description System (BDS) which is the precursor to modern BIM as we know it today [2]. In BDS, the system performs a few functions which are similar to what BIM is capable of: (1) a graphical user interface to store and display complex element shapes; (2) an interactive graphic language to arrange different elements in a design space; (3) it can produce perspective or orthographic drawings and; (4) the database can be sorted by attributes such as material type. Since then, BDS has evolved over several stages to become what BIM is today as shown in figure 1.

In 1977, Graphical Language for Interactive Design (GLIDE) had been introduced which improved on the features of BDS and offered more accurate drawings [3]. In 1989, Building Product Model (BPM) was introduced and offered more features such as estimation [4]. In 1995, the Generic Building Model (GBM) was introduced which defined the building type in the model and allowed a higher degree construction management to be achieved [5]. Beginning in the early 2000s, the concept of BIM took off among the architecture, engineering and construction (AEC) industry when parametric 3D modelling was developed [6]. It is this amalgamation of different features that spanned more than 40 years of development which defines modern BIM.

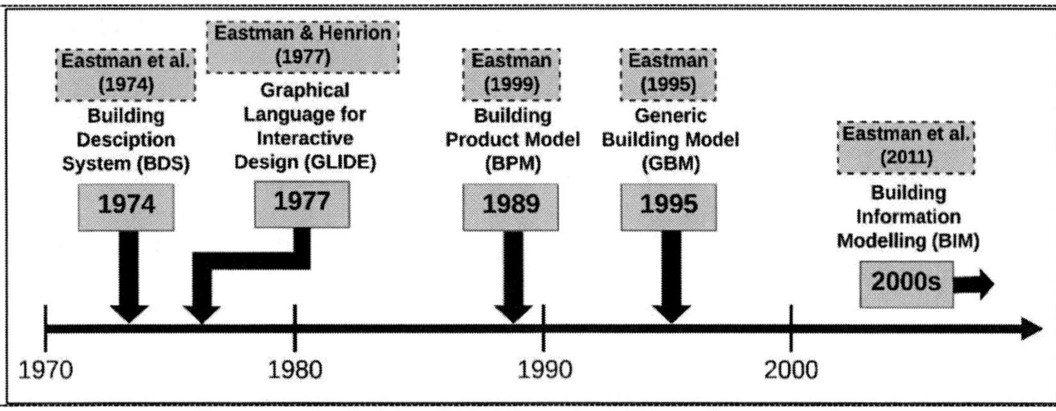

Figure 1. Evolution of BIM from the 1970s to the present day.

1.2. Definition of BIM

Many different definitions of BIM have appeared in the literature throughout the lifespan of BIM. Despite these different definitions, there exists certain common keywords which represent the essence of BIM. These keywords will become apparent by examining the different definitions of BIM. Firstly, BIM is often viewed as a computer software, a collection of software or simply as a tool. Van Nederveen and Tolman [7] defined BIM as "a model of information about a building that comprises complete and sufficient information to support all lifecycle processes and which can be interpreted directly by computer applications. It comprises information about the building itself as well as its components, and comprises information about properties such as function, shape, material and processes for the building life cycle". In early research on BIM, it was viewed as a tool to digitize the depiction of a building as a computer-generated model and allows important information to be embedded in these models.

Gu and London [8] expands on this definition further by framing BIM as a database for any parties in a project to refer to at any stage in construction as stated here: "building information modelling (BIM) is an IT-enabled approach that involves applying and maintaining an integral digital representation of all building information for different phases of the project lifecycle in the form of a data repository". By being a database, the collaborative capabilities of BIM become apparent and different parties can have a single reference point for any information related to a construction project.

This definition of BIM would be expanded to consider BIM as a methodology or process which Azhar [9] defines as "a virtual process that encompasses all aspects, disciplines, and systems of a facility within a single, virtual model, allowing all design team members (owners, architects, engineers, contractors, subcontractors, and suppliers) to collaborate more accurately and efficiently than using traditional processes". Instead of being thought of as a tool used in a project workflow, in this definition BIM is considered as the actual workflow.

This is more holistic as it considers the use of BIM throughout the entire project lifecycle. Going further, BIM can be regarded as a combination of both process and software. Indeed, for BIM to be fully effective, it must be used as a combination of technology, process and policy which is another layer added onto the BIM definition by Succar [10] who states BIM is "a set of interacting policies, processes and technologies generating a methodology to manage the essential building design and project data in digital format throughout the building's life-cycle". The definition of BIM has come around full-circle and blends the different definitions of BIM into a comprehensive one. Succar's [10] definition of BIM is widely used in various academic studies and technical reports related to BIM.

1.3. BIM maturity

BIM Maturity refers to the quality, repeatability and degree of excellence within a BIM capability [11]. In other words, it is a measure of BIM use within an organization. Two scales are frequently used to measure BIM Maturity namely the Bew-Richards model [12] and the Succar model [10]. The Bew-Richards model is more detailed compared to the Succar model, but both use similar descriptions to define the boundaries between Levels. Level 0 (i.e. Pre-BIM) is characterized by traditional building delivery methods which uses paper-based mediums for all documents and technical drawings. Level 1 represents object-based modelling with the use of 2D or 3D digital objects to represent structural elements. Level 2 is where model-based collaboration occurs between different parties involved in a construction project. Level 3 signifies network-based integration and is the step that precedes integrated project delivery as the long-term objective of BIM implementation. Figure 2 depicts the Bew-Richards model of BIM Maturity.

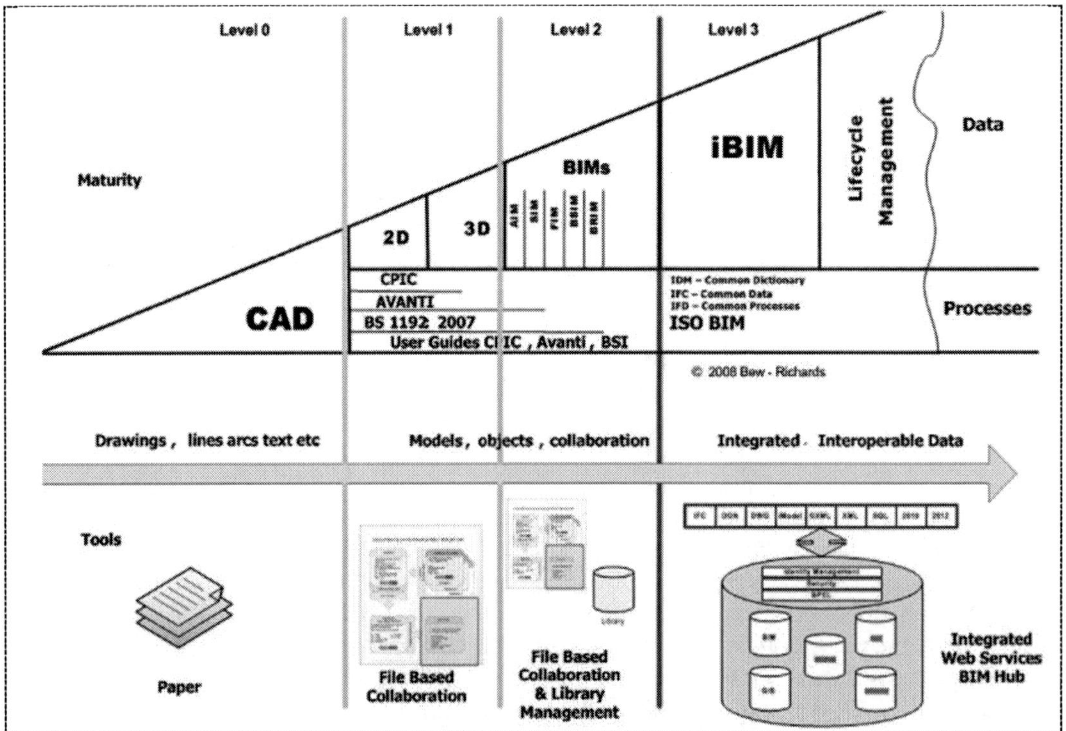

Figure 2. Bew-Richards model for BIM Maturity [12].

1.4. History of BIM in Malaysia

Malaysia is a country open to technological innovations as can be seen in the implementation of a computer chip-equipped national identification card known as MyKad and the first RFID-equipped passport in the world in 1998. Therefore, it can be said that Malaysia is also open to technological innovations in all sectors including construction. It follows that the government is very keen on adopting new technologies in the construction industry. However, the level of BIM Maturity in Malaysia has lagged those of other developed countries and its neighbors such as Singapore. In 2013, Zahrizan, Ali, Haron, Marshall-Ponting and Hamid [13] determined that BIM Maturity in Malaysia was between Level 0 and Level 1. In 2014, Enegbuma, Aligha and Ali [14] found that the Maturity level had risen to between Level 1 and Level 2. The government intends to reach a BIM Maturity of Level 2 by 2019 which is network-based integration of BIM across the construction industry [15]. In 2019, the PWD has enforced the mandatory requirement for any public construction projects valued above MYR 100 million (approximately USD 24 million) to use BIM [16]. Currently, there is no legal framework to enforce the use of BIM in private construction projects. The use of BIM in private projects mainly relies on the requirement of the client [13]. On the other hand, public construction projects fall under the jurisdiction of the PWD which is able to impose requirements for BIM. The CIDB has proposed that BIM be used in certain private construction projects beginning 2020 [17]. Though it is yet to be seen if a legal framework to support the enforcement of BIM will be created.

Private construction companies began using BIM in the early 2000s, but on a smaller scale than today [18]. BIM was formally introduced to Malaysia by the Public Works Department (PWD) in 2007 with the publication of their BIM Standard Manual and Guidelines [19]. The PWD was also responsible for establishing a BIM committee for BIM training tools and drawing a BIM Roadmap. However, the PWD is not the sole government body responsible for promoting BIM. The Construction Industry Development Board (CIDB) developed a BIM Portal and formed the first BIM Steering Committee. The Multimedia Super Corridor (MSC) initiative allocated grants for BIM training and awards BIM training certificates. The Construction Research Institute of Malaysia (CREAM) provides training and education on BIM. Figure 3 summarizes the key events in BIM implementation in Malaysia. There were 3 pilot projects which used BIM in Malaysia by the PWD which are (1) National Cancer Institute, Putrajaya, (2) Healthcare Centre Type 5, Pahang and (3) Malaysian Anti-Corruption Commission Administration Complex, Selangor. These pilot projects served to provide the PWD with the experience necessary to formulate a framework for nationwide implementation of BIM.

2. Literature Review

2.1. Global literature

Van Nederveen and Tolman [7] proposed an approach which uses so-called 'aspect models' to model building information in a simple and natural way. This was the first mention of any sort of BIM model in the literature. Since then, the number of published articles regarding BIM has increased. Zhao [20] searched all articles in the Web of Science Core Collection (WoS CC) database for any articles related to BIM and produced a comprehensive scientometric analysis of the subject. The author found that growth into BIM research grew rapidly from about 2010 onwards. In fact, it was in 2012 that global BIM research really took off and the number of BIM articles increased by approximately 50% every year. In another review of BIM literature, Santos, Costa and Grilo [21] point out that from the period between 2005 and 2015, more than 90% of all BIM-related papers included in their study were published from 2011 onwards. This shows the increasing attention that BIM has received in recent years and that the period between 2011-2012 was when BIM research grew rapidly.

A key article by Succar [10] received the most citations according to WoS. In this paper, Succar [10] laid out the theoretical framework relating to BIM application in the construction industry using a combination of ontology and visual knowledge models. This landmark paper was the influence on many subsequent papers in the field of BIM. Another authority on BIM is the previously mentioned Charles

Eastman who has produced important publications on BIM and his 'BIM Handbooks' series represent key textbooks in the study of BIM, receiving the most co-citations [20].

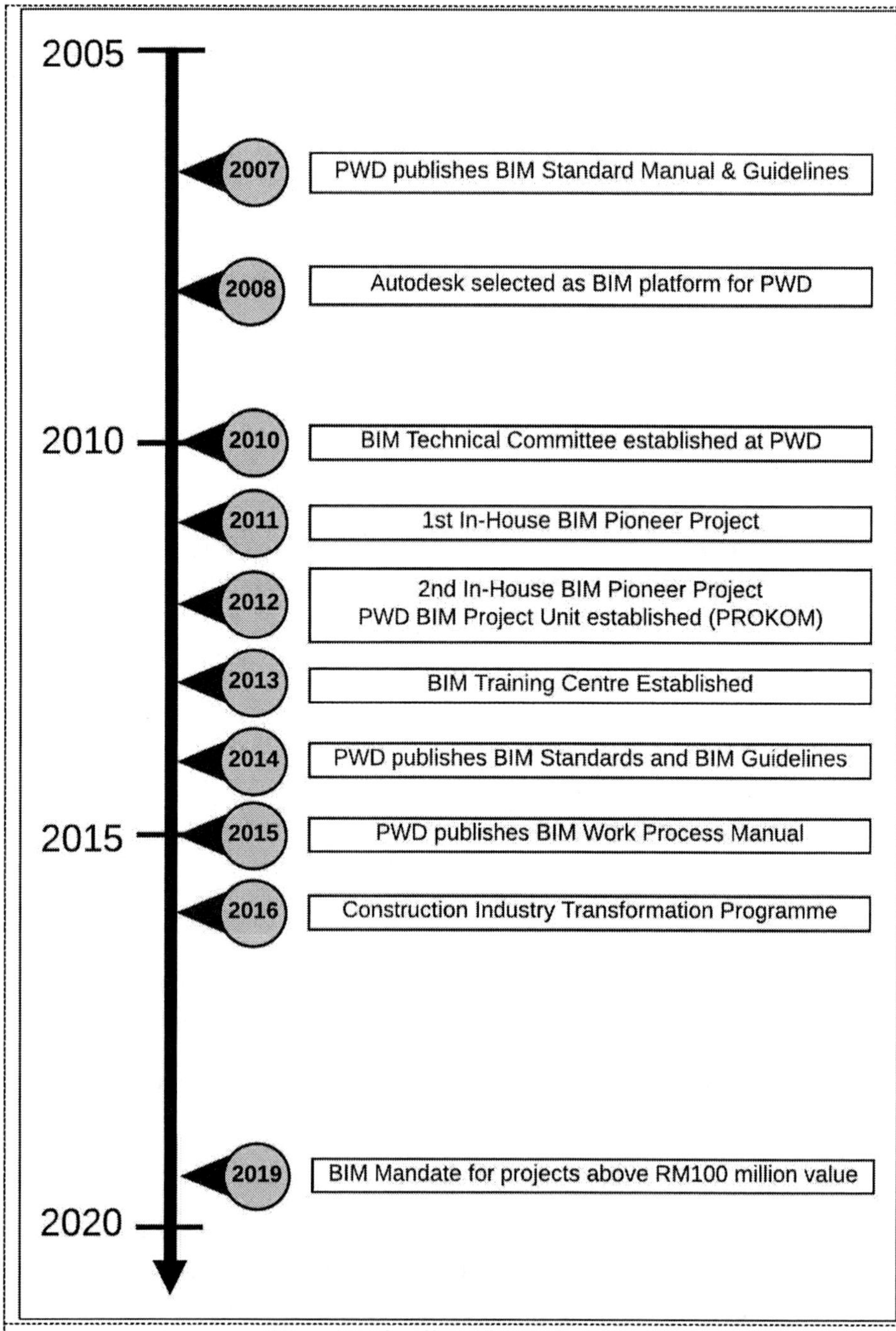

Figure 3. Timeline of BIM in Malaysia.

2.2. Malaysian literature

In the present study, a literature search was carried out in the WoS CC using a similar methodology proposed by Zhao [20]. The following code was entered into the Advanced Search field of WoS: TS =

(building information model* AND BIM* AND Malaysia*). This search was conducted in September 2019. Here, "TS" denotes an article subject and "*" represents a fuzzy search. The time span of the search was between 1999 and 2019, but the earliest paper to appear in the search results was published in 2012. A total of 69 bibliographic records were obtained from the initial search. After discarding 3 documents that were deemed unrelated to the scope, a total of 66 records were obtained.

The number of publications was the same for 2012 and 2013 but increased each year after 2013 and peaked in 2016 after which it decreased. The decrease from 2016 onwards suggests that quality BIM research is slowing down in Malaysia as shown in figure 4. In total, the publications included 39 conference papers and 27 journal articles. Table 1 shows the breakdown of subject areas of BIM literature in Malaysia. Most of the papers addressed issues related to the AEC industry in general with 28 publications. BIM studies related to quantity surveying and higher education were the second and third highest at 7 and 6 publications respectively. The 3 most cited papers in the WoS search results were Rogers, Chong and Preece [22], Takim, Harris and Nawawi [23] and Shoubi, Shoubi, Bagchi and Barough [24] with 14, 12 and 10 total citations respectively. Universiti Teknologi Malaysia (UTM) is affiliated with 24 out of 66 publications which is the highest among the higher education institutions in the country. Ali, K. N., Embi, M. R., Latiffi, A. A. and Enegbuma, W. I. were the most productive authors with 7, 6, 5 and 4 documents produced respectively.

Table 1. Subject areas of BIM literature in Malaysia between 2012 and 2019.

Subject Area	Number of Publications	Percentage
Construction industry (in general)	28	42.4%
Quantity surveying	7	10.6%
Higher education	6	9.1%
Industrialized building system	3	4.5%
Housing development	3	4.5%
Interior design	3	4.5%
Automated safety rule checking	2	3.0%
Decision support	2	3.0%
Finance	2	3.0%
Green buildings	2	3.0%
Data management	2	3.0%
BIM governance	2	3.0%
Facilities management	1	1.5%
Transportation	1	1.5%
Architecture	1	1.5%
Cultural heritage management	1	1.5%
TOTAL	66	100%

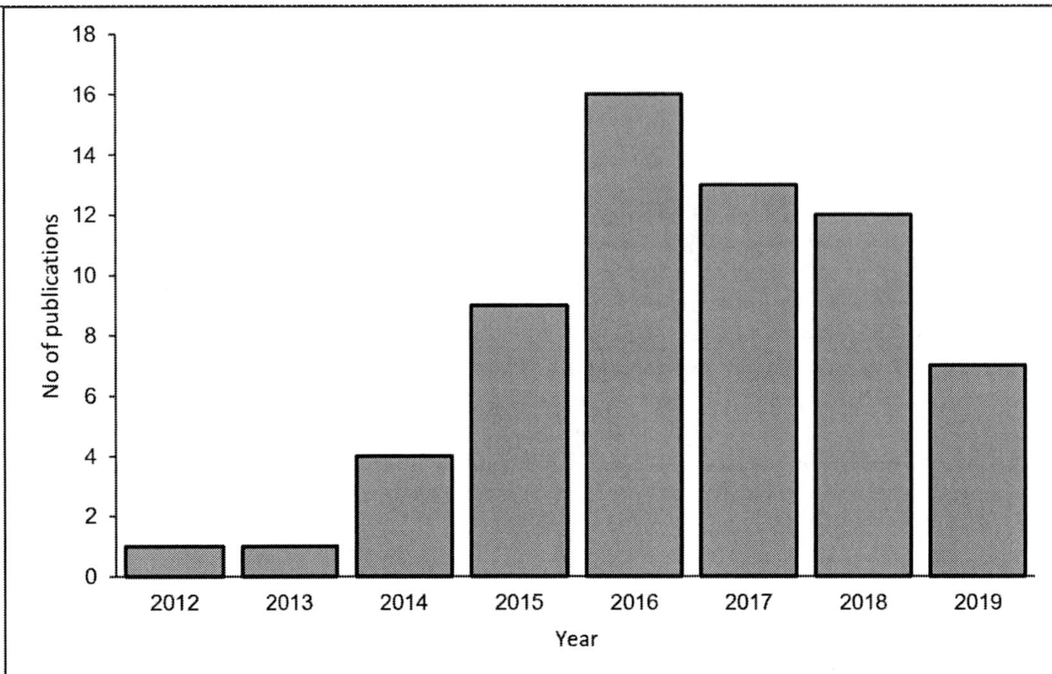

Figure 4. Number of Malaysian BIM publications in WoS between 2012 and 2019.

3. Government Efforts to Promote BIM

3.1. Increasing BIM competency

Many studies state the importance of BIM training among the construction industry for increasing BIM adoption in the country. The National Construction Industry Transformation Plan (CITP) shows BIM adoption in the country as being at only 10% [25]. Thus, they suggest increasing knowledge of BIM through training can increase BIM adoption. Mamter, Aziz and Zulkepli [26] showed that increased BIM training for new employees can lead to enhanced BIM manager capabilities of staff. Results show that software training for employees is needed to make them experts in BIM software. This situation leads to positive reinforcement in BIM manager competencies and finally lead to the creation of more BIM managers. Thus, the argument put forward is that for something to be widely adopted, information about it must be spread to as many people as possible to dispel any misconceptions.

Latiffi, Brahim and Fathi [27] found that respondents understand BIM as being a process (improving construction activities and sequence of work) and a technology (medium of communicating information in construction projects). However, the other categories of design, performance and productivity were not understood well by the respondents and they did not attribute these to BIM. The study revealed that the understanding of BIM among construction industry players was lacking. More training into BIM can clear up any misconceptions and lead to a higher adoption rate.

Another study by Latiffi, Brahim and Fathi [28] suggests that BIM competency among construction players is important for the development of BIM adoption in the country. In fact, only 2 out of the 28 respondents reported using BIM in the post-construction phase. This shows that the understanding of construction industry players towards BIM is as a pre-construction tool. Roughly half of the respondents used BIM because of their own initiative while the other half used it because of client demand. This suggests that a 'push and pull' is required from both clients and contractors for BIM to take off. They suggest that simply the use of BIM technology is not enough for its successful implementation, but also the competencies of the players involved. Thus, a tool must be used with accompanying expertise for it to be fully effective.

Qin [29] reported that companies that heavily use computer technology are more open to accepting new technologies such as BIM. In fact, the findings indicate the most important issue in implementation of BIM is the lack of training and awareness on BIM. They conclude that people are the key factors for determining the adoption of BIM. Thus, educating people on BIM may help increase uptake of BIM. As a conclusion, this situation is similar to how computer aided design (CAD) was first introduced in construction where a great deal of effort to train and prepare staff to use the new technology. Therefore, in preparation for widespread adoption of BIM, the groundwork for a new technology must be laid by educating and training staff.

3.2. Spreading BIM awareness

Since 2014, CIDB has organized a BIM Day every year except in the year 2017. This event is held as part of its annual International Construction Week (ICW) to promote and spread awareness of BIM. The first BIM Day was held on 22 September 2014 at the Putra World Trade Center and the latest one on 20 March 2019 at the Sunway Putra Hotel. Although a BIM Day was not held in 2017, the ICW for that year did include a variety of BIM-related seminars and workshops. These large-scale events are beneficial for the publicity of BIM to both the construction industry and the general public. Another effort by the CIDB is organizing regular BIM roadshows all over Malaysia and moving from state to state [30]. In fact, CIDB has also made efforts to promote BIM through digital means as well, most notably through social media.

CIDB has taken to social media to spread awareness of BIM through the hashtag "#letsBIM". A hashtag is a word or phrase preceded by a hash sign (#) used on social media to identify messages concerning a specific topic. Interestingly, "#letsBIM" was not created by CIDB, but by a Twitter user named David Delgado Vendrell with the Twitter username @daviddelven. The first Twitter post to use the hashtag was on 27 February 2015 [31]. CIDB first started using the hashtag in a Twitter post on 8 January 2016 [32]. Since this first post, CIDB has been consistently using the hashtag to identify BIM related content that it posts to social media including on Facebook and Twitter.

In conjunction with BIM Day 2018, CIDB decided to organize an Intervarsity BIM Marathon from 19 to 22 March 2018. In this competition, student representatives from 9 higher education institutes in Malaysia competed to produce a 3D BIM model. The competition lasted for 66 hours non-stop and was recognized by the Malaysia Book of Records as the longest BIM Modelling Competition [33]. Overall, CIDB has shown to produce creative efforts to promote BIM in Malaysia especially the younger generation.

CIDB has invested MYR 2.5 million (approximately USD 600,000) for the establishment of myBIMcentre on 20 November 2017 which is a one-stop referral centre that hosts training courses for developers, contractors, consultants and suppliers [34]. The centre is used to provide BIM software training and is used alongside the National BIM Library Portal. In fact, these companies can apply for BIM training courses under the CIDB Transformation Fund programme of which MYR 1 million (approximately USD 240,000) has been allocated as financial incentive for construction firms to adopt BIM [35].

3.3. BIM in higher education

Yusuf, Embi and Ali [36] emphasize the importance of competency in BIM among students in higher education. By establishing a foundation for BIM early, it allows new graduates entering the workforce to either (1) influence the company to move to BIM or (2) adapt to companies that already use BIM. The construction industry in Malaysia needs competent graduates with ready knowledge on BIM. Including BIM into the curriculum of engineering graduates can help solve this problem. However, several challenges must be overcome. First, the lack of BIM training for academic staff in BIM-related departments. Organizations such as CREAM have been providing training on BIM to industry participants for some time. Thus, it would not be difficult to adapt the curriculum used in these training programs for academic purposes. Second, the amount of resources needed to set up laboratories, purchase computer hardware and BIM software is a big investment. However, engineering faculties are equipped with computer labs for the learning of CAD software. Therefore, such labs may be able to be used for the teaching of BIM. Only the BIM software needs to be purchased for this to occur. Third, there is a lack of standard BIM curriculum. The question arises as to who the relevant authority shall be to establish such a curriculum. Feedback from students, academics and industry can help improve such a course and become the basis for a national BIM curriculum.

There is a debate between business BIM and academic BIM. Business oriented BIM is aimed at making a profit and it is appropriately named due to its business nature. Academic BIM is geared towards the knowledge development of BIM and is regulated by a governing body [37]. This also creates conflict between teaching-oriented universities and research universities. Teaching universities are more focused on preparing the graduate for the work environment and providing the skills necessary for them to operate in such environment. Research universities are more geared towards the advancement of learning and discovery of knowledge. Thus, there needs to be a balance between teaching and research as is with any subject, not just BIM. Eadie, Solan, Magee and Rice [38] studied pedagogical strategies of teaching BIM in the UK Higher Education setting and found that a constructivist approach is best used when teaching software elements of BIM. A constructivist approach is one where knowledge is constructed through experience and reflection of those experiences. Mamter, Aziz and Zulkepli [26] also found that graduates equipped with BIM knowledge and skills can increase uptake of BIM among companies in the industry. Not only that, but knowledge in BIM for graduates increases quality and employability of said graduate. Some HEIs including Universiti Tun Hussein Onn (UTHM) and Universiti Malaysia Pahang (UMP) have offered elective courses on BIM as part of a pioneer project to start the teaching of BIM in a higher education setting [33]. This is a good start for the widespread education of BIM among higher education students to produce graduates who are BIM-capable when entering the workforce.

4. Conclusions

As an agent of change, the PWD, CIDB and CREAM have been critical for the dissemination of BIM-related knowledge for the past 5 years and their efforts continue to pursue the aim of widespread BIM adoption in the country. They are more open to use new methods to promote BIM that resonate with the younger generation of today. The Malaysian government has shown effort to increase awareness of BIM among construction industry players and assist in capacity building for the use of BIM in individual firms. However, to achieve Level 2 BIM Maturity (model-based collaboration), efforts should be made to increase the collaboration between key players in the construction industry using BIM as a medium for cooperation.

Review of the literature published in WoS related to BIM in Malaysia has shown decline since 2016. Studies on BIM adoption in Malaysia have a very small number of respondents although these studies are meant to be qualitative in nature. Furthermore, these studies do not indicate which state they are conducted in. Construction activities vary greatly from state to state. Thus, different states may have different levels of BIM adoption. This is especially true with the difference in construction activities between East and West Malaysia. At the time of this writing, there is no systematic nationwide survey on BIM in Malaysia. Comprehensive nationwide studies such as by Ding, Zuo, Wu and Wang [39] for China and Hong, Hammad, Sepasgozar and Akbarnezhad [40] for Australia should also be carried out in Malaysia. Further research in Malaysia should be aimed at getting a better understanding of the current state of BIM in the country before appropriate plans can be made for the widespread implementation of BIM.

Acknowledgements

The authors would like to sincerely thank the University of Malaya Fellowship Scheme and University of Malaya Research Grant, Project No: RU019W-2017 and GPF010A-2019.

IOP Conf. Series: Materials Science and Engineering **943** (2020) 012057 doi:10.1088/1757-899X/943/1/012057

References

[1] Department of Statistics 2017 *Malaysia Economy 2017* (Kuala Lumpur: Department of Statistics Malaysia)

[2] Eastman C M, Fisher D, Lafue G, Lividini J, Stoker D and Yessios C 1974 *An outline of the building description system: Research report no. 50* (Pittsburgh, PA: Carnegie-Mellon University)

[3] Eastman C M and Henrion M 1977 Glide: A language for design information systems *ACM SIGGRAPH Computer Graphics* 11 24-33

[4] Eastman C M 1999 *Building product models: Computer environments supporting design and construction* (Boca Raton, FL, USA: CRC Press, Inc.)

[5] Eastman C M 1995 A generic building product model incorporating building type information *Automat Constr* 3 283-304

[6] Eastman C M, Teicholz P, Sacks R and Liston K 2011 *BIM handbook: A guide to building information modelling for owners, managers, designers, engineers and contractors* (Hoboken, New Jersey: John Wiley & Sons, Inc.)

[7] Van Nederveen G A and Tolman F P 1992 Modelling multiple views on buildings *Automat Constr* 1 215-24

[8] Gu N and London K 2010 Understanding and facilitating BIM adoption in the AEC industry *Automat Constr* 19 988-99

[9] Azhar S 2011 Building information modeling (BIM): Trends, benefits, risks, and challenges for the AEC industry *Leadership and Management in Engineering* 11 241-52

[10] Succar B 2009 Building information modelling framework: A research and delivery foundation for industry stakeholders *Automat Constr* 18 357-75

[11] Succar B, Sher W and Williams A 2012 Measuring BIM performance: Five metrics *Archit Eng Des Manag* 8 120-42

[12] Bew M and Richards M 2008 Bew-Richards BIM maturity model *BuildingSMART Construct IT Autumn Members Meeting*, Brighton.

[13] Zahrizan Z, Ali N M, Haron A T, Marshall-Ponting A and Hamid Z A 2013 Exploring the adoption of building information modelling (BIM) in the Malaysian construction industry: A qualitative approach *International Journal of Research in Engineering and Technology* 2 384-95

[14] Enegbuma W I, Aliagha U G and Ali K N 2014 Preliminary building information modelling adoption model in Malaysia *Construction Innovation* 14 408-432

[15] Iskandar P W 2018 *Penggunaan Level 2 BIM menjelang 2019* Retrieved on September 1, 2018 from https://www.bharian.com.my/berita/nasional/2018/07/446276/penggunaan-level-2-bim-menjelang-2019

[16] Yusof A and Shah A 2017 *MyBIM promotes modern construction technology* Retrieved September 1, 2018 from https://www.nst.com.my/business/2017/11/305294/mybim-promotes-modern-construction-technology

[17] Bernama 2019 *CIDB recommends mandatory use of BIM in certain private sector projects* Retrieved October 24, 2019 from https://www.nst.com.my/news/nation/2019/03/470468/cidb-recommends-mandatory-use-bim-certain-private-sector-projects

[18] Zainon N, Mohd-Rahim F A and Salleh H 2016 The rise of BIM in Malaysia and its impact towards quantity surveying practices *MATEC Web of Conferences* 66 1-8

[19] Latiffi A A, Mohd S, Kasim N and Fathi M S 2013 Building information modelling (BIM) application in Malaysian construction industry *International Journal of Construction Engineering and Management* 2 1-6

[20] Zhao X 2017 A scientometric review of global BIM research: Analysis and visualization *Automat Constr* 80 37-47

[21] Santos R, Costa A A and Grilo A 2017 Bibliometric analysis and review of Building Information Modelling literature published between 2005 and 2015 *Automat Constr* 80 118-36
[22] Rogers J, Chong H-Y and Preece C 2015 Adoption of building information modelling technology (BIM): Perspectives from Malaysian engineering consulting services firms *Engineering, Construction and Architectural Management* 22 424-45
[23] Takim R, Harris M and Nawawi A H 2013 Building information modeling (BIM): A new paradigm for quality of life within architectural, engineering and construction (AEC) industry *Procedia - Social and Behavioral Sciences* 101 23-32
[24] Shoubi M V, Shoubi M V, Bagchi A and Barough A S 2015 Reducing the operational energy demand in buildings using building information modeling tools and sustainability approaches *Ain Shams Engineering Journal* 6 41-55
[25] CIDB 2016 *Construction Industry Transformation Programme (CITP) 2016-2020* (Kuala Lumpur, Malaysia: Construction Industry Development Board)
[26] Mamter S, Aziz A R A and Zulkepli J 2017 Root causes occurrence of low BIM adoption in Malaysia: System dynamics modelling approach *AIP Conference Proceedings* 1903 1-6
[27] Latiffi A A, Brahim J and Fathi M S 2017 Building information modelling (BIM) after ten years: Malaysian construction players' perception of BIM *IOP Conference Series: Earth and Environmental Science* 81 012147
[28] Latiffi A A, Brahim J and Fathi M S 2016 Transformation of Malaysian construction industry with building information modelling (BIM) *MATEC Web of Conferences* 66 1-8
[29] Qin T X 2012 A study of building information modeling (BIM) In Malaysia construction industry. In: *Faculty of Engineering and Science,* (Kampar, Perak, Malaysia: Universiti Tunku Abdul Rahman) pp 1-78
[30] Zahrizan Z, Ali N M, Haron A T, Marshall-Ponting A and Hamid Z A 2014 Exploring the barriers and driving factors in implementing building information modelling (BIM) in the Malaysian construction industry: A preliminary study *The Journal of the Institution of Engineers, Malaysia* 75 1-10
[31] Vendrell D D 2015 *#DimecresTecnologics al @Coacatalunya 4 març / 16h: "Aproximació al disseny arquitectònic amb BIM" https://www.arquitectes.cat/ca/aproximacio-bim ... #LetsBIM #BIM* Retrieved September 1, 2018 from https://twitter.com/daviddelven/status/571302316508192768
[32] @BimCidb 2016 *#LetsBIM #CITPMalaysia http://fb.me/E2VwQEtHRetrieved* September 1, 2018 from https://twitter.com/BimCidb/status/773202295953711104
[33] CIDB 2018 *Universiti Malaysia Pahang wins BIM marathon competition* Retrieved September 1, 2018 from http://www.cidb.gov.my/index.php/en/media-1/latest-news/888-universiti-malaysia-pahang-wins-bim-marathon-competition
[34] Berita Nasional RTM 2017 *CIDB peruntuk RM2.5 juta bina pusat latihan myBIM* Retrieved September 1, 2018 from https://www.youtube.com/watch?v=c6c16tE2UqQ
[35] The Sun Daily 2017 *Public projects worth RM100m and more must use BIM system from 2019* Retrieved September 1, 2018 from https://www.thesundaily.my/archive/public-projects-worth-rm100m-and-more-must-use-bim-system-2019-YUARCH504849
[36] Yusuf B Y, Embi M R and Ali K N 2017 Academic readiness for building information modelling (BIM) integration to Higher Education Institutions (HEIs) in Malaysia. In: *2017 International Conference on Research and Innovation in Information Systems (ICRIIS),* (Langkawi, Malaysia) pp 1-6
[37] Yusuf B Y, Ali K N and Embi M R 2015 Building information modeling (BIM): A potential for effective building industry practice in Malaysia *Jurnal Teknologi* 77 55-62
[38] Eadie R, Solan B, Magee B J and Rice M 2016 The pedagogy of building information modelling. In: *Civil Engineering Research in Ireland (CERI) 2016,* (Galway, Ireland)
[39] Ding Z, Zuo J, Wu J and Wang J Y 2015 Key factors for the BIM adoption by architects: a China study *Engineering, Construction and Architectural Management* 22 732-48

IOP Conf. Series: Materials Science and Engineering **943** (2020) 012057 doi:10.1088/1757-899X/943/1/012057

[40] Hong Y, Hammad A W A, Sepasgozar S and Akbarnezhad A 2018 BIM adoption model for small and medium construction organisations in Australia *Engineering, Construction and Architectural Management* 26 154-83

Evaluation of practicality of virtual design and construction (VDC) with 5D building information modelling (BIM) through a case study

M L Lee[1*], W T Cheah[2], S H Lau[2], X S Lee[3], A M Abdullahi[1] and S Y Wong[1]

[1] University of Nottingham Malaysia, Jalan Broga, 43500 Semenyih, Selangor, Malaysia
[2] Universiti Tunku Abdul Rahman, Sungai Long, 43000 Kajang, Selangor, Malaysia
[3] Heriot-Watt University Malaysia, 62200 Putrajaya, Malaysia

E-mail: minlee.lee@nottingham.edu.my

Abstract. Virtual design and construction (VDC) is defined as the use of multi-disciplinary performance models for the purposes of design-construction projects. The use of VDC in construction projects enables the stakeholders to communicate and visualize the buildability and construction sequence which will impact on a project's success in terms of delivery time and cost. This case study investigated the process of integrating different dimensions of information such as time scheduling (4D) and cost estimation (5D) with a 3D model to simulate a 5D Building Information Modelling (BIM) process. The study explored the application of VDC with 5D BIM for a commercial building, namely Universiti Tunku Abdul Rahman (UTAR) Sungai Long City Campus. Through this case study, it enables comparisons of a virtual model and an actual construction in order to evaluate the advantages, practicality, and limitations of the VDC applications. The 5D model provided comprehensive information to manage the complex design-construction process by considering its timeline and cost. Six modelling stages were undertaken, namely inputting and modelling, clash detection test, time scheduling, quantification and cost estimation, virtualization, and walkthrough of the model. The results revealed that the practicability of the 5D BIM was influenced by the modelling effort, inter-operability, information output and limitations. The study also showed that VDC with 5D BIM significantly reduced the technical barrier for clients to participate in construction project, thus increasing the clients' satisfaction of the works by bridging the gap of perception and actual outcomes from the construction.

1. Introduction

Construction processes have evolved with the implementation of Virtual Design and Construction (VDC) in visualizing the 3D environment of the building. In recent years, a growing number of researchers focused on the study of the benefits, limitations and practicality of VDC in design-construction projects [1-5]. By adopting BIM technology (i.e., subset of VDC), an enhancement of the construction planning emerged from two dimensional (2D) drawing [6] to the integration of modern digital technologies with various dimensions of useful information (i.e. 3D, 4D, 5D) [7]. This systematic integrated digital information could then be communicated through VDC (i.e., visualisation) for an effective communication between the stakeholders during the project design phase.

In construction industry, the applications of BIM can be classified into preconstruction, construction and post-construction phases. Four dimensional (4D) modelling enables the users to visualize construction activities and to identify time and space constraints with the aids of relevant computing tools [4]. Based on technological sequence of scheduling the tasks of a project according to the specified technical requirements [8] and resources [9], the time management of an entire construction project can be derived [10]. Construction plans can be represented graphically by adding a layer of the time dimension to the 3D model to allow project planners to simulate and analyze any unpredicted scenarios before commencing work execution on site [11]. With the advantages of building systems analysis (e.g. clash detections test) [12] and time planning scheduling [5], stakeholders are able to track the progress of various parties working on the same project and make a better decision based on their expertise. With the supporting materials of photo-realistic representation that can be generated through VDC, it also improves collaborations among the stakeholders [13].

Recent advancement in building information modelling has seen the incorporation of the n-Dimensional (nD) information in building-construction, i.e. cost estimation. This includes the fifth Dimensional (5D) modelling that is made possible after generating the quantity takeoff for the construction components. The use of 5D modelling enables tracking of cost expenses [9] and quality level affected [14] in a construction project. This study aims to explore the applications of VDC with 5D BIM for a commercial building. Through the VDC with 5D BIM, their contributions to better project time and cost management during the construction phase are evaluated. The application of VDC could also benefit the facility management during the operation phase as the facilities and utilities which have been modelled can be tracked and monitored for maintenance purposes.

2. Methodology

This paper aims to study the visualisation of a design and construction process of a commercial building using Autodesk Revit 2016 and Autodesk Naviswork Manage. The sample project used is the building of Universiti Tunku Abdul Rahman (UTAR) Sungai Long City Campus, a building located on a commercial land title completed in 2015. The building stood 11 stories tall with 2 levels of basement. In order to enhance the effectiveness of the building information model, clash detection test, quantification of components, time scheduling for construction components, and case study of walkthrough the completed virtual building are demonstrated in this study.

2.1. Collecting, inputting and managing of building information from 2D plan

Completing the 3D building information model is the very first step to enhance the quality of a construction project by enabling a better coordination of site planning, activity sequencing, site organization, site management and clash detection test with other professional works. Thus, analysis of the building information from the 2D plan and identifying their significance levels are essential before they can be input into the 3D model. BIM promotes the representation of a building design in a 3D model incorporating geospatial data, graphical, and non-graphical data provided by any involved practitioner to allow manipulation of its database. After all the data have been analysed, the information was ready to be imported into Autodesk Revit by incorporating other disciplines e.g. civil and structural engineering and interior design. Figure 1 shows the components that were input and managed in a proper coordination with the aids of grid lines from the 2D plan.

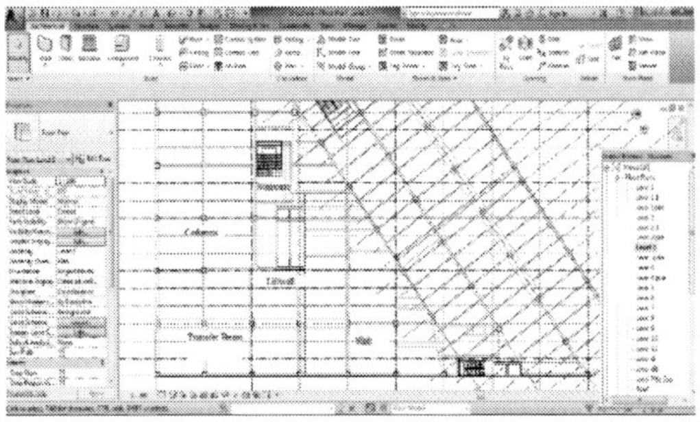

Figure 1. Structural and architectural components for the case study.

2.2. Clash detection between information models

Clash detection test enables users to identify and manage the potential impact before construction phase begins. Besides, it proves a better coordination between the construction participants.

2.3. Quantification of information components for cost estimation

Quantification of information components is also known as quantity takeoff, enabling the users to estimate the cost, volume of concrete casting, area and number of units for specific components. Quantity surveyor or cost engineers will have an advantage of performing the quantity takeoff for the materials with better accuracy and faster results. Figures 2 and 3 show the quantification of the components performed by Autodesk Revit and Autodesk Naviswork Manage, respectively.

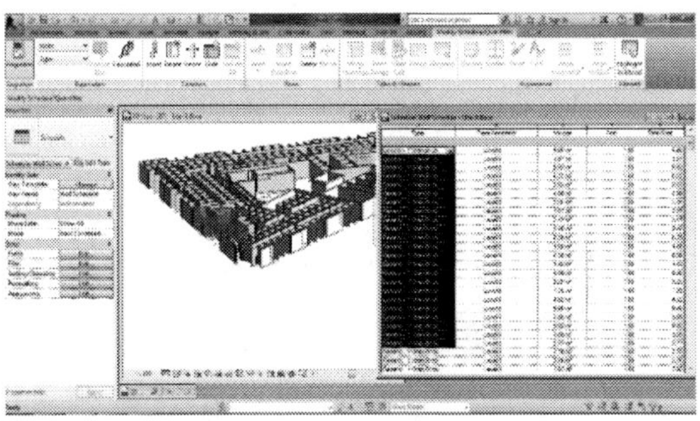

Figure 2. Quantity takeoff result from Autodesk Revit.

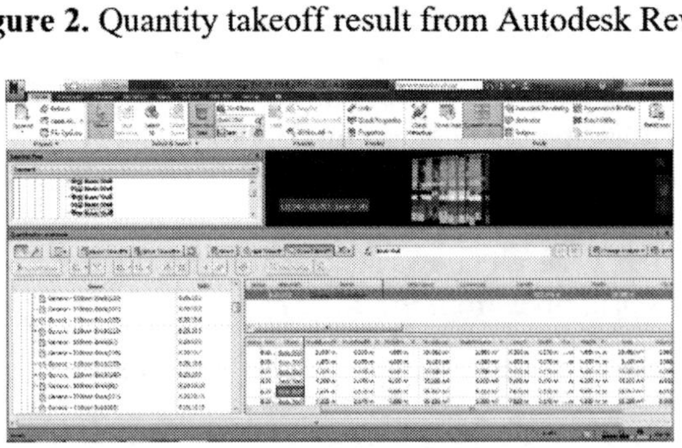

Figure 3. Quantity takeoff result from Autodesk Naviswork Manage.

2.4. Scheduling information progress

Time scheduling was performed to estimate the cost with animation features after quantification of information components. It showed a reference to match up with construction process during the planning phase of a construction project. Figure 4 shows the inputs of time scheduling with estimated cost into the time-liner function.

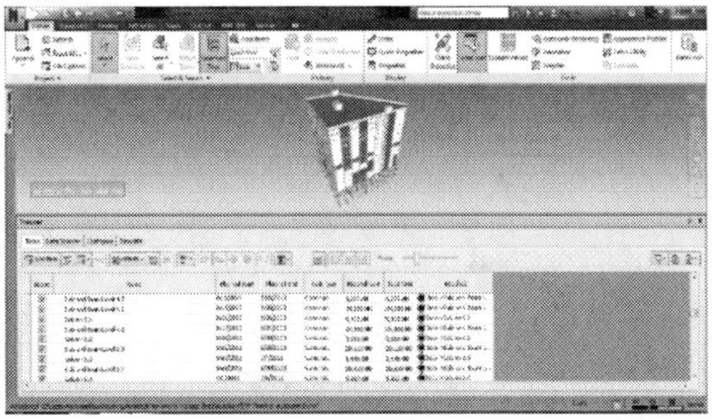

Figure 4. Inserting time and cost for information components in Naviswork Manage.

2.5. Virtualization of building information model

Animation of construction progress will be created concurrently with the input of time scheduling from the previous stage. By using the animation, it enabled the presentation of 5D Building Information Model for the needs of construction participants. Besides, for the actual construction progress, 5D BIM as-built information virtualisation can be carried out to monitor between the actual progress and the planned progress.

2.6. Third-Person walkthrough for virtual building

The final stage involved a third-person walkthrough in a virtual building. In order to further enhance and inspect the quality of a model at a closer view, the walkthrough viewpoint was generated to help clients to preview and link to his perspective since it is similar as a real person walk around the project site.

3. Results and Discussions

Figures 5a and 5b show the completed virtual building model and the actual constructed building, respectively. Apparently, the virtual model showed a close representation of the actual building. Clash detection, quantification, time scheduling, and building walkthrough were performed based on the virtual model.

(a) (b)

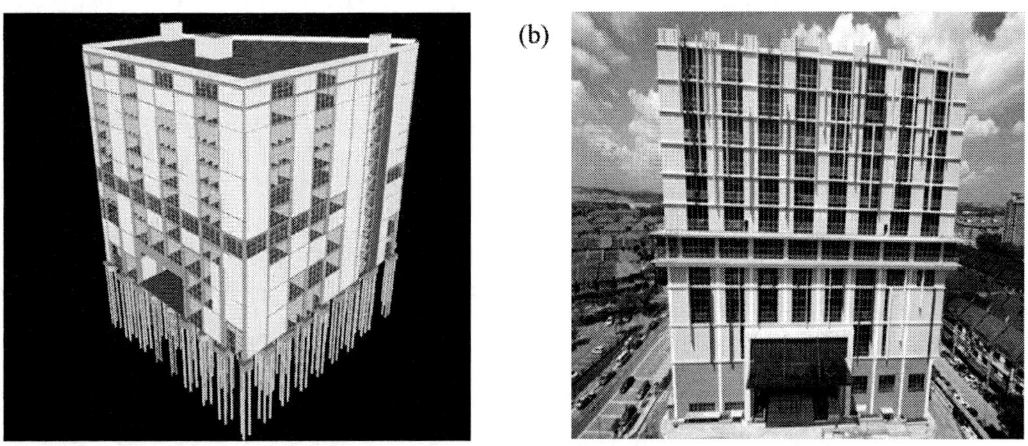

Figure 5. (a) 3-dimensional view of the virtual model, (b) Actual completed building.

3.1. Clash detection reports

Corrective measures were essential for resolving clashes of the models created. Clash report with figures and descriptions could be generated into HTML file or txt file. By performing this test, it reduced the possibility of clashes between different building components, and hence reduced the cost and time for a construction.

3.2. Quantification by Autodesk Revit and Autodesk Naviswork Manage

Quantification was performed by both Autodesk Revit and Autodesk Naviswork Manage. In Autodesk Revit, components that were taken off from the model can be clearly shown in terms of location and coordination. However, in Autodesk Naviswork Manage, it automated and separated the takeoff components into complex information while more information was presented in the model.

For residential buildings or relatively small-scaled buildings, quantity takeoff performed by Autodesk Naviswork Manage could be a preferred choice as it enables presentation of detailed results due to the lesser amount of building information compared to commercial or high rise buildings. As for the commercial building in the present case study, the quantification by Autodesk Revit is a recommended choice. Quantity takeoff in Autodesk Revit provides multiple options in taking off the building information, allowing multiple category scheduling such as analytical space, casework, ceiling and etc. Quantification of Autodesk Naviswork Manage was automatically categorized into different categories once the Revit file was appended into Naviswork Manage. For example, the components in Autodesk Revit such as brickwalls and interior partition wall were defined as "Basic Wall" type in Autodesk Naviswork manage.

Autodesk Revit enables user to choose the available parameters, or add users' own parameters or even customize a parameter's formula to present the data and values of current models. However, Autodesk Naviswork Manage does not provide flexibility to choose, decide or edit the parameters for the quantity takeoff because it has fixed the parameters such as length, width, thickness, height, perimeter, area, volume, weight, count and primary quantity. However, users can edit the formula for the variables mentioned above and also edit the SI units for the variables. Besides, cost estimation only can be performed when the quantification workbook was exported to a spreadsheet file.

3.3. Time scheduling integrated with cost estimation

By virtualising the construction progress of a building information model, construction participants are able to save cost by arranging the equipment and material for their own professional works such as scaffoldings, shipping of material, installation of service and etc. They are able to "preview" the progress of the construction work before it begins. In Naviswork Manage, it enables the input of costs for labour, machinery, and subcontractors. However, the present study only attempted on the analysis of material costs.

3.4. Advantages of walkthrough a virtual building

Some clients may have difficulties to imagine the interior views of the proposed buildings due to limited knowledge and experiences in building design and construction. By providing a walkthrough viewpoint for clients, it further ensures that the proposed design is meeting the clients' requirement and intention. In addition, the quality of the building construction can be improved owing to the better visualization of proposed building even before the construction progress begins. Figures 6 and 7 show samples of walkthrough views between the completed building and the virtual building.

(a) (b)

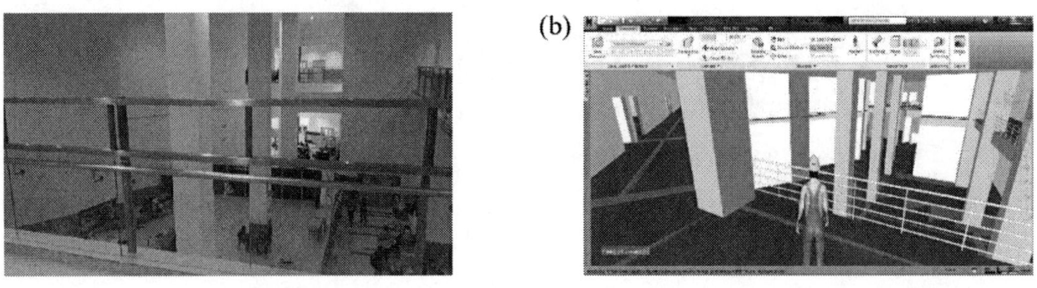

Figure 6. Viewpoints 1 of (a) Actual completed building, (b) Virtual building.

(a) (b)

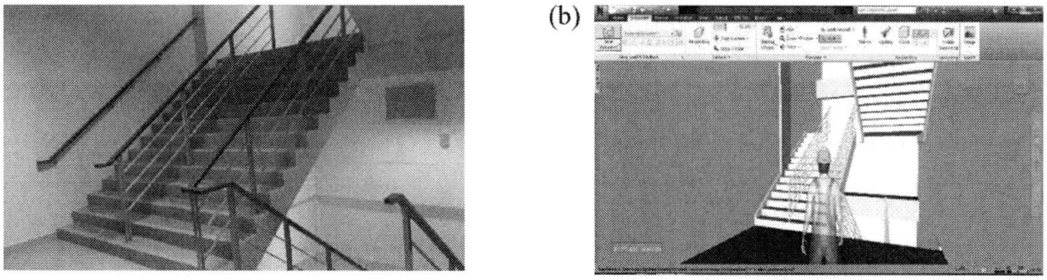

Figure 7. Viewpoints 2 of (a) Actual completed building, (b) Virtual building.

4. Practicality of 5D BIM

Based on the present case study, the practicality of the 5D BIM modelling were evaluated by the authors based on the experiences of conducting the BIM modelling. Qualitative ratings ranging from "poor" to "neutral" and "good" were given to each of the processes involved. From the first stage, it was found that the effort of modelling was highly dependent on a user's skill level. This is similar to CAD, and hence it was considered "neutral". Inter-operability was considered "poor" because different modelling tools are based on unique system and engine that are competing against one another. The information output was very comprehensive and thus it was considered as "good" when compared to 2D plan. For stakeholders who lacked of knowledge in BIM, it may deter the stakeholders to proceed to subsequent stages.

In terms of inputting the information for cost estimation, the effort was rated "good" since the process was considerably simpler compared to manual taking off. The inter-operability was considered as "neutral" as compared with other more advanced software that are normally used by quantity surveyors. However the output information was very fast and accurate. The limitation for this stage showed that suppliers cannot update the information for costing automatically.

With regards to simulated time scheduling in 3D model, the modelling effort was considered as "neutral". Unlike the first stage, collaboration that involved completed building information model was "easy" due to high inter-operability. The information output was "good" as non-professional stakeholders were empowered to easily understand 5D BIM. However, when it comes to the planning for the construction, it needed site knowledge and experience to analyse the quality of the results.

In the process of virtualisation, the effort highly depends on existing 5D BIM quality. It was determined as "neutral" in this study. However, the advancement of technology certainly will reduce the modelling effort. Both inter-operability and information output were "good" due to the direct link to the perspective of clients or end users.

5. Conclusion

The present case study shows the potential of VDC with BIM leading to a more efficient process in construction industry. Even though it is possible to integrate up more than five dimensions of information, excessive information may trigger the complexity and impede the VDC implementation process. On one hand, the integration of VDC and 5D BIM promotes the efficiency and accuracy of construction planning process. On the other hand, it also enhances construction industry stakeholders' decision making abilities which are conventionally constrained by the limitation of 2D CAD workflow. Moreover, VDC with 5D BIM significantly reduces the technical barrier for clients to participate in construction project, thus increasing the clients' satisfaction of the works by bridging the gap of perception and actual project outcomes.

IOP Conf. Series: Materials Science and Engineering **943** (2020) 012058 doi:10.1088/1757-899X/943/1/012058

References

[1] Fischer M, Hartmann T, Rank E, Neuberg F, Schreyer M, Liston K and Kunz J 2004 *Proc. of the World IT Construction Conf.* pp 167-182
[2] Kunz J and Fischer M 2012 *Center for Integrated Facility Engineering (CIFE) Working Paper* (Stanford University) pp 1-50
[3] Gilligan B and Kunz J 2007 *Center for Integrated Facility Engineering (CIFE) Working Paper* (Stanford University) pp. 1-31
[4] Haymaker J and Fischer M 2001 *Center for Integrated Facility Engineering (CIFE) Working Paper* (Stanford University) pp 1-17
[5] Pucko Z, Suman N and Klansek U 2014 *Organization, technology & management in construction: an international journal* 6(1) 958-971
[6] Mihindu S and Arayici Y 2008 *Proc. of the 2008 International Conference Visualisation, IEEE Computer Society* pp 29-34
[7] Arayici Y, Egbu C O and Coates P 2012 *Journal of Information Technology in Construction* 17 75-92
[8] McKinney K and Fischer M 1998 *Automation in construction* 7(6) 433-447
[9] Donath D, Loemker T M and Richter K 2004 *Automation in Construction* 13(2) 159-166
[10] Popov V, Juocevicius V, Migilinskas D, Ustinovichius L and Mikalauskas S 2010 *Automation in Construction* 19(3) 357-367
[11] Mallasi Z and Dawood N 2002 *Proc. of CIB w78 2002 Conference on Distributing Knowledge in Buildings* vol 2 pp 83-90
[12] Wang L and Leite F 2015 *Journal of Computing in Civil Engineering* 04015017
[13] Fischer M 2003 *Civil Engineering* 73(8) 60-67
[14] Kim G H, An S H and Kang K I 2004 *Journal of Building and Environment* 39 1235-1242
[15] Lee X S, Tsong C W and Khamidi M F 2016 *MATEC Web of Conf.* 66 00026

IOP Conf. Series: Materials Science and Engineering **943** (2020) 012059 doi:10.1088/1757-899X/943/1/012059

A BIM-Based Automated Assessment Tool for Green Building Index

G Khoshdelnezamiha[1*], S C Liew[1], V N S Bong[1] and D E L Ong[1,2]

[1] Swinburne Sarawak Research Centre for Sustainable Technologies, Faculty of Engineering, Computing and Science, Swinburne University of Technology Sarawak Campus, Jalan Simpang Tiga, 93350, Kuching, Sarawak, Malaysia
[2] School of Engineering and Built Environment, Griffith University, 170 Kessels Road, Nathan, Queensland 4111, Australia

E-mail: gkhoshdelnezamiha@swinburne.edu.my

Abstract. In Malaysia, Green Building Index (GBI) is one of the sustainability rating tools for the construction industry, focusing on six main categories including Energy Efficiency (EE), Materials and Resources (MR), Water Efficiency (WE), Indoor Environment Quality (EQ), Sustainable Site Planning & Management (SM) and Innovation (IN). Currently, the process for GBI assessment and certification is carried out manually, which requires additional time and effort for preparation of supporting documents. Since Building Information Modelling (BIM) is perceived as a holistic solution for digital integration of various construction project attributes, it is anticipated that sustainability aspects can be addressed through BIM if an appropriate approach is implemented. Therefore, GBI assessment process can potentially benefit from BIM application to transition from conventional manual approach to an automated approach. This research is aimed at developing an Automated Green Building Assessment tool (AGBIA) using visual programming (Dynamo) to run as a supplementary application to the existing BIM authoring software (Revit). The AGBIA system consists of 6 modules that automatically analyses digital BIM models in accordance to GBI requirements, and ultimately produces the necessary documentation as a final output. AGBIA is an efficient supporting BIM tool as compared to assessment and documentation of green building data manually.

1. Introduction

In Malaysia, GBI has been introduced to be the driving initiative for the construction industry to incorporate environment-friendly features that will have a positive impact in the long term. Energy Efficiency, Materials and Resources, Water Efficiency, Indoor Environment Quality, Sustainable Site Planning & Management and Innovation are the categories that must be evaluated for GBI certification. Upon assessment of the design details and supportive documents for each category, allocated points will be granted at the design stage, followed by another round of verification upon project completion. Currently, the GBI process for green building assessment and certification is carried out manually, which requires additional time and effort for preparation of supporting documents especially with large and complex projects. Combining several building features from different disciplines demands for tools and technology that can ensure well connectivity [1]. Many researchers have highlighted the need for BIM application to overcome some of the barriers in delivering green buildings since BIM is known to be an efficient tool that integrates various building aspects digitally. Ayman, Alwan and McIntyre [2] anticipated the direct impact of BIM utilisation on the sustainability barriers would be a reduction in time and cost due to better design coordination and effective decision making during the design phase. Based on the review conducted by Lu, Wu, Chang and Li [3], the application of popular BIM tools for green building assessment however, are limited to specific regions only. Although the GBI assessment process can be accelerated with the utilisation of existing tools and software, the available tools are based on sustainability assessment guidelines that are not entirely in line with GBI requirements in terms of local references and baselines, as well as method of calculation, analysis and documentation. Therefore, there is a need to identify appropriate methods for integration and automation of GBI assessment with available BIM authoring tool to address local standards for green building design. In order to propose an automated BIM-based approach for green buildings in Malaysia, the following objectives must be achieved:

- To design a BIM template addressing green property sets in accordance to GBI
- Automatic assessment of green data extracted from a standard BIM model
- Automatic production of documentation for green building assessment

To accomplish the above objectives, this research is aimed at developing an Automated Green Building Index Assessment (AGBIA) tool that automatically analyses BIM models in accordance to the GBI requirements, and ultimately produces the documentation/report as a final output. AGBIA is an efficient alternative as compared to assessment and documentation of green building data manually.

2. Literature Review

The application of BIM for green building assessment has been investigated by number of researchers. Revit as a widely used BIM authoring tool has been explored and evaluated for integration with various green building rating systems. Table 1 contains a summary of researchers who introduced technical strategies for integration of BIM with various green building rating systems. For instance, Azhar et al. [4] demonstrated that 17 LEED points and 2 prerequisites can be directly assessed via Revit and IES-VE default functionalities. Furthermore, Barnes and Castro-Lacouture [5] optimised a similar process in Revit by proposing an additional LEED toolbar for Revit users. Meanwhile, Wong and Kuan [6] utilised user defined Revit features such as custom parameters and schedules for BEAM-Plus assessment. Alternatively, a number of researchers opted for external tools to assess the information extracted from the BIM models [7-9]. From a regional perspective, the developed tools are limited to local green building guidelines, standards and baselines. Therefore, in terms of regional relevancy, the methods explored by Lim et al. and Khoshdelnezamiha et al. [8-9] are more applicable. In order to optimise the existing methods, an alternative approach is proposed to gain direct access to the building data and process the information rapidly without the use of any external tools.

Table 1. Summary of previous studies related to BIM application for green building assessment.

Author	Rating System	Region	Points	Tool Used
[4]	LEED	US	17 points and 2 prerequisites	Revit and IES-VE
[5]	LEED	US	13 points and 1 prerequisite	Revit
[6]	BEAM-Plus	Hong Kong	26 points	Revit
[7]	BREEAM	UK	14 points	ArchiCAD and GBAT
[8]	GBI	Malaysia	28 points	Revit, Excel and BEIT
[9]	GBI	Malaysia	56 points	Revit and GBI Document Assessor

3. Methodology

The AGBIA system design consists of 6 modules that automatically analyses BIM models in accordance to GBI requirements. Autodesk Revit was the selected BIM authoring tool for this methodology, as it is commonly used in Malaysia. In order to achieve an automated assessment process for the users, AGBIA was designed using Dynamo, a visual programming tool. Figure 1 illustrates the process involved with developing AGBIA. The AGBIA development process began with identifying GBI assessment subcategories. Subsequently, a set of green parameters was designed to be further assigned with values either by default (using AGBIA built-in database) or by the user. Afterwards, the green data was collected and sorted to be processed through defined calculation. The processed green information was further evaluated based on conditional statements retrieved from GBI assessment guideline. Finally, the result containing provisional GBI points and summary of green building features and design calculation was exported as supporting documents for the certification.

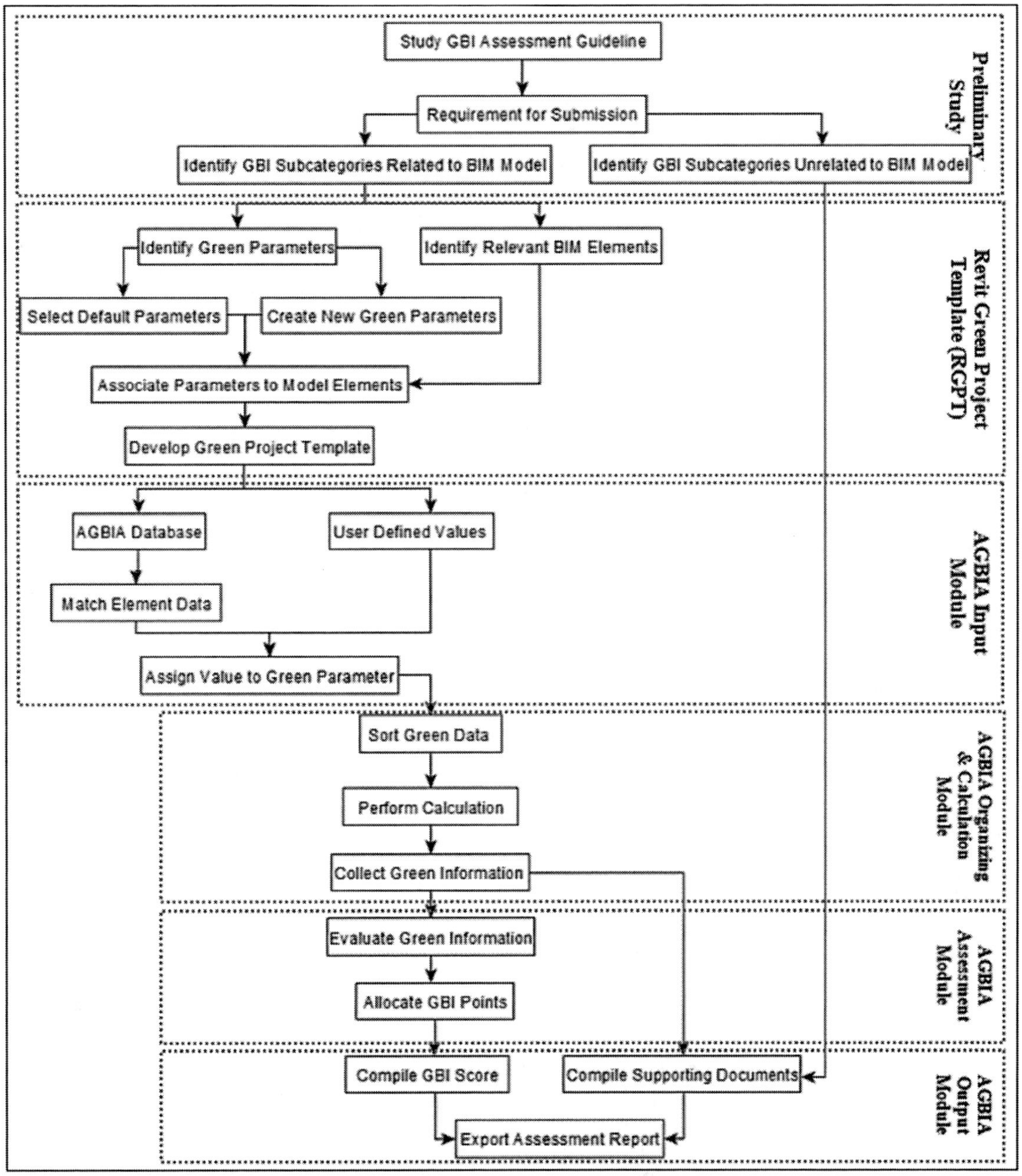

Figure 1. AGBIA system design flowchart.

4. Verification

The practicality of the developed AGBIA was evaluated by applying it on a sample BIM model. In order to verify the results obtained, similar building was assessed via the conventional method. The conventional approach involved manual extraction and calculation to obtain benchmark values for comparison. Some GBI assessment categories were verified using the available GBI excel calculators such as Building Energy Information Tool (BEIT). Although such conventional calculators can produce rapid outputs, the user is required to manually provide several design inputs as well as general project information. This by itself can be tedious and hinders the automation process of linking design data with the calculator engines. Therefore, in this paper, it is considered as a conventional approach in relation to the fully automated process proposed.

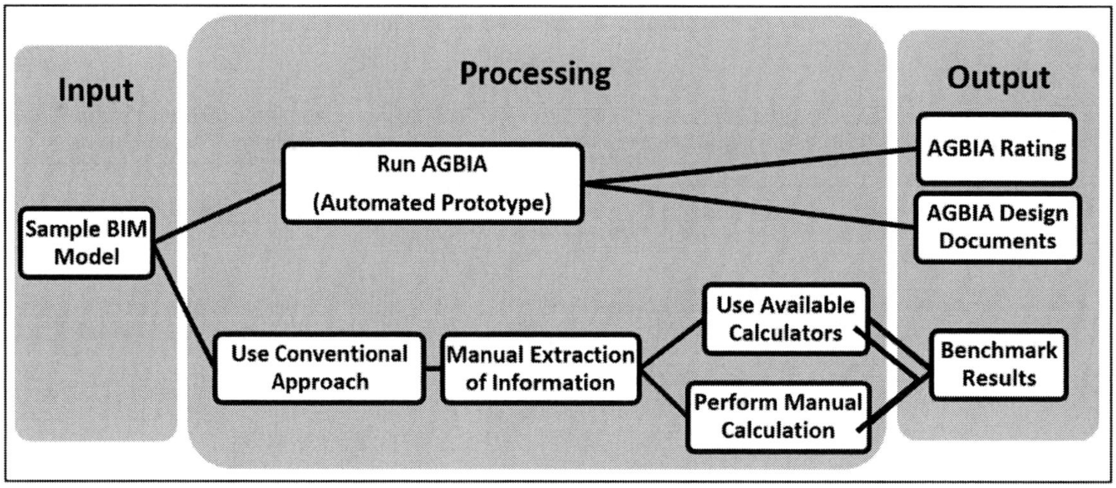

Figure 2. Proposed verification process.

5. Results and Discussion

A sample building model was analysed by AGBIA. The automated process eliminated the need for manual design data extraction and calculation. For demonstration purposes, the Overall Thermal Transfer Value (OTTV) was selected as one of the GBI EE subcategory. Figure 3 demonstrates the automated identification of relevant BIM elements such as external façade, external windows and shading devices required for the assessment of OTTV. The AGBIA report provided a summary of building data such as façade area, orientation, heat transfer coefficients and shading coefficients extracted from the sample BIM model (Figure 4). In order to verify the calculation carried out by the designed tool, the OTTV of the similar model was manually calculated using the BEIT tool. The BEIT calculation for OTTV was 53.83 W/m^2 whereby the AGBIA estimation was 51.91 W/m^2. The differential error of 3.57% is due to limitations in parametric inputs from the conventional method. The BEIT calculation input assumes constant values for each façade orientation which makes it more suitable for simplified designs. The AGBIA on the other hand, is intended to calculate parametric items individually and take the average values of the respective orientations, thus making it more practical for complex designs. For instance, the use of external shading devices for the east and west facades in the sample model were inconsistent which resulted in different shading coefficient values in each method.

6. Conclusion

The feasibility and potential for application of green BIM for various sustainability rating tools was demonstrated in the literature study. Similar to the well-known rating tools such as LEED and BREEAM, GBI as a Malaysian rating tool can automated and be integrated with BIM. Upon investigating the GBI assessment guideline and submission requirements, a set of standard property sets were identified as a green project template. Visual programming was incorporated to develop a supporting tool for Revit called AGBIA to automate the process of creating Revit green project template (RGPT) as well as to extract, interpret and process green data from a BIM model [9]. the final output from AGBIA is a report containing provisional GBI rating and a set of design calculation that can be used as supporting documents for GBI assessment. Incorporating AGBIA at the design stage can optimise the utilisation of BIM authoring tool to assist designers in automatic assessment and documentation of green buildings.

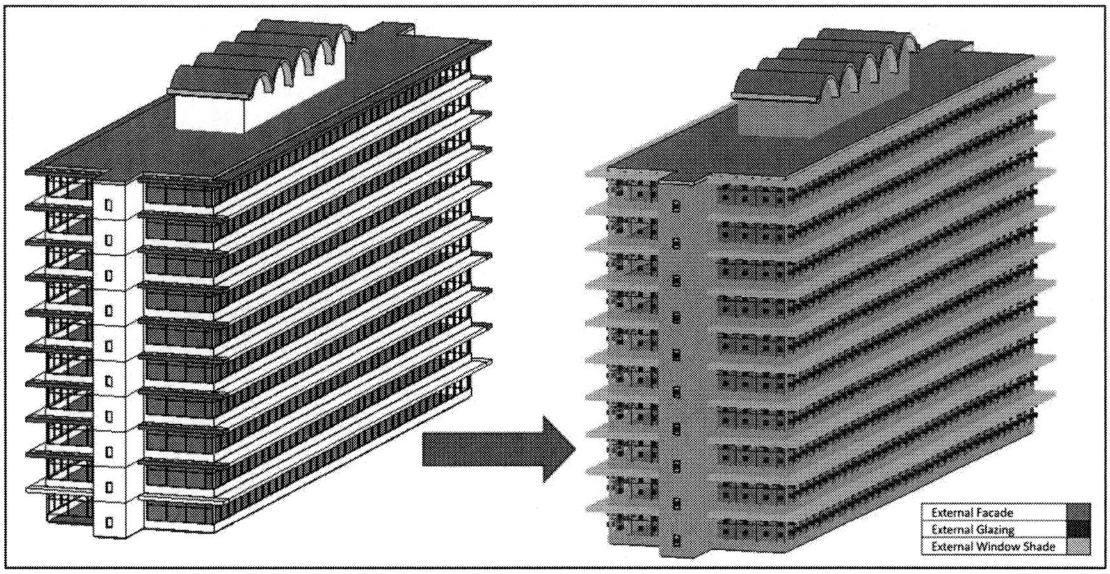

Figure 3. Sample BIM model before and after AGBIA OTTV analysis.

Heat Conduction Through Walls					
Orientation	Total Facade Area	WWR	α	Thermal Transafer Value (OTTV)	A x OTTV
N	3126.72	0.47	0.30	3.30	10322.34
E	527.17	0.47	0.30	3.35	1768.09
S	2949.27	0.47	0.30	3.36	9897.81
W	710.85	0.33	0.30	4.23	3009.38
NE	0.00	0.00	0.00	0.00	0.00
SE	0.00	0.00	0.00	0.00	0.00
NW	0.00	0.00	0.00	0.00	0.00
SW	0.00	0.00	0.00	0.00	0.00

Heat Conduction Through Windows					
Orientation	Total Facade Area	WWR	U value Window	Thermal Transafer Value (OTTV)	A x OTTV
N	3126.72	0.47	6.70	19.10	59714.76
E	527.17	0.47	6.70	18.76	9890.81
S	2949.27	0.47	6.70	18.75	55294.09
W	710.85	0.33	6.70	13.14	9338.31
NE	0.00	0.00	0.00	0.00	0.00
SE	0.00	0.00	0.00	0.00	0.00
NW	0.00	0.00	0.00	0.00	0.00
SW	0.00	0.00	0.00	0.00	0.00

Solar Heat Gain Through Windows							
Orientation	Total Facade Area	WWR	Orientation Factor (CF)	Shading Coeff of Glazing (SC1)	Shading Coeff of External Shading Device (SC2)	Thermal Transafer Value (OTTV)	A x OTTV
N	3126.72	0.47	0.90	0.59	0.67	32.78	102496.74
E	527.17	0.47	1.23	0.59	0.33	21.68	11427.78
S	2949.27	0.47	0.82	0.58	0.67	32.90	97018.01
W	710.85	0.33	0.94	0.59	0.38	13.36	9494.91
NE	0.00	0.00	1.09	0.00	0.00	0.00	0.00
SE	0.00	0.00	1.13	0.00	0.00	0.00	0.00
NW	0.00	0.00	0.90	0.00	0.00	0.00	0.00
SW	0.00	0.00	0.90	0.00	0.00	0.00	0.00

Total Solar Heat Gain
379673.02
Overall Building OTTV
51.91

AGBIA Generated Report

Figure 4. Sample report produced by AGBIA.

Acknowledgement
This research is funded by Melbourne Sarawak Research Collaboration Scheme (MSRCS) – Building Information Modelling Project supported with Strategic Research Grant (StarRG).

IOP Conf. Series: Materials Science and Engineering **943** (2020) 012059 doi:10.1088/1757-899X/943/1/012059

References

[1] Samari M, Godrati N, Esmaeilifar R, Olfat P and Shafiei M W M 2013 The Investigation of the Barriers in Developing Green Building in Malaysia *Modern Applied Science* 7 1-10

[2] Ayman R, Alwan Z and McIntyre L 2018 Factors Motivating the Adoption of BIM-based Sustainability Analysis. In: *International SEEDS Conference 2018,* (Dublin, Eire: LSIPublishing)

[3] Lu Y, Wu Z, Chang R and Li Y 2017 Building Information Modeling (BIM) for green buildings: A critical review and future directions *Automation in Construction* 83 134-48

[4] Azhar S, Carlton W A, Olsen D and Ahmad I 2011 Building information modeling for sustainable design and LEED® rating analysis *Automation in Construction* 20 217-24

[5] Barnes S and Castro-Lacouture D BIM-Enabled Integrated Optimization Tool for LEED Decisions *Computing in Civil Engineering (2009)* 258-68

[6] Wong J K-W and Kuan K-L 2014 Implementing 'BEAM Plus' for BIM-based sustainability analysis *Automation in Construction* 44 163-75

[7] Ilhan B and Yaman H 2016 Green building assessment tool (GBAT) for integrated BIM-based design decisions *Automation in Construction* 70 26-37

[8] Lim Y-W, Sediadi E, Shahsavari F and Azli N F b M N 2016 Building Information Modelling for Building Energy Efficiency Evaluation Integration with Green Building Index (GBI) in Malaysia *Annual International Conference on Architecture and Civil Engineering (ACE 2016)*

[9] Khoshdelnezamiha G, Liew S C, Bong V N S and Ong D E L 2019 BIM-based Approach for Green Buildings in Malaysia *IOP Conference Series: Earth and Environmental Science* 268 012052

A review of the green building rating systems

K H D Tang[1], C Y H Foo[1] and I S Tan[1]

[1]Faculty of Engineering and Science, Curtin University Malaysia, CDT 250, 98009 Miri, Sarawak, Malaysia

E-mail: daniel.tang@curtin.edu.my

Abstract. Green buildings have received much attention since the early 90s due to increasing concerns over the sustainability of buildings. The rise of green buildings spurred the need to systematically evaluate hence certify those buildings using a set of established criteria. This led to the emergence of numerous green building rating systems. This review compares 10 popular green building rating systems to identify their similarities, differences and the potential future studies in this area. It identifies 17 evaluation criteria as the basis of comparison and ranks the importance of the most common criteria based on their average weights across the rating systems of interest. This review shows that energy efficiency, indoor environment, health and wellbeing, sustainable siting, material efficiency, water efficiency and innovation as the most prevalent evaluation criteria in descending average weights. This study identifies the future need to probe into the implementation of the rating systems and the building performance ensuing such implementation. It is also crucial to understand the barriers in popularizing green building awards or certification. This review contributes to both the knowledge and practices of green building rating.

Keywords: Green buildings, rating, sustainability, energy, performance

1. Introduction

Conventionally, the construction and occupation of buildings are linked to numerous environmental impacts. The building sectors comprising residential and commercial structures consumed approximately 20% of the world's energy in 2018 and their energy consumption has been projected to increase by 1.3% per year to 22% in 2050 [1]. Increase in buildings' energy consumption is largely contributed by developing countries due to increasing energy demand driven by intensifying urbanization as well as better access to electricity and income [2]. Though industrial sectors were still the largest energy consumer, drawing about 40% of the global energy in 2018, the increase is projected to be in the range of 0.5% and 1.1%, leading to a decline in its overall share of energy consumption to 35% by 2050 [1].

The building and industrial sectors are in fact intricately related with building materials such as iron, steel, aluminium and paints coming from the industrial sectors [1, 2]. Taking the production and transportation of building materials into consideration, energy consumption of buildings would surpass the estimated energy consumption of 20% [1]. Electricity channelled to residential buildings are mainly used for space heating and cooling, water heating as well as refrigeration while electricity distributed to commercial buildings are used mainly to power computers and office equipment, refrigeration, space cooling system and ventilation [1, 3]. The increasing use of light emitting diode (LED) bulbs and compact fluorescent lamps has and will continue to reduce electricity intensity for lighting [1, 3].

While being energy-demanding, buildings also emit greenhouse gases [4]. It has been estimated that buildings and construction, including upstream power generation, contribute to 39% of the total CO_2 emitted [5]. In addition, buildings consume water and produce wastewater concurrently. Every year, about 13.6% of total potable water, equivalent to 15 trillion gallons, is used by buildings [5]. Construction of buildings generates demand for construction materials whose production emits greenhouse gases. In fact, greenhouse gases are emitted throughout the entire building lifecycle, from extraction and processing of raw materials, transportation of construction materials to demolition of buildings at the end of their service life [6]. Besides, demolition of building produces copious waste. The EPA estimated that the US produced 548 million tons of construction and demolition (C&D) debris in 2015 and this was double the amount of municipal solid waste produced in the same year. Demolition waste comprised 90% of the C&D waste generated [5].

The environmental impacts associated with buildings have resulted in interest to make buildings more energy efficient and environmental friendly. This sparked off the green building movement calling for integration of sustainability into the design and construction of new buildings as well as retrofitting of existing buildings [7]. Sustainable building embeds building life-cycle probing the adoption and synergy of green practices. The emergence of green buildings gives rise to practices, technology and techniques to reduce environmental and health impacts of building while tapping into renewable energy to provide electricity [8]. Not only that, green buildings also optimize the use of plants and trees, reduce rainwater run-off, utilize environmental friendly materials and promote recycling of waste materials [9]. The US green building movement culminated in the establishment of the United States Green Building Council (USGBC) in 1993 to encourage sustainable design, construction and operation of buildings [7]. The USGBC saw the need to set standards for green buildings which led to the subsequent inception of the Leadership in Energy and Environmental Design (LEED). The LEED has thence evolved into a comprehensive set of standards delving into the design, operation, maintenance, innovation as well as social functions of green buildings [10].

Though widely used today, LEED is not the first green building rating system to be developed. The Building Research Establishment's Environmental Assessment Method (BREEAM) was launched in 1990, four years before the release of LEED [11]. Since then, many other green building rating systems were introduced. For instance in 1996, the Building Environmental Assessment Method (BEAM) of Hong Kong was developed by the BEAM Steering Committee which later collaborated with other major players to form the Hong Kong Green Building Council (HKGBC). The HKGBC, in 2010, released an upgraded version of BEAM, called the BEAM Plus [12]. Subsequent to LEED, other regional green building rating systems such as the Comprehensive Assessment System for Built Environment Efficiency (CASBEE) of Japan, the Green Mark (GM) of Singapore and the Green Building Index (GBI) of Malaysia were also launched [13].

The rating systems generally cover the common attributes such as energy efficiency, water efficiency, material efficiency, indoor environmental quality enhancement, waste reduction as well as operations and maintenance optimization [9, 13]. However, the emphasis may vary in different rating systems. There have been multiple studies conducted to examine individual rating system in detail [14] or compare multiple rating systems [7, 9, 13]. Numerous studies have compared LEED and BREEAM [7, 8, 15]. Other studies extended beyond the two to include comparison with regional rating systems. Aye and Hes, for example, compared LEED, BREEAM and Green Star (GS) of Australia [16] while Geng et al. included BEAM, CASBEE, and Assessment Standard for Green Building (ASGB) of China in the comparison [17]. There are studies which compare multiple rating systems including that of Awadh which compares 6 rating systems [18] and that of Illankoon et al. comparing 8 rating systems [19]. Most studies include LEED in the comparison probably because LEED is most widely used. Though numerous studies on green rating systems have been conducted, there is an apparent lack of comparison of the popular rating systems across a comprehensive list of evaluation criteria and a lack of clarity in the weights used for comparison since different categories of the same rating system may have different weights allocated for the evaluation criteria.

2. Method

This review compares 10 green building rating systems. The rating systems were identified by running a search through the 'Google Scholar' engine on literature related to green building rating systems over the past 10 years. The keywords used in the search consisted primarily of green building rating systems, green building rating, green building certification and LEED, and secondarily of green building and green certification which yielded a wider array of search results. The rating systems that were most frequently subject to studies, hence most frequently appearing in the search were extracted. They are, not in sequence of study frequency, listed below:

- LEED
- BREEAM
- CASBEE
- GS
- ASGB
- BEAM
- Green Globes (GG) of Canada
- GM
- GBI
- Indian Green Building Council (IGBC) of India

This review provides general details of the rating systems encompassing the initiators, the countries of origin and the rating categories. This review examined the evaluation criteria of the rating systems and generated a list of common criteria as the basis of comparing the rating systems [13]. Zooming into six most common criteria, this review identified the weights allocated by each rating system on the criteria and presented the average weight for each criterion.

3. Results and Discussion

General information of the green building rating systems compared in this study are presented in table 1. Different green building rating systems provide different categories of certification, hence different rating tools. The most common categories are the certification of new buildings, existing buildings, interiors and communities, though some rating systems, particularly ASGB, GM, GBI and IGBC make a detailed distinction between the types of building (Table 1). CASBEE and IGBC differentiate communities into categories such as urban development, cities, residential societies and green townships. This demonstrates a trend of increasingly specific rating systems to purpose-fit different types of development rather than a one-size-fit-all rating system [9].

It is also noteworthy that the green building rating systems were initiated by private, member-based organizations except GM of Singapore and ASGB of China (Table 1). This could mean that governance and regulation may be a concern of the rating systems. However, in most instances, there is a third-party assessment by independent bodies on the outcomes of green building assessments to ensure objectivity and transparency [18]. More studies are needed to delve into the implementation of the rating systems particularly the governance, reliability and the effectiveness of audits encompassing objectivity and transparency to conclude unbiased implementation of the rating systems. The pursuit of green building certification also faces the challenges of resistance to change and costs, and it is worthwhile to carefully examine these challenges to promote green building certification [9]. While table 1 shows the most commonly studied green building rating systems in the past decade, many other regional rating systems have been developed for example, the Estidama Pearl Rating System (EPRS) of Abu Dhabi and the Global Sustainability Assessment System (GSAS) of Qatar.

Table 1. General description of the green building rating systems reviewed.

Green Building Rating System	Initiator	Rating Tools/ Categories	Country
LEED [10]	US Green Building Council	LEED v 4.1 BD+C (design and construction of full buildings) for new construction and major renovation, core and shell development, schools, retail, data centres, warehouses and distribution centres, hospitality as well as healthcare LEED v4.1 ID+C (interior design and construction) for commercial interiors, retail and hospitality LEED v4.1 O+M (operational and occupied buildings) for existing buildings and existing interiors LEED v4.1 Residential for single family and multifamily homes as well as multifamily homes core and shell. LEED v4.1 Cities and Communities for cities and communities	US
BREEAM [11]	Building Research Establishment (BRE)	BREEAM – Communities; BREEAM (International or UK) – New Construction; Home Quality Mark (HQM); Civil Engineering and Public Realm (CEEQUAL) Version 6; BREEAM In-Use; BREEAM Refurbishment and Fit-Out – Homes and commercial buildings	UK
CASBEE [20]	Japan Sustainable Building Consortium	CASBEE for Buildings (New Construction) 2014; CASBEE for Buildings (Existing Buildings) 2014; CASBEE for Buildings (Renovation) 2014; CASBEE for Market Promotion 2014; CASBEE for Commercial Interiors 2014; CASBEE for Temporary Construction 2007; CASBEE for Urban Development 2014; CASBEE for Cities 2013; CASBEE for Cities – Worldwide use 2015; CASBEE for Detached Houses (New Construction) 2014; CASBEE for Dwelling Unit (New Construction) 2014	Japan
GS [21]	Green Building Council Australia	Green Star – Communities, for precinct planning and development; Green Star – Design & As Built, for building design and construction ; Green Star – Interiors, for fitout design and construction; Green Star – Performance, for building operations and maintenance	Australia
ASGB [17]	Ministry of Housing and Urban-Rural Development of People's Republic of China	Evaluation Standard for Green Industrial Building (2013); Evaluation Standard for Green Office Building (2013); Assessment Standard for Green Store Building (2015); Evaluation Standard for Green Hospital Building (2016); Assessment Standard for Green Hotel Building (2016); Assessment Standard for Green Museum and Exhibition Building (2016); Assessment Standard for Green Retrofitting of Existing Building (2015); Assessment Standard for Green Building (2014); Assessment Standard for Healthy Building (2016)	China

Table 1. General description of the green building rating systems reviewed (continued).

Green Building Rating System	Initiator	Rating Tools/ Categories	Country
BEAM [22]	Hong Kong Green Building Council	BEAM Plus New Buildings; BEAM Plus Existing Buildings; BEAM Plus Interiors; BEAM Plus Neighbourhood	Hong Kong
GG [23]	Green Building Initiative	Green Globes for New Construction; Green Globes for Existing Buildings; Green Globes for Sustainable Interiors	Canada
GM [24]	Building and Construction Authority	Green Mark Scheme for the categories of 1) New non-residential buildings, 2) Super low energy, 3) New residential buildings, 4) Transit stations, 5) Existing non-residential buildings, 6) Existing residential buildings, 7) Landed houses, 8) Existing schools, 9) Healthcare facilities, 10) Laboratories, 11) Infrastructure, 12) District, 13) Healthier workplaces, 14) Office interior, 15) Restaurants, 16) Supermarket, 17) Retail, 18) Data centre, 19) Overseas projects – new developments, 20) Overseas projects – existing buildings, 21) Circular economy/ sustainable construction	Singapore
GBI [25]	Greenbuildingindex Sdn Bhd	Non-Residential New Construction (NRNC); Residential New Construction (RNC); Non-Residential Existing Building (NREB); Industrial New Construction (INC); Industrial Existing Building (IEB); NRNC – Data Centre, Retail, Hotel, Resort; NREB – Data Centre, Retail, Hotel, Resort, Historical Building; GBI Township	Malaysia
IGBC [26]	Indian Green Building Council	IGBC Green New Buildings; IGBC Green Existing Buildings; IGBC Green Homes; IGBC Green Affordable Housing; IGBC Green Residential Societies; IGBC Green Schools; IGBC Green Factory Buildings; IGBC Green Data Center; IGBC Green Campus; IGBC Green Villages; IGBC Green Townships; IGBC Green Cities; IGBC Green Existing Cities; IGBC Green SEZs; IGBC Green Landscapes; IGBC Green Mass Rapid Transit System; IGBC Green Existing Mass Rapid Transit System; IGBC Green Affordable Housing	India

Table 2 shows the criteria used in the 10 green building rating systems. The criteria represent the key areas of evaluation. There may be overlapping between areas such as sustainable siting, and land use and outdoor environment. Placing a mark on a criterion in Table 2 means the rating system puts a significant emphasis on the particular criterion in contrast to it being a small component under an overarching criterion.

IOP Conf. Series: Materials Science and Engineering **943** (2020) 012060 doi:10.1088/1757-899X/943/1/012060

Table 2. Comparison of the criteria of green building rating systems.

Criteria	LEED	BREEAM	CASBEE	GS	ASGB	BEAM	GG	GM	GBI	IGBC
Energy Efficiency	●	●	●	●	●	●	●	●	●	●
Water Efficiency	●	●	●	●	●	●	●	●	●	●
Material Efficiency	●	●	●	●	●	●	●	●	●	●
Indoor Environmental Quality, Health and Wellbeing	●	●	●	●	●	●	●	●	●	●
Sustainable Siting	●					●	●		●	●
Innovation	●	●		●	●				●	●
Land Use and Outdoor Environment		●	●	●	●					
Waste Management		●		●				●	●	
Construction Project Management		●		●	●		●			
Transport		●		●						
Integrative Process	●					●				
Pollution		●								
Sustainable Architecture and Design										●
Offsite Environment			●							
Climate Consideration								●		
Regional Priority	●									
Advanced Green Efforts								●		

Note: ● indicates that the criterion has been significantly addressed in the green building rating systems

IOP Conf. Series: Materials Science and Engineering **943** (2020) 012060 doi:10.1088/1757-899X/943/1/012060

Table 2 shows a range of criteria used in green building rating systems and the range may expand if more rating systems were to be included in this comparison. Four criteria that all the rating systems compared cover are energy efficiency, water efficiency as well as indoor environmental quality, health and welfare. Certain rating systems such as GM and BREEAM use health and wellbeing to represent the wider domain of noise, indoor air quality, spatial quality and thermal comfort while others use indoor environmental quality to capture the more specific indoor air, noise and thermal aspects. Sustainable siting and innovation are also emphasized in many of the rating systems though in rating systems such as BREEAM, sustainable siting is an extremely small component of the overarching land use and outdoor environment [11]. Besides, waste and project management also makes regular appearance in green building rating systems.

Figure 1 compares the weights of the six most prevalent green building rating criteria for new buildings. It shows GG and LEED having a comparatively higher weighting for energy efficiency than other rating systems while IGBC and ASGB allocate relatively higher weight for water efficiency. For all rating systems compared, the weights of material efficiency vary within a narrower range of 9 to 16. BEAM, BREEAM and CASBEE put considerable emphasis on indoor environmental quality, health and well-being [20, 24, 27]. GBI, however, emphasizes sustainable siting strongly (Figure 1).

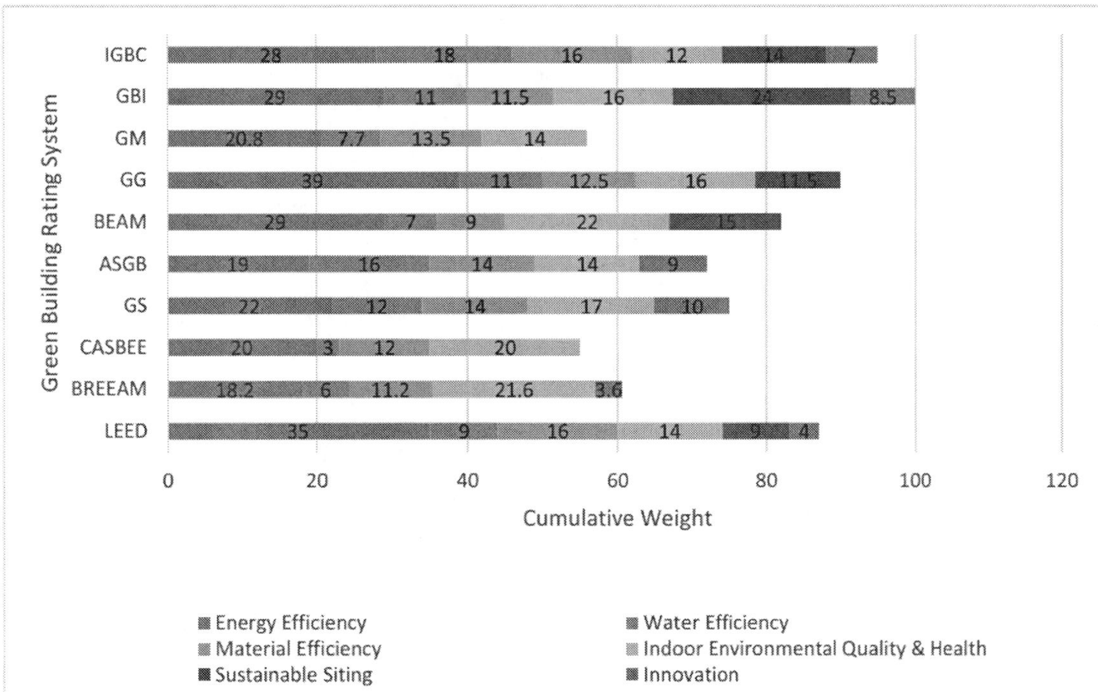

Note: GBI and GM have separate weights for new residential and non-residential buildings. The weights above are the average of the two.

Figure 1. Weights of six most prevalent green building rating criteria for new buildings.

All the rating systems compared fall short of a cumulative weight of 100% except GBI whose evaluation criteria revolve only around the predominant six. Other rating systems include criteria beyond the six most prevalent ones. This is largely in line with the findings of Shan and Hwang though the values of the weights differ as the authors did not indicate the rating categories from which the weights were sourced [13].

Figure 2 shows the average weights for each of the six rating criteria. Energy turns out to have the highest average weight (26%), followed in descending sequence by indoor environmental quality, health and well-being (16.7%), sustainable siting (14.7%), material efficiency (10.1%), water efficiency (10.1%) and innovation (7%). Shan and Hwang also revealed energy as the most highly weighted criterion in their study but the weights for other criteria in descending order are site, indoor environment, land and outdoor environment, material, water and innovation. While Shan and Hwang compared 15 green building rating systems encompassing most of those in this study, the rankings are largely in agreement with this study [13].

The six criteria in Figure 2 also aligns with the review findings of Illankoon et al. revealing seven main criteria namely energy, site, indoor environment quality, water, material, waste and pollution as well as management from eight rating systems reviewed, all covered in this study [9]. Energy is in fact the most common and predominant criterion [28].

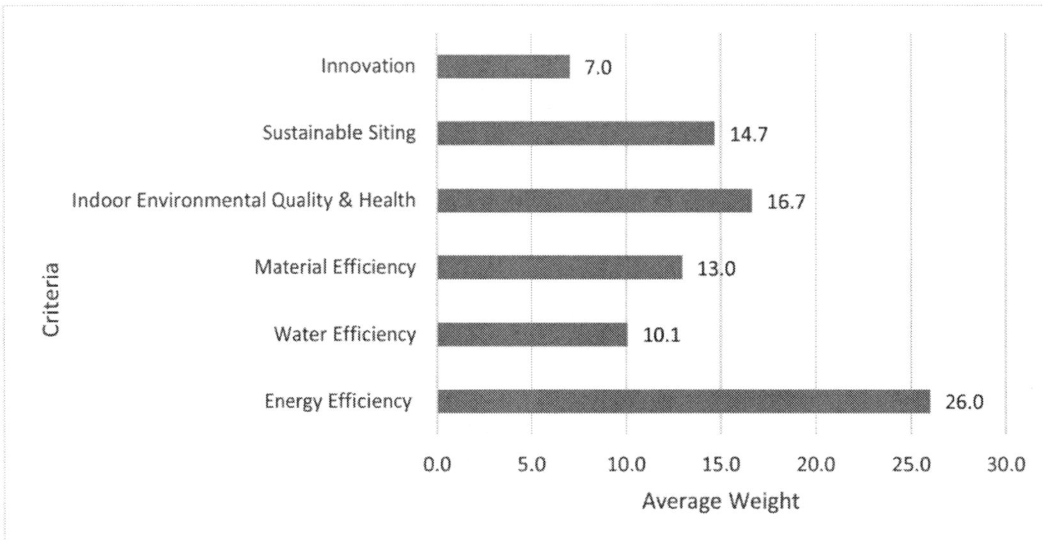

Figure 2. Average weights of the six most prevalent criteria of green building rating systems.

Nonetheless, perceived importance of the criteria may differ with types of building. Kim and Osmond, in their survey among hospital staff, patients and visitors, showed indoor environmental quality was perceived as the most important criterion for certification of green healthcare building [29]. Illankoon et al. opined that the current green building rating systems, while extensively address environmental and social sustainability, do not seem to confer equal attention to economic sustainability [19]. Besides, there is a lack of comprehensive definition for certain aspects related to design, particularly passive design strategies [30].

In addition to the limitations of green building rating systems, this study also has inherent limitations. The criteria named differently in different rating systems may address overlapping aspects for instance between sustainable siting and land use as well as between climate consideration, integrative process and construction project management. This study does not map to a detailed extent the similarities between the criteria named differently, though it does merge criteria of obvious resemblance such as indoor environmental quality, and health and wellbeing. In view of the sheer number of green building rating systems, this study is limited by only looking at the popular ones. In Malaysia alone, besides the most widely used GBI, other tools such as MyCREST and GreenRE also exist while in India, there is also Green Rating for Integrated Habitat Assessment. This may call for comparison of regional green building rating systems in the future.

4. Conclusion

This study compares 10 green building rating systems across 17 criteria and found that the most prevalent criteria used are energy efficiency, indoor environmental quality, health and wellbeing, sustainable siting, material efficiency, water efficiency and innovation in descending average weights. This study contributes to better understanding of the most common green building rating systems used around the world. It highlights the differences in the evaluation criteria and weights of the criteria to enable developers or building managers to pay due attention to the criteria most pertinent to the rating systems of interest. It recommends the examination of overlapping aspects within a rating system to continuously improve efficiency of the rating system. It also recommends the expansion of economic aspects in the rating systems. For future study, there is a need to look at how perceived importance of the green building rating criteria aligns with the allocated weights. It is also interesting to examine how the green buildings in different countries perform in terms of sustainability, in comparison to non-green buildings. This study also points to the need to look at implementation of the rating systems and the barrier towards implementing them.

5. References

[1] U.S. Energy Information Administration 2019 International energy outlook 2019 with projections to 2050 https://www.eia.gov/outlooks/ieo/pdf/ieo2019.pdf

[2] U.S. Energy Information Administration 2018 Use of energy explained – energy use in commercial buildings https://www.eia.gov/energyexplained/use-of-energy/commercial-buildings.php

[3] U.S. Energy Information Administration 2019 Electricity intensity of U.S. homes and commercial building deceases in coming decades https://www.eia.gov/todayinenergy/detail.php?id=38332

[4] KHD Tang 2018 Climate change in Malaysia: Trends, contributors, impacts, mitigation and adaptations Science of the Total Environment **650**, 1858-1871

[5] UN Environment and International Energy Agency 2017 Towards a zero-emission, efficient, and resilient buildings and construction sector – global status report 2017 https://www.worldgbc.org/sites/default/files/UNEP%20188_GABC_en%20%28web%29.pdf

[6] Zhang Y, Kang J and Jin H 2018 A review of green building development in China from the perspective of energy saving *Energies* **11** 334

[7] Haapio A and Viitaniemi P 2008 A critical review of building environmental assessment tools *Environmental Impact Assessment Review* **28** 469–482

[8] Ali HH and Nsairat ALSF 2009 Developing a green building assessment tool for developing countries – Case of Jordan *Building and Environment* **44** 1053–1064

[9] Illankoon IMCS, Tam VWY, Le KN and Shen L 2017 Key credit criteria among international green building rating tool *Journal of Cleaner Production* **164** 209–220

[10] U.S. Green Building Council 2019 LEED v4.1 https://new.usgbc.org/leed-v41?creative=340432438239&keyword=green%20building%20certification&matchtype=b&network=g&device=c&gclid=EAIaIQobChMIms_yi_e5QIVSoqPCh16GgIBEAAYASAAEgLkY_D_BwE

[11] Building Research Establishment Ltd 2019 BREEAM https://www.breeam.com/

[12] BEAM Society Limited 2019 Milestones https://www.beamsociety.org.hk/en_about_us_2.php

[13] Shan M and Hwang B 2018 Green building rating systems: Global reviews of practices and research efforts *Sustainable Cities and Society* **39** 172-180

[14] Rastogi A, Choi JK, Hong T and Lee M 2017 Impact of different LEED versions for green building certification and energy efficiency rating system: A multifamily midrise case study *Applied Energy* **205** 732-740

[15] Wu P and Low SP 2010 Project management and green buildings: Lessons from the rating systems *Journal of Professional Issues in Engineering Education and Practice* **136** 64–70

[16] Aye L and Hes D 2012 Green building rating system scores for building reuse. *Journal of Green Building* **7** 105–112
[17] Geng Y, Dong H, Xue B and Fu J 2012 An overview of Chinese green building standards. *Sustainable Development*, **20**, 211–221
[18] Awadh O 2017 Sustainability and green building rating systems: LEED, BREEAM, GSAS and Estidama critical analysis *Journal of Building Engineering* **11** 25–29
[19] Illankoon IMCS, Tam VWY and Le KN 2017 Environmental; economic; and social parameters in international green building rating tools *Journal of Professional Issues in Engineering Education and Pract*, **143**
[20] Japan Sustainable Building Consortium and Institute for Building Environment and Energy Conservation 2019 CASBEE http://www.ibec.or.jp/CASBEE/english/
[21] Green Building Council of Australia 2015 Green Star https://new.gbca.org.au/green-star/
[22] Hong Kong Green Building Council 2019 BEAM Plus – introduction https://www.hkgbc.org.hk/eng/beam-plus/introduction/
[23] Green Building Initiative 2019 Green Globes Certification https://www.thegbi.org/green-globes-certification/
[24] Building & Construction Authority 2017 About BCA Green Mark Scheme https://www.bca.gov.sg/greenmark/green_mark_buildings.html
[25] Greenbuildingindex Sdn Bhd 2019 GBI rating tools https://new.greenbuildingindex.org/how/tools
[26] Indian Green Building Council 2019 IGBC rating systems https://igbc.in/igbc/
[27] Hong Kong Green Building Council 2019 BEAM Plus – BEAM Plus new buildings https://www.hkgbc.org.hk/eng/beam-plus/beam-plus-new-buildings/
[28] KHD Tang 2019 Are we already in a climate crisis? *Global Journal of Civil and Environmental Engineering* **1** 25-32
[29] Kim S and Osmond P 2014 Analyzing green building rating tools for healthcare buildings from the building user's perspective. *Indoor and Built Environment*, **23**, 757–766
[30] Chen X, Yang H and Lu L 2015 A comprehensive review on passive design approaches in green building rating tools *Renewable and Sustainable Energy Reviews*, **50**, 1425–1436

IOP Conf. Series: Materials Science and Engineering **943** (2020) 012061 doi:10.1088/1757-899X/943/1/012061

Adoption of Building Information Modelling in Malaysia Road Construction

A T B Yung[1], E Aminudin[1,2], C N S Liat[1], M Neardey[1], R Zakaria[1], A R A Hamid[1], F Ahmad[1] and L Y Yong[3]

[1] Department of Structures and Materials, School of Civil Engineering, Faculty of Engineering, Universiti Teknologi Malaysia, 81310 Johor Bahru, Johor, Malaysia.
[2] Construction Research Centre, UTM CRC, Universiti Teknologi Malaysia, 81310 Johor Bahru, Johor, Malaysia.
[3] Department of Civil Engineering, Faculty of Engineering, Universiti Malaysia Sarawak, 94300 Kota Samarahan, Sarawak.

E-mail: eeydzah@utm.my

Abstract. Building Information Modeling (BIM) has changed the road development industry in the most recent decades. In spite of the fact that there are activities by administration of Malaysia to improve the selection of BIM in road development, yet the usage of BIM is still not true to form because of misty guide of BIM reception plan in Malaysia. Along these lines, it is essential to recognize the issues and difficulties in usage of BIM so as to improve the selection of BIM in this industry. This paper plans to analyze the street development industry in Malaysia to build up an unmistakable comprehension about BIM execution and to give the key arrangements and system for street development industry to expand the reception of BIM in Malaysia. This paper has built up broad BIM understanding which makes ready for the investigation by means of subjective and quantitative technique where the information is gathered through study poll among the street development partners in Malaysia and the gathered information is dissected with Average Index utilizing SPSS Software. The outcomes indicated the degree of current BIM usage in Malaysia street development industry is still extremely low and the issues and difficulties of BIM are distinguished in six zones incorporate specialized, process, legitimate, return of venture, government and individuals. Government has been recognized to assume a significant job in offering help to improve appropriation of BIM in Malaysia. The result of this investigation is communicated as a system for the usage of BIM in Malaysia involving issues that require thought for association to advance on BIM development stepping stool. This investigation clears a firm establishment for association to settle on choice in the appropriation of BIM in street development inside the general association structure.

1. Introduction

In highway construction, changes in the project scope will bring enormous impact to the project completion time and project cost due to the size of the project. [1] For example, minor changes in road design drawing will affect the overall road design due to its complication of project. Issues, for example, absence of assets, poor contractual worker for the board, less worker, structure deferrals and planning and scheduling inadequacy are among the reasons for delay in road development which are regularly been perceived something that is unavoidable would now be able to be illuminated effectively by utilizing Building Information Modeling. Building Information

Modelling able to solve the traditional design problems which make the all highway construction phases from inception till operation more efficient and east to be managed. Although there is a great deal of advantages of adopting BIM in road construction and design that have been proved theoretically and practically, application of BIM is still comparatively slow in developing countries such as Malaysia due to its high investment in time and cost in the early stage of construction and lack of awareness about the software. The execution of BIM in road development still cannot be ventured into broad around the world with different answer including reluctant of more seasoned age to adapt new abilities and designer's moderate nature with regards to an innovation that presently not yet to be tried and attempted. However, a proliferation of noticeable advantages, BIM has been broadened excessively application than just building.

Although there are initiatives by government of Malaysia under Highway Authority Malaysia, but the usage of BIM in road construction is still not as expected because there is no clear road map of BIM adoption plan in Malaysia. Besides, many industry players are more conservative and not prepared to embrace new innovations such as BIM into their company because there is currently no site clear for BIM market and they are not confident in using the new technology. Over the years, many pilots and live projects have been completed in Malaysia and the acknowledgment of the advantages of BIM is dependent upon an appropriate execution of BIM at a hierarchical level and its reconciliation at the business level. Study has demonstrated the guide for usage of BIM in UK development industry [2], but currently there is still no clear BIM acknowledgement for the framework has been developed for Malaysia road construction industry. Hence, this paper provides the understanding on the adoption of BIM implementation, issues and challenges in road construction besides providing the framework based on the best implementation of BIM road construction in Malaysia.

2. Literature review

2.1. Issues and Challenges of BIM Adoption

BIM in road construction has its own benefits however the barriers encounter needs to be solved in order to increase the adaptation of BIM in Malaysia construction industry. Majority of the players in construction industry see BIM as a "disruptive technology" that change the traditional construction process. Implementing BIM in road construction is never an easy task because there are a lot of things to be considered. Although BIM has been promoted as "one of the most promising recent developments in the architecture, engineering and construction (AEC) industry", equipped for lessening venture cost, improving efficiency and quality and diminishing the ideal opportunity for venture conveyance, distinguishing the difficulties of execution of BIM has been viewed as a prelude for upgrading BIM appropriation [3]. There are numerous difficulties and issues in BIM selection into road development since it is accepted as a generally new innovations for road development industry in Malaysia. To begin with, specialized difficulties that incorporate the distinctions, absence of completely embraced BIM work process for street development venture, far reaching execution cost and necessity for better equipment to deal with huge volumes of information. Among all difficulties, absence of interoperability as one of the significant specialized difficulties when comes to executing BIM in street development. Interoperability can be characterized as the capacity for a framework, for example, programming to work with different frameworks with no exertion with respect to the end client of the framework, particularly with regards to trading data between frameworks. Furthermore, process-related test is likewise one of the obstructions in BIM reception. These difficulties incorporate the requirement for the business to streamline the business procedure after the utilization of BIM, change of job in the undertaking partners, absence of standard and legally binding dialects. So, it is imperative to build up an exacting benchmarks and definition that would streamline the business procedure [4]. Thirdly, legitimate difficulties. It remembers absence of understanding for the lawful provisions about the computerized marks, stamps and expectations, uprightness of information during transmission, classified data and protection strategies to cover duties of partners. So, as to defeat this difficulties, solid arrangements, for example, expressing proprietorship and obligation of the information in contracts are essential to polish the procedure of BIM advancements reception in the street development industry [5]. Ultimately, return of investment (ROI) is additionally one of the primary worried in BIM appropriation since greater expense of BIM selection will in the end increment the expense of road development. return on initial capital investment challenges incorporate expanded of speculation for procurement of BIM empowered programming, updating current IT frameworks, teaching engineers and changing current venture conveyance technique and expectations. Despite the fact that the push-pull impact of the market will in the long run lessen the cost and increment the availability of BIM innovation, yet at the present phase of advanced upheaval, there are as yet activity that has be finished by the governing body in order to grow the modification of BIM among tremendous industry players by the organization to construct the change of BIM among colossal industry players. [2]. Table 1 has shown the identification of issues and challenges in adoption of BIM.

Table 1. Identification of issues and challenges in adoption of BIM.

Identification of challenges in adoption of BIM	Challenges in Implementation of BIM	References							
		1	2	3	4	5	6	7	8
Technical challenges	Lack of clear BIM workflow.		*	*	*			*	
	Existing hardware incapable of running basic BIM software.		*						
	Lack of ability to exchange data with other systems.	*	*		*		*		
	BIM software is complicated to use.	*	*			*	*	*	
Process-related challenges	Change of Business Model (Type of contract bid for).		*						
	Change in responsibilities of Stakeholders.		*	*	*				
	Lack of standard regarding BIM.	*	*			*	*		
	Lack of contractual language.		*		*				
Legal Challenges	Lack of agreement on the legal clauses about the digital signatures.		*						
	Lack of insurance policies to cover responsibilities of stakeholders.		*		*				
Return of Investment Challenges	High investment for acquisition of BIM enabled software	*	*	*	*	*	*	*	*
	High cost of upgrading current IT system.	*	*		*				*
	Changing current project delivery method.	*	*	*					
	High training cost.	*	*	*					*
Governance Challenge	Lack of handouts from Government.								*
	Government departments are not ready to accept 3D submission.								*
	Building plan approval is tedious as model re-working is expensive.								*
	Lack of BIM mandate by government departments.								*
People related Challenge	Insufficient availability of BIM training.							*	
	Lack of awareness of BIM benefits.		*			*		*	
	Reluctant to initiate new workflows for the implementation of BIM.							*	

**References*: 1= [3] ;2= [4] ; 3= [2]; 4= [5]; 5= [6]; 6= [7]; 7= [8]; 8= [9]*

Malaysia government specialists have effectively arranging mindfulness building exercises. Singapore and Hong Kong governments are as of now a long way ahead with BIM appropriation; notwithstanding, they embraced the ideas from Malaysia. Malaysia government is taking a gander at an approach for another economy and BIM is especially in its core. CIDB and LLM are attempting to improve the selection of BIM as they are assuming a significant job to characterize job and bearing of BIM in Malaysia development industry. There is not sufficient help for BIM by government, in contrast to Singapore. The way toward building plan endorsement is dreary in Malaysia and structures will continue changing and this impedes BIM use as changes for model is costly. Moreover, Malaysia government divisions are a long way from having the option to acknowledge 3D entry. In order to increase the adaptation of BIM in this industry, it is important for the organizations to understand that they should change the way the work currently, change in staffing needs and change in the way they use information in road construction project. According to Peter Smith in his studies on global strategies of BIM implementation has found implementation of BIM in construction has showed visible improvement by that provide consistent national standards, BIM protocols and legal contracts, quality of model, national BIM product database and libraries and changes in business model [10]. A study done by Tristan in understanding the application of BIM has suggested organization should re-evaluate and reengineer business practice, provide defined responsibility to all parties, provide consistent national standard and develop a clear adoption policy for BIM models as BIM implementation strategies [11]. Ali has suggested that implementation of BIM can be improved assigning BIM managers and lead inter-operational professional, provide BIM protocol document and provide cybersecurity for BIM tools outcome [12].

Besides, study done by Taqiadden has proposed to provide staffs with BIM training opportunity, provide consistent national standards for BIM, development of project procurement system based on Integrated Project Delivery (IPD) and provide introduction of BIM in undergraduate or graduate program in order to increase the rate of implementation of BIM in construction. There is an additionally requirements for instruction and preparing of potential BIM collaborator. A methodology to accomplish this could be through the presentation of BIM in the educational plans for undergrad and graduate projects in the fields of design, building and development. [13]. As far as government intercession in the business in connection to the selection of new advancements, there are three applicable viewpoints: data support, budgetary help and staff support [14]. An effective national procedure for advancing household advancement regularly incorporates the dynamic investment of a nation's financial improvement board. Table 2 has indicated the ID of usage procedure of BIM.

Table 2. Identification of usage procedure of BIM.

Identification of Usage Procedure of BIM	Strategy in BIM implementation	References										
		1	2	3	4	5	6	7	8	9	10	11
Skills	Provide staffs with BIM training opportunity.	*				*	*	*	*		*	
	Provide introduction of BIM in undergraduate or graduate program					*	*		*			
Process-related solution	Re-evaluate and re-engineer business practise by firms.		*	*								
	Provide defined responsibility of who owns and can act upon monitored performance data.			*	*				*			
	Provide free and easy access to online BIM building data and libraries with international consistency.		*							*		
	Provide a BIM protocol document.		*		*	*						
	Provide cyber security for BIM tools outcomes.				*				*			
	Support from top management of the organisation.						*				*	
	Provide contractual arrangement for BIM adoption.						*					
	Mandate the use of BIM in publicly funded project.		*					*		*		
Legal solution	Develop a clear adoption policy for BIM models				*				*	*		
	Optimization of upfront cost in BIM Implementation enhances the benefits associated with BIM.							*				
Return of Investment solution	Enhancements in college industry collective innovative work (R&D) connections and focused on innovation advancement programs											*
Government- related solution	Support in relation to taxation.											*
	Provide consistent global and national standards.		*	*		*		*	*	*	*	
	Increase of government incentives in BIM implementation.							*				
Awareness	Provide awareness of the benefits of BIM.							*		*		
	Provide chance to observe efficiency of BIM tools.	*										

**References: 1 = [10];2= [15] ; 3= [11]; 4= [16]; 5= [13]; 6= [17]; 7= [12]; 8= [5]; 9= [14]; 10= [18]; 11= [19]*

3. Methodology

3.1. Research instrument

This research focuses on assessing the current readiness and maturity level of Malaysia road cosntruction industry in adopting BIM and their expectation towards the development in this area. Poll review was utilized as a strategy for gathering information as this is a progressively compelling methodology in gathering objective based and quantifiable information, which assume a significant job in this examination. The survey is designed into four (4) segments involving for the most part close-finished inquiry. The essential territory (1) is to perceive the respondent's profile and nuances of their enthusiasm for BIM. Area two (2) level of execution of BIM in respondent's association, three (3) issues and difficulties in selection of BIM and finally in segment four (4) usage technique for BIM in road development. Questions in section three(3) and section four (4) adopted Likert-scale questions where the respondents are asked to choose the numerical scale from "1" to "5" where "1" indicated "Strongly disagree" , "3" indicated "Undecided" and "5" indicated "Strongly agree". Prior to information assortment, a pilot study had been directed with experienced academicians and related nearby specialists to get fundamental substance validation for the surveys. Average Index analysis was adopted in the data analysis of survey responses where the data was analyzed using SPSS Version 25 software.

3.2. Demographic profile of respondents

Aggregate of 85 polls were conveyed to the respondents in a time of (1) week, to whom has encountered in street development venture either from private or open association in Malaysia. Out of 85 surveys dispersed, 40 polls were finished with the portrayal of 47.67%. The reaction rate is proper for a development look into. Table 3 decide to recognize the respondent's job and association.

Table 3. Profile of Respondents

Profile	Category	Frequency	(%)
Number of respondents	Age 20-30	16	40
	Age 31-40	12	30
	Age 41-50	6	15
	Age more than 50	6	15
Category of organization	Government agency	8	20
	Developer	5	12.5
	Engineering consultancy	13	32.50
	Quantity Surveying firm	3	7.5
	Contractor	10	25
	Supplier and manufacturer	1	2.5
Year of experience in construction industry	Less than 1 year	6	15
	1 – 4 years	9	22.5
	5 – 9 years	5	12.5
	More than 10 years	20	50
Size of organization	3 – 19	5	12.5
	20 – 49	8	20
	59 – 199	12	30
	200 and above	15	37.5

4. Result and discussion

Respondents` organization on BIM implementation was investigated through section B of the questionnaire. The result has shown that the implementation of BIM in Malaysia road construction is still low. None of the respondent`s organization has reached BIM level 3 where full life cycle integration considering maintenance and operation. 15% of the respondents have implemented BIM up to level 2, 27.5% reached BIM level 1 where 2D and 3D design integrated with product data management and 57.5% has not implemented BIM in their organization with just Level 0 BIM. Among the respondents who have implemented BIM in their organization, all the respondents use Autodesk Revit because it does support for topography coordinate systems just that it is better to design road below 30km to prevent precision issues. BIM is still considered a new technology in Malaysia especially road construction industry.

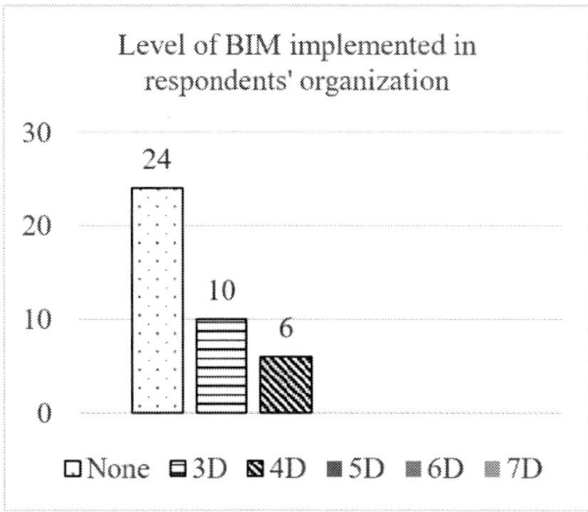

Figure 1. Level of BIM used in respondent's organization BIM tools.

4.1. Issues and challenges in adoption of BIM

As summarized in Table 4, the data collected from the questionnaire were analysed to delineate the issues and challenges in adoption of BIM in road construction. A total of twenty-one (21) variables are categorized into six BIM factors (Technical, Process, Legal, Return of Investment, Government and People). As it is, the procedure output has an overall raw alpha of 0.764 which is good considering that .70 is the cut off value for being acceptable According to the findings from the respondents, five (5) challenges that are most frequently chose by the respondents are lack of standard regarding BIM (mean = 4.325), high investment for acquisition of BIM enabled software (mean = 4.225), government departments are not ready to accept 3D submission (mean = 4.200), insufficient availability of BIM training (mean = 4.175) and lack of BIM mandate by government departments (mean = 4.15). The top challenge among all is the lack of standard regarding BIM. Although Malaysia local authorities Jabatan Kerja Raya (JKR) has built a reference document aiming at providing a unified BIM standard by JKR (Garis Panduan dan Piawaian BIM JKR and Garis Panduan Pengunaan Template JKR) that can easily be easily adopted to suit different projects, but the consultants and contractors are still fail of using those standards that creates better information exchange and communication because it is not compulsory for them to use it since currently our government has not mandate the use of BIM in construction industry yet. This study also focuses government-related factors in the issues and challenges of BIM adoption that is ignored in the development of roadmap for implementation of BIM in the UK construction industry [2]. Two out of top five challenges listed above are governance related challenges which showed that our government should come out with more initiative to boost the adoption of BIM in Malaysia. Neighboring nation, for example, Singapore had effectively actualized BIM in nearby specialists from BIM consistence to administrative prerequisite which helped in boosting the reception of BIM in Singapore. Hence, local authorities need to play a significant role in increasing the adoption of BIM in road construction.

Within the Process context, all respondent agreed that there is lack of standard regarding BIM with the mean value at 4.325. Inside the process setting, all respondent concurred that there is absence of standard in regard to BIM with the mean an incentive at 4.325. Despite BIM has been in the market for quite a long while, there is still no institutionalization BIM rule for organizations in Malaysia. BIM achievement factor is exceptionally controlled by investment of government in setting the standard for BIM so as to clear uncertainty for all partners in street development industry. In relation, respondents believe that there is lack of contractual language in BIM with mean at 4.100. Prior developing any contract language, the most important item is to consider the overall purpose of BIM implementation and there is no standard of what should be added in BIM contract language in Malaysia since it is considered a new tool in road construction industry. In terms of technical challenges, all respondents agree that there is lack of clear workflow for BIM in Malaysia which shows the same challenges faced in the implementation of BIM in UK. During the implementation of BIM, there is no clear factor like model exchange, storage, naming conventions and organization of model that allows seamless inclusion of models throughout the life of documentation process. Importantly there is a need to have a clear plan for implementation and support for organization to fully leverage the benefits of using BIM tools in road construction.

In the area of legal challenges, all the respondents agreed there is a lack of agreement on the legal clauses about the digital signatures with mean of 4.000. Under Malaysia law, although a written signature is not necessarily required for a valid contract, but the use of digital signatures in road construction is still in a grey area because there are cases that are specifically barred from digital signatures such as instrument effecting any dealing with real property under the Malaysian National Land Code. Besides, it was agreed that legal were among the least areas of understood and will represent notable risks to the project management team.

The return of investment challenges is high investment for acquisition of BIM enabled software which ranked second overall with mean of 4.225. The respondents also agreed that adoption of BIM requires high cost of upgrading current IT system (mean=4.075) and high training cost (mean=4.025). As BIM requires investment of equipment, software, hardware and training, it would greatly impact on the cost of project.

Table 4. Mean score and rank of issues and challenges in adoption of BIM.

No.	Factors	Issues and Challenges in Adoption of BIM	Mean	S.D	Rank
C1		Lack of clear BIM workflow	4.100	0.871	10
C2	Technical challenges	Existing hardware incapable of running basic BIM software	3.625	1.191	20
C3		Lack of ability to exchange data with other systems	3.925	0.997	14
C4		BIM software is complicated to use	3.850	1.026	16
C5		Change of Business Model (Type of contract bid for)	3.600	1.057	21
C6	Process-related challenges	Change in responsibilities of Stakeholders	3.675	0.997	17
C7		Lack of standard regarding BIM	4.325	0.656	1
C8		Lack of contractual language	4.100	0.778	9
C9	Legal Challenges	Lack of agreement on the legal clauses about the digital signatures	4.000	0.716	13
C10		Lack of insurance policies to cover responsibilities of stakeholders	3.625	0.868	18
C11		High investment for acquisition of BIM enabled software	4.225	0.733	2
C12	Return of Investment Challenges	High cost of upgrading current IT system	4.075	1.022	11
C13		Changing current project delivery method	3.625	1.102	19
C14		High training cost	4.025	0.831	12
C15		Lack of handouts from Government	4.100	0.672	8
C16	Government-related Challenge	Government departments are not ready to accept 3D submission	4.200	0.758	3
C17		Building plan approval is tedious as model re-working is expensive	4.150	0.735	5
C18		Lack of BIM mandate by government departments	4.150	0.770	6
C19		Insufficient availability of BIM training	4.175	0.747	4
C20	People-related Challenge	Lack of awareness of BIM benefits	3.925	1.022	15
C21		Reluctant to initiate new workflows for the implementation of BIM	4.125	0.723	7

4.2 Implementation strategy for BIM

Table 5 summaries the overall strategic solutions according to related factors. According to the findings from the respondents, five (5) strategic solutions that are most frequently chose by the respondents are provide staffs with BIM training opportunity (mean = 4.55), increase of government incentives in BIM implementation (mean = 4.425), provide cyber security for BIM tools outcomes (mean = 4.325), provide introduction of BIM in undergraduate or graduate program (mean = 4.325) and improvements in university–industry collaborative research and development (R&D) relationships and targeted technology development programs (4.300). Overall, the findings show that all the implementation strategy has a mean score more than 3.500.

Skill-related solutions are the most impactful solution as two out of top five solutions are related to this factor. All the respondents are strongly agreed that providing staffs with BIM training opportunity (mean = 4.55) is able to improve the adoption of BIM in Malaysia. There are a lot of factors that need to be considered such as size of firm and existing of expertise when it comes to planning for BIM training for staffs in order to teach right skills to the right set of people with minimal disruption to the on-going project. Provide introduction of BIM in undergraduate or graduate program (mean = 4.325) is also one of the most impactful implementation strategies as it ranked third overall. As the technology that evolved rapidly in the construction industry, it is important to constantly expose undergraduate with new construction technology in their curriculum in order to meet the future needs of the construction industry.

In terms of process related solutions, providing cyber security for BIM tools (mean = 4.325) is perceived as good solution as it is agreed by all respondents. A shared 3D model may expose to intellectual property to competitors where a fly-through visualization of a building could share sensitive information about the building design-key structural components, locations of key building services, placement of CCTV or other security equipment. A shared 4D model might expose period when assets could be susceptible to sabotage, or sites could be vulnerable to thieves.

In the area of legal-related solution, mandate the use of BIM in publicly funded project (mean = 4.225) is agreed by all respondents. Government role in implementation strategy of BIM has not been studied by the UK roadmap case study before. Currently, Malaysia government is trying to mandate the use of BIM in government project that worth more than RM100 millions by 2020 because BIM system could optimize the cost of a project in line with the Construction Industry Transformation Program (CITP) according to Construction Industry Development Board (CIDB).

Optimization of upfront cost in BIM implementation enhances the benefits associated with BIM is a significant solution for return of investment related challenges. It is important for construction industry participants to optimize the upfront cost of BIM as the gain of BIM implementation does not reflect immediately upon the point of implementation because it is considered as a long- term investment to be more cost effective and efficient in project delivery.

Improvements in university–industry collaborative research and development (R&D) relationships and targeted technology development programs (4.300) that is related to government initiative is ranked fifth in overall. This strategy has not been studied by the UK roadmap case study. Such collaborative research programs increase the importance of the research opportunities, results and knowledge transfers between the participating companies and universities. Companies are beginning to realize the importance of collaborative research in evolving their core competencies and in making sound strategic decisions on how to best deploy their R&D resources to maintain competitive advantage.

Increase of government incentives in BIM implementation (mean = 4.425) was ranked second overall. This implement strategy is not mentioned in the Roadmap for implementation of BIM in the UK construction industry case study. This strategy is vital because it shows the importance of Malaysia government initiatives in improving the adoption of BIM in Malaysia. Local road construction authorities such as Highway Authority Malaysia (LLM) has just started BIM initiative last year where they are still at data collection phase and has yet to run any campaign to create awareness about use of BIM in road construction project. Thus, with the increasing initiatives, Malaysia government will be able to produce more construction industry players that are equipped with BIM knowledge which allowed tendering for BIM project to avoid the probability of project failure. With the identified strategy in BIM implementation, framework for BIM adoption has been identified in Figure 2.

Table 5. Mean score and rank of implementation strategy of BIM

No.	Factors	Strategy in BIM implementation	Mean	S.D	Rank
S1	Technical-related solution	Provide staffs with BIM training opportunity.	4.55	0.597	1
S2		Provide introduction of BIM in undergraduate or graduate program.	4.325	0.730	4
S3	Process-related solution	Re-evaluate and re-engineer business practice by firms.	3.975	0.947	17
S4		Provide defined responsibility of who owns and can act upon monitored performance data.	4.175	0.675	12
S5		Provide free and easy access to online BIM building data and libraries with international consistency.	4.225	0.862	10
S6		Provide a BIM protocol document.	4.150	0.770	15
S7		Provide cyber security for BIM tools outcomes.	4.325	0.693	3
S8		Support from top management of the organization.	4.275	0.640	6
S9		Provide contractual arrangement for BIM adoption.	4.150	0.700	13
S10	Legal-related solution	Mandate the use of BIM in publicly funded project.	4.225	0.768	9
S11		Develop a clear adoption policy for BIM models	4.150	0.736	14
S12	Return of Investment- related solution	Optimization of upfront cost in BIM Implementation enhances the benefits associated with BIM.	4.225	0.577	7
S13	Government-related solution	Improvements in university–industry collaborative research and development (R&D) relationships and targeted technology development programs	4.300	0.791	5
S14		Support in relation to taxation	4.225	0.630	8
S15		Provide consistent global and national standards.	4.200	0.608	11
S16	Awareness- related	Increase of government incentives in BIM implementation.	4.425	0.630	2
S17		Provide awareness of the benefits of BIM.	4.025	0.800	16

S18	solution	Provide chance to observe efficiency of BIM tools.	3.950	0.638	18

Figure 2. BIM implementation framework in road construction.

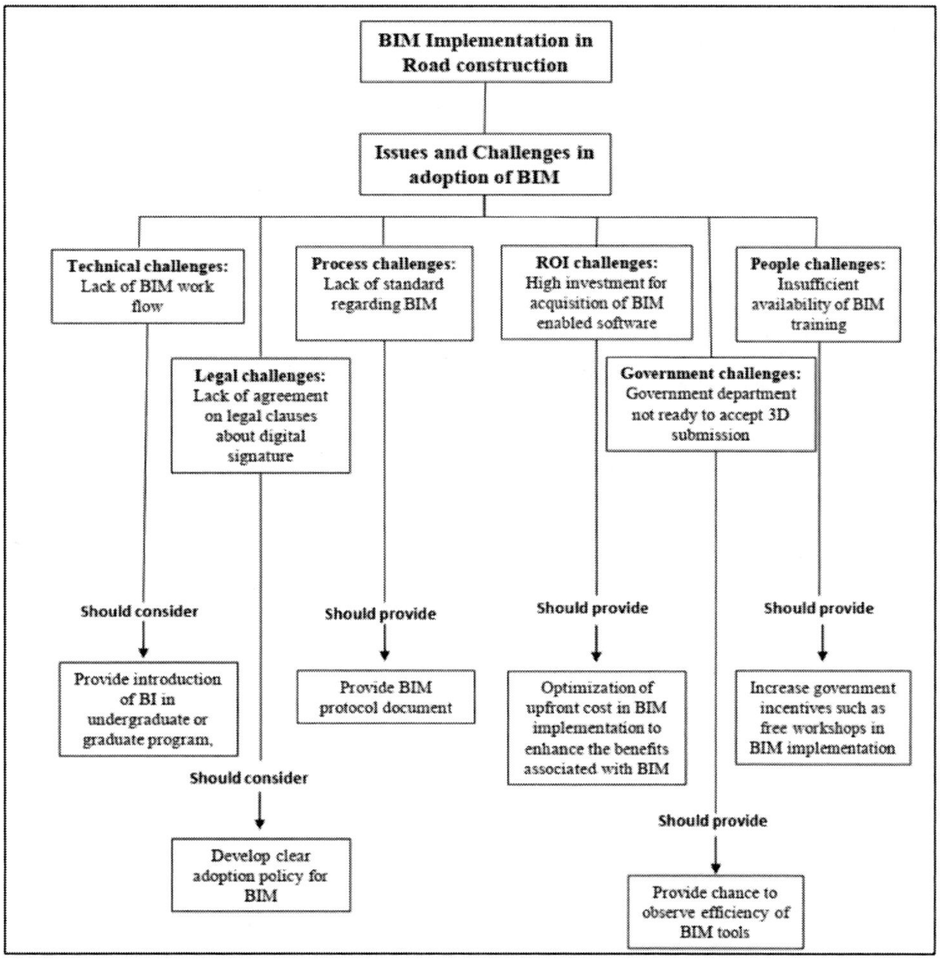

BIM has proven to be an effective technology with numerous studies had highlighted its potential capabilities. However, in the context of Malaysia road construction industry, the implementation of BIM is still in very slow pace. There are several factors that can be concluded that hinders the adoption of BIM has been identified in this research (1) lack of standard regarding BIM, (2) high investment for acquisition of BIM enabled software, (3) government departments are not ready to accept 3D submission, (4) insufficient availability of BIM training and (5) lack of BIM mandate by government departments. As Malaysia government is looking to increase adoption of BIM to prepare the industry for the future before 2020, it is important that the issues and challenges are to be addressed for the improvement of the road construction industry in Malaysia. Therefore, strategic BIM solutions should be implemented in order to increase the pace of BIM adoption in the Malaysia road construction industry. Several implementation strategies have been identified in this research (1) provide staffs with BIM training opportunity, (2) increase of government incentives in BIM implementation, (3) provide cyber security for BIM tools outcomes, (4) provide introduction of BIM in undergraduate or graduate program and (5) improvements in university–industry collaborative research and development (R&D) relationships and targeted technology development programs. Finally, several limitations need to be acknowledged. In general, the time constraint and relatively small sample size may lead to concerned on generalizations of the research.

5. Acknowledgments
Our thanks to Highway contractor PLUS Berhad Section S2 Malaysia for highly corporation with this study. AUN-Seed Net by JICA and University Technology Malaysia for supporting financial grant, Cost Centre No: R.J130000.7322.4B322, R.J130000.7322.4B3265, R.J130000.7322.4B332, R.J130000.4B334 & R.J130000.7322.4B331.

References

[1] Tyne N. U., Computer S., and Designer A. 2016. Highway infrastructure and building information modelling in UK. Proc. Inst. Civil Eng-Munic, 169:220–232.

[2] Farzad Khosrowshahi Y. A. 2012. Roadmap for implementation of BIM in the UK construction industry, J. Eng. Constr. Archit. Manag., 19:610–635.

[3] Yijun Zhou J.-B. Y. and Yu Yang. 2018. Barriers to BIM implementation strategies in China. J. Eng. Constr. Archit. Manag., 26(3):1–22.

[4] Costin A., Adibfar A., Hu H., and Chen S. S. 2018. Building Information Modeling (BIM) for transportation infrastructure – Literature review , applications , challenges , and recommendations, J. Autom. Constr., 94:257–281.

[5] Abd Jamil A. H. and Fathi M. S. 2018. Contractual challenges for BIM-based construction projects: a systematic review," Built Environ. Proj. Asset Manag., 8, no. 4, pp. 372–385.

[6] Sarhan J., Xia B., Fawzia S., Karim A., and Olanipekun A. 2018. Barriers to implementing lean construction practices in the Kingdom of Saudi Arabia (KSA) construction industry," J. Constr. Innov., 18(2):246–272.

[7] Akbar Marefat R. M. H. and Hossein Toosi. 2018. A BIM approach for construction safety : applications , barriers and solutions," J. Eng. Constr. Archit. Manag., pp. 1–24.

[8] Rogers J., Chong H.Y. and Preece C.N. 2015. Adoption of Building Information Modelling technology (BIM) Perspectives from Malaysian engineering," J. Eng. Constr. Archit. Manag., vol. 22, no. 4, pp. 424–445.

[9] Hong Y., Hammad A. W. A., Sepasgozar S. and Akbarnezhad A. 2019. BIM adoption model for small and medium construction organisations in Australia," J. Eng. Constr. Archit. Manag., vol. 26, no. 2, pp. 154–183.

[10] Shushnov M. S. and Falko A. I. 2008. Building Information Modelling (BIM) Implementation and Remote Construction Projects: Sissues, Challenges and Critiques," Telecommun. Radio Eng. (English Transl. Elektrosvyaz Radiotekhnika), vol. 67, no. 10, pp. 925–944, 2008.

[11] Gerrish T., Ruikar K., Cook M., Johnson M., Phillip M. and Lowry C. 2017.0BIM application to building energy performance visualisation and managementChallenges and potential," Energy Build., vol. 144, pp. 218–228.

[12] GhaffarianHoseini A., Doan D. T., Naismith N., Tookey J. and GhaffarianHoseini A. 2017. Amplifying the practicality of contemporary building information modelling (BIM) implementations for New Zealand green building certification (Green Star)," Eng. Constr. Archit. Manag., vol. 24, no. 4, pp. 696–714.

[13] Almuntaser T., Sanni-Anibire M. O., and Hassanain M. A. 2018. Adoption and implementation of BIM – case study of a Saudi Arabian AEC firm," Int. J. Manag. Proj. Bus., vol. 11, no. 3, pp. 608–624.

[14] Wong A. K. D., Wong F. K. W. and Nadeem A. 2011. Government roles in implementing building information modelling systems," Constr. Innov., vol. 11, no. 1, pp. 61–76.

[15] P. Smith. 2014. BIM implementation - Global strategies," Procedia Eng., vol. 85, pp. 482–492.

[16] Ghaffarianhoseini A., Tookey J., Ghaffarianhoseini A., Naismith N., Azhar S., Efimova O. and Raahemifar K. 2017. Building Information Modelling (BIM) uptake: Clear benefits, understanding its implementation, risks and challenges," Renew. Sustain. Energy Rev., vol. 75, no. September 2016, pp. 1046–1053.

[17] Ding Z., Zuo J., Wu J. and Wang J. Y. 2015. Key factors for the BIM adoption by architects: A China study," Eng. Constr. Archit. Manag., vol. 22, no. 6, pp. 732–748.

[18] Jensen P. A. and Jóhannesson E. I. 2013. Building information modelling in Denmark and Iceland," Eng. Constr. Archit. Manag., vol. 20, no. 1, pp. 99–110.

[19] Liu S., Xie B., Tivendal L. and Liu C. 2015. The Driving Force of Government in Promoting BIM Implementation The Driving Force of Government in Promoting BIM Implementation," J. Manag. Sustain., vol. 5, no. December 2015, pp. 158–164.

IOP Conf. Series: Materials Science and Engineering **943** (2020) 012062 doi:10.1088/1757-899X/943/1/012062

Simulation on Lighting Energy Consumption based on Building Information Modelling for Energy Efficiency at Highway Rest and Service Areas Malaysia

M Neardey[1], E Aminudin[1,2], L P Chung[3], R M Zin[1], R Zakaria[1], C M F H Che Wahid[1], A R A Hamid[1] and Z Z n Noor[4]

[1] Department of Structures and Materials, School of Civil Engineering, Faculty of Engineering, Universiti Teknologi Malaysia, 81310 Johor Bahru, Johor, Malaysia.
[2] Construction Research Centre, UTM CRC, Universiti Teknologi Malaysia, 81310 Johor Bahru, Johor, Malaysia.
[3] Faculty of Built Environment and Surveying, Universiti Teknologi Malaysia, 81310 UTM Johor Bahru, Johor, Malaysia.
[4] Center for Environmental Sustainability and Water Security (IPASA), Universiti Teknologi Malaysia, Skudai, Johor Bahru 81310, Malaysia.

E-mail: eeydzah@utm.m y

Abstract. Building consumes 40% of global energy consumption which contributes for such a majority amount of carbon dioxide emission. Enhancing the energy efficiency strategies in buildings by retrofitting through different approaches is essential for reducing end user on the amount of energy consumption and carbon footprint emission. Previous research shows that only a few studies have been conducted on retrofitting the energy performance on Rest and Service Area yet the result shows that lack of technical feasibility practice was reviewed. Hence, this paper explores the energy efficiency strategy by improving the lighting performance based on 24 hours service building at Rest and Service Area Ayer Keroh Northbound Malaysia. The simulation that was performed in lighting energy consumption was EnergyPLUS software, Autodesk Revit and RELUX Plug in tool in Revit. Preliminary study was conducted to identity the problem on energy performance at Rest and Service Area Ayer Keroh by determine the total energy consumption by sources and Building Energy Intensity [BEI]. The result shows lighting contributes to the most total load consumption due to lack of energy conservation management and energy efficiency strategy practices. The lighting simulation scenarios was created to enhance the energy consumption performance and reduced the irrelevant carbon footprint emission. The simulation in Revit and EnergyPLUS showed 17% and 49% reduction respectively on the end load of lighting energy consumption by reducing the overall time operation and daylight control dimming system.

1. Introduction

Situated in a tropical region, Malaysia experience a hot and humid weather which results in a huge amount of energy consumption. From the year of 2000 to 2018, the total energy consumption in Malaysia gradually increased from 53 billion (kWh) to 132 billion (kWh) [1] with the electrical power generation mainly relies on fossil fuel and it is expected to grow from 48% in 2015 to 66% of the overall consumption by 2023 [2]. Thus, energy-efficient and low-carbon energy strategies for buildings is play an important role to achieve a major reduction of carbon dioxide emission toward a sustainable nation [3].

To cope with such weather, many studied has been focusing on the energy refurbishment in existing buildings such as retrofitting on the envelopes which can compromise thermal comfort and reduce the energy load [4]. Other studies are focused on retrofitting the building mechanism system such as heating, cooling and lightings to achieve significant energy reductions by study on the pattern of occupant behavior or optimize the equipment design its self [5]. Artificial lighting is one the most energy consumer is many non-residential buildings in which energy efficient measure such as dynamic simulation should be promoted to reduce the load of consumption [6]. A recent study by [7] analyze the behavior of indoor lighting environment in a cigarette factory and energy consumption through simulation in Ecotect software. By adjusting on the size of skylight and change material on the roof material, the indoor lighting environment is enhanced with adequate illuminance and efficient lighting energy consumption is performed. [8] conducted an artificial lighting controls simulation integrates with daylight harvesting in reducing the heat energy consumption in EnergyPlus simulation program. The effect from dimming control with daylighting can save about 40% of the total energy consumption and enhance the visual comfort in indoor environment [9].

This paper simulates the artificial lighting energy consumption on a 24hours service building by implement the effect of ON/OFF control dimming integrate with natural daylighting for potential energy saving. There are a few studies has conducted at Rest and Service Areas Malaysia regarding potential energy retrofitting yet there is lack of study which consider on energy simulation for estimating the real energy consumption and saving from the energy efficiency measure [10]. The building model of Rest and Service Area is built up in Autodesk Revit before it was exported to EnergyPLUS for energy consumption simulation. The lighting profile is simulated in Autodesk Revit 2018 by using the RELUX plugin tool for analyze on the exiting lighting performance of the building.

2. Literature Review

2.1. Net-ZEB and sustainable Rest and Service Areas

The concept of Net-ZEB in buildings is the use of energy efficiency and the capability to generate electricity from renewable sources to fulfil the energy demand [11]. Various studies have conducted on the energy refurbishment of existing buildings due to a great attention has been paid on such matter in the recent EU directives [3]. [3] proposed an optimum approach for selecting the retrofitting measure and the outcome showed that modernization of energy systems is the most beneficial measure for reaching target of achieving net zero-energy. The study evaluated the energy efficiency measure in a historical building by analyse on the thermal comfort in EnergyPLUS software for potential energy saving. Cost-optimal is one of the most concerned factors for the practice of energy efficiency in order to archive the nearly zero energy building (nZEB) for energy performance [12] [13].

Rest and service areas are function as a part of highway services and facilities to provide passenger a more enjoyable journey. This type of facility offers a 24 hours service which refurbishment is essential to ensure the standard of services when the lifetime of building get to the highest productivity life. [14] explored the retrofitting measure by analysis on the variables effecting energy consumption in a non-residential building. The result from factor loading prioritized the level of significant of each variable effecting energy consumption which can provide a clear roadmap for energy retrofitting. There are numerous green building assessment tools such as BREAM, LEED, CASBEE, Green Star and GBI established for the purpose of promote the application of renewable energy sources and introduced to the passive design for reduce the operational energy use in building. On-site renewable energy sources combined with hi-tech energy efficiency application is the fundamental concept in green building toward nearly zero energy building or low carbon buildings [15]. [16] conducted a case study on rest areas at United States which adapted various energy efficiency strategies such as installing lighting sensors, enhance HVAC and window system, minimise toilet ventilation and improve the insulation enable to save the operational costs by $15,000 per year and reduce the electricity load by 500 million UK thermal Units. Another study conducted at 32 Colorado rest areas for optimize the cost saving from energy efficiency in order to consider a best decision for retrofitting options. The assessment on these rest areas is performed to determine the sustainability design and the quality of operation and maintenance activities by referred to LEED assessment criteria. The rest areas were categorized into four different types according to their function provided and each group was tested by adapted various type of energy retrofitting measure and implement the renewable energy sources. The assessment outcome showed that renewable energy sources such as solar power (PV) and wind turbines are highly potential to adapted at rest areas for generate on-site energy sources due to the potential alternative energy harvesting offered by each area. Heating and lighting load are the main dominated to electricity consumption and lighting is a primary energy consumer for some specific areas such as toilet, parking and walkway [17].

2.2. Lighting control methods

Artificial lighting contributes such a significant amount of global energy consumption. In non-residential buildings, the consumption of artificial lighting accounts about 40% of total load of consumption [18]. Therefore, control of artificial lighting and keeping the visual comfort is essential way to decrease the lighting load and reduce the building operational costs. Adapting the change of occupancy and integrated with daylight are the main methods for energy efficiency lighting control [19]. At 24 hours service building, the small change in occupancy behaviour can creates a significant effect to the energy consumption. Dimming control of luminaires and occupancy sensor used to adapt the amount of localized illuminance according to daylighting enable to provide a deep energy saving and flexible luminaires [20]. [19] categorized lighting control system into 2 primary designs namely centralized control and distributed control which is based on luminaire sensing as shown in figure 1.

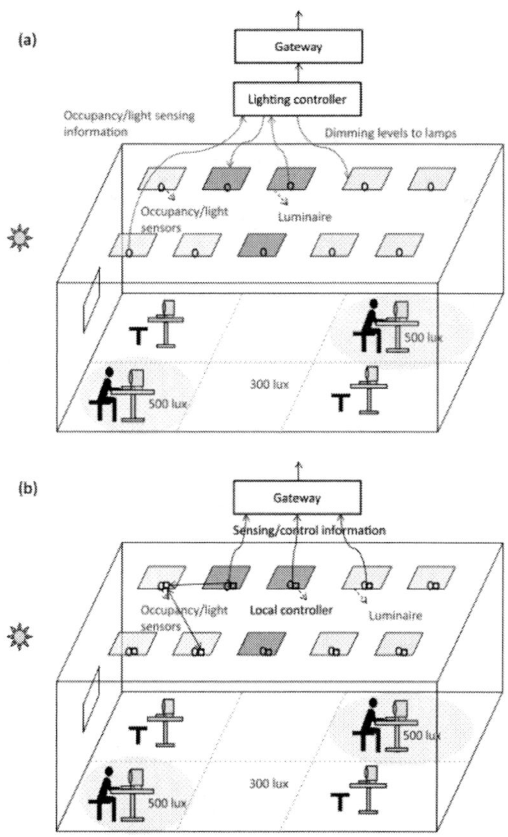

Figure 1. Smart indoor lighting system (a) centralized control and (b) distributed control [19].

IOP Conf. Series: Materials Science and Engineering **943** (2020) 012062 doi:10.1088/1757-899X/943/1/012062

3. Methodology

In this paper, a case study was conducted at Rest and Service Areas Ayer Keroh Northbound dining area building which located at west bound of Peninsular Malaysia. The aim of this study is to adapt the lighting energy efficiency to the building which is overload by the lighting energy consumption due to the lack of energy conservation and energy management measures [14]. Preliminary study revealed that the building energy consumption occupied majority by lighting load which account about half of the total energy consumption with the Building Energy Intensity [BEI] 287.88kWh/m^2/year according to the calculation. Equation 1 indicates the Building Energy Intensity based on Tahir [20] which most like being used for building.

Calculation Building Energy Intensity (BEI):

$$BEI = \frac{Total\ Energy\ Used\ per\ year}{Gross\ Floor\ Area} \left(\frac{\frac{kWh}{m^2}}{year}\right)$$ (Eq1) (Tahir *et al.*, 2015) [20]

Lighting energy efficiency was proposed with the use of dimming effect combine with natural daylight and turn off the light when there is adequate daylighting inside the building. 3 scenarios analysis was created for the simulation in Autodesk Revit and EnergyPLUS. In addition, RELUX desktop is use for testing the lighting performance of the existing scenario. The building architecture and electrical model was created in Autodesk Revit 2018 before import into EnergyPLUS simulation tool. There are total 29 light bulbs consist in the dining area of Ayer Keroh Rest and Service Area which does not include the light bulbs from the stall area or shop lot since the consumption for stall owner is separate from the building owner. The light bulbs type is high pressure sodium High bay NIKKON brand (Malaysia) with the power consumption of each bulb is 70W.

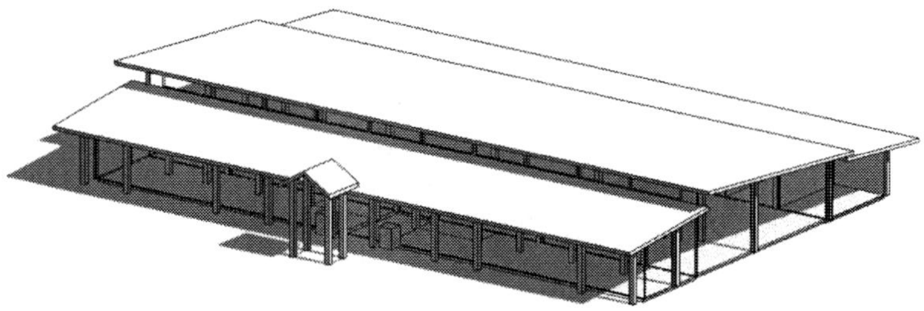

Figure 2. Ayer Keroh dining area Building Architecture model in Autodesk Revit 2018.

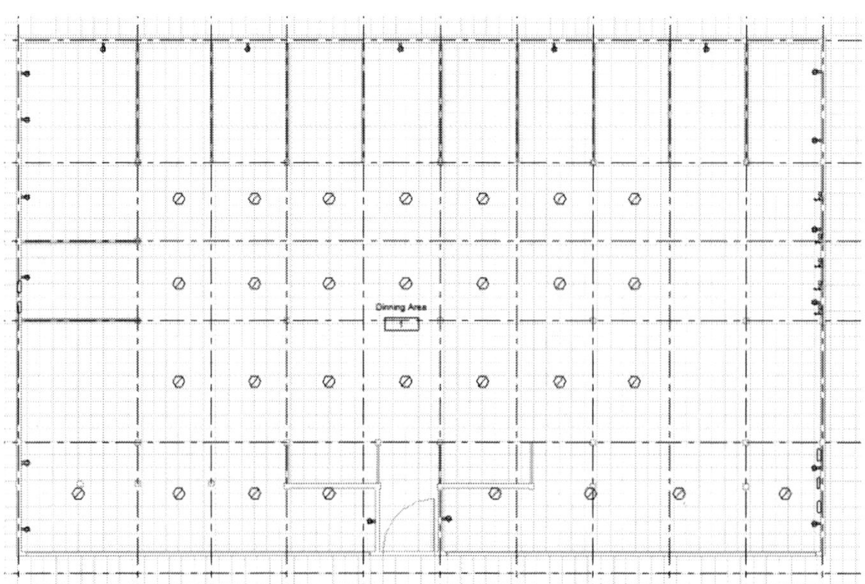

IOP Conf. Series: Materials Science and Engineering **943** (2020) 012062 doi:10.1088/1757-899X/943/1/012062

Figure 3. Ayer Keroh dining area layout model in Autodesk Revit 2018.

3.1. *Scenario analysis*

The lighting scenarios was made as control schemes to see the effect the lighting consumption of the building. Three scenarios with various lighting control measures are simulation and compare to the baseline scenario. The first scenario is considered as the baseline which the light is consumed to Turn ON for 24 hours as the practice of the existing condition of the building. The second scenario is the light turn ON when the area is dark or there is not adequate daylight in the dining area. The third scenario is the use of dimming lighting control technic by integrated daylight with artificial lighting. The artificial lighting responses to the affect from daylighting during the daytime by reducing its dimming which results in decreasing the load consumption. The illuminance will adjust accordingly to the area by the contributed from natural daylighting and it also adapt the same measure as scenario 2 which the light is turn ON when the area is dark.

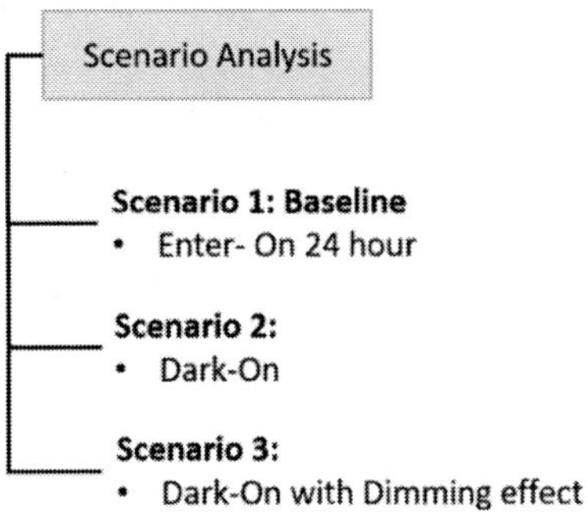

Figure 4. Scenario Analysis for Lighting simulation.

3.2. *Simulation procedure*

The architecture model electrical model was prepared inside BIM application software called Autodesk Revit 2018. The building components have to keep as simple as possible to avoid any error during the file export to EnergyPLUS simulation software. After the model is ready to export, the energy analytical model was necessarily needed before convert to gbXML format file. The gbXMLdata format was developed by Green Building Studio Inc. in 1999 for the purpose energy analysis. EnergyPLUS runs the simulation from the data transfer BIM model in the format of gbXML file [21]. In addition, RELUX plug in tool is also used to see the lighting performance of the existing scenario.

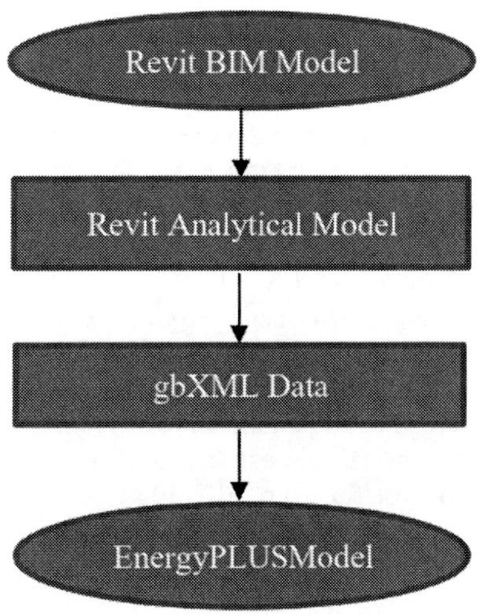

Figure 5. Data Diagram for Revit to EnergyPLUS transition.

4. Findings

4.1. Lighting profile

The lighting performance was simulation inside Autodesk Revit with the RELUX plug in tool. RELUX is software developed by RELUX Informatik AG in Switzerland with the collaboration from the various number of luminaire and lamp manufacturers provides an easy way to import the lighting manufacturers' database to the simulation [22]. This lighting simulation tool is simple and reliable for conducting lighting simulation and it is also simulated the visual building design elements such as materials, furnishing and colour [9]. [9] conducted a daylight simulation in RELUX for analysis of annual energy saving in an enclosed building in University of Nottingham, UK. With the help of RELUX, the illuminance distribution in the room is displayed.

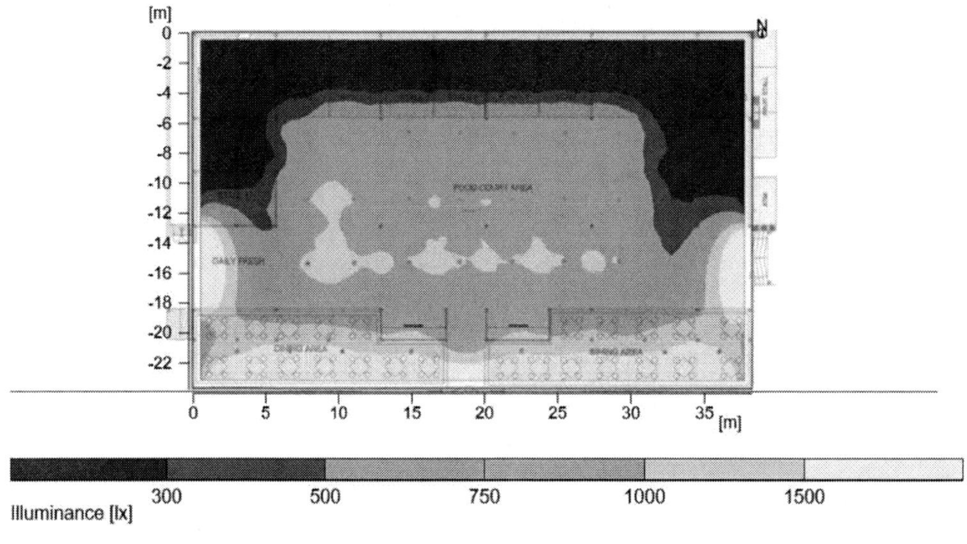

Figure 6. Artificial and daylight profile of the dining area of the Rest and Service Area.

The stall area was assumed there is no light install since there is a separate meter between the dining area and the shop lots. The electrical bill was assigned separately to the shop lots owner and the dining area lighting load was meant for PLUS Behad company which is one of the Malaysia highway concessionaires. The lighting bulb that use in the dining area is high bay lighting with the power consumption is 70W NIKKON Brand. As can be seen from the graph, some area in the dinning space is over illuminance which the colour is turn to light green. and the lighting distribution in the area is not uniform. According, Malaysia Standard MS 1525:2014 for non-residential building, the adequate illuminance for dining space is only 300 lux [23] . Due to the building is as open space building, harvesting the daylighting and decrease the demand for artificial lighting to illuminance the area is considered as an efficiency strategy for energy saving this building. Thus, replace the convention lighting bulb and the lighting control technology is compulsory to reduce the energy consumption for the area.

4.2. Lighting consumption from the Scenarios

The benchmark result of building energy performance from Autodesk Revit showed that BEI of the building is 379 kW/m^2/year which the cost for energy consumption per meter square per year is about 27.5 USD. The standard of the benchmark is followed ASHRAE 90.1 standard for energy efficiency index. ASHRAE 90.1 stands for American Society of Heating, Refrigeration and Air-Conditioning Engineers which is standard manual for rating building performance. Performance Cost Index (PCI) was modelled by ASHRAE to compare the prototype building regulated energy cost for a given building prototype, climate zone and edition [24]. This benchmark is use as future reference for the energy saving and energy efficiency measure.

$$BPF_{YearX} = \sum \frac{Prototype\ Building\ Regulated\ Energy\ Cost\ yearx}{Prototype\ Building\ Regulated\ Energy\ Cost\ 2004} / N_p \quad (2)\ [24]$$

Where:

Prototype Building Regulated Energy Cost year x = The portion of annual energy cost due to regulated energy use from PNNL prototype, climate zone and edition of Standard 90.1.

Prototype Building Regulated Energy Cost 2004 = The portion of annual energy cost due to regulated energy use from PNNL prototype buildings for a given building prototype, climate zone and the 2004 edition of Standard 90.1.

N_p = Number of prototype buildings of a particular building type

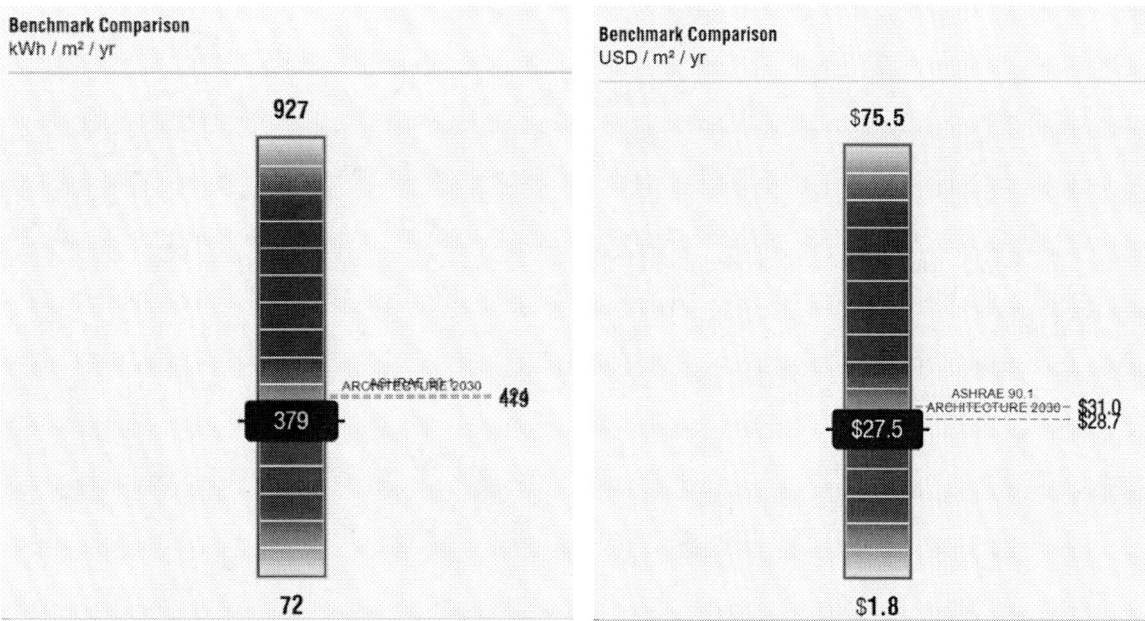

Figure 7. Building Energy Index Benchmark.

The result from EnergyPLUS software showed that the lighting consumption performed a major reduction from the lighting control measured adapted. From the first scenario, the lighting consumption of the existing condition of the building is about 25000 MWh per year. After using the control measure by switch off the light during the daylight period, a reduction of 17% of total energy load per year is achieved. The last scenario showed a major reduction in electrical consumption by save 12121.65 MWh which is about 49% compare to the baseline scenario. The dimming lighting control combine with timing sensor technic adapts in scenario 3 is more energy saving efficient compare to scenario 2 which the load of energy consumption is 7982.55 MWh difference. Harvesting the sunlight during the daytime for open space building is beneficial and efficient for building lighting refurbishment. From the simulation result, the energy saved from scenario 2 and 3 is significant. It shows an opportunity to achieved lighting energy saving by switching from conventional lighting systems to light with daylight responsive dimming functions.

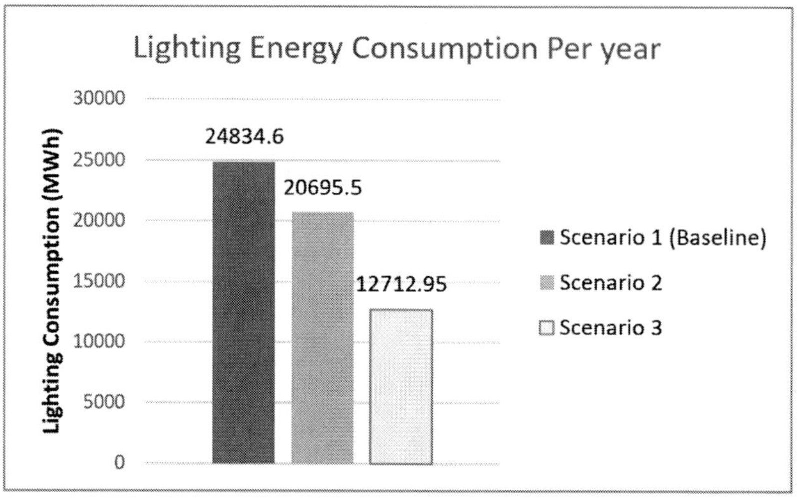

Figure 8. The lighting energy consumption from the simulation result.

As the reduction of the energy consumption is achieved, the carbon emission from fossil fuel combustion is also decreased. According to [25] the emission factor for electricity in Malaysia is 0.714 Kg per KWh based on the fuel mix input with coal and oil. Electricity consumption accounts in scope 2 (indirect emission sources) of Carbon Footprint emission which categorized under purchase energy is one of the most carbon emission from buildings (Sotos, 2014) . The reduction in the Carbon footprint emission play a major role to the climate change and sustainability. The graph shows a major reduction of 17% and 49% respectively for scenario 2 and scenario 3 compare to the baseline scenario.

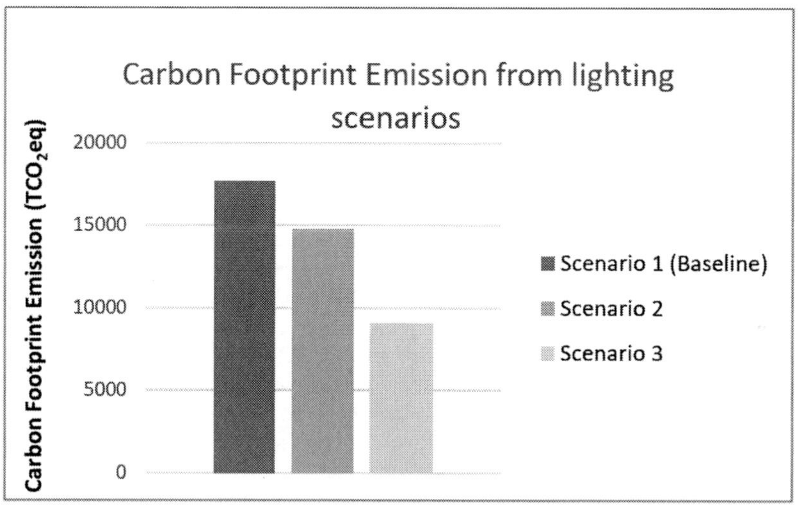

Figure 9. Carbon Emission reduction from the lighting consumption.

The cost of energy retrofitting is one of the most important factors in decision making for energy efficiency in building. Approximately 6 million Ringgit Malaysia was saved from the operational cost annually compare to the existing baseline scenario if the dimming lighting control measures is adapted at RSA Ayer Keroh Northbound Malaysia. The combine of two energy efficiency measures on lighting can save a massive amount of electrical bill annually. By only switch off the light during the daytime as adapt in scenario 2, about 2 million Ringgit Malaysia of operational cost is saved annually from the retrofitting measure.

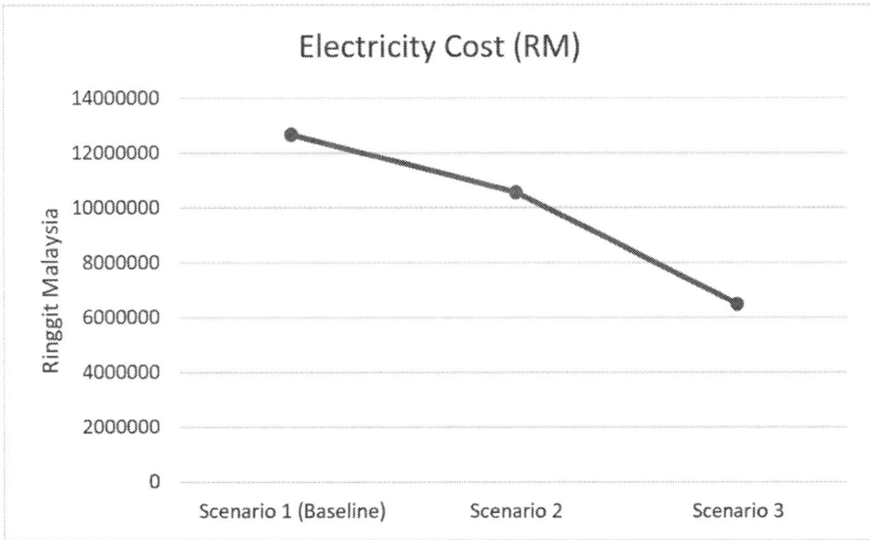

Figure 10. Electricity Cost from the lighting consumption.

5. Conclusion

Non-residential building such as Rest and Service Areas demand an efficiency refurbishment on the energy for enhance the performance of the building. Energy efficiency on the lighting control is a suitable measure for energy retrofitting in this building due to the overload of lighting consumption and high opportunity of energy saving from the simulation result. Natural lighting is one of the important sources of energy to harvest for open space building. It not only capable of improving the visual comfort but also can save the energy consumption. Deep energy retrofitting is can be achieved if there is a combine between more than two energy efficiency measures. From the results of this study, the opportunity to achieved lighting retrofitting is feasible and should be introduced for implement at Rest and Service Area Ayer Keroh Malaysia.

This study is beneficial to support the concept of nZEBs to promote the energy efficiency and Green building in Malaysia.

6. Acknowledgments

Our thanks to Highway contractor PLUS Behad Section S2 Malaysia for highly corporation with this study. AUN-Seed Net by JICA and University Technology Malaysia for supporting financial grant, Cost Centre No: R.J130000.7322.4B322, R.J130000.7322.4B3265, R.J130000.7322.4B332, R.J130000.4B334 & R.J130000.7322.4B331.

References

[1] Malaysia Energy Commission. 2017. Malaysia Energy Statistics Handbook 2017, *Malaysia Energy Commission*, pp. 1–86.

[2] Babatundea K.A., Saida F.F. and Nora N. G. M. 2018. Reducing Carbon Dioxide Emissions from Malaysian Power Sector: Current Issues and Future Directions, *Engineering Journal*, 1(6), pp. 59–69.

[3] Ascione F., Bianco N., De Masi R. F., Mauro G. M. and Vanoli G. P. 2017. Energy retrofit of educational buildings: Transient energy simulations, model calibration and multi-objective optimization towards nearly zero-energy performance, *Energy and Buildings*, 144, pp. 303–319.

[4] El-Darwish I. and Gomaa M. 2017. Retrofitting strategy for building envelopes to achieve energy efficiency', *Alexandria Engineering Journal*. Faculty of Engineering, Alexandria University, 56(4), pp. 579–589.

[5] Zhu P., Gilbride M., Yan D., Sun H. and Meek C. 2017. Lighting energy consumption in ultra-

low energy buildings: Using a simulation and measurement methodology to model occupant behavior and lighting controls', *Building Simulation*, 10(6), pp. 799–810.

[6] Li D. H. W., Lam J. C. and Wong S. L. 2002. Daylighting and its implications to overall thermal transfer value (OTTV) determinations', *Energy*, 27(11), pp. 991–1008.

[7] Zhu T., Li R. and Li C. 2017. The Analysis of Natural Lighting Simulation and Study on Energy Saving in Cigarette Factory', *Procedia Engineering*, 205, pp. 895–901.

[8] Chen Y., Liu J., Pei J., Cao X., Chen Q. and Jiang Y. 2014. Experimental and simulation study on the performance of daylighting in an industrial building and its energy saving potential', *Energy and Buildings*, 73, pp. 184–191.

[9] Yu X., Su Y. and Chen X. 2014. Application of RELUX simulation to investigate energy saving potential from daylighting in a new educational building in UK', *Energy and Buildings*, 74, pp. 191–202.

[10] Ainee F. 2012. Energy consumption and potential retrofitting of highway rest and service areas (RSAs), Master thesis, Universiti Teknologi Malaysia.

[11] Sartori I., Napolitano A. and Voss K. 2012. Net zero energy buildings: A consistent definition framework', *Energy and Buildings*, 48, pp. 220–232.

[12] Congedo P. M., Baglivo C., D'Agostino D. and Zacà I. 2015. Cost-optimal design for nearly zero energy office buildings located in warm climates, *Energy*, 91(244), pp. 967–982.

[13] Ferrara M., Fabrizio E., Virgone J. and Filippi M. 2015. Appraising the effect of the primary systems on the cost optimal design of nZEB: A case study in two different climates, *Energy Procedia*, 78, pp. 2028–2033.

[14] Neardey M., Aminudin E., Zin R. M., Zakaria R., Hafifi C. M. F. and Ramli M. R. 2019. Factor analysis on variables effecting energy consumption for retrofitting initiative at highway rest and service areas Malaysia, pp. 63–67.

[15] Ghaffarianhoseini, Amirhosein, Dahlan, N. D., Berardi, U., Ghaffarianhoseini, Ali, Makaremi, N. and Ghaffarianhoseini M. 2013. Sustainable energy performances of green buildings: A review of current theories, implementations and challenges, *Renewable and Sustainable Energy Reviews*, 25, pp. 1–17.

[16] Hirsch A. H. 2014. Sustainable Rest Areas Design and Operations, *Icsi 2014*, (October), pp. 819–830.

[17] Kreminski R., Hirsch A. and Boand J. 2011. Assessment of Colorado Department of Transportation Rest Areas for Sustainability Improvements and Highway Corridors and Facilities for Alternative Energy Source Use, (March).

[18] Mukherjee S., Birru D., Cavalcanti D., Shen E., Patel M., Wen Y.-J. and Das S. 2010. Closed Loop Integrated Lighting and Daylighting Control for Low Energy Buildings, *2010 ACEEE Summer Study on Energy Efficiency in Buildings*, pp. 9–252.

[19] Pandharipande A. and Caicedo D. 2015. Smart indoor lighting systems with luminaire-based sensing: A review of lighting control approaches, *Energy and Buildings*, 104, pp. 369–377.

[20] Tahir M. Z., Nawi M. N. M. and Rajemi M. F. 2015. Building energy index: A case study of three government office buildings in Malaysia, *Advanced Science Letters*, 21(6), pp. 1798–1801.

[21] Ivanova I., Kiesel K. and Mahdavi A. 2015. BIM-generated data models for EnergyPlus: A comparison of gbXML and IFC formats', *Building Simulation Applications*, 2015-February, pp. 407–414.

[22] Information about Relux Available at.[http://www.relux.biz]

[23] Malaysia and Standard (2012) 'Malaysian Standard', *Chrysanthemum Standard*, pp. 1–3.

[24] Pacific Northwest National Laboratory (2016) 'ANSI/ASHRAE/IES Standard 90.1-2010 Performance Rating Method Reference Manual', *Pnnl-255130*, (May).

[25] Information about SEDA Available at [http://www.seda.gov.my/]

IOP Conf. Series: Materials Science and Engineering **943** (2020) 012063 doi:10.1088/1757-899X/943/1/012063

Transformation from 2D structural drawing to building information model: Perspectives from a small-scaled company

Y H Lee[1*], L S Thien[2], Y Y Lee[3], S Y Wong[1,4], C S Tan[5] and C S Chai[6]

[1]Department of Civil and Construction Engineering, Faculty of Engineering and Science, Curtin University Malaysia, CDT 250, 98009 Miri, Sarawak, Malaysia.
[2]Adda Engineering Consultant, Bangunan USC, No. 683, Lorong Song 1A, Off Jalan Song, 93350 Kuching, Sarawak, Malaysia.
[3]Department of Civil Engineering, Faculty of Engineering, Universiti Malaysia Sarawak, Jalan Datuk Mohammad Musa, 94300 Kota Samarahan, Sarawak, Malaysia.
[4]School of Built Environment, University College of Technology Sarawak, No. 1, Jalan Universiti, 96000 Sibu, Sarawak, Malaysia.
[5]School of Civil Engineering, Faculty of Engineering, Universiti Teknologi Malaysia, 81310 Johor Bahru, Johor, Malaysia.
[6]Department of Master of Construction Management, School of Engineering, Faculty of Engineering, Computing and Science, Swinburne University of Technology Sarawak Campus, Jalan Simpang Tiga, 93350 Kuching, Sarawak, Malaysia.

Email: yeong.huei@curtin.edu.my

Abstract. Building information modelling (BIM) is one of the revolutions in construction industry. BIM is one of the long-waiting solutions for construction industry in order to solve the arisen quality and effectiveness problems. Many researchers have proved the benefits gained from BIM. In this paper, the structural package available in BIM platform is summarized and its maturity is discussed. The BIM projects in Malaysia are listed and it showed a low local BIM implementation. In the perspectives of engineers, migrating from 2D drawing to building information model is discussed with the faced problems and challenges. The technical supports such as internet supports, cloud system and etc. was lacked for small-scaled companies; and self-transformation plan is not available which is believed to minimize the lost during transformation. From structural engineers' perspective, a better visualization with building information models is critical to address these major problems occurred throughout the migration. However, more time are expected to be consumed in producing the models as it was found that there is a low level of model sharing between engineers and architects. Although the BIM implementation in Malaysia is still in a low level, it is advised that industry players and government should work together to migrating from traditional method to BIM environment.

1. Introduction

Building information modelling (BIM) comes across to minimize the problems which may potentially arise during construction, and reduce the cost through virtual construction during planning stage in architecture, engineering and construction (AEC) industry. As the virtual model is completely developed, this building information model has accurate geometries and applicable input data for design, quantity, fabrication and construction activities in order to make it tangibly exists [1].

BIM has been discussed actively and excitedly in last 10 years where AEC industry was seeking for a long period to solve the quality problem while increase the productivity [2]. It is also noticeable that BIM can alter the workflow and project delivery processes [3]. Moreover, the model also can be transferred after construction for maintenance purposes. From the study [4], BIM projects were able to reduce the project time by 7% compared to traditional method, achieve a save of 10% in contract value due to early found of clash activities before construction, and obtain up to 80% time savings in generating a cost estimation.

Questionnaire surveys were conducted in United State in order to understand the BIM effect in AEC industry [5]. A sample size of 302 was collected and it was found that architects were the BIM heaviest users. A 66% of the respondents believed BIM helped them to increase the chances for winning a project and 79% stated that BIM improved project outcomes with decreased the problems raised during construction. The value in using BIM and the factors contributing to success have been determined [6]. They believed the benefits gained from BIM but some of AEC players still retain traditional drawing method.

Malaysia, one of the developing countries that affected by the BIM wave as the AEC industry is active throughout the development process. BIM has been promoted in Malaysia through myBIM center of Construction Industry Development Board (CIDB) with provided training and incentives. There were several projects that involved BIM implementation, namely Multipurpose Hall of Universiti Tun Hussein Onn Malaysia, National Cancer Institute of Malaysia, EduCity Sports Complex Nusajaya, and Ancasa Hotel Pekan [7].

BIM has been implemented in Malaysia AEC industry. The effect on the changes is needed to be investigated especially from the perspective of structural designers, one of the AEC players. This paper addresses a study of the change to BIM environment which affecting the routine works of designers. BIM projects in Malaysia have been listed and BIM implementation has been discussed in the perspective of engineers.

2. Structural Analysis

Structural design is involved at the tendering process where the building safety should be obtained for occupants. Structural analysis is a main design stage to adequately transfer the building loads to foundations without significant changes of physical shape of the building. Stability and load transfer are concerned in design of structural members.

2.1. BIM platform

There are several platforms that allow BIM users to work with, namely Autodesk (Revit, Autodesk BIM 360, Navisworks), Bentley (AECOsim Building Designer), Graphisoft (Archicad), Tekla (Tekla BIMsight), etc. There are summarized in Table 1.

2.2. Available structural package

A good amount of structural package is available in the market to assist designers with adequate structural analysis. Table 2 summarizes the structural analysis packages that avaible in the market and their file setting.

Table 1: Available BIM platform

Platform	Package	Benefits

		Multidiscipline solution for AEC	Open BIM	Own library	Free tool	Design review, comments, marks up	Visualization of modifications	Clash detection	Building performance analysis	Mobile application
Autodesk	Revit	/	/	/				/	/	
	Navisworks		/			/		/		
	Autodesk BIM 360		/			/	/			BIM 360 field
Tekla	Tekla BIMsight		/	/	/					BIMsight note
Graphisoft	ArchiCAD		/							BIMx
Bentley	AECOsim Building Designer	/	/						/	

2.3. Maturity of design packages

Other than structural analysis, software developers are now working towards the feature of open-code file format of Industry Foundation Classes (IFC) created by buildingSMART which can be viewed in most of the platforms or software. IFC has features of 3D elements, ability to store data, filterable, reference model representation, global access by software developers, enable fast code implementation and able to localize national standards. The available structural packages listed in Table 2 are all able to import or export IFC files. Structural packages are unable to be excluded from the integration in BIM community as the model needs the inputs from structural analysis.

The accuracy of structural analysis in obtaining a safe and reliable building is the main characteristic that should be achieved before integration into BIM environment. Elimination of some structural packages from AEC industry may also happen if failure occurred in engaging with BIM atmosphere. Therefore, both accuracy and engagement with BIM should be obtained to survive in current AEC industry.

3. Review in Malaysia

There are completed or on-going projects that implement BIM which are summarized in the following contents. Other than review published projects, reanalysis of the completed project with several structural package also has been conducted to know the capability of these package in BIM environment. Malaysia government is enforcing to use BIM for those local authorities with city status by 2021 where national BIM e-submission is required [8].

3.1. Previous investigation

Review on previous studies in Malaysia is conducted and summarized accordingly. The level of implementation is discussed and the BIM projects are documented in following contents.

Table 2: Structural analysis package that available in market

Structural analysis package				**Intermediate / convertor**	**Landing environment / files / platform**
		File type			
Software	**Platform**	**Original**	**Available**		
Orion	Tekla/CSC	Orion	.cxl .dxf	Orion-Revit Integrator (add-in of Revit)	Revit
ProtaStructure	Prota	Prota	-	ProtaBIM	Revit, Archicad, Tekla structure, IFC
Tekla Structure	Tekla	Tekla	-	IFC insert IFC export	IFC2X3 IFC4
ROBOT	Autodesk	ROBOT	.RTD .STR	ROBOT, Add-in, integration	Revit
STAAD (Staad Pro, staad pro advanced, RCDC, Staad foundation, Microstran)	Bentley	STAAD	.STD	Structural Synchronizer	IFC ISM
RAM (RAM Structural system, RAM Connection, RAM Elements, RAM Concept, Structural component Visa)	Bentley	RAM	.rss	Structural Synchronizer	IFC ISM
SCIA	Nemekschek	SCIA	.esa	BIM Toolbox	IFC2X3
RFEM / RSTAB	Dlubal	-	.asf	Results Explorer (Structural Analysis Toolkit of Autodesk)	Revit IFC
FEM-Design	StrucSoft	-	-	StruXML Revit add-in Tekla StruXML	Revit, Tekla structure

3.1.1 Level of BIM implementation

Level of BIM implementation has been assessed with literature review and interview method [9] where it was found difficulties in implementing process due to the lack of national standard to comply with. Many local companies developed their own strategies to adopt BIM technology. From traditional method to BIM, small or medium scale projects may come across as the best practice in reducing the risk in migration. Some factors of the low BIM implementation and suggestions to overcome have been reported through a questionnaire survey [10]. Construction players also need to perform their roles in promoting BIM, rather than government work alone in the implementation process.

Recent publication on level of BIM implementation has been established in Melaka, Malaysia [11]. There was a number of 46 questionnaires were used for analysis for a population of 90 construction companies in Melaka. The lack of awareness was found as the main factor in delaying the BIM implementation. Hence, organizations were advised to practice BIM and initiatively develop own transformation plan to overcome the faced challenges. Hence, it could conclude that Malaysia is in low level of BIM implementation [12]. In order to successfully implement BIM, a framework has been proposed, as shown in Table 3, initiated by Jung and Joo [13] and reported by Bui et.al. [14].

Table 3: The BIM implementation framework [13]

Technical (T)	1. Data property 2. Relation 3. Standards 4. Utilization
Perspective (P)	1. Industry 2. Organization 3. Project
Construction business function (C)	1. R&D 2. General admin 3. Finance 4. HR management 5. Safety management 6. Quality management 7. Cost control 8. Contracting 9. Materials management 10. Scheduling 11. Estimating 12. Design 13. Sales 14. Planning

3.1.2 Previous building project - National Cancer Institute (NCI)

The first Malaysian project that implemented BIM was the building of National Cancer Institute [15]. The project details have been summarized in Table 4. NCI had been completed two weeks ahead from the expected completion date. There were 1800 clashes found and able to solve before construction. In this project, they utilized Revit (architecture, structure, M&E) and Navisworks.

3.1.3 Previous infrastructure project Pan Borneo Highway Sarawak (PBHS)

Pan Borneo Highway Sarawak (PBHS) is one of the BIM initiation project in infrastructure which can be referred as a national benchmark for infrastructure BIM [16]. The details of the project are tabulated in Table 5.

3.2. On-going BIM projects

There are some identified on-going Malaysian projects that utilize BIM as a working platform to integrate all relevant involved parties. There are several pilot projects that involved with BIM implementation for building construction, namely Healthcare Center Type 5 at Sri Jaya Maran, Pahang; Administration Complex Project of Suruhanjaya Pencegah Rasuah Malaysia (SPRM) at Shah Alam, Selangor; primary school of Meru Raya Ipoh, Perak; and primary school of Tanjung Minyak 2, Melaka Tengah [7,17].

For infrastructure, MMC-Gamuda is working as project delivery partner for KVMRT Sungai Buloh-Serdang-Putrajaya (SSP) line. The alignment is 52.2 km in length, 38.7 km elevated tracks and 24 elevated stations. It is divided into 19 work package contractors. This project is aimed to be completed in 2022. Phase 1 will operate as early in July 2021 which is from Sungai Buloh station to Kampung Batu Station. This project is adopting BIM level 2. Moreover, Phase 2 of PBHS on maintenance and upgrade will also use the BIM approach to conduct the project.

Furthermore, several companies are actively applying BIM in their current projects, such as UEM Sunrise, MMC-Gamuda, IJM Construction, Sunway Construction, etc. All high-rise building projects are now adopting BIM since 2017 at UEM Sunrise. MMC-Gamuda is applying BIM into their MRT project. IJM Construction won the best project award for major buildings above RM100 million during Malaysian Construction Industry Excellence Awards 2019 with Equatorial Plaza using BIM platform for integration. Sunway Construction is one of the first to introduce virtual design and construction which is equivalent to BIM platform.

Table 4: Project details for NCI [15]

Project	National Cancer Institute (NCI)
Location	Putrajaya (2.926907,101.6739021)
Scale	70,000 m^2
Duration	3 years
Client	Ministry of Health Malaysia and Public Works Department
Value	RM 690,000,000
Architect	Perunding Alam Bina
Contractor	Kiara Teratai – IJM Construction Joint Venture
BIM consultant	Precision Design Solutions Sdn Bhd
Energy consultant	IEN
Project delivery	Design and build
Description	A total of three building blocks where a hospital block comprising of one level sub-basement for radiotherapy treatment, one level of car park, four levels of oncology diagnostic and treatment services, as well as two blocks of seven levels wards.

Table 5: Project details for PBHS [16] (Akob et.al., 2019)

Project	Pan Borneo Highway Sarawak (PBHS)
Location	Sarawak
Scale	1060 km
Duration	2016 – 2021 (highway) phase 1
Client	State government and Public Works Department
Value	RM 16.5 billion
BIM resources	Using level 2 of BIM and provided up to level of development of 500. This project consists of five BIM modellers and one coordinator for each package, the total number of dedicated personnel is 55 BIM modellers, 11 BIM coordinators, 8 BIM specialists, 4 BIM lead family, 1 BIM lead coordinator and 1 BIM director.
Description	Construction of 786 km of a new 4-lane highway in 11 works packages.

3.3. *From 2D to BIM*

Shifting from traditional method to BIM affects positively as stated in previous research. The significant output should address the reduction of request of information and potential clashes during construction period. The transformation plan is essential to minimize the loss during migration from traditional method. Self-transformation plan by consultant firms may consume some budget before gaining profits. These budgets include the expenses of staff training program, purchase of BIM tools/software, upgrade internet system, increase cloud spaces, etc. The cost always becomes predominant in BIM implementation for small-scaled companies [18].

From the point of technical support, BIM technology requires more on the internet system, cloud space, live communication environment, etc. These requirements seem like to be a burden for small scaled construction companies in developing countries [14] which they need to upgrade their system to support this technology. In turn, it will reduce the gained profit. These companies usually outsource to others and this would not help in sustaining their business in BIM world. In Sarawak, one of the states in Malaysia, the developments may focus in rural areas where there is a lack of proper communication system, like internet connection that used for project integration. The up-to-date BIM models may not delivery on time or not able to be applied in this context.

However, in term of structural analysis, BIM provides a clearer visualization in 3D version and may reach 4D (time), 5D (cost), 6D (sustainability) or 7D (facility management). Structural engineers may need more detail in analysis where 3D drawings are available for checking before construction stage. The small and unnoticeable areas may need designers' attention in the design stage. Therefore, it is believed that structural engineers may take more time in the design stage where it can be catch up in later construction time. As the structural design needs more time, workload may subsequently increase and may need to hire more engineers at early stage of transformation. Draughtsperson, who assists engineers, may also need to upgrade their drawing to 3D version and training courses are needed for the in-house promotion to BIM modellers or restructure the organization. This also needed to be considered in the company self-transformation plan.

Often, engineers are wasting their time in redrawing the proposed structure as architects are fail to share their 3D models. Therefore, it requires more time in analysis and provides 3D drawings. From previous research [19], the successful of information sharing is relied on the organisational behaviour supported by the collaborative constructs. The 3D drawings that produced by engineers may not compatible with architects as minor geometric mistakes may occur due to human errors.

4. Conclusions

This paper summarized the transformation from traditional method to BIM platform. The available structural packages in BIM environment have been studied and the completed and on-going projects in Malaysia were reported. There is a proposed framework for this migration where technical, perspective and business functions are the design parameters. The level of BIM implementation in Malaysia is still in a low level that both government and industry players should work together to make the transformation successful.

It is advisable to have company own migration plan in order to minimize the risks in the transformation process where in-house training and system upgrading should be performed to overcome the BIM wave. The backup solution should be applied for BIM non-applicable zone, such as rural area where technical supports are not parallel with the technology.

BIM serves as a platform to integrate all parties in construction which has been proven beneficial to AEC industry. It might be a challenge for small-scaled companies of developing countries in the transformation process from traditional 2D model to building information model. However, it should be positively solved for building a quality AEC industry.

Acknowledgement

Authors would like to acknowledge the supports from ADDA consultant firm in Kuching in completing the initial case study and the supports from Curtin University Malaysia.

References

[1] Eastman C., Teicholz P., Sack R. and Liston K. 2008. BIM handbook: A guide to building information modeling for owners, managers, designers, engineers and contractors, Wiley, NewYork.

[2] Azhar S., Nadeem A., Mok J.Y.N. and Leung B.H.Y. 2008. Building information modeling (BIM): A new paradigm for visual interactive modeling and simulation for construction projects. Proceedings of the First International Conference on Construction in Developing Countries, Karachi, Pakistan, 435-446.

[3] Hardin B and McCool D. 2015. BIM and construction management: Proven tools, methods and Workflows, John Wiley & Sons, Inc, Indianapolis, Indiana.

[4] Azhar S., Hein M. and Sketo B. 2008. Building information modelling (BIM): Benefits, risks and challenges. Proceedings of the 44th ASC Annual Conference, Auburn, Alabama, April 2-5, 200.

[5] McGraw-Hill Construction. 2008. Building information modelling: Transforming design and construction to achieve greater industry productivity, New York.

[6] Gilligan B. and Kunz J. 2007. VDC use in 2007: Significant value, dramatic growth, and apparent business opportunity. CIFE Technical Report #TR171, Stanford University, California.

[7] Latiffi A.A., Mohd S., Kasim N. and Fathi M.S. 2013. Building information modelling (BIM) application in Malaysian construction industry. International Journal of Construction Engineering and Management, 2(4A):1-6.

[8] Bernama, 2019. All city status local authorities to use BIM eSubmission by 2021. Online news, retrieved on 8 January 2020, from http://www.bernama.com/en/news.php?id=1707184

[9] Zahrizan Z., Ali N.M., Haron A.T., Marshall-Ponting A. and Hamid Z.A. 2013. Exploring the adoption of building information modelling (BIM) in the Malaysian construction industry: A qualitative approach. International Journal of Research in Engineering and Technology, 2(8):384-395.

[10] Zahrizan Z., Ali N.M., Haron A.T., Marshall-Ponting A. and Hamid Z.A. 2014. Exploring the barriers and driving factors in implementing building information modelling (BIM) in the Malaysian construction industry: A preliminary study. The Institution of Engineers, Malaysia, 75(1):1-10.

[11] Al-Ashmori Y.Y., Othman I.B., Mohamad H.B., Rahmawati Y. and Napiah M. 2019. Establishing the level of BIM implementation – A case study in Melaka, Malaysia. IOP Conference Series: Materials Science and Engineering 601: 012024.

[12] Musa S., Marshall-Ponting A., Nifa F.A.A. and Shahron S.A. 2018. Building information modelling (BIM) in Malaysian construction industry: Benefits and future challenges. AIP Conference Proceedings 2016, 020105.

[13] Jung Y. and Joo M. 2011. Building information modelling (BIM) framework for practical implementation. Automation in Construction, 20(2):126-133.

[14] Bui N., Merschbrock C. and Munkvold B.E. 2016. A review of building information modelling for construction in developing countries. Procedia Engineering, 164:487-494.

[15] Latiffi A.A., Mohd S., Kasim N. and Brahim J. 2015. Application of building information modeling (BIM) in the Malaysian construction industry: A story of the first government project. Applied Mechanics and Materials, 733-774: 943-948.

[16] Akob Z., Hipni M.Z.A. and Rosly M.R. 2019. Leveraging on building information modelling (BIM) for infrastructure project: Pan Borneo Highway Sarawak phase 1. IOP Conference Series: Materials Science and Engineering, 512:012060.

[17] Latiffi A.A., Brahim J., Mohd S. and Fathi M.S. 2014. The Malaysian government's initiative in using building information modelling (BIM) in construction projects. Proceedings of International Structural Engineering and Construction: Sustainable Solutions in Structural

Engineering and Construction, Eds Chantawarangul K., Suanpaga W., Yazdani S., Vimonsatit V. and Singh A., ISEC Press, USA, 767-772.

[18] Ismail N.A.A., Chiozzi M. and Drogemuller R. 2017. An overview of BIM uptake in Asian developing countries. AIP Conference Proceedings 1903, 080008.

[19] Ibrahim C.K.I.C., Sabri N.A.M., Belayutham S. and Mahamadu A. 2018. Exploring behavioural factors for information sharing in BIM projects in the Malaysian construction industry. Built Environment Project and Asset Management, 9(1):15-28.

IOP Conf. Series: Materials Science and Engineering **943** (2020) 012064 doi:10.1088/1757-899X/943/1/012064

Barriers of implementing sustainable housing in Malaysian residential property sector

S Y Wong[1*], J H Loh[1], Y H Lee[1]

[1] Department of Civil and Construction Engineering, Curtin University Malaysia, CDT250, 98009 Miri, Sarawak, Malaysia.

Email: shiyee.wong@curtin.edu.my

Abstract. Designing and constructing sustainable housing by changing its design, construction and operation could help in reducing resource consumption and mitigate climate change. However, it is found that there are barriers impeding the implementation of sustainable housing. This research aims to determine such barriers in Malaysian residential property sector. A questionnaire was designed and distributed to the key stakeholders (i.e. engineers, real estate agents, contractors and home buyers) in Malaysia. This questionnaire asked the respondents to rate the relative importance of 21 barriers which were categorized into four key categories (i.e. design and technical barrier, economic barrier, institutional barrier and socio-culture barrier). The results showed that economic barrier is the most critical category among the four categories of barriers. Out of the 21 barriers, the top critical barrier is high capital/investment cost. This research identified the importance of different barriers based on the perceptions of and experiences of the respondents. While most studies focused on the professional stakeholders or home buyers, there are only few studies dedicated to the combination of the perceptions for both professional stakeholders and homebuyers in Malaysia. This research could contribute to both the knowledge and practices of sustainable housing implementation in Malaysia.

1. Introduction

Triple bottom lines requirements (i.e. social, economic and environmental aspects) are essential for sustainable housing, in the design, construction and operation stage [1, 2], in the early days. Recently, some researchers explored that sustainable housing could be defined as residential property which is ecologically suitable, flexible, long-lasting, resource efficient and can ensure occupants' comfort as well as health and safety [3-5].

Previous studies had proved on the advantages of sustainable housing in terms of the occupants comfort and energy efficiency [6]. A study supported this notion by stating that sustainable housing could have obvious outcome to the house owners by enhancing the environment and air quality of the house [7]. Despite the evidences on its advantages, houses with sustainability features are not widely adopted in Malaysia [8].

The adoption of sustainable housing in Australia, Germany, and Sweden are evidenced with the support from government and the positive return on investment [6, 9, 10]. However, there is little-to-no evidence on the adoption of sustainable housing in Malaysia. The Green Building Index (GBI), a rating tool utilized in Malaysia are mainly focusing on the commercial building instead of the residential property [11]. Therefore, it seems to be a research gap on the implementation of sustainable housing in Malaysian residential property sector.

The aim of this paper is to determine the barriers of implementing sustainable housing in Malaysian residential property sector. This research focuses on both supply chain agents and demand-side stakeholders in the property sector, to provide a complete picture on the opinions of both supply and demand stakeholders. It is worth noting that the supply chain agents recruited in this research only consist of engineers, real estate agents and contractors. These three groups of supply chain agents were recruited as engineers involved in the design of the houses, contractors involved in the construction stages, and real estate agents involved in selling houses to the home buyers.

2. Identification of potential barriers

Multiple studies had examined on the barriers that affect the implementation of sustainable housing. For instance, [12-15] discussed on the barriers of sustainable construction implementation through the interview and/or questionnaire survey methods with the supply chain agents, such as contractors, consultants, architects, quantity surveyors etc. [16] provided a thorough literature review on the key barriers of green building adoptions. However, most of the studies were carried outside Malaysia (such as Ghana and Finland) and did not combine the perspectives of supply chain agents and demand side stakeholder (i.e. home buyers). The barriers extracted from the literature review were summarized in Table 1. Twenty-one barriers were identified from the main literatures.

Table 1. List of identified barriers from literature review

Identified barriers	**Literature**							
	1	2	3	4	5	6	7	8
Lack of professional expertise in sustainability		√	√		√	√		
Lack of knowledge on sustainability technologies	√		√		√	√		√
Lack of experts in operating sustainable related technologies			√	√	√		√	
Complexity in pre-project planning								√
Lack of green building measurement rating tools/program		√						√
High capital/investment cost		√	√	√	√	√	√	√
Lack of proper property valuation system on sustainability								√
Market competition hold back sustainability adoption		√						
Lack of financial incentives	√			√	√			
Ineffective mortgage system/financial mechanism				√				√
Long period of return of investment		√	√		√			√
Lack of government incentives	√	√		√	√	√	√	√
Resistance to change	√							
Slow process in certificating and policy approval for construction sustainable housing								√
Lack of institutional cooperative consciousness collaborations				√		√	√	√
Lack of education and training	√			√				√
Lack of public awareness and interest on sustainability	√		√		√	√	√	√
Lack of understanding on the sustainability trend in the current business						√		√
Challenges in changing exiting customers' mindsets								√
Lack of end users demand	√	√			√	√	√	
Lacked of knowledge on benefits of sustainable practices	√		√					

Note: 1 = [12]; 2 = [13]; 3 = [14]; 4 = [15]; 5 = [8]; 6 = [17]; 7 = [18]; 8 = [16]

IOP Conf. Series: Materials Science and Engineering **943** (2020) 012064 doi:10.1088/1757-899X/943/1/012064

3. Research methodology

3.1. Development of questionnaire

The targeted populations were supply chain agents (i.e. engineers, real estate agents, and contractors) and demand-side stakeholder (i.e. home buyers) who play important roles in implementing sustainable housing in Malaysia. The data collection method was questionnaire survey. Due to the limitation of time, this quantitative research method was adopted to ensure that the data could be collected and analyzed completely.

The main questions in the questionnaire was requiring the respondents to rate the relative importance of 21 barriers which were categorized into four key categories (i.e. design and technical barrier, economic barrier, institutional barrier and socio-culture barrier). These questions were presented in the Likert scale, with 1 being not impacting at all to 5 being strongly impacting. Table 2 showed the classification and the abbreviations allocated for the 21 barriers.

Table 2. Classification and abbreviations of the key barriers

Code	Barriers
DT	**Design and technical barriers**
DT1	Lack of professional expertise in sustainability
DT2	Lack of knowledge on sustainability technologies
DT3	Lack of experts in operating sustainable related technologies
DT4	Complexity in pre-project planning
DT5	Lack of green building measurement rating tools/program
E	**Economic barriers**
E1	High capital/investment cost
E2	Lack of proper property valuation system on sustainability
E3	Market competition hold back sustainability adoption
E4	Lack of financial incentives
E5	Ineffective mortgage system/financial mechanism
E6	Long period of return of investment
I	**Institutional barriers**
I1	Lack of government incentives
I2	Resistance to change
I3	Slow process in certificating and policy approval for construction sustainable housing
I4	Lack of institutional cooperative consciousness collaborations
I5	Lack of education and training
I	**Socio-cultural barriers**
SC1	Lack of public awareness and interest on sustainability
SC2	Lack of understanding on the sustainability trend in the current business
SC3	Challenges in changing exiting customers' mindsets
SC4	Lack of end users demand
SC5	Lacked of knowledge on benefits of sustainable practices

3.2. Data collection method

The data collection was conducted from 1st July 2019 to 1st September 2019, and approached the targeted respondents through face-to-face visitation or email. The targeted participants were real estate agents, contractors, engineers and home buyers. These supply chain agents were identified through Construction Industry Development Board (CIDB), and the Google search. The home buyers were identified through the author's connections as well as the real estate companies.

The questionnaire was distributed to 200 targeted participants through email (180 participants) and face-to-face visitation (20 participants). There were 28 valid responses (i.e. 14% response rate) received and used in the data analysis.

4. Results and discussions

4.1. Demographic details

Fifty per cent of the respondents (14 respondents) were engineers, 39% (11 respondents) were home buyers, 7% (two respondents) were contractors and 4% (one respondent) was real estate agents. This seems to indicate that majority of the respondents were engineers and home buyers, with minority (11%) of the respondents were contractors and real estate agents. Hence, this paper had classified the respondents into three groups (i.e. Gl – Engineers, G2 – home buyers, G3 –real estate agents and contractors).

For experience in the construction industry, 82% of the respondents (23 respondents) have less than 5 years of experience and 14% (four respondents) with experience of 6-10 years. There was one respondent who worked as contractors (i.e. G3) had the working experience of more than 20 years. This seems to support the results obtained in G3 as one of the respondents is quite experienced.

4.2. Key categories of the barriers in sustainable housing implementation

The respondents were asked to rank the importance of four key categories barriers from the scale 1 to 4, with 1 being least important to 4 being most important (see Table 3).

The result showed that economic barrier (E) was ranked as the most important barrier among the four key categories of barriers, with the mean value ranging between 3.10 and 3.67 among different group of respondents. This seems to indicate that the stakeholders may presume that the cost of sustainable housing is higher compared to the other normal housing, and/or it may take longer period of time for the return of investment. According to [19], the real estate market perceived a premium price be achieved for the building with high sustainability features. However, this discussion was related to the commercial building. Research carried out by [9, 10, 20] showed that sustainability features had a positive impact on the selling and/or rental price of the residential properties, as well as comparatively shorter listing period in the real estate market. Hence, the economic barrier could be an arguable barrier if more evidence could prove that the cost of investing in sustainable housing is reasonable and the return on investment (in terms of selling/renting price) is positive.

The ranking of design and technical barrier (DT), socio-cultural barrier (SC) and institutional barrier (I) were ranked as second, third and fourth important barriers respectively by engineers (G1), home buyers (G2) and overall group respectively. This seems to imply that the respondents had the similar opinions on the level of importance on all the barriers. Real estate agents and contracts (G3), on the other hand, had the same opinions on all these three barriers, with same mean value and the standard deviation was zero. This may due to the low number of respondents in this group (G3), which leads to the similar rankings on these three barriers.

Table 3. Mean, standard deviation and ranking of key categories of barriers in sustainable housing implementation

Catego ries	Overall			G1			G2			G3		
	Mean	SD	Rank	Mean	SD	Rank	Mean	SD	Rank	Mean	SD	Rank
DT	3.04	0.53	2	3.08	0.64	2	3.00	0.47	2	3.00	0.00	2
E	3.36	0.70	1	3.50	0.67	1	3.10	0.74	1	3.67	0.58	1
I	2.70	0.95	4	2.69	0.94	4	2.56	1.13	4	3.00	0.00	2
SC	2.78	0.64	3	2.86	0.54	3	2.60	0.84	3	3.00	0.00	2

Notes: G1 – Engineers; G2 – Home buyers; G3 – Real estate agents and contractors
DT – Design and technical barrier; E – Economic barrier; I – Institutional barrier; SC – Socio-cultural barrier

4.3. Barriers of sustainable housing implementation

There were different key barriers under the four categories of barriers. The results of the relative importance of the 21 key barriers were shown in Table 4. High capital/investment cost (E1) was ranked as the most important barriers by engineers (G1) and home buyers (G2), which leads to the highest ranking in the overall group. This seems to be tallied with the findings in Table 3 that economic barrier is the most critical category of barrier. This finding was supported by the literature review finding by [13-16].

The second most important barrier was lack of professional expertise in sustainability (DT1), under the category of design and technical barrier (mean value = 3.93). There is possibility that the respondents' believed they intend in sustainability but there is a lack of expertise that could construct and/or operate sustainable housing in Malaysia. This 'vicious circle of blame' could hinder the adoption of sustainable housing [21]. The home buyers (G2), and real estate agents and contractors (G3) ranked this barrier as the third and second most important barrier respectively. The engineers (G1), interestingly, ranked this barrier in the place of tenth (mean value = 3.77). This seems to indicate that there is a discrepancy of opinions on different groups of stakeholders on the barriers that hinder the implementation of sustainable housing in Malaysia.

The barriers listed under the categories of institutional and socio-cultural barriers seems to be ranked as least important barriers compared to the other two categories. This may indicate that the respondents were not very concern on the barriers listed under institutional and socio-cultural barriers. However, the mean value of the least important barrier (SC2) was 3.30, which seems to indicate that the respondents did considered the importance of all barriers.

Table 4. Mean, standard deviation and ranking of barriers in sustainable housing implementation

	Overall			G1			G2			G3		
Barriers	Mean	SD	Rank	Mean	SD	Rank	Mean	SD	Rank	Mean	SD	Rank
DT1	3.93	1.00	2	3.77	1.01	10	4.00	1.00	3	4.33	1.16	2
DT2	3.89	1.17	4	3.93	1.21	4	3.91	1.30	5	3.67	0.58	14
DT3	3.61	0.99	13	3.57	1.09	16	3.55	1.04	12	4.00	0.00	5
DT4	3.64	1.06	12	3.43	1.02	20	3.73	1.19	8	4.33	0.57	1
DT5	3.52	1.11	16	3.43	0.94	19	3.55	1.29	13	3.67	1.53	15
E1	4.21	0.79	1	4.29	0.75	1	4.18	1.30	1	4.00	1.00	8
E2	3.41	0.84	20	3.54	0.70	17	3.09	1.33	19	4.00	1.00	8
E3	3.93	1.02	3	3.79	1.05	7	4.09	1.04	2	4.00	1.00	8
E4	3.85	1.06	6	4.00	1.00	2	3.64	1.03	10	4.00	1.73	13
E5	3.52	0.85	15	3.54	0.97	18	3.36	0.81	16	4.00	0.00	5
E6	3.88	0.99	5	4.00	1.08	3	3.60	0.84	11	4.33	1.16	2
I1	3.70	0.91	9	3.92	0.95	5	3.36	0.81	16	4.00	1.00	8
I2	3.54	1.03	14	3.62	1.04	13	3.70	1.06	9	2.67	0.58	21
I3	3.52	1.12	17	3.77	0.93	9	3.00	1.18	20	4.33	1.16	2
I4	3.50	0.76	18	3.57	0.85	14	3.44	0.73	15	3.33	0.58	16
I5	3.44	1.00	19	3.64	0.84	12	3.18	1.17	18	3.33	1.16	19
SC1	3.71	0.94	8	3.79	1.05	7	3.73	0.91	7	3.33	0.58	16
SC2	3.30	0.99	21	3.38	0.87	21	3.00	1.18	20	4.00	0.00	5
SC3	3.64	0.95	11	3.57	1.02	15	3.91	0.83	4	3.00	1.00	20
SC4	3.71	0.90	7	3.86	0.77	6	3.45	1.04	14	4.00	1.00	8
SC5	3.70	0.91	9	3.69	0.95	11	3.82	0.98	6	3.33	0.58	16

Notes: G1 – Engineers; G2 – Home buyers; G3 – Real estate agents and contractors
DT – Design and technical barrier; E – Economic barrier; I – Institutional barrier; SC – Socio-cultural barrier

5. Conclusions

This paper investigated the perspectives of supply chain agents (i.e. engineers, contractors and real estate agents) and demand stakeholder (i.e. home buyers) on the barriers of the lack of sustainable housing in Malaysia. The results proved that the most important category of barrier was economic barrier, followed by the design and technical barrier. The two most important barriers were high capital cost, and lack of professional expertise in sustainability. This paper also proved that there is a discrepancy of opinions between different stakeholders on the importance of the barriers. For example, engineers (G1) and home buyers (G2) ranked high capital cost as the most important barrier, while the contractors and real estate agents (G3) only ranked high capital cost as the 8th most important barrier.

This paper contributed to both the knowledge and practices of sustainable housing implementation in Malaysia. It could serve as the basis for the supply chain agents to look into different barriers when implementing sustainable housing in Malaysia. The home buyers could realize that there are different opinions between different stakeholders and utilize the list of barriers to make informed decision when purchasing a property. The government could come out with mitigation strategies to overcome these barriers.

6. References

[1] Choguill, C.L., 1994 *Sustainable housing programmes in a world of readjustment.* Habitat International **18**(2): p. 1-11 DOI: https://doi.org/10.1016/0197-3975(94)90047-7.

[2] Elkington, J., 1994 *Towards the sustainable corporation: Win-win-win business strategies for sustainable development.* California Management Review **36**(2): p. 90.

[3] Miller, W., J. Stenton, H. Worsley, and T. Wuersching, 2014 *Strategies and solutions for housing sustainability : building information files and performance certificates.* Queensland University of Technology Australia.

[4] Wong, S.Y., C. Susilawati, W. Miller, and M. Diaswati, 2020 *Perspectives of Australian property practitioners on sustainability features in residential property.* Journal of Housing and the Built Environment DOI: 10.1007/s10901-019-09726-3.

[5] Roshanfekr, S., N.M. Tawil, and N.A. Goh, 2016 *Investigation of Sustainable Housing Criteria.* MATEC Web of Conferences **66**: p. 00096 DOI: 10.1051/matecconf/20166600096.

[6] Romanach, L., T. Jeanneret, and N. Hall, 2015 *EnergyFit Homes Initiative working paper 4: Housing specialist and real estate industry survey results*. CSIRO Australia.

[7] Manoliadis, O., I. Tsolas, and A. Nakou, 2006 *Sustainable construction and drivers of change in Greece: a Delphi study.* Construction Management and Economics **24**(2): p. 113-120 DOI: 10.1080/01446190500204804.

[8] Samari, M., et al., 2013 *The investigation of the barriers in developing green building in Malaysia* Modern Applied Science **7**(2): p. 1-10 DOI: doi:10.5539/mas.v7n2p1

[9] Hogberg, L., 2013 *The Impact of Energy Performance on Single-family Home Selling Prices in Sweden.* Journal of European Real Estate Research **6**(3): p. 242-261 DOI: https://doi.org/10.1108/JERER-09-2012-0024.

[10] Brounen, D. and N. Kok, 2010 *On the economics of EU energy labels in the housing market.* The Royal Institution of Chartered Surveyors (RICS): London.

[11] Shan, M. and B. Hwang, 2018 *Green building rating systems: Global reviews of practices and research efforts.* Sustainable Cities and Society **39**: p. 172-180.

[12] Ametepey, O., C. Aigbavboa, and K. Ansah, 2015 *Barriers to successful implementation of sustainable construction in the Ghanaian construction industry.* Procedia Manufacturing **3**: p. 1682-1689 DOI: https://doi.org/10.1016/j.promfg.2015.07.988.

[13] Chan, A.P.C., A. Darko, A.O. Olanipekun, and E.E. Ameyaw, 2018 *Critical barriers to green building technologies adoption in developing countries: The case of Ghana.* Journal of Cleaner Production **172**: p. 1067-1079 DOI: https://doi.org/10.1016/j.jclepro.2017.10.235.

[14] Häkkinen, T. and K. Belloni, 2011 *Barriers and drivers for sustainable building.* Building Research & Information **39**(3): p. 239-255 DOI: 10.1080/09613218.2011.561948.

[15] Luthra, S., S. Kumar, D. Garg, and A. Haleem, 2015 *Barriers to renewable/sustainable energy technologies adoption: Indian perspective.* Renewable and Sustainable Energy Reviews **41**: p. 762-776 DOI: doi:10.1016/j.rser.2014.08.077

[16] Darko, A. and A.P.C. Chan, 2016 *Review of barriers to green building adoption.* Sustainable Development **25**(3): p. 167-179 DOI: https://doi.org/10.1002/sd.1651.

[17] Zhang, X.L., L.Y. Shen, Y. Wu, and G.Y. Qi, 2011 *Barriers to implement green strategy in the process of developing real estate projects.* Zhang, X. L., Shen, L. Y., Wu, Y. Zh., & Qi, G. Y. **4**: p. 33-37 DOI: http://dx.doi.org/10.2174/1876400201104010033.

[18] Mohd Nordin, R., A.H.A. Halim, and J. Yunus, 2018 *Challenges in the implementation of green home development in Malaysia: Perspective of developers.* IOP Conference Series: Materials Science and Engineering **291**(1): p. 012020 DOI: 10.1088/1757-899X/291/1/012020.

[19] Mangialardo, A., E. Micelli, and F. Saccani, 2019 *Does sustainability affect real estate market values? Empirical evidence from the office buildings market in Milan (Italy)* Sustainability **11** (11): p. 12 DOI: 10.3390/su11010012

[20] Wong, S.Y., C. Susilawati, W. Miller, and D. Mardiasmo, 2016 *Assessing the impact of sustainability-related features on residential property price*, in *22nd Annual Pacific RIM Real Estate Society Conference*: Sunshine Coast, Australia.

[21] Cadman, D., 2000 *The vicious circle of blame.* Royal Institution of Chartered Surveyor.

Is risk assessment matter in infrastructure projects on soft soils?

W W Low[1], K S Wong[1] and J W Lim[1]

[1] Department of Civil and Construction Engineering, Faculty of Engineering and Science, Curtin University Malaysia, CDT 250, Miri, Sarawak, Malaysia

E-mail: low.wai.wah@curtin.edu.my

Abstract. Infrastructure projects are not merely complicated but they require high investment costs and a considerable amount of resources and stakeholders are necessary to assure their performance. Infrastructure projects especially transportation projects like highway, bridge, etc. cover a significantly large area and involve a lot of different stakeholders as such they are high in risks and uncertainties compared to other types of projects. The large covered area has led to construction on different types of soils including soft soils. This will complicate the construction process. The results of the literature review showed that existing knowledge on risk assessment of infrastructure on soft soils is not merely insufficient. Some researchers have highlighted the importance of risk assessment in construction projects on soft soils. Hence, the aim of this research was to explore the perception of industry personnel about the significance of risk assessment in infrastructure projects. A questionnaire survey was designed and distributed to consultants and contractors in Malaysia. Data were analyzed based on the descriptive method. Results showed that industry people perceived that the level of risks and uncertainties of infrastructure projects on soft soils is higher than other types of soil. As such, the impact of infrastructure projects on soft soils on project cost performance will be more significant than other types of soil. Hence, construction industry professionals perceived that risk management is critical in mitigating risks especially in infrastructure projects on soft soils.

1. Introduction

The construction of mega infrastructure projects, namely East Coast Rail Link (ECRL), Mass Rapid Transit 2 (MRT2) and Light Rapid Transit 3 (LRT3) has shed light on the importance of risk management in infrastructure projects. In the Malaysian Budget 2020, the Malaysian government will continue providing support to the primary infrastructure to improve the competitiveness of the country [1]. Examples of infrastructure projects include development of Chuping Valley Industrial area, Kuantan port-related projects, construction of centralized sewerage treatment plant in Johor, development of infrastructure in Samalaju Industrial Park, upgrading FELDA roads and basic infrastructure, development of tourism-related infrastructure, development of basic infrastructure to increase the connectivity between urban and rural areas especially in Sabah and Sarawak and so on [1].

The large covered area has led to construction on different types of soils and these include soft soils. Soft soils such as peat soil and organic soil are problematic [2]. The low bearing capacity of soft soils tends to increase risks and uncertainties during the design and construction stages. According to the Project Management Body of Knowledge (PMBOK) guide 5th edition, *"project risk is an uncertain event or condition that, if it occurs, has a positive or negative effect on one or more project objectives such as scope, schedule, cost, and quality*"[3, 310]. It increases the challenge of assuring project performance and success. In Malaysia, peat covers over 25,000 square km and about 69% is located at Sarawak [4]. Because of the significant coverage of soft soils, those major infrastructure development projects are on soft soils [5]. In addition, each infrastructure project has its own unique risks and uncertainties.

In view of this, decision making plays a critical role in assuring the performance of infrastructure projects. Risk management is one of the ways to improve the performance of infrastructure projects on soft soils. Academic researchers have highlighted the importance of risk management process in managing and monitoring risks and uncertainties in infrastructure projects. Risk assessment is one of the processes in risk management involves the estimation of probability and severity of each identified risk on project objectives. Also, recent literature has shown that existing knowledge on risk assessment of infrastructure projects especially on soft soils is not sufficient [6, 7].

Although academic researchers have addressed the importance of risk management in infrastructure projects, the perspective from industry professionals about the importance of risk management is still in doubt. As such, the aim of this research was to explore the perception of industry personnel about the significance of risk assessment in infrastructure projects.

2. Literature Review

Risk management in infrastructure projects is gaining momentum in academia. Considerable past studies have focused on the risk assessment aspect. Risk assessment in infrastructure projects has been studied in different types of infrastructure projects, different geographical locations and different project stages. For examples, Zhi [8] focused on general infrastructure, Ghosh and Jintanapakanont [9] assessed risks of underground mass rapid-transit in Thailand, Zayed et al. [10] conducted study on risk assessment in a highway project, Wong et al. [11] studied the cross-sea route project in China and Johnsen and Veen focused on communication infrastructure in railway, just to name few. From the perspective of the academic researchers, risk assessment is critical in mitigating, monitoring and controlling risks in infrastructure projects. But what do industry professionals say about this? According to Low et al. [6], risk assessment in infrastructure projects based on soil types (e.g. soft soils) is not getting much attention among academic researchers, especially in the management context. Does risk assessment a matter in infrastructure projects on soft soils among industry practitioners?

3. Research Methodology

To explore the significance of risk assessment in infrastructure projects on soft soils, four fundamental aspects were focused on this study. The first aspect was about risk level, the second aspect was about the uncertainties level, the third aspect was about cost performance and the final aspect was concerned with the application of risk management. A questionnaire was designed and distributed to the contractors and consultants in Malaysia after getting the ethics approval from the human research ethics committee. To assure validity and reliability of the study, industry professional with at least 10 years of working experience in infrastructure projects on soft soils was targeted in this study. To speed up the data collection process, random and convenience sampling methods were used in this study. The descriptive analysis method (e.g. percentage, mean) was used to analyze the data.

4. Results and discussions

One-hundred eighty questionnaires have been sent out by post, email and hand-delivered, 72 questionnaires have been received. However, a total of 62 valid responses were used in this study. This made a response rate of 34%. About 34% of them are at the seniority level and 66% are at the management level. About 35% of them have more than 20 years and above working experience in the construction industry. Table 1 indicates the results of the data collected from the 62 respondents. It was found that the statement, "*Risk management is critical in mitigating risks and important in infrastructures projects soft soils*" has the highest mean value, 4.10. This followed by the statement, "*Infrastructure projects on soft soils will contribute significant impact on project's cost performance than infrastructures project on other types of soils*" with the mean value of 4.06. The third highest mean value's statement was "*The uncertainties level in infrastructure projects on soft soils is higher than infrastructure projects on other types of soils*" with a mean value of 3.98. The fourth highest mean value was "*The risks level in infrastructure projects on soft soils is higher than infrastructure projects on other types of soils*" with the mean value of 3.94. In general, most respondents supported that the risks and uncertainties levels in infrastructure projects on soft soils are higher than other types of soil. As such, they perceived that it will contribute a significant impact on project cost performance. Cost performance is always an element that concerned by all the project team members. This is also the reason for this study taking this aspect into consideration. Hence, they perceived that risk management is critical in mitigating risks and it is important in infrastructure projects on soft soils.

Table 1. Mean values of four different aspects (adapted from [12]).

Four aspects	Percentage (%)					Mean
	Strongly disagree	Disagree	Neither agree or disagree	Agree	Strongly agree	
• The risks level in infrastructure projects on soft soils is higher than infrastructure projects on other types of soils.	0	8.1	12.9	56.5	22.6	3.94
• The uncertainties level in infrastructure projects on soft soils is higher than infrastructure projects on other types of soils.	0	3.2	19.4	53.2	24.2	3.98
• Infrastructure projects on soft soils will contribute significant impact on project's cost performance than infrastructures project on other types of soils.	1.6	4.8	9.7	53.2	30.6	4.06
• Risk management is critical in mitigating risks and important in infrastructures projects on soft soils.	3.2	1.6	11.3	50.0	33.9	4.10

5. Conclusions

In view of the high risks and uncertainties of infrastructure projects on soft soils compared to other types of soils from the perspective of industry professionals. It can be concluded that risk management process such as risk assessment plays a vital role in assuring the performance or even success of infrastructure projects on soft soils. Results of the study also indicated that the research value of future studies focusing on this area. As such, future studies can focus on proposing a risk management methodology for infrastructure projects on soft soils or assess and compare risks of different soil types based on a longitudinal approach.

References

[1] Lim G E 2019 2020 Budge speech 16th January 2020 http://www.bnm.gov.my/documents/budget/bs2020.pdf

[2] Jafer H M, Atherton W, Sadique M, Ruddock F and Loffill E 2018 Development of a new ternary blended cementitious binder produced from waste materials for use in soft soil stabilization *Journal of Cleaner Production* **172** 516-28

[3] Rose K H 2013 A Guide to the Project Management Body of Knowledge (PMBOK® Guide)—Fifth Edition *Project management journal* **44** e1-e

[4] Construction Research Institute of Malaysia 2015 Guidelines for construction on peat and organic soils in Malaysia. (Kuala Lumpour, Malaysia

[5] Kaniraj S R and Joseph R R 2006 Geotechnical Behavior of Organic Soils of North Sarawak. In: *Fourth International Conference on Soft Soil Engineering,* (Vancouver, Canada

[6] Low W W, Wong K S and Lee J L 2019 Cost-influencing risk factors in infrastructure projects on soft soils *International Journal of Construction Management* 1-13

[7] Low W W, Wong K S and Lee J L 2018 Risk assessment framework on time impact: Infrastructure projects in soft soil during construction stage *IOP Conference Series: Materials Science and Engineering* **344** 012022

[8] Zhi H 1995 Risk management for overseas construction projects *International journal of project management* **13** 231-7

[9] Ghosh S and Jintanapakanont J 2004 Identifying and assessing the critical risk factors in an underground rail project in Thailand: A factor analysis approach *International Journal of Project Management* **22** 633-43

[10] Zayed T, Amer M and Pan J 2008 Assessing risk and uncertainty inherent in Chinese highway projects using AHP *International journal of project management* **26** 408-19

[11] Wang T, Wang S, Zhang L, Huang Z and Li Y 2016 A major infrastructure risk-assessment framework: Application to a cross-sea route project in China *International Journal of Project Management* **34** 1403-15

[12] Lim G J W 2019 Risk assessment of transportation infrastructure on soft soils Curtin University Malaysia [Undergraduate thesis]

IOP Conf. Series: Materials Science and Engineering **943** (2020) 012066 doi:10.1088/1757-899X/943/1/012066

Estimation of Individual Leaf Area by Leaf Dimension using a Linear Regression for Various Tropical Plant Species

A C Tay and J Z L Ling

Department of Civil and Construction Engineering, Curtin University Malaysia, Miri, Sarawak, Malaysia

Email tayaichen@curtin.edu.my

Abstract. Leaf area estimation is significant to assess leaf development and plant growth. A simple and efficient model (Regression model) was developed to estimate leaf area of various tropical plant species using leaf dimension (length x width). The objective of this study was to determine a relation between leaf area and leaf dimension to get the best fit line of a linear regression. A total of twenty plant species were selected at Miri, Sarawak, Malaysia. The leaf shape and plant growing environment were investigated as a background data. The leaf shapes were elliptic, acuminate, aristate, obtuse, lobed, linear, peltate and lanceolate. The recorded plant growing environment were location, soil pH and sunlight exposure. The plants grew at a slope or flat, shaded or high exposure of sunlight with the soil pH between 4.8 and 7.3. A regression equation for each leaf was established with coefficient of determination (R^2) of 0.9. It strongly proved that the leaf areas of all the plant species are well correlated with leaf dimension (length x width). The linear regression is not influenced by different leaf shape and growing environment. Generally, the plant grew at high exposure of sunlight indicated higher leaf area. There is no obvious trend of leaf area at different soil pH. From the R^2 values, it concluded that the leaf area of individual plant can be estimated by the linear regression established in this study by determining the leaf length and width.

1. Introduction

Leaf area is an important parameter to determine biophysical condition such as light interception, photosynthesis and evapotranspiration [1]. Leaf area can be estimated accurately by establishing its regression model from the measurement of leaf length and leaf width [2]. The regression model is a non-destructive analysis technique because this technique does not destroy the plant, and the plant growth can be determined periodically [3]. Leaf area estimation is useful for a phenology study which the plant development could be assessed in responding to environment. This phenology study may contribute to the field of conservation, agriculture, ecology and evolution [4]. Phenological stages responding to environmental factors provide a consistent ecological biometric identification of climate change [5]. The aim of this research was to develop a linear regression analysis models to analyse relationship between leaf area and leaf dimension (length x width) for various tropical plant species. These linear regression analysis models are able to be used for plant leaf area estimation.

2. Materials and Methods

A total of twenty plant species were selected at Miri, Sarawak, Malaysia. The leaf shape and plant growing environment were investigated as background data. The plant growing environment recorded were location, soil pH and growing sunlight condition.

2.1. Plant species

The selected plant species were adult plant with 90% fully expanded leaves which are healthy and active in photosynthesis. The scientific name of selected plant species are listed in **Table 1**. For each plant species, total of thirty (30) leaves collected were in the different sizes ranging from the smallest to the biggest. The leaf shapes collected were linear, elliptic, acuminate, obtuse, lobed, peltate, lanceolate, aristate (**Table 1**).

Table 1. Description of selected plant species.

No	Plant species	Leaf shape	Soil pH	Growing environment
1	*Annona muricate*	Elliptic	6.7	Near to the small drainage, whole-day sunlight exposure
2	*Arbutus menziesii*	Elliptic	5.9	Roadside, whole-day sunlight exposure
3	*Artocarpus integer*	Elliptic	6.5	Roadside, whole-day sunlight exposure.
4	*Bougainvillea glabra*	Acuminate	7.3	Down-slope area, expose to sunlight during afternoon time only
5	*Cassia fistula*	Aristate	6.3	Flat ground, whole-day sunlight exposure
6	*Citrus limon*	Elliptic	6.8	Flat ground, whole-day sunlight exposure
7	*Duranta erecta*	Aristate	5.3	Compete with other same species, under shaded area with less sunlight exposure
8	*Excoecaria cochinchinensis*	Elliptic	7.0	Compete with other same species, under shaded area with less sunlight exposure
9	*Gardenia jasminoides*	Elliptic	5.5	Compete with other same species, whole-day sunlight exposure
10	*Hibiscus rosa-sinensis*	Aristate	7.2	Shaded area with less sunlight exposure
11	*Hypoestes phyllostachya*	Obtuse	5.3	Compete with other same species, under shaded area with less sunlight exposure
12	*Jatropha integerrima*	Lobed	6.9	Roadside, whole-day sunlight exposure
13	*Lantana camara*	Obtuse	6.0	Compete with other same species under shaded area with less sunlight exposure
14	*Mangifera indica*	Linear	6.4	Roadside, whole-day sunlight exposure
15	*Nephelium lappaceum*	Obtuse	7.0	Compete with other same species, under shaded area with less sunlight exposure
16	*Photinia fraseri*	Elliptic	6.6	Sunlight exposure during afternoon time only

17	*Plectranthus verticillatus*	Peltate	4.8	Compete with other same species, under shaded area with less sunlight exposure
18	*Psidium guajava*	Elliptic	6.8	Roadside, under shaded area with less sunlight exposure
19	*Viburnum suspensum*	Elliptic	6.1	Flat ground, whole-day sunlight exposure
20	*Xanthostemon* sp.	Lanceolate	4.8	Flat ground, whole-day sunlight exposure

2.2. Growing environment

A pH meter was used to measure soil pH for each plant species by digging up 2 – 3 inches hole into the soil. The hole was filled with rain water to create a pool for pH measurement. The soil pH ranged between 4.8 and 7.3 (**Table 1**). The location and sunlight condition were observed for the plant which the leaves were collected. The sunlight condition of sampling areas were shaded, less exposure or whole-day exposure as shown in **Table 1.**

2.3. Leaf dimension

Leaf length and width were measured using a ruler for every single leaf. The leaf length was measured from leaf apex until the leaf edge connecting to petiole. The leaf width was measured from left to right for its wide margin. The leaf dimension was obtained by multiplying leaf length and leaf width (LW).

2.4. Linear regression

The ImageJ software was used to analyse the leaf image captured for adaxial and abaxial surfaces to measure actual leaf area. From the data obtained or the leaf area and leaf dimension, the relation between leaf area and leaf dimension was plotted to get the best fit line of a linear regression. A coefficient of determination (R^2) was identified to indicate the estimation of leaf area via leaf dimension related to the actual leaf area.

3. Results and Discussion

A regression equation for each plant species was established with coefficient of determination (R^2) of 0.9 (**Table 2**). It showed that the linear regression models with leaf dimension (LW) is accurate to be used for leaf area estimation for all the selected tropical plant species in this experiment. If R^2 closes to 1.0, it indicated a more accurate leaf area estimation [6]. Among all the plant species, *Arbutus menziesii, Cassia fistula* and *Viburnum suspensum* (**Figure 1**) were the plant species indicating the highest R^2. The leaves of both *Arbutus menziesii* and *Viburnum suspensum* were in elliptic shape. The leaf of *Cassia fistula* was in aristate shape. *Jatropha integerrima* which the leaf was in lobed shape showed the lowest R^2, 0.9436. The plant species were under different growing environment such as various soil pH and sunlight condition, however, they could perform in a strong relation in the linear regression [7, 8].

Table 2. Linear regression relation between actual leaf area (A) and leaf dimension (LW).

No	Plant species	Regression equation	Coefficient of determination, R^2
1	*Annona muricate*	A = 0.7181 LW + 1.4882	0.9869
2	*Arbutus menziesii*	A = 0.6502 LW + 0.3589	0.9952
3	*Artocarpus integer*	A = 0.7211 LW - 2.126	0.9942
4	*Bougainvillea glabra*	A = 0.636 LW - 0.7235	0.9925
5	*Cassia fistula*	A = 0.7341 LW + 0.4823	0.9945
6	*Citrus limon*	A = 0.7642 LW + 0.0628	0.9902

7	*Duranta erecta*	A = 0.6107 LW + 0.2933	0.9687
8	*Excoecaria cochinchinensis*	A = 0.7135 LW - 0.4776	0.9941
9	*Gardenia jasminoides*	A = 0.6142 LW + 0.7824	0.9864
10	*Hibiscus rosa-sinensis*	A = 0.6776 LW - 1.2945	0.9867
11	*Hypoestes phyllostachya*	A = 0.6342 LW + 0.6808	0.9769
12	*Jatropha integerrima*	A = 0.5115 LW + 3.0722	0.9436
13	*Lantana camara*	A = 0.6485 LW + 0.0771	0.9829
14	*Mangifera indica*	A = 0.9528 LW - 11.78	0.9922
15	*Nephelium lappaceum*	A = 0.6177 LW + 0.1912	0.9937
16	*Photinia fraseri*	A = 0.7545 LW - 0.7981	0.9828
17	*Plectranthus verticillatus*	A = 0.973 LW - 1.1596	0.9932
18	*Psidium guajava*	A = 0.6873 LW + 1.2865	0.9871
19	*Viburnum suspensum*	A = 0.6971 LW - 0.192	0.9948
20	*Xanthostemon* sp.	A = 0.6107 LW + 0.9614	0.9737

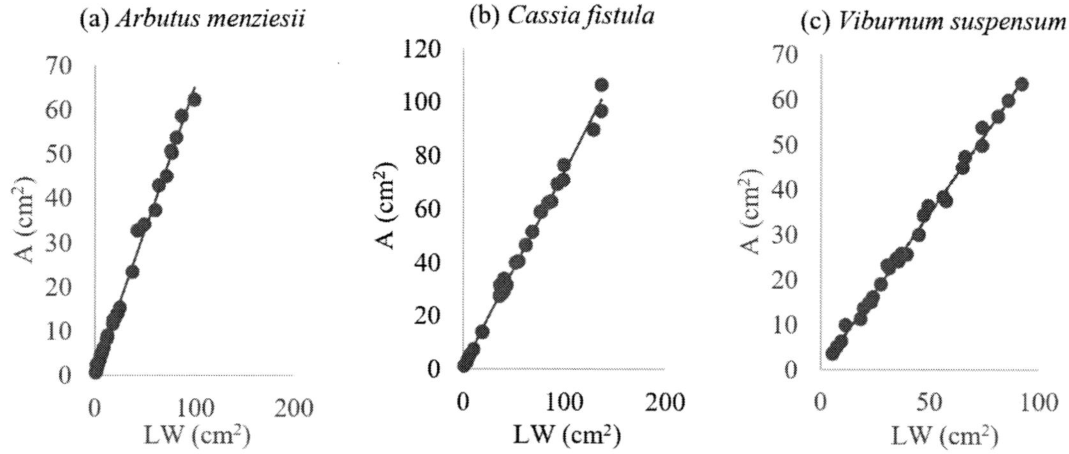

Figure 1. Relationship between leaf area (A) and leaf dimension (LW) for (a) *Arbutus menziesii*, (b) *Cassia fistula* and (c)*Viburnum suspensum*.

4. Conclusion

The equation derived can be used to estimate the actual leaf area of the plant. The best correlation in linear regression between actual leaf area with the leaf dimension (length x width) is applicable for all different plant species [9, 10]. The linear regression analysis model that using LW as independent variable is suitable for all species at any growing environment [11, 12]. Among all the leaf shapes, the elliptic leaf shape plant was more prone to develop a high efficient linear regression analysis model. Lobed leaf shape plant was observed to create a less accurate leaf area estimation models.

References

[1] Niinemets U 2010 A review of light interception in plant stands from leaf to canopy in different plant functional types and in species with varying shade tolerance *Ecological Research* **25** 693

[2] Kandiannan *et al* 2002 Allometric model for leaf area estimation in black pepper (*Piper nigrum* L.) *Journal Agronomy & Crop Science* **188** 138

[3] Rouphael *et al* 2007 Leaf area estimation of sunflower leaves from simple linear measurements *Photosynthetica* **45** 306

[4] Scheifinger *et al* 2002 Atmospheric mechansims governing the spatial and temporal variability of phenological observations in Central Europe *International Journal of Climatology* **22** 1739

[5] Ianovici N 2012 Researches on anatomical adaptations of the alpine plants - *Plantago atrata Annals of West University of Timişoara, Biology* **15** 1

[6] Rouphael *et al* 2007 Leaf area estimation of sunflower leaves from simple linear measurements *Photosynthetica* **45** 306

[7] Bianco *et al* 2008 Estimation of the leaf area of *Sida cordifolia* and *Sida rhombifolia* using linear dimensions of the leaf blade *Planta Daninha* **26** 807

[8] Brito-Rocha *et al* 2016 Regression models for estimating leaf area of seedlings and adult inviduals of Neotropical rainforest tree species *Brazilian Journal of Biology* **76** 983

[9] Leroy *et al* 2006 Practical methods for non-destructive measurement of tree leaf area *Agroforestry system* **71** 2

[10] Souza M C and Habermann G 2014 Non-destructive equations to estimate the leaf area of *Styrax pholii* and *Styrax ferrugineus Brazilian Journal of Biology* **74** 1

[11] Antunes *et al* 2008 Allometric models for non-destructive leaf area estimation in coffee (*Coffea arabica* and *Coffea canephora*) *Annals of Applied Biology* **153** 33

[12] Zhang L and Pan L 2011 Allometric models for leaf area estimation across different leaf-age groups of evergreen boardleaved trees in a subtropical forest *Photosynthetica* 49 219

AUTHOR INDEX